Handbook of Energy-Aware and Green Computing

Volume 1

CHAPMAN & HALL/CRC
COMPUTER and INFORMATION SCIENCE SERIES

Series Editor: Sartaj Sahni

PUBLISHED TITLES

Handbook of Energy-Aware and Green Computing

Volume 1

Edited by

Ishfaq Ahmad
Sanjay Ranka

CRC Press
Taylor & Francis Group
Boca Raton London New York

CRC Press is an imprint of the
Taylor & Francis Group, an **informa** business

A CHAPMAN & HALL BOOK

First published 2012 by Chapman and Hall

Published 2019 by CRC Press
Taylor & Francis Group
6000 Broken Sound Parkway NW, Suite 300
Boca Raton, FL 33487-2742

© 2012 by Taylor & Francis Group, LLC
CRC Press is an imprint of Taylor & Francis Group, an Informa business

No claim to original U.S. Government works

ISBN-13: 978-1-4398-5040-4 (hbk)

Library of Congress Cataloging-in-Publication Data

Handbook of energy-aware and green computing / edited by Ishfaq Ahmad and Sanjay Ranka.
    p. cm. -- (Chapman & Hall/CRC computer and information science series)
  Includes bibliographical references and index.
  ISBN 978-1-4398-5040-4
  1. Electronic digital computers--Power supply. 2. Computer systems--Energy conservation. 3. Green technology. 4. Low voltage systems. I. Ahmad, Ishfaq. II. Ranka, Sanjay.

TK7895.P68H36 2012
621.39'5--dc23
    2011029259

Visit the Taylor & Francis Web site at
http://www.taylorandfrancis.com

and the CRC Press Web site at
http://www.crcpress.com

Contents

PART III   Green Networking

PART IV   Algorithms

PART V   Real-Time Systems

Preface

Green computing is an emerging interdisciplinary research area spanning the fields of computer science and engineering, electrical engineering, and other engineering disciplines. Green computing or sustainable computing is the study and practice of using computing resources efficiently, which in turn can impact a spectrum of economic, ecological, and social objectives. Such practices include the implementation of energy-efficient central processing unit processors and peripherals as well as reduced resource consumption. During the last several decades and in particular in the last five years, the area has produced a prodigious amount of knowledge that needs to be consolidated in the form of a comprehensive book.

Researchers and engineers are now considering energy as a first-class resource and are inventing means to manage it along with performance, reliability, and security. Thus, a considerable amount of knowledge has emerged, as is evident by numerous tracks in leading conferences in a wide variety of areas such as mobile and pervasive computing, circuit design, architecture, real-time systems, and software. Active research is going on in power and thermal management at the component, software, and system level, as well as on defining power management standards for servers and devices and operating systems. Heat dissipation control is equally important, forcing circuit designers and processor architects to consider not only performance issues but also factors such as packaging, reliability, dynamic power consumption, and the distribution of heat. Thus, research growth in this area has been explosive.

The aim of this edited handbook is to provide basic and fundamental knowledge in all related areas, including, but not limited to, circuit and component design, software, operating systems, networking and mobile computing, and data centers. As a comprehensive reference, the book provides the readers with the state of the art of various aspects of energy-aware computing at the component, software, and system level. It also provides a broad range of topics dealing with power-, energy-, and temperature-related research areas of current importance.

The *Handbook of Energy-Aware and Green Computing* is divided into two volumes, and its 52 chapters are categorized into 10 parts:

Volume 1
    Part I: Components, Platforms, and Architectures
    Part II: Energy-Efficient Storage
    Part III: Green Networking
    Part IV: Algorithms
    Part V: Real-Time Systems

Volume 2
    Part VI: Monitoring, Modeling, and Evaluation
    Part VII: Software Systems

Part VIII: Data Centers and Large-Scale Systems
Part IX: Green Applications
Part X: Social and Environmental Issues

Readers interested in architecture, networks, circuit design, software, and applications can find useful information on the topics of their choice. For some of the topics, there are multiple chapters that essentially address the same issue but provide a different and unique problem solving and presentation by different research groups. Rather than trying to have a single unified view, we have tried to encourage this diversity so that the reader has the benefit of a variety of perspectives.

We hope the book provides a good compendium of ideas and information to a wide range of individuals from industry, academia, national labs and industry. It should be of interest not just to computer science researchers but to several other areas as well, due to the inherent interdisciplinary nature of green computing.

Ishfaq Ahmad
University of Texas at Arlington
Arlington, Texas

Sanjay Ranka
University of Florida
Gainesville, Florida

For MATLAB® and Simulink® product information, please contact:

The MathWorks, Inc.
3 Apple Hill Drive
Natick, MA, 01760-2098 USA
Tel: 508-647-7000
Fax: 508-647-7001
E-mail: info@mathworks.com
Web: www.mathworks.com

Editors

Ishfaq Ahmad received his BSc in electrical engineering from the University of Engineering and Technology, Pakistan, in 1985, and his MS in computer engineering and PhD in computer science from Syracuse University, New York in 1987 and 1992, respectively. Since 2002, he has been a professor of computer science and engineering at the University of Texas at Arlington. Prior to this, he was an associate professor of computer science at the Hong Kong University of Science and Technology. His research focus is on the broader areas of parallel and distributed computing systems and their applications, optimization algorithms, multimedia systems, video compression, and energy-aware green computing.

Dr. Ahmad has received numerous research awards, including three best paper awards at leading conferences and the best paper award for *IEEE Transactions on Circuits and Systems for Video Technology*, the IEEE Service Appreciation Award, and the Outstanding Area Editor Award from *IEEE Transactions on Circuits and Systems for Video Technology*. He was elevated to the IEEE Fellow grade in 2008.

Dr. Ahmad's current research is funded by the U.S. Department of Justice, the National Science Foundation, SRC, the Department of Education, and several companies. He is the founding editor in chief of a new journal, *Sustainable Computing: Informatics and Systems*, and a cofounder of the *International Green Computing Conference*. He is an editor of the *Journal of Parallel and Distributed Computing*, *IEEE Transactions on Circuits and Systems for Video Technology*, *IEEE Transactions on Parallel and Distributed Systems*, and *Hindawi Journal of Electrical and Computer Engineering*. He has guest edited several special issues and has been a member of the editorial boards of the *IEEE Transactions on Multimedia* and *IEEE Concurrency*.

Sanjay Ranka is a professor in the Department of Computer Information Science and Engineering at the University of Florida, Gainesville, Florida. His current research interests are energy-efficient computing, high-performance computing, data mining, and informatics. Most recently, he was the chief technology officer at Paramark, where he developed real-time optimization software for optimizing marketing campaigns. He has also held a tenured faculty position at Syracuse University and has been a researcher/visitor at IBM T.J. Watson Research Labs and Hitachi America Limited.

Dr. Ranka received his PhD and BTech in computer science from the University of Minnesota and from IIT, Kanpur, India, respectively. He has coauthored 2 books, *Elements of Neural Networks*

(MIT Press) and *Hypercube Algorithms* (Springer-Verlag), 70 journal articles, and 110 refereed conference articles. His recent work has received a student best paper award at ACM-BCB 2010, best paper runner-up award at KDD-2009, a nomination for the Robbins Prize for the best paper in the journal *Physics in Medicine and Biology* for 2008, and a best paper award at ICN 2007.

Dr. Ranka is a fellow of the IEEE and AAAS, and a member of the IFIP Committee on System Modeling and Optimization. He is also the associate editor in chief of the *Journal of Parallel and Distributed Computing* and an associate editor for *IEEE Transactions on Parallel and Distributed Computing*; *Sustainable Computing: Systems and Informatics*; *Knowledge and Information Systems*; and the *International Journal of Computing*.

Dr. Ranka was a past member of the Parallel Compiler Runtime Consortium, the Message Passing Initiative Standards Committee, and the Technical Committee on Parallel Processing. He is the program chair for the 2010 International Conference on Contemporary Computing and co-general chair for the *2009 International Conference on Data Mining* and the *2010 International Conference on Green Computing*.

Dr. Ranka has had consulting assignments with a number of companies (e.g., AT&T Wireless, IBM, Hitachi) and has served as an expert witness in patent disputes.

Contributors

S. Akoush
Computer Laboratory
University of Cambridge
Cambridge, United Kingdom

Hideharu Amano
Department of Information and Computer
    Science
Keio University
Yokohama, Japan

Lachlan L.H. Andrew
Centre for Advanced Internet Architectures
Swinburne University of Technology
Melbourne, Australia

Y. Audzevich
Computer Laboratory
University of Cambridge
Cambridge, United Kingdom

Hakan Aydin
Department of Computer Science
George Mason University
Fairfax, Virginia

Omid Azizi
Stanford University
Stanford, California

and

Hicamp Systems, Inc.
Menlo Park, California

Salman A. Baset
IBM T.J. Watson Research Center
Hawthorne, New York

Medha Bhadkamkar
Washington State University
Pullman, Washington

Puranjoy Bhattacharjee
Department of Computer Science
Virginia Polytechnic Institute and State University
Blacksburg, Virginia

Pascal Bouvry
Computer Science and Communications
    Research Unit
University of Luxembourg
Luxembourg City, Luxembourg

Ali R. Butt
Department of Computer Science
Virginia Polytechnic Institute and State University
Blacksburg, Virginia

Kevin Chang
Washington State University
Pullman, Washington

Danny Ziyi Chen
Department of Computer Science
    and Engineering
University of Notre Dame
Notre Dame, Indiana

Yiran Chen
University of Pittsburgh
Pittsburgh, Pennsylvania

Jerry Chou
Lawrence Berkeley National Laboratory
Berkeley, California

Marek Chrobak
Department of Computer Science
University of California, Riverside
Riverside, California

J. Crowcroft
Computer Laboratory
University of Cambridge
Cambridge, United Kingdom

Sujay Deb
Washington State University
Pullman, Washington

X. Dong
School of Electronic and Electrical Engineering
University of Leeds
Leeds, United Kingdom

Bernabé Dorronsoro
Interdisciplinary Centre for Reliability, Security,
    and Trust
University of Luxembourg
Luxembourg City, Luxembourg

T. El-Gorashi
School of Electronic and Electrical
    Engineering
University of Leeds
Leeds, United Kingdom

J. Elmirghani
School of Electronic and Electrical
    Engineering
University of Leeds
Leeds, United Kingdom

Amlan Ganguly
Rochester Institute of Technology
Rochester, New York

Chris Gniady
The University of Arizona
Tucson, Arizona

Ann Gordon-Ross
Department of Electrical and Computer
    Engineering
University of Florida
Gainesville, Florida

Sudhanva Gurumurthi
Department of Computer Science
University of Virginia
Charlottesville, Virginia

Mateusz Guzek
University of Luxembourg
Luxembourg City, Luxembourg

J. Kevin Hicks
Department of Computer Engineering
Rochester Institute of Technology
Rochester, New York

A. Hopper
Computer Laboratory
University of Cambridge
Cambridge, United Kingdom

Xiaobo Sharon Hu
Department of Computer Science
    and Engineering
University of Notre Dame
Notre Dame, Indiana

Jinoh Kim
Lawrence Berkeley National Laboratory
Berkeley, California

Taewhan Kim
School of Electrical Engineering and Computer
    Science
Seoul National University
Seoul, Korea

Youngjae Kim
National Center for Computational Sciences
Oak Ridge National Laboratory
Oak Ridge, Tennessee

Masaaki Kondo
Graduate School of Information Systems
The University of Electro-Communications
Tokyo, Japan

Dhireesha Kudithipudi
Department of Computer Engineering
Rochester Institute of Technology
Rochester, New York

Benjamin C. Lee
Duke University
Durham, North Carolina

Fei Li
Department of Computer Science
George Mason University
Fairfax, Virginia

Hai Li
Polytechnic Institute
New York University
Brooklyn, New York

Keqin Li
Department of Computer Science
State University of New York at New Paltz
New Paltz, New York

Minghong Lin
Computing and Mathematical Sciences
California Institute of Technology
Pasadena, California

Shaobo Liu
Binghamton University
State University of New York
Binghamton, New York

Jun Lu
Binghamton University
State University of New York
Binghamton, New York

Prabhat Mishra
University of Florida
Gainesville, Florida

Tania Mishra
Department of Computer and Information
    Sciences and Engineering
University of Florida
Gainesville, Florida

Shivajit Mohapatra
Applications Research Center
Motorola Mobility
Libertyville, Illinois

A. Moore
Computer Laboratory
University of Cambridge
Cambridge, United Kingdom

Hiroshi Nakamura
Department of Information Physics and
    Computing
The University of Tokyo
Tokyo, Japan

Mitaro Namiki
Department of Computer and Information
    Sciences
Tokyo University of Agriculture and Technology
Tokyo, Japan

Dimin Niu
Pennsylvania State University
University Park, Pennsylvania

Partha Pande
Washington State University
Pullman, Washington

Johnatan E. Pecero
University of Luxembourg
Luxembourg City, Luxembourg

R. Penty
Department of Engineering
University of Cambridge
Cambridge, United Kingdom

Xiaoke Qin
University of Florida
Gainesville, Florida

Qinru Qiu
Binghamton University
State University of New York
Binghamton, New York

M. Reza Rahimi
Department of Computer Science
University of California, Irvine
Irvine, California

Marisha Rawlins
Department of Electrical and Computer
    Engineering
University of Florida
Gainesville, Florida

Da Qi Ren
Department of Computer Science
The University of Tokyo
Tokyo, Japan

A. Rice
Computer Laboratory
University of Cambridge
Cambridge, United Kingdom

Doron Rotem
Lawrence Berkeley National Laboratory
Berkeley, California

Sartaj Sahni
Department of Computer and Information
    Sciences and Engineering
University of Florida
Gainesville, Florida

Hiroshi Sasaki
Department of Information Physics and
    Computing
The University of Tokyo
Tokyo, Japan

Henning Schulzrinne
Department of Computer Science
Columbia University
New York, New York

Bing Shi
Department of Electrical and Computer
    Engineering
University of Maryland
College Park, Maryland

Behrooz Shirazi
Washington State University
Pullman, Washington

Anand Sivasubramaniam
Department of Computer Science
    and Engineering
Pennsylvania State University
University Park, Pennsylvania

R. Sohan
Computer Laboratory
University of Cambridge
Cambridge, United Kingdom

Ankur Srivastava
Department of Electrical and Computer
    Engineering
University of Maryland
College Park, Maryland

Reiji Suda
Department of Computer Science
The University of Tokyo
Tokyo, Japan

S. Timotheou
Computer Laboratory
University of Cambridge
Cambridge, United Kingdom

Kimiyoshi Usami
Department of Information Science
    and Engineering
Shibaura Institute of Technology
Tokyo, Japan

Nalini Venkatasubramanian
Department of Computer Science
University of California, Irvine
Irvine, California

Guanying Wang
Department of Computer Science
Virginia Polytechnic Institute and State University
Blacksburg, Virginia

H. Wang
Department of Engineering
University of Cambridge
Cambridge, United Kingdom

Weixun Wang
University of Florida
Gainesville, Florida

I. White
Department of Engineering
University of Cambridge
Cambridge, United Kingdom

Adam Wierman
Computing and Mathematical Sciences
California Institute of Technology
Pasadena, California

A. Wonfor
Department of Engineering
University of Cambridge
Cambridge, United Kingdom

Qing Wu
Air Force Research Laboratory
Rome, New York

Yuan Xie
Pennsylvania State University
University Park, Pennsylvania

Yumin Zhang
Synopsys Inc.
Mountain View, California

Dakai Zhu
The University of Texas at San Antonio
San Antonio, Texas

I

Components, Platforms, and Architectures

1

Subthreshold Computing

J. Kevin Hicks
*Rochester Institute of
Technology*

Dhireesha
Kudithipudi
*Rochester Institute of
Technology*

1.1 Introduction

Historically, power consumption has been a secondary design consideration behind speed and area. However, with shrinking technology sizes and the rapid growth of portable electronic devices, power consumption can no longer be an afterthought—energy efficiency has become a critical aspect of digital circuit design. Traditionally, voltage scaling, a mechanism in which the supply voltage is varying and the threshold voltage is constant, has been an effective solution in meeting stringent energy requirements. However, voltage scaling comes at a cost of reduction in performance. The limits of voltage scaling, and therefore energy minimization, can be explored by operating a circuit at subthreshold [1], which has been garnering more attention since Vittoz et al. demonstrated the breakthrough 0.95 V subthreshold operation of a CMOS frequency divider with 2 MHz performance [2].

In subthreshold circuits, the supply voltage is reduced well below the threshold voltage of a transistor, relying on leakage currents to drive the transistor operation. By operating at a subthreshold supply voltage, the energy consumption of a circuit can be reduced by more than 20× [3]. However, the performance of the circuit suffers as a result of the weak transistor drain currents driving the circuit. When compared to transistor operation with a superthreshold supply voltage, the transient behavior of logic in subthreshold is orders of magnitude slower [4].

1.2 Subthreshold Operation

The transistor operation in subthreshold (also known as weak inversion) varies from the traditional superthreshold operation of the device, where the transistor is being driven by a supply voltage that is greater than its threshold voltage. This variation yields a substantially weaker drain current when compared to superthreshold operation.

First, consider the traditional superthreshold operation of an NMOS transistor, as shown in Figure 1.1. A positive gate–source voltage (V_{GS}) is applied to the circuit, where $V_{GS} > V_t$. Because of this condition, the holes in the source–drain channel are repelled from the positive gate voltage. Thus, the channel is inverted from a p-type region to an n-type region, where electrons are the majority carriers [5]. Electrons flow through the source–drain channel due to the electric field created by the drain–source voltage (V_{DS}).

In this scenario, the drain current under superthreshold operation is predominantly drift based, and the current can be calculated using Equation 1.1,

$$I_{D,super} = \frac{1}{2}\mu_{eff}C_{ox}\frac{W}{L}(V_{GS} - V_t)^2(1 + \lambda V_{DS}) \tag{1.1}$$

where
  W is the width of the transistor
  L is the length of the transistor
  μ_{eff} is the effective mobility
  C_{ox} is the oxide capacitance
  λ accounts for channel-length modulation

Next, consider the subthreshold operation of an NMOS transistor, as shown in Figure 1.2. In this mode, the gate–source voltage (V_{GS}) is less than the threshold voltage (V_t), $V_{GS} < V_t$. Thus, V_{GS} does not exceed the requirement for carrier inversion in the source–drain channel. However, a key

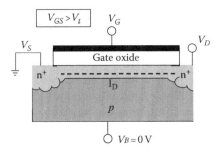

FIGURE 1.1   NMOS transistor drift current in superthreshold due to the inversion of the source–drain channel.

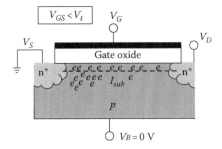

FIGURE 1.2   NMOS transistor diffusion current in subthreshold due to source leakage with minority carriers.

point to remember about subthreshold operation is that $V_{GS} < V_t$ does not result in a current of $I_D = 0$ [6]. The current does not exhibit an immediate drop to zero due the presence of subthreshold leakage currents, resulting in an exponential decrease in current strength. Hence, as long as $V_{GS} > 0$, the potential barrier created by the source-channel junction is weakened. As a result, electrons are injected from the source into the channel. The diffusion current in subthreshold is present due to these injected electrons, where minority carriers flow [5]. Thus, the transistor operates through the modulation of this leakage current, and diffusion current is the primary component of the overall drain current in subthreshold operation. The drain current in the subthreshold region is shown in Equation 1.2 [1],

$$I_{D,sub} = \frac{W}{L_{eff}} \mu_{eff} C_{ox} (m-1) V_T^2 \exp\left(\frac{V_{GS} - V_t}{mV_T}\right) \left(1 - \exp\left(\frac{-V_{DS}}{V_T}\right)\right) \qquad (1.2)$$

where
W is the width of the transistor
L_{eff} is the effective length
μ_{eff} is the effective mobility
C_{ox} is the oxide capacitance
m is the subthreshold slope factor
V_T is the thermal voltage, where $V_T = (kT/q)$

In the computation of the thermal voltage, k is the Boltzmann constant, T is the absolute temperature, and q is the magnitude of the electrical charge on an electron.

The subthreshold slope characterizes the magnitude of the subthreshold current in the transistor, measuring by how much V_{GS} has to be reduced for the drain current to drop by a factor of 10. This fact can be determined by plotting $\log I_D$ versus V_{GS} [7] or through the calculations from the equations shown in Equation 1.3 [5],

$$m = 1 + \frac{C_{dm}}{C_{ox}} = 1 + \frac{3t_{ox}}{W_{dm}} \qquad (1.3)$$

where
C_{dm} is the maximum depletion-layer capacitance
C_{ox} is the oxide capacitance per unit area

The second equivalence follows from $C_{dm} = \varepsilon_{si}/W_{dm}$, $C_{ox} = \varepsilon_{ox}/t_{ox}$, and $\varepsilon_{si}/\varepsilon_{ox} = 3$, where ε_{si} is the permittivity of silicon, W_{dm} is the maximum depletion-layer width, ε_{ox} is the oxide permittivity, and t_{ox} is the oxide thickness. Typical subthreshold slope values range from 60 to 100 mV/decade, where lower values indicate lower leakage currents in the subthreshold operation of a device.

The current in the subthreshold region is considered to be undesirable when operating a transistor in the superthreshold region. However, this current is quintessential as far as subthreshold operation is concerned. This leakage current is utilized by subthreshold circuits as their drive current. Because subthreshold operation takes place through the modulation of leakage currents, subthreshold operation results in an exponential decrease in the drain current strength. Due to the fact that the drive current is weaker in subthreshold circuits, the transistors switch slower, resulting in inferior performance when compared to traditional superthreshold designs [8]. It has been observed that transistor operation in subthreshold is 3–4 orders of magnitude slower when compared to superthreshold transistor operation.

Although the transistor operation in subthreshold is slower than superthreshold operation, there are substantial energy savings by operating in subthreshold. Total energy consumption is identified as the sum of the dynamic and static energies. The energy savings of subthreshold operation are evident by analyzing the equation for dynamic energy, as shown in Equation 1.4,

$$E_{dynamic} = \frac{1}{2} C_L V_{DD}^2 \alpha \qquad (1.4)$$

where

C_L is the load capacitance
V_{DD} is the supply voltage
α is the activity factor

Due to the quadratic reduction in power with respect to the supply voltage, subthreshold circuits are classified as ultralow power circuits.

The other component of total energy, static energy, is measured as the energy due to leakage currents and is calculated using Equation 1.5,

$$E_{static} = I_{leakage} V_{DD} t_p \qquad (1.5)$$

where

$I_{leakage}$ is the leakage current
t_p is the propagation delay

In subthreshold operation, while the supply voltage is reduced, the propagation delay increases [9]. Therefore, the benefit of utilizing subthreshold operation is primarily observed through dynamic energy savings.

1.3 Design and Implementation of Custom Subthreshold Cells

This section introduces the design process for a standard cell library operating at a subthreshold supply voltage. One of the primary reasons to form a standard cell library is to create a set of cells that can be used as basic building blocks for larger circuits. Performance enhancements to subthreshold standard cells will then be presented, resulting in improved performance through enhancements including multiple body-biasing techniques.

1.3.1 Subthreshold Standard Cell Library Design

The subthreshold and superthreshold regions of operation are highlighted in Figure 1.3. In the superthreshold region, the current is fairly linear in nature. The transistor current in the subthreshold region is exponentially dependent on V_t and the supply voltage, thus making power, delay, and current matching between two transistors exponentially dependent on these two parameters as well. This exponential dependence is a key challenge in designing circuits in subthreshold. Several of the design parameters that are affected by this challenge are process variations, noise margins, soft errors, and output voltage swings. Thus, when designing minimum-energy subthreshold circuits, these parameters play an important role.

The enabling factor for subthreshold design is energy minimization, so identifying the operating voltage range for the optimal energy forms the design basis when developing a subthreshold standard cell library. Two commonly used terms in subthreshold design are V_{min}, the voltage at which the energy of the circuit is at a minimum, and $V_{dd,limit}$, the lowest supply voltage at which the circuit can be operated. In most cases, V_{min} is greater than $V_{dd,limit}$, and V_{min} denotes the ideal supply voltage at which the circuit should be operated. The location of this minimum-energy point for any circuit is a compromise between the

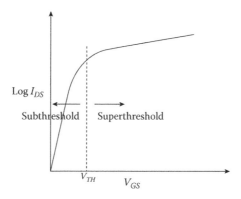

FIGURE 1.3 Transistor current characteristics in subthreshold and superthreshold.

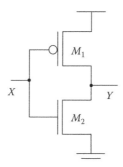

FIGURE 1.4 Schematic representation of CMOS inverter.

dynamic and leakage energies. The point of intersection of the dynamic and leakage energy curves is defined as the minimum-energy point of the circuit. Parameters including the switching activity factor α, threshold voltage V_t, effective length L_{eff}, subthreshold slope m, and drain current I_D are interdependent and should all be considered when determining the minimum-energy point of any design.

During standard cell library development, transistor sizing is performed on the fundamental building block, the base inverter, as shown in Figure 1.4. In order to identify the optimal sizing for the inverter, a seven-stage ring oscillator can be utilized. The optimal sizing is identified as the "ideal" aspect ratio (ratio of PMOS to NMOS widths) at which the charging and discharging currents are equal and a symmetrical output is observed. However, optimal sizing does not imply that a circuit operates at the minimum-energy point, so additional analysis of transistor sizing for the base inverter must be performed to determine this point.

To observe an example of the base inverter design, consider an example case with a 65 nm technology model. The base inverter operates at a $V_{dd,limit}$ of 60 mV, representing the lowest supply voltage at which the inverter can operate. To identify V_{min}, energy measurements from the ring oscillator are measured when operated at various supply voltages, and the resulting behavior can be observed in Figure 1.5 [10].

Utilizing the base inverter as a reference, additional building blocks based on the inverter aspect ratio can be designed for minimum-energy operation. By characterizing these additional cells, the necessary delay and energy consumption measurements to build a standard cell library are available. Energy consumption measurements are obtained using Equation 1.6,

$$E_{AVG} = \frac{\alpha \int_0^T i_{DD}(t)V_{DD}dt}{TransitionCount} \tag{1.6}$$

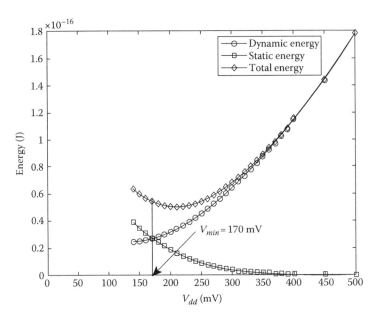

FIGURE 1.5   Inverter energy characteristics with a 5:1 aspect ratio.

where
  α is the switching activity factor
  $i_{DD}(t)$ is the instantaneous current
  V_{DD} is the supply voltage
  the *TransitionCount* is the number of transitions triggering a change in output over time period T

Example delay and energy consumption measurements of a subthreshold standard cell library in a 65 nm process operating at 0.3 V are shown in Figures 1.6 and 1.7.

The subthreshold standard cells can be used as the building blocks of a larger design. However, the simplest possible gates should be selected for use due to the characteristics of subthreshold operation and the effects of process variations. Static CMOS gates with many parallel transistors, such as wide multiplexers, do not operate well in subthreshold because the leakage through the OFF transistors can exceed the current through the ON transistor, especially considering process variation [11]. Subthreshold systems have been observed to be extremely sensitive to process variations [12,13]. Hence, it is integral to incorporate variation tolerant techniques for these designs. The effect of 3σ process variations, for the power and frequency of an inverter operating in subthreshold, is demonstrated in Figures 1.8 and 1.9. The impact of process variation is more evident at lower supply voltages, with observed variations in subthreshold gate delay as high as 300% from the nominal case [14]. For more complex gates, a supply voltage of 300 mV or higher may be required to ensure proper operation in the worst-case corners.

The exponential sensitivity to process variations makes it difficult to ensure proper operation of ratioed-logic circuits. Due to the variation in behavior, latches and registers with weak feedback devices should be avoided in subthreshold design, although conventional register design functions well. Additionally, the ratioing problem that is presented through the use of charge keepers in a dynamic circuit presents another design challenge due to process variations [11].

Several design approaches have been suggested to mitigate these process variations. For example, substrate biasing is a design technique that increases the robustness of a standard cell to process variations, while increasing performance [12]. Substrate biasing has been analyzed for a subthreshold standard

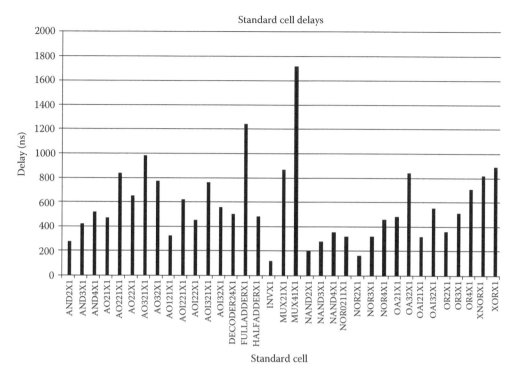

FIGURE 1.6   Gate delays of example subthreshold standard cell library.

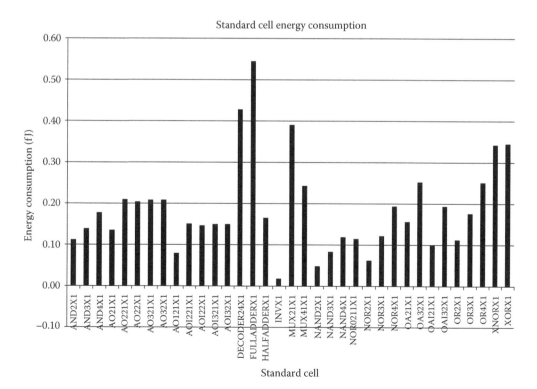

FIGURE 1.7   Energy consumption of example subthreshold standard cell library.

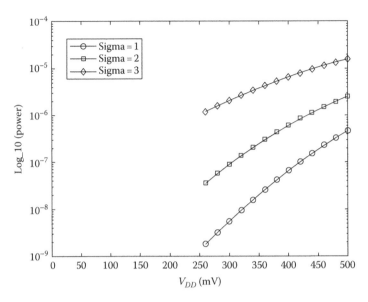

FIGURE 1.8   Effect of process variations on inverter power in subthreshold operation.

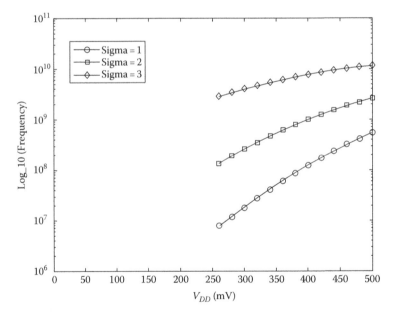

FIGURE 1.9   Effect of process variations on inverter frequency in subthreshold operation.

cell library, achieving a 75.60% reduction in process variation for the base inverter [15,16]. An alternative solution is presented by Roy et al., incorporating an adaptive β ratio modulation approach [17]. Hence, these design approaches that mitigate process variations should be utilized in the design of subthreshold circuits in order to ensure proper operation of the design under various operating conditions.

1.3.2 Performance Enhancements to Subthreshold Standard Cells

Traditional CMOS design in subthreshold operation results in performance degradation 3–4 orders of magnitude slower than traditional superthreshold CMOS designs. Thus, performance enhancement techniques to the subthreshold operation of transistors have been investigated [18].

The performance enhancement technique of substrate biasing encompasses several biasing schemes, including gate–gate, drain–drain, and supply–ground biasing. Substrate biasing is achieved by providing a bias voltage to the body of a MOS transistor. By providing a positive voltage to the body of a MOS transistor, the threshold voltage can be reduced. As the threshold voltage reduces, the drive current increases, resulting in faster charging and discharging of the load capacitances, reducing the delay of the circuit and thus improving the performance of subthreshold circuits. The threshold voltage of a four-terminal MOS transistor is given by Equation 1.7 [19],

$$V_t = V_{t0} + \gamma \left(\sqrt{\phi_0 + V_{SB}} - \sqrt{\phi_0} \right) \tag{1.7}$$

where
    V_{t0} is the threshold voltage with zero bias
    γ is the body effect parameter
    ϕ_0 is the surface potential of a MOS transistor
    V_{SB} is the source–body substrate bias

As seen in this equation, the threshold voltage can be varied by changing V_{SB}. A reduction in threshold voltage is observed when V_{SB} assumes negative values; V_{SB} becomes negative when the substrate of the device is forward biased. Thus, the threshold voltage of the device can be reduced by forward biasing the substrate of a MOS transistor.

Three substrate biasing mechanisms are shown in Figure 1.10. The biasing mechanism shown in Figure 1.10a is termed gate–gate biasing [12], where the substrate of the PMOS and NMOS are biased using a connection between respective gates and substrates. The biasing mechanism shown in Figure 1.10b is termed drain–drain biasing [20], using a connection between the respective drains and substrates. The third biasing mechanism of Figure 1.10c is termed supply–ground biasing, in which the substrate of the NMOS is biased with the supply voltage and the substrate of the PMOS is biased with ground.

Simulations based on the biasing circuits for a 65 nm technology demonstrate the frequency and power patterns for the three biasing techniques at varying supply voltages. These results are shown in Figures 1.11 and 1.12, where it can be observed that frequency and power increase exponentially with

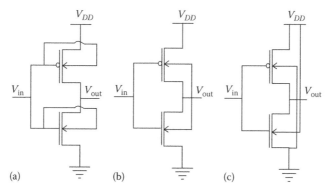

FIGURE 1.10 Inverter with various biasing schemes: (a) gate–gate biasing, (b) drain–drain biasing, (c) supply-ground biasing.

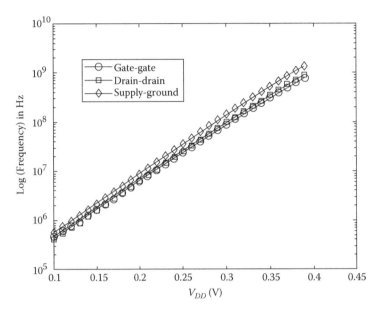

FIGURE 1.11 Frequency versus V_{DD} of an inverter for a 65 nm technology and various biasing schemes.

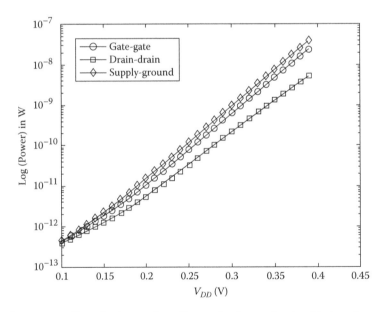

FIGURE 1.12 Power versus V_{DD} of an inverter for a 65 nm technology and various biasing schemes.

supply voltage. It can also be observed that frequency and power values are higher for supply–ground biasing compared to the other two biasing methods. This observation is present due to the fact that both the NMOS and PMOS are biased at all times in supply–ground biasing, which is not the case with the other techniques. In gate–gate biasing, either the NMOS or PMOS is biased depending on the input logic level of the circuit, while in drain–drain biasing, either the NMOS or PMOS is biased depending on the output logic level of the circuit. Furthermore, the frequency and power values are higher in the case of gate–gate biasing compared to drain–drain biasing. In the case of gate–gate biasing, the transistors are biased instantaneously when the input voltage is applied. In the case of drain–drain biasing, the output

TABLE 1.1 Substrate-Biasing Technique Measurements for 65 nm Inverter

Methodology	Delay (ns)	Energy (fJ)
Base case	29.56	0.08
Gate–gate	13.14	9.97
Drain–drain	13.94	6.48
Supply–ground	8.74	29.60

voltage gradually changes, taking the time equal to the delay of the circuit to apply to the substrate bias. The substrate bias voltage for drain–drain biasing as a function of time can be modeled as shown in Equation 1.8,

$$V_{SB}(t) = V_{SBt_d}\left(\frac{t}{T}\right) \tag{1.8}$$

where V_{SBt_d} is the substrate bias voltage at time $t = t_d$. Substituting the value of $V_{SB}(t)$ in Equation 1.8 results in the variation of threshold voltage for drain–drain biasing as a function of time, as shown in Equation 1.9.

$$V_t = V_{t0} + \gamma\left(\sqrt{\phi_0 + V_{SBt_d}\left(\frac{t}{T}\right)} - \sqrt{\phi_0}\right) \tag{1.9}$$

The expression for the drive current in subthreshold can be simplified as shown in Equation 1.10,

$$I_{ON} = A_1 \exp\left(A_2 - A_3 V_t\right) \tag{1.10}$$

where

$$A_1 = \frac{W}{L_{eff}}\mu_{eff}C_{ox}\left(m-1\right)V_T^2\left(1 - \exp\left(\frac{-V_{DS}}{V_T}\right)\right)$$

$$A_2 = \frac{V_{GS}}{mV_T}$$

$$A_3 = \frac{1}{mV_T}$$

Substituting the expression for $V_t(t)$ from Equation 1.9 into Equation 1.10 gives the expression for the drive current as a function of time, as shown in Equation 1.11:

$$I_D = A_1 \exp\left(A_2 - A_3\left(V_{t0} + \gamma\left(\sqrt{\phi_0 + V_{SBt_d}\left(\frac{t}{t_d}\right)} - \sqrt{\phi_0}\right)\right)\right) \tag{1.11}$$

The delay and energy consumption measurements for an inverter utilizing these three substrate biasing techniques are presented in Table 1.1.

The performance enhancements to subthreshold cells provide an opportunity to improve performance. However, these enhancements yield additional energy consumption in the resulting circuit when compared to traditional subthreshold design.

1.4 System-Level View of Subthreshold Design

In order to take advantage of the low-energy operation of standard cells in subthreshold, the cells must be incorporated into larger systems. These approaches utilize low-energy operation while compensating for the delay penalty incurred by subthreshold operation. Several approaches to the integration of sub-threshold cells into logic have been investigated, including the parallelization of subthreshold functional modules at the block level, as well as integration at the gate level based on the critical path method. A micropipelining approach for subthreshold circuit design has also been investigated.

1.4.1 Integration via the Parallelization of Functional Modules

In traditional designs, parallelism is used to increase the throughput of a system through the replication of circuit elements. Because there are multiple circuit elements performing the same operation, different data can be sent to each processing element simultaneously, yielding parallel output and increased throughput.

When the concept of parallelism is applied to circuits integrating subthreshold elements, it can be exploited in a different manner. In this approach, the supply voltage of the parallel processing elements is reduced to a subthreshold voltage. Thus, the parallel modules perform computations at a slower operating frequency due to the relationship between the operating frequency and the supply voltage. However, because multiple units are now working in parallel, the throughput remains unchanged from the original design, and the performance degradation resulting from subthreshold operation is mitigated. This design approach comes at the cost of an area overhead, as entire functional modules must be replicated in the final design to maintain the original throughput.

This approach to the integration of subthreshold cells has been implemented on the architecture of a 1024-bit Fast Fourier Transform (FFT) architecture [8]. In this FFT architecture, the memory elements are implemented in superthreshold while the parallel processing elements are implemented in subthreshold, similar to the block diagram shown in Figure 1.13. The serial data are distributed from the RAM to each processor, and a collection bus reads the data back into a serial stream upon completion of the operation. With this method of operation, subthreshold processing elements yield the same throughput as the superthreshold design while consuming less total power. A superthreshold FFT architecture operating at 1.0 V consumes a total power of 26.11 mW at 222 MHz for an operation, whereas a subthreshold FFT architecture operating at 0.3 V with 32 parallel processing elements consumes a total power of 8.54 mW for the same throughput as the superthreshold design. However, there is a vast area penalty with this design technique, as the superthreshold design requires a total area of 0.200 mm^2 while the parallel subthreshold design requires a total area of 1.74 mm^2.

1.4.2 Integration via the Critical Path Method

Another integration technique based on the critical path method can potentially integrate subthreshold cells with less area overhead than the replication of entire functional modules. In this approach, a design is initially synthesized utilizing high-performance superthreshold standard cells. A two-stage approach is then utilized in the integration of subthreshold cells into a high-performance design. The available slack time is determined via the critical path method. These slack times are used by the cell placement algorithm to minimize the energy of the design through the replacement of superthreshold logic, if possible.

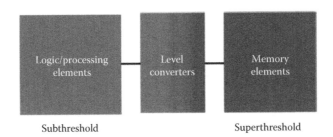

Subthreshold   Superthreshold

FIGURE 1.13   Subthreshold–superthreshold interface used throughout the hybrid design of FFT architecture.

The goal of the critical path method is to identify the slack times of each of the superthreshold cells in the high-performance design [21]. This method operates on directed acyclic graphs, where each arc on the graph has an associated delay. In order for a node to be processed, all of the node's predecessors must have completed their operation. Therefore, the earliest time that a node can start processing is equal to the length of the path with the longest delay from the start of the graph to that node; this is referred to as the early start time. Similarly, the late start time is the latest time that a node can start processing and still finish before the output is needed by its successors.

The early and late start times are used to calculate the available slack of each arc in the graph. The slack of each arc (i, j) in the graph can be calculated by using Equation 1.12, where $t_{i,j}$ is the delay of arc (i, j):

$$\text{Slack}(i, j) = \text{Late Start Time}(j) - \text{Early Start Time}(i) - t_{i,j} \tag{1.12}$$

The slack of an arc shows the additional time that the arc delay could be extended without affecting the critical path of the graph. The arc could utilize the entire slack available to it, and it would still complete its operation before the output is needed by the critical path.

The critical path method identifies the available slack time for each arc in a directed acyclic graph. In order to apply this method to digital circuits for cell placement, the circuit must first be represented as a directed acyclic graph. Because sequential logic circuits often incorporate feedback loops into the design, sequential elements must be represented as two nodes on the graph: one node for the sequential element input and one node for the sequential element output. Due to this aspect of the design, all sequential elements remain at superthreshold to ensure that the propagation delays of these cells do not change. With this design approach, standard cells are represented as the arcs of the graph, and the input and output pins of these cells are represented as the nodes of the graph, similar to the circuit and associated graph shown in Figure 1.14a and b. The delay time for each arc of the graph corresponds to the propagation delay of the respective standard cell. Thus, the superthreshold propagation delays are used during the initial slack time calculation.

With the calculated slack times, the cell placement algorithm can be used to minimize energy consumption by integrating subthreshold logic into a high-performance design while satisfying timing constraints. The cell placement algorithm utilizes a reverse-order traversal of the circuit, working from the outputs toward the inputs of the circuit. During the processing of each combinational logic block, the algorithm determines whether a level shifter would be necessary. The slack time is then compared to the additional delay that would be required to insert a subthreshold cell and a level shifter (if needed); if the available slack time exceeds the required slack time, then a subthreshold cell is inserted into the design. The cell placement algorithm is summarized in Figure 1.15.

To provide greater flexibility during the design process, a methodology to reduce the operating frequency of the circuit is provided by a performance degradation factor (PDF). When a PDF is specified, the late start time of each cell in the circuit is multiplied by this degradation factor, providing additional time for the operation to complete. This additional processing time is reflected in slack time calculations, increasing the likelihood that enough slack is present to allow for cell replacement.

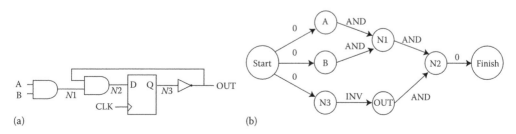

(a)   (b)

FIGURE 1.14 (a) Sequential circuit incorporating a feedback loop. (b) Directed acyclic graph representation of the sequential circuit with flip-flop input and output divided into two nodes.

Algorithm: Greedy-Subthreshold Cell Placement

Require: Completion of critical path method for slack times
for *all arcs in reverse-order traversal* **do**
        $RequiredSlack \leftarrow t_{p,subthreshold} - t_{p,superthreshold}$
        if *LevelShifterRequired* **then**
            $RequiredSlack \leftarrow RequiredSlack + t_{p,sub\_to\_gate\_levelshifter}$
        end if
        if *RequiredSlack $\geq$ AvailableSlack* **then**
            Replace superthreshold cell with subthreshold cell
        end if
        Update late start time of arc
end for

FIGURE 1.15 Greedy-subthreshold cell placement algorithm pseudocode integrating subthreshold combinational logic.

In order to interface the cells operating at different supply voltages, the use of level shifters is required. This level shifter must convert a subthreshold logic level to a superthreshold logic level. Conventional level shifter design has been labeled impractical for subthreshold design because of the contention between the pull-up network and pull-down network; the NMOS transistors, being driven by the weak subthreshold voltage, cannot overcome the drive strength of the PMOS transistors [22]. Alternative approaches to level shifter design in subthreshold have been presented, including a constant current mirror with cross-body ties [23]. While this proposed subthreshold level shifter design achieves the desired functionality, the operation comes at a high-energy cost due to the constant presence of a static current path from the supply voltage to ground; this energy overhead renders it impractical for use in the proposed design methodology.

Less energy overhead is observed in a proposed contention-mitigated level shifter, adding quasi-inverters to conventional level shifter design [24]. In this level shifter, as shown in Figure 1.16, contention between the pull-up network and pull-down network that is present in conventional level shifter design is reduced due to the fact that the logical values of node A and B are established faster by the quasi-inverter structure. This contention results in crowbar current flows during switching events in the conventional level shifter; hence, reducing this contention results in less crowbar current flow, directly reducing the

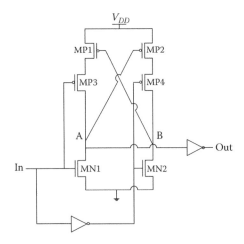

FIGURE 1.16 Transistor schematic representation of a contention-mitigated level shifter. (From Tran, C.Q. et al., Low-power high-speed level shifter design for block-level dynamic voltage scaling environment, in *International Conference on Integrated Circuit Design and Technology*, pp. 229–232, May 9–11, 2005, Austin, TX.)

energy consumption of the level shifter. The lower energy consumption makes this level shifter suitable for use in the integration of subthreshold cells at the gate level.

When the subthreshold level shifter is implemented in a 65 nm technology, the level shifter operates with a delay of 68.10 ns while consuming 3.23 fJ. A 1:10 aspect ratio was utilized for level shifters that convert a subthreshold voltage level, compensating for the weak input-gate voltage of the NMOS transistors. The necessity of level shifters adds delay, energy, and area overhead to the optimized design.

1.4.3 Micropipelining Approach for Subthreshold Circuit Design

In order to enhance the speed of subthreshold circuits, a micropipelined asynchronous network of programmable logic arrays (PLAs) has been investigated [25]. By utilizing PLAs of a fixed size, the constant delay of a dynamic PLA over all input combinations remains consistent across all PLAs in the network. Thus, there is a predictable circuit delay regardless of the circuit input. A handshaking protocol is utilized across all PLA levels, where later pipelined stages await a completion signal from previous stages before performing computations.

In order to adjust for inter-die and intra-die process, supply voltage, and temperature variations, bulk voltage adjustments are performed in a closed-loop fashion through the use of a charge pump and a phase detector [26]. The PLAs are clustered into interconnected networks, with each cluster sharing a common Nbulk node. A self-adjusting body-biasing circuit is utilized to phase lock the circuit delay of each of these PLA networks to a globally distributed beat clock.

This micropipelining architecture has been observed to provide a speed up of $7\times$ with an area penalty of 47%, yielding an energy improvement of about $4\times$ for subthreshold designs.

1.5 Applications of Subthreshold Design

Through the integration of subthreshold cells into circuit designs, low-energy computing is a reality. However, the application spectrum of subthreshold design is not only limited to low-energy computing. This section will investigate both traditional and unique applications of subthreshold computing.

Traditional applications of subthreshold design include energy-constrained applications [27]. In these designs, operating frequency is a lower priority than energy consumption, so the performance degradation resulting from subthreshold operation is tolerable. Applications such as watch battery life and hearing aids fall under this category. In fact, the first application that needed to limit the power consumption of integrated circuits at the microwatt level was the electronic watch, bringing subthreshold current to the attention of the digital design community.

Another energy-constrained application that has benefitted from subthreshold design is the microsensor node. These sensors are used for a variety of purposes, including the monitoring of habitats [28] and health [29]. Because these sensors are typically located in remote locations, extending the battery lives of these devices is critical. The typical behavior of these sensors also lend themselves to subthreshold operation; many microsensor nodes remain in an idle or sleep state for a majority of the time and become active for short bursts of time, making a measurement and then returning to the idle state. Thus, the devices do not have stringent performance requirements, so the performance degradation in subthreshold is not a significant issue.

While energy-constrained applications may be the traditional area for subthreshold computing, there are other unique applications that can incorporate the use of subthreshold devices as well. For example, cryptographic circuits provide another field of study that could benefit from the incorporation of subthreshold computing, potentially providing protection from side-channel attacks [10]. A side-channel attack is an attack on the physical implementation of a cryptographic system. Timing information, electromagnetic radiation, and power consumption are side channels that can be exploited to gain information of the cryptographic system, possibly revealing the secret keys used in the circuits.

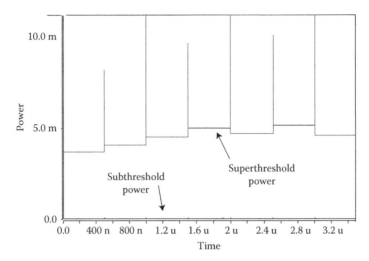

FIGURE 1.17   Subthreshold and superthreshold cryptographic multiplier power trace comparison.

Reducing the signal size and introducing noise can be an effective solution in increasing resistance against side-channel power attacks. It is effective because by reducing the signal-to-noise ratio, the attacker requires a larger amount of sample data to implement the attack. Subthreshold power consists of 80% leakage power and 20% active power. Power corresponding to useful computation (active power) in subthreshold is insignificant when compared to superthreshold where active power corresponds to 99% of the total power consumed. Inherently, the signal-to-noise ratio of a subthreshold system is significantly lower than its superthreshold counterpart. Thus, subthreshold cryptographic systems are less prone to power analysis attacks, as the reduced supply voltage and larger leakage current guarantee power variations and signal-to-noise ratios that are orders of magnitude less than those in the superthreshold case. Hence, the power variations are much more difficult to measure.

To demonstrate the susceptibility of superthreshold circuits to power attacks and the benefits of subthreshold operation, consider the power trace graphs shown in Figure 1.17 [10]. These graphs show the resulting power analysis of a cryptographic multiplier. As can be seen from the graph, the subthreshold power is very small as compared to the superthreshold case. At such low-power levels, it becomes increasingly difficult for the attacker to mount a power analysis attack. The signal magnitude becomes comparable to noise, making it very difficult for an attacker to correlate the circuit outputs to a change in inputs.

In addition to the unique applications of subthreshold computing, hybrid design methodologies that integrate subthreshold cells at the gate level open the door to a variety of opportunities. Because the gate-level integration based on the critical path method does not affect the operating frequency of the circuit, these cells can be integrated into high-performance designs. Thus, subthreshold computing could become commonplace in high-performance processors or any device that has noncritical path logic with enough slack time to allow for the insertion of a functionally equivalent subthreshold cell.

1.6 Summary

This chapter introduced the concept of subthreshold computing, discussing the design and implementation of a subthreshold standard cell library. Due to the significant performance degradation of subthreshold operation, performance enhancements to subthreshold standard cells have been investigated. In order to take advantage of the low-energy operation of subthreshold designs, an integration technique must be selected; traditional approaches have integrated subthreshold cells as the block level or gate level. The application spectrum of subthreshold computing has also been discussed. Traditional applications of subthreshold computing have restricted its use to energy-constrained applications, but other areas including cryptographic circuits can benefit from this design technique as well. With the integration of subthreshold cells now taking place at the gate level, the application spectrum of subthreshold computing will continue to expand.

Acknowledgments

The authors would like to acknowledge Hrishikesh Kanitkar from Ramtron International, Sumanth Amarchinta, Dr. James Moon, and Dr. Mike Hewitt from Rochester Institute of Technology for providing their input and support in the design of a subthreshold standard cell library used throughout the examples of this chapter.

References

1. S. Hanson, B. Zhai, K. Bernstein, D. Blaauw, A. Bryant, L. Chang, K. K. Das, W. Haensch, E. J. Nowak, and D. Sylvester. Ultralow-voltage, minimum-energy CMOS. *IBM Journal of Research and Development*, 50(4–5):469–490, 2006.
2. R. Swanson and J. Meindl. Ion-implanted complementary MOS transistors in low-voltage circuits. *Solid-State Circuits Conference. Digest of Technical Papers. IEEE International*, XV, pp. 192–193, Stanford, CA, 1972.
3. R. G. Deslinski, M. Wieckowski, D. Blaauw, D. Sylvester, and T. Mudge, Nearthreshold computing: Reclaiming Moore's law through energy efficient integrated circuits. *Proceedings of the IEEE*, 98(2):253–266, 2010.
4. B. H. Calhoun and D. Brooks. Can subthreshold and near-threshold circuits go main-stream? *IEEE Micro*, 30(4):80–85, 2010.
5. Y. Taur and T. Ning. *Fundamentals of Modern VLSI Devices*. Cambridge University Press, Cambridge, U.K., 2009.
6. J. Uyemura. *Fundamentals of MOS Digital Integrated Circuits*. Addison-Wesley Publishing Company, Boston, MA, 1988.
7. W. Wolf. *Modern VLSI Design*. Prentice Hall PTR, Upper Saddle River, NJ, 2002.
8. M. Henry and L. Nazhandali. Hybrid super/subthreshold design of a low power scalable-throughput FFT architecture. *Lecture Notes in Computer Science*, 5409:278–292, 2008.
9. J. M. Rabaey, A. Chandrakasan, and B. Nikolic. *Digital Integrated Circuits: A Design Perspective*. Prentice Hall Electronics and VLSI Series, Upper Saddle River, NJ, 2003.
10. H. Kanitkar. *Subthreshold Circuits: Design, Implementation and Application*. RIT Digital Media Library, New York, September 2008.
11. N. Weste and D. Harris. *CMOS VLSI Design: A Circuits and Systems Perspective*. Addison-Wesley, Boston, MA, 2010.
12. H. Soeleman, K. Roy, and B. Paul. Robust ultra-low power sub-threshold DTMOS logic. In *Proceedings of the International Symposium on Low Power Electronics and Design*, pp. 25–30, West Lafayette, IN, 2000.
13. E. Krimer, R. Pawlowski, M. Erez, and P. Chiang. Synctium: A near-threshold stream processor for energy-constrained parallel applications. *Computer Architecture Letters*, 9(1):21–24, 2010.
14. B. Zhai, S. Hanson, D. Blaauw, and D. Sylvester. Analysis and mitigation of variability in subthreshold design. In *Proceedings of the 2005 International Symposium on Low Power Electronics and Design*, pp. 20–25, San Diego, CA, August 8–10, 2005.

15. S. Amarchinta, H. Kanitkar, and D. Kudithipudi. Robust and high performance subthreshold standard cell design. In *52nd IEEE International Midwest Symposium on Circuits and Systems*, pp. 1183–1186, Cancun, Mexico, August 2009.

16. S. Amarchinta and D. Kudithipudi. Performance enhancement of subthreshold circuits using substrate biasing and charge-boosting buffers. *Proceedings of the 20th ACM Great Lake Symposium on VLSI (GLSVLSI 2010)*, Providence, RI, May 2010.

17. K. Roy, J. P. Kulkarni, and M. Hwang. Process-tolerant ultralow voltage digital subthreshold design. In *IEEE Topical Meeting on Silicon Monolithic Integrated Circuits in RF Systems*, pp. 42–45, Orlando, FL, January 23–25, 2008.

18. S. Amarchinta. *High Performance Subthreshold Standard Cell Design and Cell Placement Optimization*. RIT Digital Media Library, New York, June 2009.

19. Y. P. Tsividis. *Operation and Modeling of the MOS Transistor*. McGraw-Hill, New York, 1987.

20. L. A. P. Melek, M. C. Schneider, and C. Galup-Montoro. Body-bias compensation technique for subthreshold CMOS static logic gates. In *17th Symposium on Integrated Circuits and Systems Design, SBCCI 2004*, pp. 267–272, Pernambuco, Brazil, 2004.

21. R. Rardin. *Optimization in Operations Research*. Prentice Hall, Englewood Cliffs, NJ, 1998.

22. A. Hasanbegovic and S. Aunet. Low-power subthreshold to above threshold level shifter in 90 nm process. In *NORCHIP*, pp. 1–4, Trondheim, Norway, November 16–17, 2009.

23. A. Chavan and E. MacDonald. Ultra-low voltage level shifters to interface sub and super threshold reconfigurable logic cells. In *IEEE Aerospace Conference*, pp. 1–6, Big Sky, Mt, March 1–8, 2008.

24. C. Q. Tran, H. Kawaguchi, and T. Sakurai. Low-power high-speed level shifter design for block-level dynamic voltage scaling environment. In *International Conference on Integrated Circuit Design and Technology*, pp. 229–232, Austin, TX, May 9–11, 2005.

25. N. Jayakumar, R. Garg, B. Gamache, and S. Khatri. A PLA based asynchronous micropipelining approach for subthreshold circuit design. In *Proceedings of the 43rd Design Automation Conference*, pp. 419–424, San Francisco, CA, September 2006.

26. N. Jayakumar and S. Khatri. A variation-tolerant sub-threshold design approach. In *Proceedings of the 42nd Design Automation Conference*, pp. 716–719, San Francisco, CA, June 2005.

27. B. H. Calhoun, A. Wang, and A. Chandrakasan. *Sub-Threshold Design for Ultra Low-Power Systems*. Springer, Berlin, Germany, 2006.

28. A. Cerpa, J. Elson, D. Estrin, L. Girod, M. Hamilton, and J. Zhao. Habitat monitoring: Application driver for wireless communications technology. In *Proceedings of the ACM SIGCOMM Workshop on Data Communications in Latin America and the Caribbean*, pp. 20–41, San Jose, Costa Rica, 2001.

29. L. Schwiebert, S. Gupta, and J. Weinmann. Research challenges in wireless networks of biomedical sensors. In *Mobile Computing and Networking*, pp. 151–165, Rome, Italy, 2001.

30. S. Amarchinta and D. Kudithipudi. Ultra low energy standard cell design optimization for performance and placement algorithm. *Proceedings of First International Workshop on Work in Progress in Green Computing*, Chicago, IL, August 16–18, 2010.

2

Energy-Efficient Network-on-Chip Architectures for Multi-Core Systems

Partha Pande
Washington State University

Amlan Ganguly
Rochester Institute of Technology

Sujay Deb
Washington State University

Kevin Chang
Washington State University

2.1 Introduction

Modern large-scale computing systems, such as data centers and high-performance computing (HPC) clusters are severely constrained by power and cooling costs for solving extreme-scale (or exascale) problems. This increasing power consumption is of growing concern due to several reasons, for example, cost, reliability, scalability, and environmental impact. A report from the Environmental Protection Agency (EPA) indicates that the nation's servers and data centers alone used about 1.5% of the total national energy consumed per year, for a cost of approximately $4.5 billion. The growing energy demands in data centers and HPC clusters are of utmost concern and there is a need to build efficient and sustainable computing environments that reduce the negative environmental impacts. On the other hand, continuing progress and integration levels in silicon technologies make possible complete end-user systems on a single chip. This massive level of integration makes modern multi-core chips all pervasive in domains ranging from weather forecasting and astronomical data analysis to consumer electronics, smart phones, and biological applications. Consequently, designing multi-core chips for exascale computing is perceived to be a promising alternative to traditional cluster-based solutions. As an example, driven by massive parallelism and extreme-scale integration, Intel has recently created an experimental "single-chip cloud computer" targeting energy-efficient HPC applications.

Design technologies in this era of massive integration present unprecedented advantages and challenges, the former being related to very high device densities, while the latter to soaring power dissipation and reliability issues. However, given the current trends in terms of power and performance figures, it will be difficult if not impossible to design future 1000 core platforms using the existing approaches. Instead, more out-of-the-box, nature-inspired approaches (typically pursued in other areas) should be adopted to achieve the required power-performance trade-offs.

In this chapter, we elaborate on new, far-reaching, design methodologies that can help breaking the energy efficiency wall in massively integrated single-chip computing platforms. Our aim is to look into the design of massive multi-core chips from new perspectives. We conjecture that the ideas presented in this chapter will bring a paradigm shift and introduce a new era in the design of sustainable single-chip exascale computing systems.

2.2 State of the Art in Network-on-Chip Fabrics

The network-on-chip (NoC) paradigm has emerged as a communication backbone to enable a high degree of integration in multi-core system-on-chips (SoCs) [1]. Despite their advantages, an important performance limitation in traditional NoCs arises from planar metal interconnect-based multi-hop communications, wherein the data transfer between two distant blocks causes high latency and power consumption. To alleviate this problem, insertion of long-range links in a standard mesh network using conventional metal wires has been proposed [2]. Another effort to improve the performance of multi-hop NoC was undertaken by introducing ultralow-latency and low-power express channels between communicating nodes [3,4]. But these communication channels are also basically metal wires, though they are significantly more power and delay efficient compared to their more conventional counterparts. According to the International Technology Roadmap for Semiconductors (ITRS) [5] for the longer term, improvements in metal wire characteristics will no longer satisfy performance requirements and new interconnect paradigms are needed. Different approaches have been explored already, such as 3D and photonic NoCs and NoC architectures with multiband radio frequency (RF) interconnect (RF-I) [6–8]. Though all these emerging methodologies are capable of improving the power and latency characteristics of the traditional NoC, they need further and more extensive investigation to determine their suitability for replacing and/or augmenting existing metal/dielectric-based planar multi-hop NoC architectures. Consequently, it is important to explore further alternative strategies to address the limitations of planar metal interconnect-based NoCs.

In this chapter, we show how an energy-efficient NoC can be designed by replacing multi-hop wired paths by high-bandwidth single-hop long-range wireless links. Over the last few years there have been considerable efforts in the design and fabrication of miniature antennas operating in the range of tens of GHz to hundreds of THz, opening up the possibility of designing on-chip wireless links [9–11]. Recent research has uncovered excellent emission and absorption characteristics leading to dipole-like radiation behavior in carbon nanotubes (CNTs), making them promising for use as antennas for on-chip wireless communication [11]. In this chapter, the design principles of wireless network-on-chip (WiNoC) architectures are presented. The performance benefits of these WiNoCs due to the utilization of high-speed wireless links are evaluated through cycle-accurate simulations. On-chip wireless links enable one-hop data transfers between distant nodes and hence reduce the hop counts in inter-core communication. In addition to reducing interconnect delay, eliminating multi-hop long-distance wired communication reduces energy dissipation as well.

Conventional NoCs use multi-hop packet-switched communication. At each hop the data packet goes through a complex router/switch, which contributes considerable power, throughput, and latency overhead. To improve performance, the concept of express virtual channels (EVCs) is introduced in [3]. It is shown that by using virtual express lanes to connect distant cores in the network, it is possible to avoid the router overhead at intermediate nodes, and thereby greatly improve NoC performance in terms of power, latency, and throughput. Performance is further improved by incorporating ultralow-latency, multi-drop on-chip global lines (G-lines) for flow control signals [4]. In [2], performance of NoCs has been shown to improve by insertion of long-range wired links following principles of small-world graphs [12]. Despite significant performance gains, the schemes in [2–4] still require laying out long wires across the chip and hence performance improvements beyond a certain limit may not be achievable.

The performance improvements due to NoC architectural advantages will be significantly enhanced if 3D integration is adopted as the basic fabrication methodology. The amalgamation of two emerging paradigms, namely NoCs and 3D ICs, allows for the creation of new structures that enable significant performance enhancements over traditional solutions [6,13,14]. Despite these benefits, 3D architectures pose new technology challenges such as thinning of the wafers, inter-device layer alignment, bonding, and interlayer contact patterning [15]. Additionally, the heat dissipation in 3D structures is a serious concern due to increased power density [15,16] on a smaller footprint. There have been some efforts to achieve near speed-of-light communications through on-chip wires [17,18]. Though these techniques achieve very low delay in data exchange along long wires, they suffer from significant power and area overheads from the signal conditioning circuitry. Moreover, the speed of communication is actually about a factor of one-half the speed of light in silicon dioxide. By contrast, on-chip data links at the true velocity of light can be designed using recent advances in silicon photonics [19,20]. The design principles of a photonic NoC are elaborated in [6], [20], and [48]. The components of a complete photonic NoC, for example, dense waveguides, switches, optical modulators, and detectors, are now viable for integration on a single-silicon chip. It is estimated that a photonic NoC will dissipate an order of magnitude less power than an electronic planar NoC. Although the optical interconnect option has many advantages, some aspects of this new paradigm need more extensive investigation. The speed of light in the transmitting medium, losses in the optical waveguides, and the signal noise due to coupling between waveguides are the important issues that need more careful investigation. Another alternative is NoCs with multiband RF-I [21]. Various implementation issues of this approach are discussed in [8]. In this particular NoC, instead of depending on the charging/discharging of wires for sending data, electromagnetic (EM) waves are guided along on-chip transmission lines created by multiple layers of metal and dielectric stack [21]. As the EM waves travel at the effective speed of light, low-latency and high-bandwidth communication can be achieved. This type of NoC too, is predicted to dissipate an order of magnitude less power than the traditional planar NoC with significantly reduced latency.

On-chip wireless interconnects were demonstrated first in [22] for distributing clock signals. Recently, the design of a wireless NoC based on complementary metal-oxide semiconductor (CMOS) ultra wideband (UWB) technology was proposed [23]. The particular antennas used in [23] achieve a transmission range of 1 mm with a length of 2.98 mm. Consequently, for a NoC spreading typically over a die area of 20 mm × 20 mm, this architecture essentially requires multi-hop communication through the on-chip wireless channels. Moreover, the overheads of a wireless link may not be justifiable for 1 mm range of on-chip communication compared to a wired channel. We therefore propose to use relatively long-range on-chip wireless communication data links to achieve energy-efficient and low-latency wireless NoC architectures.

2.3 Wireless NoC Architectures

In a generic wired NoC, the constituent embedded cores communicate via multiple switches and wired links. This multi-hop communication results in data transfers with high energy dissipation and latency. To alleviate this problem, we propose long-distance high-bandwidth wireless links between distant cores in the chip. In the following subsections, we will explain the design of a scalable architecture for WiNoCs of various system sizes.

2.3.1 Topology

Modern complex network theory [24] provides us with a powerful method to analyze network topologies and their properties. Between a regular, locally interconnected mesh network and a completely random Erdõs-Rényi topology, there are other classes of graphs [24], such as small-world and scale-free graphs. Networks with the small-world property have a very short average path length, which is commonly measured as the number of hops between any pair of nodes. The average shortest path length of small-world graphs is bounded by a polynomial in $\log(N)$, where N is the number of nodes, which makes them particularly interesting for efficient communication with minimal resources [25,26]. This feature of small-world graphs makes them particularly attractive for constructing scalable WiNoCs. Most complex networks, such as social networks, the Internet, as well as certain parts of the brain exhibit the small-world property. A small-world topology can be constructed from a locally connected network by rewiring connections randomly to any other node, which creates shortcuts in the network [27]. These random long-range links between nodes can also be established following probability distributions depending on the distance separating the nodes [28]. It has been shown that such "shortcuts" in NoCs can significantly improve the performance compared to locally interconnected mesh-like networks [2,26] with fewer resources than a fully connected system.

Our goal here is to use the "small-world" approach to build a highly efficient NoC based on both wired and wireless links. Thus, for our purpose, we first divide the whole system into multiple small clusters of neighboring cores and call these smaller networks subnets. As subnets are smaller networks, intrasubnet communication will have a shorter average path length than a single NoC spanning the whole system. Figure 2.1a shows a subnet with mesh topology. This mesh subnet has NoC switches and links as in a standard mesh-based NoC. The cores are connected to a centrally located hub through direct links and the hubs from all subnets are connected in a second-level network forming a hierarchical network. This upper level of the hierarchy is designed to have characteristics of small-world graphs. Due to a limited number of possible wireless links, as discussed in later subsections, neighboring hubs are connected by traditional wired links forming a bidirectional ring and a few wireless links are distributed between hubs separated by relatively long distances. Reducing long-distance multi-hop wired communication is essential in order to achieve the full benefit of on-chip wireless networks for multi-core systems. As the links are initially established probabilistically, the network performance might not be optimal. Hence, after the initial placement of the wireless links, the network is further optimized for performance by using simulated

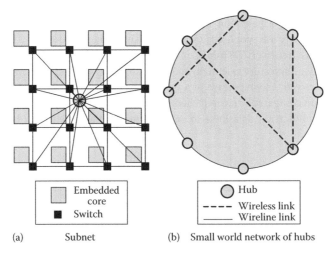

| (a)   Subnet | (b)   Small world network of hubs |

FIGURE 2.1   (a) Mesh topology of subnet with a hub connected to all switches in the subnet. (b) Network topology of hubs connected by a small-world graph with both wired and wireless links.

annealing (SA) [29]. The particular probability distribution and the heuristics followed in establishing the network links are described in Section 2.3.2. Key to our approach is establishing optimal overall network topology under given resource constraints, that is, a limited number of wireless links. Figure 2.1b shows a possible interconnection topology with eight hubs and three wireless links. Instead of the ring used in this example, the hubs can be connected in any other possible interconnect architecture. The size and number of subnets are chosen such that neither the subnets nor the upper level of the hierarchy become too large. This is because if either level of the hierarchy becomes too large then it causes a performance bottleneck by limiting the data throughput in that level. However, since the architecture of the two levels can be different causing their traffic characteristics to differ from each other, the exact hierarchical division can be obtained by performing system level simulations as shown in Section 2.4.3.

We propose a hybrid wired/wireless NoC architecture. The hubs are interconnected via both wireless and wired links while the subnets are wired only. The hubs with wireless links are equipped with wireless base stations (WBs) that transmit and receive data packets over the wireless channels. When a packet needs to be sent to a core in a different subnet, it travels from the source to its respective hub and reaches the hub of the destination subnet via the small-world network consisting of both wireless and wired links, where it is then routed to the final destination core. For inter-subnet and intra-subnet data transmission, wormhole routing is adopted. Data packets are broken down into smaller parts called flow control units or flits [30]. The header flit holds the routing and control information. It establishes a path, and subsequent payload or body flits follow that path. The routing protocol is described in Section 2.3.4.

2.3.2 Wireless Link Insertion and Optimization

As mentioned earlier, the overall interconnect infrastructure of the WiNoC is formed by connecting the cores in the subnets with each other and to the central hub through traditional metal wires. The hubs are then connected by wires and wireless links such that the second level of the network has the small-world property. The placement of the wireless links between a particular pair of source and destination hubs is important as this is responsible for establishing high-speed, low-energy interconnects on the network, which will eventually result in performance gains. Initially, the links are placed probabilistically; that is, between each pair of source and destination hubs, i and j, respectively, the probability P_{ij} of having a wireless link is given by (2.1), where h_{ij} is the distance measured in number of hops along the ring and f_{ij} is the frequency of communication between the ith source and jth destination. This frequency is expressed as the percentage of traffic generated from i that is addressed to j. This frequency distribution is based on the particular application mapped to the overall NoC and is hence set prior to wireless link insertion. Therefore, the *a priori* knowledge of the traffic pattern is used to optimize the WiNoC. This optimization approach establishes a correlation between traffic distribution across the NoC and network configuration as in [31]:

$$P_{ij} = \frac{h_{ij} f_{ij}}{\sum_{i,j} h_{ij} f_{ij}}. \tag{2.1}$$

The optimization metric μ is given by

$$\mu = \sum_{i,j} h_{ij} f_{ij}. \tag{2.2}$$

In this particular case, equal weight is attached to distance as well as frequency of communication in the metric. The probabilities are normalized such that their sum is equal to one. Once the network is initialized, an optimization by means of SA heuristics is performed. Since the subnet architectures are independent of the top-level network, the optimization can be done only on the top-level network of hubs and hence the subnets can be decoupled from this step. The optimization step is necessary as the random initialization might not produce the optimal network topology. SA offers a simple, well-established, and scalable approach for the optimization process as opposed to a brute force search.

If there are N hubs in the network and n wireless links to distribute, the size of the search space S is given by

$$|S| = \left(\begin{array}{c} \binom{N}{2} - N \\ n \end{array} \right). \tag{2.3}$$

Thus, with increasing N, it becomes increasingly difficult to find the best solution by exhaustive search. In order to perform SA, a metric has been established, which is closely related to the connectivity of the network. The metric to be optimized is the average distance, measured in number of hops, between all source and destination hubs. To compute this metric, the shortest distances between all hub pairs are computed following the routing strategy outlined in Section 2.3.4. In each iteration of the SA process, a new network is created by randomly rewiring a wireless link in the current network. The metric for this new network is calculated and compared to the metric of the current network. The new network is always chosen as the current optimal solution if the metric is lower. However, even if the metric is higher, we choose the new network probabilistically. This reduces the probability of getting stuck in a local optimum, which could happen if the SA process were to never choose a worse solution. The exponential probability shown in (2.4) is used to determine whether or not a worse solution is chosen as the current optimal:

$$P(\mu, \mu', T) = \exp\left(\frac{\mu - \mu'}{T} \right). \tag{2.4}$$

The optimization metrics for the current and new networks are h and h' respectively. T is a temperature parameter, which decreases with the number of optimization iterations according to an *annealing schedule*. In this work, we have used Cauchy scheduling, where the temperature varies inversely with the number of iterations [29] as

$$T = \frac{T_0}{k}, \tag{2.5}$$

where
   T is the temperature profile
   T_0 is the initial temperature
   k is the current annealing step

The convergence criterion is that the metric at the end of the current iteration differs by less than 0.1% from the metric of the previous iteration. The algorithm used to optimize the network is shown in Figure 2.2.

An important component in the design of the WiNoCs is the on-chip antenna for the wireless links. In Section 2.3.3, we describe various alternative on-chip antenna choices and their pros and cons.

2.3.3 On-Chip Antennas

Suitable on-chip antennas are necessary to establish wireless links for WiNoCs. In [9] and [46], the authors demonstrated the performance of silicon-integrated on-chip antennas for intra- and inter-chip communication. They have primarily used metal zigzag antennas operating in the range of tens of GHz. Design of an UWB antenna for inter- and intra-chip communication is elaborated in [32]. This particular antenna was used in the design of a wireless NoC [23] mentioned earlier in Section 2.2. The aforementioned antennas principally operate in the millimeter wave (tens of GHz) range and consequently their sizes are on the order of a few millimeters.

If the transmission frequencies can be increased to THz/optical range then the corresponding antenna sizes decrease, occupying much less chip real estate. Characteristics of metal antennas operating in the

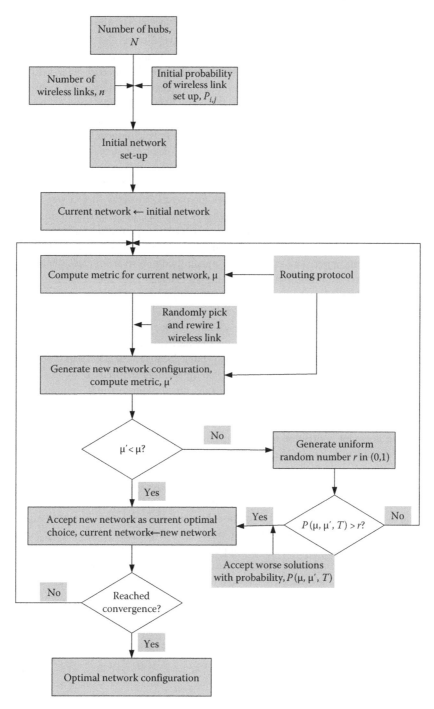

FIGURE 2.2 Flow diagram for the SA-based optimization of WiNoC architectures.

optical and near-infrared region of the spectrum of up to 750 THz have been studied [33]. Antenna characteristics of CNTs in the THz/optical frequency range have also been investigated both theoretically and experimentally [10,11]. Bundles of CNTs are predicted to enhance performance of antenna modules by up to 40 dB in radiation efficiency and provide excellent directional properties in far-field patterns [34]. Moreover, these antennas can achieve a bandwidth of around 500 GHz, whereas the antennas operating in the millimeter wave range achieve bandwidths of tens of GHz [34]. Thus, antennas operating in the THz/optical frequency range can support much higher data rates. CNTs have numerous characteristics that make them suitable as on-chip antenna elements for optical frequencies. Given wavelengths of hundreds of nanometers to several micrometers, there is a need for virtually 1D antenna structures for efficient transmission and reception. With diameters of a few nanometers and any length up to a few millimeters possible, CNTs are the perfect candidate. Such thin structures are almost impossible to achieve with traditional microfabrication techniques for metals. Virtually, defect-free CNT structures do not suffer from power loss due to surface roughness and edge imperfections found in traditional metallic antennas. In CNTs, ballistic electron transport leads to quantum conductance, resulting in reduced resistive loss, which allows extremely high current densities in CNTs, namely four to five orders of magnitude higher than copper. This enables high transmitted powers from nanotube antennas, which is crucial for long-range communications. By shining an external laser source on the CNT, radiation characteristics of multi-walled carbon nanotube (MWCNT) antennas are observed to be in excellent quantitative agreement with traditional radio antenna theory [11], although at much higher frequencies of hundreds of THz. Using various lengths of the antenna elements corresponding to different multiples of the wavelengths of the external lasers, scattering and radiation patterns are shown to be improved. Localized heaters in the CMOS fabrication process to enable localized chemical vapor deposition (CVD) of nanotubes without exposing the entire chip to high temperatures are used [47].

As mentioned earlier, the NoC is divided into multiple subnets. Hence, the WBs in the subnets need to be equipped with transmitting and receiving antennas, which will be excited using external laser sources. As mentioned in [7], the laser sources can be located off-chip or bonded to the silicon die. Hence their power dissipation does not contribute to the chip power density. The requirements of using external sources to excite the antennas can be eliminated if the electroluminescence phenomenon from a CNT is utilized to design linearly polarized dipole radiation sources [35]. But further investigation is necessary to establish such devices as successful transceivers for on-chip wireless communication.

To achieve line of sight communication between WBs using CNT antennas at optical frequencies, the chip packaging material has to be elevated from the substrate surface to create a vacuum for transmission of the high-frequency EM waves. Techniques for creating such vacuum packaging are already utilized for micro electro mechanical systems (MEMS) applications [36], and can be adopted to make creation of line of sight communication between CNT antennas viable. In classical antenna theory, it is known that the received power degrades inversely with the fourth power of the separation between source and destination due to ground reflections beyond a certain distance. This threshold separation, r_0 between source and destination antennas assuming a perfectly reflecting surface, is given by (2.6),

$$r_0 = \frac{2\pi H^2}{\lambda}.$$

(2.6)

where
   H is the height of the antenna above the reflecting surface
   λ is the wavelength of the carrier

Thus, if the antenna elements are at a distance of H from the reflective surfaces like the packaging walls and the top of the die substrate, the received power degrades inversely with the square of the distance until it is r_0. Thus, H can be adjusted to make the maximum possible separation smaller than the threshold separation r_0 for a particular frequency of radiation used. Considering the optical frequency ranges of

CNT antennas, depending on the separation between the source and destination pairs in a single chip, the required elevation is a few tens of microns only.

2.3.4 Routing and Communication Protocols

In the proposed WiNoC, intrasubnet data routing depends on the topology of the subnets. For example, if the cores within a subnet are connected in a mesh, then data routing within the subnet follows dimension order (e-cube) routing. Inter-subnet data is routed through the hubs, along the shortest path between the source and destination subnets in terms of number of links traversed. The hubs in all the subnets are equipped with a prerouting block to determine this path through a search across all potential paths between the hubs of the source and destination subnets. In the current work, paths involving only a single wireless link and none or any number of wired links on the ring are considered. All such paths as well as the completely wired path on the ring are compared and the one with the minimum number of link traversals is chosen for data transfer. For a data packet requiring inter-subnet routing, this computation is done only once for the header flit at the hub of the originating subnet. The header flit needs to have a field containing the address of the intermediate hub with a WB that will be used in the path. Only this information is sufficient as the header follows the default wireline path along the ring to that hub with the WB from its source, which is also the shortest path along the ring. Since each WB has a single, unique destination, the header reaches that destination and is then again routed via the wireline path to its final destination hub using normal ring routing. The rest of the flits follow the header, as wormhole routing is adopted in this chapter. Considering only those paths that have a single wireless link reduces computational overheads in the WB routers as it limits the search space. As the wireless links are placed as long-distance shortcuts they are always comparable in length to the diameter of the ring. Hence the probability that a path with multiple wireless links between any source/destination pair will be shorter than paths with a single wireless link is extremely low. So in order to achieve the best trade-off between the router complexity and network performance, only paths with single wireless link are considered. Also, if two alternatives have the same number of hops, the one with the wireless link is chosen, as this will have less energy dissipation. In this routing scheme, the path is predetermined at the source hub and hence, no cycles are possible. Consequently, there is no possibility of a deadlock or livelock.

An alternative routing approach is to avoid the one-time evaluation of the shortest path at the original source hub and adopt a distributed routing mechanism. In this scenario, the path is determined at each node by checking for the existence of a wireless link at that node, which if taken will shorten the path length to the final destination. If this wireless link does not exist or shorten the path in comparison to the wireline path from that node, then the default routing mechanism along the ring is followed to the next node. This mechanism performs a check at every node by computing and comparing the path lengths by using the default wireline routing or the wireless link. The adopted centralized routing performs all the checks at the original source hub, which includes all the wireless links and the wireline path from the source to the destination. We will present the comparative performance evaluation of these two schemes later in Section 2.4.3.

By using multiband laser sources to excite CNT antennas, different frequency channels can be assigned to pairs of communicating subnets. This will require using antenna elements tuned to different frequencies for each pair, thus creating a form of frequency division multiplexing (FDM) creating dedicated channels between a source and destination pair. This is possible by using CNTs of different lengths, which are multiples of the wavelengths of the respective carrier frequencies. High directional gains of these antennas, demonstrated in [11,34], aid in creating directed channels between source and destination pairs. In [37], 24 continuous wave laser sources of different frequencies are used. Thus, these 24 different frequencies can be assigned to multiple wireless links in the WiNoC in such a way that a single frequency channel is used only once to avoid signal interference on the same frequencies. This enables concurrent use of multiband channels over the chip. The number of wireless links in the network can therefore vary from 24 links, each with a single frequency channel, to a single link with all 24 channels. Assigning

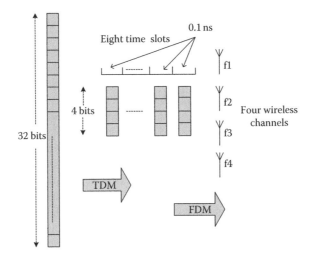

FIGURE 2.3   Adopted communication protocol for the wireless channel.

multiple channels per link increases the link bandwidth. Currently, high-speed silicon-integrated Mach-Zehnder optical modulators and demodulators, which convert electrical signals to optical signals and vice versa, are commercially available [38]. The optical modulators can provide 10 Gbps data rate per channel on these links. At the receiver a low noise amplifier (LNA) can be used to boost the power of the received electrical signal, which will then be routed into the destination subnet. As noted in [37], this data rate is expected to increase manifold with future technology scaling. The modulation scheme adopted is noncoherent on–off keying (OOK), and therefore does not require complex clock recovery and synchronization circuits. Due to limitations in the number of distinct frequency channels that can be created through the CNT antennas, the flit width in NoCs is generally higher than the number of possible channels per link. Thus, to send a whole flit through the wireless link using a limited number of distinct frequencies, a proper channelization scheme needs to be adopted. In this work, we assume a flit width of 32 bits. Hence, to send the whole flit using the distinct frequency channels, time division multiplexing (TDM) is adopted. The various components of the wireless channel viz., the electro-optic modulators, the TDM modulator/demodulator, the LNA, and the router for routing data on the network of hubs are implemented as a part of the WB. Figure 2.3 illustrates the adopted communication mechanism for the inter-subnet data transfer. In this WiNoC example, we use a wireless link with four frequency channels. In this case, one flit is divided into eight 4 bit nibbles, and each nibble is assigned a 0.1 ns timeslot, corresponding to a bit rate of 10 Gbps. The bits in each nibble are transmitted simultaneously over four different carrier frequencies. The routing mechanism discussed in this section is easily extendable to incorporate other addressing techniques like multicasting. Performance of traditional NoC architectures incorporating multicasting have been already investigated [39], and it can be similarly used to enhance the performance of the WiNoC developed in this work. For example, let us consider a subnet in a 16-subnet system, which tries to send packets to three other subnets such that one of them is diagonally opposite to the source subnet and the other two are on either side of it. In absence of long-range wireless links, using multicasting the zero load latency for the delivery of a single flit is 9 cycles whereas without multicasting the same flit will need 11 cycles to be delivered to the respective destinations. Here, the communication takes place only along the ring. However, if a wireless link exists along the diagonal from the source to the middle destination subnet then with multicasting the flit can be transferred in five cycles if there are eight distinct channels in the link. Four cycles are needed to transfer a 32 bit flit to the diagonally opposite hub via the wireless links and one more hop along the ring to the final destinations on either side. The efficiency of using multicasting varies with number of channels in the link as it governs the bandwidth of the wireless link.

2.4 Experimental Results

In this section, we analyze the characteristics of the proposed WiNoC architectures and study trends in their performance with scaling of system size. For our experiments, we have considered three different system sizes, namely 128, 256, and 512 cores on a die of size 20 mm × 20 mm. We observe results of scaling up the system size by increasing both the number of subnets as well as the number of cores per subnet. Hence, in one scenario, we have considered a fixed number of cores per subnet to be 16 and varied the number of subnets between 8, 16, and 32. In the other case, we have kept the number of subnets fixed at 16 and varied the size of the subnets from 8 to 32 cores. These system configurations are chosen based on the experiments explained later in Section 2.4.3. Establishment of wireless links using SA, however, depends only on the number of hubs on the second level of the network.

2.4.1 Establishment of Wireless Links

Initially the hubs are connected in a ring through normal wires and the wireless links are established between randomly chosen hubs following the probability distribution given by (2.1). We then use SA to achieve an optimal configuration by finding the positions of the wireless links that minimize the average distance between all source and destination pairs in the network. We followed the same optimization methodology for all the other networks. The corresponding average distances for the optimized networks with different system sizes are shown in Table 2.1. It should be noted that the particular placement of wireless links to obtain the optimal network configuration is not unique because of symmetric considerations in our setup, that is, there are multiple configurations with the same optimal performance.

In order to establish the performance of the SA algorithm used, we compared the resultant optimization metric with the metric obtained through exhaustive search for the optimized network configuration for various system sizes. The SA algorithm produces network configurations with total average hop count exactly equal to that generated by the exhaustive search technique for the system configurations considered in this chapter. However, the obtained WiNoC configuration in terms of topology is nonunique as different configurations can have the same average hop count. Figure 2.4a shows the number of iterations required to arrive at the optimal solution with SA and exhaustive search algorithms. Clearly, the SA algorithm converges to the optimal configuration much faster than the exhaustive search technique. This advantage will increase for larger system sizes. Figure 2.4b shows the convergence of the metric for different values of the initial temperature to illustrate that the SA approach converges robustly to the optimal value of the average hop count with numerical variation in the temperature. This simulation was performed for a system with 32 subnets with one wireless link. With higher values of the initial temperature, it can take longer to converge. Naturally, for large enough values of the initial temperature, the metric does not converge. On the other hand, lower values of the initial temperature make the system converge faster but at the risk of getting stuck in a local optimum. Using the network configurations developed in this subsection, we will now evaluate the performance of the WiNoC based on well-established performance metrics.

TABLE 2.1   Average Distance for Optimized WiNoCs

No. of Subnets (N)	Average Distance (Hops)		
	1 Wireless Link	6 Wireless Links	24 Wireless Links
8	1.7188	1.3125	1.1250[a]
16	3.2891	2.1875	1.5625
32	6.3301	3.8789	2.6309

[a] In case of 8 subnets, only 12 wireless links are used with 2 channels per link.

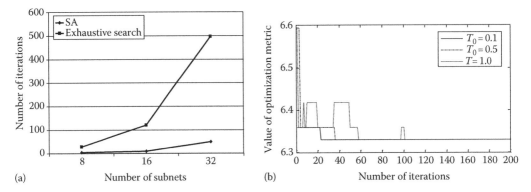

FIGURE 2.4   (a) Number of iterations required to reach optimal solution by the SA and exhaustive search methods.
(b) Convergence with different temperatures.

2.4.2 Performance Metrics

To characterize the performance of the proposed WiNoC architectures, we consider three network parameters: latency, throughput, and energy dissipation. Latency refers to the number of clock cycles between the injection of a message header flit at the source node and the reception of the tail flit at the destination. Throughput is defined as the average number of flits successfully received per embedded core per clock cycle. Energy dissipation per packet is the average energy dissipated by a single packet when routed from the source to destination node; both the wired subnets and the wireless channels contribute to this. For the subnets, the sources of energy dissipation are the inter-switch wires and the switch blocks. For the wireless channels, the main contribution comes from the WBs, which includes antennas, transceiver circuits, and other communication modules like the TDM block and the LNA. Energy dissipation per packet, E_{pkt}, can be calculated according to (2.7):

$$E_{pkt} = \frac{N_{intrasubnet}E_{subnet,hop}h_{subnet} + N_{intersubnet}E_{sw}h_{sw}}{(N_{intrasubnet} + N_{intersubnet})}. \tag{2.7}$$

In (2.7), $N_{intrasubnet}$ and $N_{intersubnet}$ are the total number of packets routed within the subnet and between subnets respectively. $E_{subnet,hop}$ is the energy dissipated by a packet traversing a single hop on the wired subnet including a wired link and switch, and E_{sw} is the energy dissipated by a packet traversing a single hop on the second network level of the WiNoC, which has the small-world properties. E_{sw} also includes the energy dissipation in the core to hub links. In (2.7), h_{subnet} and h_{sw} are the average number of hops per packet in the subnet and the small-world network.

2.4.3 Performance Evaluation

The network architectures developed earlier in this section are simulated using a cycle-accurate simulator that models the progress of data flits accurately per clock cycle accounting for flits that reach destination as well as those that are dropped. One hundred thousand iterations were performed to reach stable results in each experiment, eliminating the effect of transients in the first few thousand cycles.

The mesh subnet architecture considered is shown in Figure 2.1a. The width of all wired links is considered to be same as the flit size, which is 32 in this chapter. The particular NoC switch architecture, adopted from [40] for the switches in the subnets, has three functional stages, namely, input arbitration, routing/switch traversal, and output arbitration. The input and output ports including the ones on the wireless links have four virtual channels per port, each having a buffer depth of two flits [40]. Each packet consists of 64 flits. Similar to the intrasubnet communication, we have adopted wormhole routing in

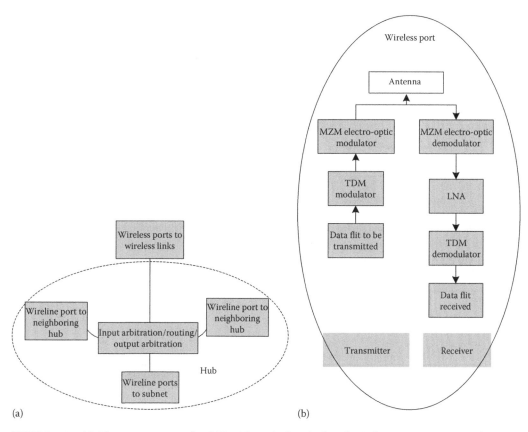

FIGURE 2.5 (a) The components of a WB with multiple wired and wireless ports at input and output. (b) A wireless port.

the wireless channel too. Consequently, the hubs have similar architectures as the NoC switches in the subnets. Hence, each port of the hub has same input and output arbiters, and equal number of virtual channels with same buffer depths as the subnet switches. The number of ports in a hub depends on the number of links connected to it. The hubs also have three functional stages, but as the number of cores increases in a subnet the delays in arbitration and switching for some cases are more than a clock cycle. Depending on the subnet sizes, traversal through these stages need multiple cycles and this has been taken into consideration while evaluating overall latency of the WiNoC. The wireless ports of the WBs are assumed to be equipped with antennas, TDM modules, and electro-optic modulators and demodulators. The various components of a WB are shown in Figure 2.5. A hub consisting of only ports to wired links is also highlighted in the figure to emphasize that a WB has additional components compared to a hub. A simple flow control mechanism is adopted uniformly for wireless links in which, the sender WB stops transmitting flits only when a *full* signal is asserted from the receiver WB. This full signal is embedded in a control flit sent from the receiver to the sender only when the receiver buffer is filled above a predefined threshold. When the full signal is asserted, flits do not move and are blocked spanning multiple switches or hubs. This in turn can block other messages in the network as in wormhole routing. In case all buffers are full, the new injected packets from the cores are dropped until new buffer space is available. A more advanced flow control mechanism could be incorporated to improve WiNoC performance further [4]. The NoC switches, the hubs, and the wired links are driven with a clock of frequency 2.5 GHz.

Figure 2.6 shows throughput and latency plots as a function of injection load for a system with 256 cores divided into 16 subnets, each with 16 cores. The delays incurred by the wired links from the cores to

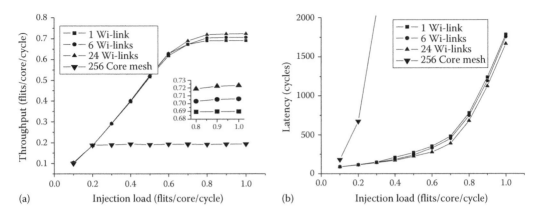

FIGURE 2.6   (a) Throughput and (b) latency of 256 core WiNoCs with different numbers of wireless links.

TABLE 2.2   Delays on Wired Links in the WiNoCs

System Size	No. of Subnets	Subnet Size	Core-Hub Link Delay (ps)	Inter-Hub Link Delay (ps)
128	8	16	96	181/86[a]
	16	8	60	86
256	16	16	60	86
512	16	32	60	86
	32	16	48	86/43[a]

[a] For 8 and 32 subnets, the inter-subnet distances are different along the two planar directions.

the hub for varying number of cores in the subnets for different system sizes are shown in Table 2.2. The delays in the inter-hub wires for varying number of subnets are also shown. As can be seen these delays are all less than the clock period of 400 ps, and it may be noted that the lengths of both core-to-hub and inter-hub wireline links will reduce with increase in the number of subnets as then each subnet becomes smaller in area and the subnets also come closer to each other. The delays incurred by the electro-optic signal conversions with the Mach-Zehnder Modulator (MZM) devices are 20 ps. When computing the overall system latency and throughput of the WiNoCs, the delays of these individual components are taken into account. This particular hierarchical topology was selected as it provided optimum system performance. Figure 2.7 shows the saturation throughputs for alternative ways of dividing the 256 core WiNoC into different numbers of subnets with a single wireless link. As can be seen from the plot all alternative configurations achieve worse saturation throughput. The same trend is observed if we vary the number of wireless links. Using the same method, the suitable hierarchical division that achieves best performance is determined for all the other system sizes. For system sizes of 128 and 512, the hierarchical divisions considered here achieved much better performance compared to the other possible divisions with either lower or higher number of subnets.

By varying the number of channels in the wireless links, various WiNoC configurations are created. We have considered WiNoCs with 1, 6, and 24 wireless links in our experiments. Since the total number of frequencies considered in this work is 24, the number of channels per link is 24, 4, and 1, respectively. As can be seen from Figure 2.6, the WiNoCs with different possible configurations outperform the single wired monolithic flat mesh architecture. It can also be observed that with increasing number of wireless links, throughput improves slightly. It should be noted that even though increasing the number of links does increase the number of concurrent wireless communication links, the bandwidth on each link decreases as the total number of channels is fixed by the number of off-chip laser sources. This causes the total bandwidth over all the wireless channels to remain the same. The only difference is in the degree of distribution across the network. Consequently, network throughput increases only slightly with

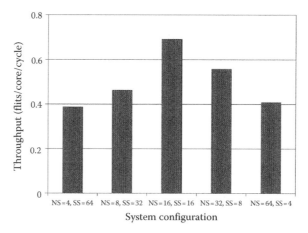

FIGURE 2.7 Throughput of 256 core WiNoC for various hierarchical configurations.

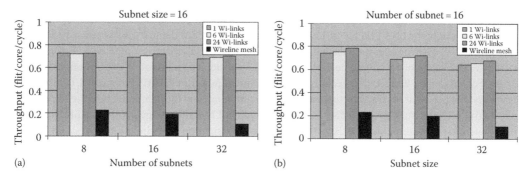

FIGURE 2.8 Saturation throughput with varying (a) number of subnets and (b) size of each subnet.

increasing number of wireless links. However, the hardware cost increases with increasing numbers of links. Thus, depending upon whether the demand on performance is critical, the designer can choose to trade off the area overhead of deploying the maximum number of wireless links possible. However, if the constraints on area overhead are really stringent then one can choose to employ only one wireless link and consequently provide more bandwidth per link and have only a little negative effect on performance.

In order to observe trends among various WiNoC configurations, we performed further analysis. Figure 2.8a shows the throughput at network saturation for various system sizes while keeping the subnet size fixed for different numbers of wireless links. Figure 2.8b shows the variation in throughput at saturation for different system sizes for a fixed number of subnets. For comparison, the throughput at network saturation for a single traditional wired mesh NoC of each system size is also shown in both of the plots. As in Figure 2.6, it may be noted from Figure 2.8 that for a WiNoC of any given size, number of subnets and subnet size, the throughput increases with increase in number of wireless links deployed.

As can be observed from the plots, the maximum achievable throughput in WiNoCs degrades with increasing system size for both cases. However, by scaling up the number of subnets, the degradation in throughput is smaller compared to when the subnet size is scaled up. By increasing the subnet size, we are increasing congestion in the wired subnets and load on the hubs and not fully using the capacity of the high-speed wireless links in the upper level of the network. When the number of subnets scales up, traffic congestion in the subnets does not get worse and the optimal placement of the wireless links makes the top-level network very efficient for data transfer. The effect on throughput with increasing system size is therefore marginal.

To determine the energy dissipation characteristics of the WiNoCs, we first estimated the energy dissipated by the antenna elements. As noted in [11], the directional gain of MWCNT antennas we propose to use is very high. The ratio of emitted power to incident power is around -5 dB along the direction of maximum gain. Assuming an ideal line-of-sight channel over a few millimeters, transmitted power degrades with distance following the inverse square law. Therefore, the received power P_R can be related to the transmitted power P_T as

$$P_R = \frac{G_T A_R}{4\pi R^2} P_T. \tag{2.8}$$

In (2.8), G_T is the transmitter antenna gain, which can be assumed to be -5 dB [11]. A_R is the area of the receiving antenna and R is the distance between the transmitter and receiver. The energy dissipation of the transmitting antennas, therefore, depends on the range of communication. The area of the receiving antenna can be found by using the antenna configuration used in [11]. It uses a MWCNT of diameter 200 nm and length 7λ, where λ is the optical wavelength. The length 7λ was chosen as it was shown to produce the highest directional gain, G_T, at the transmitter. In one of the setups in [11], the wavelength of the laser used was 543.5 nm, and hence the length of the antenna is around 3.8 μm. Using these dimensions, the area of the receiving antenna, A_R can be calculated.

The noise floor of the LNA [41] is -101 dBm. Considering the MZM demodulators cause an additional loss of up to 3 dB over the operational bandwidth, the receiver sensitivity turns out to be -98 dBm in the worst case. The length of the longest possible wireless link considered among all WiNoC configurations is 23 mm. For this length and receiver sensitivity, a transmitted power of 1.3 mW is required. Considering the energy dissipation at the transmitting and receiving antennas, and the components of the transmitter and receiver circuitry such as the MZM, TDM block and the LNA, the energy dissipation of the longest possible wireless link on the chip is 0.33 pJ/bit. The energy dissipation of a wireless link, E_{Link} is given as

$$E_{Link} = \sum_{i=1}^{m} (E_{antenna,i} + E_{transceiver,i}), \tag{2.9}$$

where

    m is the number of frequency channels in the link

    $E_{antenna,i}$ and $E_{transceiver,i}$ are the energy dissipations of the antenna element and transceiver circuits for the ith frequency in the link

The network switches and hubs are synthesized from an RTL-level design using 65 nm standard cell libraries from chip multiprocessor (CMP) [42], using Synopsys Design Vision and assuming a clock frequency of 2.5 GHz. A large set of data patterns were fed into the gate-level netlists of the network switches and hubs, and by running Synopsys^{TM} Prime Power, their energy dissipation was obtained.

The energy dissipation of the wired links depends on their lengths. The lengths of the inter-switch wires in the subnets can be found by using the formula

$$l_M = \frac{l_{edge}}{M - 1}. \tag{2.10}$$

where

    M is number of cores along a particular edge of the subnet

    l_{edge} is the length of that edge

A 20 mm $\times$ 20 mm die size is considered for all system sizes in our simulations. The inter-hub wire lengths are also computed similarly as these are assumed to be connected by wires parallel to the edges of the die in rectangular dimensions only. Hence, to compute inter-hub distances along the ring, parallel to a

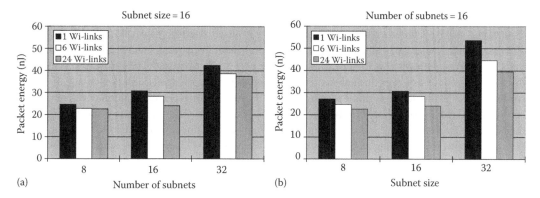

FIGURE 2.9 Packet energy dissipation with varying (a) number of subnets and (b) size of each subnet.

TABLE 2.3 Packet Energy Dissipation for Flat Wired Mesh, WiNoC, and Hierarchical G-Line NoC Architectures

System Size	Subnet Size	No. of Subnets	Flat Mesh (nJ)	WiNoC (nJ)	NoC with G-Line (nJ)
128	16	8	1319	22.57	490.3
256	16	16	2936	24.02	734.5
512	16	32	4992	37.48	1012.8

particular edge of the die, (2.10) is modified by changing M to the number of hubs along that edge and l_{edge} to the length of that particular edge. In each subnet, the lengths of the links connecting the switches to the hub depend on the position of the switches as shown in Figure 2.1a. The capacitances of each wired link, and subsequently their energy dissipation, were obtained through HSPICE simulations taking into account the specific layout for the subnets and the second level of the ring network.

Figure 2.9a and b shows the packet energy dissipation for each of the network configurations considered in this work. The packet energy for the flat wired mesh architecture is not shown as it is higher than that of the WiNoCs by orders of magnitude, and hence cannot be shown on the same scale. The comparison with the wired case is shown in Table 2.3 in the next subsection along with another hierarchical wired architecture.

From the plots, it is clear that the packet energy dissipation increases with increasing system size. However, scaling up the number of subnets has a lower impact on the average packet energy. The reason for this is that the throughput does not degrade much and the average latency per packet also does not change significantly. Hence, the data packets occupy the network resources for less duration, causing only a small increase in packet energy. However, with an increase in subnet size, the throughput degrades noticeably, and so does latency. In this case, the packet energy increases significantly as each packet occupies network resources for a longer period of time. With an increase in the number of wireless links while keeping the number of subnets and subnet size constant, the packet energy decreases. This is because higher connectivity of the network results in higher throughput (or lower latency), which means that packets get routed faster, occupy network resources for less time, and consume less energy during the transmission. Since the wireline subnets pose the major bottleneck as is made evident by the trends in the plots of Figures 2.8 and 2.9 their size should be optimized. In other words, smaller subnets imply better performance and lower packet energies. Hence, as a designer one should target to limit the size of the subnets as long as the size of the upper level of the network does not impact the performance of the overall system negatively. The exact optimal solution also depends on the architecture of the upper level of the network, which need not be restricted to the ring topology chosen in this work as an example.

The adopted centralized routing strategy is compared with the distributed routing discussed in Section 2.3.4 for a WiNoC of size 256 cores split into 16 subnets with 16 cores in each. Twenty four wireless links were deployed in the exact same topology for both cases. With distributed routing, the throughput was 0.67 flits/core/cycle, whereas with centralized routing, it was 0.72 flits/core/cycle as already noted in Figure 2.6, which is 7.5% higher. Centralized routing results in a better throughput as it finds the shortest path whereas the distributed routing uses nonoptimal paths in some cases. Hence, the distributed routing has lower throughput. The distributed routing dissipates a packet energy of 31.2 nJ compared to 30.8 nJ with centralized routing. This is because on an average, the number of path length computation with the distributed routing is more per packet, as this computation occurs at every intermediate WB. However, with centralized routing each hub has additional hardware overhead to compute the shortest path by comparing all the paths using the wireless links. This hardware area cost is discussed in Section 2.4.5.

2.4.4 Comparison with Wired NoCs

We evaluated the performance of the WiNoCs in terms of energy dissipation compared to different wired NoC architectures. As demonstrated in Section 2.4.3, with increase in system size, increasing the number of subnets while keeping the subnet size fixed is a better scaling strategy; hence, we followed that in the following analysis.

The first wired architecture considered was the conventional flat mesh architecture. Table 2.3 quantifies the energy dissipation per packet of the WiNoC and the wired architectures for various system sizes. The WiNoC configuration with 24 wireless links was chosen because it has the lowest packet energy dissipation among all the possible hybrid wired/wireless configurations. It is evident that the WiNoC consumes orders of magnitude less energy compared to the flat wired mesh network. In case of WiNoC, very small average path length due to its small-world nature and due to the low-power wireless channels the absolute value of this energy dissipation is very small.

The performance of the flat mesh NoC architectures can be improved by incorporating EVC, which connect the distant cores in the network by bypassing intermediate switches/routers. It is demonstrated that the switch/router energy dissipation of the baseline mesh architecture is improved by about 25%–38% depending on the system size by using dynamic EVCs. The energy dissipation profile is improved by another 8% over the EVC scheme by using low-swing, multi-drop, ultra low-latency global interconnect (G-lines) for the flow control signals [4]. Recently, a number of studies have shown the possibility of communicating near speed-of-light across several millimeters on a silicon substrate. Among them, low-swing, long-range, and ultralow-latency communication wires as proposed in [43] achieve higher bandwidth at lower power consumption [4]. G-lines use a capacitive pre-emphasis transmitter that increases the bandwidth and decreases the voltage swing without the need of an additional power supply. To avoid cross-talk, differential interconnects are implemented with a pair of twisted wires. A decision feedback equalizer is employed at the receiver to further increase the achievable data rate. It is evident that though introduction of EVCs improves the energy dissipation profile of a flat wired mesh NoC, the achievable performance gain is still limited compared to the gains achieved by the WiNoCs. This is because the basic architecture is still a flat mesh and the savings in energy principally arises from bypassing the intermediate NoC switches.

As a next step, we undertook a study where we compared the energy dissipation profile of the proposed hybrid NoC architecture using wireless links to that of the same hierarchical network using G-lines as long-range communication links. To do so, we replaced the wireless links of the WiNoCs by the G-lines while maintaining the same hierarchical topology with shortcuts in the upper level. Here, each G-line link is designed such that it has the same bandwidth as the wireless link it replaces. Thus, the overall throughput and end-to-end latency of the hierarchical NoC with G-line links is the same as that of the WiNoC. We performed simulations in 65 nm technology. The lumped wire resistance is 20 Ω/mm, and the capacitance is 400 fF/mm. The simulated power dissipation is found to be 0.6 mW/transmitter and 0.4 mW/receiver. In order to achieve the same bandwidth as the wireless links in our experiments, multiple G-line links are used in place of a single wireless channel between a pair of source and destination hubs.

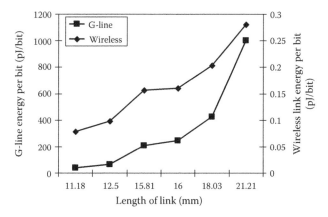

FIGURE 2.10 Energy dissipation per bit on G-line and wireless links with varying link lengths.

TABLE 2.4 Percentage of Pocket Energy Dissipation on Long-Range Links

No. of Links	1	6	24
Wireless	0.5	2.8	3.4
G-line	47.3	95.8	98.5

For example, a single G-line can sustain a bandwidth of around 2.5 Gbps for a wire length of 11 mm, whereas each wireless channel can sustain a bandwidth of 10 Gbps. Therefore, to maintain the same data rate as provided by a single wireless channel, we need four G-lines between a source and destination pair separated by 11 mm. Moreover, since each G-line works on differential signals, we will need eight wires to replace a single wireless link in this case.

The packet energy dissipation for a WiNoC and hierarchical NoC with G-line links are also shown in Table 2.3 for various system sizes. The WiNoC's energy per packet consumption is one order of magnitude less compared to the hierarchical NoC with G-line links of the same bandwidth as the wireless channels. This experiment was conducted to highlight the savings in energy dissipation due to two factors viz., the architectural innovation proposed here and the use of on-chip wireless links in place of highly optimized wired connections. The difference in energy dissipation between the flat wired mesh NoC and the hybrid NoC with G-lines arises primarily due to the architecture proposed here. The difference in energy dissipation between the WiNoCs and the hybrid NoC with G-lines is solely due to the use of wireless channels. Figure 2.10 shows the energy dissipation of a wireless link considered in this chapter and that of the G-line link as a function of communication distance between source and destination WBs considered here. This shows how high the energy dissipation of a G-line link of the same bandwidth as the wireless link is. The impact of this is reflected in the packet energy dissipation profiles shown in Table 2.3, which is obtained after full system simulation using these links. Table 2.4 shows the percentage of total packet energy dissipated on the wired and wireless links for a WiNoC with 128 cores divided into 16 subnets. The percentage of packet energy dissipated on the G-line links replacing the wireless links are also shown to signify the trade-off in energy dissipation as more wireline (G-line) links are replaced with the wireless links for a single network configuration. As shown in Tables 2.3 and 2.4, the hierarchical NoC with G-line links dissipate higher packet energy than the WiNoC and the long-distance G-line links dissipate a considerably larger proportion of that high packet energy.

Another wireline architecture developed in [2] uses long-range wired shortcuts to design a small-world network over a basic mesh topology. We considered a system size of 128 cores and 8 wireline shortcuts were optimally deployed on a basic wireline mesh following the scaling trend outlined in [2]. The chosen

WiNoC configuration was 16 subnets with 8 cores in each with 8 wireless links. The throughput of the wireline small-world NoC proposed in [2] was 0.26 flits/core/cycle, which is 18.7% more than that of a flat mesh NoC. In comparison, the WiNoC had a throughput of 0.75 flits/core/cycle. This huge gain was due to the hierarchical division of the whole NoC as well as the high-bandwidth wireless links used in creating the shortcuts. The packet energy dissipation for the NoC proposed in [2] for the configuration mentioned earlier is 984 nJ. This energy dissipation is about 25% less than the packet energy dissipation in a flat mesh. However, even this packet energy is an order of magnitude higher than that of the WiNoC for the same size of 128 cores as shown in Table 2.3.

From the previous analysis, it is clear that the proposed WiNoC architectures outperform their corresponding wired counterparts significantly in terms of all the relevant network parameters. Moreover, the WiNoC is much more energy efficient compared to an exactly equivalent hierarchical wired architecture implemented with the recently proposed high-bandwidth low-latency G-lines as the long-range communication links.

2.4.5 Comparative Analysis with Other Emerging NoC Paradigms

There are several emerging paradigms that enhance the performance of NoCs using nontraditional technology such as photonic interconnects, RF-I, and on-chip wireless communication using sub-THz links. In this subsection, we perform a comparative analysis to establish the relative performance benefits achieved by small-world wireless NoCs with respect to a subset of recently proposed photonic NoCs.

We consider two types of small-world NoC architectures. In one, the long-range links are established using on-chip wireless interconnects and in the other on-chip RF-Is are used as long-range links. Again, for the wireless interconnects, we consider two types of links, viz., THz links implemented with CNT antennas (WiNoC) and sub-THz links as used in [44]. As shown in [44], this available bandwidth can be divided into multiple frequency channels, each operating at 10–20 Gbps for the small-world NoC with sub-THz links (WiNoC_Sub-THz).

2.4.5.1 Small-World NoC with RF Interconnects (RFNoC)

Another possible way to establish the long-range link is to use multiband RF-Is . The RF NoC is designed by replacing the wireless links of the WiNoCs by RF-Is, maintaining the same hierarchical topology with shortcuts in the upper level. In 65 nm technology, it is possible to have eight different frequency channels each operating with a data rate of 6 Gbps [8]. Like the wireless links, these RF links can be used as the long-range shortcuts in the hierarchical NoC architecture. These shortcuts are optimally placed using the same SA-based optimization as used for the WiNoC.

2.4.5.2 NoCs with Optical Interconnects

NoCs with high-speed optical interconnects is another promising alternative to conventional wireline NoCs. We consider two such optical NoC architectures for the comparative study. The first one is adopted from [7] and is referred as Photonic NoC for the rest of this chapter. This NoC requires an electrical control network to configure photonic switching elements, which uses a flat wireline mesh. Photonic NoC architecture employs an optical circuit-switched network for bulk message transmission and an electronic packet-switched network for distributed control and short message exchange. The optical torus network is augmented with additional optical paths to provide path multiplicity, so that blocking can be avoided and path setup latency is accordingly reduced. Maximum path multiplicity in the Photonic NoC ensures that there is no blocking in the network. An illustrative 16 core Photonic NoC architecture with path multiplicity of 2 is presented in Figure 2.11.

The second optical NoC considered in this chapter is the Clos network proposed in [45]. Clos is a low-diameter nonblocking network. It uses multiple stages of routers to create a large nonblocking all-to-all connectivity. Clos network uses point-to-point channels instead of global shared channels resulting in

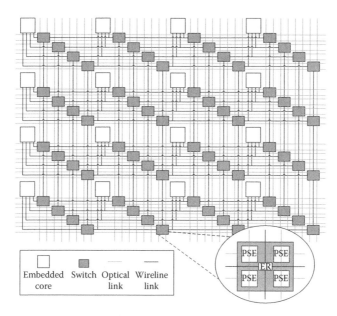

FIGURE 2.11 A representative 16 core photonic NoC with path multiplicity of 2.

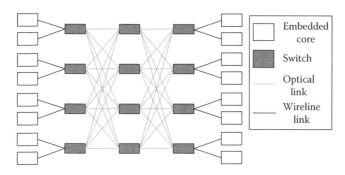

FIGURE 2.12 A 16 core photonic Clos network.

significantly less overhead compared to global photonic crossbars. A 16 core photonic Clos network is shown in Figure 2.12.

As a first step, we analyze the characteristics of the different NoC architectures in presence of a uniform random traffic and study trends in their performance with scaling of system size. For our experiments, we consider three different system sizes, namely 128, 256, and 512 cores. All the network architectures mentioned in this section are simulated using a cycle-accurate simulator, which models the progress of data flits accurately per clock cycle accounting for flits that reach their destinations as well as those that are dropped. In this work, we assume flit width of 32 bits and each packet consists of 64 flits. A self-similar traffic injection process is assumed.

For small-world architectures, the number of subnets and their sizes are chosen for optimum performance for different system sizes. A 128 core system is divided into 16 subnets with each subnet consisting of 8 cores, 256 core system is divided into 16 subnets with each subnet consisting of 16 cores, and 512 core system is divided into 32 subnets with each subnet consisting of 16 cores. For this exercise, we assumed that the cores within each subnet are connected in a mesh topology. Each core also has a direct path to the central hub. The upper level of the network is considered to be a ring with wireless shortcuts. The shortcuts are optimally deployed among the hubs following the SA methodology.

For small-world NoCs, the subnet switches and the digital components of the hubs are synthesized using 65 nm standard cell library from CMP [42] at a clock frequency of 2.5 GHz. The delays in flit traversals along all the wired interconnects that are introduced to enable the small-world NoC architecture were considered while quantifying the performance.

It is shown that 24 distinct frequency channels can be created for CNT antennas [37] used in WiNoC. In the WiNoC with sub-THz wireless links, 24 frequency channels each with a data rate of 10 Gbps is considered to provide same bandwidth as the CNT-based scheme, For WiNoC using CNT antennas, the energy dissipation of the longest possible wireless link on the chip was found to be 0.33 pJ/bit. However, the sub-THz wireless link dissipates 4.5 pJ/bit in the same topology [44]. In case of RFNoC, the energy dissipation for RF-Is in 65 nm technology is considered to be 1 pJ/bit [8].

For the Photonic NoC architecture, we have considered maximum path multiplicity for each system size for highest sustainable system throughput. The energy dissipation value for different optical components in the Photonic NoC is taken from [7].

The Clos NoC architecture uses multiple stages of small routers to create a larger nonblocking all-to-all network. As in [45], all of the Clos routers are implemented electrically and the inter-router channels are implemented with photonics. As assumed in [45], enough WDM optical channels are considered to enable a flit to be transmitted in a single cycle. The conservative energy projection values for all the optical components from [45] are used for energy calculations.

Figure 2.13 shows the achievable overall network bandwidth for all the different NoCs under consideration. The bandwidth of flat mesh is also included for comparison. It is evident that all these emerging NoCs outperform the flat mesh. Among the small-world and optical NoCs, WiNoC with both types of wireless links and Clos perform significantly better than the others for all system sizes. As the different architectures produce different peak bandwidths, the packet energy per bandwidth for different NoC architectures is considered as a relevant metric and is presented in Table 2.5. From the packet energy per bandwidth results, it can be seen that Clos NoC performs best for all system sizes followed by THz WiNoC with CNT antennas.

Figure 2.14 presents the area overhead arising out of the hubs and switches inclusive of necessary transceiver modules for the various NoC architectures. It is evident that Clos and the Photonic NoC have significantly higher area overhead compared to other NoCs. For 512 core system, the area overheads for

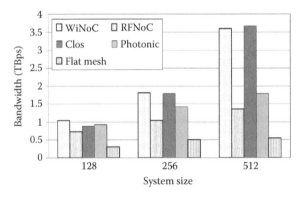

FIGURE 2.13   Achievable bandwidth for different NoC architectures with scaling of system size.

TABLE 2.5   Packet Energy per Bandwidth (nJ/TBps) for Different NoC Architectures with Scaling of System Size

System Size	Flat Mesh	WiNoC_THz	WiNoC_Sub-THz	RFNoC	Clos	Photonic NoC
128	560.0	21.0	31.3	34.1	16.6	34.1
256	721.7	13.3	19.6	25.8	9.0	29.6
512	1171	10.0	15.4	32.8	7.3	35.1

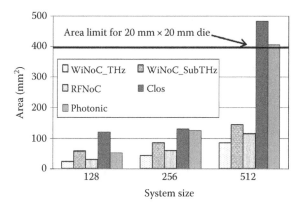

FIGURE 2.14 Area overhead for different NoC architectures with scaling of system size.

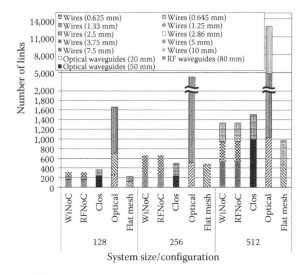

FIGURE 2.15 Total wiring, RF, and optical waveguide requirement of various lengths for a 20 mm × 20 mm die for different NoCs.

Clos and Photonic NoC even exceeds the limits of a 20 mm × 20 mm die. The area overhead of Photonic NoC can be reduced by decreasing path multiplicity, which will, however, correspondingly degrade the achievable bandwidth.

Figure 2.15 shows the total wiring, RF, and optical waveguide requirement of various lengths for the different NoC architectures considered in this work implemented in a 20 mm × 20 mm die. The wiring requirement for flat mesh architecture is also shown for comparison. It is evident that all these emerging NoCs have extra interconnect overhead with respect to a flat mesh. WiNoC and RFNoC introduce less number of extra links compared to the Photonic and Clos NoCs. In the Photonic NoC, high interconnect overhead arises mainly due to the path multiplicity, and in the Clos this happens due to the nonblocking all-to-all nature of the network.

From all the earlier analysis, it is clear that among different NoC architectures compared in this chapter, WiNoC with CNT antennas provides the best trade-off between performance and area overhead as system size scales up. This is because in WiNoC, the high-performance long-range links are optimally placed to achieve a small-world network. WiNoC with sub-THz wireless links have higher packet energy per bandwidth and area overhead compared to WiNoC with CNT antennas. Though RFNoC uses the similar small-world network architecture, the less number of links (8 links compared to 24 available in WiNoC) and limited bandwidth of each link (6 Gbps compared to 10 Gbps of WiNoC) affect the performance, especially when the system size scales up. Both Clos and Photonic NoCs are benefited by the high-bandwidth low-power optical links. The Clos NoC dissipates least packet energy per bandwidth as it is nonblocking all-to-all network, whereas the Photonic NoC is affected by the electrical path set up latency. Though the Clos network dissipates least packet energy per bandwidth, but it comes at a high price in real estate overheads compared to the small-world architectures.

2.4.6 Performance Evaluation in Presence of Application-Specific Traffic

In order to evaluate the performance of the different NoC architectures with nonuniform traffic patterns, we considered both synthetic and application-dependent traffic distributions. This result highlights the advantages of the small-world-based WiNoC and RFNoC networks, which have optimized topologies based on traffic distribution. In the following analysis, a 128 core system size is considered. We have principally considered inter-core communication patterns arising out of these traffic distributions.

We considered two types of synthetic traffic to evaluate the performance of the different NoC architectures. First, a transpose traffic pattern [2] is studied where a certain number of cores are considered to communicate more frequently with each other. We have assumed three such pairs and 50% of packets generated from one of these cores are targeted toward the other in the pair. The other synthetic traffic pattern considered is the hotspot [2], where each core communicates with a certain number of cores more frequently than with the others. We have assumed three such hotspot locations to which all other cores send 50% of the packets that originate from them. To represent a real application a 256-point fast Fourier transform (FFT) is considered on the 128 core system. Each core is assigned to perform a two-point radix-2 FFT computation. The traffic pattern generated in performing multiplication of two 128×128 matrices is also used to evaluate the performance of the NoC architectures.

Figure 2.16 shows the bandwidth for nonuniform traffic distributions for all the NoC architectures discussed in this chapter. From the bandwidth results, it is evident that WiNoC (both configurations with THz and sub-THz links) has a better bandwidth for all the different traffic patterns for 128 core system. This is precisely because WiNoC architecture is optimized taking the traffic distribution into account.

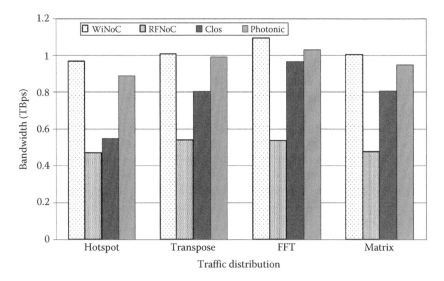

FIGURE 2.16 Achievable bandwidth for nonuniform traffic distributions for different NoCs.

The same traffic-dependent optimization is carried out in RF NoC. But the achievable bandwidth is less due to less number of available shortcuts and their lower bandwidths. From this performance analysis, it becomes clear that WiNoC provides better performance in both uniform and nonuniform traffic scenarios compared to the other alternative NoCs considered in this study.

2.5 Conclusions

This chapter demonstrates how a hierarchical NoC architecture with long-range wireless links significantly improves energy dissipation profile of multi-core computing platforms. We propose and evaluate the performance of WiNoC architectures used as communication backbones for multi-core systems. By establishing long-range wireless links between distant cores and incorporating small-world network architectures, the WiNoCs are capable of outperforming their more traditional wired counterparts in terms of network throughput, latency, and energy dissipation. The architectural innovations proposed in this work are made possible by the use of low-power and high-speed wireless links capable of communicating directly between distant parts of the chip in a single hop. The gains in network performance and energy dissipation are in part due to the architecture and the rest is due to the adopted high-bandwidth, energy-efficient wireless links.

We undertook a comparative performance evaluation between small-world wireless/RF NoC and optical NoC architectures. It is shown that small-world network employing THz wireless shortcuts provide the best performance-overhead trade-off compared to the optical NoC architectures considered in this study for both uniform and nonuniform traffic distribution. Following the findings of this chapter, we can conjecture that in the future, a hybrid NoC with wired as well as wireless links will be able to break the energy efficiency wall of single-chip massive multi-core platforms.

References

1. L. Benini and G. D. Micheli, Networks on chips: A new SoC paradigm, *IEEE Computer*, 35(1), January 2002, 70–78.
2. U. Y. Ogras and R. Marculescu, "It's a small world after all": NoC performance optimization via long-range link insertion, *IEEE Transactions on Very Large Scale Integration (VLSI) Systems*, 14(7), July 2006, 693–706.
3. A. Kumar et al., Toward ideal on-chip communication using express virtual channels, *IEEE Micro*, 28(1), January–February 2008, 80–90.
4. T. Krishna et al., NoC with near-ideal express virtual channels using global-line communication, *Proc. of IEEE Symposium on High Performance Interconnects, HOTI*, Stanford, CA, August 26–28, 2008, pp. 11–20.
5. ITRS, http://www.itrs.net/Links/2007ITRS/Home2007.htm, 2007. (Accessed 8/17/2011.)
6. V. F. Pavlidis and E. G. Friedman, 3-D topologies for networks-on-chip, *IEEE Transactions on Very Large Scale Integration (VLSI)*, 15(10), October 2007, 1081–1090.
7. A. Shacham et al., Photonic network-on-chip for future generations of chip multi-processors, *IEEE Transactions on Computers*, 57(9), 2008, 1246–1260.
8. M. F. Chang et al., CMP network-on-chip overlaid with multi-band RF-interconnect, *Proc. of IEEE International Symposium on High-Performance Computer Architecture (HPCA)*, Salt Lake City, UT, February 16–20, 2008, pp. 191–202.
9. J. Lin et al., Communication using antennas fabricated in silicon integrated circuits, *IEEE Journal of Solid-State Circuits*, 42(8), August 2007, 1678–1687.
10. P. J. Burke et al., Quantitative theory of nanowire and nanotube antenna performance, *IEEE Transactions on Nanotechnology*, 5(4), July 2006, 314–334.

11. K. Kempa et al., Carbon nanotubes as optical antennae, *Advanced Materials*, 19, 2007, 421–426.

12. D. J. Watts, *Small Worlds: The Dynamics of Networks between Order and Randomness*, Princeton University Press, Princeton, NJ, 1999, pp. xvi + 262.

13. B. Feero and P. P. Pande, Networks-on-chip in a three-dimensional environment: A performance evaluation, *IEEE Transactions on Computers*, 58(1), January 2009, 32–45.

14. D. Park et al., MIRA: A multi-layered on-chip interconnect router architecture, *IEEE International Symposium on Computer Architecture, ISCA*, Beijing, China, June 21–25, 2008, pp. 251–261.

15. W. R. Davis et al., Demystifying 3D ICs: The pros and cons of going vertical, *IEEE Design and Test of Computers*, 22(6), November–December 2005, 498–510.

16. A. W. Topol et al., Three-dimensional integrated circuits, *IBM Journal of Research & Development*, 50(4/5), July/September 2006, 491.

17. A. P. Jose et al., Pulsed current-mode signaling for nearly speed-of-light intrachip communication, *IEEE Journal of Solid-State Circuits*, 41(4), April 2006, 772–780.

18. R. T. Chang et al., Near speed-of-light signaling over on-chip electrical interconnects, *IEEE Journal of Solid-State Circuits*, 38(5), May 2003, 834–838.

19. I. O'Connor et al., Systematic simulation-based predictive synthesis of integrated optical interconnect, *IEEE Transactions on Very Large Scale Integration (VLSI) Systems*, 15(8), August 2007, 927–940.

20. M. Briere et al., System level assessment of an optical NoC in an MPSoC platform, *Proc. of IEEE Design, Automation & Test in Europe Conference & Exhibition, DATE*, Nice Acropolis, France, April 16–20, 2007, pp. 1084–1089.

21. M. F. Chang et al., RF interconnects for communications on-chip, *Proc. of International Symposium on Physical Design*, Portland, Oregon April 13–16, 2008, pp. 78–83.

22. B. A. Floyd et al., Intra-chip wireless interconnect for clock distribution implemented with integrated antennas, receivers, and transmitters, *IEEE Journal of Solid-State Circuits*, 37(5), May 2002, 543–552.

23. D. Zhao and Y. Wang, SD-MAC: Design and synthesis of a hardware-efficient collision-free QoS-Aware MAC protocol for wireless network-on-chip, *IEEE Transactions on Computers*, 57(9), September 2008, 1230–1245.

24. R. Albert and A.-L. Barabasi, Statistical mechanics of complex networks, *Reviews of Modern Physics*, 74, January 2002, 47–97.

25. M. Buchanan, *Nexus: Small Worlds and the Groundbreaking Theory of Networks*, Norton, W. W. & Company, Inc., New York, 2003.

26. C. Teuscher, Nature-inspired interconnects for self-assembled large-scale network-on-chip designs, *Chaos*, 17(2), 2007, 026106.

27. D. J. Watts and S. H. Strogatz, Collective dynamics of "small-world" networks, *Nature*, 393, 1998, 440–442.

28. T. Petermann and P. De Los Rios, Physical realizability of small-world networks, *Physical Review E*, 73, 2006, 026114.

29. S. Kirkpatrick et al., Optimization by simulated annealing, *Science. New Series*, 220(4598), 1983, 671–680.

30. J. Duato et al., *Interconnection Networks—An Engineering Approach*, Morgan Kaufmann, New York, 2002.

31. P. Bogdan and R. Marculescu, Quantum-like effects in network-on-chip buffers behavior, *Proc. of IEEE Design Automation Conference, DAC*, San Diego, CA, June 4–8, 2007, pp. 266–267.

32. M. Fukuda et al., A 0.18 µm CMOS impulse radio based UWB transmitter for global wireless interconnections of 3D stacked-chip system, *Proc. of International Conference Solid State Devices and Materials*, Yokohama, Japan, September 2006, pp. 72–73.

33. G. W. Hanson, On the applicability of the surface impedance integral equation for optical and near infrared copper dipole antennas, *IEEE Transactions on Antennas and Propagation*, 54(12), December 2006, 3677–3685.

34. Y. Huang et al., Performance prediction of carbon nanotube bundle dipole antennas, *IEEE Transactions on Nanotechnology*, 7(3), May 2008, 331–337.

35. M. Freitag et al., Hot carrier electroluminescence from a single carbon nanotube, *Nano Letters*, 4(6), 2004, 1063–1066.

36. T. S. Marinis et al., Wafer level vacuum packaging of MEMS sensors, *Proc. of Electronic Components and Technology Conference*, Lake Buena Vista, FL, May 31–June 3, 2005, Vol. 2, pp. 1081–1088.

37. B. G. Lee et al., Ultrahigh-bandwidth silicon photonic nanowire waveguides for on-chip networks, *IEEE Photonics Technology Letters*, 20(6), March 2008, 398–400.

38. W. M. J. Green et al., Ultra-compact, low RF power, 10 Gb/s silicon Mach-Zehnder modulator, *Optics Express*, 15(25), 2007, 17106–17113.

39. Z. Lu et al., Connection-oriented multicasting in wormhole-switched networks on chip, *Proc. of IEEE Computer Society Annual Symposium on VLSI*, Karlsruhe, Germany, 2006, pp. 205–210.

40. P. P. Pande et al., Performance evaluation and design trade-offs for network-on-chip interconnect architectures, *IEEE Transactions on Computers*, 54(8), August 2005, 1025–1040.

41. A. Ismail and A. Abidi, A 3 to 10 GHz LNA using a wideband LC-ladder matching network, *Proc. of IEEE International Solid-State Circuits Conference*, San Francisco, CA, February 15–19, 2004, pp. 384–534.

42. Circuits Multi-Projects. http://cmp.imag.fr

43. E. Mensink et al., A 0.28 pf/b 2 gb/s/ch transceiver in 90 nm CMOS for 10 mm on-chip interconnects, *Proc. of IEEE Solid-State Circuits Conference*, San Francisco, CA, February 2007, pp. 412–413.

44. S. B. Lee et al., A scalable micro wireless interconnect structure for CMPs, *Proc. of ACM Annual International Conference on Mobile Computing and Networking (MobiCom)*, Beijing, China, September 2009, pp. 20–25.

45. A. Joshi et al., Silicon-photonic Clos network for global on-chip communication, *Proc. of the Third International Symposium on Networks-on-Chip (NOCS-3)*, San Diego, CA, May 2009, pp. 124–133.

46. K. K. O et al., The feasibility of on-chip interconnection using antennas, *Proc. of IEEE/ACM International Conference on Computer-Aided Design*, ICCAD-2005, San Jose, CA, pp. 979–984.

47. Y. Zhou et al., Design and fabrication of microheaters for localized carbon nanotube growth, *Proc. of IEEE Conference on Nanotechnology*, Arlington, TX, 2008, pp. 452–455.

48. D. Vantrease et al., Corona: System implications of emerging nanophotonic technology, *Proc. of IEEE International Symposium on Computer Architecture (ISCA)*, Beijing, China, June 21–25, 2008, pp. 153–164.

3

Geyser: Energy-Efficient MIPS CPU Core with Fine-Grained Run-Time Power Gating

Hiroshi Sasaki
The University of Tokyo

Hideharu Amano
Keio University

Kimiyoshi Usami
Shibaura Institute of Technology

Masaaki Kondo
The University of Electro-Communications

Mitaro Namiki
Tokyo University of Agriculture and Technology

Hiroshi Nakamura
The University of Tokyo

3.1 Introduction

Leakage power dissipation in LSI chips has been increasing exponentially with device scaling [14], and has grown to be a major component in the total power dissipation today. Among existing techniques to reduce leakage power, power gating is one of the most promising approaches. Power gating is a technique to shut off the power supply and put the target circuit into sleep mode during periods of inactivity by turning off an embedded power switch transistor.

So far, power gating control has been implemented at a coarse granularity both in terms of area and time. For example, IP cores such as CPU or DSP cores in an SoC are power gated and put into sleep mode depending on applications [7,15,16], for example, in an SoC for cell phone applications. IP cores only used at video telephony are powered off when the operation is switched to the voice call. More aggressively, several run-time leakage power reduction techniques have focused primarily on caches in the microprocessor, which occupy a large area on the processor die [3,11]. It is also relatively easy to apply power gating to the caches compared to other components in the microprocessor such as the functional units in the microprocessor pipeline.

In contrast, we have been studying more aggressive techniques to power gate internal circuits in the microprocessor in much finer granularity at run-time [9,19]. In [9], the authors present a technique to power gate functional units such as a fixed-point unit and a floating-point unit in a microprocessor. In [19], an approach has been proposed to power gate a group of combinational logic gates by employing an enable signal in a gated clock design. These fine-grained power gating techniques have more opportunities to reduce leakage at run time than coarse-grained power gating techniques.

In order to effectively apply fine-grained run-time power gating to functional units in the microprocessor, it is important to detect/predict when and which unit should be put into sleep mode. This should be done with the help of both hardware and software because there exist dynamically or statically determined idle periods. Thus, we propose an architectural (hardware) technique to tackle idle periods caused by dynamic events and a compiler (software) technique to predict and handle statically determined idle periods. That is to say, the architecture should be built in a way that it can detect the chance of power gating and appropriately put the target unit into sleep mode. Also, it should offer a suitable ISA interface to the software to enable controlling power gating from the software. Compiler should decide which functional unit to be put into sleep mode by analyzing the target source code and predict how long would the idle period be of each functional unit after its use. We show our techniques are effective in reducing leakage power by designing and implementing a real chip with fine-grained run-time power gating.

The next section in the chapter describes (1) a top-down design methodology to implement fine-grained run-time power gating, (2) architecture supports and enhancements to effectively apply power gating, and (3) compiler code generation by precisely analyzing the idle time of target functional units. Based upon these fundamental techniques, we design and fabricate Geyser, an MIPS R3000–compatible CPU with fine-grained run-time power gating, and introduce it as a case study of a real chip implementation in Section 3.3. Finally, Section 3.4 summarizes this chapter.

3.2 Fine-Grained Run-Time Power Gating for Microprocessors

In this section, we first describe a top-down design methodology to implement fine-grained run-time power gating, which is composed of the proposal of a design flow to use locally extracted sleep signals. Our technique utilizes the enable signals of gated clock design to automatically partition the target circuit into power gating domains. In Section 3.2.2, we introduce an efficient architectural method to apply fine-grained run-time power gating to the computational units of the classic five-stage RISC-style CPU pipeline. Also, the ISA extension to enable the control of power gating by sleep instructions is introduced. Finally, a code generation technique that analyzes the source code of the program and appropriately inserts the sleep instructions is described in Section 3.2.3.

3.2.1 Design Approach for Fine-Grained Run-Time Power Gating

3.2.1.1 Structure and Generation of Fine-Grained Run-Time Power Gating

3.2.1.1.1 *Exploiting Enable Signals of Gated Clock*

Gated clock is a technique to reduce the dynamic power of the clock network by stop toggling the clock when data stored in flip-flops are not updated. During this period, combinational logic gates located at the transitive fan-in of the flip-flops are not required to compute new data. If outputs of the combinational logic gates are not used anywhere else, the logic gates are considered as "idle" and by detecting this period, we turn off the power switches provided to the combinational logic gates, which results in reducing active leakage power. Figure 3.1 shows the basic structure that we use for fine-grained run-time power gating. We fully exploit the enable signals of gated clock design to control both power switches provided to the

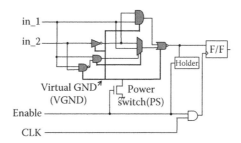

FIGURE 3.1 Basic structure used for fine-grained run-time power gating.

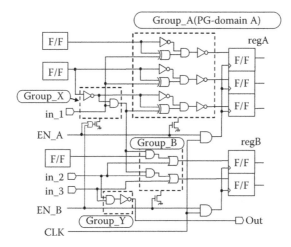

FIGURE 3.2 An example of power gating domains.

combinational logic gates and isolation cells. The isolation cell is composed of low leakage transistors (e.g., high-Vth and thicker gate oxide) and inserted between power-gated and non-power-gated circuits. When the enable signal is zero, power switches are turned off and active leakage current is cut off at the power-gated logic circuits. The isolation cells keep the input voltage of the non-power-gated circuits to avoid signal floating. When the enable signal is one, power switches are turned on and updated data are loaded into the flip-flop.

3.2.1.1.2 Power Gating Domain

In actual clock-gated designs, it is likely that more than one enable signals exist. To perform fine-grained run-time power gating for these designs, we propose an idea of "power gating domain" (PG-domain). PG-domain is defined as a group of circuits that are power gated together with a unique enable signal. We describe the PG-domain by using an example shown in Figure 3.2. In this circuit, there are two enable signals EN_A and EN_B which control clock-gating for multi-bit registers regA and regB, respectively. Combinational logic gates enclosed with a dotted line indicated as "Group_A" perform computation only for regA; logic gates in Group_A become idle if regA is not updated. This allows power gating the combinational logic gates in Group_A with the enable signal EN_A. Hence, we refer to Group_A as "PG-domain A." Similarly, logic gates indicated as Group_B can be power gated using enable signal EN_B, which we refer to as "PG-domain B."

In contrast, combinational logic gates indicated as "Group_X" influence not only regA but also regB. These logic gates become idle only when both regA and regB are not updated. Therefore, we refer to Group_X as "PG-domain AB" and power gate the domain using both EN_A and EN_B.

Logic gates indicated as "Group_Y" are not power gated because their transitive fan-outs are connected to the output pins. Data at the output pins may be used outside of this circuit, and hence should be kept updated. Thus we do not power gate the logic gates in Group_Y and they do not belong to any PG-domain. As an extension, if this scheme is applied to a coarse-grained run-time power gating where the entire circuit is put into sleep, we put the gates in Group_Y into an independent PG-domain. The PG-domain is controlled by a power switch that is turned off only when the entire circuit becomes idle.

3.2.1.1.3 Algorithms to Partition a Circuit into Power Gating Domains

We describe an algorithm to partition a given circuit into PG-domains. Let us assume the circuit depicted in Figure 3.3. First, we focus on a flip-flop and find an enable signal controlling it. For example, flip-flop FF1 is controlled by enable signal EN_A. Next, we traverse the combinational logic network backward until reaching input pins of the given circuit or an output terminal of a flip-flop from the data-input terminal of it. We put a label "A" to all the combinational logic gates that we meet during the traversal, which are the logic gates located at transitive fan-in of FF1. Then we move to the next flip-flop FF2 and find an enable signal of the flip-flop. In this case, the enable signal is identified as EN_A again, so a label "A" is put to logic gates located at transitive fan-in of FF2.

Since flip-flop FF3 is controlled by enable signal EN_B, label "B" is put to logic gates located at transitive fan-in of FF3 if unlabeled. It should be noted that we do not put a label "B" to gates G1 and G2 since they are already labeled "A." Instead, we put a new label "AB" to G1 and G2 by ripping off the old label. Next, we focus on FF4 controlled by EN_C and put a label "C" to the extracted logic gates in the same way. Because label "AB" is already put to gate G2, we update the label to "ABC." After we finish labeling logic gates located at transitive fan-in of all the flip-flops, we focus on the output pins of the circuit and perform a similar backward traversal. We put a label "N/A" to the extracted logic gates because they are not power gated in the fine-grained run-time power gating. If the extracted gates are already labeled, we update the label to "N/A."

After we complete labeling all the combinational logic gates, we create PG-domains according to the labels and a power switch is connected to each PG-domain. For example, logic gates labeled "A" are put into the PG-domain A, while those labeled "AB" are put into the PG-domain AB. Note that logic gates labeled "N/A" are not put into any PG-domain because they are not going to be power gated.

3.2.1.1.4 Generation of Control Logic for Power Switches

Since PG-domains are built based on the labels we added, each power switch connected to the PG-domain is controlled by the enable signal corresponding to it. For example, the power switch connected to PG-domain A is controlled by enable signal EN_A. In contrast, the power switch connected to PG-domain AB

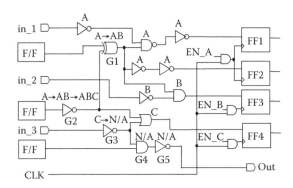

FIGURE 3.3 Example circuit to partition into power gating domains.

has to be controlled by both EN_A and EN_B. Since logic gates in PG-domain AB are idle when EN_A and EN_B are both "Low," EN_A and EN_B are OR-ed and used to control the power switch. A three-input OR gate whose inputs are EN_A, EN_B, and EN_C is added for the power switch connected to PG-domain ABC and the output of the OR gate is connected to the power switch.

3.2.1.2 Implementation Methodology

3.2.1.2.1 Local Virtual Ground (VGND) Scheme

To implement fine-grained run-time power gating, conventional global VGND rail is not effective because partitioning the global rail is extremely difficult. Instead, we use a local VGND scheme in which logic cells and power switch cells within a PG-domain are connected with a local VGND line [13]. To implement this scheme, we modified the existing technology library of logic cells such that the source of NMOS transistor within the cell is disconnected from the real ground rail and is instead connected to a newly created VGND pin. Power switch cell contains an NMOS power switch transistor whose drain and gate are connected to a VGND pin and an enable pin, respectively. The VGND pin of the logic cells are connected to those of power switch cells through a local VGND line, which is routed as an inter-cell wire at the routing stage. We describe the design flow to implement the fine-grained run-time power gating from RTL down to the layout utilizing this scheme.

3.2.1.2.2 Design Flow

Gated clock design is performed in the synthesis step in which we synthesize the gate-level Verilog netlist using the conventional low-Vth standard cell library from RTL description. For the clock-gated netlist, we build a fine-grained run-time power gating structure by using a technique described in Section 3.2.1.1. The clock-gated design is partitioned into PG-domains based on the enable signals. The decision of whether to apply power gating or not is made by considering the break-even point. Break-even point is a certain point in time at which the leakage saving energy and the overhead of switching the power switch are equal. For the PG-domains, power switches are inserted and a control logic for them is generated by adding OR gates. In this way, a fine-grained power gated netlist is generated. The netlist is fed to a placement tool and initial placement is performed. The placement result is given to the power switch optimization engine where power switch sizing is performed (we used CoolPower [4] in our design flow). Power switches are sized such that the voltage bounce at each VGND line may not exceed the user-specified upper limit. Isolation cell insertion to avoid signal floating is also performed at this step. The result of this step is sent to a router and the final layout is generated.

3.2.2 Architectural Technique to Support Run-Time Power Gating

By applying the fine-grained run-time power gating design explained in Section 3.2.1.1, functional units can be shut off in a much finer granularity. Here, we consider applying such an aggressive dynamic control of power gating into a traditional RISC CPU pipeline. For example, multipliers and dividers occupy a large area in such CPU layouts but are not used in every instruction. Such components can be woken up only when they are required by checking the fetched instruction.

However, such an aggressive control has a problem: turning the power switches on and off requires energy, and it takes a certain amount of time for the leakage power to decrease after turning the components off. Thus, in order to reduce the leakage power, the time spent in sleep mode must be more than a certain break-even point. In other words, if the sleep time is less than the break-even point, the energy is increased by applying power gating. The main concern is how to control the sleep signals so that the sleep time will be longer than the break-even point to reduce the energy consumption of the target components.

3.2.2.1 Dynamic Fine-Grained Power Gating for CPU Pipeline

3.2.2.1.1 *Target CPU and Its Functional Units*

We propose a technique to apply our proposed fine-grained run-time power gating to a classic style in-order pipelined RISC CPU, such as MIPS [10], SPARC [18], Motorola 88000 [2], and DLX [8], which is today commonly used in embedded processors. Standard five-stage pipeline structure consists of Instruction Fetch (IF), Decode (ID), Execute (EX), Memory Access (MEM), and Writeback (WB). *
We select computational units in the EX stage as the functional units to be power gated. Computational units occupy a big portion of the total CPU area, and it is easy to identify their usage based on the fetched instructions. Other leakage-consuming units specific to each processor can also be considered as targets. In our implemented real chip Geyser, we additionally selected the coprocessor 0 (CP0) to be power gated, which will also be explained in this section. CP0 supports the operating system for processing exceptions or interrupts. So it only works in special conditions when the operating system is invoked.

Here, we explain the functional units selected as the target to be power gated.

- *ALU (common arithmetic and logic unit)*: General computational units including addition and subtraction. It can be put into sleep mode when branch, node occurrence probability (NOP), or memory access instructions without address calculation is fetched.
- *Shift unit*: Since the barrel shifter occupies a considerable area but is not frequently used, it is selected as an individual unit.
- *Multiplier unit*: The multiplier unit takes several clock cycles to complete a multiplication.
- *Divider unit*: The divider unit also takes several to tens of clock cycles to complete a division.
- *CP0*: A coprocessor, which is an interface between the CPU and the operating system for handling exceptions and controlling TLB entries. It can be put into sleep mode when the CPU runs in user mode.

3.2.2.1.2 *Fundamental Control of Fine-Grained Run-Time Power Gating*

Sleep signals for the target units are generated in the sleep controller shown in Figure 3.4.

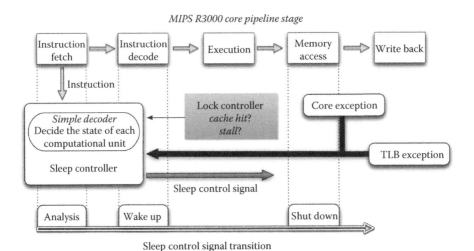

FIGURE 3.4   Sleep controller in the CPU pipeline.

---

* As will be introduced in the next section, we implemented a real chip Geyser, which is a MIPS R3000–compatible CPU, so some of the following description will be dedicated to MIPS architecture, but basically the fundamental idea can be applied to other classic style in-order CPUs.

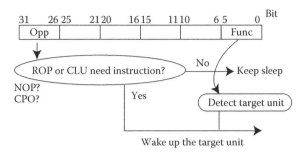

FIGURE 3.5 Detector for generating a wake-up signal.

By considering the usage of each computation unit, all units except ALU are put into sleep mode automatically after finishing the operation in EX stage. When an instruction is fetched in the IF stage, the sleep controller must check the fetched instruction and judge which computation unit is to be used for the fetched instruction. Since it takes a certain amount of time to wake (approximately 1 cycle is assumed here) the unit, the detection must be done in the IF stage. For this purpose, a high-speed simple decoder, which only detects which computational unit is to be used, is provided in the IF stage so that the unit can be available for use when required.

Figure 3.5 shows a high-speed detector for generating a wake-up signal considering the MIPS ISA. The detector checks the uppermost 6 bits of the instruction, and judges whether the instruction is a register–register operation or not. If so, the unit to be used is identified by the last 6 bits of the instruction. Otherwise, some immediate instruction or load/store instruction, which uses the ALU, is detected.

Since we are assuming an in-order processor, the pipeline is stalled during both an instruction and data cache miss (and also in the case of hit because our design takes multiple cycles to access the cache). Thus, all units are put into sleep mode when either cache miss/hit signals or a clock wake-up is detected until the data becomes ready. The sleep controller also detects the stall from the multiplier or the divider while waiting for data, and puts all units into sleep mode.

The sleep control for CP0 operates in a completely different manner; it is put into sleep mode when the CPU runs in user mode. The exception must be received in the sleep controller first, which then sends CP0 the wake-up signal, and also sends a signal to the pipeline to stall. Since the contents of registers in CP0 are lost by putting into sleep mode [12], the values of registers that need to be saved are also handled appropriately by the sleep controller.

3.2.2.2 Sophisticated Power Control Methods

3.2.2.2.1 Instruction with Power Gating Direction

The fundamental method described earlier has a potential problem that it can possibly increase the leakage power. For example, when multiple multiply operations are executed within an interval of few cycles, the multiplier unit will be woken up as soon as it is put into sleep mode. In such a case, the time spent in sleep mode might be less than the break-even point, and the power overhead of power gating increases the total power. Such a combination of instructions can be detected by the compiler, and we can save such overhead by retaining the power of the multiplier unit after the first computation is finished. For this purpose, we introduce an instruction set with power gating direction encoded in it. The technique to effectively utilize this instruction by the compiler is explained in the next section.

Figure 3.6 shows an example of instruction format with power gating direction. In the MIPS ISA, when the uppermost 6 bits (ope-code) are all zeros, the operation is applied with two registers shown in operand fields, and the type of operation is indicated with the last 6 bits. Instead of all zeros, we use "100111,"

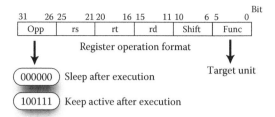

FIGURE 3.6   Instruction with power gating directive.

which is not defined in the original ISA, as the uppermost 6 bits to indicate the power gating direction. After executing such instruction, the functional unit is not put into sleep mode, but stay in active mode. The functional unit is kept awake until the next instruction which uses it with the ope-code of all zeros is executed.

3.2.2.2.2 Power Gating Policy Register

The leakage power is sensitive to the temperature and will increase drastically as the temperature rises. Thus, the break-even point is also influenced by the temperature, and tends to be short when the chip becomes hot. In such a situation, it is more beneficial to aggressively apply run-time power gating because power saving is possible in a short period of sleep. Conversely, if the chip is cool, a less-aggressive power gating policy suppresses the overhead. Thus, the policy of power gating should be changed based on the temperature that can be measured by a thermal sensor, or by the amount of leakage current itself, which can be measured by a leak monitor. Both devices can be integrated on chip and the values can be read out by the operating system.

Here we introduce three policies that can be applied:

1. *Policy-1*: The fundamental policy that always puts every unit into sleep mode dynamically after using it or after a stall, including a cache miss/hit.
2. *Policy-2*: A more conservative policy that puts every unit into sleep mode only when a cache miss/hit is detected.
3. *Policy-3*: A policy that does not apply power gating and no unit is ever put into sleep mode.

We introduce a power gating policy register that stores the data that indicates one of the aforementioned policies, and they can be written only in the kernel mode. The operating system controls the sleep mode depending on the data of the thermal sensor, as shown in Figure 3.7. Note that a policy can be selected for each functional unit, because the break-even point differs among each of them. Two bits are needed to represent the policy for computational units, and 1 bit is needed to indicate whether to sleep or not for CP0.

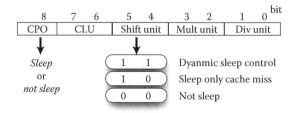

FIGURE 3.7   Power gating policy register.

3.2.3 Compiler-Assisted Run-Time Power Gating Technique

As mentioned in the previous section, compiler can predict the idle period of each functional unit by analyzing the source code of a program. According to this information, each functional unit is selected to be kept awake or put into sleep mode after execution. More precisely, if the predicted idle period of the functional unit is shorter than the target break-even point, the compiler sets the ope-code to 100111 so that the target functional unit will not be put into sleep mode after execution.

One of the reasons for long idle periods are long latency events such as cache misses that occur dynamically. As mentioned in the previous section, cache miss/hit events are detected and handled by hardware. On the other hand, compiler analysis is effective for predicting idle times that are statically determined by instruction sequences. Because there are both short and long idle periods, we need to perform a global code analysis that focuses not only inside the basic block or the procedure but also across the procedure calls. By applying a hybrid technique that combines the compiler-assisted power gating and the cache miss/hit triggered power gating, it becomes possible to capture both statically and dynamically determined idle times very effectively. The details of the compiler analysis will be described in Section 3.2.3.1.

3.2.3.1 Overview of the Compiler Analysis

In this section, we describe the compiler analysis to predict the idle time of functional units. Our analysis is analogous to data-flow analysis [1], which is conventionally used in compiler optimization techniques. We analyze a *control flow graph* (CFG) and obtain the expected idle time for each functional unit. For example, if we want to analyze the idle time of a multiplier, we count the expected number of nodes that lie between the target multiply instruction and the succeeding multiply instruction for each multiply instruction in the CFG. By assuming each instruction is executed in a single cycle, the aforementioned number of nodes give the expected idle cycles of the multiplier.

Usually, there are many procedure calls and loop structures in a program. The accuracy of the expected usage interval of functional units can be strongly affected by them. Therefore, we need to analyze across branches or procedure calls for precise prediction. We construct a *call graph* (CG), which illustrates the relation between procedure calls for interprocedural analysis.

3.2.3.2 Intraprocedural Analysis

First, we describe the analysis within a procedure that has no procedure call in it. The basic framework is similar to a data flow analysis scheme [1]; however, we define real number variables (RNV), which express the expected usage interval of functional units, for each node in the CFG instead of typical data-flow values such as reaching definitions. This analysis depends on the information that is computed in the reverse order of the control flow in a program, because we want to know where the following instruction next uses the target functional unit.

We define RNV for the nodes in a CFG as follows:

$$IN_D[s], IN_P[s], OUT_D[s], OUT_P[s]. \tag{3.1}$$

$OUT_D[s]$ expresses the expected number of instructions between the node s and the next instruction that uses the target functional unit. Therefore, $OUT_D[s]$ indicates the predicted idle time of the functional unit with the assumption that every instruction is executed in a single cycle. $IN_D[s]$ represents the same meaning value as $OUT_D[s]$ defined for the point right before node s. $OUT_P[s]$ expresses the probability of reaching to the exit point of the procedure from the point after node s without executing the instruction that uses the target functional unit. $IN_P[s]$ has the same value as $OUT_P[s]$ defined for the point right before node s.

Next, we give the data-flow equation, which gives the constraint between the variables of nodes. For preparation, we define two constant values $T_D[s]$ and $T_P[s]$. These variables are defined for each node in the CFG, and their values are determined whether the node uses the target functional unit or not.

$$T_D[s] = \begin{cases} 0 & \text{(if } s \text{ uses the target unit)} \\ 1 & \text{(otherwise)} \end{cases} \tag{3.2}$$

$$T_P[s] = \begin{cases} 0 & \text{(if } s \text{ uses the target unit)} \\ 1 & \text{(otherwise)} \end{cases} \tag{3.3}$$

$T_D[s]$ expresses the expected idle time of the functional unit from the program point right before the node s to the program point right after the node s. $T_P[s]$ expresses the probability of reaching the program point right after the node s from the program point right before the node s without executing the instruction that uses the target functional unit. When analyzing a procedure that has no procedure call in it, these two values seem to be trivial because there is always only one instruction between the program point right before the node s and the program point right after the node s. However, there can be several instructions between the two program point right before and after the node s in the interprocedural analysis, which we describe later, because of the presence of the procedure call instructions.

By using these values, the data flow equations are given as follows:

$$IN_D[s] = T_P[s]OUT_D[s] + T_D[s]$$
$$IN_P[s] = T_P[s]OUT_P[s] \tag{3.4}$$

$$OUT_D[s] = \begin{cases} q[s] * IN_D[s_{suc1}] + (1 - q[s]) * IN_D[s_{suc2}] & \text{(if } s \text{ is a branch instruction)} \\ IN_D[s_{suc}] & \text{(otherwise)} \end{cases} \tag{3.5}$$

$$OUT_P[s] = \begin{cases} q[s] * IN_P[s_{suc1}] + (1 - q[s]) * IN_P[s_{suc2}] & \text{(if } s \text{ is a branch instruction)} \\ IN_P[s_{suc}] & \text{(otherwise)} \end{cases} \tag{3.6}$$

s_{suc} indicates the following node after the node s when the node s is not a branch instruction. s_{suc1} and s_{suc2} indicate the following two nodes after the node s when the node s is a branch instruction. $q[s]$ is the probability that the branch node s jumps to node s_{suc1}. Values $q[s]$s are control parameters and can be set for each branch node respectively. $q[s]$ values can be obtained in several ways such as dynamic profiling technique or static branch prediction techniques.

Finally, we give the initial and boundary values for the iterative calculation by equations (Equations 3.4 through 3.6). Initial values for each node s except the exit node in the CFG are given as below:

$$IN_D^{(0)}[s] = T_P[s] * T_D[s], IN_P^{(0)}[s] = T_P[s],$$

$$OUT_D^{(0)}[s] = 0, OUT_P^{(0)}[s] = 0. \tag{3.7}$$

Let s_{exit} be the exit node of a procedure in the CFG, the boundary values are given as below:

$$IN_D^{(b)}[s_{exit}] = 0, IN_P^{(b)}[s_{exit}] = 1. \tag{3.8}$$

3.2.3.3 Interprocedural Analysis

Here, we describe how to handle the procedure calls in a procedure, which is often the case in a program. A pessimistic way to handle the call is to assume that the instruction that uses the target functional unit exists at the entrance of the called procedure. However, it is easy to imagine that it would result in a very conservative and rough prediction. Therefore, we need to apply interprocedural analysis.

In the following, we use a pseudo instruction "jal" as the instruction calling the procedure for explanation. Now we describe the overview of the interprocedural analysis by the following three steps: (1) preprocessing and analyzing the call graph (CG), (2) seeking the transfer constant values of each procedure (described later), and (3) passing the information from the exit point of the program to each procedure.

First, we analyze each procedure as described in the intraprocedural analysis. In steps (2) and (3), the analysis order of procedures is important because we need to exchange information properly between procedures. Therefore, we first analyze the call graph in step (1) to determine the proper analysis order.

In step (1), we decompose the CG into strongly connected components to handle the loops that are formed by recursive procedures or procedures that call each other and make a new graph CG$'$. Next, we give the post-order label to the nodes in the CG$'$ through depth first searching. We analyze each strongly connected component in the labeled order in step (2), and analyze them in reverse order in step (3). We analyze the procedures in the same component several times. For example, assume that two procedures A and B are in the same component. Then, we analyze the procedures in the order of A-B-A-B. Here we set the number of iterative times to two, which is the number of procedures in the component, to assure that the information will be passed over the components appropriately.

Next, we describe how to exchange the information between procedures in step (2). The values are given as below:

$$T_D[prc] := IN_D[s_{ent}], T_P[prc] := IN_P[s_{ent}] \tag{3.9}$$

where

prc is the procedure

s_{ent} is the entrance instruction of procedure prc

We assume that the procedure prc has been already analyzed in order to obtain the proper $IN_D[s_{ent}]$ and $IN_P[s_{ent}]$ values. We redefine the constant values T_D and T_P for the *instructions* (nodes) as below:

$$T_D[s] = \begin{cases} 0 & \text{(if } s \text{ uses the target unit)} \\ T_D[prc_{called}] & \text{(if } s \text{ is "jal")} \\ 1 & \text{(otherwise)} \end{cases} \tag{3.10}$$

$$T_P[s] = \begin{cases} 0 & \text{(if } s \text{ uses the target unit)} \\ T_P[prc_{called}] & \text{(if } s \text{ is "jal")} \\ 1 & \text{(otherwise)} \end{cases} \tag{3.11}$$

prc_{called} is the procedure that is called by the "jal" instruction. In step (2), we analyze the procedures according to the order obtained in step (1). We use the definition Equations 3.10 and 3.11 instead of Equations 3.2 and 3.3 used in the intraprocedural analysis. With the same initial and boundary values given by Equations 3.7 and 3.8, we seek the solution through an iterative calculation in each procedure. From the result, we can obtain the constant values T_D and T_P of the procedures, which will be used in the analysis of other procedures that call the procedure.

Finally in step (3), we decide the expected usage interval of functional units at each point by passing the boundary values from the exit point of the program to the exit point of each procedure in reverse order of the control flow direction. In order to analyze a certain procedure, we set OUT_D values of the "jal" instruction s'_{call}, which call the procedure to the boundary value $IN_D[s_{exit}]$ instead of setting the boundary values given by Equation 3.8. Note that the instruction s'_{call} usually exists in another procedure. Therefore, we have to analyze the procedures in the proper order to pass the boundary values to each procedure.

In summary, we seek the expected usage interval of functional units through passing the information between procedures appropriately. We analyze each procedure with iteratively using Equations 3.9 through 3.11 and the data flow equations (Equations 3.4 through 3.6).

3.3 Case Study: Geyser-1

In this section, we introduce Geyser [17], a MIPS CPU in which fine-grained run-time power gating is applied to computational units in the execution stage, and is able to control the state of each unit. Putting into sleep mode and waking up from sleep mode are controlled via architecture and software, and each unit can change their state every cycle. Although the fundamental techniques to develop Geyser is introduced in the previous section, the compiler that utilizes the sleep instruction is still under development and results are not shown in this evaluation. Geyser has applied fine-grained power gating in terms of both area and time, and has shown the ability of effectively reducing leakage power in a real chip with real-world applications.

3.3.1 Geyser-1: An MIPS R3000 CPU Core with Fine-Grained Run-Time Power Gating

3.3.1.1 Design Policy

Our first chip implementation trial with fine-grained run-time power gating Geyser-0 [17] did not work because of problems on the layout. To avoid those problems, the second prototype Geyser-1 was designed with the following policies: (1) Only the CPU core with power gating is implemented on a chip. Cache and TLB provided in Geyser-0 chip have been moved outside the chip. (2) The design flow is improved so that no manual edit on the layout is needed. (3) 65 nm Fujitsu's low power CMOS process is used instead of 90 nm standard process used in Geyser-0.

Policy (1) gives serious impact to the CPU design. Because of the pin-limitation problem, part of address/data signals must be multiplexed. Additional delay of such multiplexers, long wires, and I/O buffers to access the cache outside the chip severely degrades the operation clock frequency. Moreover, the electric characteristics of the supported package sets are another problem for high-speed operation. As a result, the maximum clock frequency of Geyser-1 was set to be 60 MHz at the layout stage, which is a very conservative value for operation frequency.

3.3.1.2 Implementation and Required Number of Gates

Geyser-1 was designed in Verilog XL, synthesized with Synopsys Design Compiler 2007.03-SP4, and generated its layout by using Synopsys Astro 2007.03-SP7, like common CPUs. The cell library CS202SN 65 nm 12-metal-layer CMOS from Fujitsu is used as a basic cell design. The core voltage is 1.2 V while the I/O requires 3.3 V. For fine-grained run-time power gating, a limited set of standard cells were modified manually to separate the VGND. Such PG cells are replaced with the standard cells between the place and route. In this implementation, we could not modify all standard cells into PG cells because of the limitation of design time.

Optimum number of sleep transistors are inserted in the post layout netlist by the Sequence Designs tool. Cool Power 2007.3.8.5. The condition of optimization is as follows: when all gates that share a sleep transistor work together in the active mode, the maximum level of VGND is less than 0.2 V. The optimized netlist is read by Astro to generate the final mask patterns. Table 3.1 shows the area of each target component.

PS shows the area of power switches while IsoCell stands for the area of isolation cells. It appears that 56% of cells are occupied with the multiplier and divider. The overhead of sleep transistors and isolation cells are about 5.4%–12.6%.

Figure 3.8 shows the layout of Geyser-1. The chip size is 2.1 mm × 4.2 mm. The black boxes are target components of fine-grained run-time power gating. The four small black boxes located near each corner are leakage monitors.

TABLE 3.1　Area Overhead of Fine-Grained Run-Time Power Gating

	Total (μm^2)	PS (μm^2)	IsoCell (μm^2)	Overhead (%)
ALU	3,752.8	296.4	79.2	10.3
Shifter	3,078.0	298.8	76.8	12.6
Multiplier	23,863.6	1762.0	153.6	8.5
Divider	27,918.4	1301.2	153.6	5.4
Others	46,304.4	—	—	—

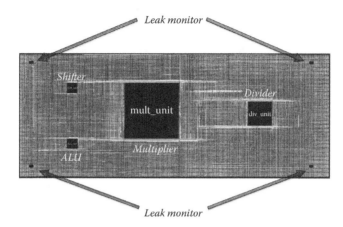

FIGURE 3.8　Layout of Geyser-1.

3.3.2　Evaluation of Geyser-1

3.3.2.1　Operation Frequency of Geyser-1

First of all, we evaluated the maximum clock frequency of Geyser-1. Here, we define "PG (power gating) mode" in which the fine-grained run-time power gating is applied in policy (1), which is the fundamental policy. By using the instructions with the direction, all units can work without being put into sleep mode. This mode is called the "ACT mode." A simple benchmark program is executed, and it appears that the maximum operation clock frequency is 60 MHz in both modes. The computation results were confirmed to be correct. This shows that the wake-up mechanism can correctly work at 60 MHz clock frequency.

3.3.2.2　Break-Even Point Analysis

In order to evaluate the effect and overhead of run-time power gating mode, we evaluated the break-even point of each component on the real chip. Figure 3.9 shows the break-even point of the multiplier unit. The test program is a simple loop that consists of a multiplication, some interval cycles, and a return to the loop entrance. The longer the interval becomes, the less frequently the state transition overhead occurs, i.e., the larger the power savings by power gating grows. So we could investigate the break-even point by changing the interval cycles and comparing the power difference between PG mode and ACT mode. The curve in the graph shows the difference. Positive value means that the state transition energy is larger than the leakage energy saved by power gating. When the operation clock frequency is 50 MHz and the temperature is 25°C, the break-even point becomes 44 clock cycles. That is, the module should be in the sleep mode if the interval of two consecutive "multiply" instructions is larger than 44 clock cycles. The compiler must use the instruction directive if the interval of two instructions is predicted to be smaller than the break-even point.

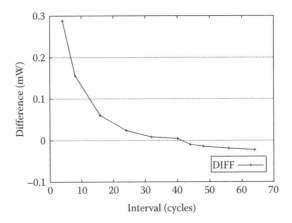

FIGURE 3.9   BET of the multiplier.

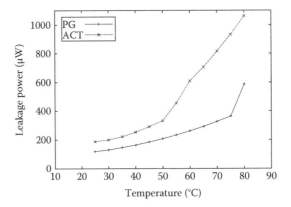

FIGURE 3.10   Leakage power reduction of Geyser-1.

3.3.2.3 Leakage Power Reduction in Power Gating Mode

First, we evaluated the power consumption with all target units set in the sleep state and ACT mode when the clock is stopped. In this case, dynamic power is not consumed and only the leakage power is consumed. The results are shown in Figure 3.10. It shows that the power gating can reduce the leakage power for about 5% at 25°C. When the temperature grows, the leakage power savings of power gating becomes larger.

3.3.2.4 Evaluation Using Benchmark Programs

To evaluate Geyser-1, we used two programs from MiBench [5] benchmark suite: Quick Sort (QSORT) from mathematics package and Dijkstra from the network package. Also, DCT (discrete cosine transform) from the JPEG encoder program was selected as an example of media processing. Unfortunately, the delay of the Block RAM equipped inside the FPGA becomes large, so the evaluation was done with a 10 MHz clock frequency.

Figures 3.11 through 3.13 show the power consumption of three benchmark programs. The leakage power savings of run-time power gating mode is the largest in Dijkstra, which does not use the multiplier and divider. The savings are 8%–24% of the total power consumption, which is more than the values shown in Figure 3.10. However, it is not strange since the computational units put into sleep mode also reduce the dynamic power that is accounted in the ACT mode. The power reduction of QSORT, which

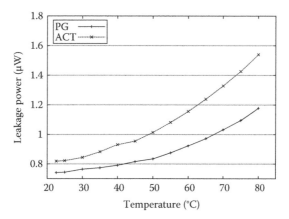

FIGURE 3.11 Power consumption of Dijkstra.

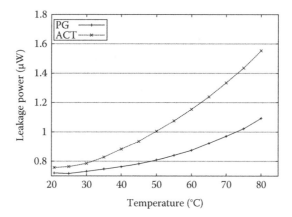

FIGURE 3.12 Power consumption of QSORT.

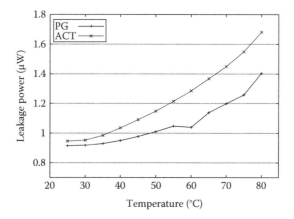

FIGURE 3.13 Power consumption of DCT.

sometimes uses the multiplier and divider, is from 4% to 29%, while that of DCT, which uses multiplier, frequently is 3%–17%. A large portion of the power reduction in DCT comes from the dynamic power reduction in run-time power gating mode, since most intervals between instructions using multiplier are not solong as the break-even point. The power reduction numbers are not as worse when compared with coarse-grained power gating technique [6], which fundamentally cannot be applied to computational units of the CPU. That is, it is demonstrated that our fine-grained run-time power gating can effectively reduce the computation power in the CPU.

As shown in this section, Geyser-1, a prototype MIPS R3000 CPU with a fine-grained run-time power gating is available. The evaluation results with the real chip reveals that the fine-grained run-time power gating mechanism works without electric problems. It reduces the leakage power by 7% at 25°C and 24% at 80°C. The evaluation results using benchmark programs show that the power consumption can be reduced from 3% at 25°C to 30% at 80°C.

3.4 Conclusion

This chapter presented fundamental techniques to enable fine-grained run-time power gating in a microprocessor: design methodology to implement fine-grained run-time power gating in the circuit level, an architectural technique to handle dynamically occurring idle periods with run-time power gating and the ISA extension to enable the control of power gating from the software, and a sophisticated compiler analysis to predict the length of statically determined idle periods together with the code generation. By utilizing the aforementioned techniques, Geyser-1, an MIPS R3000–compatible CPU with fine-grained run-time power gating is developed. The evaluation of Geyser-1 with real-world applications showed that fine-grained run-time power gating works without problems, and we can effectively reduce the leakage power of computational units in the CPU.

References

1. A. V. Aho, R. Sethi, and J. D. Ullman. *Compilers: Principles, Techniques, and Tools.* Addison-Wesley Longman Publishing Co., Inc., Boston, MA, 1986.
2. M. Alsup. Motorola's 88000 family architecture. *IEEE Micro*, 10:48–66, May 1990.
3. K. Flautner, N. S. Kim, S. Martin, D. Blaauw, and T. Mudge. Drowsy caches: Simple techniques for reducing leakage power. In *Proceedings of the 29th Annual International Symposium on Computer Architecture*, ISCA'02, Anchorage, AK, pp. 148–157, 2002. IEEE Computer Society, Washington, DC.
4. J. Frenkil and S. Venkatraman. Power gating design automation. In D. Chinnery and K. Keutzer, eds., *Closing the POWER Gap between ASIC & Custom: Tools and Techniques for Low Power Design*, pp. 269–272. Springer-Verlag, New York, 2005.
5. M. R. Guthaus, J. S. Ringenberg, D. Ernst, T. M. Austin, T. Mudge, and R. B. Brown. Mibench: A free, commercially representative embedded benchmark suite. In *Proceedings of the Workload Characterization, 2001. WWC-4. 2001 IEEE International Workshop*, Austin, TX, pp. 3–14, 2001. IEEE Computer Society, Washington, DC.
6. T. Hattori, T. lrita, M. Ito, E. Yamamoto, H. Kato, G. Sado, Y. Yamada, K. Nishiyama, H. Yagi, T. Koike, Y. Tsuchihashi, M. Higashida, H. Asano, I. Hayashibara, K. Tatezawa, Y. Shimazaki, N. Morino, K. Hirose, S. Tamaki, S. Yoshioka, R. Tsuchihashi, N. Arai, T. Akiyama, and K. Ohno. A power management scheme controlling 20 power domains for a single-chip mobile processor. In *Solid-State Circuits Conference, 2006. ISSCC 2006. Digest of Technical Papers. IEEE International*, pp. 2210–2219, 2006.

7. T. Hattori, T. Irita, M. Ito, E. Yamamoto, H. Kato, G. Sado, T. Yamada, K. Nishiyama, H. Yagi, T. Koike, Y. Tsuchihashi, M. Higashida, H. Asano, I. Hayashibara, K. Tatezawa, Y. Shimazaki, N. Morino, Y. Yasu, T. Hoshi, Y. Miyairi, K. Yanagisawa, K. Hirose, S. Tamaki, S. Yoshioka, T. Ishii, Y. Kanno, H. Mizuno, T. Yamada, N. Irie, R. Tsuchihashi, N. Arai, T. Akiyama, and K. Ohno. Hierarchical power distribution and power management scheme for a single chip mobile processor. In *Proceedings of the 43rd Annual Design Automation Conference*, DAC'06, San Francisco, CA, pp. 292–295, 2006. ACM, New York.

8. J. L. Hennessy and D. A. Patterson. *Computer Architecture: A Quantitative Approach*. 4th edn. Morgan Kaufmann Publishers Inc., San Francisco, CA, 2006.

9. Z. Hu, A. Buyuktosunoglu, V. Srinivasan, V. Zyuban, H. Jacobson, and P. Bose. Microarchitectural techniques for power gating of execution units. In *Proceedings of the 2004 International Symposium on Low Power Electronics and Design*, ISLPED'04, Newport Beach, CA, pp. 32–37, 2004. ACM, New York.

10. G. Kane. *MIPS RISC Architecture*. Prentice-Hall, Inc., Upper Saddle River, NJ, 1988.

11. S. Kaxiras, Z. Hu, and M. Martonosi. Cache decay: Exploiting generational behavior to reduce cache leakage power. In *Proceedings of the 28th Annual International Symposium on Computer Architecture*, ISCA'01, Göteborg, Sweden, pp. 240–251, 2001. ACM, New York.

12. S. Kim, S. V. Kosonocky, D. R. Knebel, K. Stawiasz, and M. C. Papaefthymiou. A multi-mode power gating structure for low-voltage deep-submicron CMOS ICS. *IEEE Transactions on Circuits and Systems II: Express Briefs*, 54(7):586–590, 2007.

13. T. Kitahara, N. Kawabe, F. Minami, K. Seta, and T. Furusawa. Area-efficient selective multi-threshold CMOS design methodology for standby leakage power reduction. In *Proceedings of the Conference on Design, Automation and Test in Europe—Volume 1, DATE'05*, Munich, Germany, pp. 646–647, 2005. IEEE Computer Society, Washington, DC.

14. D. E. Lackey, P. S. Zuchowski, and J. Koehl. Designing mega-asics in nanogate technologies. In *Proceedings of the 40th Annual Design Automation Conference, DAC'03*, Anaheim, CA, pp. 770–775, 2003. ACM, New York.

15. T. Luftner, J. Berthold, C. Pacha, G. Georgakos, G. Sauzon, O. Homke, J. Beshenar, P. Mahrla, K. Just, P. Hober, S. Henzler, D. Schmitt-Landsiedel, A. Yakovleff, A. Klein, R. Knight, P. Acharya, H. Mabrouki, G. Juhoor, and M. Sauer. A 90 nm CMOS low-power GSM/EDGE multimedia-enhanced baseband processor with 380 MHz ARM9 and mixed-signal extensions. In *Solid-State Circuits Conference, 2006. ISSCC 2006. Digest of Technical Papers. IEEE International*, San Francisco, CA, pp. 952–961, 2006.

16. P. Royannez, H. Mair, F. Dahan, M. Wagner, M. Streeter, L. Bouetel, J. Blasquez, H. Clasen, G. Semino, J. Dong, D. Scott, B. Pitts, C. Raibaut, and U. Ko. 90 nm low leakage SoC design techniques for wireless applications. In *Solid-State Circuits Conference, 2005. Digest of Technical Papers. ISSCC. 2005 IEEE International*, Vol. 1, pp. 138–589, 2005.

17. N. Seki, Lei Zhao, J. Kei, D. Ikebuchi, Yu. Kojima, Yohei Hasegawa, H. Amano, T. Kashima, S. Takeda, T. Shirai, M. Nakata, K. Usami, T. Sunata, J. Kanai, M. Namiki, M. Kondo, and H. Nakamura. A fine-grain dynamic sleep control scheme in MIPS R3000. In *2008. IEEE International Conference on ICCD 2008. Computer Design*, Labe Tahoe, CA, pp. 612–617, 2008.

18. CORPORATE SPARC International, Inc. *The SPARC Architecture Manual: Version 8*. Prentice-Hall, Inc., Upper Saddle River, NJ, 1992.

19. K. Usami and N. Ohkubo. A design approach for fine-grained run-time power gating using locally extracted sleep signals. In *International Conference on Computer Design, ICCD 2006*, San Jose, CA, pp. 155–161, 2006.

4

Low-Power Design of Emerging Memory Technologies

Yiran Chen
University of Pittsburgh

Hai Li
New York University

Yuan Xie
Pennsylvania State University

Dimin Niu
Pennsylvania State University

4.1 Introduction

The mainstream memory technologies, that is, static random-access memory (SRAM), dynamic RAM (DRAM), and Flash, have achieved tremendous accomplishments in the history of semiconductor industry. However, all these technologies are facing significant scaling challenges when the technology node enters 32 nm and beyond [1], such as high-leakage power consumption, degraded device reliability, and large process variations. Thus, many memory technologies have been invented to provide not only the better performance but also the improved reliability and scalability. Some examples include resistive RAM (RRAM), phase change RAM (PCRAM), and magnetic RAM (MRAM). Generally, they can be categorized as "universal memory," which features high integration density, nonvolatility, zero standby power, and nanosecond access speed. Because of the small memory cell area, RRAM and PCRAM primarily target storage class memory applications, for example, replacing DRAM as main memory [2–4] and replacing Flash as solid-state drive (SSD) [5], respectively. MRAM, however, primarily focuses on embedded memory applications such as L2 and L3 caches [6,7] for its nanosecond read access speed (e.g., 11 ns for 64 Mb MRAM [8]).

Similar to all nanoscale electronic devices, emerging memory technologies also suffer from some general technical limitations such as process variations, limited endurance, and retention time. Overcoming these technical limitations, however, incurs significant overheads on design, performance, and power.

In this chapter, we will give a short introduction to some promising emerging memory technologies and then present some examples of emerging memory low-power design techniques and their applications in on-chip cache. The unique data storage mechanisms of emerging memories introduce new design perspectives on power and reliability enhancements, which may potentially change the landscape of memory low-power designs.

4.2 Fundamentals of Emerging Memory Technologies

Figure 4.1 shows the overview of some popular emerging memory technologies. In this section, we will briefly introduce the physical mechanisms of MRAM, PCRAM, RRAM, and the newly proposed memristor devices.

4.2.1 Magnetic Memory

The data storage component of MRAM—magnetic tunneling junction (MTJ)—usually includes two ferromagnetic layers (FLs) and one oxide barrier layer, for example, MgO. MTJ resistance is determined by the relative magnetization directions (MDs) of the two FLs: when the MDs are parallel (antiparallel), the MTJ is in a low (high)-resistance state, as shown in Figure 4.2. The MD of one FL (called the "reference layer") is fixed by coupling to a pinned magnetization layer; the MD of the other FL (called the "free layer") can be changed under magnetic or electrical excitations [9].

In conventional MRAM designs, the MTJ resistance is changed by using a polarized magnetic field to switch the magnetization of the MTJ [10,11]. When the size of the MTJ scales down, the required amplitude of the switching magnetic field increases accordingly. The incurred high write power consumption severely limits the scalability of conventional MRAM.

Recently, a new write mechanism was introduced to MRAM design. The new write mechanism uses spin polarization current to switch the magnetization of the MTJ: when applying a positive voltage on point A in Figure 4.2, the MD of the free layer is flipped to be parallel to one of the reference layers, and the resistance of MTJ changes to low. When applying a positive voltage on point B, the MD of the free layer is flipped to be antiparallel to that of the reference, and the resistance of MTJ changes to high. The corresponding new MRAM design, called spin-transfer torque random-access memory (STT-RAM), is believed to have a better scalability than conventional MRAM [9,12].

Figure 4.1 shows the popular one-transistor-one-MTJ (1T1J) STT-RAM structure, where one MTJ is connected to one NMOS transistor in series [9]. The interconnects attached to the MTJ, the source/drain, and the gate of the NMOS transistor are called the bit line (BL), the source line (SL), and the word line (WL), respectively. The MTJ is usually modeled as a variable resistor in the equivalent circuit schematic.

4.2.2 Phase Change Memory

Figure 4.1 also shows a conceptual cross section of a PCRAM cell. A layer of phase change materials, that is, chalcogenide, is integrated between a top electrode and a bottom electrode. A heating element (normally a resistor) is grown on the bottom electrode and contacts the chalcogenide layer. The programming of a PCRAM cell can be divided into two different operations, as shown in Figure 4.3 [13]. During SET operation, a moderate but long current is injected into the heater and raises the temperature of chalcogenide layer through Joule heating. Then the chalcogenide layer is crystallized by heating it above its crystallization temperature T_x. The set pulse lasts for a sufficiently long duration (t_2) so as to maintain the device temperature in the rapid crystallization range for sufficient long for crystal growth. The Phase Change Random Access Memory (PCRAM) cell shows a low-resistance state. During RESET operation, a high but short current is injected and raises the temperature of the programmed volume in the chalcogenide layer above the melting point. The polycrystalline order of the material is eliminated,

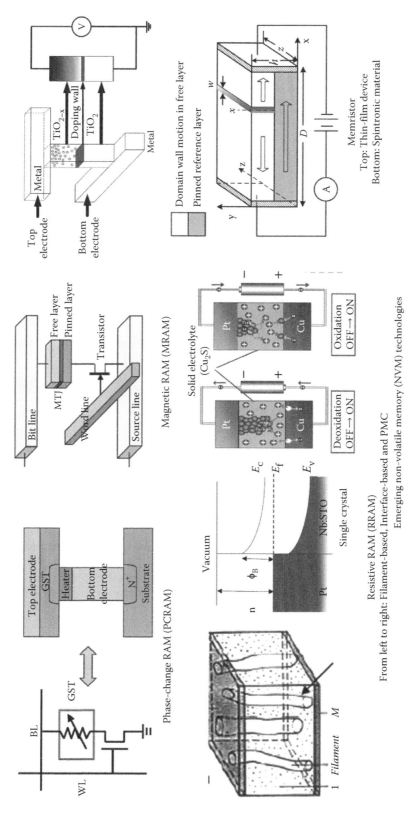

FIGURE 4.1 Overview of some emerging nonvolatile memory technologies, including MRAM (particularly STT-RAM), PCRAM, RRAM, and memristor.

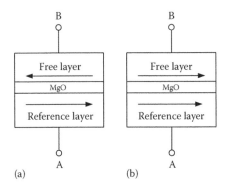

FIGURE 4.2   MTJ structure. (a) Antiparallel (high resistance). (b) Parallel (low resistance).

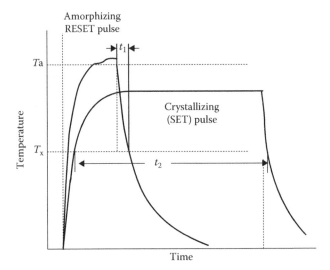

FIGURE 4.3   Programming procedures of a PCRAM cell. (From Lai, S. and Lowrey, T., OUM—A 180 nm nonvolatile memory cell element technology for standalone and embedded applications, in *International Electron Devices Meeting*, pp. 36.5.1–36.5.4, 2001.)

and the volume returns to the amorphous state. When the reset pulse is sharply terminated, the device quenches to "freeze in" the disordered structural state. The device shows a high-resistance state. Because of the change in reflectivity, the amorphous volume appears as a mushroom cap–shaped structure. This quench time (t_1) is usually determined by the thermal environment of the device and the falling time of the RESET pulse.

4.2.3 Resistive Memory

Although RRAM technology involves many different storage mechanisms, there are only two "conventional" operation types in RRAM design: *unipolar switching* and *bipolar switching*. Within this context, unipolar operation executes the programming/erasing by using short and long pulse, or by using high and low voltage with the same voltage polarity. In contrast, bipolar operation is achieved by applying short pulses with opposite voltage polarity.

One typical unipolar switching example is filament-based RRAM devices [14]. A filament or conducting path is formed in an insulating dielectric after applying a sufficiently high voltage. Once the filament is formed, it may be set (leading to a low resistance) or reset (leading to a high resistance) by appropriate voltages.

One typical bipolar switching example is programmable metallization cell (PMC) devices, which are composed of two solid metal electrodes, one relatively inert the other electrochemically active. A thin electrolyte film is allocated between two electrodes. When a negative bias is applied on the inert electrode, the metal ions that are originally in the electrolyte and from the positively charged active electrode flow into the electrolyte. Finally, the ions form a small metallic "nanowire" between the two electrodes. As a result, the resistance between two electrodes is dramatically reduced. When erasing the cell, a positive bias is applied on the inert electrode. Metal ions will migrate back into the electrolyte and eventually to the negatively charged active electrode. The nanowire is broken and the resistance increases again.

4.2.4 Memristor

Recently, RRAM technology was also extended to build memristors [15]—the fourth fundamental passive circuit element predicted by Chua in 1971 [16]. Memristors have a unique feature to remember the historic profile of the current/voltage applied to itself [15,17]. After HP reported the first TiO_2-based memristor device in 2008 [15], many types of memristors and their applications have been identified, including nonvolatile memory, signal processing, neuromorphic computing, and control systems [18].

4.3 Multilevel Cell MRAM

Switching current density (J_{cr}, unit: A/cm^2) and tunneling magnetoresistance (TMR) ratio are two major performance criteria of STT-RAM cells. TMR ratio is defined as $(R_{high} - R_{low})/R_{low}$, where R_{high} and R_{low}, respectively, denote the high and low resistances of an MTJ. Generally, a small switching current density is desired to reduce the write power consumption of STT-RAM. A high TMR make it easier to differentiate the two resistance states of MTJs. Many magnetic engineering techniques, that is, introducing the perpendicular anisotropy in out-of-plane direction [19] and perpendicular MTJ technology [20], have been explored to reduce switching current. Also, new materials, that is, Fe(001)/MgO(001)/Fe(001) [21] and FeCo(001)/MgO(001)/FeCo(001) [22], have been successfully adopted in the oxide layer of MTJs to raise TMR ratio to more than 200%.

High TMR ratio motivated the research on multilevel cell STT-RAM. Multilevel cell (MLC) technology can effectively improve the integration density of memory by storing n-bit ($n > 2$) information in a single memory device as 2^n states. MLC technology has achieved great commercial success in NAND flash [23] and has been explored in RRAM [24].

Very recently, we reported a 2 bit MLC MTJ device in [25]. Two-digit information—00, 01, 10, and 11—is represented by four MTJ resistance states. The transitions among different MTJ resistance states are realized by passing the spin-polarized currents with different amplitudes and/or directions. In this section, we discuss the access (read and write) scheme of the MLC MTJ device proposed in [25] and its optimization. The trade-offs of different write schemes on power, performance, and hardware cost are also analyzed from both circuit and architectural perspectives.

4.3.1 Structure of MLC MTJ

Two types of MLC MTJs were proposed in [25]: parallel MLC MTJs and serial MLC MTJs. Because parallel MTJs have relatively higher TMR ratio, smaller switching current, and better reliability than serial MTJs, here we mainly discuss the structure and working mechanism of parallel MTJs.

Figure 4.4a shows the structure of a conventional single-level cell (SLC) MTJ. One oxide layer, that is, MgO, is sandwiched by two FLs. The MD of one FL (reference layer) is fixed by coupling to a pinned magnetization layer (which is not shown here) while the MD of the other FL (free layer) can be changed by passing a switching current polarized by the magnetization of reference layer. When the MDs of the two FLs are parallel (antiparallel), the MTJ is in low (high)-resistance state.

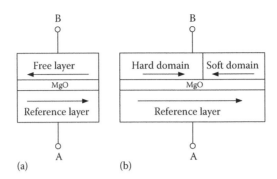

FIGURE 4.4 MTJ structure. (a) SLC MTJ. (b) Parallel MLC MTJ.

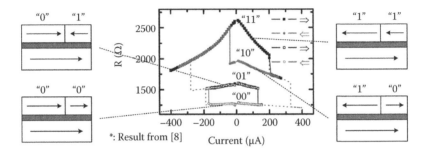

FIGURE 4.5 Four resistance states of an MLC MTJ and R–I swap curve.

Figure 4.4b shows the structure of a parallel MLC MTJ. The free layer of the MLC MTJ has two magnetic domains of which the MDs can be changed separately. The MD of one domain (soft domain) can be switched by a small current while that of the other domain (hard domain) can be switched by only a large current. Four combinations of the MDs of the two domains in the free layer correspond to the four resistance states of MTJs, as shown in Figure 4.5. The first and the second digits of the 2 bit data refer to the MDs of the hard domain and the soft domain, respectively.

4.3.2 Access Schemes of MLC STT-RAM

4.3.2.1 Read Circuitry of MLC STT-RAM Cells

The read schemes of MLC memory cells have been well studied in MLC NAND flash design [26–28]. Here, we designed a current-based read circuitry of MLC STT-RAM based on "Dichotomic" scheme [28] because it provides a well-balanced solution between the long operation latency of "sequential" scheme [26] and the high hardware cost of "parallel" scheme [27].

As shown in Figure 4.6a, a read current is applied to the MTJ and generates the BL voltage V_{BL}. V_{BL} is then compared to three reference voltages V_{REF1}, V_{REF2}, and V_{REF3} based on the Dichotomic search (or binary search) algorithm, as shown in Figure 4.6b. Here, $V_{BL00} < V_{REF1} < V_{BL01} < V_{REF2} < V_{BL10} < V_{REF3} < V_{BL11}$. $V_{BL00} \sim V_{BL11}$ denotes the corresponding BL voltage when the value of the MLC STT-RAM bit is 00–11, respectively. The first and the second digits of 2 bit data are read out respectively in the two comparisons in sequence.

In this read circuitry, only one sense amplifier is required. The reference voltages can be generated from dummy cells. The read operation takes two steps to obtain the first and second digits separately. The first digit is used to choose V_{REF1} or V_{REF3} during the second digit detection.

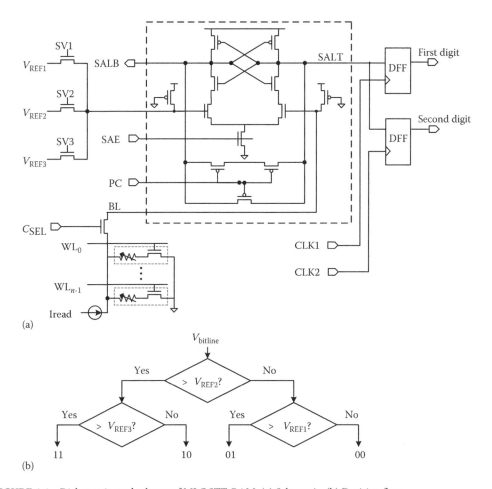

FIGURE 4.6 Dichotomic read scheme of MLC STT-RAM. (a) Schematic. (b) Decision flow.

TABLE 4.1 Switching Currents of MLC STT-RAM Cell (μA)

		To			
		00	01	10	11
From	**00**	0	−189	X	−280
	01	130	0	X	−280
	10	328	X	0	−45
	11	328	X	197	0

4.3.2.2 Write Mechanism of MLC STT-RAM Cells

In the write operation of MLC STT-RAM cells, writing currents with different amplitudes and directions are applied to the MTJ by adjusting the biases on the WL and BL in Figure 4.6a.

Table 4.1 shows the switching currents required by the transitions between the different resistance states of an MTJ with a 100 nm × 200 nm elliptical shape. A positive current denotes the current direction from free layer to reference layer. Although the MD of soft domain can be switched alone, the switching of the MD of hard domain is always associated with the switching of the MD of soft domain. In general, switching the MD of soft domain requires a smaller current than the one required to switch the MDs of both hard and soft domains.

Transitions between some pairs of resistance states cannot complete directly (shown as "X" in Table 4.1). For example, the transition $11 \rightarrow 01$ has to go through two steps: the first step is from 11 to 00 by applying a switching current of 328 μA; the second step is from 00 to 01 by applying a switching current of -189 μA. "0" denotes the switching currents required by the transitions between the same states.

Without loss of generality, the transitions of the MTJ resistance states can be summarized as the following four types:

1. Zero transition (ZT): MTJ stays at the same state.
2. Soft transition (ST): Only the MD of soft domain is switched in the transition, that is, $00 \leftrightarrow 01$ and $10 \leftrightarrow 11$.
3. Hard transition (HT): The MDs of both soft and hard domains are switched in the transition, that is, $00 \rightarrow 11, 01 \rightarrow 11, 10 \rightarrow 00$, and $11 \rightarrow 00$.
4. Two-step transition (TT): Transition completes with two steps, including one HT followed by one ST, that is, $00 \rightarrow 10, 01 \rightarrow 10, 10 \rightarrow 01$, and $11 \rightarrow 01$.

4.3.2.2.1 Simple Write Scheme of MLC STT-RAM Cells

Depending on the bit that is being written, a simple write scheme of MLC STT-RAM can be proposed as follows:

- Writing 00 and 11: the MLC STT-RAM bit is directly programmed to the state by one HT.
- Writing 01 and 10: a TT is executed—the MLC STT-RAM bit is first programmed to 00 or 11 by one HT, and then programmed to 01 or 10 by one ST.

We refer to this write scheme as "simple write scheme."

4.3.2.2.2 Complex Write Scheme of MLC STT-RAM Cells

In fact, many unnecessary transitions are introduced in simple write scheme. For example, for the transition $10 \rightarrow 11$, a TT is executed in the simple write scheme even it can be completed by an ST only. To solve this issue, we proposed a complex write scheme: A read operation is conducted first. Based on the values of the new data being written and the original data stored in the MLC STT-RAM bit, a ZT, ST, HT, or TT will be executed exactly according to the transition procedure shown in Table 4.1.

4.3.2.2.3 Hybrid Write Scheme of MLC STT-RAM Cells

The power dissipation of an HT is significantly higher than that of an ST. We noticed that the transitions between the two resistance states with the same first digit can be completed by only one ST, that is, $00 \leftrightarrow 01$ and $10 \leftrightarrow 11$. Also, the first digit of a MLC STT-RAM bit value can be read by only one comparison between V_{BL} and V_{REF2} (see Figure 4.6). Based on these two observations, we propose a hybrid write scheme to reduce the power dissipation by minimizing the write operations that require an HT. The write operation types of hybrid write scheme can be summarized as follows:

- Soft hybrid (SH) write: If the first digit of the new data and the original data is the same, only one ST is executed to complete the transition, that is, $00 \leftrightarrow 01, 10 \leftrightarrow 11, 00 \leftrightarrow 00$, and $11 \leftrightarrow 11$.
- Hard hybrid (HH) write: If the first digit of the new data and the original data is different, and the new data is either 00 or 11, only one HT is needed to complete the transition, that is, $00 \rightarrow 11, 01 \rightarrow 11, 10 \rightarrow 00$, and $11 \rightarrow 00$.
- Two-step hybrid (TH) write: If the first digit of the new data and the original data is different, and the new data is either 01 or 10, transition completes with a TT, that is, $00 \rightarrow 10, 01 \rightarrow 10, 10 \rightarrow 01$, and $11 \rightarrow 01$.

For random inputs, the probabilities of SH, HH, and TH are 1/2, 1/4, and 1/4, respectively. The state transition graphs of all three write schemes are summarized in Figure 4.7.

Without loss of generality, the longest delay of each write schemes can be expressed as $L_{RD} + L_{ST} + L_{HT}$. Here, L_{RD}, L_{ST}, and L_{HT} denote the latencies of the corresponding read operation, an ST, and an HT,

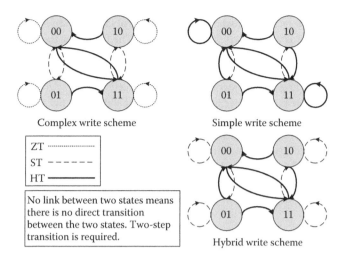

Complex write scheme   Simple write scheme

ZT	----------
ST	- - - - - -
HT	————

No link between two states means
there is no direct transition
between the two states. Two-step
transition is required.

Hybrid write scheme

FIGURE 4.7 State transition graphs of write schemes.

respectively. For simple write scheme, $L_{RD} = 0$. The average energy consumption of each write schemes can be calculated by $\sum_{all\,type} Pr_i \cdot (E_i)$. Here, Pr_i and E_i denote the occurrence probability and the energy consumption of each write operation types, respectively. E_i includes the total energy consumed in the read operation, HT and ST if any. In STT-RAM design, different write schemes may be chosen for various requirements of timing, power, and hardware cost, as well as the probabilities of different write operations types.

4.3.2.2.4 Resistance-Logic State Encoding

The four resistance states of an MLC MTJ (denoted as 00, 01, 10, and 11 before) from low to high are not necessarily assigned to the combinations of two data bits—00, 01, 10, and 11, respectively. In fact, there are total of 4! = 24 encoding schemes to map the four resistance states to the four two-digit data—"00," "01," "10," and "11." As we shall show later on, without considering the cases of staying at the same value, writes of data 00 contribute to the most significant portion of the total write operations. As shown in Table 4.1, the transitions to the resistance state 11 require smaller energy consumption than other states (note "X" means the transition requires two steps, which indicate more energy consumptions). Therefore, assigning resistance state 11–00 may be power efficient.

4.3.2.2.5 Endurance of MLC STT-RAM Cells

Although there are no known endurance issues of the magnetic materials, the lifetime of an MTJ is limited by the time-dependent dielectric breakdown (TDDB) of the MgO layer. Based on the 1/E model, the TDDB time to failure (TTF) can be modeled as follows: $\ln(TTF) \approx 1/E$, where E is the electric field applied [29]. Compared to SLC STT-RAM, the increased switching current of the transitions from 10 or 11 to 00 exponentially degrades the lifetime of the MTJ though the smaller switching current of other transitions alleviates the TDDB issues.

4.4 Experiments

4.4.1 Circuit-Level Simulation

To evaluate the functionality and performance of our proposed read and write schemes, we generate the netlists of an STT-RAM cell, the Dichotomic read scheme, and the three write schemes with PTM

90 nm technology [30]. All necessary logic and timing control circuitry are included in the netlists as well. The size of MLC MTJ is assumed as 90 nm × 180 nm. The four resistance states of MLC MTJ are 1560 Ω (00), 1955 Ω (01), 2350 Ω (10), and 3030 Ω (11), respectively. The read current is set to 60 μA. Figure 4.8 shows the simulation results when a data of "00" is read out. The whole read operation takes two steps that read out the first and the second digits, respectively.

In order to meet the write performance requirement of embedded systems, the writing pulse width is set to 10 ns [6]. To simplify the design of write scheme, the switching current of ST and HT is respectively set to ±160 and ±266 μA, which are carefully scaled from the measurement in [25]. The circuit-level simulations were conducted by using Cadence Virtuoso Spectre simulator. The read operation of an MLC STT-RAM bit can finish within 1.924 ns. The write energies of one HT or one ST without considering the peripheral circuit are 1.92 and 3.192 pJ, respectively. The design parameters of all three write schemes are shown in Table 4.2.

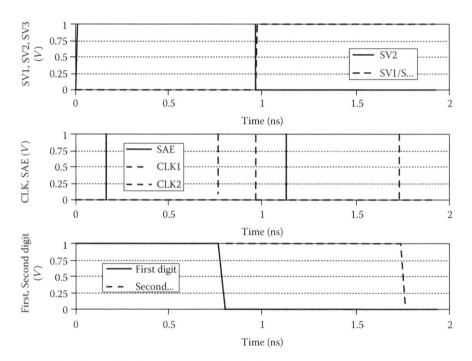

FIGURE 4.8 Simulation results when a data of "00" is read out.

TABLE 4.2 Design Parameters of Three Write Schemes

	Simple	Complex	Hybrid
Transistor number of control logic[a]	56	108	80
Control logic energy (pJ)	0.127	0.268	0.158
Average write energy per bit[b] (pJ)	4.152	2.556	3.036
Total latency (ns)[c]	20	21.924	20.962

[a] Per bit.

[b] For random inputs, not including the peripheral or control logic circuit energy.

[c] Including two 10 ns write pulse width (HT + ST) and read operation (if applicable).

4.4.2 Architecture Level Comparison

We also conducted architecture level simulations to evaluate the application of each write scheme in an MLC STT-RAM-based cache of a microprocessor. The working frequency is set to 400 MHz. The peripheral circuit delay of cache write access (not including the latencies of write schemes shown in Table 4.2) is 1.063 ns, which is calculated by CACTI 5.3 [31]. The write latencies of complex, simple, and hybrid write schemes are set to 10, 9, and 9 cycles, respectively. The precompiled Alpha SPEC2000 binaries [32] were simulated by using SimpleScalar [33].

Figure 4.9a shows the transition distribution between the different values of MLC STT-RAM bits. Most of the transitions occur between the same values, and the MTJ resistance state does not need to change at all. As a result, simple write scheme saved more than 80% write energy reduction compared to complex write scheme, as shown in Figure 4.9b. Here, the dynamic power of the peripheral circuit has been included. On average, the one-cycle additional latency associated with complex write scheme leads to only 0.58% instruction per cycle (IPC) degradation. Obviously, complex write scheme can significantly reduce the write energy of an MLC STT-RAM-based cache in an embedded microarchitecture with minimized performance overhead.

Figure 4.10 compares the normalized write energy under 24 different encoding schemes (E1–E24) of the cache on average for all benchmarks. We found that the following encoding is optimal, that is, {R00 → 11, R01 → 10, R10 → 01, R11 → 00}. Here, we use "R" to denote the resistance levels, which

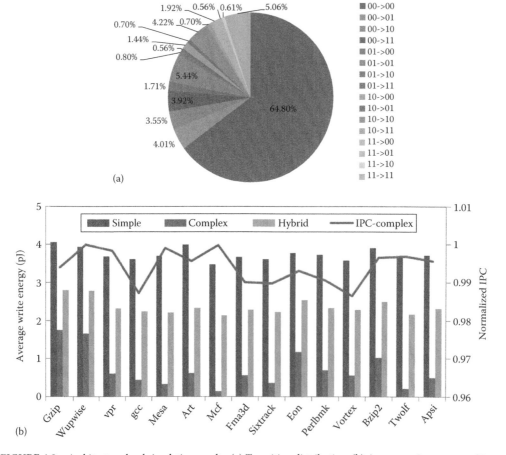

(a)

(b)

FIGURE 4.9 Architecture-level simulation results. (a) Transition distribution. (b) Average write energy and IPC.

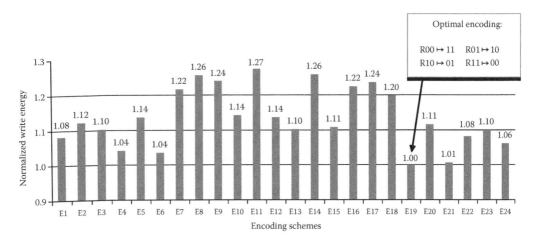

FIGURE 4.10   Write energy comparison of the MLC STT-RAM cache under different encoding schemes on average.

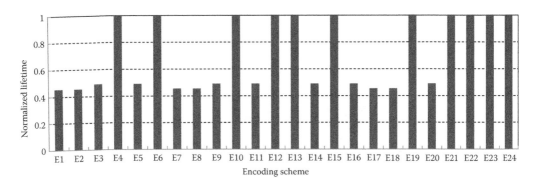

FIGURE 4.11   Relative lifetime of a 16 MB MLC STT-RAM cache with different encoding schemes for benchmark *zeusmp*.

are different from logical levels. The write energy consumption can differ by as much as 27.5% among various encoding schemes.

Figure 4.11 shows the simulated normalized lifetime of the MLC STT-RAM cache with different encoding schemes for benchmark *zeusmp*. The large lifetime improvement at encoding schemes E19, etc., is due to the minimization of the state transitions that require large switching current, which is also the primary reason for the improvement in energy consumption of the same encoding scheme in Figure 4.10.

4.5 Memristor-Based Memory Design

For memory design, reducing the energy consumption is one of the most important design metrics for both high-end computing and low-end embedded applications. In this section, we perform a comprehensive study to analyze the power and energy consumption of memristor-based memory design, which adopts HP's thin-film memristor technology [15], and propose a novel dual-element memristor-based memory structure.

4.5.1 Preliminaries

4.5.1.1 Memristor Model

The memristor is formally defined as a two-terminal element in which the magnetic flux between the terminals is a function of the amount of electric charge q that passes through the device, which is explicitly expressed by the equation $d\varphi = Mdq$, where M is called memristance. If the relation is linear, memristance becomes constant and the memristor acts as a resistor. By contract, if it is nonlinear, the memristance varies with the charge, which determines the hysteretic behavior of current/voltage profile, and can be defined as $M(q)$. Generally, a memristive system can be described by the following relations:

$$\begin{cases} V(t) = M(\omega, i, t)I(t) \\ \omega = f(\omega, i, t) \end{cases}, \tag{4.1}$$

where

ω is the variable that indicates the internal state of the memristive system

$V(t)$ and $I(t)$ denote the voltage and current across the memristive system

Note, M is the memristance and depends on the system state, current, as well as time. If the memristance in Equation 4.1 only depends on the cumulative effect of current, it becomes a charge-dependent device and is called a current-controlled memristor.

The structure of HP's memristor cell is shown in Figure 4.12a [15]. In this metal–oxide–metal structure, the top electrode and the bottom electrode are two thick metal wires on Pt, and two titanium dioxide films are sandwiched by the electrodes. The upper titanium oxide is doped with oxygen vacancies, noted as TiO_{2-x}, and has relatively high conductivity as a semiconductor. The lower oxide, which is perfect titanium oxide without dopants, has its natural state as an insulator. When a biased voltage is applied cross the thin film, the oxygen vacancies are driven from the doped oxide to the undoped oxide and consequently lower the memristance of the whole cell. Contrarily, with a reversed bias voltage, the dopants migrate back to the doped oxide.

Based on the dopant drifting characteristics, the memristor can be modeled as a two-terminal device, which is made up of three resistors in series. Figure 4.12b shows the schematic of memristor model. The thicknesses of the upper and bottom oxides are L_{top} and L_{btm}, respectively. The upper oxide has a constant resistance of R_{top}. But the resistance of the bottom oxide depends on the status of the oxide. We

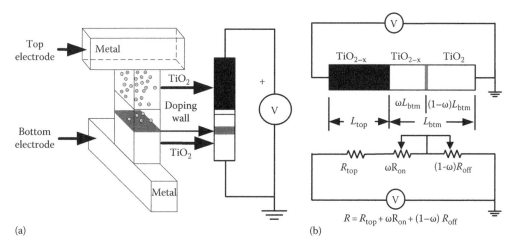

(a)   (b)

FIGURE 4.12 Structure and model of memristor cell.

can denote the full undoped resistance as R_{off} and the full doped resistance as R_{on}. Obviously, R_{top}, R_{on}, and R_{off} have the relationship as follows: $R_{off} \gg R_{on}, R_{off} \gg R_{top}$, and $R_{top}/R_{on} = L_{top}/L_{btm}$.

Since R_{top} is a constant and relatively small, the total resistance is predominantly determined by the status of the bottom oxide. The state variable w is defined as the proportion of the length of the doped region to the total length of the bottom oxide, as shown in Figure 4.12b. Thus, the mathematical model for the memristance of the cell can be described by the resistances and the internal state w as [15]

$$M(w) = R_{top} + wR_{on} + (1 - w)R_{off}. \tag{4.2}$$

Under a biased voltage $v(t)$, the length of the doping region will change, which results in the change of the state variable w. For example, a positive voltage causes the dopants (oxygen vacancies) to drift to the undoped region and therefore result in a lower memristance of the whole memristor. In the linear drift model, the boundary between the doped and undoped regions drifts at a constant speed as $v_{drift} = \mu_v R_{on} i(t)/L_{btm}$. Based on this fact, the memristor can be modeled as [15]

$$\begin{cases} v(t) = (R_{top} + wR_{on} + (1 - w)R_{off})i(t) \\ \frac{dw}{dt} = \frac{\mu_v \rho_{TiO_{2-x}}}{LS} i(t) \end{cases} \tag{4.3}$$

and the internal state (w) can be deduced as

$$w(t) = \frac{\xi_1}{\xi_2} - \sqrt{\left(\alpha_0 - \frac{\xi_1}{\xi_2}\right)^2 - \frac{\lambda}{\xi_2}\phi(t)}, \tag{4.4}$$

where

$\xi_1 = R_{off} + R_{top}$
$\xi_2 = R_{off} - R_{on}$
$\lambda = 2\mu_v \rho/LS$
α_0 denotes the initial state of the memristor

Since the parameters ξ_1, ξ_2 and λ are constants and only depend on the material and manufacture process, the internal state w can be considered as a flux-driven variable. Accordingly, the memristance is determined by the flux applied to it, regardless the waveform of the input voltage. Considering the flux is the integral of voltage on time, if a square wave pulse is applied to the cell, the change of the memristance consequently depends on the pulse width of the input voltage. In this case, these properties make it suitable to employ memristor as a voltage-driven memory cell.

4.5.1.2 Memristor-Based Memory

Attractive properties of the memristor such as nonvolatility, high distinguishability, low-power, scalability, and fast accessing speed make it a promising candidate for the next-generation high-density and high-performance memory technology [34,35]. Consequently, memristor-based memory structure has been studied recently [34]. The basic structure of memristor-based memory is very similar to a typical memory array such as SRAM-based and DRAM-based memory, which consists of the word line decoder, memory array, sense amplifier, and output multiplexer (MUX). However, due to the special electronic characteristics of memristor, several additional peripheral circuits, such as pulse generator and R/W selector, should be added to implement the write/read operations. With the nonvolatility and the flux-driven properties, the key features for the memristor-based memory can be summarized as follows:

1. *Information storage*: It has been reported that the off-to-on resistance ratios of HP's memristor can achieve as much as 1000 [36], which implies a good distinguishability of the memristor-based memory. Since the memristance can reflect the internal state w of the memristor, we can define

the high resistance as logic "0" and the low resistance as logic "1." In addition, in the presence of noise, a safety margin should be defined to ensure the reliability of the cell.

2. *Write operation*: The write operation is to change the internal state of the memristor and therefore change the information stored in the cell. A simple way to write the memristor is to apply enough positive (writing logic "1") or negative (writing logic "0") voltage to the memristor cell. By applying a square wave pulse to the cell, the net flux is only determined by the polarity and pulse width of the input voltage. Thus, one can carefully adjust the input voltage to ensure the correct write operation.

3. *Read operation*: The read operation is more complicated than the write. To read a memristor cell, a voltage is applied across the cell and the current is converted to voltage output by a sense amplifier. Besides, in order to avoid the disturbing read and ensure the stabilization of the memristor's internal state, a two-state read operation is proposed in [34]. The read pattern contains a negative pulse and a positive pulse with the same magnitude and duration, making sure zero net flux is imported into the memristor cell.

4. *Refresh scheme*: If the waveform of the read operation is not perfectly symmetrical, the net flux applied to the memristor becomes nonzero. After several read operations, the accumulative net flux disturbs the internal state significantly and results in soft errors. Thus, given a boundary of the mismatch between negative and positive pulse widths, a refreshing operation is needed to refresh the internal state after every specific number of read operations.

The recently proposed memristor-based memory array structure design used the crossbar structure [34]. In such crossbar-based memory structure, a diode in series with data storage cell is used as the selector of the cell. Such structure can work well for the unipolar memory cell. However, for a bipolar device as memristor, it is difficult to provide a method to select a memristor in a crossbar array without disturbing the adjacent memristors, and such method can result in significant leakage current for the memory array [37,38]. Consequently, in our design, we follow the traditional memory structure and use MOSFET as the selector of the memory cells. However, this design methodology can be easily used in crossbar-based memristor memory array as long as the appropriate nonohmic selector is discovered.

4.5.2 Memristor-Based Dual-Element Memory Design

In this section, we first characterize the power/energy consumption of a memristor cell. Based on our derivations, we propose a low-power dual-element memristor-based memory structure. The circuitry for the dual-element memristor-based memory is also described.

4.5.2.1 Concept of Dual-Element Memristor Cell

One of the most distinguished features of memristor is that its internal state is only determined by the initial state (ω_0) and the accumulated flux (φ) applied to it, which is shown in Equation 4.4. Besides, Equation 4.2 shows that there is a one-to-one correspondence between the internal state (ω) and the memristance (M). Therefore, from Equations 4.2 and 4.4, the memristance of a cell is solely dependent on the flux as

$$M(\phi) = \sqrt{(\xi_1 - \alpha_0 \xi_2)^2 - \lambda \xi_2 \phi}. \tag{4.5}$$

Based on this observation, if a constant voltage, V_{in}, is applied to the memristor, the current–voltage relationship can be described as

$$i(t) = \frac{V_{in}}{\sqrt{(\alpha_0 \xi_2 - \xi_1)^2 - \lambda \xi_2 V_{in} t}}. \tag{4.6}$$

From the model presented in Equation 4.2, the lower bound and upper bound of the memristance are $R_{top} + R_{on}$ and $R_{top} + R_{off}$. Thus, Equation 4.6 is valid during the time interval of

$$\frac{-2\alpha_0\xi_1 + \alpha_0^2\xi_2}{\lambda V_{in}} < t < \frac{-2(1-\alpha_0)\xi_1 - (1-\alpha_0^2)\xi_2}{\lambda V_{in}}. \tag{4.7}$$

Otherwise, the currents are $i_{max} = V_{in}/(R_{top} + R_{on})$ and $i_{min} = V_{in}/(R_{top} + R_{off})$, respectively.

Assuming that the programming time stays in the valid interval defined in Equation 4.7, the power and energy consumption can be calculated as

$$\begin{cases} P(t) = \dfrac{V_{in}^2}{\sqrt{(\alpha_0\xi_2 - \xi_1)^2 - \lambda\xi_2 V_{in}t}} \\ E(t) = \int_0^t P(x)dx = \dfrac{2V_{in}}{\lambda\xi_2}\left[(\xi_1 - \alpha_0\xi_2) - \sqrt{(\xi_1 - \alpha_0\xi_2)^2 - \lambda\xi_2 V_{in}t}\right] \end{cases}. \tag{4.8}$$

Note that in Equation 4.8, the second part in the brackets has the same expression as the memristance shown in Equation 4.5, meaning that from the same state α_0, the energy consumption is only determined by the final memristance after the write operation. Therefore, if we do not use the whole range of the resistance, and only program a part of the memristor, the energy consumption can be saved remarkably. With this partial programming, the write pulse width is also reduced, and hence the write speed is increased. As shown in Figure 4.13, the time for programming the cell from R_{high} to $(R_{high} - R_{low})/2$ is reduced to 75% of the original time. On the other hand, if we program the cell from R_{low}, the programming time is reduced to 25%. Intuitively, programming the low-resistant part of the memristor is faster than programming the high resistance part. Therefore, we can take advantage of this kind of "partial programming" to optimize the memristor-based memory.

One obvious limitation of partial programming is that the distinguishability of the memristor cell is reduced. The distinguishability of a memory cell is the ability to distinguish logic "0" and logic "1" during the read operation. For a memristor-based memory cell, the read operation is realized by applying

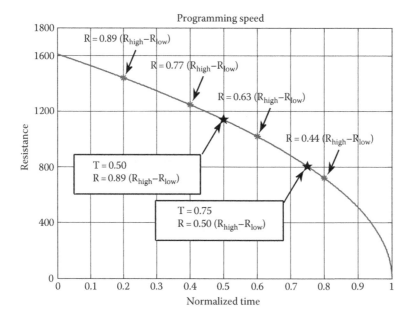

FIGURE 4.13 Programming time versus resistance.

a small voltage across the cell and then sensing the current through the cell. Since the magnitude of current is determined by the resistance of the memory cell, to ensure the reliability of the read operation, current accessing the cell should be distinguishable enough. As a result, the ratio of high resistance to low resistance can reflect the distinguishability between logic 0 and logic 1. For our memristor model, this ratio, $Ratio_M$ can be calculated as

$$Ratio_M = \frac{R_{high}}{R_{low}} = \frac{\xi_1}{\xi_1 - \xi_2}. \tag{4.9}$$

It is important to notice that $Ratio_M$ is the physical parameter of a memristor cell and only determined by the storage material. If we also consider the sense scheme of the cell, which is shown in Figure 4.14a, the distinguishability of the cell is determined by the voltage gap applied to the sense amplifier. Under an input voltage V, the voltage applied to the sense amplifier is $V_{out} = V_{in}R_X/(R_x + R_{cell})$. Given the high resistance R_{high} and low resistance R_{low}, the series resistor R_X should be selected carefully to achieve the maximum voltage difference. This "optimal" value of R_X is

$$R_x = \sqrt{R_{high}R_{low}}.$$

Therefore, the maximum voltage difference applied to the sense amplifier, defined as Λ, can be determined by

$$\Lambda = \frac{V_{in}}{2} \times \frac{R_{high} - R_{low}}{(R_{high} + R_{low}) + 2\sqrt{R_{high}R_{low}}}. \tag{4.10}$$

If partial programming is applied to the cell, both $Ratio_M$ and Λ are reduced, which results in a degradation of the distinguishability of the cell. To overcome this disadvantage, a dual-element memristor-based memory is proposed. The basic idea of the dual-element memory cell is to use a pair of memristors to store the differential signals of the input data. During the write operation, the differential signals are saved in the cells separately. One of the cells, named the positive cell, stores the data with the same polarity of the input data, while the other one, negative cell, stores the complementary data. For example, in order to write logic "0" into the cell, the positive cell is written to high resistance. At the same time, the negative cell is written to low resistance. The read scheme is different from the general memristor memory proposed in [34]. As shown in Figure 4.14b, during the read operation, instead of comparing the voltage to a reference resistance, the voltages from the two memristors are compared. Therefore, the voltage gap input to the sense amplifier is also increased. In order to obtain the same voltage gap, the memristor should be programmed to the state that

$$R_{int} = \sqrt{R_{high}R_{low}}.$$

In this way, we can partially program each of the cell separately and use the associated information to achieve enough distinguishability.

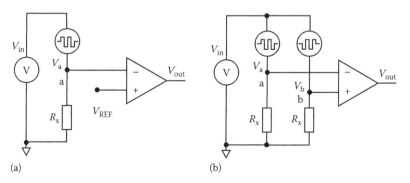

FIGURE 4.14 Sense scheme for single-/dual-element memristor-based memory cell.

4.5.2.2 Design Space Exploration

As mentioned earlier, the partial programming is faster than full programming. Also, the change of memristance is only determined by the flux applied to the memristor. For energy optimization under a specific performance constraint, we can apply a low-magnitude long-width voltage pulse to program the cell. According to Equation 4.8, the energy consumed in the write operation is reduced as the input voltage decreases. On the other hand, one can also optimize the performance of the memory array under a specific energy constraint. Consequently, to design a dual-element memristor-based memory cell, the trade-offs among speed, energy, and $Ratio_M$ should be carefully explored. In order to explore the design space for dual-element memristor-based memory, the following constraint should be considered:

- $Ratio_M$ or Λ should be above a minimum value to guarantee enough distinguishability for the whole cell.
- The programming voltage should not exceed a maximum voltage.
- For memristor, the read speed is much faster than write speed. Therefore, the operation speed mainly depends on the write speed. Thus, the write speed should be lower-bounded.
- Finally, the energy consumption for read and write a memory cell should not exceed certain power budget. Note that in this work, we assume the energy consumption is dominated by the write operation.

Consequently, three different design optimization strategies can be applied for the proposed dual-element memristor-based memory: *energy-driven*, *speed-driven*, or R_{ratio}-*driven* design optimizations. In this chapter, due to the limited space, we focus on the energy-driven design optimization to implement a low-power memory cell. However, under other specifications such as high speed or high distinguishability, the *speed-driven* or R_{ratio}-*driven* optimization can be performed as well.

For the energy-driven optimization, the objective is to optimize the programming voltage V_{in} and pulse width t_{pulse} to minimize the write energy consumption E_{write}, given the design requirements of $Ratio_{min}$, $Speed_{min}$, and V_{max}. The design space of V_{in} and t_{pulse} for the energy-driven optimization is shown in Figure 4.15. Note, the partial programming from low-resistant and high-resistant side of the memristor consumes different amount of energy. Particularly, at the same speed and R_{ratio}, programming from the high resistance side consumes more energy than from the low resistance. Also, the energy consumption increases with the increase of write speed. We should notice that the upper and lower boundaries of write speed are restricted by the voltage constraints and noise immunity issues, which are

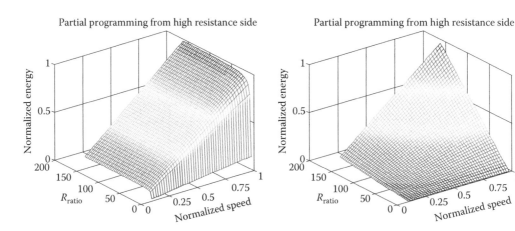

FIGURE 4.15   Design space of dual-element cell.

denoted as V_{max} and V_{min}. Thus, for a low-power design, we should partially program the low resistance side of the memristor and program at the speed of *Speed*$_{min}$.

Based on the energy-driven optimization, a low-power dual-element memristor-based memory cell is proposed, which has the same speed and distinguishability as one cell memristor-based memory and consumes considerably less energy.

4.5.2.3 Memristor-Based Memory Circuitry Design

In this section, a dynamic memristor-based memory is proposed. This design can switch easily between a high-capability normal mode and a low-capability low-power mode, with very small area and energy overhead compared to general memristor-based memory.

Figure 4.16 shows the overall structure of the proposed memory array with peripheral circuits. Generally, the memory structure is based on the design proposed in [34]. As mentioned, in order to avoid the problems due to the sneak paths, the MOSFET is used as the selector in our design. Moreover, we add several additional circuits to the design, ensuring that the memory can switch between the two modes. The memory is divided into two banks, bank 0 and bank 1. The positive cells are located in bank 0, and the negative cells are located in bank 1. Each bank has its own pulse generator. In Mode 0, two banks work independently, whereas in Mode 1, two banks work together to realize the dual-element memristor-based memory, and each pair of the associated cells are located at the same positions in every bank.

The pulse generator provides different voltage levels to the memory cell. A mode bit is assigned to each pulse generator to choose the voltage level. If mode bit is logic "1," indicating the memory works in low-power mode, the low voltage V_L is applied to memristor cells. If mode bit is logic "0," the normal mode uses the high voltage V_H for the circuit.

The read operation is similar to the single-cell memristor memory. As shown in Figure 4.14, to extract the information of the cell, a voltage V_{in} is applied across the whole cell and the current is sensed. In Figure 4.17, if the mode bit is set to 0, the circuit works as two independent memory cells. The voltage V_1 and V_{REF} are compared in sense amplifier 1 (SA$_1$) while V_2 and V_{REF} are compared in sense amplifier 2 (SA$_2$). However, if this bit is set to 1, the circuit works in low-power dual-element mode. The resistors used to generate reference voltage are useless and are isolated. Additionally, the voltages V_1 and V_2 are compared to extract the data.

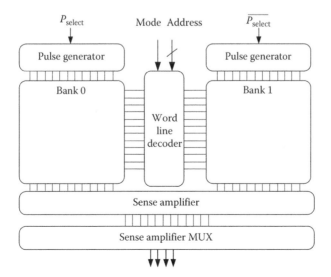

FIGURE 4.16 Overall structure of the memory array.

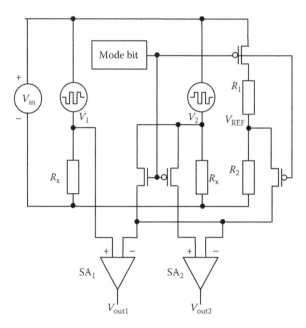

FIGURE 4.17 Circuit structure for dual-element memristor-based memory cell.

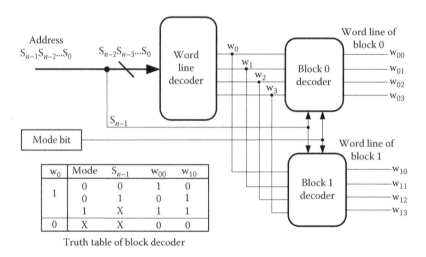

Truth table of block decoder

FIGURE 4.18 Word line decoder for dual-element memristor-based memory cell.

The modified word line decoder is illustrated in Figure 4.18, which contains one traditional decoder and two block decoders. As mentioned, memristors in the dual-element cell locate in the same location of bank 0 and bank 1, respectively. For example, in Figure 4.16, the memristors of address "0000" is associated with the memristors at "1000." For each pair of the associated memristor, only the first bit of the address is different. Thus, to decode the $n-$ bit address, an $n-1$ bit decoder is firstly used to decode the low $n-1$ bits of the address. After that, two word lines are selected at the same time. Then the highest bit and the mode bit are used together to generate the control signal for the two block decoders, deciding which line or both lines are activated. The truth table for the block decoder is shown in Figure 4.18. If the mode bit is 0, only one line is activated at one time, indicating the memory works at normal mode. Contrarily, if the mode bit is 1, two word lines are activated simultaneously.

TABLE 4.3 Experiment Parameters

Parameters	Description	Value
R_{on}/R_{off}	Low/high resistance	50/10 KΩ
R_{top}	Resistance of upper oxide	50 Ω
μ_v	Equivalent dopants mobility	10^{-7} m^2/V · s
L	Thin film thickness	5 nm
S	Cross section area	30 nm × 30 nm
N_1/N_2	Memory size	32/16 MB
W	Word width	8 bits
V_{w0}	Write level voltage at Mode 0	1 V
V_{w1}	Write level voltage at Mode 1	1/3 V
V_r	Read voltage	1 V

4.6 Experimental Results

4.6.1 Experiments Setup

In our experiments, we model the memristor-based memory structure with 45 nm CMOS technology (i.e., the memory storage cells are memristors while the peripheral circuitry is built with 45 nm CMOS technology). The Synopsys HSPICE tool with 45 nm technology library is employed to get the delay, energy, and area values of the peripheral circuits of our memory design. Also, we use Cadence Spectre Analog environment to simulate the design of sense amplifier. The memristor parameters and other memory parameters used in the experiments are shown in Table 4.3 [34,36].

4.6.2 Energy Consumption Optimization

As mentioned earlier, to design the dual-element memory cell, a trade-off should be made among the energy, speed, and distinguishability considerations. Three different kinds of designs can be realized by using the dual-element memristor-based memory: energy-driven, speed-driven, and R_{ratio}/Λ -driven design. Our experiments mainly focus on the energy-driven optimization. However, we also show some "intermediate" schemes of the optimization, which demonstrate the speed and distinguishability improvements of the dual-element cell. For the energy-driven optimization, the problem can be formulated as follows:

Objective: Optimize the programming voltage V_{in} and pulse width t_{pulse} to minimize write energy E_{write}.

Constraints:

$$\begin{cases} \Lambda > \Lambda_{min} \\ Speed_{write} > Speed_{min} \\ V_{min} < V_{in} < V_{max} \end{cases}. \tag{4.11}$$

According to the energy-driven optimization, four schemes of dual-element memristor-based memory cell are proposed, which are shown in Table 4.4. Baseline is the single-element memristor-based memory, in which the full range of the memristor is used. The other four schemes, schemes 1–4, are the proposed dual-element memory with different resistance ranges and write strategies. Especially, scheme 4 shows the most aggressive optimization of the design, which has the least resistance range and the equivalent distinguishability compared to the baseline scheme. Note that we use Λ and t_{write} as the design constraints in this paper to ensure the dual-element design has the comparable performance to the baseline design. Thus, the design constraints are $\Lambda > \Lambda_{baseline} = 0.41$ V, $t_{write} < t_{baseline} = 25.38$ ns, $V_{max} = 1$ V, and $V_{min} = 1/3$ V. From this table, we can see that under these constraints, the proposed dual-element memristor-based memory can gain speed, distinguishability, and energy improvement with different design schemes. For example, scheme 1 has the similar energy consumption as the baseline design.

TABLE 4.4   Different Optimization Schemes

Scheme	Resistance	$\Lambda(V)$	V_{write}	t_{write}(ns)	Energy/Cell (pJ)
Baseline	$R_{low} \sim R_{high}$	0.41	V_{w0}	25.38	3.20
Scheme 1	$R_{low} \sim R_{high}/2$	0.74	V_{w1}	19.02	3.16
Scheme 2	$R_{low} \sim R_{high}/3$	0.68	V_{w1}	8.85	2.10
Scheme 3	$R_{low} \sim R_{high}/4$	0.63	V_{w1}	4.98	1.54
Scheme 4	$R_{low} \sim \sqrt{R_l R_{high}}$	0.41	V_{w1}	0.78	0.60

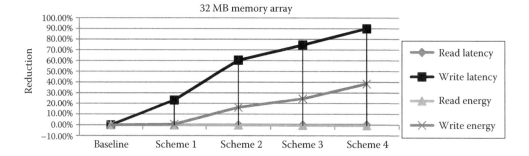

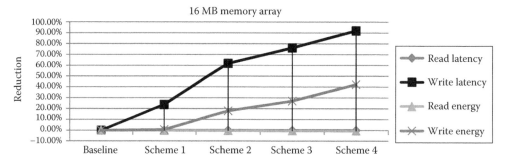

FIGURE 4.19   Comparison of latency/energy reduction.

However, the distinguishability, denoted as Λ, increases from 0.41 to 0.74 V. On the other hand, in the most aggressive scheme (scheme 4), the Λ is equal to $\Lambda_{baseline}$. But the energy consumption is only about 18% of the baseline.

To study the efficiency of the proposed design, we compare the latency and energy consumption for two different memory arrays, which are shown in Figure 4.19. It is important to notice that the numbers of memristor used in the same size of memory are different. For example, the baseline design of the 32 MB memory has totally 32 M × 8 memristors. However, the other schemes have 64 M × 8 memristors. The results show that, although the proposed schemes use doubled number of memristors, the write energy and latency are improved significantly. The energy consumption can be reduced by 40%, and the write latency can be improved by up to 90%.

4.7 Conclusion

In this chapter, we illustrate the applications of two emerging memory technologies, MRAM and memristor in on-chip cache and memory designs, respectively. Our analysis shows that emerging memories can introduce significant power and area benefits with enhanced reliability and lifetime. The innovations of some design technologies to overcome the intrinsic drawbacks of emerging memories, for example, the large write power and long write latency, are essential to ensure the successful adoption of these new technologies.

References

1. The International Technology Roadmap for Semiconductors. http://www.itrs.net, 2008 (accessed Jan, 2011).
2. R. F. Freitas and W. W. Wilcke. Storage-class memory: The next storage system technology. *IBM Journal of Research and Development*, 52:439–447, 2008.
3. W. Zhang and T. Li. Exploring phase change memory and 3D die-stacking for power/thermal friendly, fast and durable memory architectures. *18th International Conference on Parallel Architecture and Compilation Technology*, Raleigh, North Carolina, pp. 101–112, 2009.
4. P. Zhou, B. Zhao, J. Yang, and Y. Zhang. A durable and energy efficient main memory using phase change memory technology. *36th International Symposium on Computer Architecture*, Austin, Texas, pp. 14–23, 2009.
5. G. Sun, Y. Joo, Y. Chen, D. Niu, Y. Xie, Y. Chen, and H. Li. A hybrid solid-state storage architecture for the performance, energy consumption, and lifetime improvement. *15th International Symposium on High-Performance Computer Architecture*, Bangalore, India, pp. 1–12, 2010.
6. X. Dong, X. Wu, G. Sun, Y. Xie, H. Li, and Y. Chen. Circuit and microarchitecture evaluation of 3D stacking magnetic RAM (MRAM) as a universal memory replacement. *Design Automation Conference*, Anaheim, California, pp. 554–559, 2008.
7. G. Sun, X. Dong, Y. Xie, J. Li, and Y. Chen. A novel architecture of the 3D stacked MRAM L2 cache for CMPs. *14th International Symposium on High-Performance Computer Architecture*, Raleigh, North Carolina, pp. 239–249, 2009.
8. K. Tsuchida et al. A 64 Mb MRAM with clamped-reference and adequate-reference schemes. *IEEE International Solid-State Circuits Conference*, San Francisco, California, pp. 258–259, 2010.
9. M. Hosomi et al. A novel nonvolatile memory with spin torque transfer magnetization switching: Spin-RAM. *IEEE International Electron Device Meeting*, Washington, DC, pp. 459–462, 2005.
10. M. Motoyoshi et al. A study for 0.18 μm high-density MRAM. *IEEE Symposium on VLSI Technology*, Honolulu, HI, pp. 22–23, 2004.
11. Y. K. Ha et al. MRAM with novel shaped cell using synthetic anti-ferromagnetic free layer. *IEEE Symposium on VLSI Technology*, Honolulu, HI, pp. 24–25, 2004.
12. T. Kawahara et al. 2 Mb spin-transfer torque RAM (SPCRAM) with bit-by-bit bidirectional current write and parallelizing-direction current read. *IEEE International Solid-State Circuits Conference*, San Francisco, California, pp. 480–617, 2007.
13. S. Lai and T. Lowrey. OUM—A 180 nm nonvolatile memory cell element technology for standalone and embedded applications. *International Electron Devices Meeting*, Washington, DC, pp. 36.5.1–36.5.4, 2001.
14. I. H. Inoue, S. Yasuda, H. Akinaga, and H. Takagi. Nonpolar resistance switching of metal/binary-transition-metal oxides/metal sandwiches: Homogeneous/inhomogeneous transition of current distribution. *Physical Review B*, 77(3):7, 2008.
15. D. B. Strukov, G. S. Snider, D. R. Stewart, and R. S. Williams. The missing memristor found. *Nature*, 453(7191):80–83, May 2008.
16. L. Chua. Memristor—The missing circuit element. *IEEE Transactions on Circuit Theory*, 18(5):507–519, September 1971.
17. L. Chua and S. M. Kang. Memristive devices and systems. *Proceedings of the IEEE*, 64(2):209–223, February 1976.
18. Y. Chen and X. Wang. Compact modeling and corner analysis of spintronic memristor. *IEEE/ACM International Symposium on Nanoscale Architectures*, San Francisco, California, pp. 7–12, 2009.

19. H.-J. Suha and K.-J. Lee. Reduction in switching current density for current-induced magnetization switching without loss of thermal stability: Effect of perpendicular anisotropy. *Current Applied Physics*, 9(5):985–988, September 2009.

20. M. Yoshikawa et al. Tunnel magnetoresistance over 100% in MgO-based magnetic tunnel junction films with perpendicular magnetic $L1_0$ − FePt electrodes. *IEEE Transaction on Magnetics*, 44(11):2573–2576, November 2008.

21. S. Yuasa, T. Nagahama, A. Fukushima, Y. Suzuki, and K. Ando. Giant room-temperature magnetoresistance in single-crystal Fe/MgO/Fe magnetic tunnel junction. *Nature Materials*, 3:868–873, 2004.

22. S. S. P. Parkin et al. Giant tunneling magnetoresistance at room temperature with MgO (100) tunnel barriers. *Nature Materials*, 3:862–867, 2004.

23. J.-H. Park et al. 8 Gb MLC (multi-level cell) NAND flash memory using 63 nm process technology. *IEEE International Electron Devices Meeting*, Washington, DC, pp. 873–876, December 2004.

24. I. G. Baek et al. Multi-layer cross-point binary oxide resistive memory (OxRRAM) for post-NAND storage application. *IEEE International Electron Devices Meeting*, Washington, DC, pp. 750–753, December 2005.

25. X. Lou, Z. Gao, D. V. Dimitrov, and M. X. Tang. Demonstration of multilevel cell spin transfer switching in MgO magnetic tunnel junctions. *Applied Physics Letter*, 93:242502, 2008.

26. M. Horiguchi, M. Aoki, Y. Nakagome, S. Ikenaga, and K. Shimohigashi. An experimental large-capacity semiconductor file memory using 16-levels/cell storage. *IEEE Journal of Solid State Circuits*, 23(1):27–33, February 1988.

27. C. Calligaro, R. Gastaldi, A. Manstretta, and G. Torelli. A high-speed parallel sensing scheme for multi-level non-volatile memories. *IEEE International Workshop on Memory Technology, Design and Testing*, San Jose, CA, pp. 96–101, 1997.

28. C. Calligaro, V. Daniele, R. Gastaldi, A. Manstretta, and G. Torelli. A new serial sensing approach for multistorage non-volatile memories. *IEEE International Workshop on Memory Technology, Design and Testing*, San Jose, CA, pp. 21–26, 1995.

29. K. Croes and Z. Tokei. E−and $\sqrt{E}$−model too conservative to describe low field time dependent dielectric breakdown. *International Reliability Physics Symposium*, Anaheim, CA, pp. 543–548, 2010.

30. W. Zhao and Y. Cao. New generation of predictive technology model for sub-45 nm early design exploration. *IEEE Transactions on Electron Devices*, 53(11), 2816–2823, 2006.

31. Hewlett Packard. Research. http://www.hpl.hp.com/research (accessed June, 2008).

32. D. Weaver. Pre-compiled little-endian Alpha ISA SPEC2000 binaries. http://www.eecs.umich.edu/~chriswea/benchmarks/spec2000.html (accessed June, 2008)

33. D. Burger and T. M. Austin. The simple scalar tool set, version 2.0, University of Wisconsin. Technical Report TR1342, June 1997.

34. Y. Ho, G. M. Huang, and P. Li. Nonvolatile memristor memory: Device characteristics and design implications. *International Conference on Computer-Aided Design*, San Jose, CA, pp. 485–490, 2009.

35. J. M. Tour and T. He. Electronics: The fourth element. *Nature*, 453(9) 42–43, 2008.

36. R. Williams. How we found the missing memristor. *IEEE Spectrum*, 45(12):28–35, December 2008.

37. M. Dong and L. Zhong. Challenges to crossbar integration of nanoscale two-terminal symmetric memory devices. *IEEE Conference on Nanotechnology*, Arlington, TX, pp. 692–694, August 2008.

38. H. Li and Y. Chen. An overview of non-volatile memory technology and the implication for tools and architectures. *Design, Automation & Test in Europe Conference and Exhibition*, Dresden, Germany, pp. 731–736, 2009.

Energy-Efficient Storage

5

Reducing Delays Associated with Disk Energy Management

Puranjoy
Bhattacharjee
*Virginia Polytechnic
Institute and State
University*

Ali R. Butt
*Virginia Polytechnic
Institute and State
University*

Guanying Wang
*Virginia Polytechnic
Institute and State
University*

Chris Gniady
University of Arizona

5.1 Introduction

Research on energy conservation has traditionally been focused on battery-operated devices. However, recent works have also highlighted the positive financial and environmental implications of energy conservation for stand-alone servers and workstations [2,4,7,15]. For example, large organizations often require shutting down workstations, unneeded servers, and cooling systems overnight [20] to reduce energy costs. Setups such as academic institutions and businesses, where users frequently work remotely and at all hours, also employ dynamic energy management.

Dynamic management saves energy by identifying periods of inactivity for a device, and then keeping the device in a low-power state during such periods. Accurately predicting such idle periods [8,12] is critical for energy reduction and minimization of exposed delays. However, these mechanisms still expose powering-on delays (e.g., disk spin-up delays), even if the predictions are correct and provide energy savings. Delays can significantly impact system performance, irritate users, and also reduce the energy savings since the system has to operate longer to satisfy user requests. Furthermore, excessive delays may irritate users to the point where they simply disable energy management techniques. "Therefore, the challenge lies in realizing the energy savings, by keeping the system powered down for as long as possible, yet reducing the performance impact associated with energy management delays, e.g., response latencies on powering the device up when it is needed."

Current approaches for reducing energy management delays rely on predicting when I/O requests will arrive and powering on the device ahead of time [5]. Alternatively, energy management delays can be reduced by utilizing surrogate sources that may be available [14]. The implication here is that the requests destined for a given device are somehow satisfied by a lower-power and higher-performance alternative source. Thus, the delays associated with energy management are avoided.

In this chapter, we focus on reducing disk energy management delays that are significantly longer than any other system component because disks contain mechanical platters and require significant amount of time to spin up from a low-power mode. Moreover, disks are significant energy consumers [18,21,25], and disk energy management, e.g., shutting down idle disks [8], is a common practice present on almost every system in some form.

We do a survey of currently prevalent disk energy management techniques and the associated delays inherent in such algorithms. We focus on whether such techniques have explicit mechanisms to handle delays or not; and if they do, how they try to reduce them. We consider these mechanisms with an example of a user working in an enterprise environment.

Subsequently, we look at SARD [22] which treats delay as a first-class concern in disk energy management and tries to reduce the delay exposed to users from energy management techniques. SARD presents an alternate way of satisfying I/O requests destined for a typical desktop or workstation disk in low-power mode in enterprise environments. While hiding latency might not seem directly related to saving energy, it is crucial in enabling higher rate of adoption of power-saving techniques. Servicing I/Os from alternate sources provides opportunities for keeping the local disks in low-power mode, and may reduce energy consumption as an additional bonus.

The rest of the chapter is organized as follows. In Section 5.2, we survey a number of current disk energy management techniques. Following that, we explore the design of SARD and look at some of the results in Section 5.3. Finally, we conclude in Section 5.4.

5.2 Disk Energy Management Techniques and Delays

In this section, we survey various energy management techniques in use today and look into the delays imposed by these techniques. Table 5.1 gives a brief summary of how current techniques strive to manage the delay exposed to the users. We note here that this survey is by no means exhaustive; rather, it is meant to be representative of the wide flavor of energy management techniques prevalent today and gives us a spectrum of approaches in order to understand how delay management is undertaken.

5.2.1 No Explicit Management

Some of the disk energy management techniques do not address potential delays explicitly. In the following, we discuss a few of these techniques.

5.2.1.1 Adaptive Spin-Down

The paper by Douglis et al. [6] provides a theoretical framework to decide the thresholds for spin-down time and wait time intelligently. The key idea presented by the authors is that it is better to avoid close calls rather than run the risk of a quick spin-up. To achieve that, the authors use the insight that different users have different priorities when it comes to the trade-off between energy conservation and performance. In addition, even for a single user, the workload and usage pattern vary over time. Based on these factors, the threshold is changed dynamically to adjust to current usage conditions. Therefore, there is no explicit delay management in adaptive spin-down; rather, the goal is to reduce the number of undesirable delays.

TABLE 5.1 Table Showing the Delay Management Approaches Taken by the Current Energy Management Techniques

Project	Explicit Delay Management?	Discussion
Adaptive spin-down	No	• Theoretical framework to decide spin-down time and wait-time • Observes that it is better to avoid close calls rather than run the risk of a quick spin-up
PCAP	No	• Tries to reduce the number of mispredictions of wake-ups • Uses call-site signatures for prediction • Waits for the window time before actually shutting down the disk
Best shifting	No	• Dynamically selects the best approach to power management based on the workload • No single power-saving policy • Delay imposed related to the particular algorithm chosen
SRCMap	No	• Works on virtual machines • Replicates the active datasets of various volumes and consolidates them on a single machine; I/Os that would have gone to the various volumes are satisfied at the single machine • Synchronization is undertaken offline
Hibernator	Yes	• Proactively monitors the performance of the disk array • Runs disks at full speed if the performance falls below threshold • Migrates blocks so that the hottest files are on top-tier disks
PARAID	Yes	• Enables disk arrays to be operated in multiple gears • Higher gears having higher disk parallelism for peak performance • Reduces average rotational latency which can be significant for small file accesses
PDC	Yes	• Dynamically migrates popular disk data to a subset of disks • Load skewed toward a few disks, remaining spun down • Works because network servers workloads have files with varying popularity
BlueFS	Yes	• Reads first few blocks of a file proactively from server • Targets mobile environment with typically small files that fit in the pre-read blocks • Performance improvement and reduced delay
System shutdown	Yes	• Introduces pre-wake-up to reduce delay overhead • Upcoming idle period predicted using exponential average approach • Used to predict the arrival of the next wake-up signal and disk woken up before that
Self-learning	Yes	• Machine learning–based approach to identify task-specific power management policies • Allows users to specify performance requirements of applications • Trains the system to choose the appropriate spin-down policy at runtime • Delay managed with the specified configuration file

5.2.1.2 PCAP

With program-counter access predictor (PCAP) [9], the delay is equal to the time taken to spin-up a disk. However, the number of such delays is reduced by PCAP's policies. PCAP uses call-site-based prediction to accurately predict when the disk needs to be shut down. Higher accuracies eliminate unnecessary shutdowns and therefore eliminate the delays associated with wrongful shutdowns and resulting spin-ups. The key idea in this approach is that applications have repetitive behavior that is strongly correlated to the particular call-site responsible for a given access activity. Spin-down predictions are made based on past history of disk accesses and the program counter paths that describe a particular call-site responsible for given accesses.

5.2.1.3 Best Shifting

The best shifting [17] prefetching strategy changes a disk access workload by adapting to the current workload and leads to more consistent good performance under varying workload conditions. This turns out to be more energy efficient than fixed prefetching policies and traditional disk spin-down strategies. Best shifting is useful to guard against spikes in energy consumption when faced with pathological or problem workloads. Best shifting is therefore a dynamically self-optimizing algorithm and leads to longer disk idle periods and higher energy savings.

5.2.1.4 SRCMap

Sample-replicate-consolidate mapping (SRCMap) [21] creates an energy proportional storage system. SRCMap optimizes the storage virtualization layer by consolidating the I/O workload on a few physical volumes proportional to the workload. This is done by sampling the volumes and replicating blocks from the working set on other volumes. It then spins down the unused volumes while servicing the workload from the consolidated volumes.

5.2.2 Data Migration

Some techniques strive to reduce energy consumption in disk arrays rather than in individual disks. The target environment usually has a cluster of disks with data replication to ensure reliability, and these techniques take advantage of the fact that there are more than one disks. In this section, we look at a few such techniques.

5.2.2.1 Hibernator

Hibernator [25] tackles the problem of energy consumption for disk arrays in data centers. Data centers have strict performance goals under which they have to operate, so unacceptable delays are not an option. Hibernator proactively monitors the performance metrics to manage possible delays explicitly. It takes advantage of the fact that different disks in an array operate, and can operate, at different speeds. It uses a coarse-grained approach to determine which disks to spin at what speeds, and this information is used to migrate data blocks to disks spinning at appropriate speeds. Based on the monitors, hibernator can spin up all the disks to operate at full speed if required to meet performance agreements.

5.2.2.2 PARAID

RAID is a commonly used technique in storage systems to protect against data loss. Power-aware RAID (PARAID) [23] reduces the power consumption in server-class RAIDs without sacrificing performance or reliability. It uses a skewed striping pattern to determine the number of powered disks based on the system load pattern. It provides gears to satisfy the demand of different system loads. The skewed striping pattern creates an organization like hierarchical overlapping sets of RAIDs, which serve requests via the data blocks or the replicated blocks. Delay is addressed by changing the system to operate at higher gears if necessary. In lower gears, fewer disks are spinning thus reducing the probability of incurring a full rotation wait from a single disk thus reducing delay as well.

5.2.2.3 PDC

The popular data concentration (PDC) [16] is an energy conservation technique for disk array–based network servers. It migrates frequently accessed data to a subset of the disks. This leads to a skewed I/O access load to a subset of the disks. The remaining disks can be spun down to conserve energy. PDC takes advantage of the fact that network server workloads have been observed to have files with different access rates.

5.2.3 Predictive Models

Many disk energy conservation techniques are predicated on accurate prediction of disk accesses and consequently disk idle times. In this section, we look at such techniques.

5.2.3.1 BlueFS

The blue file system (BlueFS) [14] targets energy conservation in portable storage devices such as mobile disks and USB thumb drives. It predicts the disk access demands and fetches the data from the most energy efficient source, which can be a local hard drive, flash drive, or a network server. If BlueFS decides that the most energy efficient device is the hard drive that is in the off state, it fetches the file from alternate devices that contain the file while the disk is spinning up. Once the disk starts spinning, it takes over serving the file, thus minimizing delays.

5.2.3.2 System Shutdown

Hwang and Wu [11] proposed a system shutdown technique to conserve energy by exploiting sleep mode operations. They introduce a technique called pre-wake-up to manage delays. In this technique, disks are woken up based on the prediction of the next wake-up signal. Thus, when the signal actually arrives, the disk is already spinning leading to a reduced delay. It uses an exponential average–based method to model upcoming idle periods.

5.2.3.3 Self-Learning

Self-learning [24] is a machine learning–based approach to power management. It identifies task-specific policies to shut down the disk to conserve energy. Users are allowed to specify performance requirements of certain applications. The system is thus trained to choose an appropriate spin-down policy at runtime. The operating system automatically chooses one of a set of policies that are optimized for specific workload, application scenario or computing platform. Delay is managed with the help of the configuration file specified by the user.

5.2.4 Discussion

Approaches to handle delay cover a wide range from explicit absolute delay management to an implicit reduction in the average delay by reducing the number of delays exposed to the users. Different disk energy management schemes take advantage of the particular characteristics and delay expectations of their target environments to provide different services. We summarize the discussion of different techniques in Table 5.1.

From this discussion, we observe that different energy management techniques approach delay handling in different ways. However, delay is still not a first-class concern for these mechanisms. In the next section, we look at one technique that strives explicitly to reduce delay springing from applying energy management techniques in an enterprise environment.

5.3 Reducing Energy Management Delays in Disks

System-wide alternative retrieval of data (SARD) [22] targets the delay prevalent in enterprise environments when energy management techniques are used. The key insight exploited by SARD is that the machines in an enterprise environment are centrally managed and administrators install the same versions of operating systems and other system- and user-level software. Therefore, all the binaries running on various machines are the same. SARD reduces delay associated with waking up a shutdown disk by retrieving the requested binary page over the network from an already awake machine.

5.3.1 Design of SARD

In SARD, all machines join a p2p overlay network, which enables them to interact with each other in a decentralized and dynamic fashion. Participants run a software that advertises their in-memory applications to others via the overlay. Advertisements enable participants to learn what applications (or

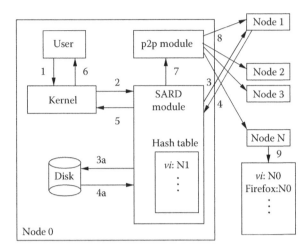

FIGURE 5.1 SARD architecture.

parts thereof) are available in memory of peers. When an application is executed on a node, it can use the remote availability information and decide whether to retrieve the application from the local disk or remote memory. Servicing requests from remote memory helps avoid spin-up delays of powered down disks, and can also improve energy savings by keeping disks in a low-power mode longer.

The impact of SARD is evident by the observation that energy management delays are crucial and all energy management schemes try to avoid unnecessary spin-ups exactly for this reason. Our approach is no different, except that it makes reducing such delays the main objective to encourage adoption of energy management techniques. As stated earlier, up to 10 s of delay per disk spin-up following an energy-related spin-down can be avoided using SARD.

Figure 5.1 shows the architecture of SARD. The only kernel modification is to intercept and reroute disk I/O requests to the SARD module. The module consists of a UDP server, a /proc entry, a hash table, and interfaces required to route the intercepted I/O calls. The hash table contains information about file availability on remote nodes. The content of the hash table is provided by the p2p system (described later). The /proc entry (/proc/SARD) is used to communicate the files currently available in local memory to the p2p system for advertising to remote nodes.

After intercepting an I/O call (in the read_pages() function of standard Linux kernel), SARD checks the hash table to determine alternative sources for serving it. If a remote source is found, a UDP message requesting the image is sent to that node. The corresponding SARD UDP server on the remote node receives and serves the request. Once a reply containing the requested image is received back at the requester, the image is returned to the kernel just as if the request was serviced from the local disk.

Figure 5.1 also shows how an example call is serviced by SARD. Consider the case where a user wants to load *vi*. The user types *vi* at the command path and the *execv* initiates execution by entering the kernel (1) and mapping the image of *vi* into virtual memory. SARD intercepts the I/O requests to the disk (2) and looks up *vi* in the hash table. SARD finds *vi* in its table and that Node 1 has the image in memory. Then, SARD sends out a UDP packet to Node 1 with a request for *vi*'s image (3). The UDP server at Node 1 retrieves the image from local memory and sends it back (4) to SARD at Node 0. Once the reply is received, SARD on Node 0 replies to the kernel I/O request (5). In this case, a disk access is avoided. In case SARD cannot find the requested image in memory of any known remote machine, it routes the request to the disk for servicing (3a, 4a), similarly as in the original kernel.

Additionally, after loading the pages, the kernel module invokes the p2p module (in user space) (7), and sends out a broadcast message (8), announcing to everyone in the overlay that the node (Node 0 in this case) has the application image available in memory, and is willing to share. Other nodes then save this information (9) for later use.

5.3.2 SARD Results

5.3.2.1 Remote Binary Serving

Modern operating systems load portions of applications from disk on-demand. In a typical system that runs many different applications, on-demand accesses essentially translate to reading random pages from the disk. This random access behavior is modeled in the controlled experimental setting by using an application that performs random I/O.

For this purpose, the PostMark [13] benchmark is used, which supports many knobs that essentially allow us to serve files of increasing sizes, in essence emulating on-demand random page loads of varying lengths.

For each case, the time it would take to service the request locally from disk or from remote memory is measured. Figure 5.2 shows the ratio of the time used for servicing a request locally compared to that served remotely. Note that for these measurements the disks were spinning and in ready state, which is the best case scenario if SARD is absent. We observe that for smaller files, the disk performance is poor compared to remote retrieval—servicing from disk takes orders of magnitude longer compared to over the network. The comparative benefit from remote retrieval is somewhat reduced for larger file sizes because the time to retrieve data from the disk improves significantly with increasing file sizes, i.e., large sequential accesses.

Overall, SARD provides improved performance mainly because of the fact that while random accesses have poor I/O performance for disks, the difference between random and sequential accesses is immaterial for SARD, which retrieves contents from memory and does not require any disk movement.

5.3.2.2 Impact of SARD on Remote Machines

In the next set of results, let us study the impact of SARD on remote node performance.

First, how a node's overall performance is impacted when servicing varying rates of page requests is determined. For this purpose, a benchmark was designed that generates a controlled number of remote page requests at one of the test machines. On the other test machine, the Linux kernel was compiled and the compilation time observed for each case as the number of requests generated was increased per second from 1 to the extreme case of 65,536. Figure 5.3 shows the results. The horizontal line shows the average time it takes to compile the kernel on a standard setup without any remote load.

We observe that up to 256 requests per second are serviced without any observable performance degradation, and only 5.59% degradation is observed when as much as 16,384 requests are serviced per second.

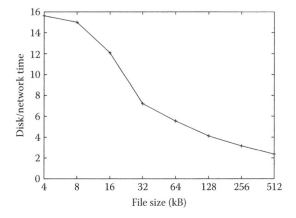

FIGURE 5.2 The ratio of local access time compared to serving the binaries remotely.

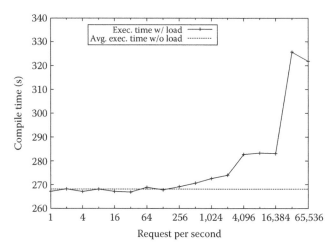

FIGURE 5.3   Impact of servicing remote memory requests.

TABLE 5.2   Requests per Second for Various Applications, and Their Estimated Impact on Remote Node Performance

Application Name	Requests per Second	Impact on Remote Node (%)
cscope	112	0
`make`	6.25	0
PostMark (8 kB)	530	0.90
PostMark (32 kB)	1073	1.63
PostMark (128 kB)	2064	2.14
PostMark (512 kB)	2541	2.94

Also note that the dip in the curve as the requests-per-second are increased to 65,536 is due to lost requests from either network congestion or kernel queue overflow at such large rate.

Second, using the aforementioned information, how several test applications will affect remote nodes was determined. For this experiment, the following applications were used: a *cscope* [19] query on Linux 2.6.22.9 source code, `make` to compile the same kernel, and PostMark [13] with different file sizes. We observe the average rate of remote page requests issued by these applications.

Table 5.2 shows the request rates. We then used the load impact numbers of Figure 5.3 to estimate the impact of the studied applications on a remote node serving the requests. Here, it is assumed that all requests from an application are serviced at a single remote node. In particular, observe that both *cscope* and `make` incur negligible overhead, and PostMark (8 kB) that models on-demand application loading incurs less than 1% overhead. Furthermore, requests may be sent to multiple nodes to reduce the performance impact on individual nodes. By distributing a node's requests across multiple locations in the system, the overall rate of requests serviced at each node can be maintained within acceptable limits.

5.3.3 Simulation Results for SARD's Energy Impact

SARD can reduce the delays associated with energy management without using any additional hardware and avoid the increase in energy consumption that additional hardware would cause. SARD also reduces the overall system energy consumption by servicing I/O requests from remote machines. This is illustrated by the following simulation results.

TABLE 5.3   Disk Energy Consumption Specifications for WD2500JD

State	
Read/write power	10.6 W
Seek power	13.25 W
Idle power	10 W
Standby power	1.8 W
Spin-up energy	148.5 J
Shutdown energy	6.4 J
State transition	
Spin-up time	9 s
Shutdown time	4 s

TABLE 5.4   The Number and Duration of Traces Collected for the Studied Applications

Appl.	Trace Length [h]	Number of Reads	Writes	Referenced [MB] Reads	Writes	User Files (%)	Application Files (%)
Mozilla	45.97	13,005	2483	66.4	19.4	17.92	82.08
Mplayer	3.03	7,980	0	32.3	0	96.37	3.63
Impress	66.76	13,907	1453	92.5	40.1	43.45	56.55
Writer	54.19	7,019	137	43.8	1.2	3.50	96.50
Calc	53.93	5,907	93	36.2	0.4	5.98	94.02
Xemacs	92.04	23,404	1062	162.8	9.4	0.15	99.85

5.3.3.1 Methodology

Detailed traces of user-interactive sessions for each application were obtained by a `strace`-based tracing tool [3] over a number of days. A Western Digital Caviar WD2500JD was used for the simulation with specifications shown in Table 5.3. The WD2500JD has a spin-up time of about 9 s from a sleep state, which is common in high-speed commodity disks.

Table 5.4 shows six desktop applications that are popular in the enterprise environments: *Mozilla* web browser, *Mplayer* music player, *Impress* presentation software, *Writer* word processor, *Calc* spreadsheet, and *Xemacs* text editor. The table also shows trace length and the details of I/O activity. Read and write requests satisfied in the buffer cache are not counted, since they do not cause disk activity. Finally, Table 5.4 illustrates read activity in the applications by separating it into user file accesses and application file accesses. Accesses of application files dominate the read activity for interactive applications. This is true for all traced interactive applications except *Mplayer* traces, which show that a majority of reads are targeting user files, reflecting the primary function of this particular application. It is clear that peer nodes mirroring common application files would meet most of the demand for file reads from the remaining applications, and that the demand for files other than application files represents a significantly smaller fraction of the total observed read requests. Finally, it is assumed that user files are stored on a central file server, which is a common practice in an enterprise environment.

5.3.3.2 Energy Consumption

Serving the I/O from remote machines increases the length of idle periods by eliminating spin-ups required to serve the I/O requests from the local disk. Table 5.5 illustrates the impact of serving I/O requests on the length of idle periods. It shows the number and average length of idle periods for varying fractions of requests served by the remote machines. The case of 100% of requests served locally illustrates the stand-alone workstation that serves all requests from the local disk. By serving more and more requests from other workstations the number of idle times is reduced since the idle periods are concatenated

TABLE 5.5 Number and Average Length of Application Idle Periods as Increasing Number of Requests Are Serviced Remotely

% of Reqs.	Mozilla		Calc		Impress		Writer		Mplayer		Xemacs	
Served Locally	Idle Prds	Length [s]	Idle Prds	Length [s]	Idle Prds	Length [s]	Idle Prds	Length [s]	Idle Prds	Length [s]	Idle Prds	Length [s]
100	165	985	150	1283	227	1048	136	1423	4	2712	95	3477
15	102	1601	89	2170	122	1959	88	2206	4	2713	59	5604
10	88	1858	77	2511	110	2174	80	2427	4	2712	56	5906
5	82	1995	70	2763	87	2752	70	2776	4	2713	49	6751
2	58	2825	52	3724	66	3631	51	3814	4	2713	41	8070
1	49	3346	34	5701	45	5330	40	4867	2	5435	38	8708

resulting in fewer and longer periods. In the case of 1% of requests served locally, the average number of idle periods is reduced by 73.2% and the average length is extended by 205.5%.

In addition, results are shown for serving 2%, 5%, 10%, and 15%, which may be encountered for a small number of workstations in the network.

Reduction in number of periods and lengthening the duration of the idle periods has a twofold impact on energy efficiency. First, fewer number of periods indicates that there are fewer spin-ups required to serve the I/O requests resulting in lower energy spent on powering up the devices and shutting them down. Second, longer idle periods allows the disk to remain in a power saving state also reducing energy consumption. These can be seen in Figure 5.4, which shows distribution of the local disk energy consumption among three categories: Busy—due to serving the I/O requests, Idle—due to waiting for more requests to arrive during timeout interval, and Power-Cycle—due to shutting down and spinning up the disk. We show numbers normalized to the case when a standard energy saving mechanism is used in a stand-alone system, i.e., with 100% requests served locally. All of the states are impacted by SARD. The energy spent serving I/O requests is not significant since most of the applications are interactive with long user think times or they are accessing user files that are mounted on a remote file server. The two largest components are power-cycle and idle energy. The average fraction of energy spent on spinning up and shutting down the disks in the case of local disk only is 69.3%. The energy is reduced as we serve more and more from remote machines and reaches the average fraction of 22.2% for the case of only 1% of requests served locally. Similarly, the time spent in idle is reduced due to fewer timeout periods

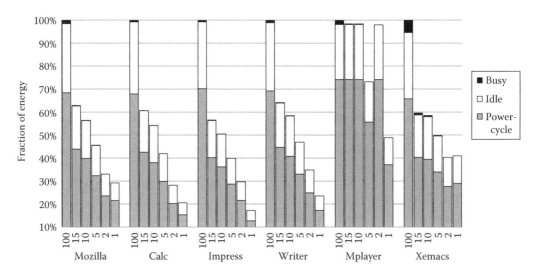

FIGURE 5.4 Breakdown of energy savings as more and more requests are serviced from remote nodes normalized to the case of a stand-alone system (100% local requests).

encountered as we increase fraction of requests served remotely. The average fraction of energy consumed at idle is 28.8% for all requests served locally and is reduced down to 7.9% for when serving only 1% of requests locally. Moreover, compared to an always-on scheme with no energy saving mechanism, the 100% approach provides an average savings of 80.6%, which is further improved to an average of 81.7% for the case with 1% local requests.

It is intuitive that if the system requires fewer disk spin-ups, the overall delays exposed to the users will also be shortened. To verify this, an experiment was designed where network latencies of a 1 Gbps Ethernet connection with at most 20% degradation due to contention modeled randomly were assumed. Table 5.6 illustrates the total delay exposed to the user as the fraction of I/O requests served locally is varied. The average delay across applications is reduced from 1165.5 s for all requests served locally to 312 s, i.e., a 73% reduction, when only 1% of requests is served locally. Reduction in delay has two benefits: First, the user experience is improved since the user will see fewer lags due to disk spinning up. As a result, the user is more likely to use energy management techniques as opposed to turning the energy management off to prevent the irritating delays. Second, the shorter delays will allow the user to accomplish the task quicker, which increases the efficiency of the system.

The increase in energy efficiency of the system can be illustrated by the energy-delay product (EDP) [10]. Figure 5.5 shows the EDP for the studied applications, normalized to the case of a stand-alone system with 100% requests served locally. Each point is calculated by multiplying the total execution time (Table 5.5) of a trace and the corresponding energy consumed (Figure 5.4). On average, the energy-delay product is reduced by 5.8% for the system serving only 1% of requests locally as compared to a stand-alone system with 100% requests serviced locally.

TABLE 5.6 Delay due to Disk Spin-Up as More and More Requests Are Serviced from Remote Node

Local %	Total Delay [s]					
	Mozilla	Calc	Impress	Writer	Mplayer	Xemacs
100	1485	1350	2043	1224	36	855
15	918	801	1098	792	36	531
10	792	693	990	720	36	504
5	738	630	783	630	36	441
2	522	468	594	459	36	369
1	441	306	405	360	18	342

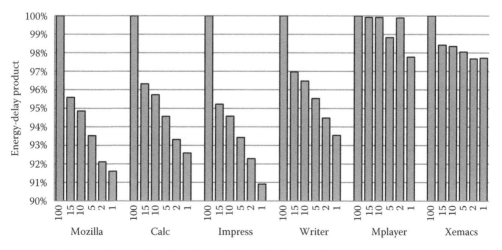

FIGURE 5.5 Energy-delay product for the studied applications under various remote servicing conditions, normalized to case of a stand-alone system (100% local requests).

An observation here is that while the use of p2p overlay may lead to increased CPU energy consumption for processing the network stack, the power impact of managing the overlay is encapsulated by the overall power consumed by SARD. Therefore, there is no need to determine the finer-grained power consumption of the overlay.

We note that longer delays prevent users from adopting energy-saving measures. SARD can remedy this by reducing the exposed delays to the user, who will otherwise opt for no energy management: SARD reduces energy-delay product by 81.63% on average, compared to the always on case. Consequently, "SARD can potentially hasten the adaptation of energy-saving mechanisms."

5.3.4 SARD Case Study

In the next experiment, energy saving and performance impact of SARD are studied using 10 departmental machines used for typical desktop use. Each machine has an Intel Pentium 4 3.0 GHz processor, 1 GB RAM, 40 GB hard disk, and is connected via 1 Gbps Ethernet. The network interface card (NIC) on all the participating nodes in experiment remains in active power-up mode. Therefore, no NIC wake-up latency is observed when retrieving data remotely. If the NIC is put to sleep as well, techniques such as Somniloquy [1] can help mitigate any resulting delays. Nonetheless, given the nodes in question are actively processing, though not from disk, the assumption about the NIC being awake is reasonable.

For a period of 2 weeks, system usage of the workstations were traced. Subsequently, these traces on the systems were replayed, measuring the energy consumed by the machines (using Watts up? PRO power meters). Next, the machines were configured to run SARD, the traces replayed, and the energy measured. The energy consumed by displays is not taken into account as a simple timeout mechanism will work equally well for them. Also, long idle periods of system inactivity are factored out, e.g., 6 PM to 6 AM, when SARD has no benefit over a standard energy-saving mechanism. This focuses on cases where in the absence of SARD, no energy savings are possible. The total energy consumed by the machines over the duration of the week reduced from 165,312 to 156,895 Wh, a saving of 5.1%. This study shows the potential for SARD to provide energy benefits. Moreover, as long as a copy of an application is available in-memory at some machine in the system, disks at other machines can stay off. Thus, with a large number of machines as in enterprise environments, SARD has potential to provide even larger savings by keeping more disks in low-power state.

5.3.5 Discussion

We have looked at SARD, a p2p-based system that mitigates delays associated with disk energy management by allowing sharing of in-memory application images across peers. SARD transparently retrieves application images from nodes on which the applications are already loaded, thus minimizing the need for costly disk accesses and hiding the spin-up delays from the users. Remote application retrieval also provides opportunities for keeping the local disks off, thus saving energy. The evaluation of SARD shows an average performance improvement of 33.5% for typical desktop applications, and a case study with real usage shows average energy savings of 5.1%. Most importantly, a 71% reduction in delays associated with energy management is observed. Finally, the effect of remote application retrieval on remote nodes is shown to be minimal (less than 3%), and demonstrates that SARD can serve as a practical and effective tool for mitigating energy management delays, which also improves energy efficiency in enterprise environments.

5.4 Summary

In this chapter, we have explored the relationship between energy-management techniques for disks and the wake-up latency exposed to the user. We argue that delay management should be considered as a first-class concern for any power-saving technique lest they remain unused by users not willing to accept the disk wake-up latency. To this end, we have presented a survey of currently used power management techniques and their associated delays. We have also presented SARD, a project with the primary concern of reducing hiding disk wake-up latency from users. Thus, SARD is useful as a foil for other power-saving techniques because it encourages users to adopt such techniques without having to face wake-up delays. We believe the techniques used in SARD can lead to the design of more advanced energy management techniques with better integrated delay management policies.

Acknowledgments

This work is supported by the U.S. National Science Foundation through grants CNS-1016408, CNS-1016793, and CNS-1016198.

References

1. Y. Agarwal, S. Hodges, R. Chandra, J. Scott, P. Bahl, and R. Gupta. Somniloquy: Augmenting network interfaces to reduce PC energy usage. In *Proceedings of the NSDI*, Boston, MA, 2009.
2. R. Bianchini and R. Rajamony. Power and energy management for server systems. Technical Report DCS-TR-528, Department of Computer Science, Rutgers University, Piscataway, NJ, June 2003.
3. A. R. Butt, C. Gniady, and Y. Charlie Hu. The performance impact of kernel prefetching on buffer cache replacement algorithms. *IEEE Transactions on Computers*, 56(7):889–908, 2007.
4. J. Chase, D. Anderson, P. Thackar, A. Vahdat, and R. Boyle. Managing energy and server resources in hosting centers. In *Proceedings of the SOSP*, Banff, Alberta, Canada, 2001.
5. I. Crk and C. Gniady. Context-aware mechanisms for reducing interactive delays of energy management in disks. In *Proceedings of the USENIX ATC*, Berkeley, CA, 2008.
6. F. Douglis, P. Krishnan, and B. Bershad. Adaptive disk spin-down policies for mobile computers. In *Proceedings of the USENIX Symposium on Mobile and Location-Independent Computing*, Ann Arbor, MI, 1995.
7. M. Elnozahy, M. Kistler, and R. Rajamony. Energy conservation policies for web servers. In *Proceedings of the USITS*, Seattle, WA, 2003.
8. C. Gniady, A. R. Butt, Y. Charlie Hu, and Y.-H. Lu. Program counter-based prediction techniques for dynamic power management. *IEEE Transactions on Computers*, 55(6):641–658, 2006.
9. C. Gniady, Y. Charlie Hu, and Y.-H. Lu. Program counter based techniques for dynamic power management. In *Proceedings of the Sixth Symposium on Operating Systems Design and Implementation*, San Francisco, CA, December 2004.
10. R. Gonzalez and M. Horowitz. Energy dissipation in general purpose microprocessors. *IEEE Journal of Solid-State Circuits*, 31(9):1277–1284, September 1996.
11. C.-H. Hwang and A. C. H. Wu. A predictive system shutdown method for energy saving of event driven computation. *ACM Transactions on Design Automation of Electronic Systems*, 5(2):226–241, April 2000.
12. Y.-H. Lu, E.-Y. Chung, T. Simunic, L. Benini, and G. De Micheli. Quantitative comparison of power management algorithms. In *Proceedings of the DATE*, Paris, France, 2000.
13. Network Appliance Inc. PostMark file system performance benchmark, January 2008. http://www.netapp.com/tech_library/3022.html
14. E. B. Nightingale and J. Flinn. Energy efficiency and storage flexibility in the blue file system. In *Proceedings of the OSDI*, San Francisco, CA, 2004.
15. E. Pinheiro, R. Bianchini, E. V. Carrera, and T. Heath. Load balancing and unbalancing for power and performance in cluster-based systems. In *Proceedings of the Workshop on Compilers and Operating Systems for Low Power*, Barcelona, Spain, 2001.

16. E. Pinheiro and R. Bianchini. Energy conservation techniques for disk array-based servers. In *ICS '04: Proceedings of the 18th Annual International Conference on Supercomputing*, Saint Malo, France, pp. 68–78, 2004 ACM, New York.

17. J. P. Rybczynski, D. D. E. Long, and A. Amer. Expecting the unexpected: Adaptation for predictive energy conservation. In *Proceedings of the StorageSS*, Fairfax, VA, 2005.

18. P. Sehgal, V. Tarasov, and E. Zadok. Evaluating performance and energy in file system server workloads. In *FAST2010: Proceedings of the Eighth USENIX Conference on File and Storage Technologies*, San Jose, CA, February 2010.

19. J. Steffen and H.-B. Bröker. CSCOPE, January 2008. http://cscope.sourceforge.net/

20. Tufts. Computers and energy efficiency. *Online Specification*, 2008. http://www.tufts.edu/tie/tci/Computers.html

21. A. Verma, R. Koller, L. Useche, and R. Rangaswami. Srcmap: Energy proportional storage using dynamic consolidation. In *FAST2010: Proceedings of the Eighth USENIX Conference on File and Storage Technologies*, San Jose, CA, February 2010.

22. G. Wang, A. R. But, C. Gniady, and P. Bhattacharjee. A light-weight approach to reducing energy management delays in disks. In *Proceedings of the International Green Computing Conference*, Chicago, IL, August 2010.

23. C. Weddle, M. Oldham, J. Qian, A.-I. Andy Wang, P. Reiher, and G. Kuenning. Paraid: A gear-shifting power-aware raid. In *Proceedings of the USENIX FAST*, San Jose, CA, 2007.

24. A. Weissel and F. Bellosa. Self-learning hard disk power management for mobile devices. In *Proceedings of the IWSSPS*, Seoul, Korea, 2006.

25. Q. Zhu, Z. Chen, L. Tan, Y. Zhou, K. Keeton, and J. Wilkes. Hibernator: Helping disk arrays sleep through the winter. *SIGOPS Operating Systems Reviews*, 39(5):177–190, 2005.

Power-Efficient Strategies for Storage Systems: A Survey

Medha Bhadkamkar
Washington State University

Behrooz Shirazi
Washington State University

6.1 Introduction

Data in the digital world are increasing at a tremendous pace. A report by the International Data Corporation (IDC) estimates that by 2011, the data storage requirement would grow up to 2 zettabytes (20^{21} bytes). By 2011, every household can have over 1 TB of data footprint, which can be stored on personal computers (PCs) or laptops, external storage devices, and digital data from smart phones or cameras. This increase is attributed to widely used applications that are becoming increasingly popular such as media and Internet applications including e-mail, search engines, social networking sites and photo (Flickr) and video sharing (YouTube), database applications that generate structured data, applications such as Office tools including presentations, spread sheets, or documents that create unstructured data, data-intensive applications such as animation rendering and scientific computing, or even digitalized data created by personal devices such as cell phones, smart phones, or personal digital assistants (PDAs). Since storage capacities are cheap ($0.21/GB for hard disks), once created, most of these data are never destroyed, even after their utility ends. People continue to archive their personal outdated data on the local disk drives on their PCs or on the cloud. Organizations transfer outdated data on archival storage for future reference if the need arises. Further, generating duplicates of existing data is fairly common. For instance, copies of data are maintained, possibly in distant geographical locations, for backups to avoid loss of data due to system failures or natural disasters. Data can also be duplicated frequently by e-mail servers that handle content such as e-mail messages or attachments, and several versions of such content that are vastly similar may exist.

Numerous storage devices are required to store and access such massive amounts of data that are generated every day, while maintaining the necessary throughput. This requires a significant amount of energy. It is estimated that the storage component of a data center alone attributes to approximately 25% of the overall energy consumption. The Environment Protection Agency (EPA) report [1] indicates that in 2011, the annual cost of energy consumption for storage in data centers is approximately $7.4 billion. Further, current trends indicate that these numbers are likely to double in the next 5 years.

These numbers indicate that the growing energy demands for data storage are of utmost concern and there is a need to build efficient and sustainable data centers that reduce the environmental impact. A substantial amount of recent work has focused on improving the power efficiency of storage in all aspects—data consolidation, migration, reduction in the data footprint, and tiered storage. The proposed strategies minimize the power consumption for both single-disk environments such as laptops, where the goal is to increase the battery life, or PCs, where the goal is to achieve cost benefits, as well as multidisk environments, such as data centers and servers, where the goal is to reduce the operational costs. However, it should be noted that single-disk power management strategies cannot be applied to multidisk environments [2].

This chapter provides a comprehensive overview of the power-efficient storage strategies incorporated in enterprise level, multidisk environments such as data centers as well as single-disk systems, with a greater emphasis on the former. This chapter is organized as follows. It first identifies the sources of power consumption in storage systems. Next, it discusses how the various storage devices are organized in the storage hierarchy followed by a description of the storage devices. It then discusses the existing popular architectures used in enterprise systems. Power-saving strategies using these devices and the architectures are described in the following. Finally, we discuss some research directions that can aid the development of power-aware systems.

6.2 Sources of Power Consumption in Storage Systems

Power consumption is directly proportional to the hardware that includes the storage devices, interconnects, and the cooling infrastructure required to maintain the reliability of the devices. In a typical data center that drives hundreds of thousands of disk drives that are actively spinning at any given time, the power consumption is over 27% of the overall power used in data centers [3]. The heat generated by these components can affect the reliability [4] of the drives where for every 10°C increase in temperatures, the failure rate for disk drive doubles [5]. Hence, cooling mechanisms are installed to maintain the desired temperatures.

Several data management operations such as encryption, replication, and backups and archival need to be carried out on the data content to ensure reliability, data protection against failures and natural disasters, or data recovery after failure. Additional processing power, hardware, and networking resources required for these operations add to the overall power usage.

6.3 Storage Devices

Disk drives are the most ubiquitous secondary storage devices. Flash-based storage devices such as solid-state drives (SSDs) are now being used higher in the storage stack as nonvolatile read caches or write buffers to improve performance, reliability, or power efficiency and hold the potential to replace disk drives in the future. These block-based devices export the same logical block addressing to hide the underlying physical device characteristics from the file system interface. The data blocks are accessed by the device controller based on the geometry, for instance, hard disk drive controllers use CHS (Cylinder, Head, Sector) addressing scheme. In this section, we compare the device power and performance characteristics for SSDs and hard drives.

6.3.1 Hard Disk Drives (HDDs)

Hard disk drives are ubiquitous as the secondary storage device due to their cost-effectiveness and performance. Enterprise level, high-end disks are used for *online* storage while low-end, power-efficient disks are used in *nearline* or *offline* storage of data.

Hard drives are electromechanical devices that consist of several platters or disks coated with a magnetic material. These platters rotate around a spindle, typically at speeds of 7,200 or 10,000 rpm. Data is recorded on these platters and read by the disk heads. Binary values of the data are recorded on the platters by magnetizing the material to represent "0" or "1." Every platter surface has a corresponding disk head that is controlled by the actuator arm assembly. The actuator assembly is responsible for moving the disk head to the desired location and ensuring that the disk head does not make direct contact with the disk platter surface. If the heads make direct contact with the platters, it results in a "head crash" and can permanently damage the disk drive. Several optimizations discussed later in the chapter have been implemented to reduce the power consumption of disk drives, but there still remains room for further improvement.

Depending on its state of operation and the components engaged, disk drives have several modes of power consumption and corresponding modes of operations explained as follows. It should be noted that the number of modes may vary based on the manufacturer and some drives may have intermediate power modes as well.

1. *Active*: The disk drive is in active mode and its power consumption depends on the type of workload—sequential or random, reads or writes—which determines how the various components of the drive are engaged.
2. *Idle*: In this mode, the disk drive is active and spinning at the specified rpm, but no requests are being serviced, thus there is no power consumption by the actuator motors to move the disk heads.
3. *Standby*: Disk drives can be spun down in periods of low activity to conserve energy consumed by the spindle motor. Additional power savings can be achieved if the actuator assembly is disengaged and the disk heads are parked away from the disk platters. The heads can be loaded back on the platters in less than a second. Power continues to be consumed by the onboard components in this mode.
4. *Sleeping*: In the mode, the drive is completely shut down. No power is consumed by the spindle motors, the arm assembly, or the onboard electronics. It can take several seconds for the disk drive to start up before it can start servicing incoming requests.

6.3.2 Solid-State Drives

Most SSDs store persistent data using nonvolatile NAND flash memory and can be designed for energy efficiency, high performance, or reliability. They provide numerous advantages such as higher bandwidth, random read and write performance orders of magnitude higher than disk drives, and greater reliability and energy efficiency due to the absence of moving electromechanical components. They can be organized in as 4–10 banks of flash chips that can be accessed concurrently for high performance. It has been shown that for sequential reads and writes, SSDs are up to 10× and 5× faster than disk drives, respectively. For random reads and writes they are 200× and 135× faster, respectively [6]. However, they are not cost-effective when compared to disk drives and thus their popularity remains restricted. Several storage device vendors sell "hybrid" drives that combine hard drives with SSDs or other nonvolatile flash-based memory in a single device to leverage the advantages of both devices.

For flash-based SSDs, writes to a nonempty page incur an erase to this page before data are written. SSD's have limited erase cycles (of the order hundreds of thousands) depending on the device manufacturer and for every block and erasures beyond this limit can lead to the device being unreliable. The SSD device controller is responsible for performing SSD-specific data management optimizations such

TABLE 6.1   Comparison of Disk Drives and SSDs

Device Type	Vendors and Model	Capacity (GB)	Power Idle (W)	Performance/W		$/GB
				Read	Write (BW/W)	
Consumer SATA SSD	Western Digital SiliconEdge Blue	256	2/0.6	125	125/70	3.43
Enterprise SATA SSD	Intel X-25M	80	2.5/0.06	125	35	2.11
Enterprise SATA SSD	Intel X-25E	64	2.4/0.06	125	35	9.5
PATA SSD	Solidata P1	256	2/0.5	55	40	3.5
PATA SSD	Solidata P2	64	2/0.5	55	40	9.3
Enterprise SATA HDD	Western Digital RE4-GP	2000	6.8/3.7	16	16	0.11
Enterprise SATA HDD	Western Digital RE4	2000	10.7/8.2	12.8	12.8	0.10
Enterprise SCSI HDD	Seagate ST3300007LW	300	16.4/10.1	19.5	19.5	0.58

as wear leveling, handling write amplification, and garbage collection to improve the write performance and reliability. Since SSDs do not have power hungry motors and electromechanical components, their power consumption is at least 3× lower than traditional disk drives.

There are two implementations for SSDs, multilevel cell (MLC) and single-level cell (SLC). The MLCs have larger access times and are less reliable than the SLCs but are cost-effective. The shortcomings of MLCs can be overcome by optimizations such as smart writing algorithms and wear leveling.

Table 6.1 compares some of the characteristics of SSDs and disk drives manufacture by different vendors for both enterprise and consumer level devices. The values are obtained from the data sheets for each of the devices. While some of these devices have recently been introduced, others are popular choices and are widely available. While the characteristics of Intel X-25M and Intel X-25E appear to be the same, these two devices are based on the MLC and the SLC architecture, respectively and differ in their reliability. X25-M has a mean time before failure (MTBF) of 1.2 million hours while the X25-E has an MTBF of 2 million hours. Similarly, the Solidata P1 and P2 drives are based on the MLC and SLC architectures, respectively. Western Digital RE4-GP hard drive is a green drive optimized for greater power efficiency while Western Digital RE4 is optimized for greater performance. RE4-GP is more recent than the Cavier Green drive and exhibits greater power savings and performance.

6.4 Power-Efficient Strategies

We identify three major categories for implementing power-saving strategies for storage systems as follows. Each of these is explained in detail in the following sections:

1. *Hardware optimizations*: Strategies in this category achieve power savings by managing and controlling the underlying hardware, which includes powering down devices or switching them to low-power modes during idle times in the workload. Optimizations can also be incorporated in hardware to improve power savings, such as optimizing the head seek operation in disk drives or increasing the density of data stored on drives. In addition to hardware designed for power savings, *tiered storage* where devices with different power-consuming characteristics are incorporated in the storage hierarchy for maximum power savings.

2. *Data management optimizations*: This category includes power management techniques that manage data storage and retrieval. This includes managing data layout, intelligent caching and

prefetching, reducing the overall data footprint, data consolidation, and resource allocation. Data layout management can be done by reorganizing data on storage devices for power-efficient accesses. Caching and prefetching techniques reduce the number of accesses to the hard drive and thus result in power savings. Reducing the data footprint directly translates to power savings since as the amount of data to be stored reduces, it in turn decreases the number of devices, hardware components, networking resources, and cooling requirements used to store and access the data. The overall data footprint can be reduced by two techniques: *data compression*, where different compaction techniques are used to compress the data and *data de-duplication*, where duplicate copies of the data are eliminated. Data management also includes *consolidation* of data on a smaller number of devices or *resource allocation techniques* that direct requests to the underlying devices such that it maximizes the power savings by optimizing the utilization and increasing idle periods for devices, so that they can be switched to low-power consuming modes.

3. *Storage virtualization*: Storage virtualization is the pooling of heterogeneous physical storage devices to appear as a single storage device. It offers several advantages such as centralized management of resources, aggregation of resources for cost efficiency, and increased utilization and energy efficiency.

6.4.1　Trace Analysis

In this section, we also present some results from our analysis of traces obtained from different servers at the Microsoft Research Lab, Cambridge (MSR) [7] to gain further insight into different workload characteristics such as the areas of frequent access, the randomness of accesses, and the device idle time. The goal of this trace analysis is twofold. First, it identifies important workload characteristics that show that there is ample scope for implementing power-efficient strategies. Second, it implies that certain characteristics depend of the type of workload and hence the optimum power-savings strategies, such as spinning disks down or colocating the frequently accessed data, specific to the workload should be implemented with some knowledge of workload characteristics to derive maximum power.

For the sake of brevity, in this chapter, we show results for all the volumes for four different workloads: *usr*, which consists of accesses to the home directories of all the users, *web*, which consists of accesses to the web server, *rsrch*, which contains files related to research projects, and *src1*, which is one of the source control repository. The I/O traces used in this study were collected over a 7 day period. A snapshot of the *usr* trace on disk 0 is given in Figure 6.1. The traces comprise the following per-request I/O information. Timestamp of the request in Windows time, workload name, volume or disk number, the type of request (read or write), the block offset accessed in a request followed by the length of the request in bytes, and the per-request response time.

Table 6.2 summarizes the relevant characteristics of these traces for every disk (volume) included in the server. It indicates the number of disks used in each workload. The table also shows the total data accessed in disk blocks and the ratio of reads and writes. The table also shows the total data accessed in disk blocks and the ratio of reads and writes. As can be observed, the selected workloads are a mix of read

```
128166372002993263, usr, 0, Read, 10995532800, 16384, 30123
128166372010284761, usr, 0, Write, 3207667712, 24576, 82327
128166372010327568, usr, 0, Write, 3154132992, 4096, 39521
128166372010484633, usr, 0, Write, 3154124800, 4096, 38705
128166372026523960, usr, 0, Write, 9823862784, 4096, 92960
128166372026593224, usr, 0, Write, 2037374976, 4096, 179945
128166372046503999, usr, 0, Write, 9823862784, 4096, 112794
128166372046550633, usr, 0, Write, 3131813888, 4096, 222409
128166372046730497, usr, 0, Write, 3131846656, 4096, 42544
```

FIGURE 6.1　Snapshot of I/O traces used in this study.

TABLE 6.2 Workload Characteristics

| Workload | Disk No. | Data Accessed (No. of Blocks) | | | Total No. of Requests | No. of Seeks |
		Reads	Writes	Total		
	0	9,273,455	3,857,714	12,764,783	2,237,889	1,803,980
usr	1	14,737,649	545,998,912	560,736,561	45,283,980	4,317,691
	2	6,958,640	109,101,787	116,060,427	1,057,0046	202,444
	0	3,186,736	4,575,092	7,761,828	2,029,945	1,132,923
web	1	181,147	1,009,484	1,190,631	160,891	32,132
	2	216,502	69,193,260	69,409,762	5,175,368	3,001,151
	3	111,591	188,252	299,843	31,380	18
	0	364,955	2,888,684	3,253,639	1,433,655	837,240
rsrch	1	150	42,251	42,401	13,780	190
	2	136,364	75,699	212,063	207,587	44,764
	0	192,475,082	213,435,847	405,910,929	37,415,613	230,447
src1	1	390,791,272	7,963,243	398,754,515	45,746,222	945,964
	2	2,338,547	11,615,326	13,953,873	1,907,773	97,633

and write intensive workloads with disk 1 on *src* showing maximum read traffic of 98% while disk 2 on the *web* workload showing maximum write accesses of 99.6%. The next column shows the total number of requests in the workload. The final column indicates the number of seek operations the head performs for the given workload. A seek operation is incurred when a nonsequential access is performed. The length of the seek operation determines the seek latencies incurred and varies based on the disk vendor.

The following sections present our analysis of some of the relevant workloads characteristics.

6.4.2 Hardware Optimizations

Several optimizations can be incorporated in the hardware to reduce power consumption. Since disk drives are the most popular power-consuming devices, numerous studies have explored hardware solutions to reduce their power consumption.

6.4.2.1 Optimizing Disk Drive Technology

First, energy consumption can be reduced in disk drives by controlling the acceleration of the spindle motor. In a conventional disk drive, for every random access, the disk drive head typically locates the destination track and reaches it with maximum acceleration, incurring seek latency. It also incurs a rotational latency until it waits for the desired sector to position under the disk head so that data can be read or written. Considering the frequent number of random accesses in most real-life workloads where several thousand blocks are accessed from a single-hard drive, the rapid acceleration of the disk heads results in increased energy usage. Power-efficient disk drives optimize this operation by controlling the acceleration of the heads during the seek operation, so that the disk head reaches the destination sector just in time before the desired sector reaches the disk head. Since the heads now accelerate at a slower pace, power savings are achieved. However, controlling disk head acceleration increases the seek latencies that in turn degrades the system response time.

The contact start/stop (CSS) or the load/unload strategy enables turning off the power to the actuator assembly and the disk heads thereby increasing power savings. In standard drives, these components require power even during idle times since the actuator assembly that controls the disk heads are still enabled and continue to consume energy while ensuring that the disk heads do not crash into the platters. For instance, an enterprise level disk (Cheetah 15 K.7 SAS) from Seagate consumes approximately 12 W when idle (16 W when busy) [8]. Further, in some drives, the heads are periodically moved to random locations during disk idle periods. This location is determined by several factors such as wear or power reduction by minimizing the effect of air resistance, head cycling, or even accumulation of particles on the drive heads. CSS-based drives park the disk heads on a "landing zone," an area on the innermost periphery of the platter containing no data, when the drive is spun down. However, greater power is required to overcome the *stiction* forces, attributed to the heads sticking to the platters when the disk was spun down. Stiction force cause components with smooth surfaces to "stick" or adhere to each other requiring significant power to separate them. To overcome this drawback, CSS drives use a start/stop zone, which is a region on the drive that contains no data and is roughened or texturized. However, this limits the magnetic spacing on the platters thereby reducing the storage density of the drives. Maximum density can be achieved when the platters are smooth. Further, reliability was also compromised in the CSS strategy due to the additional wear arising from the direct contact of the disk heads and the platters thus limiting the number of start/stop cycles. These drawbacks were overcome by the load/unload technology proposed by IBM [9] and first implemented by Hitachi [10] in their disk drives. The load/unload strategy enables greater power savings and also improves reliability and capacity in disk drives. In the load/unload technology, the disk heads and I/O channels are disengaged when the disk is spun down. The disk heads are lifted off the platters and are parked in a safe location referred to as "ramps," adjacent to the outer periphery of the disk platters. When the disk becomes active, the heads are loaded back on the platters once the drive reaches the appropriate rotational velocity. This technology can be extended to offer greater power savings by parking the disk heads in their ramps when the disk is in idle mode. The CSS and the load/unload strategy are shown in Figure 6.2. It should be noted that this strategy of parking the heads and spinning down the drives can degrade the overall system I/O performance, since some latency is incurred in spinning the disks back up (~15 s) and loading the heads on the platters (~1 s) [11].

Green hard drives [12,13] use the aforementioned techniques to achieve power savings and are now being widely used. These drives can have a capacity of up to 2 TB and can deliver power savings as much as 50% (a difference of approximately 5 W) over the standard disk drives. This makes them more cost-effective since they can handle increased storage capacities without exceeding the power budget.

There has been a consistent increase in the areal density for disk drives, from a modest 1 kB per square inch, when magnetic recording was first introduced in the 1960s, to 1 TB per square inch in 2010. Consequently, physical space, energy, and cooling requirements were reduced as the drives became smaller. The areal density has increased and power consumption is reduced since the load that is driven by the spindle motor is reduced. As mentioned earlier, the spindle motor is the highest power-consuming component of the disk drive since it is responsible for maintaining the spinning speed of the platters and the disk heads. If the number of platters and the corresponding disk heads of the disk drive are reduced, it results in direct power savings since it reduces the number of platters and disk heads thereby reducing the power consumed by the spindle motor to start and continuously spin the motor. Samsung drives [13] employ this principle and increase the recording density for the disk drives. This also increases

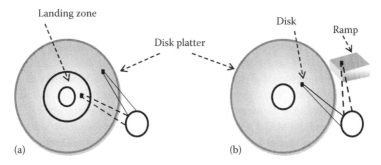

FIGURE 6.2   (a) CSS and (b) load/unload technology. The position of the disk head arm during idle periods is shown with dotted lines.

reliability as the number of disk heads are reduced, lowering the probability of disk-head-related failures. However, disk drives have reached their maximum limit of 1 TB per square inch for perpendicular recording and a new technique called *shingled writes* [14,15] has been proposed to further increase the areal density of disk drives. It has been expected that shingled drives can increase the density by a minimum of 2.3× [16]. In shingled writing, a track that contains data can be partially overwritten by another track through a specialized writing technique. Only a small portion of the previous track does not contain overlapping data. However, this small portion is readable by heads using the GMR (Giant Magneto-Resistive) technique. Random access reads can be serviced in drives that use shingling using the traditional technique. Appending writes to a track is straightforward; however, if data have to be overwritten, it potentially results in rewriting several adjacent tracks due to the shingling, thus making it inefficient. Thus, shingled write disks can be used in archival systems where writes are sequential and reads may be random.

6.4.2.2 Managing Device Power Usage

Another power-saving strategy that has been explored in several studies is to dynamically modulate the spinning speeds (rpms) in disk drives to conserve energy [17–21]. The spindle motor is the highest power-consuming component in a disk drive and has been known to account for 81% of the energy consumed by the disk when idle for server class disks [22]. Power consumption by the spinning motor is directly proportional to its spinning speeds and hence power savings can be achieved by controlling the rotational speed of the motor. The rotational speed of the disk can be dynamically controlled by turning on and off the power supply to the spindle motor at a specific frequency based on the desired rpm required [23]. Previous studies [17] have shown that dynamically modulating the rotational speed of disk drives has more scope for power savings than completely spinning down disks since a much larger torque, which consumes a significant amount of energy, is required to overcome the stiction force to start spinning the disk again. However, this has been vastly improved by the advent of the load/unload technology as mentioned earlier. Dynamically controlling the rotational speeds in drives is particularly useful when the idle times between workloads are low and the benefits of powering down disks or switching to low-power modes are outweighed by the energy and performance overheads incurred in spinning them back up. However, slower spinning speeds can affect the rotational latencies and the transfer rate when servicing requests, thereby affecting performance.

Redundant Array of Independent Disks (RAID) systems that enable power management by spinning down disks that are inactive are called massive array of idle disks or MAID [24]. These arrays can be constructed using low cost disks such as Serial ATA (SATA) disks. First-generation MAIDs provided significant power savings by limiting the number of active devices at a given instance. However, if I/Os were requested to an array that was powered down, that array had to be powered up to service the request while another array had to be spun down to maintain the power savings resulting in large overheads unacceptable in some workloads. This restricted the workloads that could use MAIDs for power efficiency such as offline data storage where performance is of low priority. Second-generation MAIDs provide intelligent power management and can be applied for most workloads with minimal impact on performance. Second-generation MAIDs operate in three power saving levels and the appropriate level can be chosen based on the application workload access patterns and performance requirements. Level 1 provides modest power savings of 15%–20% by unloading the disk heads. In addition to this, Level 2 also slows the spinning speeds of disks, which result in 30%–50% power savings. Level 3 exhibits maximum power savings up to 70% by completely spinning down disk drives. Level 0 indicates that power management is disabled and that the array operates at the peak RAID performance. The performance penalty is less than a second for MAID 1 to move the heads back in position on the platters to actively service requests, it is between 12 and 15 s in Level 2 to bring the disks up to the regular spinning speed and as high as 45 s in Level 3 required to power up the devices. To ensure that the disks are turned off for the maximum possible duration, MAIDs employ one of the two techniques: *migration*, where data are moved to the active arrays, and *caching*, where the most recently used data are cached. Both techniques have their trade-offs. Migration is space efficient since there is no duplication of data and can have acceptable overheads if the data to be migrated is small. Caching creates duplicate copies of data on a small number of disks and has no overheads of moving data and maintaining the data structures that contain the mapping across all the devices. Caching also maintains mapping though the size of the map is proportional to the size of the disks used for caching. There are several design choices that can be made based on the usage patterns of the MAID systems and the type of devices used when caching is used. For instance, if the caching policy should vary for reads and writes or if the data should be separated from the metadata operations.

While Table 6.2 presents the ratio of total reads and writes, the following Figure 6.3 shows the ratio of all the unique read and write accesses for each of the workload used in this study. Unique accesses refer to all the data that are accessed at least once. The number of unique accesses is important to determine policies for caching read data or buffering write data and this may differ from the overall read and write accessed. For instance, while the table shows for volume 1 on workload *usr*, the writes are 97% of the overall accesses making it a write intensive workload, the Figure 6.3 shows that the unique writes are just about 5% of the overall unique accesses. Thus, for this workload, by using a buffer that can store all the 5% unique writes, or migrating all the writes, the disks in MAID can be powered down for longer durations.

Several studies in reducing energy consumption focus on powering down disk drives, or switching to a low-power mode during idle times. This strategy is workload dependent and can have significant power savings when the workloads are bursty with adequate idle times that can be used to power down the devices. However, for workloads that do not exhibit any idle periods, the observed power savings are modest, if not worse. This is because disk drives consume up to $3\times$ more energy when spinning back up than when rotating with constant acceleration. Hence, a significant amount of energy is consumed if the drives are powered down and turned back on several times. Further, when an idle drive is powered up, a spin-up latency is incurred while the spindle motor reaches its rotational velocity and begins servicing requests. I/O requests waiting for data from the drives can thus experience higher than usual latencies before the request is serviced. For instance, for the *usr* and the *web* workload, if the idle time is distributed

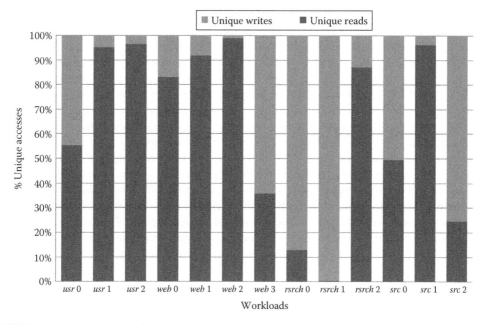

FIGURE 6.3 Ratio of unique reads and writes.

such that all the requests are serviced without any idle times and disk operates in the sleep mode for the remainder of the time, power savings up to 71% can be observed for the *usr* workload and as much as 80% for the *web* workload. However, the distribution of idle times is workload specific and it can be utilized to power down the disk only if it is sufficiently large enough. If for the given workloads, the disk is powered down only if the idle period exceeds the time to power down and to power the device back up, the ratio changes to 17% for the *web* and 29% for the *usr* workload.

To avoid performance degradation and maximize power savings, it is important to know when to put the drives in a low-power state and how long it should remain in that state. However, since future idle-time characteristics of a workload are unknown, answering these questions are a challenge. Riska et al. [11] explore when and how long-disk drives can be turned off without affecting performance. They propose monitoring the length of idle time intervals in a workload and the user-perceived request response time to control the performance degradation within a predefined target. Given a predefined target and a power-saving mode, their model can determine the idle time interval after which the system can switch to the specified power-saving mode and the total time the system remains in that power-saving mode.

For "bursty" workloads where requests arrive in batches or "bursts" and exhibit adequate idle periods, switching disk drives to low-power modes is straightforward and can be carried out transparent to the user. This is better elucidated from the following Figure 6.4, which shows the disk accesses over time for a 7 day period for the *rsrch* workload. The Y-axis shows the logical block addresses (LBA's) accessed in the volume while the X-axis shows the timeline for the week. Figure 6.4a shows high disk utilization for volume 0. This implies that the power-saving gains achieved by powering down disks can be relatively modest. Figure 6.4b for volume 2, however, shows high periods of inactivity and is an example of a bursty workload. Clearly, powering down disk drives is a reasonable approach to achieve power savings here. It should be noted that for volume 0, due to the scale used for the X-axis, idle periods that are several seconds long and can be used for powering down devices, are not visible. However, since the devices are powered on several times in comparison to volume 2, the relative power savings remain low for volume 0.

However, not all workloads exhibit these characteristics and several optimizations have been proposed to increase idle periods in disk drives. Idle periods can be increased by employing several techniques that exhibit low overheads as follows. First, caching and prefetching techniques [25,26] ensure that the data are available in cache and reduce the number of accesses to the disks. It has been previously shown before that while maintaining an optimal cache hit ratio ensures performance efficiency, it results in energy inefficiency [27]. For improved energy efficiency, data from the inactive storage devices need to be maintained longer in cache to ensure the idle times of these devices are prolonged. Partition Based-Least Recently Used (PB-LRU) cache replacement algorithm [28] is an optimized algorithm that dynamically estimates the partitions size in cache for each disk in the system to improve energy efficiency.

Next, concentrating popular data on limited number of disks [29–33] can limit the number of active disks at a given time. The drives in low-power modes are powered up only if requests to nonpopular content not present in the cache are made. However, this can negatively impact the I/O performance since limiting the number of disks reduces the benefits of parallelism. Figure 6.5 shows the percentage of the top 10% most frequently accesses data of the total accesses in the different volumes in the workloads. Each column in the graphs represents a volume for the given workload as labeled in the X-axis. The Y-axis indicates the percentage of data that comprise the top 10% frequent accesses. The higher numbers in the graphs indicate that a small portion of the data is accessed over and over again. Thus, workloads such as *usr*_using volumes 0 and 1 can benefit greatly if the highly popular data is moved to a limited number of disks.

Finally, disk burstiness refers to clustering of the disk accesses in time by the operating system to increase the idle periods and the device utilization and can be implemented to improve power savings. Rather than clustering requests, operating systems deliberately try to maintain uniformity in the temporal access patterns for disk drives to prevent any I/O bottlenecks. While this improves I/O performance,

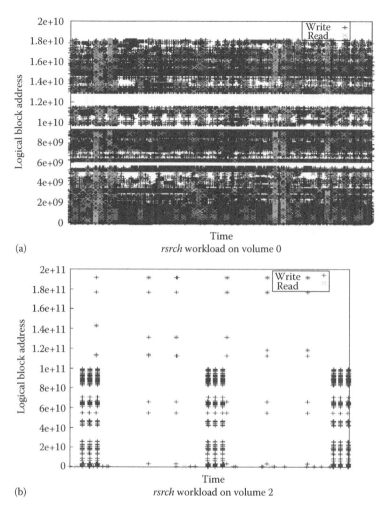

(a) *rsrch* workload on volume 0

(b) *rsrch* workload on volume 2

FIGURE 6.4 Disk access patterns for *rsrch* workload using volumes (a) 0 and (b) 2, respectively.

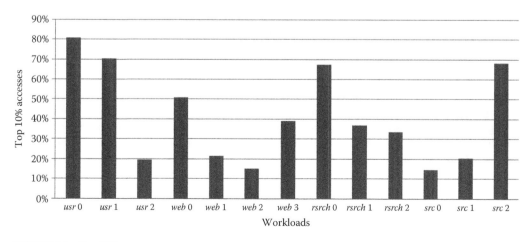

FIGURE 6.5 Top 10% accesses.

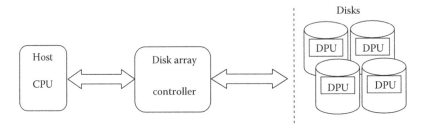

FIGURE 6.6   Processors in the I/O stack.

it decreases the energy efficiency as the idle periods are shortened. Clustering requests to improve the burstiness has shown to improve the power savings up to 55% [34].

Workloads can also be dynamically adaptive and can be executed during active periods [35]. Studies show that idle times based on workload characteristics can be predicted using modeling techniques and experimentation [36]. This is helpful to determine when the disks can be powered down. Prediction-based techniques can also be used to predict idle times [37].

Other studies have made a case for active storage devices [38,39], where the storage systems are extended to perform some processing. Storage centric computation can be off-loaded to the storage controllers that are much closer to the storage devices thereby improving both performance and power savings. Smullen et al. [39] propose that computation can be off-loaded in two specific levels in the I/O stack (Figure 6.6), the disk drive processor (DPU) or the disk array controller.

The host CPU, the disk array controller, and the DPU have heterogeneous power and performance characteristics. The host CPU typically operates at frequencies in the GHz range, while the array controller and DPU operate at lower frequencies in the MHz range, with DPUs having the lowest frequency. Consequently, the host CPU consumes the highest amount of energy, as much as 40% of the power in a system with eight disks. When the computation from the host CPU is offloaded, it can transition to a lower operating frequency with dynamic frequency scaling. Smullen et al. show that this results $3\times$ reduction in the power consumption. Offloading the computation closer to the storage increases the ability to efficiently use parallelism to achieve $3\times$ to $6\times$ increase in performance in most cases.

6.4.2.3 Tiered Storage

Moving data to storage device that can best serve its power and performance requirements is called tiering [40]. Devices with low-power consumption characteristics are used at various levels in the storage hierarchy for power efficiency where data can be cached, prefetched, or migrated. Storage class memories (SCM) such as SSDs or DRAM are based on semiconductor technology and promise devices that are much faster, reliable, and power efficient than disk drives, due to the absence of electromechanical components. However, they are cost inefficient with restricted capacity and hence cannot be used as a replacement to disk drives. Thus, several studies [7,41–43] have proposed using a tiered architecture with devices such as Dynamic Random Access Memory (DRAM), a nonvolatile memory device, or SSDs [44] as storage devices for selected, popular data. Depending on the tiering architectures, tiered storage offers two significant advantages. First, data can be accessed at much lower latencies and at a much lower power than disk drives. Second, the underlying disk drives can be spun down to maximize power savings. This is the principle behind augmenting RAID with SSDs [44], which shows that using a naïve approach where the most recent reads are cached in the SSDs, and writes are first written to the SSDs before they are flushed to the RAID disks to increase the idle time, results in power savings of up to 17%. Greater power savings can be achieved with optimizations such as prefetching data on the SSDs or piggybacking reads and write requests.

Workloads that exhibit temporal locality where the same data is accessed over time (Figure 6.11) and when the footprint of unique data accessed is small enough to be stored on the SSD cache can benefit the

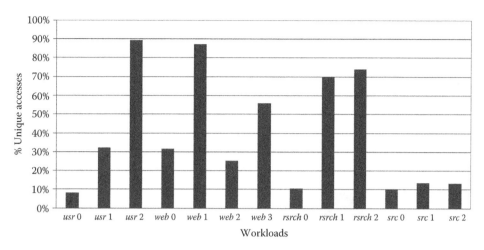

FIGURE 6.7 Percentage of unique data accesses.

most from such a strategy. Figure 6.7 shows the percentage of unique data accesses of the total accesses on the Y-axis for every workload and its corresponding volume on the X-axis. The lower number here indicates that the same data is accessed over and over again. Thus, volumes 0 for the *usr* or the *src* workload would benefit most by storing the unique data on the external caching device such as the SSD.

6.4.3 Data Management Techniques

Reducing the number of active devices directly reduces the power consumption. Techniques such as data consolidation, or data compression, and de-duplication to reduce the overall physical data footprint can be employed to reduce the number of active devices.

Data are usually striped across several disks to achieve high performance through parallelism. However, this implies that all the devices need to be powered up to service requests, which results in power inefficiency. Data allocation strategies can optimize the data placement on a minimum number of devices, so that the inactive devices can be turned off. This also results in greater space efficiency. However, there is a trade-off between power and performance when consolidating data. Studies such as [45–47] show that the number of devices used can be reduced by consolidating data on a smaller number of devices.

Figure 6.8a through d shows the spatial locality as heat maps for all the volumes for the workloads. The entire volumes are divided into 100 cells to represent the volume layout in a 10×10 matrix, where each cell represents an LBA range in the volume. Frequency of disk accesses is represented in six levels (0–5) with a different grayscale shading as shown. The lowest level 0 shown in white indicates a region in the volume, which is never accessed, while the highest level 5 indicates a region of highest activity and is shown in black. These heat maps clearly indicate that the spatial locality is workload specific, where volume 0 for the *web* workload shows minimal activity while volume 1 for the *usr* workload shows highest activity throughout the volume. Further, in 9 of the 13 volumes considered here, more than 50% of the entire volume remains unaccessed, while in 2 others, at least 20% remains unaccessed. This shows that there is ample scope for power savings by consolidating the data for the volumes that have lower utilization. For instance, for the *web* workload, the data in all the four volumes can be consolidated on just two volumes, thus halving the power consumption.

Reducing the physical data footprint or the amount of data that has to be stored also reduces the number of devices and hence results in power savings. It is necessary that the data footprint reduction techniques are effective and do not impact the performance. There are two popular techniques used—*data compression* and *data de-duplication*. Data compression encodes the data in such a way that it can be stored

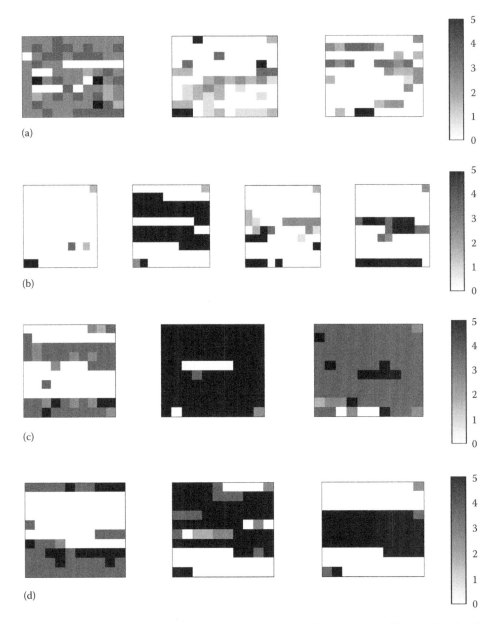

FIGURE 6.8 (a) Spatial locality for *rsrch* workload, volumes 0, 1, and 2, respectively. (b) Spatial locality for *web* workload, volumes 0, 1, 2, and 3, respectively. (c) Spatial locality for *usr* workload, volumes 0, 1, and 2, respectively. (d) Spatial locality for *src* workload, volumes 0, 1, and 2, respectively.

on the device using a fewer number of bits than the original data. Examples of popular data compression tools include WinZip or Gzip utilities. Data de-duplication techniques identify duplicate content in the data sets and eliminate all but one copy of the date. These techniques are known to improve the space efficiency by up to 20× [48].

Data compression is commonly used to reduce the size of data for improved storage space efficiency or network transmission. Data compression can be implemented in two ways: real-time and time-delayed.

In the real-time implementation, there is no performance impact on the applications since the data does not have to be decompressed to make any updates. This implementation is used in applications that are time sensitive and require hard real-time deadlines such as online transaction processing (OLTP) applications and audio and video streaming applications. For the time-delayed implementation, the data needs to be decompressed completely before it can be modified. In delta differencing, files are examined for updated blocks or bytes, and only the changed data are transferred over LAN or WAN to the targeted storage.

While all the copies of data are maintained in data compression, data de-duplication reduces the occurrence of redundant data. The redundant data here refers to the portions or segments of files that are common. Data redundancy can occur when server backups are performed. Backups are generally written to virtual tape libraries (VTLs). VTLs are arrays of disk drives where data can be written in parallel at high performance, but which have the same interface as tape libraries to ensure compatibility across heterogeneous backup services.

Backups usually examine a file for the timestamp to ensure if a file has been updated. Thus, even if a small segment of a file is updated, the entire file is backed up. This results in creation of multiple, common segments of the same files on heterogeneous storages devices. Data de-duplication techniques typically examine segments of files for similarities. If common segments are found, only a single copy is retained. The other copies of file segments that are deleted are replaced with pointers to the existing one. It has to be noted that duplicate data segments in different data streams are not necessarily aligned and can be at different offsets in the stream and the de-duplication algorithm should be able to track the duplicates irrespective of the data offset.

These common segments can be identified through hash-level comparison. Hash algorithms such as MD5 or SHA-1 are commonly used. These algorithms generate a unique identification number for every segment that is stored in an index. Every time a file is updated, only the bytes or bits that are changed are updated. While bit- or byte-level hashing is more efficient, it calls for greater processing power to build the hashes and search the indexes and greater memory to store the indexes. Hash collisions are also a problem that can lead to unique data being deleted if the algorithm produces the same hash for two unique segments. There are several ways to overcome this problem. Multiple hashing algorithms can be combined to reduce the probability of collisions. A byte-level comparison can also be done in addition to hash-level comparison to ensure unique data is always protected [48].

De-duplication can either be inline or postprocessing. In inline de-duplication, data is first examined for commonality and only unique data are written to the disk. In postprocessing de-duplication, data are first written to the disk and is then examined for commonality at a later time. Both approaches have their pros and cons and there are several factors that can be considered before making a choice. While inline de-duplication is more space efficient, postprocessing de-duplication requires enough capacity to store the duplicated data as well as to store the data once it has been de-duped. Inline de-duplication, however, has greater overheads when writing data, which can cause a bottleneck. While inline de-duplication may take longer to write the data, the total duration of the backup operation including writing the data and de-duplicating is larger in postprocessing de-duplication. Due to this reason, disaster recovery sites typically use the inline de-duplication strategy. The size of redundant data is also an important factor governing the choice of de-duplication strategy. If the size of redundant data is large and does not change much with every backup, an inline de-duplication can be more efficient in identifying common data in the different streams. If the size of redundant data is expected to be large, the processing and memory capacity needs to be sufficient to improve the performance in inline de-duplication. Also, the memory and CPU powers should be easily scalable in the event that the size of the data grows.

A hybrid approach can be used to benefit from the advantages of both the techniques. Such strategies monitor the throughput requirements of the applications and can dynamically switch between the two types.

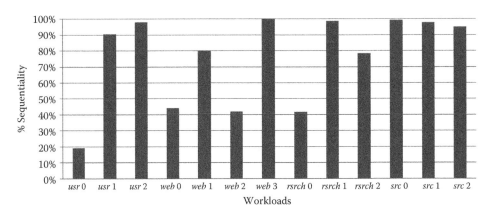

FIGURE 6.9   Percentage sequentiality.

Data de-duplication strategies can be implemented at the file, block, or the bit level. File de-duplication removes duplicate files and since this is not very efficient, de-duplication strategies generally operate at a level lower than the file system in the I/O subsystem and are oblivious to the file names, the file system, or the physical hosts in which the file segments reside. Data de-duplication requires a significant amount processing and memory power and should be used only when the benefits outweigh this additional cost.

It has been observed in the past that the number of seeks or random accesses is directly proportional to the power dissipation [49]. Reducing the number of random accesses in turn reduces the disk arm movement, thereby leading to power savings. Figure 6.9 shows the percentage for sequential accesses on the Y-axis for all the workloads and the volumes considered in this study on the X-axis. The traces show that if the percentage of random accesses is reduced by just 1%, power savings of 0.6% for *usr* and 0.2% for the *web* workload can be achieved. If the workload is made entirely sequential, the percentage improvement for the *usr* workload is up to 36.6%, while it is 15% for the *web* workload.

6.4.4 Storage Virtualization

Storage virtualization refers to the pooling of centrally managed physical storage and its abstraction in to a logical storage to hide the complexity of the underlying physical resources and making it independent of the applications [50]. The virtual infrastructure manager, also referred to as the virtual machine manager (VMM) or the hypervisor, is responsible for managing the virtualized resources and the mapping between the virtual and the underlying physical resources.

There are several advantages to virtualization. The most significant advantage is that it enables aggregation of underutilized resources including storage, network, and processing capacities, as shown in Figure 6.10. In the figure, the hypervisor layer lies between the virtual machine and the underlying physical hardware and controls all the virtual machines and their accesses to the physical hardware. Every virtual machine can execute its own operating system. Virtual machines provide isolation [51] to the applications where each application can be executed inside a dedicated virtual machine without sharing any data between them. Thus, although the virtual machines share physical resources between them, the behavior of one virtual machine and its resource consumption does not impact the performance of another. Consequentially, virtualization leads to increased utilization of resources. For instance, the amount of data stored on a physical hard drive is approximately 60% of the entire disk capacity, thus leaving a considerable amount of space free. With virtualization, the logical disk is assigned the capacity as required by the user. If the user requirement for storage capacity increases in the future, the system administrator can simply allocate another logical disk from the pool.

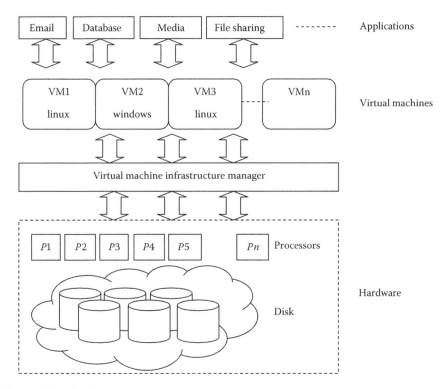

FIGURE 6.10 Virtualized environment.

Further, the hypervisor also provides a centralized point of management of the underlying heterogeneous physical hardware for system administrators. System administrators can assign the necessary storage capacity from the available pool by simply allocating a logical disk to the user, instead of the traditional approach of locating and allocating a physical device.

Data migration in virtualized storage can be carried out transparent to the user. Since the user is unaware of the physical location of the data, the actual data can be moved to a different physical location with minimal user-perceived overheads. Once the data is moved, the mapping from the logical to the physical data is updated to point to this new location. Data migration is typically carried out to move data to an underutilized device, or a device with better performance or power characteristics, or even to archival storage to minimize the number of active devices.

Using virtualization, entire virtual machines can be migrated from one physical server to another. VMware's VMotion [52] enables moving live virtual machine between servers without any noticeable impact on the user performance. This enables server consolidation that reduces the associated costs and increases energy efficiency and also provides increased reliability through decrease in downtime in case of system failure, hardware maintenance, or disasters. VMotion first transfers the memory and the system state from the source to the to the destination server. The entire system state is captured in a set of files that is stored on shared storage, where multiple servers can access the same files concurrently. The memory and the execution state are copied over high-speed network. To minimize user-perceived latencies during transfer, VMotion maintains a bitmap of ongoing memory transactions. In the final step, VMotion suspends the source virtual machine and copies this bitmap to the destination on a high-speed gigabit Ethernet network incurring a latency of less than 2 s. Since the network used by the virtual machine is also virtualized, the network identity and connections can be preserved even after the virtual machine migration.

6.5 Directions for Future Research

While the aforementioned strategies show significant power and performance benefits, there is still ample scope for improvement in ultrascale systems. With the emerging exascale platform, the challenges for power-optimization strategies lie in the scalability of existing strategies with minimal impact on performance. We identify a few more strategies here.

6.5.1 Replacing Disk Drives with Alternate Low-Power Consuming Devices

While the capacities of disk drives double every year, with 2 TB drives now becoming the trend, the performance increase is only 10% per year. The performance gains are attributed to the increased spindle speed as well as improvement in the transfer and positioning latencies. However, further improvements are restricted due to the electromechanical components used. Several classes of persistent data storage devices exist such as tapes, SSDs, and flash-based memory, which exhibit better performance, power, or energy characteristics. In spite of the emerging SCM such as SSDs or DRAM, which are based on semiconductor technology and promise devices that are much faster, reliable, and power efficient than disk drives, disk drives still remain the ubiquitous choice for data storage because of the higher costs or lower storage capacities and the likelihood that they will be replaced in the near future is bleak. The most significant reason for this is that disk drives have been used over several decades. Their access characteristics have been extensively studied and several popular applications and operating systems use data allocation policies based on the known geometrical and physical characteristics for disk drives. In order to replace the disk drive with any other device, its characteristics need to be known and exploited for better performance. For instance, data is stored on disk drives on electromagnetic platters, while SSDs are flash based and comprise various interconnected ICs. Since SSDs do not use any moving parts, they consume much lower power and do not incur any seek or rotational latencies. However, writes require block erasures and the number of erase cycles per device are limited. A device can survive up to 5 million erase cycles for a single block, which can be a problem for workloads with frequent writes. Because of this drawback, in order to use SSDs, the file system needs to be modified to alleviate the problem by using techniques such as wear leveling. In wear leveling, the erase cycles are distributed throughout the entire block device instead of over writing the data in place, as is done in disk drives, and thus the allocation policies need to be revisited in case of SSDs as a possible device replacement for hard drives. There are further considerations to be accounted for before replacing disk drives with any other persistent storage media:

1. It needs to be ensured that introducing new devices will not disrupt the existing service. Also, it is necessary that the new devices are compatible with the existing hardware (e.g., in bandwidth, interface, etc.) to avoid any bottlenecks.
2. In addition to the hardware and any software tools required for system administration to replace existing devices, the costs incurred for the necessary hardware or interconnects should be considered. If the costs outweigh the performance and power benefits, it is not practical to disrupt legacy systems.
3. The new hardware should replace the older one with ease since complex installations can increase the chances of errors and bugs. Further, the system downtime should be minimal when installing the new hardware, and hot-swappable devices should be used when possible.
4. The devices should be easily scalable. Several thousand disks are currently employed in peta-scale systems and with the increasing number of applications and their data, the number of storage devices are further expected to increase.
5. The file system used with the older devices should be compatible with the new device. This is feasible if the new device exports the same block-level interface to the file system irrespective of the device geometry or the access characteristics. Further, the device should have a standard interface

that is compatible with the applications that are executed. Also, it should permit the execution of native legacy optimizations that are typically implemented by the operating system.

6.5.2 Configurable Test Bed for Power-Optimizing Strategies

In the existing work for the aforementioned power-optimizing strategies, researchers either implement their system that may require "reinventing the wheel" each time for the same set of basic operations, or build a simulator to evaluate their strategy. Further, the evaluation platform differs with every research group that makes comparison across such strategies unfeasible. Thus, there is a need for a configurable test bed, which researchers can use to implement and evaluate their strategies. The test bed should enable configuration of system hardware parameters. The platform can also export an API to aid the development of power- or performance-efficient strategies. The API can support the interface between the power-optimizing strategy and the underlying physical resources while the developer can implement their own algorithms using the interface. The API can also enable the developer to monitor and access system characteristics such as power and performance measurements and control system resource allocation. This platform can have the following goals:

1. Build configurable systems such as server farms or data centers by choosing various computing resources including the type of processor, storage media, memory, and network.
2. Enable the execution of traces or benchmarks on the system.
3. Provide information about the usage characteristics of the system components.
4. Control the allocation of the resources.
5. Measure and observe power usage and impact on performance.

6.5.3 Autonomous, Self-Optimizing, Intelligent Systems

As is shown by our workload analysis, most characteristics are workload specific and for maximum power savings, the power-optimization strategies mentioned here should be able to select the most appropriate strategies and fine-tune its parameters based on workload characteristics. Most workloads exhibit temporal locality, where the access patterns in the workload are somewhat deterministic over certain time intervals. This is indicated in Figure 6.11, which shows the temporal locality for volume 1 executing the *src* workload. All the other workloads used in this study show similar results. The graph shows the percentage of data accesses, on the Y-axis, which overlapped for every day of the week, shown

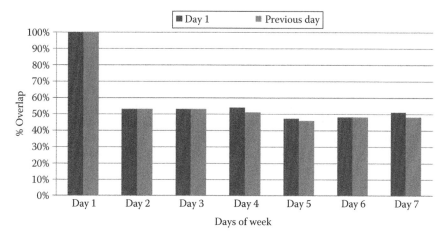

FIGURE 6.11 Temporal locality.

on the X-axis. The first bar indicates overlap with the first day of the week, while the second bar indicates the percentage of accesses that are common with those made on the previous day.

The figure indicates that on an average, the accesses common with Day 1 or with the previous day are about 50%. This indicates that the set of data accessed over the 7 day period that the traces were collected remained considerably deterministic. Such workload characteristics such as those presented here can be used by power-saving strategies to identify and predict workload characteristics and efficiently execute the most effective power-optimizing strategy for the given workload. Such a system should also be able to efficiently handle any changes in workload patterns. The system can periodically monitor the workload characteristics and can enable changes using an autonomous feedback loop to tailor the most efficient strategies for the current workload.

References

1. U.S. Environmental Protection Agency. Report to Congress on server and data center energy efficiency Public Law 109-431. s.l.: http://www.energystar.gov/
2. Gurumurthi, S. et al. Interplay of energy and performance for disk arrays running transaction processing workloads. *ISPASS'03: Proceedings of the 2003 IEEE International Symposium on Performance Analysis of Systems and Software*, 2003, Washington, DC.
3. Freeman, L. Reducing data center power consumption through efficient storage. Netapp White Paper, July 2009.
4. Kim, Y., and Gurumurth, S. Understanding the performance-temperature interactions in disk I/O of server workloads. s.l.: *Proceedings of HPCA*, 2006, Austin, TX.
5. Pinheiro, E., Weber, W.-D., and Barroso, L.A. Failure trends in a large disk drive population. *Proceedings of the 5th USENIX Conference on File and Storage Technologies*, 2007, San Jose, CA.
6. Polte, M., Simsa, J., and Gibson, G. Comparing performance of solid state devices and mechanical disks. *3rd Petascale Data Storage Workshop Held in Conjunction with Supercomputing'08*, 2008, Austin, TX.
7. Narayanan, D., Donnelly, A., and Rowstron, A. Write off-loading: Practical power management for enterprise storage. *ACM Transactions on Storage*, Vol. 4, 2008, New York.
8. Seagate. Product manual Cheetah 15 K.7 SAS. s.l.: http://www.seagate.com/staticfiles/support/disc/manuals/enterprise/cheetah/15K.7/SAS/100516226b.pdf
9. Suk, M., and Albrecht, T.R. The evolution of load/unload technology. s.l.: *12th Annual Symposium on Information Storage and Processing Systems*, 2001, Santa Clara, CA.
10. Kim, P., and Suk, M. Ramp load/unload technology in hard disk drives. s.l.: Hitachi White Paper, 2007.
11. Riska, A. et al. Feasibility regions: Exploiting trade-offs between power and performance in disk drives. *ACM SIGMETRICS*, Vol. 37, No. 3, 0163-5999, 2009, New York.
12. Western Digital. WD green power technology. s.l.: http://www.wdc.com/en/products/greenpower/
13. Samsung. Samsung Eco Green hard drives. s.l.: http://www.samsung.com/global/business/hdd/
14. Amer, A. et al. Design issues for a shingled write disk system. s.l.: *26th IEEE Symposium on Massive Storage Systems and Technologies: Research Track*, 2010, Incline Village, NEvada.
15. Gibson, G. and Polte, M. Directions for shingled-write and two-dimensional magnetic recording system architectures: Synergies with solid state disks. s.l.: Carnegie Mellon University Parallel Data Lab Technical Report, Pittsburgh, PA 2009.
16. Tagawa, I. and Williams, M. High density data-storage using shingle write. s.l.: *Proceedings of the IEEE International Magnetics Conference*, 2009, Sacramento, CA.
17. Gurumurthi, S. et al. DRPM: Dynamic speed control for power management in server class disks. s.l.: *Proceedings of the International Symposium on Computer Architecture (ISCA)*, 2003, San Diego, CA.

18. Pinheiro, E. and Bianchini, R. Energy conservation techniques for disk array-based servers. *ICS'04: Proceedings of the 18th Annual International Conference on Supercomputing*, 2004, Malo, France.

19. Sereinig, W. Motion-control: The power side of disk drives. s.l.: *International Conference on Computer Design*, 2001, Austin, TX.

20. Carrera, E.V., Pinheiro, E., and Bianchini, R. Conserving disk energy in network servers. *ICS'03: Proceedings of the 17th Annual International Conference on Supercomputing*, 2003, San Francisco, CA.

21. Elnozahy, E.N., Kistler, M., and Rajamony, R. Energy-efficient server clusters. *PACS'02: Proceedings of the 2nd International Conference on Power-Aware Computer Systems*, 2003, Cambridge, MA.

22. Harris, E. et al. Technology directions for portable computers. s.l.: *Proceedings of IEEE*, 1995.

23. Cameron, S. and Carobolante, F. Speed control techniques for hard disc drive spindle motor drivers. s.l.: *Proceedings of the Annual Symposium on Incremental Motion Control Systems and Devices*, 1993, San Jose, CA.

24. Colarelli, D. and Grunwald, D. Massive arrays of idle disks for storage archives. *Proceedings of the 2002 ACM/IEEE Conference on Supercomputing*, 2002, Baltimore, MD.

25. Papathanasiou, A.E. and Scott, M.L. Energy efficient prefetching and caching. s.l.: *Usenix Annual Technical Conference*, 2004, Boston, MA.

26. Son, S.W. and Kandemir, M. Energy-aware data prefetching for multi-speed disks. s.l.: *Proceedings of the 3rd Conference on Computing Frontiers*, 2006, Ischia, Italy.

27. Zhu, Q. et al. Reducing energy consumption of disk storage using power-aware cache. s.l.: *Proceedings of the 10th International Symposium on High Performance Computer Architecture*, 2004, Madrid, Spain.

28. Zhu, Q., Shankar, A., and Zhou, Y. PB-LRU: A self-tuning power aware storage cache replacement algorithm for conserving disk energy. *ICS'04: Proceedings of the 18th Annual International Conference on Supercomputing*, 2004, Malo, France.

29. Son, S.W., Chen, G., and Kandemir, M. A compiler-guided approach for reducing disk power consumption by exploiting disk access locality. *Proceedings of the International Symposium on Code Generation and Optimization*, 2006, Washington, DC.

30. Weddle, C., Oldham, M., and Qian, J. et al. PARAID: A gear-shifting power-aware RAID. s.l.: *ACM Transactions on Storage*, Vol. 3, No. 3, 2007, New York.

31. Zhu, Q., Chen, Z. et al. Hibernator: Helping disk arrays sleep through the winter. *Proceedings of the Twentieth ACM Symposium on Operating Systems Principles*, 2005, Brighton, U.K.

32. Ganesh, L. et al. Optimizing power consumption in large scale storage systems. *Proceedings of the 11th USENIX Workshop on Hot Topics in Operating Systems*, 2007, San Diego, CA.

33. Otoo, E., Rotem, D., and Tsao, S.C. Analysis of trade-off between power saving and response time in disk storage systems. *Proceedings of the 2009 IEEE International Symposium on Parallel and Distributed Processing*, 2009, Washington, DC.

34. Papathanasiou, A.E. and Scott, M.L. *Increasing Disk Burstiness for Energy Efficiency*. University of Rochester, Rochester, NY, 2002.

35. Li, X., Li, Z., David, F. et al. Performance directed energy management for main memory and disks. *Proceedings of the 11th International Conference on Architectural Support for Programming Languages and Operating Systems*, 2004, Boston, MA.

36. Elnozahy, E.N., Kistler, M., and Rajamony, R. Energy-efficient server clusters. *Proceedings of the 2nd International Conference on Power-Aware Computer Systems*, 2003, Cambridge, MA.

37. Gniady, C. et al. Program counter-based prediction techniques for dynamic power management. *IEEE Transactions on Computers*, Vol. 55, No. 6, 0018-9340, 2006, Washington, DC.

38. Son, S.W., Chen, G., and Kandemir, M. Power-aware code scheduling for clusters of active disks. *ISLPED'05: Proceedings of the 2005 International Symposium on Low Power Electronics and Design*, 2005, San Diego, CA.

39. Smullen, C.W. IV et al. Active storage revisited: The case for power and performance benefits for unstructured data processing applications. *CF'08: Proceedings of the 5th Conference on Computing Frontiers*, 2008, Ischia, Italy.

40. Sankar, S., Gurumurthi, S., and Stan, M.R. Intra-disk parallelism: An idea whose time has come. s.l.: *ISCA*, 2008, Beijing, China.

41. Narayanan, D., Donnelly, A. et al. Everest: Scaling down peak loads through I/O off-loading. s.l.: *Proceedings of OSDI*, 2008, San Diego, CA.

42. Yao, X. and Wang, J. RIMAC: A novel redundancy-based hierarchical cache architecture for energy efficient, high performance storage systems. *SIGOPS Operating System Review*, 2006, New York.

43. Narayanan, D. et al. Migrating server storage to SSDs: Analysis of trade-offs. *Proceedings of the 4th ACM European Conference on Computer Systems*, 2009, Nuremberg, Germany.

44. Lee, H., Lee, K., and Noh, S. Augmenting RAID with an SSD for energy relief. s.l.: *HotPower'08 Proceedings of the 2008 Conference on Power Aware Computing and Systems*, 2008, San Diego, CA.

45. Storer, M.W. et al. Pergamum: Replacing tape with energy efficient, reliable, disk-based archival. *Proceedings of the 6th USENIX Conference on File and Storage Technologies*, 2008, San Francisco, CA.

46. Strunk, J.D. et al. Using utility to provision storage systems. *Proceedings of the 6th USENIX Conference on File and Storage Technologies*, 2008, San Jose, CA.

47. Sankar, S. and Vaid, K. Addressing the stranded power problem in datacenters using storage workload characterization. *Proceedings of the First Joint WOSP/SIPEW International Conference on Performance Engineering*, 2010, San Jose, CA.

48. Brown, K. Deduplicating backup data streams with the NetApp VTL. http://www.netapp.com/us/communities/tech-ontap/vtl-dedupe.html

49. Essary, D. and Amer, A. Predictive data grouping: Defining the bounds of energy and latency reduction through predictive data grouping and replication. s.l.: *ACM Transactions on Storage*, Vol. 4, No. 1, 2008.

50. Bunn, F. et al. Storage virtualization. *SNIA Tutorial*, 2004.

51. Gupta, D. et al. Enforcing performance isolation across virtual machines in Xen. *7th USENIX Middleware Conference*, 2006, Melbourne, Victoria, Australia.

52. VMware VMotion. Live migration for virtual machines without service interruption, 2009.

7

Dynamic Thermal Management for High-Performance Storage Systems

Youngjae Kim
Oak Ridge National Laboratory

Sudhanva Gurumurthi
University of Virginia

Anand Sivasubramaniam
Pennsylvania State University

7.1 Introduction

Thermal awareness is becoming an integral aspect in the design of all computer system components, ranging from micro-architectural structures within processors to peripherals, server boxes, racks, and even entire machine rooms. This is increasingly important due to the growing power density at all granularities of the system architecture. Deeper levels of integration, whether it be within a chip, components within a server, or machines in a rack/room, cause a large amount of power to be dissipated in a much smaller footprint. Since the reliability of computing components is very sensitive to heat, it is crucial to drain away excess heat from this small footprint. At the same time, the design of cooling systems is becoming prohibitively expensive, especially for the commodity market [14,28]. Consequently, emerging technologies are attempting to instead build systems for the common case—which may not be subject to the peak power densities, and thereby operate at a lower cooling cost—and resort to dynamic thermal management (DTM) solutions when temperatures

129

exceed safe operational values. This chapter explores one such technique for implementing DTM for disk-drives.

Disk-drive performance is highly constrained by temperature. It can be improved by a combination of higher rotational speeds of the platters (called RPM), and higher recording densities. A higher RPM can provide a linear improvement in the data rate. However, the temperature rise in the drive enclosure can have nearly cubic relation to the RPM [6]. Such a rise in temperature can severely impact the reliable operation of the drive. Higher temperatures can cause instability in the recording media, thermal expansion of platters, and even outgassing of spindle and voice-coil motor lubricants, which can lead to head crashes [16]. One way of combating this generated heat is by reducing the platter sizes, which reduces the viscous dissipation by the fifth power. However, a smaller platter leads to a smaller disk capacity, unless more platters are added (in which case the viscous dissipation increases again by a linear factor). Moreover, a higher number of bits are necessary for storing error correcting codes to maintain acceptable error rates due to lower signal-to-noise ratios in future disk-drives. All these factors make it difficult to sustain the continued 40% annual growth that we have been enjoying in the data rates until now [14]. This makes a strong case for building drives for the common cause, with solutions built in for dynamic thermal management when the need arises. DTM has been already implemented in Seagate Barracuda ES drive in the industry [25].

There is one other important driving factor for DTM. It is not enough to consider individual components of a computing system in isolation any more. These components are typically put together in servers, which are themselves densely packed in racks in machine rooms. Provisioning a cooling system that can uniformly control the room so that all components are in an environment that matches the manufacturer specified "ambient" temperatures can be prohibitively expensive. With peak load surges on some components, parts of a room, etc., there could be localized thermal emergencies. Further, there could be events completely external to the computer systems—HVAC/fan breakdown, machine room door left open, etc.—which can create thermal emergencies. Under such conditions, today's disk-drives could overheat and fail, or some thermal monitor software could shut down the whole system. The disk is, thus, completely, unavailable during those periods. The need to sustain 24/7 availability and growing power densities lead to the increased likelihood of thermal emergencies. This makes it necessary to provide a "graceful" operation mode for disk-drives. During this "graceful" mode, even if the disk is not performing as well as it would have when there was no such emergency, it would still continue to service requests, albeit slowly. This graceful mode would essentially be a period during which certain dynamic thermal management actions are carried out in the background, while continuing to service foreground requests.

Multispeed disk operation [4,13] has been proposed as a solution to reduce disk-drive power, and can thus be a useful mechanism for thermal management as well. This mechanism is based on the observation that it is faster to change the rotational speed of a disk, rather than spinning it all the way down/up. DRPM allows the disk to service requests at a slower rate even at lower RPM. During a thermal emergency, we can not only reduce the speed to reduce temperature, but we can continue to service requests at the lower speed. A multispeed disk with two rotational speeds is commercially available [17]. Since the heat dissipated during the operation at a lower RPM is also much lower, the temperature within the drive can be lowered by employing this option during a thermal emergency. While a Hitachi's multispeed disk does provide a smaller window of time when the disk cannot service requests compared to a disk that only provides on-off modes, it still does not serve requests when it is at a lower RPM.

In this chapter, we explore two options for temperature management during a thermal emergency. We first consider disks that are tuned for maximum performance with the ideal/constant ambient temperature. We then introduce thermal emergencies—by adjusting the external ambient temperature of the drive— which pushes the drive into the emergency regions. We then investigate these two multispeed drive options, and show that it is indeed possible for some regions of external ambient temperature variation to service disk requests even though such situations would have caused the drive to completely shut down in a non-multispeed drive. As is to be expected, the performance during those periods is not as good as it would be when there are no emergencies. Between the two multispeed options, not servicing the requests at the lower RPM causes frequent switches between RPMs, thus not faring as good as the DRPM disk in its availability.

7.2 Background and Related Works

7.2.1 Basic Disk-Drive Components and Behavior

Figure 7.1 shows the geometry of the mechanical components of a disk-drive from vertical and horizontal views. The disk platters are attached to the hub, which is empty inside. The hub connects to the base through spindle motor. An arm actuator motor (voice-coil motor) is placed on the right end of the base. The arm assembly is composed of one or more arms and each head for data read and write from/to the platter is placed at the end of each arm. The unit of data read-write is sector. The sector is represented in Figure 7.1b as a thick curve. As shown in Figure 7.1b, a set of sectors composes a track and a disk platter is composed of multiple tracks. A cylinder is a group of tracks that are vertically grouped as shown in the figure. All these components are enclosed by the disk cover and are closed to the ambient air.

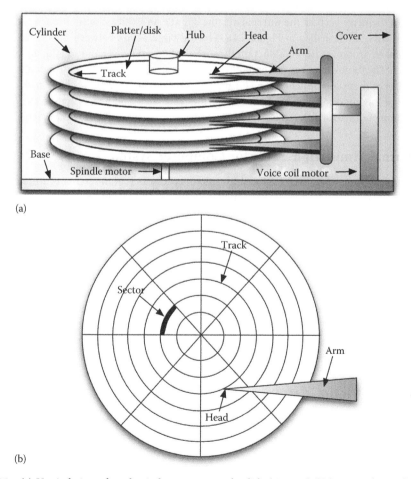

FIGURE 7.1 (a) Vertical view of mechanical components of a disk-drive and (b) horizontal view from the top. (Adapted from Ruemmler, C. and Wilkes, J., *Computer*, 27, 17, 1994.)

When an I/O request is sent to the disk-drive from the host, disk operation behaves differently according to the request type (read or write). When the request is a read, the disk controller first decodes the request. Then the controller disconnects from the bus and starts a seek operation that moves the arm assembly toward the target track of data on the platter. The disk head at the end of the arm will properly settle down (head positioning time). Then, the data transfer occurs from the disk media to the host through SCSI bus. Since reading the data of the disk media is slower than sending it over the bus, partial buffering is first required before sending it over the bus. Moreover, during this data transfer, head switch operation (to move the head to the next track) might be involved if necessary. When data transfer finishes, the complete status message is sent to the host. When the request is a write, data transfer to the disk's buffer is overlapped with head positioning time. If the head positioning time finishes, the data transfers to the disk media from the buffer. Also head switching could be involved on the needs as in read requests. If this data-recording on the media finishes, the complete status message is sent to the host.

7.2.2 Computational Model of Thermal Expansion in a Disk-Drive

The thermal simulation model is based on the one developed by Eibeck and Cohen [7]. The sources of heat within the drive include the power expended by the spindle motor (to rotate the platters) and the voice-coil motor/arm actuator motor (for moving the disk arms). The thermal model evaluates the temperature distribution within a disk-drive from these two sources by setting up the heat flow equations for different components of the drive such as the internal air, the spindle and voice-coil motor assemblies, and the drive base and cover as described in Figure 7.1a. The only interaction between the device components and the external environment is by conduction through the base and cover and subsequent convection to the ambient air. The finite difference method [20] is used to calculate the heat flow. It iteratively calculates the temperature of these components at each time step until it converges to a steady state temperature. The accuracy of this model depends on the size of the time step. The finer the time step is, the more accurate the temperature distribution is over the disk-drive, but the simulation time is large.

7.2.3 Thermal Simulation Tools for Computer Systems

Temperature-aware design has been explored for microprocessors [28], interconnection networks [26], storage systems [14], and even for the rack-mounted servers at machine rooms [5,27], because high temperature can lead to reliability problems and increase cooling costs. There have been various thermal simulation tools proposed to evaluate temperature-aware design. HotSpot [28] is a thermal simulator for microprocessors using thermal resistance and capacitance derived from the layout of microprocessor architecture. A performance and thermal simulator for disk-drives [12,19] uses the finite difference method similar to that proposed by Eibeck and Cohen [7] to calculate the heat flow and to capture the temperatures of different regions within the disk/storage system. Mercury [15] is a software suite to emulate temperatures at specific points of a server by using a simple flow equation. In addition, *ThermoStat* [5] is a detailed simulation tool for rack-mounted servers based on computational fluid dynamics (CFD) [21], which provides more accurate thermal profiles by generating three-dimensional (3D) thermal profiles.

7.2.4 Dynamic Thermal Management

Dynamic thermal management has been adopted for individual components of the systems such as microprocessors [3,30] and disk-drives [19] or distributed environments such as distributed systems [34] and rack-mounted servers at data centers [5,27]. Of all these approaches, DTM for disk-drives has already been addressed [14].

A delay-based DTM has been applied to prevent this situation from happening [11]. When the temperature of disk-drive reaches close to the thermal envelope, DTM is invoked by stopping all the requests issued, hence all the seek activities stop and the service resumes only after the temperature is sufficiently reduced. However, even if this delay-based throttling (by controlling the seek activities) is feasible, many requests cannot be issued during thermal emergencies and thereby the performance is greatly affected by them. Today's Seagate's Barracuda ES drives have a similar DTM feature by adjusting the workloads for thermal management [25]. The other possible approach is to modulate the RPM speed in a multispeed disk. Since RPM has nearly cubic power relation to the viscous dissipation [6], it can be more effective to manage the temperature of the disk-drive. This technique of dynamic RPM modulation for thermal management is discussed in the rest of this chapter.

7.3 Framework for Integrated Thermal-Performance Simulation

In order to analyze the thermal behavior of applications, we need a framework that can relate activities in the storage system to their corresponding thermal phenomena as the workload execution is in progress. In a real system, this can be achieved by instrumenting the I/O operations and leveraging the thermal sensors [16] that are commonplace in most high-end disks drives today. However, since the objective of this study is to investigate the effect of disk configurations that are not yet available in the market today, we use a simulation-based approach. In this section, we describe the design simulation framework that we have developed to study performance and thermal behavior of storage systems in an integrated manner.

The simulator consists of two components, namely, a performance model and a thermal model. As the name suggests, the performance model is used to simulate all activities in the storage system that could potentially affect workload performance, such as, interconnect latencies, disk and cache accesses, and multidisk organizations (such as RAID). The thermal model, on the other hand, considers the effect of parameters that can affect the temperature of the storage system components. Some of the parameters of a thermal model are typically not considered in a performance model, for instance, the external air temperature or nature of the material used for making a platter. On the other hand, there are parameters that affect both performance and temperature and are thus shared by both models, such as the drive RPM and the seek behavior. Therefore a thermal-performance simulator would have to communicate the states of all shared parameters appropriately between the two models.

In our simulator, the performance model we use is Disksim [9], which models the performance aspects of the disk-drives, controllers, caches, and interconnects in a fairly detailed manner. Disksim is an event-driven simulator and thus the simulated time is updated on an event rather than by maintaining an explicit (periodic) clock. The thermal model that we use is the one developed by Eibeck and Cohen [7]. This model evaluates the temperature distribution of a disk-drive by setting up the heat equations for different components of the drive such as the internal air, the spindle and voice-coil motor assemblies, and the drive base and cover. It uses the finite difference method [20] to calculate the heat flow. At each time step, the temperatures of all the components and the air are calculated and this is iteratively revised at each subsequent time step until it converges to a steady state temperature. Thus, the thermal model uses an explicit notion of time.

Each of the models maintains internal state information (program variables), some of which is shared between the models. In our implementation, the two models exchange state information via checkpoints. Here, the performance model invokes the thermal model in order to simulate the thermal behavior for a certain amount of time. After this, all the variables that capture the current thermal state are explicitly checkpointed. This allows the thermal state to be accurately restored the next time the thermal model is invoked. We maintain the checkpoints on a per-disk basis.

When doing the thermal-performance simulation, one of the components that needs to be modeled accurately is the dynamics of a physical disk seek operation. Although the time taken for a seek is already accounted for by the performance model, the mechanical work involved to effect the seek operation has a strong influence on temperature and needs to be accounted for.

7.3.1 Modeling the Physical Behavior of Disk Seeks

The seek-time depends on two factors, namely, the inertial power of the actuator voice-coil motor (VCM) and the radial length of the data band on the platter [10]. The VCM, which is also sometimes referred to as the arm actuator, is used to move the disk arms across the surface of the platters. Physically, a seek involves an acceleration phase, when the VCM is powered, followed by a coast phase of constant velocity where the VCM is off, and then a deceleration to stop the arms near the desired track when the VCM is again turned on but the current is reversed to generate the braking effect. This is then followed by a head-settling period. For very short seeks, the settle time dominates the overall seek-time whereas for slightly longer seeks, the acceleration and deceleration phases dominate. Coasting is more significant for long seeks. We capture the physical behavior of seeks using a bang-bang triangular model [18]. In this model, for any physical seek operation, the time taken for acceleration and the subsequent declaration are equal. Using the equations of motion, we now calculate time taken during the acceleration, coast, and deceleration of a physical disk seek operation.

In order to calculate the acceleration that is required of a VCM, we make the following assumptions:

- The settle time of a head is approximated as the track-to-track seek-time.
- The seek operation across 1/3 of the data-zone takes an amount of time equal to the average time for a large number of random length seeks [2]. Therefore, if you calculate the time that it takes to traverse half this distance (acceleration), we can merely double its value to obtain the average seek-time of the disk.
- There are no mechanical limitations in the VCM assembly that would prevent us from accelerating beyond a certain value. As we shall shortly show, this model is able to capture the characteristics of real disk quite closely.

Let the outer radius of the disk-drive be denoted as r_o. We set the inner radius to be half that of the outer radius, i.e., $r_i = r_o/2$ [14]. Let D_{avg} and T_{avg} denote the distance of 1/3 of the data-zone and the average seek-time. In other words, T_{avg} is the time spent for seek operation for the distance of D_{avg} according to the second assumption. Since we are calculating the time only during the movement of the disk arm and not the settling period, T_{avg} is adjusted by subtracting the settle time of a head (i.e., the track-to-track seek-time) from the average seek-time. Let V_{max} denote the maximum velocity that is permissible. This value is dictated both by the characteristics of the VCM assembly and the bandwidth of the underlying servo system (which is used to accurately position the disk arm over the desired track). We use a V_{max} value of 120 in./s, which reflects many modern disk-drive implementations.

7.3.1.1 Equations of Motion for Constant Acceleration

In order to calculate the different components of a disk seek, we make use of one-dimensional (1D) equations of motion for constant acceleration. These equations are

$$v = v_0 + a \cdot t \qquad (7.1)$$

$$d = v_0 \cdot t + \frac{1}{2} \cdot a \cdot t^2 \qquad (7.2)$$

where
  v_0 is the initial velocity
  a is the acceleration
  v is the final velocity
  t is the time interval

For a seek operation that needs to traverse a distance of D_{avg}, we need to consider three cases as shown in Figure 7.2. First, assume that in order to traverse this distance in time T_{avg}, we need to accelerate and subsequently decelerate without having to coast (as depicted in Figure 7.2a and b). In

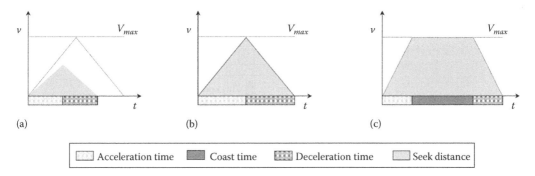

FIGURE 7.2 Different possibilities for a physical seek operation based on the values of the maximum acceleration (Acc_{max}), which is used to reach and the maximum permissible velocity (V_{max}) within the first half of the seek-time. (a) $Acc < Acc_{max}$, (b) $Acc = Acc_{max}$, and (c) $Acc > Acc_{max}$.

that case, we accelerate for time $T_{avg}/2$, traversing a distance of $D_{avg}/2$ and then decelerate for the same amount of time and then wait for the settling period. We now need to calculate the acceleration that is required to satisfy this timing requirement. Using Equation 7.2, we can calculate the numerical value of the acceleration, given that $v_0 = 0$, $d = D_{avg}/2$, and $t = T_{avg}/2$. Let us denote this calculated acceleration as Acc. After the acceleration is complete, we need to calculate the terminal velocity of the disk arms. For this, we use Equation 7.1, where $v_0 = 0$, $a = Acc$, and $t = T_{avg}/2$. If this velocity is found to be greater than the maximum permissible velocity, V_{max}, then we need to reduce the acceleration and introduce a period of time when the disk would coast at a velocity of V_{max}, before decelerating to the target track. This situation is depicted in Figure 7.2c. Therefore, we need to recalculate the acceleration value.

Let T_{new} be the time when the velocity of the arms reaches the value V_{max}. Note that, in this case, $T_{new} < T_{avg}/2$. Then, the new acceleration value, Acc_{new}, can be calculated by solving the following two simultaneous equations:

$$V_{max} = Acc_{new} \cdot \left(\frac{T_{new}}{2} \right) \tag{7.3}$$

$$\left(\frac{D_{avg}}{2} \right) = V_{max} \cdot \left[\left(\frac{T_{new}}{2} \right) + \left(\frac{T_{avg}}{2} - T_{new} \right) \right] \tag{7.4}$$

Here, $D_{avg}/2$ is essentially half the area of the equilateral trapezium shown in Figure 7.2c, which is calculated using Equation 7.4. The coast time is equal to $T_{avg} - 2 \cdot Acc$.

Thus, given the value of V_{max}, which depends on the design of the arm assembly, and a particular value of the seek-time, this procedure can be used to calculate the acceleration, deceleration, and coast times for the seek operation.

7.3.1.2 Validation

In order to validate this model, we calculated the acceleration that is computed by our model, under all the stated assumptions for a Fujitsu AL-7LX disk-drive, which is a 2.6″ 15,000 RPM disk-drive, and compared it to its measured mechanical seek characteristics, which is published [1]. It has an average seek-time of 3.5 ms with a track-to-track seek-time of 0.4 ms [8]. The reported value for the acceleration to satisfy the seek-time requirement is 220 G (2150 m/s²), whereas our model calculates it to be 253.5 G (2488.1 m/s²), which is a 15% error. The differences are mainly due to our assumptions regarding the settle time (which could be smaller than the track-to-track seek-time depending on the track layout) and also the assumption about the value for V_{max}. In fact, the computed acceleration value is very sensitive to the value of the settle time. For instance, even if the settle time was just

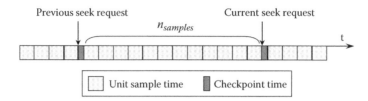

FIGURE 7.3 Sampling between I/O requests for accurate temperature profiling.

0.2 ms lesser than the track-to-track seek-time (0.2 ms), then the acceleration that is computed is 218 G (0.1% error).

7.3.2 Sampling for Speed vs. Accuracy Trade-Offs

The second issue that has to be addressed is the reconciliation of the event-driven performance model with the time-based thermal one. In order to accomplish this, we use a sampling approach. Here, we invoke the thermal model for a specific number of times (denoted as $n_{samples}$) between the beginning of any two I/O events in the performance model. This situation is depicted in Figure 7.3. This methodology is also employed for other events such as changes in the RPM of the disk, which we shall be describing in this chapter. For every sampling-event, care is taken to pass information about the state of the performance simulator (such as whether the VCM is on or off based on which phase of a disk seek is being simulated) to the thermal model.

In order to choose a reasonable value for $n_{samples}$ and also verify the correctness of our implementation, we did a workload-driven study. We ran a workload whose average inter-seek-time (the time between two successive seeks that require a movement of the arm) was 4.5 ms. (The actual value that is used for this experiment is not very important. We merely need to ensure that there are a sufficient number of disk seeks.) Each such seek operation invoked the thermal model and we recorded the simulated time and the temperature. We then compared this temperature to the output by a stand-alone version of the thermal model for the simulated time. To make sure that the stand-alone model is capturing the time-varying thermal behavior accurately, we used a very fine-grained time-step of 60,000 steps/min. Also, in order to perform the controlled experiment, we did not model the power behavior of the seek operations themselves and set the VCM power to 0 W. The stand-alone model was run under the same conditions as well. We varied the value of $n_{samples}$ from 100 to 2500 samples for the integrated model. We found that a choice of 400 samples between requests gives a thermal profile whose accuracy is within 0.1% of the stand-alone simulation, and we use this value in our evaluation.

7.3.3 Simulation "Warmup"

At the beginning of the simulation, all the disks in a cold state, have the same temperature as that of the outside air. When the simulation begins, heat is generated by the rotating platters and some of the heat is conducted through the base and cover and convected to the outside air. After a long period of time (roughly 50 min of simulated time), the temperature finally reaches a steady state. In order to prevent this behavior from skewing our results, we perform the experiments only after the system has reached the steady state idle temperature, i.e., it is assumed that the workload is run on a system that has been running for a long period of time. To simulate this physically warmed-up state, we use a method that is similar to "simulation-warmup," which is common in computer architecture studies.

In computer architecture simulation, warmup is used to study workload behavior after structures that maintain state information, such as caches and branch-predictor tables, have been initialized. This is typically accomplished by simulating the workload using a "fast-mode," instructions are simulated only functionally, and updating the state information in the structures. After this, detailed micro-architectural simulation is performed using the warmed-up state. In our experiments, we use a similar methodology to warmup the simulation. We run the stand-alone model (whose simulation time is faster) for the first 150 min of simulation assuming that the disks are idle (i.e., the disks are spinning but there are no arm movements). After this, we take a checkpoint of the thermal state and then start the performance simulation of the workload.

7.4 Handling Thermal Emergencies in Disk-Drives

Thermal emergencies are generally caused by unexpected events, such as fan-breaks, increased inlet air temperature, etc. These unexpected events threaten the reliability of disk by causing data corruption on disk. Unfortunately, predicting when such thermal emergencies happen in real time is a big challenge. In this section, we understand the impact of the external ambient temperature variation to the disk temperature with micro-benchmark tests.

7.4.1 Thermal Variation over Disk-Drive

In order to understand the heat distribution over all the components of a drive enclosure while it is in operation, we used STEAM to model an Ultrastar 146Z10 disk-drive [32] installed in a 42U computer system rack. The Ultrastar 146Z10 is composed of two 3.3″ platters and rotates at 10,000 RPM. The power curve of the spindle motor (SPM) due to its rotational speed change is obtained using the equation describing the change in the rotation speed of the disk has a quadratic effect on its power consumption [13]. And the VCM power (which is dependent on the platter dimensions) is obtained by applying the power-scaling curve from [29]. The power values of SPM and VCM are set to be 10 and 6.27 W, and all other required parameters such as disk geometry were supplied as inputs into the model.

From Figure 7.4, we see that the hottest component over the drive enclosure is the arm assembly (which has heads at its end) whose temperature is around 68°C at maximum (i.e., the disks are spinning and the arms are moving back and forth with VCM on all the time) and the lowest temperature is for

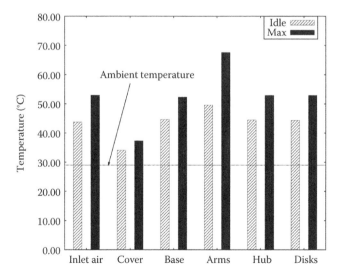

FIGURE 7.4 Temperature distribution over IBM Ultrastar 146Z10. The dotted line denotes the ambient temperature (29°C). Each label in the *x*-axis indicates each component in the disk-drive. Inlet air denotes external inlet air temperature. The description about all other labels can be seen Figure 7.1a.

the disk-cover surrounding disk-drive (around 37°C at maximum). This is because the heat is directly drained away to the ambient air through the convection process. When the disk-drive is idle (the disks are spinning without any arm movement), the disks have a similar temperature distribution to when the disk-drive is maximum (the disk-drive consumes maximum power while the disks are spinning and arms are always moving back and forth).

7.4.2 Impact of Increased External Ambient Temperature to Disk-Drive's Temperature

We performed a micro-benchmark evaluation to understand the impact of variation in ambient temperature to the disk temperature and the feasibility of dynamic throttling by RPM modulation in a multispeed disk. Figure 7.4 shows that there is high thermal variation in temperature over the disk-drive, which means that thermal sensor location is very critical in applying DTM to a multispeed disk. It is hard to decide the location that should be selected for detecting emergencies. The base temperature of disk is chosen for DTM mechanism, because a thermal sensor is mounted on the back side of the electronics card close to the base of the actual disk-drive [16]. The highest RPM speed of a multispeed disk is restricted to 20,000 and the baseline is 10,000 RPM because a 10,000 RPM disk-drive is one of the most popular server disk-drives and 20,000 RPM is known as the possible rotational speed of the disks for disk-drive's design until now. *Thermal slack* is defined as the temperature difference between the current operating temperature and the thermal envelope. We modeled two different disks for the experiments: one is a 3.3″ 1 platter disk-drive used in HPL Openmail and the other is a disk-drive with 3.3″ 4 platters used for other workloads shown in Table 7.1.

We have measured the base temperature of the disk-drive with different ambient temperatures at the steady state in STEAM. We varied the ambient temperature from 29°C to 42°C for the 3.3″ 1 platter disk and from 29°C to 33°C for the disk of 3.3″ 4 platters. In the experiment, the thermal envelope was set to be 60°C because the possible operating temperature range of a disk-drive suggested in manuals is 5°C–60°C [24,32].

From Figure 7.5a, it is observed that thermal emergencies never happen with even 20,000 rotational speed of platters and the VCM on all the time for 29°C ambient temperature. However, if the ambient temperature is increased further to 42°C, it could exceed thermal envelope (60°C). For example, a multispeed disk operating at 20,000 under 42°C is above the thermal envelope at both idle and maximum.

TABLE 7.1 Description of Workloads and Storage Systems Used and Thermal Emergency Situations for Real Workloads

Workload	HPL Openmail [31]	OLTP Application [33]	Search-Engine [33]	TPC-C
Workload description and storage systems				
Number of requests	3,053,745	5,334,945	4,579,809	6,155,547
Number of disks	8	24	6	4
Per-disk capacity (GB)	9.29	19.07	19.07	37.17
RPM	10,000	10,000	10,000	10,000
Platter diameter (in.)	3.3	3.3	3.3	3.3
Number of platters (#)	1	4	4	4
Thermal emergencies				
T_{init_amb} (°C)	29	29	29	29
T_{emg_amb} (°C)	42	33	33	33
Emg_start (s)	500.000	500.000	2,000.000	2,000.000
Emg_end (s)	2,500.000	30,000.000	12,000.000	10,000.000
Simulated Time (s)	3,606.972	43,712.246	15,395.561	15,851.512

T_{init_amb} denotes the initial ambient air temperature. *Emg_amb* denotes the increased ambient air temperature due to thermal emergencies. *Emg_start* and *Emg_end* respectively denote simulated thermal emergency starting and ending time.

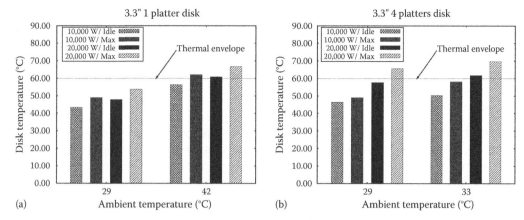

FIGURE 7.5 Each bar denotes the average temperature of each component at the steady state for maximum (where VCM is on all the times while the disk platters are spinning) and idle (VCM just turns off). The horizontal line in each graph is the thermal envelope (60°C). (a) and (b) are the steady state base temperatures of disk for different disk dimensions (such as the size of platter and rotational speed of platter) and different power consumption modes under various ambient temperatures. The horizontal line in each graph is the thermal envelope (60°C). (a) One 3.3 in. platter disk-drive and (b) four 3.3 in. platters disk-drive.

However, if RPM drops down to 10,000, it comes below thermal envelope at idle, while it exceeds 60°C at maximum. This result shows that the thermal slack would be around 4°C at the maximum. We also observe from Figure 7.5b that, if the disk-drive has a larger number of platters in a similar disk geometry, it is more prone to thermal emergencies even for small increases in ambient temperature because the heat dissipation inside the disk-drive is proportional to the number of platters [6]. Figure 7.5b shows that 33°C external ambient temperature introduces thermal emergency when it is operating at the higher speed. Thermal slack becomes larger around 10°C at the maximum. Moreover, even if the disk-drive is operating at the lower speed, the thermal emergency could exceed the thermal envelope for even 29°C ambient temperature depending on the request patterns. Similarly, since the heat generated from the disk-drive is proportional to the 4.6th power of disk platter size [6,23], the disk-drive with the larger size of platters is more sensitive to variations in the ambient temperature.

Within these thermal slacks, a dynamic throttling mechanism can be applied to avoid thermal emergencies. It can be achieved by pulling down the rotational speed of the disks (when it reaches thermal emergencies). And then once the temperature is lower than a given thermal envelope, it brings up the disk to full rotational speed after the cooling period.

7.5 Designing Dynamic Thermal Management Technique for Disk-Drives

7.5.1 Multispeed Disk-Drive

In order to study the effect of multispeed disk-drive when thermal emergencies happen, we have simulated the temperature behavior of RPM transitions in a multi-speed disk-drive. We consider the operable rotational speeds of the platters for this multispeed disk are 10,000 and 20,000. From Figure 7.5, we see how much the disk-drive's temperature can vary with different RPM speeds and external ambient temperatures. The maximum transition time taken between different rotational speeds of disk is assumed to be 7 s (from the lower to the higher and vice versa as in the commercial multispeed disk-drive of Hitachi [17]).

We used four commercial I/O traces for the experiment, whose characteristics are given in Table 7.1, and we consider two kinds of multispeed disk-drives as follows:

1. *DRPM$_{simple}$*: This is the same approach as Hitachi's multispeed disk, where the lower RPM is just used for cooling the hot disk, rather than servicing the requests.
2. *DRPM$_{opt}$*: This is the technique that was proposed in [13], where the disk-drive still performs I/O at the lower RPM.

Note that *DRPM$_{opt}$* disk-drive is a disk-drive serviceable at any rotational speed of the platters while *DRPM$_{simple}$* utilizes the lower rotational speed of the platters only for disk-drive's cooling effect without servicing any request. We evaluate two possible DTM policies (called time based and watermark based) with these multi-speed disk-drives.

7.5.2 Time-Based Throttling Policy

The time-based policy is based on a predefined period for cooling time before resuming to service the requests under thermal emergencies. The thermal sensor of the disk-drive periodically checks the temperature as a decent disk-drive does [24]. Once the disk's temperature reaches thermal emergency, the RPM drops down and the drive waits for a predefined period before resuming I/O operation by ramping up the RPM to full speed. Since *DRPM$_{simple}$* is not available to service during the cooling and transition times, the performance is constrained by these two values. However, most server workloads generally have many requests issued with short inter-arrival times and they should be processed as quickly as possible. In addition, 7 s of delay for each RPM transition is not negligible to the performance of a multispeed disk.

Figure 7.6 shows the performance degradation by *DRPM$_{simple}$*, compared to the disk without any dynamic thermal management technique under thermal emergencies. Each graph shows the CDF (cumulative distribution function) of the average response time at I/O driver across different disks. The response time is the time a disk-drive takes to finish a given input request. The solid curve in each graph shows the disk operating at the maximum speed of 20,000 RPM without DTM technique and others reflect a multispeed disk-drive with *DRPM$_{simple}$*. As is to be expected, many requests suffer from large delays (due to non-serviceable cooling time) more than 200 ms in the multispeed disk-drive of *DRPM$_{simple}$*. In Figure 7.6a, even 30%–50% requests are serviced with their response times more than 200 ms while

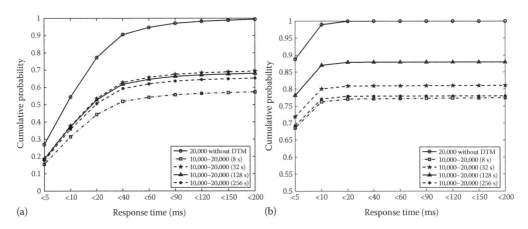

FIGURE 7.6 Performance degradation of *DRPM$_{simple}$* for the server workloads. The value in parenthesis at each graph denotes a cooling unit time (which is given as a delay time, once it becomes close to thermal envelope [60°C]). (a) HPL Openmail and (b) TPC-C.

in Figure 7.6b about 13%–25% requests suffer from large response times more than 200 ms. Even if we varied cooling unit times to compensate for the performance degradation, none of them is effective for both workloads. $DRPM_{simple}$ might be desirable for a DTM solution, because such a straightforward policy does not only require significant additional complexity to the disk-controller design but also after reasonable delays, it could still overcome thermal emergencies by resuming the service below thermal envelope.

$DRPM_{opt}$ has been designed to minimize the performance drawback caused by long delays of $DRPM_{simple}$ required for disk-drive's cooling effect. Figure 7.7 shows different thermal profiles for the workloads under thermal emergency situations described in Table 7.1. Since each disk-drive of the disk arrays of TPC-C, OLTP, and Search-Engine has the same disk dimension/characteristics and they have similar temperature profiles, we focus on the results for HPL Openmail and TPC-C. The upper curve for each graph is when DTM is not applied while operating at the maximum speed, while the lower curve (going up and down) is the result from $DRPM_{opt}$ with 400 s of cooling unit time during which it operates at the lower rotational speed of the platters. As shown in Figure 7.7, operating only at 20,000 RPM exceeds thermal envelope under thermal emergencies unless DTM is applied. However, $DRPM_{opt}$ avoids emergencies by dynamically modulating the rotational speed between high and low RPMs at need. However, even if $DRPM_{opt}$ could be available to service the requests during the cooling time, many RPM transitions (e.g., as shown from many transitions for OLTP in Figure 7.7) increase the overheads due to non-serviceable RPM transition time. Any arriving request during RPM transition should wait until it completes.

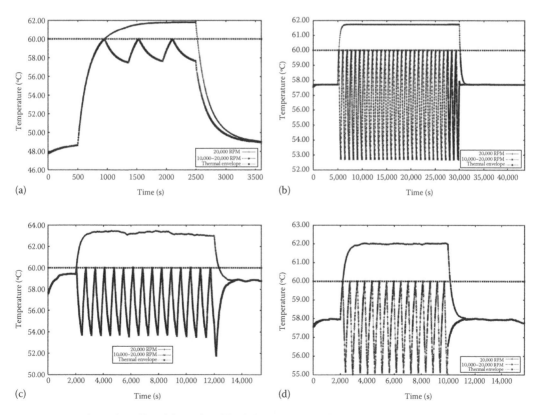

FIGURE 7.7 Thermal profiles of the real workloads for $DRPM_{opt}$ under the scenarios of Table 7.1. They are all for the disk0 of disk arrays each of which is a 10,000–20,000 multispeed disk with 7 s of RPM transition time and 400 s of a cooling unit time. (a) HPL Openmail, (b) OLTP, (c) search-engine, and (d) TPC-C.

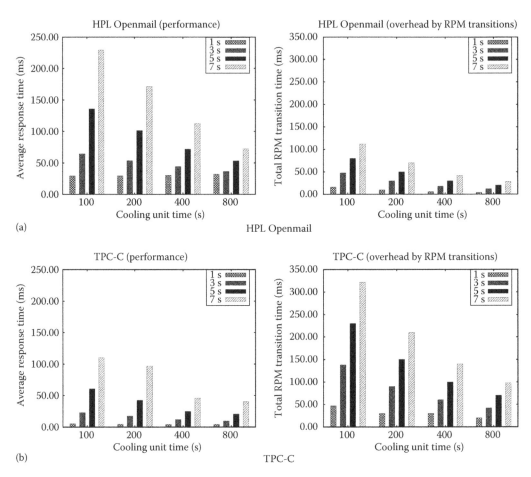

FIGURE 7.8 Correlation between cooling unit time, RPM transition time, and performance (i.e., response time). Each bar denotes an average value across the disks at disk array in the unit of millisecond.

The time taken for RPM transitions greatly affects the performance, and, thus, to study the impact of nonnegligible RPM transition to the performance, we have experimented $DRPM_{opt}$ for different RPM transitions ranging from 1 to 7 s in steps of 2 s. In Figure 7.8, cooling unit time is a predefined period during which the platters rotate at the lower rotational speed. The first column for each workload is to understand the correlation between cooling unit time and RPM transition time and the second column for each workload shows the relationship between average total time taken for RPM transitions and cooling unit times. From Figure 7.8a, we see that a small RPM transition time shows better performance for a given constant cooling unit time. In addition, it is shown in Figure 7.8b that high cooling times can hide the overhead of RPM transitions by reducing the number of RPM transitions and sparing more time for I/O disk operations. In the DTM option of $DRPM_{opt}$, still servicing the requests at the lower rotational speed of the platters works positively in the performance improvement; however, it still has an upper bound in performance improvement. This is because more cooling implies that more requests should be serviced at the lowest speed of RPMs in a multispeed disk.

7.5.3 Watermark-Based Throttling Policy

Watermark-based policy uses two thresholds, T_{high} and T_{low}. As in time-based policy, the thermal sensor of disk-drive periodically checks the temperature. If the thermal sensor detects that the temperature is close to thermal emergency (T_{high}), which is the temperature at which DTM is invoked, thermal management is applied to cool down the disk until the temperature gets down to the predetermined threshold (T_{low}). After this point, the disk controller comes to know that the emergency has been resolved.

Figure 7.9 shows the experimental results of $DRPM_{opt}$ where T_{high} is 60°C (which is set to be the same as the thermal envelope in this experiment) and T_{low} is obtained by subtracting a few degree Celsius

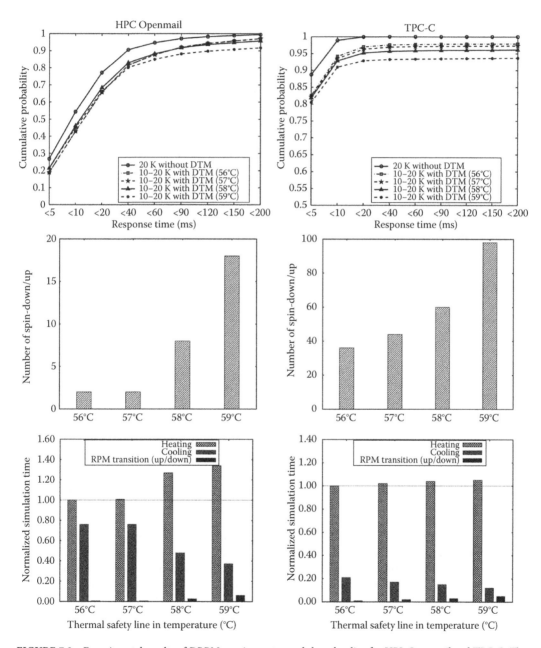

FIGURE 7.9 Experimental results of $DRPM_{opt}$ using watermark-based policy for HPL Openmail and TPC-C. *The thermal safety line* in the graphs denotes temperature at which disk-drive sufficiently cools down to operate. Note that plots on the left column are test results with HPL Openmail traces whereas those on the right column are test results with TPC-C.

from T_{high}. The graphs in the first row of Figure 7.9 shows the CDF of the average response time across disk-drives. The solid curve in each graph represents the performance of the baseline system without any DTM and others are for DTM with different lower thresholds (T_{low}) (which is denoted by a *thermal safety line*). In this experiment, the RPM transition (up and down) time is assumed to be 7 s. The lower thermal safety line of T_{low} helps the performance of a multispeed disk. This is because the lower value of T_{low} allows more relaxation in throttling and reduces the number of RPM transitions. As shown from the graphs in the second row of Figure 7.9, the lower values of T_{low} result in fewer number of RPM transitions. The graphs in the last row of Figure 7.9 shows the time breakdown for the total simulated time composed of heating/cooling times and RPM transition time for different lower thresholds. The larger fraction of heating time implies better performance, because it implies that more requests could be serviced at the maximum speed. However, it is to be noted that it is not absolutely better than the small portion of heating time, because the RPM transition time (in the order of seconds) offsets this benefit.

7.6 Conclusions

This chapter has presented graceful operation of multispeed disk to handle thermal emergencies in large disk arrays. We studied several DTM policies (i.e., time based and watermark based) for different multispeed disk techniques (i.e., $DRPM_{simple}$ and $DRPM_{opt}$) executing real workloads and observed that the DRPM technique is one of the best solutions to avoid thermal emergencies. The $DRPM_{simple}$ technique overcomes thermal emergencies by dynamically modulating the rotational speed of disks and providing predefined delays. But such delays cause poor performance (such as response time), compared to a normal disk-drive without any DTM technique. However, the $DRPM_{opt}$ technique further improves the performance by continuously servicing the requests at the lower speed. *Time-* and *watermark-based policies* have been evaluated for thermal management and they showed that the time taken for an RPM transition in a multispeed disk plays a critical role in the performance of thermal management.

References

1. K. Aruga. 3.5-Inch high-performance disk drives for enterprise applications: AL-7 Series. *Fujitsu Science and Technology Journal*, 37(2):126–139, December 2001.
2. K.G. Ashar. *Magnetic Disk Drive Technology: Heads, Media, Channel, Interfaces, and Integration.* IEEE Press, New York, 1997.
3. D. Brooks and M. Martonosi. Dynamic thermal management for high-performance microprocessors. In *Proceedings of the International Symposium on High-Performance Computer Architecture (HPCA)*, Nuevo Leone, Mexico, pp. 171–182, January 2001.
4. E.V. Carrera, E. Pinheiro, and R. Bianchini. Conserving disk energy in network servers. In *Proceedings of the International Conference on Supercomputing (ICS)*, San Francisco, CA, June 2003.
5. J. Choi, Y. Kim, A. Sivasubramaniam, J. Srebric, Q. Wang, and J. Lee. Modeling and managing thermal profiles of rack-mounted Servers with thermoStat. In *Proceedings of the International Symposium on High Performance Computer Architecture (HPCA)*, Phoenix, AZ, pp. 205–215, February 2007.
6. N.S. Clauss. A computational model of the thermal expansion within a fixed disk drive storage system. Master's thesis, University of California, Berkeley, CA, 1988.
7. P.A. Eibeck and D.J. Cohen. Modeling thermal characteristics of a fixed disk drive. *IEEE Transactions on Components, Hybrids, and Manufacturing Technology*, 11(4):566–570, December 1988.
8. Fujitsu MAM3184MC—Full Specifications. http://www.fel.fujitsu.com/home/v3__product.asp?inf=dsc&pid=301

9. G.R. Ganger, B.L. Worthington, and Y.N. Patt. *The DiskSim Simulation Environment Version 2.0 Reference Manual.* http://www.ece.cmu.edu/~ganger/disksim/

10. E. Grochowski and R.D. Halem. Technological impact of magnetic hard disk drives on storage systems. *IBM Systems Journal,* 42(2):338–346, 2003.

11. S. Gurumurthi. The need for temperature-aware storage systems. In *Proceedings of the Intersociety Conference on Thermal and Thermomechanical Phenomena in Electronic Systems,* San Diego, CA, pp. 387–394, May 2006.

12. S. Gurumurthi, Y. Kim, and A. Sivasubramaniam. Thermal simulation of storage systems using STEAM. *IEEE Micro Special Issue on Computer Architecture Simulation and Modeling,* 26(4):43–51, 2006.

13. S. Gurumurthi, A. Sivasubramaniam, M. Kandemir, and H. Franke. DRPM: Dynamic speed control for power management in server class disks. In *Proceedings of the International Symposium on Computer Architecture (ISCA),* San Diego, CA, pp. 169–179, June 2003.

14. S. Gurumurthi, A. Sivasubramaniam, and V. Natarajan. Disk drive roadmap from the thermal perspective: A case for dynamic thermal management. In *Proceedings of the International Symposium on Computer Architecture (ISCA),* Madison, WI, pp. 38–49, June 2005.

15. T. Heath, A.P. Centeno, P. George, Y. Jaluria, and R. Bianchini. Mercury and freon: Temperature emulation and management in server systems. In *Proceedings of the International Conference on Architectural Support for Programming Languages and Operating Systems,* San Jose, CA, October 2006.

16. G. Herbst. IBM's drive temperature indicator processor (Drive-TIP) helps ensure high drive reliability. In *IBM Whitepaper,* October 1997.

17. Hitachi Power and Acoustic Management—Quietly Cool. In *Hitachi Whitepaper,* March 2004. http://www.hitachigst.com/tech/techlib.nsf/productfamilies/White_Papers

18. H. Ho. Fast servo bang-bang seek control. *IEEE Transactions on Magnetics,* 33(6):4522–4527, November 1997.

19. Y. Kim, S. Gurumurthi, and A. Sivasubramaniam. Understanding the performance-temperature interactions in disk I/O of server workloads. In *Proceedings of the International Symposium on High-Performance Computer Architecture (HPCA),* Austin, TX, February 2006.

20. H. Levy and F. Lessman. *Finite Difference Equations.* Dover Publications, New York, 1992.

21. M.K. Patterson, X. Wei, and Y. Joshi. Use of computational fluid dynamics in the design and optimization of microchannel heat exchangers for microelectronics cooling. In *Proceedings of the ASME Summer Heat Transfer Conference,* San Francisco, CA, 2005.

22. C. Ruemmler and J. Wilkes. An introduction to disk drive modeling. *Computer,* 27:17–28, March 1994.

23. N. Schirle and D.F. Lieu. History and trends in the development of motorized spindles for hard disk drives. *IEEE Transactions on Magnetics,* 32(3):1703–1708, May 1996.

24. Seagate Cheetah 15K.3 SCSI Disc Drive: ST3734553LW/LC Product Manual, Vol. 1. http://www.seagate.com/support/disc/manuals/scsi/100148123b.pdf (accessed March, 2003).

25. Seagate Workload Management for Business-Critical Storage. http://www.seagate.com/docs/pdf/whitepaper/TP555_BarracudaES_Jun06.pdf (accessed July, 2006).

26. L. Shang, L.-S. Peh, A. Kumar, and N.K. Jha. Thermal modeling, characterization and management of on-chip networks. In *Proceedings of the International Symposium on Microarchitecture (MICRO),* Portland, OR, pp. 67–78, December 2004.

27. R.K. Sharma, C.E. Bash, C.D. Patel, R.J. Friedrich, and J.S. Chase. Balance of power: Dynamic thermal management for internet data centers. *IEEE Internet Computing,* 9(1):42–49, January 2005.

28. K. Skadron, M.R. Stan, W. Huang, S. Velusamy, K. Sankaranarayanan, and D. Tarjan. Temperature-aware microarchitecture. In *Proceedings of the International Symposium on Computer Architecture (ISCA),* San Diego, CA, pp. 1–13, June 2003.

29. M. Sri-Jayantha. Trends in mobile storage design. In *Proceedings of the International Symposium on Low Power Electronics*, San Diego, CA, pp. 54–57, October 1995.

30. J. Srinivasan and S.V. Adve. Predictive dynamic thermal management for multimedia applications. In *Proceedings of the International Conference on Supercomputing (ICS)*, San Francisco, CA, pp. 109–120, June 2003.

31. The Openmail Trace. http://tesla.hpl.hp.com/private_software/

32. Ultrastar 146Z10 hard disk drives specifications. http://www.hitachigst.com/tech/techlib.nsf/techdocs/6BB69573F9D1537687256BD600697AAA/$file/u146z10_sp22.pdf (accessed February, 2002).

33. UMass Trace Repository. http://traces.cs.umass.edu

34. A. Weissel and F. Bellosa. Dynamic thermal management for distributed systems. In *Proceedings of the First Workshop on Temperature-Aware Computer Systems (TACS)*, Munich, Germany, June 2004.

8

Energy-Saving Techniques for Disk Storage Systems

Jerry Chou
Lawrence Berkeley National Laboratory

Jinoh Kim
Lawrence Berkeley National Laboratory

Doron Rotem
Lawrence Berkeley National Laboratory

8.1 Introduction

Enterprise applications and large-scale scientific computing research are generating and exploring information contents of extremely large datasets most of which are maintained online for easy accessibility. These large datasets are analyzed by a large community of researchers and must be retained and managed on thousands of traditional rotating disks and sometimes supported by mass storage systems when the data is to be migrated to deep archive. As prices of disks are getting cheaper in terms of dollars per gigabyte, the prediction is that the energy costs for operating and cooling these rotating disks will eventually outstrip the cost of the disks and the associated hardware needed to control them. Currently it is estimated that disk storage systems consume about 25%–35% of the total power used in data centers. This percentage of power consumption by disk storage systems will only continue to increase, as data-intensive applications demand fast and reliable access to online data resources. This in turn requires the deployment of power hungry faster (high rpm) and larger capacity disks. As a result, there are now many research programs in industry, government, and academia, which address reducing storage energy costs at data centers. Examples of such initiatives and programs for energy-efficient computing currently underway include:

- Green Grid Consortium that include such companies as IBM, Microsoft, Google, NetApp, EMC[2], etc.
- DiskEnergy at Microsoft Research
- GreenLight project at UC San Diego
- Leadership in energy efficient computing (LEEC) at LBNL
- Green-NET Project in INRIA

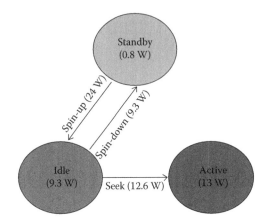

Description	Value
Idle power	9.3 W
Active power	13 W
Standby power	0.8 W
Spin-up power	24 W
Spin-down power	9.3 W
Spin-up time	15 s
Spin-down time	10 s

FIGURE 8.1   Disk power state and power consumption.

In addition, disk storage vendors are introducing more energy-efficient hard disks drives (HDD) with rotating platters, multispeed HDD, solid state drives (SSDs) whose energy consumption is only a fraction of that of HDDs, and hybrid disk drives that employ a combination of HDD and SDD in a single unit. Unfortunately the price of SSDs in terms of dollars per byte is currently an order of magnitude higher than that of HDDs, thus using them as a replacement for HDDs in large data centers does not seem practical in the near future. More information about this topic is given in Section 8.5.

Several energy saving techniques for disk-based storage systems have been introduced in the literature. Most of these techniques revolve around the idea of spinning down the disks from their usual high-energy mode (idle or active mode) into a low-energy mode (sleep or standby mode) after they experience a period of inactivity whose length exceeds a certain threshold (idleness threshold). The reason for this is that typical disks consume about one tenth of the power in sleep mode as compared with their power consumption in idle mode (see Figure 8.1).* It is well known that the optimal idleness threshold should be set to about B/p seconds, where B is the amount of energy (in joules) needed to spin up the disk from sleep to idle mode and p is the power (in watts) required to keep the disk spinning in idle mode [12]. There are problems associated with this spin-down technique as listed in the following:

- *Energy and response time penalty*: Disks can only service requests while they are in idle mode, in case a request arrives when the disk is in sleep mode there is a response time penalty of several seconds (10–15 s) before the request can be serviced. In addition, as mentioned earlier, B joules are required to spin up the disk; in some cases this can exceed the energy saved by transitioning the disk to sleep mode.
- *Expected length of inactivity periods*: Under many typical workloads found in scientific applications, disks do not experience long enough periods of inactivity (longer than the idleness threshold) thus limiting the opportunities to save energy by transitioning to sleep mode.

The research literature attempts to solve the above problems using several techniques. The idea is to reshape the workload so that I/O requests are directed to a small subset of the disks allowing the other disks to enjoy relatively long periods of inactivity. Several of these techniques are listed below. We will give more details about each of these in Section 8.3.

- *Data placement techniques*: These techniques use intelligent data placement on disks where the most popular files are packed into the fewest possible disks subject to response time constraints. This placement is based on some knowledge of data access patterns. A second approach tries

---

* We interchangeably use "sleep" and "standby" for lower energy mode of disks.

to achieve the same goal by using dynamic data migration based on observed workloads. This approach does not require an a priori knowledge of the access patterns but results in a larger timescale to adapt.

- *Write off-loading*: These techniques are directed toward minimizing energy consumed due to write requests. Newly written data is diverted to disks which are currently spinning (anywhere in the data center) thus enabling disks in sleep mode to remain in their low-energy mode for longer periods.

- *Batching I/O requests*: Aggregating I/O requests into batches where each batch is directed to a single disk can help saving energy by satisfying multiple requests with fewer transitions from sleep to idle mode.

- *Space reduction*: This class of techniques uses very efficient representations of the data in order to reduce the amount of space required to store data. This results in reducing the number of spinning disks leading to energy savings. Techniques in this class include data compression and data de-duplication.

- *Power aware caching and pre-fetching*: Due to the different energy cost of accessing data from disks in their idle or sleep modes, caching for energy savings must employ different policies than caching for performance. The idea behind such techniques is to always prefer evicting blocks from idle disks rather than from disks in sleep mode. It also seems beneficial to allocate more cache space for disks in sleep mode than disk in idle mode. Another observation is that pre-fetching data from an idle disk is beneficial in order to avoid the need to access it later when the disk is placed in sleep mode.

- *Using replication for energy saving*: Many modern file systems such as HDFS (Hadoop Distributed File System) use replication for enhancing fault tolerance and supporting data recovery. This replication can be used to save energy by accessing data copies from spinning disks while transitioning disks that contain redundant data to sleep mode.

The rest of this chapter is organized as follows: In Section 8.2, we survey some theoretical results on dynamic power management (DPM) related to competitive algorithms. In Section 8.3, we describe both static and dynamic data placement methods and show several results on energy savings achieved by these placements. We also describe how data replication may be used beneficially to save energy by diverting read requests to replicas residing on spinning disks whenever possible. Other techniques that enhance energy savings such as write off-loading are also described in this section. Storage products that support energy savings in various stages of maturity are described in Section 8.4. Several prototypes developed in academic institutions as well as commercial products are surveyed in this section. Finally, in Section 8.5, we present what we think are important commercial future trends in software and hardware as well as directions for future research.

8.2 Theoretical Background

The theory of dynamic power management of disks has also drawn a lot of attention recently from the theoretical computer science community (see [12] for an extensive overview of this work). Most of this work considers a single disk only and attempts to find an optimal idle waiting period (also called idleness threshold time) after which a disk should be moved to a state which consumes less power. This power management problem is an online problem and is similar to the *ski-rental* problem where a person on a ski vacation must make a buy-rent decision for his/her ski equipment with no knowledge of the duration of the vacation period. A storage device can serve requests when it is in a high-power state. At any given time a storage device is not aware of the arrival time of future requests. The device experiences idle periods, i.e., periods when no requests arrive. When experiencing an idle period, the problem is when to move to

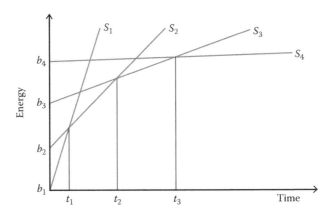

FIGURE 8.2   Energy states of a disk.

a lower-power state and benefit from the reduced energy consumption, given that the device must finally be transitioned to high-power mode again (when a request arrives) at a substantial cost.

In a more general setting, the disk may be placed in several power consumption states (see Figure 8.2). The assumption is that the disk can be transitioned among n power consumption states, S_i, where the ith state consumes more power than the jth state for $i < j$. The disk can serve file requests only when it is in the highest power state (S_1 state) which is also called the active state. The system must pay a penalty b_i if a request arrives when the disk is in the ith state, the penalty is proportional to the power needed to spin up from state S_i to the active state S_1. The penalty is increasing with the state number, i.e., $b_j > b_i$ for $j > i$, and $b_1 = 0$. The problem is that of devising online algorithms for selecting optimal threshold times, based on idle periods between request arrivals, to transition the disk from one power state to another. The problem is illustrated in Figure 8.2, where the different power states are represented by lines. The slope of each line represents the power consumption at this state and b_i represents the penalty to move from state S_i to the active state S_1. Optimally the system should transition between states at the intersection points t_i and operate along the lower envelope of the diagram. The most common case has only two states namely, active state (full power) and standby (sleep) state.

The quality of these algorithms is measured by their competitive ratio which compares their power consumption to that of an optimal offline algorithm that can see the entire request sequence in advance before selecting state transition times. An online algorithm is called c-competitive if for every input and for any idle period, the total energy consumption of that algorithm is at most c times that of the optimal one that has complete knowledge. Competitive analysis provides a strong worst-case performance guarantee. In [12], Irani et al. presented a two-competitive algorithm for both the two state and the multiple state power management problems. It is also known that this is the best possible competitive ratio that can be achieved by any deterministic algorithm.

8.3 Survey of Technology

Most data centers have low utilization with a typical average utilization in the range of 20%–30%. However, these lightly loaded servers still consume a substantial fraction of the energy as they consume at peak loads. Therefore, several techniques were proposed to exploit the opportunities of saving energy by powering down underutilized nodes (either disks or servers) and redistributing their load to the remaining powered-on nodes. In the following, we introduce some of these techniques and discuss the related performance issues or limitations of their approaches.

8.3.1 Data Placement

As observed from network server loads, file popularity in terms of access frequency vary greatly. Based on this observation, data placement strategies attempt to change the workload structure such that the disks are split into groups, with the most frequently accessed files placed in the first group of disks and the least frequently accessed files placed in the last group of disks. This creates longer idle periods on the group of disks having less frequently accessed files with an opportunity to shut down these disks for power saving. The trade-off however, is that the frequent accesses to the disks having popular files, may result in longer response times. Thus, these disks should be loaded maximally, but without exceeding some guaranteed acceptable response times. Some approaches based on such data placement technique are described in this section.

8.3.1.1 PDC: Popular Data Concentration

PDC [20] dynamically migrates frequently accessed data to a subset of the disks in the system, so that the load becomes skewed such that a few disks take a high percentage of the load while the other disks can stay in low-power mode. Specifically, the goal of PDC is to store the most popular files on the first disk, the next set of the most popular files on the second disk, and so forth. To avoid performance degradation, PDC estimates access rate of each data and migrates data onto a disk until the expected load on the disk reaches its maximum bandwidth. PDC may also need to be applied periodically to the system to accommodate the change of data popularity over time. PDC was evaluated with a broad set of parameters including file popularity, request rate, request temporal locality, number of disks, migration frequency, etc. The results showed PDC achieves consistent and robust energy savings for most of the settings. However, the energy gains of PDC does degrade substantially for long migration intervals. On the other hand, shorter migration intervals could significantly increase the percentage of delayed requests, due to the intense file migration traffic.

8.3.1.2 DiskPack

DiskPack proposed in [19] splits disks into two groups. The files accessed frequently are placed in the first group of disks and the remaining files are placed in the second group. In this work, the trade-offs between energy savings and response time in static file allocation strategies are discussed. It is shown that packing the files on disks with maximal energy savings subject to response time restrictions is equivalent to the two-dimensional vector packing problem (2DVPP). This problem is known to be *NP-Complete*, however several approximation algorithms for it are known. The authors use one such approximation algorithm and show greater than 50% energy saving with real-life workloads. The results of [19] can also be used as a tool for obtaining reliable estimates on the size of a disk farm needed to support a given workload of requests while satisfying constraints on I/O response times. The simulation results indicate that power saving decreases with greater arrival rates, while it increases with higher allowable constraints on disk loads.

8.3.1.3 Summary

Data placement techniques reduce energy consumption by reorganizing data locations in a storage system. Previous studies have shown such techniques can effectively reduce energy consumption, but performance overheads and penalties can also occur during data reorganization. Since the effectiveness of these techniques heavily relies on workload characteristics and data popularity, and such statistics can also dynamically changed, data re-organization may need to be applied periodically or on-demand to the system. As a result, the performance overhead and penalty of data placement techniques could become a concern for more dynamic systems or workloads. Therefore, it remains an interesting research problem to explore the possibility of minimizing the overheads and penalties for this class of techniques.

8.3.2 Replication

Most data centers or clusters servicing data-intensive workloads widely use data replication to ensure data availability and reliability in the presence of failures. Thus, many techniques [13,15,21] have been proposed to use these replicated data to redirect load away from the deliberately powered-down nodes for energy saving, without losing access to the data on these nodes. In the following, we describe two techniques that utilize data replication to save energy while maintaining system performance, such as load-balancing, response time, etc.

8.3.2.1 Diverted Accesses

Diverted Accesses [21] is proposed based on an observation that replicated data only needs to be accessed in the following scenarios: (1) during periods of high demand for disk bandwidth, to increase performance; (2) when disk failures occur, to guarantee reliability and availability; and (3) periodically to handle data changes. Since replicated data need not be readily accessible at all times, Diverted Accesses segregates original data from the associated replicated data onto two different subsets of disks, original disks and redundant disks. Reads are always directed to the original disks, unless redundant disks are needed to share the load or handle disk failures. Writes are performed on all disks when the load is high, while they are directed to the original disks immediately, and propagated to the redundant disks periodically afterward, when load is light or moderate. By doing so, disks for replicated data can be kept in low power mode and only be powered on when needed for the purpose of performance, fault tolerance or writes propagation.

Diverted Accesses is applicable to any storage systems with data replication, as long as the replicated data could be segregated from the original data onto different subsets of disks. The evaluation results showed that Diverted Accesses was able to reduce significant amount of energy (20%–61%) under real workload traces. The technique can also be used along with other energy saving approaches to further reduce their energy consumptions. For example, Diverted Accesses can be combined with PDC [20] by applying data migration technique on the original disks, and it can be combined with MAID [3] by using the additional cache disks in MAID as the redundant disks. The energy consumption of both techniques (PDC and MAID) were substantially reduced when they are combined with Diverted Accesses as shown in [21].

8.3.2.2 Dissolving Chains and Blinking Chains

The work in [14] investigates the interaction between replication and power down schemes to provide an energy management approach that can maximize energy efficiency by powering down some nodes while ensuring data availability and the utilization of the remaining nodes does not exceed a targeted peak utilization. Specifically, the authors propose two node power down techniques *Dissolving Chains* and *Blinking Chains* based on the *Chained Declustering* data replication scheme [10], which strips the partitions of a data set two times across the nodes of a system. In Chained Declustering, the logical ordered arrangement of all nodes is commonly referred as a *ring*. When a node in a ring is removed (or powered down in the context of the energy saving technique), the ring is broken into *segments*. Chained Declustering guarantees data availability when multiple nodes are removed from the chain if node removal does not occur on any adjacent nodes. In addition, the workload can be evenly distributed within a segment, while nodes in a longer segment of a ring have lower loads compared to a node in a shorter segment.

Based on the properties of Chained Declustering, Dissolving Chains simply powers down (up) nodes as system utilization decreases (increases). To minimize the load imbalance on the remaining nodes, it walks along segments of the ring and powers down nodes at the halfway point of the given segment, such that the length of all segments in the ring could be more close to equal. As illustrated by Table 8.1, if the target peak utilization is 4/3, two nodes can be powered down while the system still has the capacity to run the workload. Let n_0 be the first node powered down, then the second node that can be powered

TABLE 8.1 Dissolving Chains with Two Nodes Powered Down

Nodes	n_0	n_1	n_2	n_3	n_4	n_5	n_6	n_7
Original	—	B_1 (1)	B_2 (2/3)	B_3 (1/3)	—	B_5 (1)	B_6 (2/3)	B_7 (1/3)
Replica	—	b_2 (1/3)	b_3 (2/3)	b_4 (1)	—	b_6 (1/3)	b_7 (2/3)	b_0 (1)
Load	0	4/3	4/3	4/3	0	4/3	4/3	4/3

TABLE 8.2 Dissolving Chains with Three Nodes Powered Down

Nodes	n_0	n_1	n_2	n_3	n_4	n_5	n_6	n_7
Original	—	B_1 (1)	—	B_3 (2)	—	B_5 (1)	B_6 (2/3)	B_7 (1/3)
Replica	—	b_2 (1)	—	b_4 (2)	—	b_6 (1/3)	b_7 (2/3)	b_0 (1)
Load	0	2	0	2	0	4/3	4/3	4/3

TABLE 8.3 Blinking Chains with Three Nodes Powered Down

Nodes	n_0	n_1	n_2	n_3	n_4	n_5	n_6	n_7
Original	—	B_1 (1)	B_2 (1/2)	—	B_4 (1)	B_5 (1/2)	—	B_7 (1)
Replica	—	b_2 (1/2)	b_3 (1)	—	b_5 (1/2)	b_6 (1)	—	b_0 (1)
Load	0	3/2	3/2	0	3/2	3/2	0	2

down is n_4 because it is the halfway point of segment from n_1 to n_7. As a result, the two segments in the ring have the equal length and all remaining nodes reaches the peak utilization 4/3. However, Dissolving Chains has some limitations, one of which is that it can only reach a balanced state when the number of powered down nodes is 2^i and 2^i divides N, where N is the number of nodes in the system. For example in Table 8.2, if the target peak utilization decreases and one more node is allowed to be deactivated, node n_2 would be powered down but a balanced state cannot be reached.

In contrast to Dissolving Chains, the other proposed power down method, Blinking Chains, allows to power up some nodes before powering down the desired number of nodes in order to reduce load imbalances. For example, comparing to the results of Dissolving Chains in Table 8.2, Blinking Chains in Table 8.3 achieves less load variation across nodes by first powering up node n_4 then powering down node n_3 and n_6. Blinking Chains is also able to reach a balanced state whenever the number of powered down nodes exactly divides the number of nodes in the system. For example, when we attempt to powering down 3 out of 12 nodes in a system, Blinking Chains can achieve a balance state by powering down nodes n_0, n_4, and n_8, while Dissolving Chains cannot achieve a balance state due to its simply binary cut strategy. However, Blinking Chains may induce significant transition overhead as multiple nodes might need to be powered up or down.

The proposed techniques are evaluated by a simulated storage system with 1000 nodes. The energy consumption and response time were measured as a function of node utilization from a real server. The results show both proposed methods can effectively reduce 50% energy in a system with only 50% utilization by powering down half of the nodes. The results also confirm Blinking Chains achieves better load balancing than Dissolving Chains, but also has higher transitioning energy costs. However, given the same peak node utilization requirement and workload, Blinking Chains could be able to run the workload on fewer nodes than Dissolving Chains if the load is more balanced across nodes. Therefore, Blinking Chains could still overcome the transition penalty as the running time of a workload increases.

To sum up, [14] addresses the performance consideration of load balancing when designing an energy saving technique based on replication. The proposed energy management techniques based on the Chained Declustering replication scheme are able to gracefully adopt to overall system utilization. It would be interesting to further look into the possibility of developing such power down techniques with the consideration of other data placement algorithms and replication schemes as well as dealing with node failures.

8.3.2.3 Summary

Replication techniques exploit existing data replications available in storage systems in order to reduce energy consumption. The main goal of such techniques is to power down disks/nodes while maintaining system performance, such as data availability, response time, and load balancing. Unlike the data placement techniques, replication techniques are more sustainable to dynamic changes of workload characteristics and data access frequency, and they do not require periodic data placement or re-organization. However, with replicated data, when a write request for data arrives to the system, to ensure data consistency, the updated data must be written to all the disks/nodes holding a replica of the data. As a result, potentially multiple disks/nodes need to be powered on to service a single write request. Therefore, replication techniques could become less effective for write-intensive workloads. However, several techniques have been developed to minimize such inefficiency, one of them is the write off-loading technique that will be described in Section 8.3.5.

8.3.3 Caching and Prefetching

Large storage caches are commonly used in modern storage systems to improve performance by reducing the number of disk accesses [5,11]. As the number of disk accesses is reduced, disk idle time increases, thus providing more opportunities to power down the disks and save energy. Storage systems that integrate caching with disk power management have been developed and are available in commercial products [2,4]; we will introduce some of these systems in Section 8.4.2. Thus, here we focus only on the energy-aware cache management policy that could be used in these systems.

Cache management policy could affect energy consumption, because it influences the order in which requests access disks and therefore changes the average idle time of disks. For example, if most of the data from a particular disk is kept in cache by the cache replacement policy, that disk can stay in a low power mode longer and reduce its energy consumption. The relation between energy and cache management policy, specifically the cache replacement policy, is discussed by Zhu et al. [29,30]. Their goal is to design a power-aware cache replacement algorithm, such that for a given request sequence the algorithm generates a cache-miss sequence for which the disks consume the least energy, their technique is based on the multispeed disks. Their proposed algorithms are described next.

8.3.3.1 OPG: Offline Power-Aware Greedy

An off-line energy-optimal cache replacement algorithm was developed in [28], which runs in polynomial time by using dynamic programming. However, the energy-optimal algorithm is still too complex to be practical for implementation and evaluation, thus in [30] the authors proposed a heuristic off-line power aware greedy algorithm, called OPG. The OPG algorithm iteratively computes the energy penalty of evicting every resident block, and evicts the block with the minimum energy penalty. The energy penalty is computed based on the information of *deterministic misses*, which are the bound-to-happen cache misses observed from the cache content and request sequence regardless what replacement algorithm is used. The algorithm is a sub-optimal heuristic as it looks at only the current set of deterministic misses.

8.3.3.2 PA-LRU: Power-Aware LRU

PA-LRU [30] is a power-aware online cache replacement algorithm, which dynamically keeps track of workload characteristics for each disk, and classifies disks into two categories, *regular* and *priority*. The priority disks are the ones with small percentage of cold misses and large request inter-arrival lengths, and the other disks belong to the regular class. Intuitively, a cache hit from one of the priority disks has higher probability to increase disk idle time than a cache hit from a regular disk, thus data blocks from priority disks should be considered to have higher energy penalty and kept in the cache. Therefore, PA-LRU simply keeps the accessed data blocks from two classes of disks in two separate LRU stacks and always replaces the least recently used block from the regular LRU stack. However, PA-LRU does have the drawback that a parameter configuration is required for disk classification, and no parameter setting works well for all workloads.

8.3.3.3 PB-LRU: Partition-Based LRU

PB-LRU [29] is a self-tuning heuristic cache replacement algorithm. It dynamically adapts to workload changes without tedious parameter tuning. The basic idea of PB-LRU is to divide the cache into separate partitions, one for each disk, and manage each partition separately using a basic replacement policy such as LRU. The goal of PB-LRU is to find the partitioning, such that the total energy consumption of all disks is minimized. It is shown that the partitioning problem is a variant of the famous 0-1 knapsack problem [16] with respect to a list of input variables $E(i, s)$, where $E(i, s)$ is the estimated energy consumption of disk i with partition size s. PB-LRU computes the energy estimation for different partition sizes at runtime based on the results of Mattson's Stack algorithm, which determines if a request would result in a cache hit or miss for different partition sizes. Then PB-LRU solves the 0-1 knapsack problem in pseudo-polynomial time by dynamic programming with time complexity $O(nm^2)$, where n is the number of disks and m is the number of potential partition sizes. Finally, PB-LRU computes the energy-optimal partitioning solution periodically. Setting the periodicity is the only parameter that PB-LRU requires.

8.3.3.4 Summary

The above power-aware cache replacement algorithms were evaluated with both real and synthetic workloads using the Disksim simulator [7]. The power model used in the simulation is similar to the one proposed by Gurumurthi et al. [8] for multispeed disks. Major results from the evaluation are as follows: (1) OPG, the off-line heuristic algorithm, consumes even less energy than the Belady's algorithm for real-system workloads, even though Belady's algorithm is optimal in terms of the number of cache misses. (2) Comparing to LRU, online heuristic algorithms, PA-LRU and PB-LRU, could save 16% more disk energy and provide 50% better average response time. (3) PB-LRU without any parameter tuning is able to achieve similar or even better performance and energy savings than PA-LRU for every evaluated workload. One limitation of these power-aware cache replacement algorithms is that their effectiveness is depending on the workload pattern or characteristics. As a result, these techniques would be effective if sufficient amount of caching space is available or multispeed disks are used.

8.3.4 Multispeed Disks

Although not piratical yet due to technical difficulties, multispeed disks have been considered for energy saving. Similar to DVS (dynamic voltage scaling) that can increase or decrease voltage for power management, multi-speed disks provide a means of changing disk rotational speed—a greater speed for higher performance with a greater power requirement, and vice versa.

Hibernator [27] considers energy management in a storage system with multispeed disks. Figure 8.3 illustrates the disk-based storage system model in Hibernator. Disks are grouped into tiers, and disks in a single tier are assumed to be run with the same rotational speed. Each tier is in different "temperature" from the coldest to the hottest. The hottest tier disks contain the most active data, while the coldest tier contains the least active data. Hibernator determines disk speed for each tier based on the associated temperature. Naturally, the hottest tier would be run in the greatest disk speed to maximize performance, while the coldest tier would be set with the lowest disk speed to save energy.

The core problem in Hibernator is how to identify the right speed for each tier for both energy and performance. Hibernator defines an optimization problem to solve this as in the following:

$$\text{minimize} \sum_{i=1}^{N} E_{ij}, \quad \text{subject to} \sum_{i=1}^{N} \frac{n_i \cdot T_{ij}}{n} \leq T_{max} \tag{8.1}$$

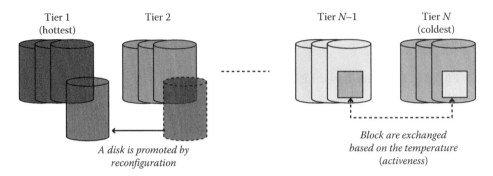

FIGURE 8.3   Hibernator tier-based architecture.

where
  N is the number of disks
  n is the total number of requests in the system
  n_i is the number of requests in disk i for a given period of time
  E_{ij} refers to a predicted energy consumption for disk i at speed j
  T_{ij} refers to the predicted response time in that case
  T_{max} defines the required average response time as the performance constraint

Hibernator periodically performs computations to determine disk speed based on measurements in the current time window for the next period of time. Thus, the basic assumption here is the stability of workload patterns. Conversely, if the workload characteristic is substantially different from the past window, it may not guarantee the performance requirement. If Hibernator detects performance unsatisfaction against the constraint, it disables the energy management function, and sets the entire disks in the system to perform at full speed. The energy management function will be enabled again when it observes that average response time satisfies the given threshold (T_{max}).

Returning back to tier configuration, tiers are reconfigured periodically for adaptability to workloads. That is, Hibernator adapts to workloads by changing the number of assigned disks for each tier. For example, when the load becomes heavier, it assigns more disks to hotter tiers to meet the performance constraint (as in Tier 1 and 2 in Figure 8.3). In contrast, under lighter load, Hibernator moves disks to colder tiers, thus saving energy. A problem arises due to the assumption of RAID-5 organization for Hibernator tiers. For example, suppose m disks are currently in a tier, and n disks ($n < m$) in the new configuration for the next window. The reconfigured tier should have parity blocks uniformly distributed across n disks. Hibernator uses *randomized shuffling* that shuffles blocks in the added disks with the ones in the existing disks randomly; this gives a fairly uniform distribution of parity blocks with ($2 \times m \times s$) block migration, where s is the number of stripes in a disk.

In addition to tier reconfiguration, Hibernator relocates data blocks between tiers, as shown in the right-side tiers in Figure 8.3. A data block that is frequently and recently accessed can be promoted by relocating itself to a hotter tier, if there is any block that has been less inactive than that. This enables Hibernator to keep maintaining hotter data in hotter tiers and colder data in colder tiers.

8.3.5 Write Off-Loading

While the two-competitive approach based on idleness threshold is attractive with its simplicity for disk power management, conventional wisdom is that the approach is not appropriate for server workloads due to short idle periods. However, Narayanan et al. have made several interesting observations, the most important of which is that the enterprise data center has a significant amount of idle time, and it is further extended if write requests are removed from consideration. According to their observations, the average amount of time that a volume is active is 21% without write requests, while it is 60% for both reads and writes. This implies that harvesting idle periods is still feasible for power management in an enterprise data center, and further saving of energy can be possible if we handle writes in a more efficient way.

The basic idea of *write off-loading* is to store write blocks into disks other than their original locations, so that it enables a subset of disks to keep spun down during a write-intensive workload period. Write blocks during such a period are "off-loaded" to any available disk elsewhere in the data center. Inactive disks (due to perhaps to the two-competitive algorithm) stay in power-saving mode, unless any read request is made to the inactive disks. Hence, it is possible to extend energy saving by keeping disks spun down under write-dominant workloads.

For write off-loading, there are two entities, *managers* and *loggers*. A manager is responsible for a set of disks (or *local disks*). All requests to the disks are intercepted by the manager, and based on the workload characteristic, the manager performs power management for the disks. A logger maintains small persistent storage, keeps off-loaded blocks temporarily, and serves requests from managers it communicates with. The operations that a logger provides are *read, write, invalidate,* and *reclaim*. Read and write operations are for RW accesses of off-loaded blocks, while invalidate operations mark associated blocks as invalid. Reclaim operations are for synchronization of updated blocks; upon the reception of a reclaim request from a manager, the logger returns valid blocks for that manager. By doing so, the manager can maintain the most updated blocks in its local disks.

Figure 8.4 illustrates a conceptual model with a manager and loggers. A manager can communicate with multiple loggers (by configuration), and the list is maintained in a persistent storage space (*Logger View* in the figure). It also manages two caches for redirection of requests (accessing off-loaded blocks) and for garbage collection (for already synchronized blocks, hence not necessary to keep any more). For each redirection block, version information is maintained for consistency, in addition to its location information.

Let us discuss read and write operations in more detail. For a read request, the corresponding manager intercepts the request, and first checks if the requested block is kept in any logger it communicates with. If so, the manager sends a read request to that logger; otherwise, the request is served by accessing its local disks that may need to be spun up if not active at that time. If the read request includes multiple blocks located different locations, the request is split as required. A write request is redirected to one of the loggers in the list: (1) if the local disks are in power-saving mode, or (2) if there already exist logged blocks for the request. The case (1) is straightforward. The case (2) is to maintain the new off-loaded

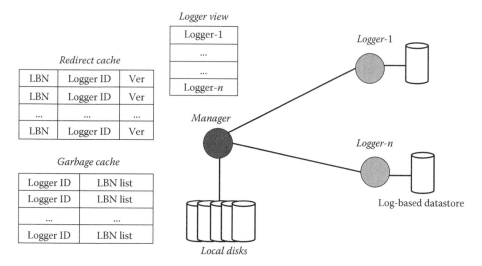

FIGURE 8.4 A conceptual model of write off-loading.

blocks as the latest version. Otherwise, the manager simply writes the blocks to the corresponding local disks.

A manager simply deactivates its local disks whenever the length of the idle period exceeds a given idleness threshold. However, if there is no logger available, deactivation is postponed until the manager finds any active logger. As described, the manager activates local disks for any read request to non-off-loaded data or the number of off-loaded blocks reaches a certain threshold for synchronization.

The trace-driven evaluation in a real testbed with 56 disks the authors conducted shows that write off-loading achieves significant energy saving (up to 60%). One (known) limitation of this approach is that by its concept read requests can suffer from large delays needed to spin up disks when demands occur, and thus maybe unsuitable for time-critical applications.

8.4 Prototypes and Products

In this section, we introduce several interesting prototypes and commercial products that have been developed focusing on energy-saving requirements. We surveyed five prototypes: PARAID for reconfigurable RAID arrays, FAWN for energy-optimized DHT (Distributed Hash Table), Pergamum for disk-based archival storage systems, and Covering Set and All-in Strategy for MapReduce cluster energy management. We then briefly introduce two families of commercial products, COPAN from SGI and AutoMAID from Nexsan, both are based on the MAID (Massive Array of Idle Disks) principle.

8.4.1 Prototypes

8.4.1.1 PARAID: Power-Aware RAID

PARAID [26] has been developed to allow opportunities to spin down disks (hence to save energy) in a RAID array. PARAID uses data replication that happens in a skewed fashion across the RAID disks. Thus, each disk may maintain a different amount of replicated data. Under light loads, it would be possible to spin down some disks without suffering from data unavailability. The main motivation for this work came from several important observations in data server systems. One observation is a load fluctuation over time, severely varying ranges from light to heavy loads. However, conventional RAIDs need to keep spinning entire disks even under very light loads. Another important observation is underutilization of storage (30%–60%), giving an opportunity to utilize unused storage for data replication.

In responding to the above first observation, PARAID uses *gear-shifting* based on load changes. PARAID gear-shifting is similar to the concept in automobiles: operating with a low gear implies downgraded performance, whereas a high gear yields a relatively better performance. Perhaps unlike automobiles, however, PARAID achieves smaller energy consumption in a low gear mode by deactivating a subset of the disks in the array. From this perspective, a conventional RAID can be considered as a single-geared device (i.e., only with the highest gear). PARAID gear-shifting is based on disk utilization periodically measured. An upshift takes place based on current disk utilization, while PARAID considers expected disk utilization (derived from current load and the number of active disks after downshifting) for a downshift. By comparing the estimated disk utilization with a predetermined threshold, a gear-shifting decision can be made.

When a downshift condition is met, a disk goes into power-saving mode. To preserve data availability at any time, PARAID uses data replication, in responding to the second observation of storage underutilization. Instead of choosing randomly a disk to spin down, PARAID makes its decisions based on its replication rule called *skewed striping*. Figure 8.5 illustrates an example of skewed striping in a RAID array. In the figure, blocks residing on disk 4 are replicated to the other disks (i.e., disk 1–3). Hence, the entire data set can still be available in case of disk 4 deactivation. In contrast, blocks of disk 3 are replicated only to disk 1 and 2. Additionally, blocks of disk 4 that are replicated on disk 3 are also replicated to disk

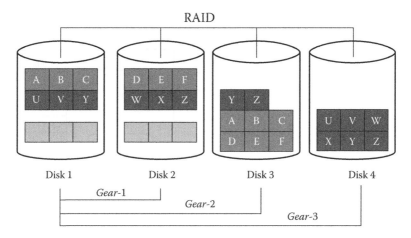

FIGURE 8.5 PARAID replication and gear-shifting.

1 and 2. Hence, disk 3 can be spun down next, after disk 4 deactivation. Based on this skewed replication, PARAID safely moves the gear level without the fear of data unavailability. As shown in the figure, in the highest gear level (i.e., gear level 3), entire disks are activated, while only disk 1 and 2 are active in the lowest gear (i.e., gear level 1) and disk 3 and 4 are in power-saving mode. Since disk 1 and 2 are always active, it is not necessary to replicate their local blocks.

There can be a consistency problem between original blocks and replicas due to updates. PARAID provides two schemes for consistency, *full synchronization* copying the entire modified blocks to the associated original blocks when a disk is powered on, and *on-demand synchronization* where stale blocks are updated only when they are accessed.

PARAID is developed as a device layer in the conventional RAID multi-device drivers, and a prototype has been implemented in a Linux kernel. According to the evaluation reported with a 5-disk prototype, PARAID saves up to 34% power compared to conventional RAIDs without significant performance loss.

8.4.1.2 FAWN: A Fast Array of Wimpy Nodes

FAWN [1] is an energy-efficient architecture for a *key-value* storage cluster with low-power embedded CPUs and flash storage. It is designed particularly for I/O-intensive workloads accessing small-sized data, e.g., thumbnail images. For such workloads, conventional disks might not be suitable due to inefficient random access with long seek latencies. Memory-based clusters may also be poorly suited since high-performance memory requires a significant amount of power in general. The FAWN architecture is optimized for small-sized data store and retrieval with less energy requirements than conventional clusters but with acceptable performance.

FAWN consists of *front-end* and *back-end* nodes. Front-end nodes receive queries from users, forward them to the associated back-end nodes (based on query keys), and give responses back to users. Back-end nodes sit in a distributed hash table (DHT) based on consistent hashing, and each one can be identified by the key value in the given query. Instead of using DHT routing, each front-end node maintains the entire membership list. Thus, a query can be forwarded to the associated back-end node by looking up the list. Figure 8.6 illustrates a DHT ring with back-end nodes, each of which has responsibility for a range of key. In the figure, front-end nodes are not included. In a physical configuration, nodes are connected each other via an Ethernet.

A back-end node is specialized with an embedded CPU and flash storage (as shown in Figure 8.6). Flash is attractive for random-access, read-intensive workloads and it is two orders of magnitude faster than magnetic disks. In addition, flash has significantly smaller power requirements than hard disks. One shortcoming of flash memory can be expensive random writes since updating requires erasing an entire

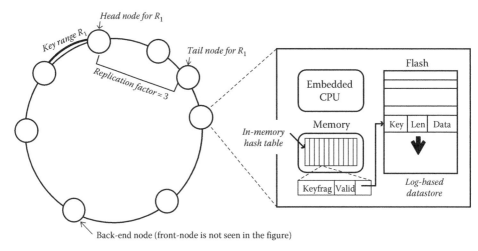

FIGURE 8.6   FAWN in a DHT ring and back-end node architecture.

block (typically 128 kB) and rewriting the modified block. Due to this intrinsic characteristic, FAWN uses it as an appending-only log-structured datastore.

To locate data objects in the back-end node, a hash table for metadata is maintained in the (back-end's) memory, each entry of which points the actual data item in the flash memory, as shown in Figure 8.6. When a lookup query comes (from a front-end node), the back-end node first locates a hash table entry based on the key, and tests if the entire key in the flash pointed by the hash table entry matches the given key. If true, then it returns the located data; otherwise, it attempts to locate a hash entry with another hash key configured in the system. For inserting a data object, it simply appends the object in the flash store, and updates the corresponding hash table entry. If it is an update, the data block originally pointed at is freed and will be reclaimed by later garbage collection (since no hash entry points to it). Deletion of an object invalidates the hash entry by the given key, and appends a special "Delete" entry (with 0-byte value) to the datastore to explicitly indicate that the old data block is not valid.

In the case of back-end failure, the hash table is reconstructed by scanning the log datastore. For this, FAWN employs a checkpoint technique that periodically stores hash table contents to the datastore. In the reconstruction phase, the in-memory hash table is restored by reading the checkpoint.

For fault tolerance, data objects are replicated to multiple back-end nodes. In the FAWN design, the successor node for the key (or *head node*) has a primary responsibility for the data, and replicates them by using a variant of *chain replication*. Along the key space, the $R - 1$ successor nodes of the head node will maintain the data objects for which it is responsible, where R is the replication factor. The last successor in the replication chain is called *tail node*. When a new node is joined between the head node and the tail node, the current tail node has no more responsibility to keep the head node's replicated data because it becomes the Rth successor. Instead, the new node should maintain replicated data for the head node. The tail node transfers the entire replicas for the head node to the newcomer. If a new node is right ahead of the head node in the key space, the datastore in the head node is split based on the keys of the new node and the head node, and the new node takes over the responsibility for the data objects belonging to its key space. When a node leaves, the successor of the departed node takes over the responsibility by merging the key spaces. Adding replicas to the following nodes are similar with the node join procedure. Table 8.4 summarizes the functions provided by back-end nodes for data handling and maintenance.

A FAWN prototype was constructed with a single front-end and 21 back-end nodes. The front-end node is running on an Intel Atom-based machine, and is connected with back-end nodes via Ethernet switches. The back-end node is implemented with a single-core 500 MHz AMD processor, 256 MB SDRAM, and 4 GB flash storage. The power requirements are 3 W for idle and 6 W for peak. The evaluation results on

TABLE 8.4 Back-End Functions

Function	Description
Store	Inserts a data object to the datastore: the new data is appended to the log, and the corresponding hash entry is created in the in-memory hash table to point the data block in the log
Lookup	Retrieves a data object from the datastore: a corresponding hash entry is retrieved by the key, and the data block is returned
Delete	Deletes a data object from the datastore: the corresponding hash entry is invalidated in the in-memory hash table by removing the pointer to the log
Split	Splits a datastore into two key ranges: the datastore is scanned and the writes data objects belonging to the new key range to the new datastore. The new datastore will be transferred to the newly joined successor for node addition
Merge	Merges two adjacent key ranges in a single datastore; two datastores are merged in case of node departure
Compact	Cleans up the log and collects garbages

the prototype show around 350 queries per Joule of energy, which is two orders of magnitude better as compared with a conventional disk-based cluster.

8.4.1.3 Pergamum: A Disk-Based Archival Storage

The Pergamum prototype has been developed for archival storage with goals of persistence and accessibility of archived data [23]. Characteristics of archival data are different from backup data in some aspects. For example, archived digital photos would not be frequently accessed, but still require fast accessibility with acceptable latencies, whereas backup data are generally accessed for fault tolerance reasons and read performance would be less important. In this regard, tape media is acceptable for storing backup data, but is ill-suited for archived data due to poor random-access performance.

Another important challenge for archival storage, in addition to reasonable data access performance, is energy efficiency. Even if initial set-up cost is cheap, expensive operational cost can be a serious burden for archival storage systems, and it is known that energy makes up a substantial fraction of the operational cost. Pergamum handles these challenges (i.e., random access performance and energy efficiency) by constructing a storage system with commodity hard disks.

Pergamum is organized with *tomes*, each of which is equipped with a low-power ARM processor, a SATA hard drive, an on-board flash memory (NVRAM), and an Ethernet port. A tome consumes less than 13 W in active mode including the network power. Figure 8.7 shows the tome architecture (the Ethernet port is omitted in the figure).

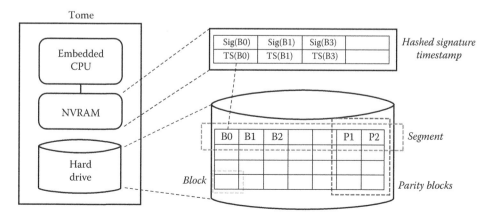

FIGURE 8.7 The architecture of a tome in Pergamum.

An interesting feature of Pergamum is the two-level redundancy architecture for data error correction: intra-tome redundancy and inter-tome redundancy. In Pergamum, the basic access unit for data is called a *block*, which can be sized from 128 KB to 1 MB. For verification of data validity, the tome computes a 64-bit hash value (*signature*) for each data block, and stores the computed hash value to NVRAM with an associated timestamp. Thus, it is possible to recognize corrupted blocks by comparing computed hash values from blocks with the corresponding ones stored in NVRAM. A fixed number of blocks are grouped into a *segment*, that also maintains parity blocks for error correction in a tome. Figure 8.7 also illustrates blocks, segments, parity blocks for segments in the disk, and signatures and timestamps stored in NVRAM.

Additionally, inter-tome redundancy is provided to deal with tome-level errors. The above intra-disk redundancy would be sufficient to handle block errors, but entire tome errors (e.g., due to hard-disk crashes) cannot be addressed. For a set of segments residing in different tomes, a special segment (called redundancy segment) is maintained in a separate tome. The redundancy segment contains erasure correcting codes for the data segments, and is used for recovery from any tome-level failure.

Pergamum can keep 95% of the disks inactive in power-saving mode. In that mode, the read operation is performed in a straightforward manner: The corresponding disk gets spun up first if deactivated, and then the associated blocks are read from the disk. Write operations are more complicated since it is necessary to update the redundancy segment contained a separate tome for consistency. For a write request, the target tome having the requested data block computes "deltas" by comparing the old and new blocks, and the deltas are sent to the tome that contains the associated redundancy segment. The receiving tome computes erasure code based on the existing code and the received deltas, replaces the segment with the newly computed code, and sends an acknowledgment to the target tome as a confirmation. Finally, the target tome updates the local disk for the new write upon the reception of confirmation.

Summarizing, a Pergamum prototype has been developed with commodity disks for energy efficiency as well as acceptable random-access performance for archival services. Energy saving can be harvested by spinning down the vast majority of disks in the system. Pergamum's design that uses disks provides acceptable performance with the benefits of fast random access (compared to tape-based media that have poor random-access performance). Pergamum also provides intra-tome and inter-tome redundancy to give a high degree of data persistence.

8.4.1.4 Covering Set and All-In-Strategy

Some prototype systems have also been developed for distributed processing frameworks in data centers. One of such frameworks getting more attention is the MapReduce-style framework, such as Yahoo!'s Hadoop [9], Google's MapReduce [6], and some others [25]. A MapReduce-style cluster is commonly built with a group of commodity hardware, and thus, node failure is treated as norm rather than exception. Thus, data must be replicated and distributed across nodes to accommodate common failures, namely, single-node and whole-rack failures. For example, Hadoop's file system (HDFS) ensures that (1) no two replicas of a single data item are stored on any one node, and (2) replicas of a data item must be found on at least two different racks. We introduce two prototype systems for energy-efficient MapReduce frameworks: *Covering Set* [15] and *All-In Strategy (AIS)* [13].

To ensure the data availability when powering down nodes, Covering Set proposes to add an additional rule into the existing replication principles, that is at least one replica of a data must be stored in a subset of nodes referred as a *covering subset*. With the new replication scheme, when the system load is low, all the nodes not in the *covering subset* can be powered down without affecting the availability of data or interrupting the normal operation of the clusters. For example, given a cluster of four racks of nodes, the first rack can be labeled as the *covering subset*, so that every data has at least one replica on the rack. When the system is under-utilized, the other three racks of nodes can be powered down to save energy, and data is still available to the users on the first rack. On the other hand, when an application cannot meet its performance requirement from users, all nodes can be powered up again to run at the full system capacity.

In contrast to Covering Set, AIS always attempts to run workload as quickly as possible at the full capacity, then power down all nodes at the end of the workload execution to reduce the idle energy cost. Thus, AIS does not require storage resource over-provisioning (i.e., the size of the *covering subset*) or modification to the existing replication scheme. AIS also has a very predictable degradation in the workload response time based on the time to power up or down a node, while there are efforts [17] to reduce this cost dramatically. More importantly, depending on the workload complexity, the response time of workload could increase nonlinearly (e.g., exponentially) to the number of powering down nodes. Therefore, it could be much more energy efficient to run complex workload at full system capacity rather than partial capacity. As shown by the AIS experiments, with larger complex workload, the energy benefit from using AIS increases more rapidly comparing to other approaches, such as Covering Set.

Both Covering Set and AIS were implemented and evaluated with real MapReduce workloads and clusters. While both of Covering Set and AIS showed promising results on energy conservation, many issues and questions remain to be addressed. For example, Covering Set runs either in energy saving mode with the covering subset nodes, or it runs in full power mode with all the nodes. Thus, a more sophisticated power down technique could be developed to manage the noncovering subset nodes and to provide more gracefully transitioning. Also as mentioned, the effectiveness of Covering Set relies on an intelligent resource provisioning to determine the size of the covering subset. On the other hand, to exploit the power saving opportunity, AIS must batch jobs and periodically power up the entire system when the system is under-utilized. Thus, a workload prediction technique might be needed to decide how long jobs schedule be batched. Nevertheless, both Covering Set and AIS serve as the first step to examine the relation between energy and MapReduce-style workloads. Given the unique computation environment and growing interest, it is expected that more efforts will be made on improving energy efficiency of MapReduce-style frameworks in the near future.

8.4.2 Commercial Products

There are quite a few commercial products on the market that implement the MAID (Massive Array of Idle Disks) principle. Early versions of MAID products, also called MAID 1.0 or "old MAID," suffered from performance problems due to time delays in spinning up the sleeping disks. Also, MAID 1.0 products supported only a limited percentage of active disks at any given time typically as little as 25%. MAID 2.0 provides more energy-saving options by adding multiple energy modes instead of the simple binary on/off approach of MAID 1.0. The idea is that different types of data have different response time requirements and therefore can be maintained different energy consumption modes. Products such as AutoMAID and Dynamic-MAID described below are examples of this approach.

There are many storage products in the market with energy-saving features, due to space limitations we will describe in this section only two such product families, one from SGI and the other from Nexsan. SGI (formerly called COPAN Systems) has a family of products called COPAN Native MAID 300M/400M. The COPAN family of products are all based on what is called Enterprise MAID platform, whose main goal is to support storage requirements of write-once/read-occasionally (WORO) data. The product comes with Disk Aerobics software that supports fault tolerance and reliability by monitoring disk "health."

The primary goals of this software (as specified by SGI [22]) are:

- Provide ongoing and continuous data verification on a drive-by-drive basis
- Proactively monitor and manage drive health by periodically exercising all disks and detecting potential drive failures before they occur
- Copy data from a "suspect," potentially failing drive to a new "healthy" spare drive, avoiding lengthy RAID rebuild times and data loss

Nexsan [18] has a product called AutoMAID (Automatic Massive array of idle disks) that has the ability to spin down drives to lower energy consumption between data references. AutoMAID is a representative of the MAID 2.0 generation that does not have the shortcomings of MAID 1.0 in terms of slow response time. AutoMAID delivers sub-second response times to the first I/O request and remains at full speed for every subsequent I/O request until enough idle time has elapsed to activate AutoMAID energy savings once again. AutoMAID disks have four power management states that incrementally increase energy savings while providing slower response times:

- AutoMAID 0: The disks are fully powered and run at peak performance without restrictions.
- AutoMAID 1: The heads are parked and powered down by policy or command. However, the actual drive continues to spin at full speed. If a request for an I/O is received, the heads wake up and load data with sub-second response times and remain at full speed for every subsequent I/O request. The overall energy savings is approximately 20%.
- AutoMAID 2: The heads are parked and the rotation speed of the disk is slowed from 7200 to 4000 RPM. If a request for an I/O is received, the drive cycles up to full speed and loads the heads resulting in up to 15 s response times. The disk remains at full speed for every subsequent I/O request. The overall energy savings is approximately 40%.
- AutoMAID 3: Heads are parked and the drive motor is turned off. If an I/O request is received, the drive spins up and loads the heads resulting in up to 30 s response time delays. The disk remains at full speed for every subsequent I/O request. The overall energy savings is approximately 60%.

Other noteworthy products and vendors with energy savings features that include MAID and data de-duplication are briefly described below. For each of them we list its MAID features and de-duplication features.

DataDirect Networks S2A9900: Dynamic-MAID (D_MAID) has a feature called variable spin down that supports user-configured disk idling periods and automatically manages disk arrays for optimum power efficiency.

EMC Disk Library 5000 Series: All-drive SATA disk spin down. Data de-duplication is performed by an add-on option that provides policy-based data de-duplication that reduces storage capacity requirements, has no impact on backup performance, lowers energy consumption, and reduces replication costs.

Clarion CX4 Series: Has a policy-driven spin down feature.

Fujitsu Storage Systems Eternus Series: Has a policy-driven MAID/spindown feature. Disk drives have Eco-mode support using MAID technology. This stops disk rotation at set times based on customer's usage patterns. Eco-mode reduces power consumption, as disks are powered down during periods of inactivity. This time-controlled mode, based on scheduled use of specific disks, can be set up for individual RAID groups and backup operations.

GreenBytes GB-X Series: Has a MAID/spin down feature. De-duplication is built into the operating system creating a true single instance storage appliance capable of on-the-fly de-duplication of data.

8.5 Summary and Future Directions

With rising concerns about energy costs and CO_2 emissions, energy use in large data centers is under growing scrutiny. A large fraction of the power consumed by data centers is attributed to their storage systems and associated cooling. The main reason for that is that storage systems in data centers typically employ thousands of spinning disk drives that are needed in order to meet peak demands. There is also a continuous pressure to store more data due to regulatory retention requirements and the need for increased availability. According to many industry observers, the energy consumption problem is also exacerbated by the inefficient use of storage assets as data centers typically utilize only a small fraction of their available storage capacity.

In this chapter, we summarized important techniques and commercial products that were developed specifically to mitigate this problem. As we outlined in the preceding sections, minimization of energy usage does not lend itself to simple solutions as it involves multidimensional trade-offs with metrics such as I/O performance, energy consumption, hardware costs, and reliability that all play a role in achieving optimal solutions. Future trends include several solutions both in hardware and software. In terms of hardware, the replacement of 3.5 in. drives with smaller 2.5 in. hard disk drives can result in significant energy savings. These smaller drives consume much less energy and also have a faster transition from standby to idle mode resulting in better response time. Technologies based on SSDs that were formerly out of reach for many enterprises due to cost per gigabyte will undergo improvements and reductions in cost leading to their commoditization. This will allow enterprises to employ these in their data centers finding them a proper place in the storage/memory hierarchy. For example, EMC is now shipping FAST (Fully Automated Storage Tiering) Cache, which uses industry standard Flash drives as a non-volatile read/write cache. FAST Cache works with Block Data Compression as part of the FAST Suite to automatically move data between high-speed SSDs and high-capacity SATA drives.

In terms of software, improvement in storage techniques such as data de-duplication and compression will result in reduction in storage requirements and improve storage utilization. In terms of academic research, more investigation is needed along the directions listed below:

- *Using workload characterization and patterns*: As we showed in the previous sections, data placement is a crucial element in the design of energy saving disk scheduling algorithms. There is a need to perform more studies of typical workloads and application codes to construct benchmarks in order to identify patterns that will allow better placement of data that enhances energy savings. To the best of our knowledge, no such storage energy-efficiency benchmark for individual products exists.
- *Dynamic data allocation*: Much more research and experimental work is needed in this important area as it requires development of data structures for fast identification of hot files and/or data blocks and dynamic reorganization techniques that converge quickly to a state that enhances energy savings while satisfying response time constraints.
- *Modeling and experimentation with SSDs*: SSDs are currently too expensive to serve as a replacement for HDD, but can be very effectively used as caches. Correct modeling of the behavior of such devices in the presence of read/write/delete operations and new energy aware caching algorithms need to be developed for such systems. Current HDD simulation programs such as Disksim will need to be extended and modified as they currently do not allow to easily capture the behavior of SSD products.
- *Data replication schemes for energy saving*: As mentioned in the previous sections, replication that is currently used for fault tolerance can be also exploited by energy saving algorithms as only one copy of the data is needed for read operations and write operations can be delayed by write-off loading or log-structured files techniques. Much more research is needed into finding energy-efficient replication schemes and incorporating such schemes into existing file systems such as HDFS so they are transparent to the users.
- *Storage virtualization*: Storage virtualization can greatly improve utilization as it provides a transparent I/O redirection layer that can be used to consolidate fragmented storage. Research and measurements are needed to establish how such virtualization can benefit energy savings in data centers. A first step in this direction is described in a recent paper [24] where a system called SRCMap (Sample-Replicate-Consolidate Mapping) is proposed for enhancing energy proportionality of storage systems.

References

1. D. G. Andersen, J. Franklin, M. Kaminsky, A. Phanishayee, L. Tan, and V. Vasudevan. FAWN: A fast array of wimpy nodes. In *Proceedings of the ACM SIGOPS 22nd Symposium on Operating Systems Principles*, SOSP'09, Big Sby, MT, pp. 1–14, 2009. New York: ACM.
2. AutoMAID, http://www.nexsan.com/products/automaid.php
3. D. Colarelli and D. Grunwald. Massive arrays of idle disks for storage archives. In *Supercomputing '02: Proceedings of the 2002 ACM/IEEE Conference on Supercomputing*, Baltimore, MD, pp. 1–11, 2002. Los Alamitos, CA: IEEE Computer Society Press.
4. COPAN, http://www.sgi.com/products/storage/maid/
5. EMC Corporation. *Symmetrix 3000 and 5000 Enterprise Storage Systems Product Description Guide*, Hopkinton, MA, 1999.
6. J. Dean and S. Ghemawat. MapReduce: Simplified data processing on large clusters. In *OSDI'04: Proceedings of the Sixth Conference on Symposium on Operating Systems Design & Implementation*, San Francisco, CA, pp. 10–10, 2004. Berkeley, CA: USENIX Association.
7. Disksim, http://www.pdl.cmu.edu/disksim/
8. S. Gurumurthi, A. Sivasubramaniam, M. Kandemir, and H. Franke. DRPM: Dynamic speed control for power management in server class disks. In *ISCA'03: Proceedings of the 30th Annual International Symposium on Computer Architecture*, San Diego, CA, pp. 169–181, 2003. New York: ACM.
9. D. Borthakur. The Hadoop distributed file system: Architecture and design, http://hadoop.apache.org/core/docs/current/hdfsdesign.pdf
10. H.-I. Hsiao and D. J. DeWitt. Chained declustering: A new availability strategy for multiprocessor database machines. In *Proceedings of the Sixth International Conference on Data Engineering*, Istanbul, Turkey, pp. 456–465, 1990. Washington, DC: IEEE Computer Society.
11. IBM Corporation. ESS-the performance leader, 1999.
12. S. Irani, G. Singh, S. K. Shukla, and R. K. Gupta. An overview of the competitive and adversarial approaches to designing dynamic power management strategies. *IEEE Transactions on VLSI Systems*, 13(12):1349–1361, 2005.
13. W. Lang and J. M. Patel. Energy management for MapReduce clusters. In *VLDB'10*, 2010, Singapore.
14. W. Lang, J. M. Patel, and J. F. Naughton. On energy management, load balancing and replication. *SIGMOD Record*, 38(4):35–42, 2009.
15. J. Leverich and C. Kozyrakis. On the energy (in)efficiency of Hadoop clusters. *SIGOPS Operating Systems Review*, 44(1):61–65, 2010.
16. S. Martello and P. Toth. *Knapsack Problems: Algorithms and Computer Implementations*. John Wiley & Sons, Inc., New York, 1990.
17. D. Meisner, B. T. Gold, and T. F. Wenisch. Powernap: Eliminating server idle power. In *Proceedings of the 14th International Conference on Architectural Support for Programming Languages and Operating Systems*, ASPLOS'09, Washington, DC, pp. 205–216, 2009. New York: ACM.
18. http://www.nexsan.com/products/automaid.php
19. E. J. Otoo, D. Rotem, and S.-C. Tsao. Analysis of trade-off between power saving and response time in disk storage systems. In *IPDPS*, Rome, Italy, pp. 1–8, 2009.
20. E. Pinheiro and R. Bianchini. Energy conservation techniques for disk array-based servers. In *ICS'04: Proceedings of the 18th Annual International Conference on Supercomputing*, Saint Malo, France, pp. 68–78, 2004. New York: ACM.
21. E. Pinheiro, R. Bianchini, and C. Dubnicki. Exploiting redundancy to conserve energy in storage systems. In *SIGMETRICS'06/Performance '06: Proceedings of the Joint International Conference on Measurement and Modeling of Computer Systems*, Saint Malo, France, pp. 15–26, 2006. New York: ACM.
22. http://www.sgi.com/products/storage/maid/400m/

23. M. W. Storer, K. M. Greenan, E. L. Miller, and K. Voruganti. Pergamum: Replacing tape with energy efficient, reliable, disk-based archival storage. In *Proceedings of the Sixth USENIX Conference on File and Storage Technologies*, FAST'08, San Jose, CA, pp. 1:1–1:16 2008. Berkeley, CA: USENIX Association.

24. A. Verma, R. Koller, L. Useche, and R. Rangaswami. SRCMap: Energy proportional storage using dynamic consolidation. In *FAST*, San Jose, CA, pp. 267–280, 2010.

25. VMotion, http://vmware.com/products/vi/vc/vmotion.html

26. C. Weddle, M. Oldham, J. Qian, A.-I. A. Wang, P. Reiher, and G. Kuenning. PARAID: A gear-shifting power-aware RAID. *Transactions on Storage*, 3(3):13, 2007.

27. Q. Zhu, Z. Chen, L. Tan, Y. Zhou, K. Keeton, and J. Wilkes. Hibernator: Helping disk arrays sleep through the winter. In *SOSP'05: Proceedings of the Twentieth ACM Symposium on Operating Systems Principles*, Brighton, U.K., pp. 177–190, 2005. New York: ACM.

28. Q. Zhu, F. M. David, C. F. Devaraj, Z. Li, Y. Zhou, and P. Cao. Reducing energy consumption of disk storage using power-aware cache management. In *Proceedings of the 10th International Symposium on High Performance Computer Architecture (HPCA)*, Madrid, Spain, 2004.

29. Q. Zhu, A. Shankar, Y. Zhou, A. Shankar, and Y. Zhou. PB-LRU: A self-tuning power aware storage cache replacement algorithm for conserving disk energy. In *Proceedings of the 18th International Conference on Supercomputing*, Saint Malo, France, pp. 79–88, 2004.

30. Q. Zhu and Y. Zhou. Power-aware storage cache management. *IEEE Transactions on Computing*, 54(5):587–602, 2005.

9

Thermal and Power-Aware Task Scheduling and Data Placement for Storage Centric Datacenters

Bing Shi
University of Maryland

Ankur Srivastava
University of Maryland

9.1 Introduction

The primary objective of large-scale storage centric datacenters is to provide data. The advent of applications such as YouTube, etc., has inspired the development of such datacenters. Storage centric datacenters are comprised of large-scale storage devices and CPUs that support data access. People produce and access a tremendous amount of data everyday or even every minute. For example, YouTube serves up to 100 million videos a day [1]; Facebook has 400 million active users and 3 billion photos uploaded each month [2]. These videos and images are stored in datacenters. Therefore, large storage centric datacenters have become an essential component of modern information and communication technology.

The tremendous amount of data storage and access to storage datacenters nowadays result in a huge amount of power consumption on servers. As we will see later, the power consumption of disks on datacenter servers has a nearly cubic relationship with the rotational speed of disks. Therefore, when the disk runs for a longer time due to the huge amount of data access requests, the power consumption of datacenters will increase dramatically. Moreover, an increase in disk power consumption will lead to a rise in disk temperature. In order to maintain the reliability of datacenters, the air conditioner needs to consume more power so as to cool the disks down and maintain the disk temperature within an acceptable level. The cooling consumes 50% of the total energy in large-scale datacenters [3]. Therefore, dynamic thermal and power management for storage datacenters, which is also a major issue for green computing, has become an active topic of research.

Hadoop is a software framework usually used to manage data file access in large-scale storage centric datacenters whose primary job is to deliver data to clients. In such systems, the primary job is to associate each data request to a specific data replica among many available replicas. This assignment impacts the workload and power distribution across the storage servers. Different from usual datacenters, each file in the Hadoop system is divided into a set of blocks of equal size. Data delivery in Hadoop is always with respect to blocks. Therefore the data access requests are block accesses and have the same workload and execution time.

There are a number of works on thermal and power management of datacenters [4–13]. Reference [4] explores three task scheduling method on blade server–based datacenters: temperature balancing, workload balancing, and computing energy minimization. The work in [5] tries to minimize the total power consumption of datacenters by formulating the optimization problem as an integer linear programming problem and solves it with heuristics. In [6], the authors try to increase the cooling efficiency by reducing the amount of heat recirculation in the datacenter. In [7,8], the authors propose task scheduling methods that save cooling power by minimizing the peak inlet air temperature and heat recirculation. A software architecture for making datacenter thermal-aware job scheduling decision proposed in [9,10] explores thermal-aware job scheduling under both constant and linear cooling models. However, to our knowledge, there are few works on storage centric datacenter thermal management.

In this chapter, we focus on Hadoop-based storage centric datacenters whose primary purpose is to provide block-based file storage and access, other than general datacenters. Specifically, we work on thermal- and power-aware task scheduling for such datacenters and explore the characteristic of data access tasks in such datacenters under the Hadoop framework. We try to decide which data node on the storage server should be used to source each data access request, so that the total power consumption of the A/C system is minimized (that equals maximizing the COP of A/C system), and at the same time, all disk temperature is kept within some acceptable thermal constraint. We investigate two problem instances: (1) all tasks have the same deadline and (2) each task has a different deadline. We formulate the problems under the ILP framework. Since it is very complex and time consuming to solve the ILP problem, we develop an efficient, minimum cost flow–based heuristic to solve the problems.

The experimental result shows that when the tasks have the same deadline, our method is very close to the optimal solution achieved by using an ILP solver. Our method forces the A/C system to output air temperature only 0.38–1.08 K lower than the optimal ILP solution. However, the runtime of our method is only 1%–2.5% of the runtime using an ILP solver, so our method is more suitable for online scheduling. On the other hand, random selection of data replica for each data request results in the required A/C output air temperature to be 6.35 K lower than our method, which forces the A/C system to work harder. When tasks have different deadlines, compared with random selection of data replica, our method needs the A/C system to output air temperature 2.5–8.0 K higher, which results in A/C power savings.

We also explore thermal- and power-aware data replica placement. Specifically, we decide which data node in the datacenter is used to store each data block so that the system is more effective from the point of view of temperature and power. The experimental results show that properly allocating the data block replicas could result in further A/C power savings.

The rest of the chapter is organized as follows: Section 9.2 introduces the thermal and power model for storage centric datacenters and the Hadoop distributed file system. In Section 9.3, we explore thermal- and power-aware task scheduling for storage centric datacenters. We first explore the problem assuming all tasks have the same deadline, and then we extend the problem to incorporate different task deadlines.

We then investigate the thermal- and power-aware data placement in the datacenter in Section 9.4. The experimental result is given in Section 9.5.

9.2 Hadoop System and Thermal Modeling of Datacenters

9.2.1 Hadoop Distributed File System (HDFS)

In this chapter, we focus on thermal and power management for storage centric datacenters. Different from usual datacenters, storage centric datacenters primarily provide data storage and data access services. The tasks on such datacenters are generally data intensive, such as file storage and file access, rather than computational intensive.

Hadoop is a software framework used to manage data file access in such systems. It is a framework widely used by many organizations like Yahoo, Facebook, IBM, etc. [14].

The Hadoop architecture is shown in Figure 9.1 and it has the following features:

1. A Hadoop system stores large files on multiple machines/racks. The size of a typical file in Hadoop ranges from gigabytes to terabytes. Each file is divided into blocks of the same size and replicated several times to ensure reliability.
2. It is composed of a number of data nodes that store data and process data accessing requests, and each data block replica is stored in a separate data node. The data nodes are placed on several racks.
3. It has a master server called NameNode, which executes operations like opening, closing, and renaming files. It also determines the placement of block replicas on data nodes.
4. Tasks that run on Hadoop are usually data accessing requests. Large file accessing requests by clients are partitioned into smaller data block accessing requests and scheduled to data nodes. The scheduling is managed by NameNode. So here we use task to denote the smaller data block accessing requests.
5. Data delivery in Hadoop is always with respect to blocks. Therefore the tasks are actually block accesses. That is, each task requests one data block. So the tasks have the same workload and execution time.
6. Data blocks are replicated for fault tolerance. The replication factor is configurable. The common data replication policy is as follows: when the replication factor is three, Hadoop stores two of them in two different data nodes on the local rack, and the other is stored in a different data node on a

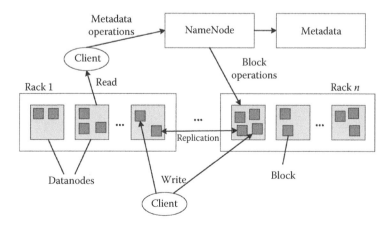

FIGURE 9.1 Hadoop architecture.

different rack. In most cases, the network bandwidth between data nodes on the same rack is larger than that for data nodes on different racks, as communication between two data nodes on different racks has to go through switches.

9.2.2 Thermal Model of Datacenters

In this chapter, we deal with datacenters that are distributed sources. In a typical datacenter, data nodes are placed on the raised floor, and the air conditioner is located under the raised floor. The cooling air produced by the air conditioner, whose temperature is called *supply temperature* T_{sup}, enters the nodes from one side and exits from the other side, taking away the heat in the nodes (as shown in Figure 9.2). The air conditioner then extracts the heated air flowing out of nodes, and pumps the cooling air back into the datacenter.

Part of the heated air coming out of the nodes will recirculate and flow back into other nodes. For example, in Figure 9.2, the air flowing out from node 1 enters node 2 with a ratio of a_{12}. Similarly, the air flowing out from node 2 will also enter node 1 with a ratio of a_{21}. Here a_{ij} is the cross interference coefficients between nodes i and j, which can be decided using the method in [15]. Therefore, the inlet air of node i whose temperature is denoted as T_{in}^i is a mix of (1) the cooling air from air conditioner and (2) the recirculated outlet air from other nodes. The hot air flowing out of node i will (1) go back to air conditioner and (2) recirculate into other nodes.

In the datacenter as modeled in Figure 9.2, the relationship between power consumption of a node P_i and its inlet and outlet air temperature (T_{out}^i and T_{in}^i) is [4,15]

$$P_i = K_i \left(T_{out}^i - T_{in}^i \right) \tag{9.1}$$

where K_i is a constant associated with each node i:

$$K_i = \rho f_i C_p \tag{9.2}$$

where
   ρ is the air density
   C_p is the specific heat of air
   f_i is the flow rate of air through node i

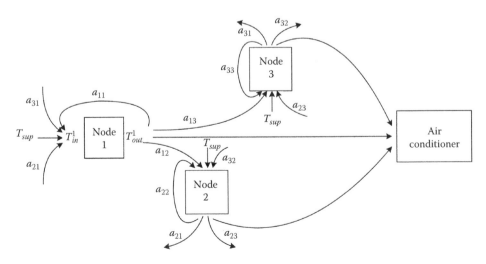

FIGURE 9.2 Thermal model of datacenters.

The work in [15] formulated the thermal model for node i as

$$K_i T_{out}^i = \sum_{j=1}^{N} a_{ji} K_j T_{out}^j + \left(K_i - \sum_{j=1}^{N} a_{ji} K_j \right) T_{sup} + P_i \qquad (9.3)$$

Here N is the total number of data nodes in the datacenter. Interested readers can refer to [4,15] for details.

9.2.3 Power Consumption of Storage Datacenters

The power consumption of disks on data nodes can be modeled as [16]

$$P_i \propto n \omega^{2.8} D^{4.6} \qquad (9.4)$$

where

n is the number of disks
ω is the rotational speed of the disks
D is the diameter of disks

The power consumption of disks has a nearly cubic relationship with the disk speed. Therefore, in this chapter, we approximate that the power consumption of two disks is the same if they work on the same speed. Besides, due to this cubic relationship, the increase in disk speed will lead to a dramatic power consumption increase of the storage datacenter.

The power consumption of the A/C system depends on its coefficient of performance (COP), which is defined in Equation 9.5.

$$\begin{aligned} \text{COP} &= \frac{\text{Amount of heat removed by air conditioner}}{\text{Energy consumed by air conditioner}} \\ &= \frac{\text{Total power consumption of all datanodes}}{P_{AC}} \\ &= \sum_{i=1}^{N} \frac{P_i}{P_{AC}} \end{aligned} \qquad (9.5)$$

where

P_{AC} is the power consumption of A/C system
P_i is the power consumption of each data node

So the power consumption of A/C system is

$$P_{AC} = \sum_{i=1}^{N} \frac{P_i}{\text{COP}} \qquad (9.6)$$

The total power consumption of Hadoop is then [4]

$$P_{total} = P_{AC} + \sum_{i=1}^{N} P_i = \left(1 + \frac{1}{\text{COP}} \right) \sum_{i=1}^{N} P_i \qquad (9.7)$$

For a detailed explanation of the power and thermal model of datacenters, interested readers can refer to [4]. Generally, as we can see from Equation 9.7, increasing COP can lead to a decrease in power consumption in A/C system and therefore reduce total power consumption. Also, COP always increases when the supply temperature T_{sup} (the output temperature of A/C system with which the thermal constraint is satisfied) is high. For example, the model used in [4,17], obtained from a water-chilled CRAC unit in HP Utility Data Center, describes the relationship between COP and T_{sup} as

$$\text{COP} = 0.0068 T_{sup}^2 + 0.0008 T_{sup} + 0.458 \tag{9.8}$$

Therefore, increasing supply temperature can result in power savings.

9.3 Thermal- and Power-Aware Storage Centric Datacenter Task Scheduling

Due to the tremendous number of access tasks nowadays, storage centric datacenters are consuming a huge amount of power, which then results in rapid disk temperature increase. In order to maintain the reliability of such datacenters, A/C systems need to consume more power so as to maintain the disk temperature within an acceptable level. So we need to consider both temperature and power at the same time when scheduling the tasks on storage datacenters. In our power- and thermal-aware storage centric datacenter task scheduling problem, we would like to make sure that each node in the storage datacenter operates at a temperature below a certain temperature threshold (we call it T_{max}). On the other hand, we also would like to minimize the total power consumption of the A/C system.

In this section, we explore this problem under the widely used Hadoop framework. We explore two problems: The first problem assumes all the tasks have the same deadline (or deadline independent), and in the second problem, tasks have different deadlines.

9.3.1 Case 1: With Same Deadline

9.3.1.1 Problem Formulation

Given

1. A Hadoop system with N data nodes. The disks on all the nodes run on the same speed. So the power consumption of all active data nodes (nodes that source some tasks) are the same. Data blocks (of the same size) are stored in the nodes as described in Section 9.2. That is, each block is replicated three times and each replica is stored in a separate data node. Two of the replicas are stored in different data nodes on the same rack and the other in a different data node on a different rack.
2. L data-intensive tasks, each of which reads a specific block. Assuming each block is replicated C times, we would like to choose one from the C replicas of this block to source the data request task. Hadoop assumes the workload of all tasks are the same, since the size of data blocks are the same. That is, each task reads a whole block on a node. Since all the disks run on the same speed, the execution time of all tasks is the same.
3. At any given time, each node can only source one data requesting task.

We would like to assign each data access task to a data node so that each task can read its required data block locally. Given the task-node assignment, we can decide the power consumption distribution on data nodes. This power distribution then decides the workload of the A/C system. We would like to maximize the COP of the A/C system so that its total power consumption is minimized. Also T_{out} of all nodes should never exceed the maximum temperature constraint T_{max}.

The thermal- and power-aware task scheduling problem is therefore formulated as follows:

$$max \quad COP$$

$$s.t. \; K_i T_{out}^i = \sum_{j=1}^{N} a_{ji} K_j T_{out}^j + \left(K_i - \sum_{j=1}^{N} a_{ji} K_j \right) T_{sup} + P_i, \quad \forall \text{ node } i$$

$$T_{out}^i \leq T_{max}, \quad \forall \text{ node } i$$

$$\sum_{c=1}^{C} x(l, c) = 1, \quad \forall \text{ task } l$$

$$\sum_{l=1}^{L} \sum_{c=1}^{C} x(l, c) g(l, c, i) \leq 1, \quad \forall \text{ node } i$$

$$P_i = p \sum_{l=1}^{L} \sum_{c=1}^{C} x(l, c) g(l, c, i), \quad \forall \text{ node } i \tag{9.9}$$

p is the power consumption of an active node when it is sourcing one data block and $g(l, c, i)$ is a function that specifies the placement of block replica on data nodes:

$$g(l, c, i) = \begin{cases} 1, & \text{if node } i \text{ stores } c\text{th replica of block task } l \text{ needs} \\ 0, & \text{otherwise} \end{cases} \tag{9.10}$$

The problem tries to maximize COP, which is equivalent to minimizing the power consumption of the A/C system. As we can see from Equation 9.8, COP is a monotonically increasing function of T_{sup}, and maximizing COP equals maximizing T_{sup}. So the objective equals maximizing T_{sup}. That is, we would like to find the task-data node mapping specified by $x(l, c)$ in Equation 9.11 so that T_{sup} is maximized.

$$x(l, c) = \begin{cases} 1, & \text{if task } l \text{ uses the } c\text{th replica of its requested block} \\ 0, & \text{otherwise} \end{cases} \tag{9.11}$$

The first constraint specifies the thermal model of the datacenter. The second constraint requires that the outlet air temperature of each node T_{out} should not exceed the maximum temperature constraint T_{max}. The third constraint ensures that each task is assigned to one and only one node among all the nodes that contain the data block this task requests, while the fourth constraint ensures that each node sources at most one task. The last constraint specifies that the power consumption of each node is either p when this node is active or 0 when it is idle.

This problem is NP-hard, but it can be solved optimally with any integer linear programming (ILP) tool. Here we use a minimum cost flow–based heuristic to solve the problem. We first come up with a reasonably good solution using minimum cost flow, and then further improve it with task rescheduling.

9.3.1.2 Formulate Minimum Cost Flow Problem

The first step of this problem is basically assigning a data node to each task so that (1) each task can get its required data, (2) each data node is assigned to at most one task, and (3) each data node has an associated cost and the total cost of selected data nodes is minimized. The notion of *cost* will be explained later.

To form the minimum cost flow problem, each task is represented by a source node with one unit of flow available. Each data node is represented by a transshipment node. There is also a sink node with demand equal to the number of tasks L. Edges connecting the sources and transshipment nodes are decided by function $g(l, c, i)$ as in Equation 9.10. For each pair of task and data node whose corresponding $g(l, c, i)$ value equals to 1, there is an edge connecting them. That is, each pair of task and the node storing a block replica this task needs is connected by an edge. These edges have capacity of 1 and cost 0. Each data node is also connected to the sink by an edge with a capacity of 1 and each data node has an associated cost whose assignment will be discussed soon. Since each data node has an associated edge connecting it to the sink, this node cost can be converted to the associated edge cost. For example, if data node i has node cost $cost(i)$, then the edge connecting node i to the sink is assigned an edge cost whose value equals $cost(i)$.

Figure 9.3 gives an example of the minimum cost flow formulation with 5 tasks and 12 data nodes, and each block is replicated three times. Table 9.1 specifies the locations of three replicas of the block each task needs. In Figure 9.3, the five nodes on the top represent five tasks, while the nodes in the middle represent data nodes. The problem is to send the flow from sources to sink through selected edges so that the total cost of selected edges is minimized. This problem can be solved with any minimum cost flow tools.

Now we decide the assignment of node cost. Intuitively, nodes that are more sensitive to the air conditioner or have greater impact on other nodes should be assigned higher cost. From Equation 9.3, we can see coefficient $a_{ji}K_j$ represents the impact of the temperature of node $j(T_{out}^j)$ on node $i(T_{out}^i)$. By summing up $a_{ji}K_j$ for all the nodes i, (i.e., $\sum_{i=1}^{N} a_{ji}K_j$), this measures node jth's impact on all the other nodes through recirculation. So if this value is high for node j, its cost should also be high as it has more impacts on other nodes. Also, from Equation 9.3, nodes with higher $K_i - \sum_{j=1}^{N} a_{ji}K_j$ tend to be more sensitive to the change of T_{sup}. That is, when T_{sup} increases, the value on the right side of Equation 9.3 increases more if $K_i - \sum_{j=1}^{N} a_{ji}K_j$ is large. So T_{out}^i on the left side will also increase more, and vice versa.

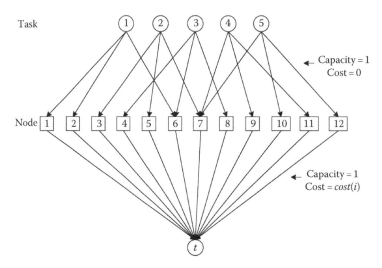

FIGURE 9.3 Example of minimum cost flow for the case where all tasks have the same deadline.

TABLE 9.1 Data Replica Location

	Task 1	Task 2	Task 3	Task 4	Task 5
Replica 1	1	3	4	7	7
Replica 2	2	5	6	9	10
Replica 3	6	7	8	11	12

Therefore this coefficient represents node i's position with respect to the air conditioner. If this value for node i is high, that means node i is close to the air conditioner and will be easier to cool down. So this node should be assigned a smaller cost. Therefore the overall cost for node i can be formulated as

$$cost(i) = k \cdot K_i \sum_{j=1}^{N} a_{ij} - \left(K_i - \sum_{j=1}^{N} a_{ji} K_j \right) \tag{9.12}$$

Here k is a constant that adjusts the weight of both components of the node cost.

After solving the minimum cost flow problem, each task is assigned to a data node. That is, we get the solution for $x(l, c)$ in Equation 9.9. So now, we can calculate the power consumption of each node P_i: the power consumption of each active data node is p and that of the idle node is 0. Given P_i for each data node i, we can then easily calculate the maximum supply temperature T_{sup} (and maximum COP) satisfying the following constraints:

$$\begin{cases} K_i T_{out}^i = \sum_{j=1}^{n} a_{ji} K_j T_{out}^j + \left(K_i - \sum_{j=1}^{n} a_{ji} K_j \right) T_{sup} + P_i, \quad \forall \text{ node } i \\ \\ T_{out}^i \leq T_{max}, \quad \forall \text{ node } i \end{cases} \tag{9.13}$$

Next, we will incorporate task rescheduling to further improve the solution.

9.3.1.3 Task Rescheduling

Because minimum cost flow formulation is a heuristic to the original ILP method, we can further improve it by task rescheduling to balance the power consumption and outlet temperature of the data nodes. The process works as follows:

1. Pick the node with highest T_{out}, say node i, and move the task on this node to another idle data node that stores a replica of the same block. If both of the other two replicas are idle, choose the one with the lower T_{out}.
2. Assuming we have moved the task on node i to node j in step 1, we then recalculate the power consumption of node i and j. That is, $P_i = 0$ and $P_j = p$.
3. Calculate the maximum T_{sup} satisfying the constraints in Equation 9.13 with the new power consumption distribution.
4. Compare the new T_{sup} $\left(T_{sup}^{new} \right)$ with the old one $\left(T_{sup}^{old} \right)$. If $T_{sup}^{new} > T_{sup}^{old}$, then we accept this task movement; else we reject the movement and undo the previous steps 1, 2, and 3.

We repeat this for several times to move the tasks from hot nodes to cool nodes so that the temperature of all the nodes is more balanced.

9.3.2 Case 2: With Different Deadlines

We now extend this problem to a more general case where tasks have different deadlines.

9.3.2.1 Problem Formulation

Given

1. A Hadoop system with N data nodes, which are similar as described in Section 9.3.1.1.
2. L data-intensive tasks with different deadlines. Assuming each task takes 1 unit of time to execute, the deadline for the tasks range from 1 to Q. Data blocks are replicated C times.

We would like to assign each task to a data node at a particular time so that each task can get the required data before its deadline. At the same time, we would like to maximize the cumulative COP for all time steps between 1 to Q $\left(\sum_{t=1}^{Q} \text{COP}(t) \right)$. Still T_{out} of all the nodes should never exceed the maximum temperature constraint T_{max} at any time.

The problem can be formulated as follows:

$$max \quad \sum_{t=1}^{Q} \text{COP}(t)$$

s.t.

$$K_i T_{out}^i(t) = \sum_{j=1}^{N} a_{ji} K_j T_{out}^j(t) + \left(K_i - \sum_{j=1}^{N} a_{ji} K_j \right) T_{sup}(t) + P_i(t), \quad \forall \text{ node } i \text{ at time } t$$

$$T_{out}^i(t) \leq T_{max}, \quad \forall \text{ node } i \text{ at time } t$$

$$\sum_{c=1}^{C} \sum_{t=1}^{deadline(l)} x(l, c, t) = 1, \quad \forall \text{ task } l$$

$$\sum_{l=1}^{L} \sum_{c=1}^{C} x(l, c, t) g(l, c, i) \leq 1, \quad \forall \text{ nodes } i \text{ at time } t$$

$$P_i(t) = p \sum_{l=1}^{L} \sum_{c=1}^{C} x(l, c, t) g(l, c, i), \quad \forall \text{ nodes } i \text{ at time } t \qquad (9.14)$$

where $g(l, c, i)$ is still the function that represents placement of data block replicas. Our purpose is to find the task-node-time mapping specified by Equation 9.15.

$$x(l, c, t) = \begin{cases} 1, & \text{if task } l \text{ uses the } c\text{th replica of its requested block at time } t \\ 0, & \text{otherwise} \end{cases} \qquad (9.15)$$

In Equation 9.14, COP(t) is the coefficient of performance at time step t. The objective is to minimize the cumulative COP during the time steps $1 - Q$, where Q is the maximum task deadline. The first constraint still specifies the thermal model of the datacenter and the second constraint places a maximum temperature constraint on the outlet air temperature of each node T_{out}. The third constraint ensures that each task is assigned to one and only one node, and executed before its deadline. The fourth constraint ensures that each node sources at most one task, while the last constraint specifies the power consumption of each node. This is a complex optimization problem, but we can still use minimum cost flow–based heuristic to solve it.

9.3.2.2 Formulate Minimum Cost Flow Problem

We use a similar method to formulate the minimum cost flow problem.

Each task is represented by a source node with one unit of flow available. Data nodes are represented by transshipment nodes. Here each data node is replicated Q times, where Q denotes the maximum task deadline. Each set of data nodes in a given replica indicates the data nodes at that specific time. Each task is connected to all the replicas of data nodes that store the data this task needs and whose corresponding

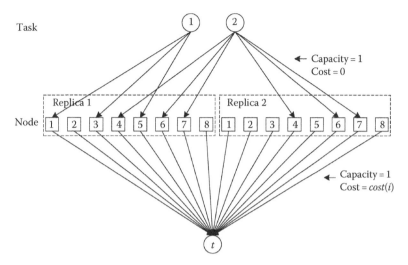

FIGURE 9.4 Example of minimum cost flow for the case where tasks have different deadlines.

deadline is not larger than the deadline of this task. That is, if the deadline of task l is q, then task node l is connected to all data node replicas storing the data task l needs and these data node replicas correspond to deadlines from 1 to q. Each edge connecting a task and a data node has capacity 1 and cost 0. There is also a sink node with demand equal to the number of tasks L. Each data node replica is connected to the sink node also with a capacity of 1, and each data node has an associated cost. We can also convert the node cost to the associated edge cost using the method in Section 9.3.1.2.

Figure 9.4 gives an example of two tasks with two different deadlines, and blocks are replicated three times. The deadline of task 1 is 1 and the deadline of task 2 is 2. There are eight data nodes. The data nodes are replicated twice as the maximum deadline is 2. The left eight data nodes correspond to deadline 1, while the right eight nodes correspond to deadline 2. Since task 1 has deadline of 1, it is only connected to the corresponding data nodes belonging to the left replica. Task 2 has deadline of 2, so it is connected to data nodes in both left and right replicas, as it can be executed at either time 1 or time 2.

The cost of data nodes is same as that specified in Section 9.3.1.2. After solving the minimum cost flow problem and calculating the power consumption of each node, we then find the maximum cumulative COP $\left(\sum_{t=1}^{Q} \text{COP}(t)\right)$ satisfying the following constraints:

$$
\begin{cases}
K_i T_{out}^i(t) = \displaystyle\sum_{j=1}^{N} a_{ji} K_j T_{out}^j(t) + \left(K_i - \sum_{j=1}^{N} a_{ji} K_j\right) T_{sup}(t) + P_i(t), & \forall \text{ node } i \text{ at time } t \\[4mm]
T_{out}^i(t) \le T_{max}, & \forall \text{ node } i \text{ at time } t
\end{cases}
\tag{9.16}
$$

9.3.2.3 Task Rescheduling

We can use a similar task rescheduling method as in Section 9.3.1.3 to further improve the solution. The process is as follows:

1. Pick the node with the highest T_{out}, say node i, and move the task on this node, say task l, to another idle data node that stores the same block and whose corresponding deadline is not larger than the deadline of task l. If there is more than one option, choose the node with lowest $T_{out}(t)$.
2. Execute the same operations as steps 2,3,4 in Section 9.3.1.3, except that when calculating maximum cumulative T_{sup}, we use the constraints in Equation 9.16.

9.4 Data Replica Placement

In the previous thermal- and power-aware task scheduling, we assume that the data block placement is already fixed. That is, the node into which each data block replica is stored is already decided, and we investigate the thermal- and power-aware task scheduling based on the given block replica placement. However, in the Hadoop system, allocating data block replicas to nodes is also flexible, and properly allocating these replicas could achieve further A/C power savings.

Basically, data blocks are replicated for fault tolerance. The replication factor is configurable. The placement of data replicas is of critical importance to the reliability and performance of the Hadoop system. The Hadoop system develops data replica placement policies that consider data reliability, availability, and network bandwidth utilization at the same time. A simple but nonoptimal policy is to evenly distribute replicas in the racks, which makes it easy to balance load on component failure. The problem with this policy is that the cost of writes is high since a write needs to transfer data blocks to multiple racks. This is because, as illustrated in Section 9.2, network bandwidth between data nodes on the same rack is larger than that on different racks, as communication between two data nodes on different racks has to go through switches. A more common data block placement policy in Hadoop systems is to place data block replicas only on a few racks. For example, when the replication factor is three, Hadoop puts one replica on one node in the local rack, another replica on a different node in the local rack, and the last replica on a node in a different rack. Here the local rack means the rack that this block is originally written into. In this policy, each data block is placed in only two racks rather than three, this reduces the inter-rack data block writes, which reduces the data writing overhead.

9.4.1 Problem Formulation

In our work, we adopt this common data replica placement policy to reduce the overhead of writes and develop the thermal- and power-aware data replica placement based on it. We would like to decide the placement of each data block replica, (i.e., which node is used to store each data block replica), so that the expected cooling cost is minimized. Here we assume that tasks have the same deadline, the problem formulation when tasks have different deadlines and methods solving the problem are similar. The problem can be formally defined as follows:

Given

1. A Hadoop system with R racks and N data nodes, so each rack contains N/R data nodes. The system is the same as that described in Section 9.3.1.1.
2. S sets of training tasks along with the probability that each task set would occur $\mathscr{A} = \{\alpha^1, \alpha^2, \ldots, \alpha^S\}$. All the tasks have same deadline.
3. B data blocks. Assuming we are given vectors of frequency that each data block would be requested in each task set s: $\mathscr{B}^s = \{\beta_1^s, \beta_2^s, \ldots, \beta_B^s\}$ $(s = 1, \ldots, S)$, then the overall probability that each data block would be requested for all task sets $\mathscr{P} = \{\pi_1, \pi_2, \ldots, \pi_B\}$ can be estimated using $\mathscr{A}$ and $\mathscr{B}^s, \forall s$. Also, each block is replicated C times. For illustration, we assume that each block is replicated three times, among which two of them are stored on two different nodes in the local rack and the other is stored in a different rack.
4. The capacity of each node is E. That is, each node could store at most E data blocks.

We would like to assign each block replica to a data node so that the data replica placement policy is satisfied. Based on this data replica placement, we then perform thermal- and power-aware task scheduling using the methods proposed in Section 9.3 for each set of tasks and decide the minimum workload of the A/C system (i.e., the maximum COP) required to maintain the outlet air temperature of all nodes within constraints. We would like to maximize the expected COP of the A/C system for all task sets.

When all tasks have the same deadline, the overall block replica placement and task scheduling problem is formulated as follows:

$$max \quad E(\text{COP}) = \sum_{s=1}^{S} \alpha^s \cdot \text{COP}^s$$

$$s.t. \quad \sum_{b=1}^{B} \sum_{c=1}^{C} h(b, c, i) \leq E, \quad \forall \text{ node } i$$

$$K_i T_{out}^{i,s} = \sum_{j=1}^{N} a_{ji} K_j T_{out}^{j,s} + \left(K_i - \sum_{j=1}^{N} a_{ji} K_j \right) T_{sup}^s + P_i^s, \quad \forall \text{ node } i, \text{ task set } s$$

$$T_{out}^{i,s} \leq T_{max}, \quad \forall \text{ node } i, \text{ task set } s \tag{9.17}$$

$$\sum_{c=1}^{C} x^s(l, c) = 1, \quad \forall \text{ task } l, \text{ task set } s$$

$$\sum_{l=1}^{Ls} \sum_{c=1}^{C} x^s(l, c) g^s(l, c, i) \leq 1, \quad \forall \text{ node } i, \text{ task set } s$$

$$P_i^s = p \sum_{l=1}^{Ls} \sum_{c=1}^{C} x^s(l, c) g^s(l, c, i), \quad \forall \text{ node } i, \text{ task set } s$$

Here COP^s is the COP achieved for task set s, $T_{in}^{i,s}$, $T_{out}^{i,s}$, and P_i^s are the inlet, outlet air temperature, and power consumption of node i for task set s, and T_{sup}^s is the A/C supply temperature for task set s. Ls is the number of tasks in each task set s. $h(b, c, i)$ is the function specifying the data replica placement:

$$h(b, c, i) = \begin{cases} 1, & \text{if node } i \text{ stores the } c\text{-th replica of block } b \\ 0, & \text{otherwise} \end{cases} \tag{9.18}$$

Function $g^s(l, c, i)$ is similar to $g(l, c, i)$ defined in Equation 9.10, which specifies that for task set s, the cth replica of the data block that task l needs is stored in node i. This could be easily calculated using $h(b, c, i)$. The definition of $x^s(l, c)$ is the same as $x(l, c)$ described by Equation 9.11, and it decides which block replica c is used to source each task l in task set s.

This problem tries to find the data block replica to node mapping $h(b, c, i)$ and also the task to block replica mapping $x^s(l, c)$ for each task set s that results in maximum expected COP for all task sets, given the probability that each task set would occur. The first constraint specifies the capacity of each node, which could accommodate at most E blocks. The following constraints are similar to the constraints in Equation 9.9.

The problem formulation when tasks have different deadlines is similar.

This problem is NP-hard, therefore, we adopt a heuristic to solve it. In Section 9.3, we have explored the methods to find task to block replica mapping $x(l, c)$ when data replica placement $g(l, c, i)$ is given (for both cases where tasks have the same deadline and tasks have different deadlines), so here we focus on finding the data block replica to node mapping $h(b, c, i)$. We could still use the minimum cost flow–based heuristic to solve the problem.

9.4.2 Formulate Minimum Cost Flow Problem

To form the minimum cost flow problem, each data block replica is represented by a source node with one unit of flow available. Since we assume each block is replicated three times, there are $B * 3$ source nodes, where B is the number of blocks. Each data node is represented by a transshipment node. There is also a sink node with demand of $B * 3$.

Each source node is connected to all the transshipment nodes. Each edge connecting source node to transshipment node has a capacity of 1 and an associated edge cost (we will discuss the edge cost assignment soon). Also, all transshipment nodes are connected to the sink node by edges with capacity of E and cost 0, where E is the capacity of each data node.

Figure 9.5 shows an example of the minimum cost flow with two data blocks and three racks. Each block is replicated three times, so there are a total of six source nodes, each of which has one unit of flow available. Each rack contains three data nodes; therefore there are nine transshipment nodes. The sink node has a demand of six units of flow. In the minimum cost flow in Figure 9.5, each of the six source nodes is connected to all nine transshipment nodes, with edges of capacity 1 and associated edge cost described later. Also, each transshipment node is connected to the sink t with edges of capacity E and cost 0. Our objective here is to send the flow from source nodes to the sink node through selected edges, so that the total cost of selected edges is minimized.

However, the resultant block replica placement solution might violate the data replica placement policy described at the beginning of this section. We use replica reallocation, which will be described in Section 9.4.3, to address the violation.

9.4.2.1 Cost Assignment

The edge cost should be decided by both the thermal characteristic of data nodes and also the frequency each block is requested. Basically, we would like to assign data blocks that are requested more frequently

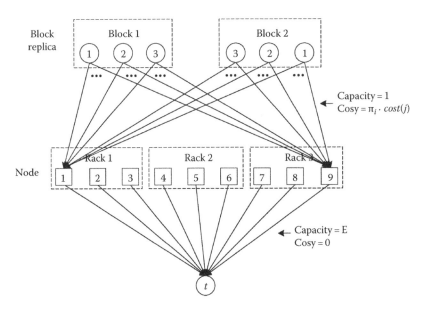

FIGURE 9.5   Example of minimum cost flow for data replica placement.

to nodes with lower node cost (node cost is defined in Section 9.3.1.1). Therefore, the cost of edges connecting the replicas of ith block to the jth node is assigned as

$$cost(i, j) = \pi_i \cdot cost(j) \tag{9.19}$$

Here $cost(j)$ is the cost of node j defined in Equation 9.12.

By solving this minimum cost flow problem, we will get the block replica placement on the data nodes. That is, we get $h(b, c, i)$ defined in Equation 9.18. After we get the data replica placement, for each set of tasks, we can decide the assignment of each task to data nodes using the methods described in Section 9.3.

9.4.3 Data Replica Reallocation

The data replica placement achieved by using the minimum cost flow described in the previous section might result in violation of the data replica placement policy, which requires the following: assuming each block is replicated three times, then among the three replicas of each data block, two of them are stored on two different nodes in a same rack, and the other replica is stored on a node in a different rack. The solution achieved using the previous minimum cost flow might lead to two types of violation: (a) all the three replicas of a block might be stored on three different nodes in the same rack or (b) the three replicas are stored in three different racks.

Case a: All three replicas stored on three different nodes in the same rack

In this case, we should move one replica to another rack. When the three replicas of a block b are stored on nodes i_1, i_2, and i_3 in the same rack, say rack r, then we choose the node with the highest node cost (node cost is defined in Equation 9.12), say node i_H, and move block b's replica on node i_H to an available node on a different rack whose node cost is the closest to the node cost of i_H. Here available node means that the number of replicas stored in the node has not reached the node capacity E, so it could still accommodate more block replicas.

Case b: Three replicas stored in three different racks

In this case, we should move a replica to one of the racks where the other two replicas are stored. The steps to address this problem is given in Algorithm 9.1.

Basically, steps 2–9 try to move a replica to one of the racks where the other two replicas are stored. We start from the replica stored in the node with the highest node cost, and then move it to an available node on the racks where the other two replicas are stored (if such an available node exists). If we cannot

---

Algorithm 9.1 Replica reallocation when three replicas are stored in three different racks

---

Assuming the three replicas of block b are stored in nodes i_c in racks r_c ($c = 1 \ldots 3$ and $r_1 \neq r_2 \neq r_3$).:
1. Sort the three nodes in descending order of node cost, s.t. $cost(i_1) \geq cost(i_2) \geq cost(i_3)$.
2. Set $H = 1$.
3. Repeat until $H > 3$:
4.   Find an available node (other than nodes $i_c, \forall c \neq H$) in racks $r_c, \forall c \neq H$, whose node cost is the closest to the node cost of i_H, denote this node as i_d.
5.   If we could find such node i_d:
6.   Move block b's replica on node i_H to node i_d, and then stop.
7.   Else:
8. $H = H + 1$.
9. End repeat.
10. Move the replicas on nodes i_1 and i_2 to two available nodes belonging to the same rack (other than rack r) and whose average node cost is the closest to the average node cost of i_1 and i_2.

---

find such a node, we try to move the replica stored in the node with the second highest node cost. If we could not find any available node for any of the three replicas in steps 2–9, we will move two replicas simultaneously to a rack with two available nodes as described in step 10.

9.5 Experimental Results

We obtained the thermal model parameters a_{ij} and K_i from [4,15]. The maximum temperature constraint of the data nodes $T_{max} = 318.15$ K. The power consumption is 10 W for each active data node and 0 for idle data node.

9.5.1 Task Scheduling

We first test our thermal- and power-aware task scheduling method assuming the data replica placement is already fixed. That is, we assume the locations of data block replicas on nodes are fixed. We first set the parameter k in the node cost function (Equation 9.12) to be 1. We will investigate the influence of different k in the next subsection.

9.5.1.1 With Same Deadline

For the problem where all tasks have same deadline, assuming there are a total of 25,000 data nodes. We tested four task sets with 5,000, 10,000, 15,000, and 20,000 tasks and compared the results achieved by our method with (a) the optimal solution obtained by solving the ILP problem in Equation 9.9 using *lp_solve* tool and (b) a solution that randomly selects a feasible data replica for each task without considering the thermal effects (we call the method *random*). The comparison is shown in Table 9.2.

As we can see from Table 9.2, the solution obtained by our method is very close to the optimal solution calculated by *lp_solve*. For example, for task set 1 containing 5000 tasks, the supply temperature T_{sup} required by our method is only about 0.38 K lower than the optimal solution obtained by *lp_solve*, while the COP in our method is only 0.24% lower. Assuming all task sets have equal probability to occur, the

TABLE 9.2 Comparison of Our Method (without Thermal- and Power-Aware Data Replica Placement) with *lp_solve* and *Random* Method When Tasks Have Same Deadlines ($k = 1$)

Task Set	Number of Task	Algorithm	With Rescheduling	Runtime(s)	COP	T_{sup}(K)	$Mean(T_{out})$(K)	$Stdev(T_{out})$
1	5,000	Our method	No	<1	628.7199	303.9008	315.9500	1.5659
		Our method	Yes	3.283	635.2791	305.4834	317.6022	0.7841
		lp_solve		134.078	636.8691	305.8658	318.1242	0.0532
		Random		<1	621.2212	302.0814	315.0697	2.4424
2	10,000	Our method	No	<1	573.7454	290.2978	314.5412	2.8201
		Our method	Yes	8.603	584.3221	292.9640	317.2313	1.4651
		lp_solve		361.555	586.7863	293.5817	318.1329	0.0827
		Random		<1	556.6016	285.9234	311.6821	5.3128
3	15,000	Our method	No	<1	520.4627	276.4756	313.2924	3.8661
		Our method	Yes	8.451	535.0867	280.3371	316.9589	1.8947
		lp_solve		789.793	539.2365	281.4232	317.6929	0.6240
		Random		<1	500.9395	271.2348	309.6551	7.1763
4	20,000	Our method	No	<1	480.8865	265.7442	312.9278	4.4476
		Our method	Yes	4.018	482.9767	266.3218	313.5804	4.2783
		lp_solve		>6 h				
		Random		<1	462.9052	260.7226	310.9223	5.7527

expected T_{sup} required by our method is merely 0.69 K lower than that required by *lp_solve*, and the expected COP in our method is 0.46% lower. However, the runtime of our method is much smaller than *lp_solve* (only about 1%–2.5% of the runtime used by *lp_solve*). Moreover, the runtime using *lp_solve* increases dramatically that when task number is 20,000, the runtime is more than 6 h. So our method is more suitable for online scheduling in large-scale storage datacenters.

On the other hand, compared with the simple method that randomly selects a feasible data replica for each task, our method achieves higher COP and T_{sup}. As we can see from Table 9.2, the supply temperature T_{sup} required by our method is about 6.3 K higher than that obtained by *random* method on average, and the COP in our method is 4.77% higher than *random* method, which means less power consumption in the A/C system. Also, the average T_{out} of the nodes are closer to T_{max} and the fluctuation of T_{out} (measured by standard deviation $stdev(T_{out})$) in our method is always smaller, which means the system is more efficiently used and the task assignment is more balanced.

9.5.1.2 With Different Deadlines

For the problem where tasks have different deadlines, we still use 25,000 data nodes and test three task sets with 50,000, 100,000, and 150,000 tasks. The task deadlines are from 1 to 10. The simulation result in shown in Table 9.3.

This problem where tasks have different deadlines is a complex optimization problem and cannot be solved with ILP. So we only compare our method with the *random* method. As we can see from Table 9.3, the T_{sup} obtained by our method exceeds that obtained by the *random* method by about 2.5–8.0 K, and the COP in our method exceeds the *random* method by 4.75% on average. Similarly, the average T_{out} of the nodes are closer to T_{max} and the standard deviation of T_{out} is usually smaller in our method. Since the *random* method should randomly assign each task to a data node containing the block replica this task needs and then get rid of infeasible assignments, the runtime of the *random* method is similar to our method. Therefore, our method has a runtime similar to the *random* method, while achieving higher COP.

9.5.2 Impact of Node Cost

We then investigate the impact of parameter k in the node cost function in Equation 9.12 by solving the thermal- and power-aware task scheduling problem using different values of k and comparing the maximum T_{sup} achieved. Here we focus on our method without task rescheduling because the performance of task rescheduling is not influenced by the assignment of node cost (so independent of k). Figure 9.6 shows the result achieved by using different values of k ($k \in \{0.2, 0.5, 1, 1.2, 1.5\}$). The x-axis denotes the number of tasks in each testing task set. The y-axis is the difference between the maximum T_{sup} achieved using our method (without task rescheduling) and the optimal result achieved by using *lp_solve* (that is, $dT_{sup} = T_{sup}^{out} - T_{sup}^{lp\_solve}$), which shows how far away our solution is from the optimal result. As we can see

TABLE 9.3　Comparison of Our Method (without Thermal- and Power-Aware Data Replica Placement) and *Random* Method When Tasks Have Different Deadlines ($k = 1$)

Task Number	Algorithm	Rescheduling	Runtime(s)	Total Power(W)	T_{sup}(K)	$Mean(T_{out})$(K)	$Stdev(T_{out})$
50,000	Our method	No	5.949	5.6178e4	305.6005	317.2310	1.1657
	Our method	Yes	35.947	5.5737e4	306.9471	317.8895	0.7927
	Random		5.477	5.6593e4	304.4631	316.6095	2.0848
100,000	Our method	No	10.347	1.3071e5	293.3775	313.7472	5.7069
	Our method	Yes	36.984	1.2396e5	296.4670	314.8426	5.7005
	Random		23.972	1.4091e5	290.1834	314.4067	5.5720
150,000	Our method	No	21.7324	3.2904e5	280.5644	313.0872	5.5704
	Our method	Yes	62.544	2.5748e5	284.8338	313.3713	4.5626
	Random		24.349	4.3561e5	276.2348	311.6551	5.1763

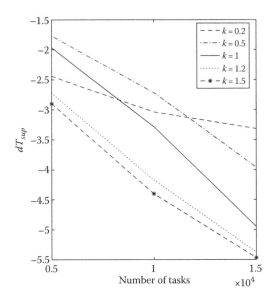

FIGURE 9.6   Impact of using different values of k.

from the figure, properly setting the value of k can result in further improvement of the solution achieved by our method. For example, by setting $k = 0.5$, we could improve our solution with respect to setting $k = 1$ by about 0.2–1 K. Usually $k = 0.2$–0.5 gives the best result. When the number of tasks L is relatively small compared with the number of nodes in the system N (e.g., $L \leq N/2$), setting k to 0.5 usually gives a better solution, while if the number of tasks L is large (e.g., $L > N/2$), setting k to 0.2 might result in a better solution.

9.5.3 Data Replica Placement

We then test the thermal- and power-aware data replica placement. Assuming the datacenter has 500 racks and each rack contains 50 data nodes, then there are a total of 25,000 data nodes. We still start with the case where all tasks have the same deadline. We generate four sets of tasks as the training tasks. Each task set contains 5,000, 10,000, 15,000, and 20,000 tasks, and the block requesting frequency distribution in each data set $\mathcal{B}^s$ follows uniform distribution or Gaussian distribution. Assuming each task set has equal probability to occur, we then estimate the probability each data block will be requested, $\mathcal{P}$, using the method described in Section 9.4. We solve the thermal- and power-aware data replica placement problem based on this probability distribution $\mathcal{P}$. Then using this block replica placement, we perform the task scheduling using the method in Section 9.3.1.1 for each task set.

Table 9.4 and Figure 9.7 show the comparison of T_{sup} achieved with and without thermal- and power-aware data replica placement. We compare four methods in Table 9.4 and Figure 9.7.

1. Method 1: Our method without task rescheduling or thermal- and power-aware data replica placement
2. Method 2: Our method without task rescheduling, but with thermal- and power-aware data replica placement
3. Method 3: Our method with task rescheduling, but without thermal- and power-aware data replica placement
4. Method 4: Our method with both task rescheduling and thermal- and power-aware data replica placement

TABLE 9.4 Comparison of T_{sup} for Methods with and without Thermal- and Power-Aware Data Replica Placement When Tasks Have the Same Deadline

Number of Task	T_{sup}			
	Method 1	Method 2	Method 3	Method 4
5,000	303.9008	304.5712	305.4834	306.0728
10,000	290.2978	291.2821	292.9640	293.8754
15,000	276.4756	277.6798	281.4232	281.9821
20,000	265.7442	267.3244	266.3218	268.7629

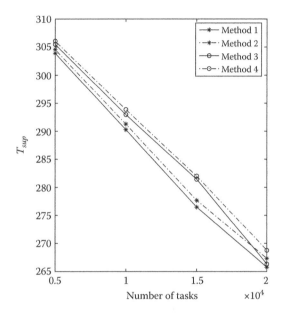

FIGURE 9.7 Impact of thermal- and power-aware data replica placement when tasks have the same deadline.

As we can see, using thermal- and power-aware data replica placement (methods 2 and 4) could further improve T_{sup} for about 1 K, compared to the method where data replica placement is fixed (methods 1 and 3).

We use the same method to test the case where tasks have different deadlines. We use three sets of training tasks, each of which contains 50,000, 100,000, and 150,000 tasks and the task deadlines range from 1 to 10. We compare the results achieved with and without using thermal- and power-aware data replica placement. The result is shown in Table 9.5 and Figure 9.8. The figure and table show that, compared with our method using a fixed data replica placement (methods 1 and 3), using our thermal- and power-aware data replica placement (methods 2 and 4) will result in about 1–1.5 K T_{sup} improvement.

TABLE 9.5 Comparison of T_{sup} for Methods with and without Thermal- and Power-Aware Data Replica Placement When Tasks Have Different Deadlines

Number of Task	T_{sup}			
	Method 1	Method 2	Method 3	Method 4
50,000	305.6005	306.4311	306.9471	307.9231
100,000	293.3775	294.8559	296.4670	297.8892
150,000	280.5644	282.3267	284.8338	286.4621

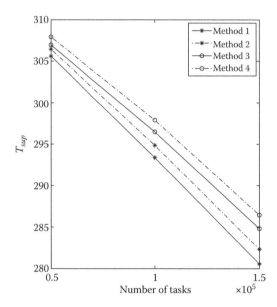

FIGURE 9.8   Impact of thermal and power aware data replica placement when tasks have different deadlines.

9.6 Conclusion

This chapter investigates thermal- and power-aware task scheduling and data placement for storage centric datacenters based on the Hadoop framework. We tried to minimize the power consumption by the datacenter, while ensuring that each data node in the datacenter operates at a temperature below a certain temperature threshold. We formulate the optimization problem as an ILP problem and then solve the problem heuristically using minimum cost flow and use task rescheduling to further balance the workload of the system. Then, based on the data replica placement rules specified by the Hadoop system, we also explore the thermal- and power-aware data replica placement in the storage datacenter so that the power consumption could be further reduced.

Acknowledgment

We would like to thank NSF grant CCF 0937865 for supporting part of this research.

References

1. Youtube serves up 100 million videos a day online, *USA Today*, 2006.
2. Facebook statistics, http://www.facebook.com
3. R. Sawyer, Calculating total power requirements for data centers. White paper, American Power Conversion, 2004.
4. Q. Tang, S. K. S. Gupta, D. Stanzione, and P. Cayton, Thermal-aware task scheduling to minimize energy usage of blade server based datacenters, in *Proceedings of the Second IEEE International Symposium on Dependable, Autonomic and Secure Computing*, Indianapolis, IN, pp. 195–202, 2006.
5. E. Pakbaznia and M. Pedram, Minimizing data center cooling and server power costs, in *Proceedings of the 2003 International Symposium on Low Power Electronics and Design (ISLPED'09)*, San Francisco, CA, pp. 145–150, 2009.

6. J. Moore et al., Making scheduling "cool:" Temperature-aware resource assignment in data centers, in *Usenix Annual Technical Conference*, Anaheim, CA, 2005.

7. Q. Tang, S. K. S. Gupta, and G. Varsamopoulos, Energy-efficient, thermal-aware task scheduling for homogeneous, high performance computing data centers: A cyber-physical approach, *Transactions on Parallel and Distributed Systems, Special Issue on Power-Aware Parallel and Distributed Systems (TPDS PAPADS)*, 19:1458–1472, 2008.

8. Q. Tang, S. K. S. Gupta, and G. Varsamopoulos, Thermal-aware task scheduling for data centers through minimizing heat recirculation, in *Proceedings of the 2007 IEEE International Conference on Cluster Computing*, Austin, TX, 2007.

9. T. Mukherjee, Q. Tang, C. Ziesman, S. K. S. Gupta, and P. Cayton, Software architecture for dynamic thermal management in datacenters, in *International Conference on Communication System Software and Middleware (COMSWARE'07)*, Bangalore, India, 2007.

10. G. Varsamopoulos, A. Banerjee, and S. K. Gupta, Energy efficiency of thermal-aware job scheduling algorithms under various cooling models, in *International Conference on Contemporary Computing (IC3)*, Noida, India, 2009.

11. P. Ranganathan, P. Leech, D. Irwin, and J. Chase, Ensemble-level power management for dense blade servers, in *Proceedings of the 33rd Annual International Symposium on Computer Architecture*, Boston, MA, 2006.

12. T. Heath, B. Diniz, E. V. Carrera, W. Meira Jr., and R. Bianchini, Energy conservation in heterogeneous server clusters, in *Proceedings of the 10th ACM SIGPLAN Symposium on Principles and Practice of Parallel Programming*, Raleigh, NC, 2005.

13. Y. Chen, A. Das, W. Qin, A. Sivasubramaniam, Q. Wang, and N. Gautam, Managing server energy and operational costs in hosting centers, in *Proceedings of the 2005 ACM SIGMETRICS International Conference on Measurement and Modeling of Computer Systems*, Berkeley, CA, 2005.

14. Hadoop wiki, http://wiki.apache.org/hadoop/PoweredBy

15. Q. Tang, T. Mukherjee, S. K. S. Gupta, and P. Cayton, Sensor-based fast thermal evaluation model for energy efficient high-performance datacenters, in *International Conference on Intelligent Sensing and Information (ICISIP2006)*, Bangalore, India, pp. 203–208, 2006.

16. N. Schirle and D. K. Lieu, History and trends in the development of motorized spindles for hard disk drives, *IEEE Transactions of Magnetics*, 32:1703–1708, 1996.

17. J. Moore, J. Chase, P. Ranganathan, and R. Sharma, Making scheduling "cool": Temperature-aware workload placement in data centers, in *Proceedings of the Annual Conference on USENIX Annual Technical Conference*, Anaheim, CA, pp. 5–5, 2005.

Green
Networking

10

Power-Aware Middleware for Mobile Applications*

Shivajit Mohapatra
Motorola Mobility

M. Reza Rahimi
University of California, Irvine

Nalini Venkatasubramanian
University of California, Irvine

10.1 Introduction

The next generation of the internet and intranets will see a wide variety of small low-power devices operating on board with high-end systems. With progressive improvements in technology and wide-ranging user demands, these computers are now being exploited to perform diverse tasks in an increasingly distributed fashion. Mobile systems come in different flavors, including laptops, cellular phones, webphones, PDAs, pocket computers, sensors to name a few. All such devices have wireless communication capabilities enabling them to operate in distributed environments. Due to their modest sizes and weights, these systems have inadequate resources—lower processing power, memory, I/O capabilities, storage, and battery life as compared to desktop systems. These portable devices are mostly battery driven and oftentimes have to run on batteries for considerable time periods. Therefore, power management is a critical issue for these systems, and tackling power dissipation and improving service times for these systems have become crucial research challenges.

The growing popularity of distributed middleware systems coupled with their scalability, flexibility, and affinity to mobile and wireless architectures has made them a dominant methodology for supporting distributed applications. *We refer to "middleware" as an abstraction layer that sits between the operating system (OS) and the application and provides a level of abstraction for distributed applications.* The primary advantage of having middleware is that it runs on heterogeneous environments shielding applications

---

from the individual underlying technologies. Today, wireless-enabled end-to-end middleware frameworks find application in a variety of areas: gaming, chat systems, and other similar peer-to-peer (P2P) applications are already available. Other applications like mobile banking, business-to-employee (B2E) systems like mobile workforce automation, machine-to-machine (M2M) systems like remote administration, manufacturing control, etc., are quickly becoming a reality.

Applications in distributed environments require a multiplicity of services in order to accomplish their tasks. A distributed middleware system can provide important additional services like reliable messaging, distributed snapshots, clock sync services, location management services, CPU scheduling, persistent data management services, system security(encryption/decryption), directory management among others, the complex details of which remain hidden from applications. In a reconfigurable middleware framework, one or more of these component services can be independently started, stopped, or moved by a meta-level entity. This plug-and-play approach obviates the need for all middleware components to be running on a low-power device at all times. Additionally, new services may need to be installed/uninstalled to cope with dynamic requirements. Customizable middleware frameworks can be considered as important candidates for energy optimization as they can be pruned depending on the workload and residual power of the device. For example, a cache management service can be off-loaded to a nearby proxy or cloud saving on both power and storage, while still providing appreciable performance benefits of caching [CM09,CBC10].

In this chapter, we present as a case study an adaptive, reconfigurable middleware framework (PARM) that can significantly improve energy uptake of low-power devices operating in distributed environments. It provides for runtime middleware customizations on a low-power device, while still maintaining the semantics of the middleware services. Furthermore, the adaptation decisions are shifted to an external entity with cross-application knowledge and overall system information. As the residual energy (ER) on the device diminishes, the PARM middleware dynamically reconfigures the component distribution on the low-power device to improve the time of service (TOS) of the device. We develop a graph theoretic solution [AMO93,CGK97] for determining the optimal middleware component reconfigurations and empirically identify when and how often reconfigurations should occur for optimal energy gains. The reconfiguration decisions are driven by comprehensive profiling of the energy usage patterns of various applications and middleware components. Finally, we simulate and extensively test and analyze the performance of our approach using three different classes of applications that form a representative workload for all low-power devices.

The rest of the chapter is organized as follows: Section 10.2 discusses the current state of the art in support for energy efficiency for mobile applications. Section 10.3 argues for a proxy-based approach where an in-network node assumes some of the processing and storage capabilities required for mobile applications and introduces a PARM for this purpose. Section 10.3 characterizes the communication/computation trade-offs as a PARM component reconfiguration problem, models it using a source parametric flow network, and presents an optimal polynomial time solution to the component reconfiguration problem. In Section 10.4, we present the performance evaluation and the prototype implementation of the PARM framework. We conclude by discussing future research direction in Section 10.5.

10.2 State of the Art in Power-Aware Mobile Computing

A tremendous amount of research has already been done for achieving power savings in low-power and embedded devices. The research efforts have ranged from defining measurement techniques [MZ95,MSS03], software strategies [LS96,IBM02], monitoring, and modeling [FLS01,FS99,FPS02] for power management to more system-level approaches like slackening or powering down components (e.g., CPU) when not in use. Modern mobile devices also present several opportunities to achieve power savings by providing capability to operate various hardware components in multiple low duty-cycle operating modes. Consequently, several interesting solutions for energy optimization have been proposed at various computational levels. At the architectural level, solutions include disk spin-down policies [DKB95,DKM94], turning of the displays during periods of inactivity, effective battery management [CR00], minimizing power by managing the mobile host's communication device [IGP95,KK98,SK97], system cache and external memory access optimization [FS96,HSA01,HKA01], dynamic power management of disks and network interfaces [C02,CV02], and efficient compilers and application/middleware-based adaptations [MV03,NSN97,FLS01].

Dynamic voltage scaling (DVS; [BSM04,LS96,LS98,PS01,WWD94,YN02,YN04,YN03]) tries to address the trade-off between performance and battery life, by adjusting CPU speed (and therefore power) depending on the system workload. It uses the fact that for most computers, the peak computing rate needed is much higher than the average throughput that needs to be sustained. DVS scales the operating voltage of the CPU along with the frequency to achieve power gains. Most voltage scaling initiatives are undertaken at the OS (or lower) level and require OS support for adjusting CPU speeds. Hughes et al. [HKA01] have analyzed architectural variability in the frame-level execution time for a number of multimedia applications on general-purpose architectures. Hughes et al. [HSA01] propose a technique for combining two hardware adaptations (architecture adaptation and DVS) for reducing energy in multimedia workloads. The study concludes that DVS alone gives most of the energy benefits, while the more aggressive architectures are more energy efficient in the presence of DVS. In [SG01,ISG02], Gupta et al. present "online" power management strategies for systems with multiple idle power states. They compare the effectiveness of the online algorithms against the optimal algorithm (competitive analysis) and present a generalized analysis of dynamic power management strategies for systems with multiple sleep states. Jejurikar and Gupta [JG02] extend the DVS approach by calculating static slowdown factors for tasks that need to synchronize for access to shared resources.

There is also a rapidly growing body of work on HW–SW cosynthesis under energy constraints [SAE05, W04]. Typical work in design space exploration proposes ad hoc scheduling policies to minimize the energy [SAE05] in the context of multimedia applications. However, such an approach adds the overhead of customized scheduling tables for each individual application. Prior work has also explored a cosynthesis methodology that guarantees real-time operation with low area/energy requirements [KBD08]; our techniques exploit the existing heterogeneity in MPSoCs by choosing from a library of well-known scheduling policies such as rate monotonic (RM), earliest deadline first (EDF), and permits extensive design space exploration multiple trade-off points can be evaluated.

There have been several research efforts to dynamically modify application behavior dynamically to conserve energy, with some help from the OS. Flinn and Satyanarayanan demonstrated that a collaborative relationship between the OS and applications can be used to meet user-specified goals for battery duration [FLS01,FS99]. Odyssey [NSN97] presents an applications aware adaptation scheme for mobile applications. In this approach, the system monitors resource levels, enforces resource allocation, and provides feedback to the applications. The applications then decide on the best possible adaptation strategy. The Milly Watt project [E99] explores the design of a power-based API that allows a partnership between applications and the system in setting energy use policy. In the context of this project, a currency model that unifies energy accounting over diverse hardware components and enables fair allocation of available energy among applications and a prototype energy-centric operating system (ECOSystem) that implements explicit energy management techniques from the system point of view have been proposed [ZEL02]. Their goal is to extend battery lifetime by limiting the average discharge rate and to share this limited resource among competing tasks according to user preferences. PowerScope [FS99] is an interesting tool that maps energy consumption to program structure. It first profiles the power consumption and system activity of a computer and then generates an energy profile from this data.

At the middleware level, the focus has been on reducing the communication and computation required at the mobile device end while providing adequate QoS for the executing mobile applications. One of the primary directions in this context has been to optimize network interface power consumption [C02,CV02, FN01]. In [SAS01], an energy-aware EDF packet scheduling scheme for real-time traffic called modulation scaling is presented. In effect, this work describes a technique for integrating network communication into a dynamic power management scheme on a radio. Chandra [C02] compares the energy consumed by network interface cards for three different video streaming formats. They suggest the traffic shaping at a proxy for sending multimedia data, so that network cards can be transitioned to sleep states during periods of inactivity for saving power. An initial study of energy consumption characteristics of network interfaces was made in [SK97]. Radkov and Shenoy [SR03] present a methodology for estimating video decoding times and also provide two mechanisms for intelligent transmission of multimedia data. Feeney and Nilsson [FN01] make a comprehensive evaluation of energy consumption of network interface cards and formulates formulas for relating energy consumption with data transfer sizes. Farkas et al. [FFB00] have characterized the energy consumption of a pocket computer and evaluated the energy and power trade-offs in the design of the JVM for the machine.

Given the high compute/communication overheads of rich multimedia traffic, much of the research in network level adaptations has focused on dealing with streaming multimedia content to the mobile device. A thorough analysis of power consumption of wireless network interfaces has been presented in [FN01]. Chandra and Vahdat [CV02] have explored the wireless network energy consumption of streaming video formats like Windows Media, Real media and Apple Quick Time via techniques such as energy-aware traffic shaping close to the mobile device [CV02]. In [SR03], Shenoy suggests performing power-friendly proxy-based video transformations to reduce video quality in real-time for energy savings. Caching streams of multiple qualities for efficient performance has been suggested in [FPS02]. In [YN00], a resource-aware admission control and adaptation is suggested for multimedia applications for optimal CPU gains. In [ABE00], the authors propose a server-controlled power management scheme much like ours where the server exploits the workload, traffic-related information, and feedback from the mobile device to minimize the energy consumption of the network interface card. Puppeteer [FLS01] presents a middleware framework that uses transcoding to achieve energy gains. Using the well-defined interface of applications, the framework presents a distilled version of the application to the user, in order to draw energy gains. In [ANF03,ANF04], Anand et al. present a middleware that allows applications to provide "ghost hints" to OS level power managers for implementing power management strategies (for disk cache management). This is in effect similar to the approach of exposing and exchanging state information between system layers for coordinated adaptations.

Power optimization techniques developed at single level, incognizant of the other abstraction hierarchy levels, potentially miss opportunities for substantial improvements achievable through cross-level integration. One of the primary arguments for a cross-layer approach was the realization that advances in battery and hardware technology cannot, by themselves, meet the energy demands of next generation mobile computers, that is, higher levels of the system must also be involved [KWW02,E99]. The cumulative power gains achievable by coordinating techniques at each stage can be potentially significant, but this requires a study of the trade-offs involved and the customizations required for unified operation. Ellis [E99] presents a case for higher level power management and states that applications/middleware can contribute significantly toward energy savings in smaller devices.

In this chapter, we argue for a middleware-based approach whose goal is to control and direct the middleware component workload on a system to effect energy gains. For example, feedback from on-device middleware environment can complement the advantages of DVS resulting in higher energy gains. Weiser et al. [WWD94] compare and contrast the characteristics of wide area and embedded applications that influence the design of quality object middleware. [ABE00,ATS06,BC00,BSM04,GNM03, HMV05,PAL09] present a principled approach toward supporting adaptation by using reflection, in a component-based middleware framework. Other techniques from compilers have been used for power-aware mobile applications [KHR03]. The GRACE project [ANJ02,ZEL03,YN04] proposed the use of cross-layer adaptations for dynamic CPU optimizations. They suggest both coarse-grained and fine-grained tuning of the CPU through global coordination and local adaptation of hardware, OS, and application layers. In [YN00], a middleware framework for integrating soft real-time scheduling and DVS is presented. The idea is to apply DVS as far as possible, but while meeting resource reservation requirements of soft real-time applications.

Our approach is realized in DYNAMO [MDN07], a cross-layer middleware framework for evaluating power/performance/QoS trade-offs, we adopt the strategy of exchanging state information dynamically at all system levels of the system hierarchy (architecture, network, OS, middleware, and applications) and driving simultaneous adaptations at each layer based on these exchanges for optimized performance and energy benefits. This work targets coordinated optimizations for all the three most power hungry components of a mobile system—the CPU, the LCD display, and the wireless network interface card. The other key feature of the DYNAMO framework is that it utilizes intermediate servers in close proximity of the mobile device to perform end-to-end adaptations such as admission control, intelligent network transmission, and dynamic video transcoding. The knowledge of these adaptations are then used to drive "on device" adaptations, which include CPU voltage scaling through OS-based soft real-time scheduling, LCD backlight intensity adaptation, and network card power management. Our experimental results show that such joint adaptations can result in energy savings as high as 54% over the case where no optimization is used while substantially enhancing the user experience on handheld systems.

The idea of task off-loading onto a remote machine is not new [OH98,RRP99]; Rudenko et al. [RRP99] shows that the task off-loading can deliver significant energy savings over noiseless wireless network channels, while the gains are offset over noisy communication channels. More recently, there have been efforts exploring the use of cloud computing to off-load tasks [GRJ09,CM09,CBC10]; the use of proxies (or near clouds) to aid mobile devices is being explored. In Clone cloud [CM09,GRJ09], a dynamic task partition scheme is presented for off-loading application subtasks onto a remote machine or virtual machine for achieving optimal performance according to the memory and processing time. MAUI [CBC10] uses the same technique for achieving the optimal power management and extending the portable device battery life. In the following section, we discuss such a proxy-based approach to effect power efficiency for mobile applications and formulate a methodology for addressing computation and communication trade-offs in such architecture.

10.3 Proxy-Based Approach to Power-Aware Computing

A typical distributed system architecture using low-power devices is envisioned in Figure 10.1. The model environment consists of several distributed servers (service providers), proxy servers, meta-data repositories (e.g., directory service), and mobile clients. The servers are high-end machines on the network that provide a variety of services (e.g., streaming video and web services). A set of proxy servers are available on the network and are used to perform an assortment of tasks. For example, a server can replicate data onto a proxy for load balancing and a client can off-load expensive tasks onto a proxy in order to conserve local resources. End clients are mobile devices that communicate with other entities via an access points (e.g., base station) in their geographical proximity. The mobile clients route their

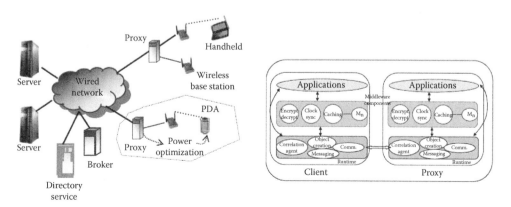

FIGURE 10.1 System architecture and PARM framework. (Based on Mohapatra, S. and Venkatasubramanian, N., PARM: Power aware reconfigurable middleware, in *ICDCS*, Providence, RI, 2003. © 2003 IEEE.)

requests to the servers through a proxy server associated with its base station. Depending on the system design, a device may either be bound to a proxy permanently or might disassociate and reconnect to another (possibly nearer) proxy. In the latter case, a proper hand of state information might be necessary between the proxies. For simplicity, we currently bind a device to a predetermined proxy. The directory service stores the overall system state information and forms an important focal point for storing and retrieving data efficiently. The architecture is further bolstered by the presence of a high-end server called the "power broker." The power broker can be looked upon as the crux of the system, making intelligent decisions and adaptations based on either the global system energy state or individual device power states. In a large-scale distributed environment, there would be many such brokers distributed over the network. In PARM, the power broker determines the set of middleware components that can be off-loaded from a device onto a proxy, to minimize the energy consumption at the device. For simplicity, we assume that the client is bound to a proxy. A more complicated scenario would include a mobility model for the clients and the state handoffs involved therein. Additionally, we define policies which dictate the instants (when), as well as the frequency (how often) at which the broker reconfigures the devices. Based on the current system state, information from the device/proxy and one or more PARM policies, the broker runs the PARM algorithm to determine a new configuration (if any) for reassignment of middleware components between the device and proxy. All low-power devices update their current state information (e.g., running applications/middleware components, residual power, and mobility information) periodically to the power broker. An optimal algorithm at the broker would ensure the most beneficial configuration of components at the device and hence maximal energy savings. For the rest of the chapter, we refer to any battery-operated computer as a device and use the terms "components" and "services" interchangeably to represent middleware services.

The implicit role of middleware in dealing with system heterogeneity, scalability, resource management, etc., makes it ideal for the above infrastructure, as it effectively shields applications from the intricacies of the underlying distribution. To make middleware technology more hard-hitting on low-power devices, it has to be both expeditious and energy efficient. Section 10.3.1 presents a middleware framework that can contribute to significant power downsize on low-power devices. For the rest of the chapter, we refer to any battery-operated computer as a device and use the terms "components" and "services" interchangeably to represent middleware services.

10.3.1 PARM: A Power-Aware Reflective Middleware Framework

In this section, we present a flexible, reflective middleware framework that is suitable for low-power devices operating in distributed environments. The goal is to realize a middleware solution that is implicitly capable of performing power optimizations. We purport that application-based adaptation in conjunction with our middleware can further substantiate the gains achieved implicitly through the middleware. Therefore, guidelines can be established for applications developed for the middleware, such that maximum power benefits can be achieved. Figure 10.1 gives a succinct depiction of the PARM middleware framework driving the applications on any computer in the system. PARM applications are developed using a set of programming interfaces exported by the middleware. Each application can either be stand alone or could be a collection of separate tasks that collectively constitute the application. The middleware framework comprises a core runtime that is a part of every system within the environment. The core runtime services include the skeletal constructs for object creation, simple message passing, and a communication framework for message routing. Any application can be developed by using these simple components. A "correlation agent" is included with the core runtime and performs various adaptations and housekeeping activities, while coordinating the various middleware components.

A set of "enrichment" middleware components are provided to impart a fuller abstraction level, which enable applications to achieve more complex tasks through a simple interface. These components are independent and can be transparently plugged into the runtime by a system-level entity. The middleware carries the burden of seamlessly adapting to the changing environment and providing an acceptable QoS to the application. For example, the core middleware runtime does not include a real-time scheduling service. A soft real-time application (e.g., video streaming) requiring a certain level of QoS may not be able to meet its resource/timing demands using the skeletal services. On the other hand, a video application linked to the soft real-time scheduler provided by the framework would be guaranteed certain resources on the machine by the scheduler component and would be able to meet its deadlines. In the process, certain extra amount of work is done by the middleware, and this might tally to a considerable cost in terms of energy consumed at the devices. Similarly, the other enrichment components carry out complex activities that would incur some energy cost on low-power devices. We target these components as possible candidates for energy optimizations. Our approach would enable low-power devices to make use of these services (whenever possible) while cutting down on the energy costs.

By "reconfigurable middleware framework," we advert to a middleware model in which the enrichment components (not the runtime) can be dynamically started and stopped/migrated by some meta-entity, without effecting the execution of the other components. Middleware services can be discontinued on a device in a number of ways. Whenever it is not feasible to migrate a service, we can decide to either stop the service locally or degrade the service, such that it does less work, albeit at a loss of performance but with power savings. Otherwise, we can move the middleware component to a proxy (with an abundance of resources) and simply use the results of the remote execution locally. In the process, a communication overhead is incurred that includes the cost of transferring state information and responses from the proxy. Additionally, off-loading/stopping the individual components should be transparent to the applications. In PARM, this is achieved by leaving a component stub on the local machine as the component migrates to the proxy. The stub provides the transparency to the application and handles all the communication with the now remotely executing component. The lightweight component stub is designed to have a very small computation overhead; it has an existence outside of the component and can be considered to be a first class object.

In most low-power devices, CPU operations and the network communication result in the highest energy costs. An optimization of the energy consumption costs of the CPU and communication would therefore optimize the overall energy consumption of the device. For simplicity, we model the energy costs for display (LCD, front-light) other than the CPU and the network adapter, as a constant fraction of the initial available energy of the device.

The basic premise of using a reconfigurable middleware framework is that auxiliary services can be added and stopped by a meta-level middleware entity, while maintaining transparency to the applications. A power efficient model could use this capability to its advantage by stopping redundant components and off-loading expensive components to a proxy, which has both the power and the resources to execute the components. The problem then becomes that of identifying the components that can be migrated away from the device, such that the energy costs of computation and communication at the device are minimized. This section describes the mechanics of the PARM framework and presents an optimal algorithm for component distribution by modeling the PARM component reconfiguration problem as a source parametric flow network.

As the residual power on the device diminishes, executing the integrated framework of all the applications and components rapidly reduces the remaining TOS for the device. The PARM framework addresses this issue by dynamically tailoring itself in a manner such that the device remains functional for as long as possible. This is achieved by moving expensive middleware services transparently to a proxy server and using the residual power of the device as a principal factor in determining the component distribution. Furthermore, the PARM framework dynamically adapts to the decreasing residual power at the device, by trying to push more and more components to the proxy as the power on the device decreases. This is achieved by inflating the costs associated for executing the components on a device, causing the middleware to reevaluate the distribution and making appropriate alterations (possibly choosing to assign the now expensive component to execute at the proxy). In the following sections, we introduce the parameters used in characterizing the problem and present an optimal graph theoretic algorithm for ascertaining the component distribution.

10.4 Characterizing Computation versus Communication in PARM

We use the following parameters to characterize the energy costs at the device for the various computation and communication operations. It is important to note that all energy costs are incurred at the device. The values are quantified by careful experimentation and profiling. Let BW_t and BW_r be the maximum bandwidth available to the device for transmitting and receiving data, respectively. $P_{transmit}$ and P_{recv} are the average power consumption rates at the device while it transmits and receives data. $P_{runtime}$ represents the power consumption rate of the PARM runtime. Let T_i be the length of the ith time interval (i.e., time interval between two consecutive executions of the PARM algorithm at the broker), and let R_i represents the residual power on the device after the ith time interval. $Size_t^k$ and $Size_r^k$ are the average sizes of the messages transmitted and received, respectively, by the kth middleware component. We characterize the some of the other parameters as follows:

- PC_k: Average rate at which the kth component consumes power due to computation
- PS_k: Average rate at which the kth component stub consumes power
- NS_{di}^k: Average number of messages transmitted by the kth component during the ith time interval, when component is executing at the device
- NR_{di}^k: Average number of messages received by the kth component during the ith time interval, when component is executing at the device
- NS_{pi}^k: Average number of messages transmitted by the kth component during the ith time interval, when component is executing at the proxy
- NR_{pi}^k: Average number of messages received by the kth component during the ith time interval, when component is executing at the proxy

Using the above characterization, we derive the following energy costs for computation and communication:

1. Computation cost:
   a. When middleware component "k" executes on the proxy during time interval "i,"

   $$EC_{proxy}^k = PS_k T_i$$

   b. When middleware component "k" executes on the device during time interval "i,"

   $$EC_{device}^k = E_0 + \lambda T_{const}$$

   c. where

   $$E_0 = PC_k T_i$$

   $$\lambda = \frac{1}{R_i} \quad if\ R_{i-1} - R_i > 0$$

   $$= \frac{1}{R_{max}} \ (\text{otherwise})$$

   T_{const} = some scaling factor determined from profiling information

2. Communication cost:
   a. When middleware component "k" executes on the proxy during time interval "i,"

   $$CC_{proxy}^k = \frac{NS_{pi}^k Size_t^k}{BW_r} P_{recv} + \frac{NR_{pi}^k Size_r^k}{BW_t} P_{transmit}$$

b. When middleware component "*k*" executes on the device during time interval "*i*,"

$$CC_{device}^k = \frac{NS_{d_i}^k Size_t^k}{BW_r} P_{transmit} + \frac{NR_{d_i}^k Size_r^k}{BW_t} P_{recv}$$

In order to optimize the energy in the system, we now have to find an allocation scheme that distributes the components between the device and proxy such that the overall energy cost at the device is minimized. Let *U* be the entire set of components that we consider. Let *X* be the set of components mapped to the device. Then *U* − *X* is set of components mapped to the proxy. Our problem now becomes that of minimizing the computation and the communication costs at each interval. For each *T*, we need to minimize the following term:

$$\sum_{k \in X} EC_{device}^k + \sum_{k \in X} CC_{device}^k + \sum_{k \in U-X} EC_{proxy}^k + \sum_{k \in U-X} CC_{proxy}^k .$$

10.4.1 Network Flow Representation of the PARM Problem

To achieve the optimal distribution of components between the device and the proxy, we cast our problem as a source parametric flow graph problem [AMO93]. The minimum cut of the flow graph then gives us an optimal mapping of components. We incorporate the energy costs of network communication and CPU computation into an energy flow network. To create the flow network (Figure 10.2), *we distinguish two special nodes in the network: a source node D and a sink node P*. Additionally, we define two conceptual nodes R_d and R_p, which represent the core runtime frameworks on device and the proxy, respectively (Figure 10.2). We associate the source node (*D*) with the low-power device and the sink node (*P*) with a proxy. In Figure 10.2, B_d and A_p represent the energy costs of the PARM runtime executing at the device and the proxy, respectively. All the other nodes in the graph correspond to the PARM middleware components M_i. The arc capacities are assigned as follows: Each A_i denotes the energy costs of computation incurred at the device, when component M_i is executing at the proxy.

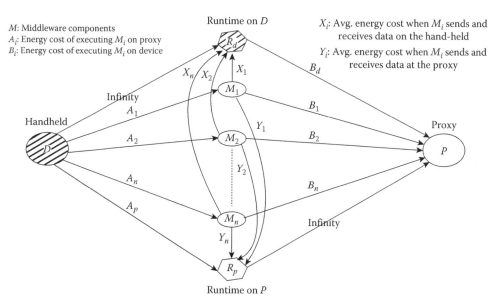

FIGURE 10.2 Flow network. (Based on Mohapatra, S. and Venkatasubramanian, N., PARM: Power aware reconfigurable middleware, in *ICDCS*, Providence, RI, 2003. © 2003 IEEE.)

In the PARM framework, it is the cost of execution of the component stub at the device. B_i denotes the energy cost of computation when component M_i executes at the device. Each B_i is defined as a non-decreasing linear function of the ER on the device. This makes the flow graph a source parametric graph, where the computation cost capacity of every source arc (B_i) increases with time. X_i represents the energy cost of network communication when component M_i is executing at the device and sending/receiving data to/from the proxy. Y_i represents the energy cost of network communication when component M_i is executing at the proxy and sending/receiving data to/from the device. The device runtime (R_d) is bound to the device (D) by assigning infinite energy cost to the arc from D to R_d. The proxy runtime is similarly bound to the proxy.

With this representation, there now exists a one-to-one correspondence between a minimum cut of the graph and the assignment of components to the source (device) and the sink (proxy). Let P_1 and P_2 be the assignment of components to the device and the proxy, respectively. The minimum cut corresponding to this assignment is then ($[\{D, R_d\} \cup P_1], \{P, R_p\} \cup P_2$). The cut of the graph effectively represents the minimum energy cost assignment of the components to the device and the proxy. Moreover, as the costs of the arcs B_i increase (ER at the device reduces), the minimum cut would tend to exclude these arcs, thereby assigning more and more components to the proxy. In our framework, the energy costs of executing the components on the device (B_i) are represented as nondecreasing linear functions of the residual power on the device. As the power drains from the device, the cost of executing components on the device increases. The PARM algorithm now tends to push more and more components toward the proxy. We solve the above problem by using a modified FIFO preflow push algorithm, as described in Section 10.4.2. Our algorithm has a worst case execution time of $O(n^3)$.

10.4.1.1 PARM Algorithm

We now describe the PARAM reconfiguration algorithm (Figure 10.3), which uses a modified FIFO pre-flow push algorithm [AMO93], generally applied to solve maximum flow problems. The reconfiguration algorithm forms an integral part of the PARAM algorithm and determines the minimum cut of the flow graph created in Figure 10.3, by first resolving the maximum flow of the graph. From the above flow graph, we construct a residual graph that tells us how much more flow (energy cost) can be sent along each arc (edge) of the graph. In effect, it represents the ER flow capacities of each arc in the graph. For every unit of flow sent along an arc, a reverse arc is added to the residual graph with the same number of units. The algorithm is initially executed on the residual graph of the graph initially generated, and subsequently, the algorithm works on the residual graph generated at the previous step. The main concept behind the algorithm is to initially push the maximum possible flow from the device (node D) to the proxy (node P, see PARM-INITIALIZE() in Figure 10.3). Depending on the capacities of the Intermediate $arcs(A_i, X_i, B_i, Y_i)$, the maximum allowable flow gets routed to the proxy (P). Once a saturation point is reached at the intermediate nodes M_i (i.e., the residual graph does not contain any forward arc to the node P), the surplus flow initially sent flows back to the device (node D). At this point, the flow in the network (from D to P) is the maximum flow between the device and the proxy. However, our interest is restricted to finding the arcs that constitute a minimum cut of the graph. To achieve this, we exploit the two phase nature of the preflow push algorithm. As stated earlier, the algorithm first tries to push all the excess energy flow toward the proxy (P), till either all the energy reaches it or the intermediate arc capacities are saturated. In the second phase, the algorithm sends the surplus flow (that could not reach node P) back to the source node D. An interesting observation here is that the algorithm typically establishes a maximum preflow (end of phase-1) long before it establishes a maximum flow (end of phase-2). In order to get a minimum cut of the graph, we keep track of the intermediate nodes (M_i, R_d, R_p) that get saturated (i.e., cannot send any more flow to node P) during the first phase of the algorithm. At the end of the first phase, we get a set of nodes that would eventually send their excess flows back to the source node (D). The minimum cut would partition the graph such that these nodes are grouped with nodes D and R_d. In effect, this is the set of nodes (components M_i) that would be assigned to the device. The other

```
Procedure PARM_INITIALIZE(Graph G)
BEGIN
  + Compute exact distance labels for each note: d(i);
  + Send max. possible flow on all arcs emanating from note 'D';
    F(D,Mi) = U(D,Mi); (for each arc (D,Mi); Enqueue(List,Mi);
  + Update energy cost flows on the residual flow graph.
  + Set d(D) = n;
  + Add node D to N'
END;


Procedure PARMRECONFIGURATION()
BEGIN
  - Synchronize time with the device;
  - Get the list of active components on the device;
  - Generate the network flow graph & the residual graph (G);
  BEGIN          //Parm reconfiguration algorithm
    PARM_INITIALIZE(G);
    WHILE(LIST not empty)
      note = DeQueue(LIST);
      ADJUST_ENERGYFLOW(node);
    ENDWHILE;
  END;
  - Determine whether reconfiguration is required at device;
  - Send RECONFIG request to the device;
END
```

```
Procedure ADJUST_ENERGYFLOW(Node i)
BEGIN
  + WHILE (excess(i) > 0)
  BEGIN
    IF(residual graph contains an admissible arc(i,k))
    BEGIN
      Send flow = min{excess(i),Ri->k} on arc(i,k);
      Update excess() values for i,k;
      Update the energy cost flows residual graph;
      ADD newly active nodes to LIST;
    ELSE
      d(i) = min{d(j)+1 }, FOR all set of arcs emanating from m
                                  all with Ri->j greater than 0.
      IF(d(i) >= n)
        ADD i to N';   //component allocated to device
        REMOVE i from LIST;
      ELSE
        Enqueue(LIST, i);
      ENDIF;
      BREAK;
    ENDIF
  END;
END;
```

FIGURE 10.3 The PARM algorithm. (Based on Mohapatra, S. and Venkatasubramanian, N., PARM: Power aware reconfigurable middleware, in *ICDCS*, Providence, RI, 2003. © 2003 IEEE.)

TABLE 10.1 Terms Used in the PARM Algorithm

n	The Number of Nodes in the Graph
N'	A *SET* of all the components that send their excess flow back to D
	The nodes in this set will be assigned to the low-power device
$F(n1,n2)$	The energy flow from node "$n1$" to node "$n2$"
$U(n1, n2)$	The maximum energy capacity of the arc from node "$n1$" to node "$n2$"
$R_{i\rightarrow j}$	The ER flow available on arc from node "i" to node "j"
Excess (n)	The excess flow at node "n" ($=$ flow into "n" $-$ flow out of "n")
Distance label $d(i)$	Distance of the node i from the sink node (node P)
Admissible arc (i, j)	The arc between node "i" and node "j" is considered an admissible arc if and only if $\mathbf{d(j)} = \mathbf{d(i)} - 1$
Active nodes	The nodes in the graph that have a positive *excess*
LIST	A queue that maintains a list of the currently active nodes

Source: Based on Mohapatra, S. and Venkatasubramanian, N., PARM: Power aware reconfigurable middleware, in *ICDCS*, Providence, RI, 2003. © 2003 IEEE.

intermediate nodes (components) would be assigned to the proxy. Table 10.1 explains the notation that has been used in PARM algorithm.

10.4.1.2 PARM Policies

We investigate the performance of our strategy under a set of policies that dictate when and how often the PARM algorithm is executed for reconfiguring components on a device. The purpose is to determine which policy returns the best results in terms of energy savings at the device and ascertain the optimal times for executing the algorithm for different classes of applications and components. Note that the PARM algorithm is executed at the power broker based on the device power state and the chosen policy. We categorize the policies based on the host (broker or device) that triggers the reconfiguration.

- *Random*: The broker performs the component reconfigurations at random intervals. The importance of having a random trigger policy is that often randomly chosen policies do an excellent job in dealing with complex and contentious test cases, outperforming more intuitive policies. Moreover, the results from the random trigger policy provide an interesting case for comparing the performances of more intuitive and predictable policies.
- *Periodic*: The broker performs the component reconfigurations at periodic intervals determined by the system administrator. It solves the following purpose: given a group of applications and a set of associated middleware components, we use this policy to learn how periodic execution of the PARM algorithm at the broker impacts the power savings at the device. Moreover, it tells us how often the reconfiguration of components needs to happen in order to obtain optimal gains. These results are of a more predictable nature and can be cataloged for future reference if the outcomes are favorable. The following two policies are device driven.
- *Application triggered*: The PARM algorithm is triggered at the broker, whenever a new application/component is started up at the device. The device sends a signal to the broker, triggering the reconfiguration algorithm. This is a "reactive" policy, where the system responds to the change in load by affecting the reconfiguration algorithm to execute. Using this policy, we determine whether this causative behavior is more prolific (in terms of power savings) than the other "proactive" policies.
- *Threshold*: A reconfiguration is triggered by the device whenever the ER of the device drops below a certain threshold, determined as a percentage of the initial energy of the device. A set of threshold values are predetermined to fix the number of reconfigurations. The idea here is more simple-minded. We choose to reconfigure the components only when we feel that the power level of the device has dropped below a certain boundary and needs attention. The value of the threshold determines the frequency of the reconfigurations.

10.4.2 Handling Component Migration

We now briefly discuss how component migration is handled within the PARM framework. The interplay between components and applications on the same machine is handled through local messaging and/or IPC mechanisms. The costs of these interactions implicitly get included in the energy costs for CPU operations. We now address the more intricate issue of communication between components and applications when components are moved to the proxy. Note that all services running on the device are replicated on the proxy. So migration costs do not involve transfer of the process image over the network. Consider a common scenario in our approach where a reconfiguration results in moving a component from the device to the proxy. The migrated service is started (if not already running) at the proxy, and the state information is transmitted from the device to the proxy. This incurs energy overhead due to communication that needs to be accounted for in our optimization. The component stub is retained at the device and handles all messages received from the remotely executing component and routes them back to the application. Interaction between components on different machines is modeled as communication from one component to the runtime of the other machine. Note here that some middleware components cannot be migrated. In such a case, the service is either stopped or degraded. We include the cost of maintaining the state information for the various components within the cost for executing our framework. We identify three different situations for which component migration needs to be addressed:

1. *Case A: Application using a single middleware service:* This is the simplest of the three cases where an application on the device is using a single middleware service. As an example, think of a navigator application using a location management component where a GPS system is used to identify its current location. If exact precision is not a necessity for the application, the computationally intensive middleware component can be executed on the proxy and the results transmitted to the device, with possible energy savings.

2. *Case B: Application using multiple middleware services:* Consider an application that requires to replicate its messages, order them, and send them reliably to applications on other hosts. Furthermore, the application might require encrypting messages that are sent to some distant hosts (outside its domain). Each of the above functions is performed separately by distinct "enrichment" middleware components. The flow of the application message through the middleware components is as follows: create replicated messages, tag the messages for ordering, encrypt the messages using some algorithm implemented by the middleware component, and then use the reliable messaging service to transmit the message. In such a scenario, migrating one or more components to a proxy might result in a condition where messages are sent back and forth between the device and the proxy. However, if the computation costs are much greater than the communication overhead, this arrangement of the components might still prove to be energy efficient.

3. *Case C: Multiple applications using a middleware component:* A more complicated situation arises when there are multiple concurrent applications use a single middleware component. When the components are shared by the applications, moving the component to a remote machine will require individual application-related state to be transferred to the remote machine. The component stub now performs more work to maintain and transfer the required data and now has to be modeled as a more power devouring unit. Additionally, if the computation overhead of such a component is not too high, it can be bound to the device and never migrated.

10.5 Performance Evaluation

In this section, we lay out the details of our simulation model and analyze the performance of our model under various system conditions. We present the results of our experiments and examine the effects of the different policies on power conservation. To simulate our system, we separately model the low-power device, the proxy server, the power broker, the PARM middleware framework, and the applications. The core middleware runtime framework executes on the device, the proxy server, and the server running the power broker. Applications are developed using the APIs exported either by the PARM runtime or by one or more of the "enrichment" components that are available with the middleware framework. Extensive profiling is used to record the power consumption pattern of individual components and applications. The system model is as follows: the low-power device is modeled after a Compaq iPaq 3650. The proxy and the broker are assumed to run on high-end ethered workstations with substantial resources. Upon start-up, the device registers itself with the nearest proxy and the PARM runtime on the device records the residual power of the device, the current set of active applications and components and updates this information with the directory service. A constant energy cost is incurred for communicating with the directory service, the proxy, and the broker. We are currently building a prototype using the CompOSE|Q [VDM01] middleware framework to integrate the PARM reconfiguration algorithm for power optimizations in a distributed environment.

10.5.1 Experimental Setup

The type of application we choose to execute on the device has a significant bearing on the results of our experiments. We therefore opt for the types of applications that are currently regarded as suitable for handheld computers and some applications that we think would be popular as the devices evolve. Moreover, we target applications that consume appreciable amounts of power either through computation, communication, or both. To generalize the experiment a bit further, we divide our set of representative applications into three classes: computation intensive (class-1, e.g., image processing applications, games, and 3D graphics applications), communication intensive (class-2, e.g., web browsing, real-time chatting, and network monitoring), and both computation and communication intensive (class-3, e.g., multimedia streaming, text to speech, and GIS/navigation). [CV02] presents the energy consumption characteristics of MPEG video playback on handheld computers. We borrow some of the results presented in this chapter to model a typical video playback application for our simulation. In [KWW02], power usage rates of some typical computation intensive applications are empirically determined. We draw on some of the results illustrated in these chapters to model the power usage patterns of some of our applications. Table 10.2 illustrates the different power utilization values for a typical application from each of the above classes. The energy consumption patterns of some of the applications and components (e.g., message ordering) are obtained from our profiling results as discussed earlier.

We choose typical applications in each category and separately profile the various computation and communication costs for the applications and the associated PARM components. An application typically communicates either by routing messages through a component or by sending independent messages via the runtime, both of which have power overheads. We regard the power consumed by the independent messages as the communication overhead for the application. The cost of messages through the

TABLE 10.2 Application Model

Application	Class-1	Class-2	Class-3
Components linked	Adaptive scheduler Encryption-DES Decryption-DES	Reliable messaging Clock synchronization Message ordering	Adaptive scheduler MM message service Clock synchronization
Average power (J/s)	0.65	0.23	0.38
Average power (J/s) (communication)	0.21 (W)	0.40	0.45
Average message size	64 (bytes)	128 (bytes)	1024 (bytes)
Average messages via component (per second)	300,150,150	610,530,480	755,830,670

Source: Based on Mohapatra, S. and Venkatasubramanian, N., PARM: Power aware reconfigurable middleware, in *ICDCS*, Providence, RI, 2003. © 2003 IEEE.

components gets accounted for, in the communication costs associated with the component. The costs of computation are recorded separately for each application. For each application, the average rate at which the application sends messages via components, its memory, and CPU usage, and other details that help in completely representing an application are also registered.

10.5.1.1 PARM Component Model

To model the PARM components, we profile the energy pattern of each component while using it with a different number of applications from each class. In particular, we record the average power consumption rate of the component running on the device, and the power overhead of running the component stub on the device while the component is executing at the proxy. We also store the average number of control messages the component uses to maintain state information when used in conjunction with different classes of applications. For example, an encryption mechanism that uses the current machine time might require the remote stub to regularly send the time of the remote machine for clock synchronization. To even out the variance introduced by single applications, an average message size per application class is used. We control the communication through a component by varying the size of the messages used as well as the number of messages routed through the component by applications from each class. Here are a few examples of the "enrichment" components that we use for our simulation: replication services, message ordering services, encryption/decryption services, reliable messaging service, scheduling, clock synchronization, location management service, watermarking, and user authentication service.

10.5.1.2 Execution Model

The simulation is carried out in two phases. In the first phase, the applications are executed on the device without the PARM algorithm. The energy consumed and the remaining TOS for the device are recorded. In the second phase, the same set of applications is executed with the PARM algorithm executing at the broker. The broker implements the PARM algorithm and policies. It retrieves the individual state information for each device from the directory service and determines the optimal reconfiguration of components for the device. It then announces the new configuration to the device and the proxy. The device then initiates a routine that performs the off-loading/downloading of the components to/from the proxy based on the new configuration. The power consumed in the transfer of components is profiled and used as a part of the PARM algorithm. The energy consumed and the remaining TOS of the device are recorded. The proxy server in our simulation is a workstation that simply receives the middleware components from a device and executes them on behalf of the device. In our simulation, we have a maximum of 10 applications each of which link to two or more "enrichment" PARM components. For each application class, we assess the gains using a set of sporadic-start applications: applications that start and stop irregularly over time. Nonsporadic applications: applications that run continuously till the device runs out of power.

10.5.2 Experimental Results

We analyze the performance of the PARM framework by evaluating its execution under different application loads and reconfiguration times. Our primary metric of evaluation is the gain achieved in the

- *ER:* showing the unexpended energy left in the device
- *Remaining TOS:* indicating the remaining time for which the device can be operational under the current operational load

By "gain," we mean the TOS/ER saved by running the applications on the middleware framework with and without the PARM algorithm. When the algorithm is in use, the power broker determines an optimal set of components that need to run on the device. The device then moves the other components to the proxy. We study how the TOS/ER gains are impacted as the number of applications on the device is scaled up. Next we examine how often the broker should reconfigure the PARM components for a given set of applications to achieve optimal energy benefits. Finally, we compare the energy savings for each class of applications and ascertain the optimal reconfiguration frequency for achieving the maximum energy profits. A default reconfiguration time of 5 min is specified for the simulations.

10.5.2.1 Performance of Class-1 Applications

Figures 10.4 through 10.9 depict the gains achieved over time for the class-1 applications. As shown in Figure 10.4, there is a clear improvement in the TOS at the reconfiguration points as the components are moved onto the proxy. The initial loss is due to the fact that the device has not been reconfigured, and the overheads for updating the directory service and registering with the proxy and the broker have been incurred. Figure 10.5 shows the TOS saved for a set of applications as the broker reconfigures the

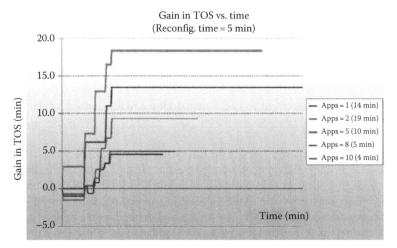

FIGURE 10.4 Effect of applications on TOS (class-1, nonsporadic). (Based on Mohapatra, S. and Venkata-subramanian, N., PARM: Power aware reconfigurable middleware, in *ICDCS*, Providence, RI, 2003. © 2003 IEEE.)

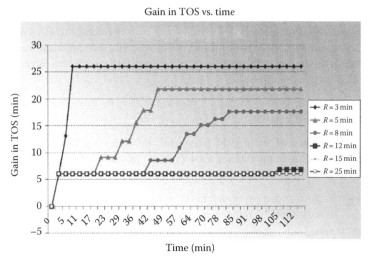

FIGURE 10.5 Effect of reconfiguration time on TOS (class-1, nonsporadic). (Based on Mohapatra, S. and Venkata-subramanian, N., PARM: Power aware reconfigurable middleware, in *ICDCS*, Providence, RI, 2003. © 2003 IEEE.)

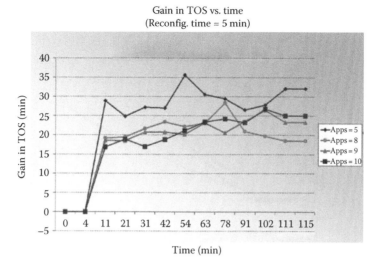

FIGURE 10.6 Effect of applications on TOS (class-1, sporadic). (Based on Mohapatra, S. and Venkatasubramanian, N., PARM: Power aware reconfigurable middleware, in *ICDCS*, Providence, RI, 2003. © 2003 IEEE.)

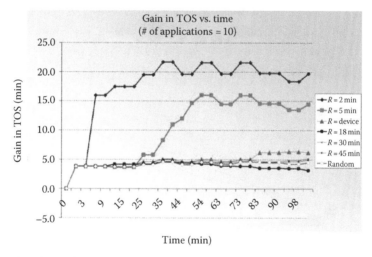

FIGURE 10.7 Effect of reconfiguration time on TOS (class-1, sporadic). (Based on Mohapatra, S. and Venkata-subramanian, N., PARM: Power aware reconfigurable middleware, in *ICDCS*, Providence, RI, 2003. © 2003 IEEE.)

components at different times. Frequent reconfigurations result in higher gains, and as the reconfiguration frequency reduces, the gains drop to almost zero (for reconfigurations at every 30 min or more). Figures 10.8 and 10.9 compare the energy gains achieved over time as the number of applications is scaled for different sets of sporadic and nonsporadic applications. As the reconfigurations occur, a steady gain in ER is noted. Moreover, it is noticed that the gains become more pronounced as the number of components used increases.

Figures 10.6 and 10.7 display the TOS gains for sporadic-start class-1 applications with different number of applications and reconfiguration times, respectively. The curves indicate irregular gains (as compared to the nonsporadic case) at different times, both when applications are scaled and when the reconfiguration times are different. The flat regions (const. gain) on the graph are the times between the configurations

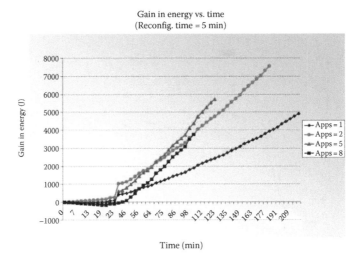

FIGURE 10.8 Effect of applications on ER (class-1, nonsporadic). (Based on Mohapatra, S. and Venkata-subramanian, N., PARM: Power aware reconfigurable middleware, in *ICDCS*, Providence, RI, 2003. © 2003 IEEE.)

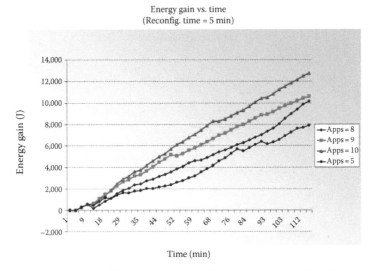

FIGURE 10.9 Effect of applications on ER (class-1, sporadic). (Based on Mohapatra, S. and Venkatasubramanian, N., PARM: Power aware reconfigurable middleware, in *ICDCS*, Providence, RI, 2003. © 2003 IEEE.)

that resulted in a redistribution of components between the device and the proxy. A comparison between the two cases shows that there is a significant energy saving as we scale the number of applications.

10.5.2.2 Performance of Class-2 Applications

Figures 10.10 and 10.11 demonstrate the TOS and the energy gains, respectively, for nonsporadic class-2 applications as the number of applications scales from 1 to 10. The broker executes the PARM algorithm every 3 min. The graphs indicate that the maximum TOS gain is achieved when the number of applications is small. A comparison with the corresponding class-1 graphs indicates that most of the gains achieved in this case are due to an initial reconfiguration of components (dependent on the nature of the components and applications started) after which the gains become steady. The explanation for this is that most of the energy overheads in this case are due to the communication. The increase in the cost of the components

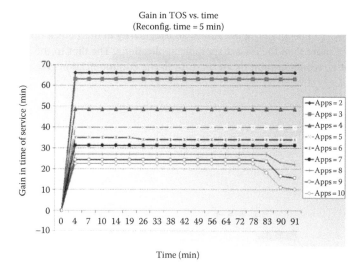

FIGURE 10.10 Effect of applications on TOS (class-2, nonsporadic). (Based on Mohapatra, S. and Venkata-subramanian, N., PARM: Power aware reconfigurable middleware, in *ICDCS*, Providence, RI, 2003. © 2003 IEEE.)

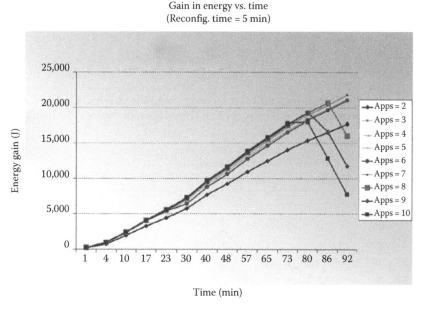

FIGURE 10.11 Effect of applications on ER (class-2, nonsporadic). (Based on Mohapatra, S. and Venkata-subramanian, N., PARM: Power aware reconfigurable middleware, in *ICDCS*, Providence, RI, 2003. © 2003 IEEE.)

with time does not seem to result in frequent reconfigurations as the computation costs are small and the device runs out of power before the costs become high enough to cause a reconfiguration. The higher gain when compared to the class-1 case is due to the different energy models used for the applications and components in this application class. It is also observed that the there is a drop in the gains toward the end when the number of applications is increased. This is caused due to the overhead of reconfiguration when the device power is depleting and happens sooner when a greater number of applications are executing.

The next two graphs (Figures 10.12 and 10.13) show the TOS gains for sporadic-start applications. As in the case of the class-1 nonsporadic applications, an irregularity in the gains is observed. However, the gains in this case are more than the class-1 sporadic applications. The dips in the curve in Figure 10.12 could be a result of a reconfiguration or an increase in the communication resulting in an additional overhead at the device.

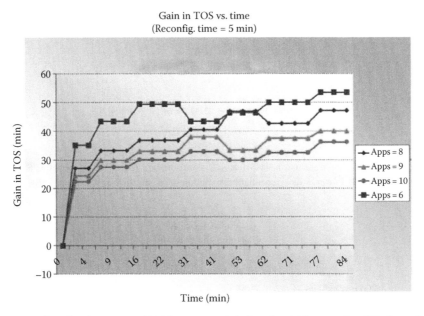

FIGURE 10.12 Effect of applications on TOS (class-2, sporadic). (Based on Mohapatra, S. and Venkatasubramanian, N., PARM: Power aware reconfigurable middleware, in *ICDCS*, Providence, RI, 2003. © 2003 IEEE.)

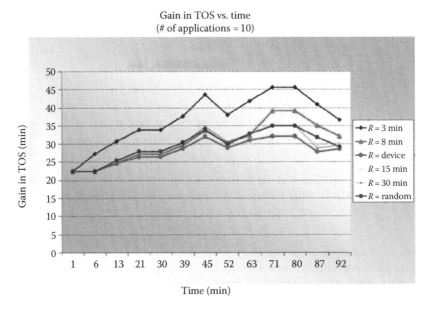

FIGURE 10.13 Effect of reconfiguration time on TOS (class-2, sporadic). (Based on Mohapatra, S. and Venkata-subramanian, N., PARM: Power aware reconfigurable middleware, in *ICDCS*, Providence, RI, 2003. © 2003 IEEE.)

10.5.2.3 Performance of Class-3 Applications

Class-3 applications present a more realistic representation of the workload in handheld devices. Most multimedia applications fall into this category. Owing to the resource scarcity in small devices, a number of applications that use moderate communication and computation would also fall into this category for smaller devices. The modeling of the components and applications for this application class is different from the models used in the previous cases. Figures 10.14 and 10.15 show the graphs for TOS and ER

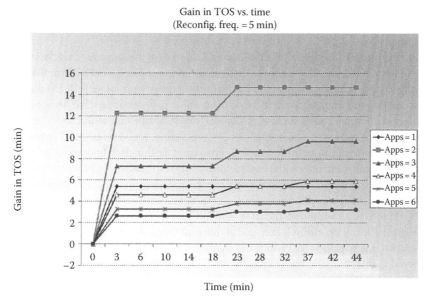

FIGURE 10.14 Effect of applications on TOS (class-3, nonsporadic). (Based on Mohapatra, S. and Venkatasubramanian, N., PARM: Power aware reconfigurable middleware, in *ICDCS*, Providence, RI, 2003. © 2003 IEEE.)

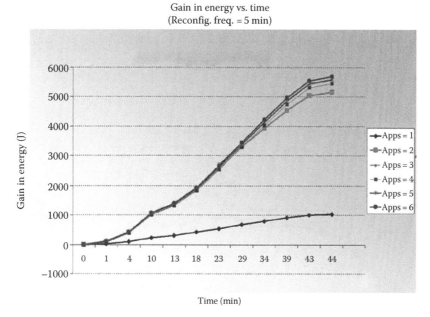

FIGURE 10.15 Effect of applications on ER (class-3, nonsporadic). (Based on Mohapatra, S. and Venkatasubramanian, N., PARM: Power aware reconfigurable middleware, in *ICDCS*, Providence, RI, 2003. © 2003 IEEE.)

gains as we scale the nonsporadic-start applications for this class. As seen from the graphs, there is a reduction in the TOS and energy gain as we scale the number of applications. An interesting observation is that with a single application, gains stay low (Figure 10.14), showing that reconfigurations with a small number of components do not result in significant gains. Figure 10.15 indicates that the gains do not change significantly as the numbers of applications are increased. A comparison with the class-1 and class-2 applications reveals that the gains in this case are not as pronounced as in the previous cases. This is because both the communication and the computation costs now play a role in determining the component reconfiguration, and it becomes hard to predict the outcome of a reconfiguration. One reason for the modest gains could be that the energy saved by off-loading components is slightly offset by the cost of communication.

Figures 10.16 through 10.19 show the TOS/ER gains achieved for the sporadic-start class-3 applications. As shown in Figure 10.16, the TOS gain for these applications is jerky. The TOS gain decreases as the number of applications scales up. Figure 10.17 presents the energy gains as we scale the number of applications. The energy gains increase as we increase the number of applications, but become unpredictable when the number of applications is close. A comparison indicates that with class-3 applications, the changes are smoother and less frequent than the corresponding class-1 and class-2 applications. Moreover, the gains are also less as compared to the other two cases. In this class, both middleware components and the applications consume substantial energy. Since, only the middleware components are considered for migration, the device drains out of power very quickly showing smaller gains. A possible inference is the with class-3 applications, application migration can significantly improve energy gains.

Figures 10.18 and 10.19 represent the TOS and ER gains achieved over six sporadic-start class-3 applications as the PARM algorithm is executed at different times. An interesting observation here is that there is a negative energy gain as we execute the algorithm at a frequency of over 8 min. A steady energy gain is observed when the algorithm is executed every 3 min. However, all the other cases seem to result in an overall loss in both TOS and saved energy. The reason for this is that the broker determines the distribution of components assuming that the applications (and the linked components) would be active

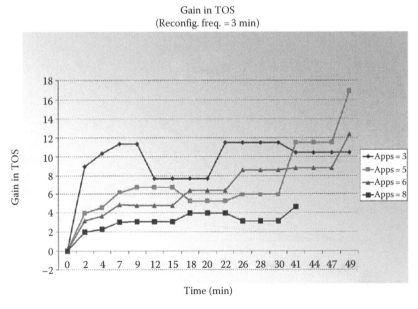

FIGURE 10.16 Effect of applications on TOS (class-3, sporadic). (Based on Mohapatra, S. and Venkata-subramanian, N., PARM: Power aware reconfigurable middleware, in *ICDCS*, Providence, RI, 2003. © 2003 IEEE.)

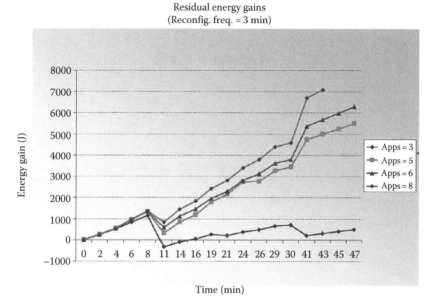

FIGURE 10.17   Effect of applications on ER (class-3, sporadic). (Based on Mohapatra, S. and Venkatasubramanian, N., PARM: Power aware reconfigurable middleware, in *ICDCS*, Providence, RI, 2003. © 2003 IEEE.)

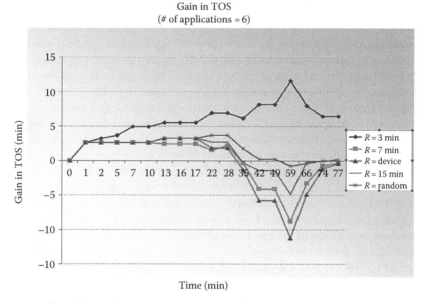

FIGURE 10.18   Effect of reconfiguration time on TOS (class-3, sporadic). (Based on Mohapatra, S. and Venkatasubramanian, N., PARM: Power aware reconfigurable middleware, in *ICDCS*, Providence, RI, 2003. © 2003 IEEE.)

over the next time interval (till the broker executes the reconfiguration algorithm again). However, in the sporadic-start case, the unpredictability of the timings of the application start/stop times seems two set the effectiveness of the PARM algorithm resulting in a loss in this particular case. This argument is supported by the fact that there is a gain when the reconfiguration frequency is high (3 min), where the random nature of the application start/stop times has the least effect.

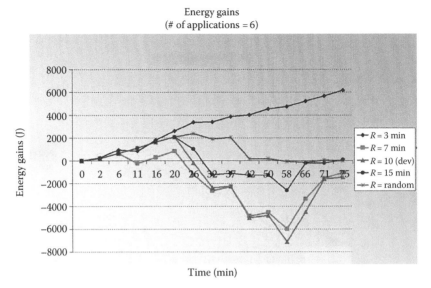

FIGURE 10.19 Effect of reconfiguration time on ER (class-3, sporadic). (Based on Mohapatra, S. and Venkatasubramanian, N., PARM: Power aware reconfigurable middleware, in *ICDCS*, Providence, RI, 2003. © 2003 IEEE.)

10.5.2.4 Comparative Performance of the Three Classes of Application

We compared the impact of the periodic broker-triggered reconfiguration and the device-triggered threshold reconfiguration policies for the three classes of applications, for both sporadic-start and nonsporadic-start applications. Figures 10.20 and 10.21 compare the impact of different reconfiguration times (3 and 10 min) for each class of applications. Both graphs show improvements in TOS gains when the reconfiguration occurs more frequently. Figures 10.22 and 10.23 illustrate that a simple-minded threshold triggered reconfiguration performs significantly worse than the periodic reconfiguration, especially for class-2 and class-3 applications. Class-3 applications show a negative gain for both the sporadic and nonsporadic-start cases. For nonsporadic class-3 applications, the threshold triggered reconfiguration performs similar to case where reconfigurations occur every 10 min (Figure 10.19) or more. We believe that the results will show marked improvements if the thresholds were set to trigger reconfigurations more frequently. Figures 10.24 and 10.25 study the impact of reconfiguration time on a mixed workload of applications. It is observed that the reconfiguration time has little impact on the particular selection of applications. Only a reconfiguration every 3 min provides a noticeable gain.

We studied the impact of scaling the number of applications and varying the rate of execution of our proposed algorithm, on the gain in TOS and ER of a handheld device, for three classes of applications. We observe that for class-1 applications, maximum service time gains of 15%–25% were achieved when the PARM algorithm was executed frequently. However, the gains were negligible when the algorithm executed once every 10 min or later. In case of class-2 applications, the gains were more moderate than the class-1 applications. On an average, there was a 7%–25% increase in the TOS of the device. In the case of the class-3 applications, both the computation and communication costs played a significant role in determining the reconfiguration of components. The gains followed a pattern similar to the gains in class-2 applications, but were more moderate as the applications drained the device power faster than the other two classes. Interestingly, it was observed that the gains in the case of sporadic-start applications in this class were negative when the reconfigurations happened less frequently (once every 8 min or more). A reconfiguration rate of around 3 min was found to give a significant gain in both energy and TOS. Our experiments illustrated that a simplistic device-triggered threshold driven policy performed inferior to the broker-driven periodic reconfiguration, especially for class-2 and class-3 applications.

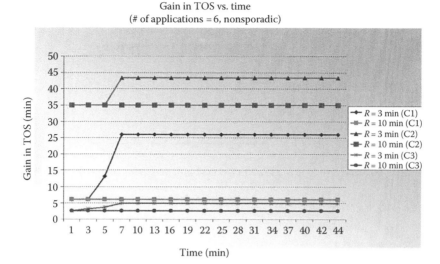

FIGURE 10.20 Impact of reconfiguration time on TOS. (Based on Mohapatra, S. and Venkatasubramanian, N., PARM: Power aware reconfigurable middleware, in *ICDCS*, Providence, RI, 2003. © 2003 IEEE.)

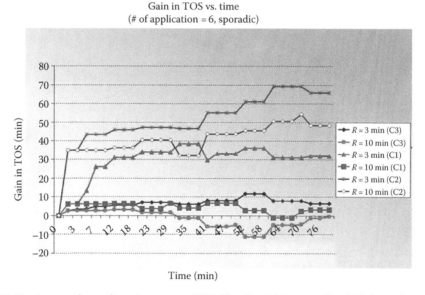

FIGURE 10.21 Impact of reconfiguration time on TOS. (Based on Mohapatra, S. and Venkatasubramanian, N., PARM: Power aware reconfigurable middleware, in *ICDCS*, Providence, RI, 2003. © 2003 IEEE.)

10.5.3 Prototype Middleware Framework

We are currently implementing a prototype of the PARM model using the CompOSE|Q middleware framework developed by us. CompOSE|Q is a message-oriented reflective middleware framework that supports a concurrent active object model. It provides mechanisms to dynamically start and stop middleware components and has an application program interface that can be used to develop applications. The entire runtime is implemented using Java and the LDAP directory service interface. A detailed discussion of the design and implementation of the CompOSE|Q runtime and PARM model are outside the scope

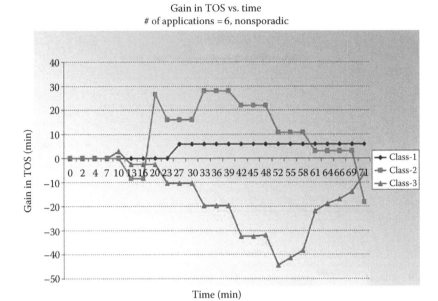

FIGURE 10.22 Threshold triggered reconfiguration (nonsporadic). (Based on Mohapatra, S. and Venkatasubramanian, N., PARM: Power aware reconfigurable middleware, in *ICDCS*, Providence, RI, 2003. © 2003 IEEE.)

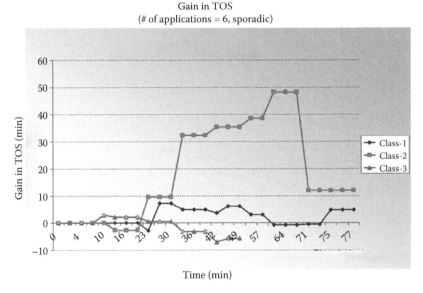

FIGURE 10.23 Threshold triggered reconfiguration (sporadic). (Based on Mohapatra, S. and Venkatasubramanian, N., PARM: Power aware reconfigurable middleware, in *ICDCS*, Providence, RI, 2003. © 2003 IEEE.)

of this chapter. So we simply present the outline of the details. Figure 10.26 shows the integration of the PARM framework with the existing architecture. The core runtime consists of a node manager that administrates the runtime components and performs functions like object creation, bootstrapping of the various meta-level runtime modules, and communication with remote middleware components. The runtime has a skeletal communication framework that includes the router, the postman, and the message

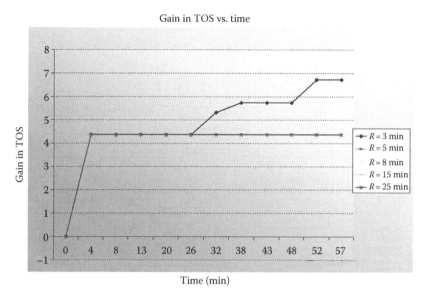

FIGURE 10.24 Impact of reconfiguration time on TOS (nonsporadic). (Based on Mohapatra, S. and Venkatasubramanian, N., PARM: Power aware reconfigurable middleware, in *ICDCS*, Providence, RI, 2003. © 2003 IEEE.)

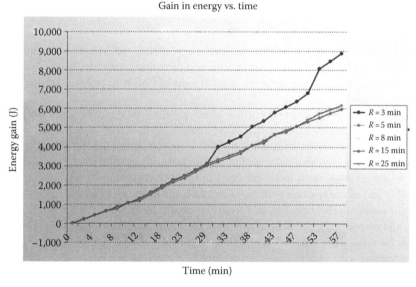

FIGURE 10.25 Impact of reconfiguration time on ER. (Based on Mohapatra, S. and Venkatasubramanian, N., PARM: Power aware reconfigurable middleware, in *ICDCS*, Providence, RI, 2003. © 2003 IEEE.)

receiver modules. The PARM-specific runtime agents are the correlation agent, the signaling agent, the registration agent, and the reconfiguration agent. Each PARM middleware component contains a stub through which all communication between the component and the runtime is routed. The stub always executes locally and provides the transparency to the applications when components are migrated. The correlation agent manages the individual component stubs on the device when the components are moved to the proxy. Incoming messages are also routed to the individual applications using the correlation agent. The agent identifies the component stub for which the messages are intended and delivers the messages

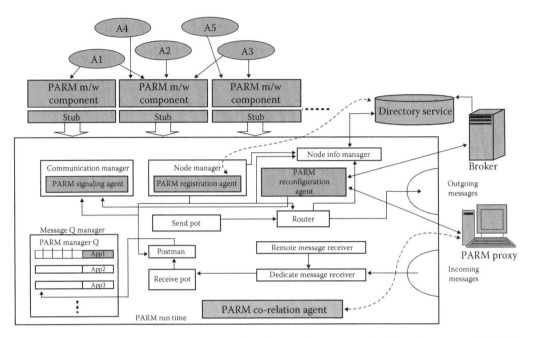

FIGURE 10.26 PARM prototype implementation. (Based on Mohapatra, S. and Venkatasubramanian, N., PARM: Power aware reconfigurable middleware, in *ICDCS*, Providence, RI, 2003. © 2003 IEEE.)

to the stub. The stub then submits the messages to the appropriate applications. The runtime generates unique identifiers for each middleware component, and the components are designed to keep track of applications that link to them. The signaling agent is used in PARM to signal either the broker or the proxy whenever required

For example, a signal can be sent to the broker to execute the PARM reconfiguration algorithm, every time new application is started at the device. A registration agent is used to register device state information with the directory service. At start-up, the device registers itself with the nearest proxy using this agent. The registration agent can also be used to recover state information after abnormal terminations. The PARM reconfiguration agent forms the heart of the PARM runtime. It regularly receives updates from the broker and in conjunction with the correlation agent manages the migration of PARM components (and possibly applications) to and from the proxy. Moreover, it maintains a list of current applications and components and actively monitors the ER of the device. The broker uses the profiled energy values for the components and applications to generate and execute the PARM reconfiguration algorithm, using a specific policy. Using the prototype, we intend to measure the performance of the proposed framework in an actual distributed environment for different classes of applications. Of particular interest is to study the impact of the reconfiguration on the performance of applications and the determination of a policy that generates the best power-performance trade-off in a truly distributed environment.

10.6 Concluding Remarks

With new emerging technologies such as cloud computing and increasing popularity of low-power mobile devices, pervasive use of these devices within distributed frameworks is inevitable. In future, distributed environments have to be cognizant of the widespread presence of these devices and confront the new challenges introduced by these resource-deficient computers. Distributed middleware frameworks hide the heterogeneity of underlying distributed environments when there are many cloud providers. These middlewares could hide and optimize many different types of mobile applications complexities while different cloud providers provide different services with diverse QoS parameters. This will provide new challenges for computer scientist in mobile computing area.

In this chapter, we presented a distributed power-aware reflective middleware framework, state of art literature and explored the viability of applying such architecture to achieve significant power gains in low-power devices. The explored framework PARM dynamically adapted to the diminishing power availability of the devices over time, by dynamically off-loading expensive middleware components to a proxy, thereby increasing the service times of the device. An optimal polynomial time graph theoretic algorithm was formulated to determine the dynamic.

References

AMO93   Ahuja, R. K., Magnanti, T. L., and Orlin, J. B. *Network Flows: Theory, Algorithms, and Applications*. Prentice-Hall, Englewood Cliffs, NJ, 1993.

ABE00   Andersen, A., Blair, G. S., and Eliassen, F. A reflective component-based middleware with quality of service management. In *PROMS*, Cracow, Poland, October 2000.

ANF03   Anand, M., Nightingale, E. B., and Flinn, J. Self-tuning wireless network power management. In *MOBICOM*, San Diego, CA, 2003.

ANF04   Anand, M., Nightingale, E. B., and Flinn, J. Ghosts in the machine: Interfaces for better power management. In *MOBISYS*, Boston, MA, 2004.

ANJ02   Adve, S. V., Nahrstedt, K., Jones, D. et al. The Illinois grace project: Global resource adaptation through cooperation. In *SHAMAN-02*, New York, 2002.

ATS06   Ashwini, H. S., Thawani, A., and Srikant, Y. N. Middleware for efficient power management in mobile devices. In *Proceedings of the Third International Conference on Mobile Technology, Applications (Mobility '06)*, Bangkok, Thailand, 2006.

ASD03   Acquaviva, A., Simunic, T., Deolalikar, V., and Roy, S. Server controlled power management for wireless portable devices, HP Labs, Technical Report, 2003.

B05   Buttazzo, G. C. Rate monotonic vs. EDF: Judgment day. *Real-Time Systems*, Vol. 29, No.1, 2005, pp. 5–26.

BSM04   Biermann, D., Sirer, E. J., and Manohar, M. A rate matching-based approach to dynamic voltage scaling. In *Proceedings of the First Watson Conference on the Interaction between Architecture, Circuits, and Compilers*, Yorktown Heights, NY, October 2004.

BC00   Blair, G. S., Coulson, G., Andersen, A. et al. A principled approach to supporting adaptation in distributed mobile environments. In *PDSE*, Limerick, Ireland, 2000.

BSO03   Balan, R. K., Satyanarayanan, M., Park, S., and Okoshi, T. Tactics-based remote execution for mobile computing. In *Proceedings of the Third International Conference on Mobile Systems, Applications, and Services (MobiSys)*, San Francisco, CA, 2003.

CY02   Chakraborty, S. and Yau, D. K. Y. Predicting energy consumption of mpeg video playback on hand helds. In *Proceedings of IEEE International Conference on Multimedia and Expo*, Lausanne, Switzerland, August 2002.

C02   Chandra, S. Wireless network interface energy consumption implications of popular streaming formats. In *MMCN*, San Jose, CA, January 2002.

CGK97   Chekuri, C., Goldberg, A. V., Karger, D. R., Levine, M. S., and Stein, C. Experimental study of minimum cut algorithms. In *Symposium on Discrete Algorithms*, New Orleans, LA, 1997, pp. 324–333.

CR00   Chiasserini, C. F. and Rao, R. Energy efficient battery management. In *IEEE Infocom*, Tel-Aviv, Israel, March 2000.

CM09   Chun, B. G. and Maniatis, P. Augmented smartphone applications through clone cloud execution. In *Proceedings of the Eighth Workshop on Hot Topics in Operating Systems (HotOS)*, Monte Verità, Switzerland, 2009.

CBC10   Cuervo, E., Balasubramanian, A., Cho, D. K., Wolman, A., Saroiu, S., Chandra, R., and Bahl, P. MAUI: Making smartphones last longer with code offload. In *ACM MobiSys 2010, Association for Computing Machinery, Inc.*, San Francisco, CA, 2010.

CV02   Chandra, S. and Vahdat, A. Application-specific network management for energy-aware streaming of popular multimedia formats. In *USENIX Annual Technical Conference*, Monterey, CA, June 2002.

DKB95   Douglis, F., Krishnan, P., and Bershad, B. Adaptive disk spin down policies for mobile computers. In *Second USENIX Symposium on Mobile and Location-Independent Computing*, Ann Arbor, MI, April 1995.

DKM94   Douglis, F., Krishnan, P., and Marsh, B. Thwarting the power hungry disk. In *WINTER USENIX Conference*, San Francisco, CA, January 1994.

ED00   Efstratiou, C., Cheverst, K., Davies, N., and Friday, A. Architectural requirements for the effective support of adaptive mobile applications. In *Middleware*, New York, 2000.

E99   Ellis, C. The case for higher level power management. In *Proceedings of HotOS*, Rio Rico, AZ, March 1999.

FFB00   Farkas, K., Flinn, J., Back, G., Grunwald, D., and Anderson, J. Quantifying the energy consumption of a pocket computer and a java virtual machine. In *ACM Conference on Measurement and Modeling of Computer Systems*, Santa Clara, CA, 2000.

FN01   Feeney, L. and Nilsson, M. Investigating the energy consumption of a wireless network interface in an ad hoc networking environment. In *IEEE Infocom*, Anchorage, AK, April 2001.

FLS01   Flinn, J., de Lara, E., Satyanarayanan, M., Wallach, D. S., and Zwaenepoel, W. Reducing the energy usage of cell applications. In *IFIP/ACM International Conference on Distributed Systems Platforms*, Heidelberg, Germany, 2001.

FS96   Feng, W. C. and Sechrest, S. Improving data caching for software mpeg video decompression. In *IS&T/SPIE Digital Video Compression: Algorithms and Technologies*, San Jose, CA, 1996.

FS99   Flinn, J. and Satyanarayanan, M. Powerscope: A tool for profiling the energy usage of mobile applications. In *Proceedings of the Second IEEE Workshop on Mobile Computing Systems and Applications*, New Orleans, LA, 1999.

FPS02   Flinn, J., Park, S., and Satyanarayanan, M. Balancing performance, energy, and quality in pervasive computing. In *Proceedings of the 22nd International Conference on Distributed Computing Systems (ICDCS)*, Vienna, Austria, July 2002.

GBW95   Golding, R., Bosch, P., and Wilkes, J., Idleness is not sloth. In *Proceedings of USENIX Winter Conference*, New Orleans, LA, 1995.

GNM03   Gu, X., Nahrstedt, K., Messer, A., Greenberg, I., and Milojicic, D. Adaptive offloading inference for delivering applications in pervasive computing environments. In *Proceedings of the First IEEE International Conference on Pervasive Computing and Communications*, Fort Worth, TX, 2003.

GRJ09   Giurgiu, I., Riva, O., Juric, D., Krivulev, I., and Alonso, G. Calling the cloud: Enabling mobile phones as interfaces to cloud applications. In *Proceedings of the 10th ACM/IFIP/USENIX International Conference on Middleware (Middleware '09)*, Urbanna, IL, 2009.

HMV05   Huang, Y., Mohapatra, S., and Venkatasubramanian, N. An energy-efficient middleware for supporting multimedia services in mobile grid environments. In *International Conference on Information Technology (ITCC): Coding and Computing*, Las Vegas, NV, 2005.

HKA01   Hughes, C. J., Kaul, P., Adve, S., Jain, R., Park, C., and Srinivasan, J. Variability in the execution of multimedia applications and implications for general-purpose architectures. In *ISCA*, Göteborg, Sweden, 2001.

HSA01   Hughes, C. J., Srinivasan, J., and Adve, S. V. Saving energy with architectural and frequency adaptations for multimedia applications. In *MICRO*, Austin, TX, 2001.

IBM02   IBM and Monta Vista Software. Dynamic power management for embedded systems, http://www.research.ibm.com/arl/projects/dpm.html (November 2002).

IGP95   Imielinski, T., Gupta, M., and Peyyeti, S. Energy efficient data filtering and communications in mobile wireless computing. In *Proceedings of Usenix Symposium on Location Dependent Computing*, Ann Arbor, MI, April 1995.

ISG02   Irani, S., Shukla, S., and Gupta, R. Competitive analysis of dynamic power management strategies for systems with multiple power saving states. In *DATE*, Paris, France, 2002.

JG02   Jejurikar, R. and Gupta, R. Energy aware task scheduling with task synchronization for embedded real time systems. In *CASES*, Grenoble, France, 2002.

KBD08   Kim, M., Banerjee, S., Dutt, N., and Venkatasubramanian, N. Energy-aware co-synthesis of real-time multimedia applications on MPSoCs using heterogeneous scheduling policies. *ACM Transactions on Embedded Computing Systems*, Vol. 7, No. 2, 2008.

KK98   Kravets, R. and Krishnan, P. Application-driven power management for mobile communication. In *Proceedings of MobiCom*, Dallas, TX, 1998.

KHR03   Kremer, U., Hicks, J., and Rehg, J. M. A compilation framework for power and energy management on mobile computers. In *Springer Languages and Compilers for Parallel Computing*, Lecture Notes in Computer Science, Springer-Verlag, Berlin, Germany, 2003.

KWW02   Krintz, C., Wen, Y., and Wolski, R. Application-level prediction of program power dissipation, Technical Report, University of California, San Diego, CA, 2002.

LWX02   Li, Z., Wang, C., and Xu, R. Computation offloading to save energy on handheld devices: A partition scheme. In *Proceedings of International Conference on Compilers, Architectures and Synthesis for Embedded Systems*, Atlanta, GA, November 2002.

LS96   Lorch, J. and Smith, A. J. Reducing processor power consumption by improving processor time management in a single-user operating system. In *Proceedings of MOBICOM 96*, Rye, NY, November 1996.

LS98   Lorch, J. and Smith, A. J. Software strategies for portable computer energy management. In *IEEE Personal Communications Magazine*, Vol. 5, No. 3, June 1998, pp. 60–73.

MZ95   Marsh, B. and Zenel, B. Power measurements of typical notebook computers. In *Proceedings of the USENIX Conference*, New Orleans, LA, January 1995.

MSS03   Martin, T. L., Siewiorek, D. P., Smailagic, A., Bosworth, M., Ettus, M., and Warren, J. A case study of a system-level approach to power-aware computing. *ACM Transactions on Embedded Computing Systems*, Vol. 2, No. 3, 2003, pp. 255–276.

MV03   Mohapatra, S. and Venkatasubramanian, N. PARM: Power-aware reconfigurable middleware. In *ICDCS-23*, Providence, RI, 2003.

MDN07   Mohapatra, S., Dutt, N., Nicolau, A., and Venkatasubramanian, N. DYNAMO: A cross-layer framework for end-to-end QoS and energy optimization in mobile handheld devices, *IEEE Journal on Selected Areas in Communication*, Vol. 25, No. 4, May 2007, pp. 722–737.

NSN97   Noble, B. D., Satyanarayanan, M., Narayanan, D., Tilton, J. E., and Flinn, J. Agile application-aware adaptation for mobility. In *Proceedings of the 16th ACM Symposium on Operating Systems and Principles*, Saint-Malo, France, October 1997.

OH98   Othman, M. and Hailes, S. Power conservation strategy for mobile computers using load sharing. In *Mobile Computing and Communications Review*, Vol. 2, No. 1, January 1998.

PAL09   Paolo, B., Antonio, B., and Luca, F. An IMS-based middleware solution for energy-efficient and cost-effective mobile multimedia services. In *Mobile Wireless Middleware, Operating Systems, and Applications*, Springer, Berlin, Germany, 2009.

PS01   Pillai, P. and Shin, K. G. Real-time dynamic voltage scaling for low-power embedded operating systems. In *Proceedings of the 18th ACM Symposium on Operating Systems Principles*, Chateau Lake Louise, Banff, Canada, 2001.

MZ95   Marsh, B. and Zenel, B. Power measurements of typical notebook computers. In *Proceedings of the USENIX Conference*, New Orleans, LA, January 1995.

RRP99   Rudenko, A., Reiher, P., Popek, G., and Kuenning, G. Portable computer battery power saving using a remote processing framework. In *Mobile Computing Systems and Application Track of the ACM SAC*, San Antonio, TX, February 1999.

SAE05   Schmitz, M. T., AL-hashemi, B. M., and Eles, P. Co-synthesis of energy efficient multimode embedded systems with consideration of mode-execution probabilities. *IEEE Transactions on Computer-Aided Design of Integrated Circuits and Systems*, Vol. 24, No. 2, 2005, pp. 153–169.

SAS01   Schurgers, C., Aberthorne, O., and Srivastava, M. B. Modulation scaling for energy aware communication systems. In *ISLPED*, Huntington Beach, CA, 2001.

SR03   Shenoy, P. and Radkov, P. Proxy-assisted power-friendly streaming to mobile devices. In *MMCN*, Santa Clara, CA, January 2003.

SG01   Shukla, S. and Gupta, R. A model checking approach to evaluating system level power management policies for embedded systems. In *Proceedings of IEEE Conference on High Level Design Validation and Test*, Munich, Germany, 2001.

SK97   Stemm, M. and Katz, R. Measuring and reducing energy consumption of network interfaces in hand-held devices. In *IEICE (Special Issue on Mobile Computing)*, Vol. E80-B, No. 8, August 1997, pp. 1125–1131.

VDM01   Venkatasubramanian, N., Deshpande, M., Mohapatra, S. et al. Design and implementation of a composable reflective middleware framework. In *ICDCS-21*, Phoenix, AZ, 2001.

W04   Wolf, W. The future of multiprocessor systems-on-chips. In *DAC '04: Proceedings of the 41st Annual Conference on Design Automation*, San Diego, CA, 2004.

WKP00   Wang, N., Kircher, M., Parameswaran, K., and Schmidt, D. C. Applying reflective middleware techniques to optimize a QoS-enabled CORBA component model implementation. In *COMPSAC*, Taipei, Taiwan, 2000.

WWD94   Weiser, M., Welch, B., Demers, A., and Shenker, S. Scheduling for reduced CPU energy. In *Symposium on Operating Systems Design and Implementation*, Monterey, CA, 1994.

YN00   Yuan, W. and Nahrstedt, K. A middleware framework coordinating processor/power resource management for multimedia applications. In *IEEE Globecom*, San Francisco, CA, November 2000.

YN02   Yuan, W. and Nahrstedt, K. Integration of dynamic voltage scaling and soft real-time scheduling for open mobile systems. In *Proceedings of 12th International Workshop on Network and Operating Systems Support for Digital Audio and Video (NOSSDAV '02)*, Miami Beach, FL, May 2002.

YN03   Yuan, W. and Nahrstedt, K. Energy-efficient soft real-time CPU scheduling for mobile multimedia systems. In *Proceedings of 19th ACM Symposium on Operating Systems Principles (SOSP '03)*, Bolton Landing, NY, October 2003.

YN04   Yuan, W. and Nahrstedt, K. Practical voltage scaling for mobile multimedia devices. In *Proceedings of ACM Multimedia 2004*, New York, October 2004.

ZEL02   Zeng, H., Ellis, C., Lebeck, A., and Vahdat, A. Ecosystem: Managing energy as a first class operating system resource. In *ASPLOS-02*, San Jose, CA, 2002.

ZEL03   Zeng, H., Ellis, C., Lebeck, A., and Vahdat, A. Currency: Unifying policies for resource management. In *Usenix Annual Technical Conference*, San Antonio, TX, 2003.

11

Energy Efficiency of Voice-over-IP Systems*

Salman A. Baset
IBM T.J. Watson Research Center

Henning Schulzrinne
Columbia University

11.1 Introduction

Voice-over-IP (VoIP) systems are increasingly prevalent in our lives. These systems come in a wide variety of flavors such as desktop-based software applications (e.g., Skype [26]), systems that replace public switched telephony network (PSTN) as the primary line voice service (e.g., Vonage [29]), and more recently, VoIP over smart phones. In November 2010, the number of concurrent VoIP users on Skype exceeded 25 million [2]. A study estimates that as of February 2010, there are approximately 110 million VoIP hard phone subscribers in the world [11]. Another study estimates that the number of mobile VoIP users will exceed 100 million by 2012 [15]. With such a large existing user base of VoIP and expected user growth, and with constantly increasing costs of energy, we ask ourselves what is the energy efficiency of these systems. To answer this question, we gather information about existing VoIP systems and architectures, build energy models for these systems, and evaluate their power consumption and relative energy efficiency through analysis and a series of experiments.

The core function of a VoIP system is to provide mechanisms for storing and locating the network addresses of user agents and for establishing voice and video media sessions, often in the presence of restrictive network address translators (NATs) and firewalls. These systems also provide additional

---

* An earlier version of this chapter titled "How green is IP-telephony?" appeared in the *Proceedings of SIGCOMM 2010 Green Networking Workshop*.

functionality such as voicemail, contact lists (address books), conferencing, and calling circuit-switched (PSTN) and mobile phones. From the perspective of energy efficiency, a VoIP system can broadly be classified according to two criteria: whether it is a primary-line phone service replacing PSTN and whether it uses a client–server (c/s) or a peer-to-peer (P2P) architecture. Vonage [29] and Google Talk [10] are examples of c/s architectures, while Skype [26] is an example of a P2P architecture. Of these, only Vonage is a primary-line phone service replacing PSTN.

We begin the chapter by describing the common configurations of deployed c/s and P2P VoIP systems (Section 11.2). We then devise a simple model for analyzing the energy efficiency of these common configurations (Section 11.3). This model enables a systematic comparison of c/s and P2P configurations of VoIP systems. We then present measurements for the c/s and P2P VoIP components of these systems (Section 11.4), which we apply to the model developed for identifying the sources of energy wastage in these systems and the incurred economic costs (Section 11.5). Based on our analysis, we provide recommendations to improve the energy efficiency of VoIP systems (Section 11.6). Finally, we present the related work (Section 11.7).

11.2 VoIP System Architecture

In this section, we briefly explain the main functionalities of VoIP systems and describe how they are typically implemented in c/s and P2P VoIP systems. We then describe in more detail the architecture of a typical Internet telephony service provider (ITSP), an enterprise VoIP system, a softphone-based c/s VoIP system, and Skype. The first three are representative of a client–server VoIP architecture, and the latter is representative of a P2P VoIP architecture.

11.2.1 Functionalities of a VoIP System

The main functionalities of a VoIP system are:

Signaling: storing and locating the reachable address of the user agents, and routing calls between user agents.

NAT keep-alive: sending and processing user agent traffic to maintain state at NAT devices for receiving incoming requests and calls.

Media relaying: sending VoIP traffic directly between two user agents or through a relay. Relaying is necessary when one or both of the user agents are behind a restrictive NAT or a firewall that prevents establishment of a direct VoIP connection.

Authentication, authorization, accounting: verifying that a user agent is permitted to use the system and tracking usage for billing purposes.

PSTN and mobile connectivity: establishing calls between VoIP clients, and PSTN and mobile phones using managed gateways.

Other services: such as voicemail, contact list storage, video calls, and multiparty audio and video conferencing.

Of the aforementioned services, signaling, NAT keep-alive, and media relaying lend themselves most easily to a P2P implementation. Consequently in the VoIP systems (including Skype) of which we are aware, all but signaling, NAT keep-alive, and media relaying functionality are implemented on centralized servers. As we will see in Section 11.5, the relative energy consumption of c/s and P2P VoIP systems will be determined by the relative efficiency of c/s and P2P implementations of signaling, NAT keep-alive, and media relaying.

11.2.2 Client–Server VoIP Architecture

We consider three types of client–server VoIP systems. The first type is an Internet telephony service provider (ITSP) that provides telephony service to residential and business customers as their primary voice service. The second type is representative of VoIP system deployment in an enterprise. The third type represents softphone-based VoIP systems like Google Talk.

11.2.2.1 Typical ITSP (T-ITSP)

We surveyed three c/s ITSPs in February 2010 to obtain information about their server systems, subscriber populations, and characteristics of the network traffic. Based on this survey, we present an overview of the largest of these whose architecture is typical for an ITSP. We refer to this ITSP as T-ITSP in order to preserve its anonymity.

T-ITSP uses an infrastructure based on open protocols, namely, SIP [20] for signaling and RTP [21] for media. It uses a SIP proxy and registrar implementation based on SIP Express Router (SER) [24]. The SIP registrar stores the reachable address of user agents (phones), whereas the proxy server forwards signaling requests between user agents. Users place calls predominantly through hardware SIP phones. Most such phones are audio-capable only, although some also support video. The vast majority of hardphones are connected to broadband Internet through a broadband modem, which in turn is connected to a home or an office router. The router is typically configured to act as a NAT/firewall. Over 90% of SIP signaling is carried over UDP. User agents connect to SIP servers, perform SIP digest authentication, and register their reachable address every 50 min to receive incoming calls, a process we refer to as a *registration* event.

Because most existing NAT devices maintain UDP bindings for a short period of time [9], hardphones behind NATs need to periodically refresh the binding in order to reliably receive incoming calls. The hardphones achieve this by sending a SIP NOTIFY request [18] every 15 s to the SIP server, which replies with a 200 OK response. While wasteful, this method proved to be the only reliable way of maintaining NAT bindings.

To establish a call, the user agents send the SIP INVITE requests to the SIP proxy servers, which then forward these requests to the destination user agents. The vast majority of hardphones are behind NATs/firewalls and a large proportion of these devices use default settings that prevent them directly exchanging voice packets. Consequently, T-ITSP needs to operate RTP relay servers to relay these calls, thereby consuming additional energy and network bandwidth. T-ITSP also maintains a number of PSTN servers for calling phones in the traditional telephone network. T-ITSP does not encrypt signaling or media traffic. Figure 11.1 illustrates the architecture of T-ITSP.

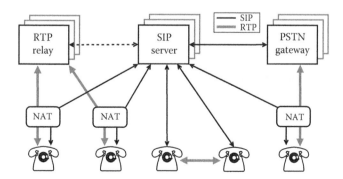

FIGURE 11.1 Client–server Internet telephony service provider (ITSP) architecture.

11.2.2.1.1 Traffic

T-ITSP has a total subscriber base of approximately 100,000 users. The peak call arrival rate is 15 calls per second (CPS) and the systems see no more than 8000 calls at any instant. Approximately 60% (or 4800) of the peak calls are to subscribers within the ITSP; the rest are being routed to PSTN/mobile phones. Hardphones register their network address with T-ITSPs SIP registrar every 50 min and send a SIP NOTIFY message every 15 s to maintain the NAT binding. For 100,000 k subscribers, these statistics imply that the SIP registrar needs to process 33 registration events and 6667 NOTIFY events per second. In Section 11.4, we extrapolate these peak numbers for a large subscriber base.

11.2.2.2 Enterprise VoIP Systems

The enterprise VoIP system comprises of SIP proxy and registrar servers, hardphones, and enterprise Ethernet switches for connecting hardphones to the proxy server. In addition to the VoIP phones, office computers are also connected to the same Ethernet switch. In some installations, the enterprise switches also provide power to the hardphones through Power-over-Ethernet (PoE) [17]. The enterprise VoIP system is connected to the other VoIP, PSTN, or mobile telephony systems through gateways. Typically, the IP address space in an enterprise is flat and the NAT devices are sporadic. Consequently, unlike T-ITSP, the hardphones do not need to periodically send SIP NOTIFY messages to keep the NAT bindings. Further, the enterprise VoIP system does not need to maintain media relay servers. When the IP address space is not flat, the VoIP systems in different departments are typically connected via gateways or call managers [5].

11.2.2.3 Softphone-Based VoIP Systems

Softphone-based client–server VoIP systems such as Google Talk are similar in their functionality to T-ITSP, except that the user agents mostly run as a software application on a desktop or a mobile device. Such systems typically do not replace PSTN as the primary phone service.

11.2.3 P2P VoIP Architecture: Skype

In this section we present an overview of Skype [26], which is representative of a P2P VoIP system. Skype is not advertised as a primary-line phone service. There are two types of nodes in a Skype network, super nodes and ordinary nodes. The super nodes form the Skype overlay network, with ordinary nodes connecting to one or more super nodes. Super nodes, which are chosen for their unrestricted connectivity and high-bandwidth, are responsible for signaling, NAT keep-alive, and media relaying. Skype encrypts signaling and media traffic to prevent super nodes from eavesdropping. Skype-managed servers provide functionality for authentication, contact list and voicemail storage, and calling PSTN and mobile phones. Figure 11.2 shows an illustration of a P2P VoIP system. Table 11.1 compares the distributed and centralized features of the T-ITSP, enterprise VoIP systems, Google Talk, and Skype.

11.3 Power Consumption Model

We present a model for understanding the power consumption of c/s and P2P VoIP system architectures. We focus on signaling, NAT traversal, and media relaying as they are accomplished using managed servers in the c/s but through super nodes in P2P VoIP systems. Let N be the total number of online subscribers of a VoIP system and let λ_{INV} be the peak rate of calls per second these subscribers make and d be the average call duration. These calls are either to other subscribers of the VoIP provider or to PSTN or mobile phones. Let p_v be the percentage of VoIP calls. Of these, let p_{relay} be the proportion of calls that need a relay.

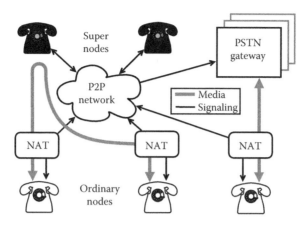

FIGURE 11.2 P2P VoIP architecture.

TABLE 11.1 Comparison of T-ITSP, Enterprise VoIP, Google Talk, and Skype Features

Feature	T-ITSP	Enterprise	Google Talk	Skype
User agents (UA)	Hardphone	Hardphone/softphone	Softphone	Softphone
UAs always on	Yes	Yes	No	No
Signaling	Centralized	Centralized	Centralized	P2P + centralized
NAT keep-alive	Centralized	None	Centralized	P2P
Media relaying	Centralized	None	Centralized	P2P
PSTN connectivity	Centralized	Centralized	Centralized	Centralized
Voicemail	Centralized	Centralized	Centralized	Centralized
Contact list	Centralized	Centralized	Centralized	Centralized

The value of "None" in the Enterprise column indicates that the user agents typically do not send NAT keep-alives, nor do they require media relays for establishing calls with user agents within the same enterprise.

11.3.1 Client–Server

As discussed in Section 11.2.2.1, a c/s VoIP architecture has dedicated servers for handling the signaling, NAT traversal, and media relaying traffic. Signaling traffic includes registration of user agent network addresses with the SIP registrar and call signaling for establishing media sessions. Let λ_{REG} and λ_{INV} denote the peak number of SIP registration events and calls per seconds, respectively, that N user agents generate. The NAT traversal traffic (SIP NOTIFY in T-ITSP) is sent by the user agents to refresh NAT bindings and ensuring reliable receipt of incoming calls. Let λ_{NAT} be the rate of these NAT traversal messages per second. λ_{NAT} will be significantly lower for signaling over TCP than over UDP. In most c/s VoIP systems, signaling and NAT traversal are handled on separate servers from those of media relaying.

Let $S(\lambda_{REG}, \lambda_{INV}, \lambda_{NAT}, PROTO)$ represent the number of signaling servers needed to handle the peak signaling and NAT traversal load under a particular transport protocol *PROTO*. The *PROTO* may be UDP, TCP, or TLS. An advantage of using permanent TCP connections between user agents and SIP servers is that it reduces the frequency of the traffic to maintain NAT bindings. However, maintaining hundreds of thousands of TCP or TLS connections on a server is costly in terms of the memory needed [23]. Let $M(\lambda_{INV}, d, p_v, p_{relay})$ represent the number of media relay servers needed to relay calls. Let w_s and w_m denote, respectively, the wattage consumed by signaling and media servers at the peak load. Let c be the system's PUE and r_s and r_m be the redundancy factor used for signaling and media servers. Then the power consumed by the signaling and media relay servers is given as follows:

$$w_{c/s} = (Sw_s r_s + Mw_m r_m)c \tag{11.1}$$

11.3.2 Peer-to-Peer

Recall from Section 11.2.3 that there are two types of nodes in a P2P communication system, namely, super nodes that forward signaling and routing traffic from other nodes and relay a call between nodes with restrictive network capacity, and second, ordinary nodes that do not participate in the overlay routing and connect to one or more super nodes. Let N_S be the number of super nodes in the P2P system with a total population of N subscribers. In contrast to c/s systems, where it is easy to attribute the energy consumption of signaling, NAT traversal and relaying, it is non-trivial to do so for super nodes in P2P systems. We consider two reasonable accounting strategies that apply as well to energy accounting on phones and network devices:

- *Delta*: count only the additional power drawn by the signaling and relaying functions of the super node machine above that of the baseline power consumption of the machine.
- *Prop*: in addition to delta, attribute to P2P VoIP a fraction of the system baseline power consumption that is proportional to the time the CPU is woken up to handle signaling, NAT traversal, and media relaying traffic.

For simplicity, assume that each super node sends and receives λ_{MAINT} messages to maintain the overlay, and receives $1/N_S$ of the total registration, call invites, and NAT traversal. Each super node relays at maximum one call at a time. A node may use a secure transport protocol such as TLS or DTLS for non-media relaying traffic. Let w_{base} denote the baseline wattage drawn by the super node machine. Let w_Δ denote the wattage drawn by the overlay maintenance, registration, signaling, NAT traversal, and media relaying functionality. Let p be the proportion of time the CPU is woken up to serve super node requests if *prop* accounting policy is chosen or zero for the *delta* policy. Then the power consumed by P2P super nodes is

$$w_{P2P} = (w_\Delta + w_{base}p)N_S \qquad (11.2)$$

11.3.3 Comparison Issues in C/S and P2P VoIP Systems

In this section, we highlight the broader issues in comparing c/s and P2P VoIP systems.

11.3.3.1 PSTN Replacement

The most important consideration for our comparison is whether the VoIP system replaces the always-on PSTN system (e.g., Vonage). For such a system, the user agents must always be reachable (or powered on) to receive incoming calls. The total energy consumed by such systems is the sum total of the energy consumed by always-on user agents and servers, if any.

In contrast, systems like Google Talk and Skype run as a software application on a desktop, laptop, or a mobile device and do not replace PSTN as the primary-line voice service. Therefore, it is not possible to directly compare them with systems like Vonage. Moreover, Google Talk uses a c/s architecture whereas Skype uses a P2P architecture. When comparing these two architectures, it is important that we examine the power consumed by the machines providing the core functionality (servers in c/s, super nodes in P2P) and not the difference in energy consumed by the user agents.

11.3.3.2 Network Costs

C/S and P2P communication systems have a different network footprint as in the latter, nodes have to exchange data to maintain the P2P network. Edge and core routers likely incur an energy cost for forwarding traffic for P2P and c/s communication systems. However, these costs are harder to quantify as the edge and core routers are always on. Although an analysis similar to [16] can be used, we focus on quantifying the energy usage of the system itself and not the network. However, we do incorporate the energy costs of broadband modems and network switches to which VoIP user agents are directly connected and that otherwise cannot be powered down without disconnecting the user agent.

11.4 Measurements and Results

In this section, we describe a set of experiments for measuring the power consumption of signaling and media relay servers, broadband modems and home routers, enterprise Ethernet switches, user agents (hardphones and softphones), and Skype super nodes. Our power measurements were taken using a Watts-up .NET power meter [31]. The meter provides 0.1 W precision and claims an accuracy of 1.5% of the measured value.

11.4.1 Signaling and Media Relay Servers

Based on the architecture and load information of T-ITSP, we set up a test bed consisting of two servers, the first for handling signaling and NAT traversal workload and the other for handling media relaying. Our goal was to measure the power consumption of these servers under peak load, and extrapolate the number of servers needed and the power consumed based on peak workload, using the model developed in Section 11.3.1. Although, this extrapolation may be considered an over simplification, it still provides useful insights into the energy consumed by large-scale c/s VoIP systems.

11.4.1.1 Test Bed Overview

In our test bed, the SIP server machine was a Dell PowerEdge 1900 server [7] with two quad-core 2.33 GHz Intel Xeon X5345 processors and 4 GB of memory. It was connected to load generators with two Intel 82545GM Gigabit Ethernet controllers. The machine had six fans. It ran Debian Squeeze (snapshot from February 26, 2010) with Linux kernel 2.6.32. We installed the latest version of SIP-Router, an open source SIP server [24], on the machine and configured it with all the features an ITSP operating in the public Internet would need to use. The SIP server was configured to use 2.5 GB of memory and 16 processes (two per core). We used MySQL 5.1.41–3 (from a Debian package) configured with 2 GB of query cache. We used SIPp [25] version 3.1.r590-1 to generate SIP traffic according to the model described in Section 11.2.2.1.

For RTP relay tests, we used an IBM HS22 blade server [12] with five blades installed. One of the blades was used as an RTP relay server; the remaining four blades and two desktop-class PCs were used as RTP load generators. Each blade had two Intel Xeon quad-core CPUs running at 2.9 GHz and a 10 GigE Intel NIC with multiple hardware transmission and receive queues and ran a Linux 2.6.31 kernel. We used the latest version of iptrtpproxy [13], a kernel-level RTP relay. The software relays RTP packets using iptables rules. We used a modified version of SEMS [22] to generate a large number of simultaneous RTP sessions.

11.4.1.2 SIP Server Measurements

We performed a number of measurements to figure out the maximum number of subscribers that our SIP server can support. We wanted to determine the maximum load on this server in three configurations: (1) signaling and NAT keep-alive (SIP NOTIFY) traffic carried over UDP as described in Section 11.2.2.1, (2) signaling traffic over UDP but without any SIP NOTIFY traffic, (3) signaling traffic over permanent TLS connections. The first configuration allowed us to reason about the maximum ITSP-like workload a server can handle. The second configuration provided insights into peak ITSP-like signaling workload a server can handle, assuming there were no NATs. The third configuration was helpful from the perspective of comparing T-ITSP to Skype, as Skype uses a TLS-like protocol to encrypt signaling and media traffic.

Before running any tests, we provisioned the database of the SIP server with 1 million unique subscribers. The baseline consumption of the server was 160 W. The machine had six fans; each fan consumed 10 W when running at full speed. The power consumption when all fans were removed and the machine was idle was 145 W. To see how CPUs contributed to the overall power consumption of the machine, we ran 8 cpuburn [6] processes (one per core). The machine consumed 332 W when all cores were fully utilized.

For the first configuration, we found out that our server could handle T-ITSPs traffic mix for approximately half a million users. Under this load, the number of calls (λ_{INV}), registrations (λ_{REG}), and NAT keep-alives (λ_{NAT}) events per second were 75 k, 166 k, 33 k, respectively, and the server consumed (w_s) 210 W. For the second configuration, in which there was no NAT traversal traffic, we found that our server could handle load for approximately 1 million subscribers. w_s was 190 W.

For the third configuration (signaling over TLS) there was no need to exchange frequent keep-alive messages over TCP connections to keep NAT bindings open, so λ_{NAT} was 0. With SIP over TLS, the SIP server used 61 kB of memory per connection and one connection was needed per user agent. Consequently, memory became our bottleneck and a maximum of 43 k simultaneously connected user agents could be supported on a single SIP server. w_s was 209 W.

Based on these measurements, we extrapolate the number of servers needed for these configurations in Table 11.2. Compared to the first configuration, observe that eliminating the keep-alive traffic reduces the number of servers by half in the second configuration. Although the number of signaling servers needed for the third configuration increases approximately by a factor of 12 as compared to the first configuration, we believe that such limitation can be addressed by (1) tuning the SSL buffer, (2) increasing memory in our server, and (3) using hardware SSL accelerators.

11.4.1.3 Media Relay Server

We managed to saturate the IBM blade with 15,000 simultaneous calls. Each call had a bit rate of 64 kb/s for an aggregate bit rate of 960 Mbit/s. At this rate, the resource bottleneck appeared to be a single CPU core overloaded by the ksoftirqd kernel thread. It is likely that even greater call volumes could be relayed by optimizing the multi-core scheduling of this machine using techniques such as [8]. At this workload, the media relay server consumed approximately 240 W (w_m). In Table 11.3, we extrapolate the number of relay servers needed as a function of user population and the number of calls that need relaying.

11.4.2 Broadband Modems, Middleboxes, and Ethernet Switches

A typical residential broadband user is connected to the Internet through a home router (Ethernet switch + WiFi), which in turn is connected to the broadband modem (cable, DSL, or fiber). Our measurements indicate that the recent models of WiFi routers with four Ethernet switches consume, on average, 4–6 W of power. Similarly, a broadband modem also consumes 4–6 W of power. In our calculations, we use 5 W as an estimate for broadband modem and home router power consumption.

TABLE 11.2   Signaling Servers Needed by Configuration

Transport	NAT Keep-Alive	100 k	1 M	10 M	100 M
UDP	YES NOTIFY/s	1	2	20	200
UDP	NO	1	1	10	100
TLS	NO	3	25	250	2500

TABLE 11.3   Media Servers Needed When Relayed Calls Are 0%, 30%, and 100% of ITSP-ITSP Calls

% Relayed Calls	100 k	1 M	10 M	100 M
0	0	0	0	0
30	1	2	10	96
100	1	4	32	320

In an enterprise, the VoIP hardphones are connected to an Ethernet switch that is typically PoE enabled. A 48 port Cisco switch model C2960S-48LPD-L consumes 70 W of power at 5% throughput [4] and has 370 W of available PoE power or 7.70 W per port.

11.4.3 User Agents

We performed measurements to determine the power consumption for a variety of user agents that included hardware SIP phones and softphones. We also performed power measurements for Skype super nodes.

11.4.3.1 Hardware SIP Phones

For a variety of SIP-based hardphones, we found that phones consume between 3 and 6 W of power. We also observed that the phone power consumption does not change when the user is in a voice call.

11.4.3.2 Softphones

We used Skype and Google Talk as representative of softphones. For several desktop machines running Windows XP and Windows 7, we did not observe any discernible change in the machine baseline power consumption when Skype and Google Talk were idle. The non-discernible change in the power draw when these softphones were idle is partially attributed to the power meter we used, which can only measure power up to a tenth of a watt with an accuracy of 1.5%. When placing a voice call, we found that on average Skype and Google Talk consumes between 6 and 8 W on a Windows XP and Windows 7 desktop machine. Similarly, for a video call, Skype and Google Talk consumed between 10 and 20 W. For laptop machines running Windows XP and Max OS X, we found that Skype and Google Talk, on average, consumed between 1 and 2 W when placing a voice call. As with the desktop machines, Skype and Google Talk did not cause any discernible power increase when idling. We observed similar power draw behavior for other SIP-based software clients.

11.4.3.3 Skype's Energy Consumption as a Super Node

Measuring Skype's energy consumption as a super node is not straightforward. First, we need a machine to transition to super node status. Since the Skype client itself decides whether to become a super node, we can only encourage this decision to be made by ensuring that the node has a public IP address, has sufficient bandwidth, and is lightly loaded, which we desired anyway given that we were trying to isolate what we assumed Skype's relatively low power consumption amid the noise of the machine's hardware and operating system. To this end, we ran a Skype client for a few hours on a machine with a public IP address and good network connectivity. To determine if the Skype is relaying a call, we performed measurements using a traffic sniffer running on another machine that was connected to the same hub as the Skype machine. We assumed a call is being relayed if the bit-rate was above a threshold [27]. Although, our meter readings indicated that there was a nonzero power increase, the difference measured was smaller than the measurement error reported by the power meter. Determining when a super node is handling signaling traffic is even harder, and the power draw per event lasts for a shorter interval and is likely smaller in magnitude. We did find that the machine can go to sleep when Skype is acting as a super node and relaying the call. The calls were either dropped or transferred to another relay; however, it is impossible for us to ascertain the status of those calls due to the closed nature of the Skype network.

11.5 Discussion

Our model and measurements allow us to answer the following questions: (1) What is the total energy consumed by a VoIP system that may or may not replace PSTN as the primary line phone service?

(2) Where is energy consumed in such a system? (3) Are P2P VoIP systems more energy efficient than c/s?

To answer questions (1) and (2), we consider T-ITSP (Section 11.2.2.1), enterprise (Section 11.2.2.2), and softphone-based c/s VoIP deployments (Section 11.2.2.3). Recall that for the T-ITSP workload that includes signaling and NAT keep-alive traffic over UDP, our SIP server can handle this workload for 500,000 subscribers, and consumes 209 W (w_S) under peak load. The RTP relay server under test consumed 240 W (w_M) and can relay 15,000 calls, with each call having a bit-rate of 64 kb/s. The number of active calls in the system for 500 k users is 24 k (by extrapolating the number of active calls for 100 T-ITSP users), requiring two relay servers to handle this load (one server can handle 15 k calls). Depending on the actual deployment, not all calls need relaying. Our conversations with various VoIP system providers suggest that using NAT traversal techniques like ICE [19] will likely bring down the relayed sessions under 30%. When relaying 30% of the 24 k calls, only one relay server is needed. We compute $w_{c/s}$ for both 100% and 30% relaying using our c/s model (Equation 11.1). We plug c (PUE) as 2, and $r_S = 1$ and $r_M = 1$ in our model. For 100% and 30% relaying, the computed w_S is 1.378 and 0.89 kW, respectively. Observe that these numbers are approximate for the peak load and will be higher if the servers are underutilized.

Table 11.4 shows the energy consumed in kilowatts for running the servers, broadband modems, home routers, and hardphones. Based on our measurements, we assign 5 W for running the broadband modem and 5 W for the WiFi router with four Ethernet ports. These numbers will be higher for a WiFi router with more than four ports. Nevertheless, the energy consumed by these devices cannot be solely attributed to VoIP because both VoIP and non-VoIP traffic share the same router. A reasonable assumption is that on average, such sharing occurs only for 12 h in a day. The rest of the time, these devices must remain powered on so that a VoIP user can receive incoming calls. Using this conservative assumption, we calculate the approximate power required to run a 100 million VoIP system to be 1000.129 MW. The number is calculated by using plugging 500 MW for phones, 500 MW for broadband modems and home routers (discounted by 50% because of our usage assumption), and 129.68 kW for running servers. The monthly cost of running such a system, at 11 cents per kWh [1] is 79.2 million dollars or 80 cents per user per month (rounded up). The energy cost per month of running the servers is $10270 or less than one thousandth of a cent per user per month.

In enterprise VoIP systems, there are typically minimal or no NATs. Consequently, the hardphones do not need to send SIP NOTIFY packets to the SIP proxy server for keeping the NAT bindings alive nor will they likely require any media relay servers. However, VoIP hardphones must be connected to the Ethernet switches. A 48-port PoE-enabled Ethernet switch when connected with hardphones that require 5 W per phone consumes 310 W. For an organization with 100,000 hardphones, the total number of such switches needed are at least 2084. If only one half of the ports in each switch are used for VoIP phones and the rest for non-VoIP usages such as the Internet, then the number of switches increases to 4168. Assuming that switches solely serve VoIP traffic for one half of the day (ignoring idle time on weekends and holidays), the monthly power consumption and economic cost of an enterprise system with 100,000

TABLE 11.4 T-ITSP Energy Consumption (kW) as a Function of Number of Users

Users	10 k	100 k	1 M	10 M	100 M
Servers (NATs)	0.90	0.90	1.78	13.16	129.68
Servers (no NATs)	0.42	0.42	0.84	4.20	40.20
Broadband modems	50	500	5,000	50,000	500,000
Home routers	50	500	5,000	50,000	500,000
Hardphones	50	500	5,000	50,000	500,000

The wattage for servers includes the PUE factor "c" of two.

users is approximately 465,033 kWh and $51,153, respectively. The latter number when rounded up is 52 cents per user per month.

These results indicate always on VoIP phones are a major source of energy waste in T-ITSP and enterprise VoIP systems. Further, the always-on broadband modems, home routers, and enterprise switches significantly add to the energy bill. In contrast, the servers only consume a tiny fraction ($<0.02\%$) of the total power consumed by a VoIP system replacing PSTN. Table 11.4 also illustrates that restrictive NATs and firewalls are wasteful in terms of server power consumption as they increase the total energy consumption of servers by a factor of two and three for number of users below and above one million, respectively.

For softphone-based c/s systems such as Google Talk that do not replace PSTN as the primary-line phone service, their servers incur the same server energy usage for an equivalent load as for the servers in VoIP systems that replace PSTN. However, the softphone energy consumption is harder to quantify in these systems. This is because the softphones typically run on PCs that are powered on anyway. If the softphones consume a small fraction of the power consumed by the PC, it is likely that they will still dominate the total power consumption of such a system; however, the relative power fraction of servers will increase. On the other hand, if the users leave their PCs powered on solely for the purpose of receiving calls (such as magicJack [14]), then the power consumption of running these softphones will be much higher than hardphones, making such systems highly energy inefficient. As such, a user study is needed to determine how long the users keep their PCs idle but powered on for receiving incoming calls.

To answer the third question whether P2P system is more energy efficient than c/s or vice versa, we note this will only hold if the power consumed by all the super nodes assuming a delta accounting policy is less than the total power consumed by the servers in c/s systems, i.e.,

$$w_\Delta N_S < w_{c/s} \qquad (11.3)$$

Observe that this equation does not include the power consumed by user agents, broadband modems, home routers, or Ethernet switches because we assume that they consume the same amount of power in c/s and P2P VoIP systems. To solve (11.3) for w_Δ, we need to estimate the total number of super nodes in the system that can process signaling, NAT keep-alive, and media relaying traffic. We estimate the number of super nodes to be 1% of the total user population, meaning that in a population of 500 k user agents, 5 k are super nodes. This assumption is reasonable since if 30% of the 24,000 active calls (7,200) need a relay, a super node roughly relays one complete call at any instant. Further, in Skype, a super node does not relay more than one call at any instant. Thus, the power consumption per super node, w_Δ, is $0.89k/5k = 0.178$ W in order for c/s and P2P systems to be equivalent in terms of energy efficiency. When the servers are underutilized, say 50%, w_Δ is twice its original value (0.356 W). The small value of w_Δ suggests that if the super nodes were to consume more power than this value in order to handle the signaling, NAT keep-alives, and media relaying workload, a P2P system using super nodes will become energy inefficient as compared to a c/s VoIP system.

Due to the low precision of our power meter, we are not able to ascertain if Skype super node and relaying power consumption is close to w_Δ. However, we speculate that the power consumed by super nodes and relays running on desktop machines may likely be close to the w_Δ calculated earlier. The reason is that the CPU of a relatively unloaded machine running a Skype super node or relay may be woken often to service these requests, thus incurring the small power draw to cause it to go above w_Δ. On the contrary, handling an additional job on a loaded server causes almost no additional CPU wakeups.

The analysis reveals that in a VoIP system replacing PSTN, hardphones and switching equipment consume 99.98% of the total energy consumed by the VoIP system. Thus, in order to make VoIP system more energy efficient, we need to take advantage of techniques that allow powering down these devices when idle. In Section 11.6, we discuss the use of these techniques.

11.6 Recommendations for Reducing Power Consumption of VoIP Systems

In this section, we discuss using a number of existing techniques that can potentially reduce the energy consumption of hardphones, switches and middleboxes, and servers when these devices are idle. As a result, the devices in a VoIP system will potentially only draw power when making or receiving VoIP calls. Observe that unlike cloud-based systems where services can be aggregated on a smaller number of servers to improve utilization and reduce energy wastage, it is not possible do so in a VoIP system. The reason is simple: the users want to receive and make calls through their telephones and it is simply not possible to aggregate phones similar to aggregating compute jobs on a server.

Our analysis showed that hardphones, broadband modems, home routers, and enterprise switches comprise the biggest chunk of the total energy consumed in a VoIP system. To reduce the energy consumption of hardphones, the various components of the phone including LCD display, processor, and Ethernet jack should be powered down when not in active use. The former two can be accomplished by turning off the LCD display and by making use of energy-efficient processors, whereas the latter can be accomplished using energy-efficient Ethernet [3]. If the phones were only used for 8 h a day and were powered down or ran on minimal power (<0.1 W) during the remaining 16 h, it will bring down the per user per month energy bill from 79 cents to 53 cents in T-ITSP-like systems, and from 52 cents to 32 cents in enterprise VoIP systems.

In T-ITSP-like systems, the hardphones must send keep-alive messages over UDP every 15 s to keep the NAT bindings alive. Such wasteful traffic prevents the phones and home routers from taking advantage of any sleep modes available on the device. To eliminate such wasteful traffic, the phones can establish a permanent TCP connection with the SIP server. Further, the ISPs can setup a SIP phone on the broadband modem, which is typically not behind an NAT device. When the SIP user agent on the broadband modem receives an incoming call, it can wake up the home router and the phone using techniques such as Wake-on-LAN [30] to receive the incoming call. This technique can further bring down the per user per month energy bill for running VoIP phones.

Our analysis also indicated the number of servers needed to support a large VoIP user base is fairly small; one SIP server can handle registration events and NAT keep-alive traffic for 500,000 users, and RTP relay server can relay calls for 15,000 calls. By setting up the VoIP user agents on cable modems, the NAT keep-alive traffic can potentially be eliminated. By using advanced NAT traversal techniques, such as ICE [19] to allow user agents to detect network conditions, the use of RTP relay server can be further minimized. Together, these techniques can significantly reduce the power consumption of VoIP servers.

11.7 Related Work

Nedevschi et al. [16] have developed models describing the relative power efficiency of c/s and P2P architectures for generalized network applications (e.g., file-sharing), and conclude that P2P approaches use system energy more efficiently than the c/s ones. Similarly, Valancius et al. [28] argue that building P2P nano data centers on the Internet gateway devices provides energy savings over traditional centralized data centers. In both papers, the energy savings argument boils down to data center servers (1) needing cooling, network, and other overheads measured by a multiplicative factor called *power utilization efficiency* (PUE), and (2) having significant baseline power consumption (i.e., power consumption when idling). Typical data center PUEs range from 1.2 to 2, while the PUE of a peer is 1 (e.g., home air-conditioning is already running) and peers are on anyway, so processes running on peers escape this baseline cost.

We examined the relative energy efficiency of c/s and P2P VoIP systems, and found, intriguingly, that the energy consumption of a peer does not need to be very large in order for a P2P architecture to be *less* energy efficient than a c/s one.

11.8 Conclusion

We identified the key components that are implemented on servers in a c/s VoIP system and by super nodes in a P2P VoIP system. We presented a model for understanding power consumption of c/s and P2P VoIP systems. We performed a number of experiments to determine the power consumption of different components of c/s and P2P VoIP systems. Our model, analysis, and measurements indicate that for VoIP systems used as a replacement for always-on PSTN system, the power consumed by hardphones and connected network devices (broadband modems, home routers, and enterprise switches) overwhelmingly dominate the total power consumed by the VoIP system and the per user per month cost is less than a dollar in such systems. Moreover, when comparing c/s and P2P VoIP systems, our results show that even when super nodes consume relatively small power for system operation, the P2P VoIP system can be less energy efficient than a c/s VoIP system. Further, we demonstrated the presence of NATs as the main obstacle to building energy-efficient VoIP systems.

References

1. Average retail price of electricity to ultimate customers by end-use sector, by State (URL). http://www.eia.doe.gov/electricity/epm/table5_6_a.html (accessed November 2010).
2. Celebrating 25 million concurrent users (URL). http://blogs.skype.com/en/2010/11/25_million.html (accessed November 2010).
3. K. Christensen, P. Reviriego, B. Nordman, M. Bennett, M. Mostowfi, and J. A. Maestro. IEEE 802.3az: The road to energy efficient ethernet. *Communications Magazine, IEEE*, 48(11):50–56, November 2010.
4. Cisco Catalyst 2960-S and 2960 Series Switches with LAN Base Software (URL). http://www.cisco.com/en/US/prod/collateral/switches/ps5718/ps6406/product_data_sheet0900aecd80322c0c.html (accessed November 2010).
5. Cisco Unified Communications Manager (CallManager) (URL). http://www.cisco.com/en/US/products/sw/voicesw/ps556/index.html (accessed November 2010).
6. CPU burn (URL). http://pages.sbcglobal.net/redelm/ (accessed June 2010).
7. Dell Power Edge 1900 Server (URL). http://www.dell.com/downloads/emea/products/pedge/en/PE1900_Spec_Sheet_Quad.pdf (accessed June 2010).
8. M. Dobrescu, N. Egi, K. Argyraki, B.-G. Chun, K. Fall, G. Iannaccone, A. Knies, M. Manesh, and S. Ratnasamy. RouteBricks: Exploiting parallelism to scale software routers. In *Proceedings of SOSP*, Big Sky, MT, October 2009.
9. B. Ford, P. Srisuresh, and D. Kegel. Peer-to-peer communication across network address translators. In *Proceedings of the USENIX Annual Technical Conference*, Anaheim, CA, April 2005.
10. Google Talk. http://www.google.com/talk/ (accessed June 2010).
11. Hard VoIP users top 100 million (URL). http://www.theinquirer.net/inquirer/news/1593216/hard-voip-users-100-million (accessed November 2010).
12. IBM Blade (URL). http://ibm.com/systems/bladecenter/ (accessed June 2010).
13. iptrtpproxy (URL). http://www.2p.cz/en/netfilter_rtp_proxy/iptrtpproxy (accessed June 2010).
14. magicJack (URL). http://www.magicjack.com/6/index.asp (accessed November 2010).
15. Mobile VoIP users to exceed 100 million by 2012 (URL). http://juniperresearch.com/viewpressrelease.php?pr=187 (accessed November 2010).

16. S. Nedevschi, J. Padhye, and S. Ratnasamy. Hot data centers vs. cool peers. In *Proceedings of the USENIX HotPower Workshop*, San Diego, CA, December 2008.

17. Power over Ethernet (PoE) (URL). http://en.wikipedia.org/wiki/Power_over_Ethernet, (accessed November 2010).

18. A. Roach. Session initiation protocol (SIP)-specific event notification. RFC 3265, June 2002.

19. J. Rosenberg. Interactive connectivity establishment (ICE). RFC 5245, April 2010.

20. J. Rosenberg, H. Schulzrinne, G. Camarillo, A. Johnston, J. Peterson, R. Sparks, M. Handley, and E. Schooler. SIP: Session initiation protocol. RFC 3261, June 2002.

21. H. Schulzrinne, S. L. Casner, R. Frederick, and V. Jacobson. RTP: A transport protocol for real-time applications. RFC 3550, July 2003.

22. SEMS (URL). http://iptel.org/sems (accessed June 2010).

23. C. Shen, E. Nahum, H. Schulzrinne, and C. Wright. The impact of TLS on SIP server performance. In *Proceedings of the IPTCOMM*, Munich, Germany, August 2010.

24. SIP Router Project (URL). http://sip-router.org/ (accessed June 2010).

25. SIPp (URL). http://sipp.sourceforge.net/ (accessed June 2010).

26. Skype (URL). http://www.skype.com/, (accessed June 2010).

27. K. Suh, D. R. Figuieredo, J. Kurose, and D. Towsley. Characterizing and detecting relayed traffic: A case study using Skype. In *Proceedings of the IEEE INFOCOM*, Barcelona, Spain, April 2006.

28. V. Valancius, N. Laoutaris, L. Massoulie, C. Diot, and P. Rodriguez. Greening the Internet with nano data centers. In *Proceedings of the CoNEXT*, Rome, Italy, December 2009.

29. Vonage (URL). http://www.vonage.com/ (accessed June 2010).

30. Wake-on-LAN (URL). http://en.wikipedia.org/wiki/Wake-on-LAN (accessed June 2010).

31. Watts up .NET power meter (URL). https://www.wattsupmeters.com/ (accessed June 2010).

12

Intelligent Energy-Aware Networks

Y. Audzevich
University of Cambridge

A. Moore
University of Cambridge

A. Rice
University of Cambridge

R. Sohan
University of Cambridge

S. Timotheou
University of Cambridge

J. Crowcroft
University of Cambridge

S. Akoush
University of Cambridge

A. Hopper
University of Cambridge

A. Wonfor
University of Cambridge

H. Wang
University of Cambridge

R. Penty
University of Cambridge

I. White
University of Cambridge

X. Dong
University of Leeds

T. El-Gorashi
University of Leeds

J. Elmirghani
University of Leeds

12.1 Introduction

Energy-efficient processes are increasingly key priorities for information and communication technology (ICT) companies with attention being paid to both ecological and economic drivers. Although in some cases the use of ICT can be beneficial to the environment (e.g., by reducing travel and introducing more efficient business processes), countries are becoming increasingly aware of the very large growth in energy consumption of telecommunication companies. In particular, the predicted future growth in the number of connected devices and the Internet bandwidth of an order of magnitude or two is not practical if it leads to a corresponding growth in energy consumption. Regulations may therefore come soon, particularly if governments mandate increasing moves toward carbon neutrality, as has already begun to occur in the public sector in British Columbia and New Zealand. Indeed the United Kingdom has highlighted this as a priority [1]. This represents a significant departure from accepted practices where ICT services are provided to meet the growing demand, with little regard for the energy consequences of relative location of supply and demand. It also departs from existing attempts to constrain energy demand as it has been shown that increased equipment power efficiency leads to more consumption, the Khazzoom–Brookes postulate [2].

Internet power usage has continued to increase over the past decade due to (i) an ever increasing number of connected devices, (ii) higher per device power consumption, and (iii) growing device usage (per day) [3]. Drivers for future bandwidth expansion and power demand include streaming and video on demand, cloud computing, and proposals to locate user computing and data within the network. New services such as Internet Protocol Television (IPTV) will increase capacity demand in the backbone, meaning that the power consumption in the backbone will become more important relative to the access network. Internet traffic growth of between 40% and 300% is expected in the next 3 years [4], fueled mainly by these broadband and video applications. In addition to the ecological impact, the economic impact is compounded by the increasing cost of electricity, which is expected to rise by 200% over the next 7–8 years [4].

A report issued by the Ministry of Internal Affairs and Communications in Japan concluded that ICT equipment [routers, servers, PCs, and network systems] consumed 4% of the total electricity generated in Japan in 2006, a figure of 45 TWh. Over the past 5 years, the figure has grown by more than 20% [5]. Similar trends are observed in Europe. In Italy, Telecom Italia is the second largest consumer of electricity (2 TWh/year) after railways, consuming about 1% of the total country demand growing by 8% and 12% with respect to 2005 and 2004, respectively [4,6]. British Telecom (BT) used 0.7% of the total U.K. energy in 2007 making it the largest single energy consumer in the nation [7]. In 2007, about 10% of the total U.K. power consumption was estimated to be related to ICT equipment [8] and ICT has been observed to produce more greenhouse gases than aviation [9]. In the United States, ICT accounts for 3% of the total countrywide electrical energy consumption [10,11].

The power consumption of telecommunication equipment has rapidly grown over recent years. The typical router capacity increased from 100 Gbps in 2000 to 10 Tbps in 2008 with its power consumption increasing from 1.7 to 50 kW [12]. Furthermore, the rate of increase in router and telecom equipment power density is starting to accelerate and air-cooling solutions are coming to the end of their capabilities with calls for the consideration of expensive liquid cooling solutions. ICT and telecommunications equipment continue to consume significant power even in the idle (not sleep) state. The power wastage is significant given that typical router utilization figures are 20%–50% in the Internet service provider (ISP) backbone, 8%–25% in enterprise networks, and less than 1% in local area networks (LANs). In ICT equipment, about 20% of the lifetime energy consumption is used in manufacturing and 80% during the use of the product in its lifetime [13]. Therefore, it is important to reduce network power consumption. It is worth observing that current estimates are that 37% of energy in ICT is due to transport of bits (telecommunication), while 63% is due to processing of bits (in data centers and in telecom equipment) [13]. Therefore, addressing network operating power, both in transmission and signal processing, are essential steps in reducing the overall carbon footprint of the Internet.

Efforts to reduce the environmental impact of ICT have mostly concentrated on improving the energy efficiency of individual ICT components or whole ICT systems. In communication networks, equipment level techniques are based on using power-saving modes when the equipment is underutilized or improving the energy efficiency of hardware parts and operations. Additionally, network-level techniques manipulate the energy profile of components by consolidating traffic, achieving energy savings of the system as a whole.

Substantial advances can be achieved through the innovative use of renewable sources and the development of new architectures, protocols, and algorithms operating on hardware, which will allow significant reductions in energy consumption. BT has announced plans to develop wind farms to generate 25% of its electricity needs by 2016. The BT wind farms will generate a total of 250 MW of electricity (about 2 MW per wind turbine), sufficient power for 122,000 homes or a city the size of Coventry. This will save 500,000 tons of CO_2 each year, the equivalent of a quarter of a million return air trips to New York. To ensure scalability, this problem has to be tackled at a national and European scale and with the activity also examining control plane mechanisms that can provide quality of service (QoS) guarantees in this novel renewable energy ICT scenario, this benefiting from newly differentiated QoS and differentiated resilience schemes [14,15] (where nodes with low renewable power are treated as failing nodes).

In data centers, the primary goal is usually to reduce the ratio of the facility to the IT equipment power usage, called power usage effectiveness (PUE), by increasing the efficiency of the cooling and power distribution systems. Nevertheless, the PUE measures neither the energy efficiency of IT components' operation nor their level of utilization. The former is achieved by technological advancements at the equipment level, while the latter requires the consolidation of resources. Virtualization is a key tool for achieving this goal through the incorporation of multiple operating environments on the same machine and the migration of workloads inside or between data centers [16]. In this way, a significant amount of energy can be saved by putting a fraction of servers to sleep, while increasing the utilization of the active servers, as illustrated in Figure 12.1.

Apart from improving the energy efficiency of ICT, its environmental impact can be further reduced by locating communication networks and data centers near clean energy generation sites, so that they can be supplied with both renewable and grid energy (Figure 12.1). Supplying remotely generated renewable energy to ICT services rather than to grid workloads in the cities is beneficial for two main reasons. First, renewable sources, such as photovoltaic and wind farms, occupy large areas and are usually built

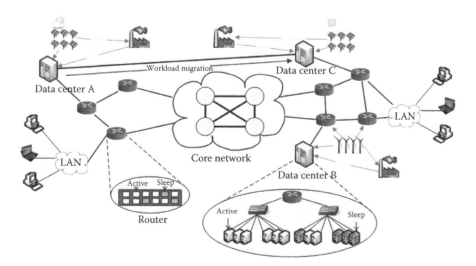

FIGURE 12.1 Example illustrating different opportunities to reduce the energy consumption and environmental impact of ICT.

at appropriate remote locations; hence, shipping the generated energy to the cities incurs electricity transmission overheads. On the contrary, information transmission over long distances is highly efficient, allowing the collocation of communication nodes or data centers and renewable energy sites. Locating data centers away from the cities is also beneficial due to the availability of affordable land, low-cost electricity, and possibly free cooling [17]. Second, the intermittent nature of renewable sources combined with the inflexibility of grid workloads and the inability to efficiently store energy, results in substantial energy waste. In contrast to grid workloads, the agility of data potentially allows their processing to be delayed or performed at different geographical locations. For example, the maximization of renewable energy usage in a communication network can be achieved by appropriately routing traffic according to the expected utilization of each node, its energy profile, and renewable energy availability. Similarly, data centers utilization can keep up with renewable energy availability by either scheduling noninteractive workloads at later times or migrating appropriate interactive workloads to data centers with abundance of renewable energy. For example, in Figure 12.1 the reduction in the solar energy generation at data center A induces workload migration to data center C via a dedicated communication link.

Transporting data instead of energy imposes a number of challenges. First, interactive workload migration has to abide by service level agreements (SLAs), which may be violated due to the migration downtime. Therefore, it is essential to characterize workloads and develop migration prediction models and even employ dedicated high-speed low-latency networks to achieve transparent migration. Second, the relationship between the energy savings due to workload migration in data centers and the energy consumption of communication networks employed in the migration of resources needs to be established in order to derive conclusions about the efficiency of the approach and the optimal frequency of reoptimization. Third, the minimization of nonrenewable energy consumption in accordance with the dynamically varying renewable energy generation and satisfaction of the migration and data center capacity requirements is a hard optimization problem that needs to be dynamically solved in near real time. Finally, efficient optimization can only be achieved by building accurate power consumption models of data centers, especially given the large gap between actual and theoretical energy consumption [18].

The remainder of the chapter is organized as follows: In Section 12.2, we discuss in detail the techniques that have been proposed to reduce the energy consumption of communication networks. Section 12.3 introduces the hybrid-power IP-over-WDM network architecture. A linear programming (LP) model is developed to minimize nonrenewable energy consumption of the network and a heuristic is proposed for improving renewable energy utilization. In order to identify the impact of the number and the location of nodes that employ renewable energy on the nonrenewable energy consumption of the network, we construct another LP model. We also investigate the additional energy savings that can be gained through adaptive link rate (ALR) techniques where different load-dependent energy consumption profiles are considered. In Section 12.4, we expand the discussion of data centers and agile workload migration using virtualization technologies. In Section 12.5, we consider some of the photonic systems advances that have the potential to reduce significantly the energy consumption within Ethernet switches and Internet Protocol (IP) routers in the data center, showing how integrated photonic switch fabrics are starting to have the performance required for energy-efficient high-switching applications. Finally, the chapter is concluded in Section 12.6.

12.2 Energy Efficiency in Communication Networks

In this section, we discuss techniques proposed in the literature for the reduction of energy in communication networks. Specifically, we classify these techniques into two main categories: (a) equipment level and (b) network level. *Equipment level* techniques aim at reducing the power consumption of individual network devices (e.g., switches, routers) by adjusting the power mode of various components of the devices (e.g., network interfaces, line cards, memories, processors) based on locally collected information. On the contrary, the objective in *network-level* techniques is to minimize power consumption of a network using global state information including the topology, the state of links, traffic demands, QoS requirements, and power consumption models of all equipment. This can be achieved either in a centralized manner by collecting all information at a single controller or in a distributed manner through node cooperation.

12.2.1 Equipment Level

Driven by the idea of reducing power consumption in network components, the research community and industry have made first attempts in the investigation and realization of power-saving technologies. A great deal of work has been done in reengineering conventional network equipment and network protocols. There are three main methods to reduce the energy consumption of *equipment level* techniques based on local information. The *adaptive tuning* approach allows the reduction of energy by allocating the optimal amount of equipment resources with respect to the current demand. *Efficient operations* present another alternative to energy saving by reengineering specific functions of individual components from a performance perspective. Finally, with *proxying techniques*, operations of single or multiple network nodes are temporarily performed by a *proxy* while the nodes themselves are sleeping.

12.2.1.1 Adaptive Tuning

Adaptive tuning considers a set of approaches that are designed to dynamically adjust the characteristics of the network equipment in correspondence to the current service demands. "Adaptation" is achieved by an extensive management of the onboard system's hardware and brought about by putting various components into a sleep mode or slowing down the hardware operation. These measures result in the reduction of the overall power consumption of the system and are beneficial for energy conservation inside the network.

In particular, adaptive tuning for switches and routers has been intensively studied within the last decade. These utilize multiple processors, memories, and interface cards for data handling and forwarding, which require high energy expenditure. A number of power consumption solutions based on adaptive tuning of these components is presented in this section.

12.2.1.1.1 Sleep/Shutdown Approach

Recently, a great amount of attention has been devoted to the minimization of power consumption in network equipment [19,20]. The sleep/shutdown scheme, being a dominating approach explored in the literature, has been broadly applied in the network interface cards (NICs) of desktop computers and switching architectures in Ethernet networks. The key principle of sleep/shutdown is to power off the system's network cards during the periods of network inactivity. Since the energy demand of an idle interface is almost identical to the one being fully loaded, shutting the device down during the periods of network inactivity can be power beneficial.

The pioneering study on power minimization in network switches is presented by Gupta et al. in [21]. In particular, the authors did the shutdown of the ports and line cards based on a prediction algorithm, which uses the history of packet inter-arrival times. The maximal power savings of the scheme are achievable only when the packet arrivals to the system are infrequent and transition times between the power states are minimal. However, this is rarely achievable in real networks, where the packet arrivals can be spontaneous and, therefore, hardly predictable for the future. In fact, being in a sleeping mode, the interface cards are not capable of either detecting or receiving arriving traffic, which may introduce additional packet loss and/or delays to the system.

In order to avoid the drawbacks of the previous approach, the schemes with modified NICs functionality were proposed in [22–30]. These schemes introduce additional hardware blocks that are powered on at all times and capable of buffering arriving packets when the system is sleeping. Packet loss minimization is achieved by the combination of different *traffic prediction algorithms*, which trigger a set of timers forcing the system to sleep [23,24,29–31], *buffer policies* that set up occupancy thresholds to reinitialize the wake-up state [23,24,26,28,29] and *shadow ports* that cluster multiple ports of the network switches to handle ingress traffic [30]. The coordination of these techniques allows in theory achieving good balance between the performance characteristics of the NIC and its energy savings (40%–60%) in underutilized networks (5%–10%) [24].

A set of energy-conservation schemes for passive optical networks was introduced in [32–34]. Due to the fact that the optical line terminal (OLT) plays the role of traffic distributor and arbitrator in a network, it is possible to switch the modes of optical network units (ONU) from active to sleep, via extensive communication and modifications on the multiple access protocol deployed at upstream direction. To achieve the maximal energy savings under light traffic loads [34], the OLT is capable of storing arriving traffic for inactive ONUs and redistributing the traffic upon the end-system wake-up (similar to [22]).

The majority of sleep/shutdown approaches presented at different levels of network equipment can achieve effectiveness in power saving only at low network utilizations. To provide reliable network service, the approaches explored use additional blocks that are powered on continuously to avoid possible losses. With the increase in the traffic load, the energy conservation becomes insufficient due to frequent transitions between the power states. Frequent state transitions, in their turn, cause undesirable in-rush currents that can significantly degrade the energy saving of a scheme (due to energy spikes) as well as damage the circuitry [25]. Additionally, the state transition times between the "power on" and "power off" states may be long enough for a packet loss to appear, which will result in retransmissions from the network protocol side. All these factors should be taken into account in the practical realization of power-aware network interfaces.

12.2.1.1.2 Adaptive Link Rate

The ALR approach [35] is mostly studied and applied in Ethernet technologies. Contemporary Ethernet provides up to 10 Gbps link rate, which consumes a considerable amount of energy irrespective of traffic demand. The ALR scheme dynamically tunes the link's data rate in correspondence to the varying traffic loads and, as a result, minimizes the energy use.

The energy-efficient Ethernet (EEE) task force [36] introduced enhancement to the traditional standard by suggesting three main link rates (10 Gbps, 1 Gbps, and 100 Mbps) for operation. The rate switching is performed by a rapid PHY selection (RPS) algorithm that uses a sophisticated frame exchange policy for the negotiation of link speeds. In practice, tuning the transmission rate may require a considerable amount of time due to adjustments performed in transmission and synchronization circuits of the Ethernet card. The solution to this issue is a challenge that faces the practical implementation of the ALR approach, since large transition times may introduce performance degradation of the system. A set of theoretical studies shows the application of the ALR technique in networks with high utilization. The authors in [22,37] proposed to shape arriving traffic into bursts and, while considering coordination of both ALR and sleep/shutdown approaches, to achieve energy-proportional power consumption [38]. Avoidance of packet loss and packet delay was achieved by over-dimensioning of interface buffers based on the idea that bursts are assembled only at the last hop of network switches. To address the problem of frequent switching between the transmission rates, the authors of [39–43] suggested control of the buffer occupancy of an interface card, while introducing various threshold policies. The thresholds controlling the link utilization were proposed in [42,43]. The key principle of the approach is to stay at the fixed lower rates (10 or 100 Mbps) until the network utilization falls below a certain value, for instance 70%. The implementation of the mentioned threshold policies allows preventing link rate oscillations both for low and high traffic loads as well as for various traffic types [40].

The practical (FPGA-based) implementation of the ALR shown in [44] emphasizes the fact that real physical layer selection times can be almost two orders of magnitude higher than proposed for EEE (1 ms). The results show that the power consumed by the device is strongly dependent on the transition type and the direction of transition. For instance, the power consumption of an interface that changes the rate directly from 1 Gbps to 10 Mbps is four times larger than in a step-by-step transition. In a similar way, the authors show the asymmetric power needs for 1 Gbps–10 Mbps (which requires higher energy) and 10 Mbps–1 Gbps transitions.

The sleep/shutdown scheme presented in the previous section proved to be energy efficient only at low network loads. The ALR scheme, in contrast, can be energy beneficial also in networks with high utilization, provided that the rate adaptation time is sufficient to maintain the network services. The latest advances in Ethernet technology show that the standards with higher data rates implement incompatible functionalities to the older designs. Therefore, the realization of the ALR approach in practice will imply the complex interaction between the incompatible layers (physical layers) of a device and can be a limiting factor for QoS guarantees.

12.2.1.1.3 Dynamic Frequency and Voltage Scaling

Dynamic frequency and voltage scaling (DVFS) [45,46] is a technique used for the power management of integrated circuits. DVFS exploits the fact that the switching power in a chip decreases in proportion to $CV^2f + P_{stat}$, where V is a core voltage applied, f denotes the clock frequency, and P_{stat} is a static leakage power. According to [45], the main contribution to the power consumption in a circuit is attributed to the first term of the formula, so the second term can be usually skipped.

The DVFS scheme proved to be extremely energy efficient in traditional high-speed CPUs, where the reduction in operating frequency allows a decrease in the supplying voltage level [45,46]. Modern processors do not show significant power reduction with DVFS due to a set of new advances introduced to the processor's architecture [47]. Since the introduction of DVFS, the transistors' size has been considerably reduced to tens of nanometers. In relation to this, the static leakage power inside the chip has become almost identical to the switching power. The impact of DVFS is further reduced by a small switching power range (transistors being smaller require a lower core voltage to be applied) and more efficient sleep states. The DVFS scheme can be widely applied in the components of network switching and routing architectures, reducing the power consumption of the system during low traffic demands.

12.2.1.2 Efficient Operations

Efficient operations consider a set of approaches that implicitly impact the power consumption of components by improving the performance of network-related operations. For instance, line coding can result in energy savings if there is a small transmission overhead and processing requirements. A second example is related to IP lookup in routers. In this case, energy savings are achieved by applying traffic classification techniques that allow the minimization of the computational and storage requirements of IP tables, resulting in the implicit reduction of energy.

12.2.1.2.1 Efficient Encoding

The traditional channel coding techniques implemented in current Ethernet standards consider delivery of the prescribed messages through the network with the maximum possible reliability. Unfortunately, the sophisticated implementation of such codes usually requires a considerable amount of energy for the interface operation. The primary investigation on the power consumption of various encoding schemes was presented in [48]. The authors suggest that the encoding circuitry makes the main contribution to the power usage inside the system. While examining the most popular line coding schemes, the authors conclude that simplified versions of the codes can be more energy beneficial but are not always reliable.

Although EEE encoding schemes are not well explored, this direction has high potential for power conservation at the network equipment level. The Ethernet encoding schemes proposed currently tend to ameliorate the bit-level clock recovery at the receiver side and increase the chances of error detection. On the other hand, the recent implementations of 10 Gbps Ethernet suffer from physical layer compatibility issues with respect to the previous versions of the standard. Therefore, the implementation of advanced encoding schemes should balance the energy consumption of the circuit, the reliability of data transmission, and performance. Efficient implementation of the encoding scheme can improve the performance of some of the approaches presented earlier. For instance, the encoding scheme can complement the ALR approach reducing the switching time between the transmission rates. In a similar manner, the encoding schemes can provide robustness of data transmission, taking advantages of the photonic networks, while dispensing with the latencies of the transport layer protocols (like Transmission Control Protocol [TCP]). Additionally, the redesign and refinement of the physical line codes can promote changes in error detection and correction techniques at the receiver side. While currently the end hosts utilize the continuous ARQ-based transport protocols to provide correctness in data delivery, fountain/raptor codes can provide data recovery without causing network retransmissions. These techniques form a good basis for low-latency packet transmissions.

12.2.1.2.2 Efficient IP Lookup

In high-end routers, the speed of IP routing becomes a major performance degradation point that limits the increase in link rates. The traditional IP lookup engines use ternary content addressable memories (TCAMs) that implement a fully parallel search during a single clock cycle. Unlike random access memories (RAM), TCAMs have more sophisticated onboard functionality featuring comparison circuits for each memory bit. This reduces their scalability in terms of power, storage, and density [49,50]. A number of studies deal with these shortcomings. Some of them demonstrate that the reduction in power consumption may be achieved by modification of the TCAM structures. The others utilize alternative low-power memory solutions, like static random access memories (SRAMs).

A set of modifications to the TCAM's structure was proposed in [51,52]. TCAM is divided into multiple blocks by equally distributing the number of packet classification rules between each block. Upon a packet arrival, only a set of blocks is searched, while the remaining blocks are disabled.

In contrast, the IPStash approach [53] implements the full associativity of TCAMs with "limited-associativity" of SRAMs. This provides good performance and lower power consumption. Unfortunately, this approach suffers from extensive power consumption when the routing table is large.

The implementation of SRAM-based lookup engines in the form of a trie can be used as an alternative [54]. Tries can be further classified into uni-bit tries and multi-bit tries [55,56]. The packet classification is performed by traversing the trie levels, where nodes represent packet prefixes. Uni-bit tries make correspondence of only one bit to a node and therefore suffer from frequent memory accesses. On the other hand, multi-bit tries, assigning multiple bits to a single node, may have larger memory occupied due to redundancy.

The large number of prefixes (hundreds of thousands) in a trie may require a large amount of memory to be allocated. In addition, the packet classifications that are performed simultaneously can lead to an extremely large memory access time. The problem is usually resolved by memory pipelining in order to perform multiple lookups in parallel [57–59]. In this case, memory arrays are subdivided into sub-arrays, with a possibility to access a specific sub-array directly. The lookup table trie entries can be assigned to the memory sub-arrays in a different way with unbalanced or balanced trie nodes distribution. The unbalanced distribution can degrade the access time, while the balancing approach can influence the system's throughput [58–60]. In contrast, the combination of SRAM lookup with external/internal caching can be beneficial in packet classification due to a reduced number of memory accesses for similar traffic arrivals [50,61].

12.2.1.3 Proxying

Currently, desktop computers receive a large amount of network management information, even when they are idle, which prevents them from sleeping. This shortcoming is overcome with the introduction of energy-efficient *proxying*, which is capable of managing the arriving network traffic on behalf of sleeping computers and maintaining network presence. Being competent in making traffic classification, the proxy decides the appropriate time for the system to wake up [3,62,63]. Proxying is implemented either locally at the computer (*internal proxying*) or globally within a LAN (*external proxying*).

The onboard proxying adapter adds extra functionality to the NIC and is usually implemented by extra hardware blocks (like microprocessors). The controller implementation presented in [61,64] shields and processes ARP, Internet Control Message Protocol (ICMP), and Dynamic Host Configuration Protocol (DHCP) requests without any interaction with the rest of the system. On the other hand, due to the limited implementation of the protocol stack, TCP segments are processed by the computer and require the system's wake-up. The packet inspection system implemented in [64] performs hardware-based content processing of packets by utilizing partitioned TCAMs and cache memories, which apart from energy savings also improves the performance characteristics of the system. The application support of the power proxy was implemented in [65,66]. The combination of secondary processor and flash memory embedded in the computer's NIC runs the full functionality of the network protocol stack plus a set of applications implemented in the forms of "stubs" (reduced functionality). These allow the support of more levels of functionality and save a considerable amount of energy during the infrequent traffic arrivals.

As the size of a LAN increases, deploying internal proxies at each computer is inefficient due to the complexity and cost of hardware. On the other hand, implementing external proxying in each subnet is capable of managing the arriving traffic, sleeping intervals, and transmission rate for a large number of computers simultaneously [65,67–69]. Two notification routines should be implemented to perform power management of a network: (a) a sleep management program implemented at the proxy and (b) a sleep notification program that runs at each client. The sleep management program informs the end hosts about sleeping possibilities and shows the network presence of sleeping devices for the outer networks. The approaches presented in [67,68] are capable of responding to some packets on the client's behalf or waking up the sleeping client for certain specified traffic (by means of wake-on-LAN packets [63]). The possibility of handling user-destined traffic presents a trade-off between the implementation complexity of the proxy and the amount of energy saved. The SleepServer approach suggests implementing virtual machine instances of the hosts at a proxying machine that provides network presence of devices and efficiently uses the protocol stack deployed at virtual machines [70]. This approach yields significant network power savings and keeps the connectivity of devices while being in sleeping mode.

12.2.1.4 Problems and Challenges

A great deal of research has been done to improve the power efficiency of communication networks at various component levels. These studies have contributed to the development of power-saving solutions in simulated environments; however, their practical application in existing networks is still challenging for two main reasons. First, power mode adaptation is not supported by current hardware, and second, compatibility has to be maintained with current standards.

Implementing energy-efficient and highly reliable encoding schemes is also challenging. Reliability requires the realization of highly complex encoding schemes, which consume a large portion of energy for encoding/decoding with respect to the actual transmission energy. Thus, designing highly robust line codes that require low processing are needed. Having robust codes also improves the energy efficiency of reliable transport protocols (e.g., TCP/IP), as the number of required retransmissions is minimized.

According to the discussion in the previous sections, energy-aware equipment may have a large response time and energy consumption during power state transitions. These may result in performance degradation in terms of extensive packet loss and delay. Additionally, it is important not only to minimize the overall transmission energy but also consider the overhead energy due to state transitions. In parallel, due attention is required in building hardware components that efficiently deal with the transition time and energy.

Finally, one of the biggest challenges is to achieve proportionality between energy consumption and utilization of components [38]. Achieving proportionally will not only be beneficial in reducing the energy consumption of network devices, but it will also eliminate the need for sophisticated energy saving approaches.

12.2.2 Network Level

In Section 12.2.1, we have discussed techniques that can be used for the energy-efficient operation of various network components such as network interfaces, switches, and routers, based on locally collected information and without any node coordination. However, equipment level techniques are not sufficient to guarantee the minimization of Internet energy consumption. Currently, deployed networks are over-provisioned to accommodate more than the maximum expected traffic demand and overredundant to deal with link and node failures. As a result, many links are underutilized and all devices are constantly in operation, which provides opportunities for energy reduction. This can be accomplished by disseminating the traffic in a way that minimizes the network's energy consumption by putting specific nodes and links to a power-saving mode. Network-level optimization requires cooperation between the nodes to collect information about the global network state, which includes the capabilities and energy consumption of all network elements, as well as full information about the traffic demand between nodes and the network topology. These techniques can be employed both at the design stage (*network design*) and during its operation either periodically (*traffic engineering*) or in real time (*distributed routing*). In these ways, we can achieve energy efficiency and meet network operation requirements such as user request satisfaction, QoS, and reliability. In this section, we discuss research undertaken in these categories in both the electrical (e.g., IP/MPLS routers) layer, as well as multilayer approaches where nodes are comprised of electrical components on top of optical components (e.g., optical cross-connects [OXCs] and reconfigurable optical add-drop multiplexer [ROADMS]).

12.2.2.1 Network Design

Traditionally, network design has focused on the minimization of the network capital expenditure (CapEx), which accounts for the equipment and installation costs of the network infrastructure [71]. However, as Internet traffic and energy costs are exponentially rising, energy consumption is becoming a major issue for network operators for several reasons: (a) the operational expenditure (OpEx) due to the energy cost is significant, (b) ICT is an important contributor to the global energy consumption and it is essential to cut down CO_2 emissions,* and (c) energy-efficient operations alleviate the heat dissipation problem [72]. As a result, energy minimization has become a main goal in designing communication networks.

Before reviewing the research undertaken in this area, we briefly describe the problem examined. The main goal of a network design problem is to find the design parameters that optimize an objective function associated with the minimization of CapEx (or energy consumption), while satisfying the provisioned network traffic demand and other design specifications (e.g., maximum link utilization, end-to-end delay, reliability). Traffic demand information is usually given in the form of a traffic matrix, which represents the peak demands between each source–destination node pair [73]. The topology of the network can either be known or the points of presence (PoPs) can be chosen from a set of candidate locations. Candidate equipment information is also available such as different router models that can be used, chassis specifications, number, and types of line cards supported by a chassis, as well as capacity, cost, and power consumption information of router components. Hence, the goal is to find the type and quantity of equipment to be installed at each location, in order to satisfy the design specifications and minimize the objective function. Mathematical formulation of the network design problem usually results in mixed integer linear programs (MILP), which fall into the class of multicommodity flow problems and are nondeterministic polynomial-time hard (NP-hard).

In the electrical layer, the problem of energy-minimized network design has been investigated in [70,74]. Chabarek et al. [70] assume a known topology and formulate the problem to minimize the total power consumption of the network when the type/number of line cards and chassis can be selected. A general model of power consumption for a router is proposed, which considers the power consumption of

---

* The UK government is already operating a carbon reduction commitment (CRC) mandatory scheme for large energy consumers.

chassis and line cards in base configuration and the extra power consumption due to the router traffic utilization, while an empirical measurement study is conducted to obtain values for these costs. Results on realistic and random topologies of small size show power reduction of up to 65%. In [74], the relationship between network power minimization and robustness is investigated using an objective function that combines power minimization with mean network delay. Network robustness is assessed with respect to survivability measures, by simulating link failures on topologies derived from the solution of the network design problem for different weighted versions of the two objective metrics. The results indicate that a good trade-off between the two metrics can lead to better network designs.

In IP-over-WDM networks, energy-minimized network design has been introduced in [72]. The authors assume that the topology is known and formulate the problem by encompassing the energy consumed by the node ports, the erbium-doped fiber amplifiers (EDFAs), and the transponders. In this case, the optimal design parameters of the physical layer (e.g., number of fibers and wavelengths on each physical link) should also be found. Additionally, the problem is very challenging as the flow conservation and capacity constraints need to be simultaneously satisfied at the IP and optical layers, resulting in an NP-hard model that has $O(N^4)$ integer variables and $O(N^3)$ constraints, where N is the number of nodes. As a result, the optimal solution can be obtained in reasonable time only for small problems, while for larger problems the authors proposed heuristics techniques. Their results on various real topologies indicated that (a) the lightpath bypass is better than nonbypass (25%–45% less energy consumption), (b) the network consumption is mostly at the IP routers (more than 90%), and (c) CapEx-minimized design is also energy efficient. The latter result was also supported by other independent studies on different networks [75,76]. Nevertheless, if the energy consumption of the nodes is assumed to be mostly dependent on its load, rather than the number of active elements, then cost-efficient design is not also energy efficient [76]. Moreover, the use of diversified-capacity lightpaths and the architecture of the node (e.g., IP-over-WDM or IP-over-OTN-over-WDM) result in energy savings [75,76] as well. The effect of the node architecture in network design is also considered in [77] for a ring network topology. It is shown that depending on the energy consumption ratio between the electrical and optical layer of one node, different node architectures can be more energy efficient.

12.2.2.2 Traffic Engineering

The objective of traffic engineering (TE) is the control of network traffic flows, in order to optimize resource utilization and network performance under specific QoS requirements [78]. Traditionally, TE has focused on balancing the traffic among links in order to avoid congestion due to traffic bursts. On the contrary, the main goal of power-aware traffic engineering (PATE) is to reduce the energy consumption of the network by consolidating traffic to a few links so as to put idle network components into a power-saving mode. Therefore, performing PATE is an even more challenging problem as energy saving has to be achieved on top of any QoS requirements. As will be discussed in this section, significant energy savings can be achieved due to the over-provisioning and high-redundancy of the network, as well as by periodically performing PATE to exploit periods of low traffic activity.

Modeling the PATE problem usually relies on multicommodity flow MILP techniques, similar to energy-efficient network design. In fact, the two problems have similar mathematical formulations, differing mainly on the decision variables. In the network design problem, the emphasis is placed on selecting the type and amount of equipment to be placed at each PoP; in the PATE problem, the decision variables are related to the power mode that has to be adopted for each network component, in order to minimize the total power consumption of the network for the provisioned traffic demand at the considered time period.

Several approaches have been adopted for modeling the power consumption of network components, which differ in the type and energy profile of subcomponents considered at each node. In [79], it is assumed that nodes cannot be put to sleep because they are either source or destination nodes; hence, the power consumption metric considered is the number of sleeping links. In [80], the authors proposed to minimize the number of active cables rather than the number of links, by highlighting that each logical link between two nodes is actually a bundle of cables that provide higher availability and linear capacity increase. A common approach adopted in the literature is based on the observation that the power consumption of current Internet routers is not significantly affected by the load [70]. As a result, an *on/off model* is adopted that accumulates the power consumption of the active components of each node. Different on/off models have varying degrees of detail; in [81] and [82] the consumption of nodes and links is taken into account, in [83] the base consumption of line cards and their integrated ports is considered, while in [84], the power consumption model includes the contribution from chassis, line cards, and ports at different line speeds. An important disadvantage of the on/off model is that it does consider the *energy profile* of a component, that is, the dependency between its power consumption and its utilization; although this dependency is weak for current communication equipment, future energy-aware components are expected to consume energy proportional to their utilization [38]. In [85,86], a simple energy-profile model is considered where each component has a base consumption (when in idle state) that increases linearly with its load utilization. Finally, the effect of various energy profiles on the energy consumption of a network is considered in [87].

As already mentioned, PATE targets the consolidation of traffic to a few links and nodes so that many network elements can sleep. Nevertheless, traffic aggregation may create violation of QoS requirements, and for this reason, formulations incorporate QoS constraints. To put a threshold on the average link queuing delay, the approach usually taken is to impose an upper bound on the link utilization, which is a fraction of the nominal link capacity (50%–70%) [81–84,86,87]. Other measures to reduce the overall delay include the consideration of paths that are smaller than a certain length [83] or the average delay in the network [85]. Load balancing has also been suggested to be performed on top of the PATE solution to provide an even distribution of the traffic among the non-asleep network components [83]. Because PATE routing removes most of the redundant paths, examining the availability of the network is also important. Nevertheless, none of the PATE approaches have incorporated the particular QoS metric into its mathematical formulation; network availability has only been examined in derived simulation results [79,84].

Because different PATE formulations are NP-hard [79,81], a number of heuristic approaches have been proposed for obtaining fast and close to optimal solutions. Pruning heuristics have been proposed in [79,81,82]. The idea is that network elements (e.g., nodes and links), that are considered for switching off, are initially ordered according to a desirability criterion such as their power consumption, aggregate traffic, or ratio of total to remaining capacity; then, sequentially these elements are orderly switched off, given that the network traffic can be routed with the remaining elements. On the contrary, the heuristics proposed in [80] start from the integer solution of the relaxed MILP problem and in the process recompute the desirability criterion in every iteration. These heuristic approaches are of polynomial time complexity and result in fast and relatively accurate solution of the problem. In [83], a nonpolynomial heuristic is proposed, which is based on constructing a path flow formulation of the considered PATE problem for the k shortest paths for each source–destination pair, instead of all the paths. Although the resulting formulation is still MILP, the number of integer variables is considerably smaller so that larger problems can be solved with very high accuracy.

Although an important amount of research has been undertaken in PATE-based routing, little attention has been paid in discussing implementation details using existing protocols and mechanisms. To the best of our knowledge, only Zhang et al. [83] investigated this matter in detail. Specifically, the use of a centralized controller node is proposed that is responsible for the aggregation of information, as well as for the computation and dissemination of the solution, under the assumption that the network runs both Open Shortest Path First (OSPF) and Multiprotocol Label Switching (MPLS) protocols. The authors describe how the controller can passively collect all necessary information and distribute the periodic solution of the problem using TE variations of the OSPF and MPLS protocols. The implementation of a state transition between on and off and the traffic splitting in flows are also discussed.

A number of interesting results about the energy saving that can be achieved via PATE routing have emerged. However, before starting the discussion, we should emphasize that these results are indicatory, due to the lack of practical implementation in all studies. Additionally, these results are based on assumptions about numerous factors such as the network topology, the traffic demand, the energy model, and power consumption values of the network components. This is also indicated by the fact that the energy reduction achieved for different studies ranges between 10% and 70% for realistic network topologies and traffic matrices [84,86]. The first main result is that the yearly energy reduction that can be achieved for the real topology and traffic data of a specific ISP is around 23%, when an on/off power model is adopted [82]. Another important result is that future energy-aware proportional network devices can result in an order of magnitude energy reduction [86,87]. Key results can also be inferred about the trade-off between energy minimization and QoS satisfaction. Several studies have shown that for low and moderate traffic volumes, energy reduction can be achieved without significantly affecting delay-related QoS metrics (e.g., maximum link utilization, route length increase) [79,83,85]. However, the availability of the network, in terms of the average number of disjoint paths between pairs of nodes, drops very quickly to its minimum value, which means that our network is prone to failures [79,84]. Thus, it is essential to develop network equipment that can switch power mode very fast. In terms of the solution stability, it has been illustrated in [83] that no abrupt changes occur between successive PATE solution computations.

12.2.2.3 Distributed Online Energy-Aware Routing

The PATE approaches discussed in Section 12.2.2.2 rely on global information about the system state, including the traffic demand matrix that is not readily available, and require the solution of NP-hard combinatorial optimization problems; hence, they are centralized and cannot be used for the online dynamic traffic rerouting. In this section, we discuss techniques that only rely on information readily available at the servers and can be used for online distributed rerouting of traffic that achieves energy savings.

Energy-aware traffic engineering (EATe) is the first scalable online technique that has been proposed for energy saving; it is based on ALR, as well as node and link sleeping [88]. The authors describe techniques that can handle two different power consumption models: (i) when the ratio of idle to maximum component power is low, the authors propose to remove traffic from underutilized links, in order to switch to a lower link rate, and redistribute it to links that are not underutilized but can support more traffic; and (ii) when the ratio of idle to maximum component power is high, the authors propose to switch off as many links and nodes as possible by redistributing the traffic of underutilized links. Stability is achieved by explicit feedback from the nodes receiving the redistributed traffic so that only feasible changes are accepted.

Green OSPF protocol relies on aggregating traffic at a neighborhood of routers in order to put links to sleep [89]. In this protocol, there are two sets of routers. "Exporter routers" compute their shortest path tree (SPT) normally, while "importer routers" take as reference a slightly modified version of the SPT of "exporter router," rather than computing their own. In this way, a number of routers share the same SPT so that several idle links can sleep. To avoid inconsistencies in routing, a final phase is required in which each router identifies the new topology (through received link state advertisements [LSAs]) and reoptimizes its routing paths. Nevertheless, this protocol does not address how link overloading can be avoided.

Contrary to green OSPF, general distributed routing protocol for power saving (GDRP-PS) considers the possibility of switching of nodes during off-peak hours [90]. In this protocol, node sleeping is coordinated by a central node, so that not many nodes sleep concurrently; nonetheless, a node's decision to ask for sleeping permission is based only on its aggregate link utilization. Periodically, the nodes wake up, reconnect to the network, and examine whether they should go back to sleep. Performance evaluation indicates that 20% power savings can be achieved; however, no simulation results or analysis is presented to support whether the network stability is maintained.

12.2.2.4 Problems and Challenges

Despite the large number of network-level studies undertaken, a number of issues needs to be further explored. As already mentioned, network design and TE problems are NP-hard and researchers often employ heuristic solution techniques. Although a number of different approaches has been proposed in this area, no comparative studies have been performed, while performance evaluation of individual studies is not compared against other similar approaches. For this reason, it is important to build a library of representative test instances for specific problem variations so that benchmarking will be possible. An example of such a library is SNDlib 1.0 for survivable fixed telecommunication in traditional network design [91].

Additionally, the available literature lacks practically implemented approaches; most studies do not even discuss any deployment details about how to collect the require information, when and how often to solve the TE problem, how to disseminate information, and how to implement the sleeping of nodes in practice. Furthermore, as TE approaches are performed offline, there is a need for more distributed online routing techniques; so far online approaches have focused on providing intuitive solutions, which provide no optimal results or rigorous guarantees for stability and QoS satisfaction.

Current routing protocols such as OSPF and intermediate system to intermediate system (IS-IS) are designed to handle a small number of failures and do not support the simultaneous shutting of a large number of links and nodes, which is the case with energy-aware techniques [82]. As a result, modified versions of these protocols need to be designed that will be capable of dynamically adjusting to the constantly changing topology. Moreover, the PATE approaches proposed so far are only suitable for intradomains, as ISPs are unwilling to share information between them. Consequently, it is important to study interdomain protocols where energy-efficient routing is performed among ISP providers. To this direction, ISPs need to be given incentives to cooperate, for example, by guaranteeing fairly distributed energy consumption and by not requiring exchange of sensitive information.

Finally, it is imperative to investigate the interdependency between equipment level and network-level energy-efficient approaches in order to avoid potential problems. For example, if a PATE technique redirects a large volume of traffic on a link that is running on a lower data rate than usual, this will cause significant delay or packet loss. Additionally, the incorporation of PATE techniques into communication networks will alter the statistical characteristics of traffic, making the efficiency of equipment level techniques doubtful.

12.3 IP-over-WDM Networks Employing Renewable Energy

An IP-over-WDM network is composed of two layers, the IP layer and the optical layer. In the IP layer, IP routers aggregate traffic from access networks. IP routers are connected to optical switches in the optical layer. The optical layer provides large capacity for the communication between IP routers. Optical switches are connected to optical fiber links where a pair of wavelength multiplexers/demultiplexers is used to multiplex/demultiplex wavelengths on each fiber. Transponders are used to provide optical–electrical–optical (OEO) processing for full wavelength conversion at each switching node. For long distance transmission, the EDFAs are used to amplify the optical signal on each fiber. IP-over-WDM networks are implemented by either lightpath *non-bypass* or *bypass*. Under the lightpath nonbypass, all the lightpaths are terminated, processed, and forwarded by IP routers in all intermediate nodes. On the other hand, under the lightpath bypass, all lightpaths are directly bypassed through intermediate nodes, that is, lightpaths are treated as virtual links in the IP layer. Lightpath bypass can significantly save the total number of IP router ports required and as IP routers are the major energy consumption components in an IP-over-WDM network, minimizing the number of IP router ports can potentially minimize the energy consumption of IP-over-WDM networks. In [72], the multihop-bypass heuristic was proposed where the bandwidth utilization is improved by allowing traffic demands between different source–destination pairs to share capacity on common virtual links (lightpaths). Improving the wavelength bandwidth utilization results in fewer virtual links, and therefore, fewer IP router ports and lower energy consumption.

As mentioned in Section 12.1, introducing the use of renewable energy in ICT networks will allow significant savings in energy consumption and CO_2 emissions. However, many challenges need to be addressed to develop and deploy renewable energy in ICT networks. In this section, we focus on reducing the CO_2 emissions of backbone IP-over-WDM networks by introducing renewable energy sources to the network. A hybrid-power IP-over-WDM network architecture is proposed where the power supply is composed of nonrenewable energy and renewable energy [92]. In this case, the total CO_2 emission of an IP-over-WDM network will be reduced if a portion of the nonrenewable energy consumption of the network is replaced by renewable energy. Therefore, the problem becomes that of minimizing the nonrenewable energy consumption of the hybrid-power IP-over-WDM network. A LP optimization model and a heuristic are set up to minimize the nonrenewable energy consumption.

12.3.1 A LP Model for Hybrid-Power IP-over-WDM Networks

The lightpath bypass was implemented in [72] to reduce energy consumption of the IP-over-WDM network by reducing the number of required IP router ports. An MILP optimization model to minimize energy consumption was developed. In this section, we develop a model focusing on minimizing the nonrenewable energy consumption in the hybrid-power IP-over-WDM network. We consider a network with the topology $G = (N, E)$ with N nodes and E physical links. The nodes that have access to renewable energy can also have access to nonrenewable energy to guarantee QoS when the renewable energy is not available. The renewable energy can be used to power the ports, transponders, optical switches, multiplexers, and demultiplexers in a node. The total nonrenewable power consumption of the network consists of [92]

1. The nonrenewable power consumption of ports without access to renewable energy

$$\sum_{i \in N} PR \left(Q_i^e + \sum_{p \in P} \delta_{ip} W_p \right).$$

2. The nonrenewable power consumption of EDFAs

$$\sum_{e \in E} PE \cdot E_e \cdot f_e.$$

3. The nonrenewable power consumption of router ports with access to renewable energy

$$\sum_{i \in N} PRS \left(Q_i^s + \sum_{p \in P} \delta_{ip} WS_p \right).$$

4. The nonrenewable power consumption of transponders with access to renewable energy and that of the transponders without access to renewable energy

$$\sum_{\theta \in E} (PT\omega_\theta + PTS\omega s_\theta).$$

5. The nonrenewable power consumption of optical switches with access to renewable energy and that of the optical switches without access to renewable energy

$$\sum_{i \in N} [PO_i (1 - y_i) + POS_i y_i].$$

6. The nonrenewable power consumption of multiplexers and demultiplexers with access to renewable energy and that of the multiplexers and demultiplexers without access to renewable energy

$$\sum_{i \in N} (PMD \cdot DMe_i) + (PMDS \cdot DMs_i).$$

The LP model is developed to minimize the total nonrenewable power and is defined as follows: Objective: minimize [92]:

$$
\sum_{i \in N} PR \left(Q_i^\theta + \sum_{p \in P} \delta_{ip} W_p \right) + \sum_{\theta \in E} PE \cdot E_\theta \cdot f_\theta + \sum_{i \in N} PRS \left(Q_i^s + \sum_{p \in P} \delta_{ip} Ws_p \right)
$$

$$
+ \sum_{\theta \in E} [(PT\omega_\theta) + (PTSws_\theta)]
$$

$$
+ \sum_{i \in N} \left[PO_i(1 - y_i) + POS_i y_i \right] + \sum_{i \in N} [(PMD \cdot DMe_i) + (PMDS \cdot DMs_i)] \tag{12.1}
$$

subject to [92]:

$$
\sum_{p \in P} x_p^d = h^d \quad \forall d \in D, \tag{12.2}
$$

$$
\sum_{d \in D} x_p^d \le (W_p + Ws_p)B \quad \forall p \in P, \tag{12.3}
$$

$$
\sum_{p \in P} (\delta_{ip} W_p + \delta_{ip} Ws_p) + Q_i \le \nabla^i \quad \forall i \in N, \tag{12.4}
$$

$$
\sum_{\theta \in E} \delta_{\theta p} \omega_\theta^p = W_p + Ws_p \quad \forall p \in P, \tag{12.5}
$$

$$
PR^s \left(Q_i^s + \sum_{p \in P} \delta_{ip} \cdot Ws_p \right) + \sum_{\theta \in E} PT^s \cdot ws_\theta \cdot \delta_{i\theta} + PMD^s \cdot DMs_i + PO_i^s \cdot y_i \le S_i, \forall i \in N, \tag{12.6}
$$

$$
\sum_{p \in P} \omega_\theta^p \le Wf_\theta \quad \forall e \in E, \tag{12.7}
$$

$$
Q_i^\theta + Q_i^s = Q_i \quad \forall i \in N, \tag{12.8}
$$

$$
\sum_{p \in P} \omega_\theta^p = \omega_\theta + ws_\theta \quad \forall e \in E, \tag{12.9}
$$

$$
DMe_i + DMs_i = DM_i \quad \forall i \in N. \tag{12.10}
$$

The variables and parameters in the previous equations are defined as follows [92]:

$E(Pa)$	Physical link set in optical layer
$P(Pa)$	Virtual link set in IP layer
$D(Pa)$	Traffic demand set between node pairs
$\delta_{ip}(Pa)$	If node i belongs to virtual link p, δ_{ip} is '1', otherwise it is '0'
$\delta_{ie}(Pa)$	If node i belongs to physical link e, δ_{ie} is '1', otherwise it is '0'
$\delta_{ep}(Pa)$	If virtual link p starts or ends at physical link e, δ_{ep} is '1', otherwise it is '0'
$PR(Pa)$ *and* $PE(Pa)$	Nonrenewable power consumption of a router port and an EDFA respectively both use non-renewable energy
$PRS(Pa)$	Nonrenewable power consumption of a router port that has access to renewable energy
PR^s	Renewable power consumption of a router port that has access to renewable energy
$PO_i(Pa)$ *and* $POS_i(Pa)$	Nonrenewable power consumption of an optical switch that has access to nonrenewable energy only or has access to renewable energy in node i, respectively

PO_i^s	Renewable power consumption of an optical switch that has access to renewable energy
$PMD(Pa)$ and $PMDS(Pa)$	Nonrenewable power consumption of a multi/demultiplexer that has access to nonrenewable energy only or has access to renewable energy, respectively
PMD^s	Renewable power consumption of a multi/demultiplexer that has access to renewable energy
$DM_i(Pa)$	The total number of multiplexers and demultiplexers in node i
$DMe_i(Ve)$	Number of multiplexers and demultiplexers in node i which use nonrenewable energy
$DMs_i(Ve)$	Number of multiplexers and demultiplexers in node i which use renewable energy
$y_i(Ve)$	If the optical switch in node i has access to renewable energy $y_i = 1$, otherwise $y_i = 0$
$\omega_e(Ve)$ and $\omega s_e(Ve)$	Number of wavelength channels on physical link e in the optical layer which use nonrenewable energy and renewable energy respectively
$W_p(Ve)$ and $Ws_p(Ve)$	Number of wavelength channels on virtual link p in the IP layer which use nonrenewable energy and renewable energy respectively
$Q_i^e(Ve)$ and $Q_i^s(Ve)$	Number of ports which are powered by nonrenewable energy or renewable energy for data aggregation in node i
$x_p^d(Ve)$	Traffic demand d between node pairs on virtual link p
$\omega_e^p(Ve)$	Number of wavelength channels of virtual link p on physical link e
$f_e(Ve)$	Number of fibers on physical link e
$E_e(Pa)$	Number of EDFAs on each fiber on physical link e
$PT(Pa)$ and $PTS(Pa)$	Nonrenewable power consumption of a transponder that has access to nonrenewable energy only or has access to renewable energy respectively
PT^s	Renewable power consumption of a transponder that has access to renewable energy
$W(Pa)$	Number of wavelengths in a fiber
$Q_i(Pa)$	Number of ports for assembling data
$\nabla^i(Pa)$	Maximum number of ports in node i
$S_i(Pa)$	The maximum output power of the renewable energy source in node i
$h^d(Pa)$	Traffic demand d between node pairs
$B(Pa)$	Capacity of each wavelength

The objective function (Equation 12.1) aims to minimize the nonrenewable power consumption of the hybrid-power IP-over-WDM network. Constraints (12.2) and (12.5) give the flow conservation constraint in the IP layer and the optical layer, respectively. Constraint (12.3) ensures that the capacity of each virtual link is not exceeded. Constraint (12.4) ensures that the maximum number of router ports in each node is not exceeded. Constraint (12.6) ensures that the renewable power consumption of router ports and transponders at each node is not larger than the renewable energy available at that node. Constraints (12.7) and (12.9) give the maximum number of wavelength channels in each physical link. Constraint (12.8) ensures that the total number of ports assembling data in each node is equal to the number of router ports powered by nonrenewable energy and the number of ports powered by renewable energy. Constraint (12.10) gives the maximum number of multiplexers and demultiplexers at each node.

12.3.2 Heuristic Approach

The multihop-bypass heuristic proposed in [72] improves the bandwidth utilization by allowing traffic demands between different source–destination pairs to share capacity on common virtual links, which results in fewer virtual links, and hence fewer IP router ports and lower power consumption. However, implementing the multihop-bypass heuristic, which is based on the shortest path routing in the hybrid-power IP-over-WDM network where renewable energy sources are only available to a limited number of nodes, will only minimize the total power consumption without considering whether this power comes from renewable or nonrenewable sources. We propose a heuristic, renewable energy optimization hop (REO-hop) that minimizes the utilization of nonrenewable power by allowing traffic flows to traverse as many nodes as possible that use renewable energy to ensure that in addition to reducing the total number of IP router ports and transponders, the nonrenewable power consumption is minimized. To maintain QoS, only the two shortest path routes are considered so as not to increase the propagation delay. As the traffic pattern is changing and the output power of renewable energy sources varies during different times of the day, the routing paths are dynamic.

The pseudo code of the REO-hop heuristic is given in Figure 12.2. The heuristic starts by ordering all the node pairs based on their traffic demands from highest to lowest and creating an empty virtual link

---

Algorithm: Renewable Energy Optimization hop (REO-hop)

Input: Traffic demand (D_n), Renewable energy (R_s)

Output: Nonrenewable power consumption of network

1	**Initialization:** Reorder node pairs according to their traffic demands and create empty virtual topology G
2	Retrieve a node pair from the ordered list
3	**if** enough capacity is available over G **then** Route D_n through the maximum number of nodes using renewable energy
4	Update G and R_s
5	**else**
6	**if** enough capacity is available over G **then**
7	Route D_n using shortest path
8	Update G and R_s
9	**else**
10	Create two new virtual links V_{n1} (maximum number of nodes using renewable energy) and V_{n2} (shortest path route), calculate non-renewable power consumption (E_{n1} and E_{n2})
11	**if** $E_{n1} < E_{n2}$ **then**
12	Route D_n on V_{n1}
13	**else**
14	Route D_n on V_{n2}
15	**end if**
16	**end if**
17	**end if**
18	**if** all the traffic demands are routed on the virtual topology *G* **then**
19	Calculate result
20	**else**
21	Continue loop
22	**end if**

FIGURE 12.2 The REO-hop heuristic pseudo code.

topology G. A node pair is retrieved from the ordered list, and its traffic demand is routed over virtual topology G by comparing the two shortest path routes and the route with the maximum number of nodes that use renewable energy is selected. If sufficient free capacity is available on virtual topology G, the selected route is accommodated and the remaining capacity on all the virtual links and renewable energy of each node are updated. If the selected route is not available, the other route is selected. If the virtual topology cannot accommodate either route, a new direct virtual link is established between the source and destination by comparing the nonrenewable power consumption of the route with the maximum number of nodes that use renewable energy and the shortest path route. The route with lower nonrenewable power consumption is selected. If the two paths have the same nonrenewable power consumption, the shortest path route is selected. The new virtual link is added to the virtual topology G, and the remaining capacity on all the virtual links and the renewable energy of each node are updated. The aforementioned steps are repeated for all the node pairs. After routing all the traffic demands on the virtual topology G, the objective function (Equation 12.1) is used to calculate the total nonrenewable power consumption of the network.

12.3.3 Simulation and Results

To evaluate the nonrenewable power consumption of the architecture and test the REO-hop heuristic, the National Science Foundation Network (NSFNET) network, depicted in Figure 12.3, is considered as an example of a real-world network. The NSFNET network consists of 14 nodes and 21 bidirectional links. Solar energy is used as the renewable energy source. Nodes in the NSFNET will experience different levels of solar energy and traffic demands at any given point in time as different parts of the network fall in different time zones. There are four time zones, Eastern Standard Time (EST), Central Standard Time (CST), Mountain Standard Time (MST), and Pacific Standard Time (PST), with an hour time difference between each time zone and the next. We use EST as the reference time.

Real sunrise and sunset data are used to determine the solar energy available at each node during different hours of the day. Table 12.1 shows the solar energy available to a node [93]. The solar energy is available from 6:00 to 22:00 and the maximum output power is at 12:00.

The average traffic demand during different hours of the day is shown in Table 12.2 [94]. The average traffic demand between each node pair ranges from 20 to 120 Gbps with the peak occurring at 22:00.

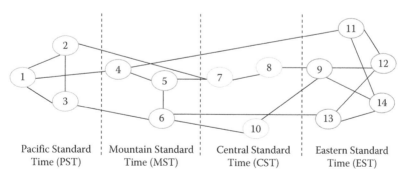

FIGURE 12.3   The NSFNET network with time zones.

TABLE 12.1   Average Traffic Demand in Different Time Zones

Time Zone \ Time (EST)	0	2	4	6	8	10	12	14	16	18	20	22
EST	40	30	35	20	40	110	90	90	110	100	70	120
CST	80	35	30	25	30	75	100	90	100	105	85	95
MST	120	40	30	35	20	40	110	90	90	110	100	70
PST	95	80	35	30	25	30	75	100	90	100	105	85

TABLE 12.2 Output Power of Solar Energy at Different Nodes in Different Time Zones

Time (EST) Node ID	0	2	4	6	8	10	12	14	16	18	20	22
Node 1	0	0	0	0	0	1.5	10	18	18	13	8	2.5
Node 5	0	0	0	0	2	6	13	20	18	13	8	2
Node 9	0	0	0	0	6	13	18	18	13	8	3	0
Node 14	0	0	0	0.5	6	13	20	13	7.5	5	2	0

The traffic demand between each node pair in a time zone is random with a uniform distribution and no lower than 10 Gbps.

12.3.3.1 Nonrenewable Power Consumption of the Network

To maximize the reduction in the nonrenewable power consumption, nodes located at the center of the network with a high nodal degree are selected to use renewable energy as these are expected to consume more power compared to nodes at the edge of the network as more traffic demands passes through them. Nodes 4, 5, 6, 7, and 9 are initially selected to employ solar energy. The impact of the location of nodes using solar energy is studied in Section 12.3.3.2.

We consider the maximum solar energy available to a node to be 20 kW. Typically a 1 m^2 silicon solar cell can produce about 0.28 kW of power [95]. Therefore, we require a total solar cell area of about 100 m^2, which can be accommodated in a typical core node. Later, we study the impact of higher solar energy output power per node.

We consider a suitable size optical switch (the Glimmerglass's 192 × 192 optical switch) based on the maximum number of wavelengths at each node. As the difference in power consumption between different optical switch sizes is negligible compared to the power consumption of a router port, we assume the same power consumption data for optical switches in the multihop-bypass heuristic and REO-hop heuristic. We found that under the multihop bypass (the heuristic requiring fewer wavelengths), the maximum number of wavelengths needed is 109.

Table 12.3 gives the simulation scenario parameters [92]. It shows the number of wavelengths, wavelength capacity, distance between two neighboring EDFAs, and power consumption of different components in the network [72,96–98]. The LP problem is solved using the AMPL/LPSOLVE software.

Figure 12.4 shows the reduction in the total nonrenewable power consumption of the NSFNET network throughout the day compared to non-bypass without solar energy case. The curve "LP optimal with solar energy" gives the lower bound on the nonrenewable power consumption. Both the multihop-bypass and REO-hop heuristics have reduced the nonrenewable power consumption. Reductions introduced by the REO-hop heuristic increase between 6:00 and 22:00, that is, when the solar energy is significant. The REO-hop heuristic still introduces more savings in the nonrenewable power consumption compared to the multihop-bypass heuristic when there is no solar energy in the network (from 0:00 to 4:00) as

TABLE 12.3 Simulation Scenario Parameters

Capacity of each wavelength	40 Gbps
Number of wavelength in a fiber	16
Distance between two neighboring EDFAs	80 km
Nonrenewable power consumption of a router port	1000 W
Nonrenewable power consumption of an optical switch	85 W
Nonrenewable power consumption of a multiplexer or a demultiplexer	16 W
Nonrenewable power consumption of a transponder	73 W
Nonrenewable power consumption of an EDFA	8 W

Source: Dong, X. et al., *J. Lightw. Technol.*, 29(1), 3, 2011.

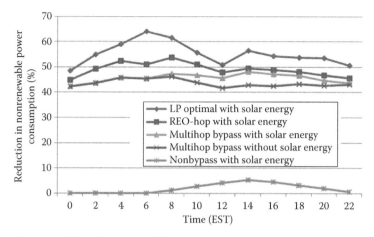

FIGURE 12.4 Reduction in the total nonrenewable power consumption (compared to nonbypass without solar energy) under different heuristics with and without solar energy.

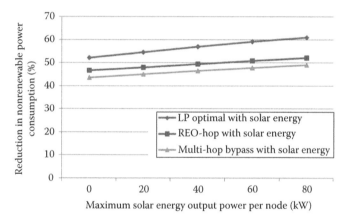

FIGURE 12.5 Reduction in the total nonrenewable power consumption (compared to nonbypass without solar energy) in 24 h period under different heuristics.

the REO-hop heuristic tries to route demands on virtual links with sufficient capacity rather than using shortest path routing as with the multihop-bypass heuristic.

The impact of different levels of the maximum solar energy (40, 60, and 80 kW) per node is examined in Figure 12.5. A solar cell area of up to 300 m^2 is needed to generate such values. Solar cell cladding with such surface area can be practically built in a typical core routing node location. Increasing the maximum solar energy output linearly reduces the total nonrenewable power consumption under different heuristics. Compared to the non-bypass without renewable energy case, the total nonrenewable power during a 24 h period has been reduced by about 47%–52% and 43%–49% under the REO-hop and multihop-bypass heuristics, respectively.

While the propagation delay of the multihop-bypass heuristic based on the shortest path is constant, the propagation delay of the REO-hop heuristic varies throughout the day as the REO-hop heuristic routes demands dynamically based on the solar energy available at nodes and the available network capacity. Figure 12.6 shows the increase in the average propagation delay of the REO-hop heuristic compared to the multihop-bypass heuristics based on the shortest path. The increase in the propagation delay is limited as the REO-hop heuristic routes demands using only the two shortest path routes (the increase is less than 0.3 ms, i.e., less than 10%) maintaining the QoS.

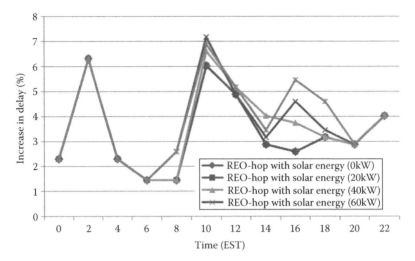

FIGURE 12.6 Increase in the average propagation delay under the REO-hop compared to the multihop-bypass heuristics.

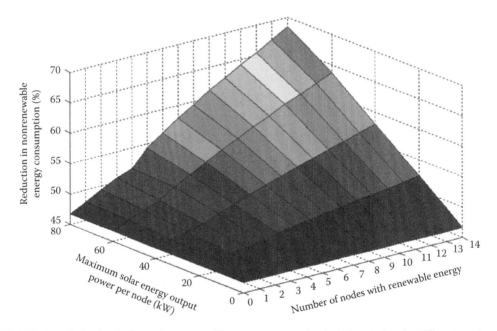

FIGURE 12.7 Reduction in the total nonrenewable power consumption in 24 h period under different maximum solar energy per node and different number of nodes employing solar energy.

12.3.3.2 Number and Location of Nodes That Use Renewable Energy

In Section 12.3.3.1, the five nodes with the highest nodal degree in the NSFNET network core were selected to use renewable energy. In this section, we investigate the impact of the number and location of nodes that use renewable energy.

In Figure 12.7, we investigate how the reduction in the nonrenewable power consumption changes with the maximum output power of renewable energy sources and the number of nodes that use renewable energy. The set of nodes selected to use renewable energy is as follows: {1}, {1,2}, {1,2,3}, {1,2,3,4}... {1,2,3... 14}, and the maximum output power of renewable energy sources ranges from 0 to

80 kW. Increasing the number of nodes with renewable energy and increasing the maximum available output power results in reducing the total nonrenewable power consumption of the network. However, the relation is not linear between the number of nodes with renewable energy and the nonrenewable power consumption, which implies that choosing some nodes to use renewable energy will result in further reductions in the nonrenewable power consumption compared to other nodes.

We investigated the impact of the location of nodes that use renewable energy on the total nonrenewable power consumption by developing another LP model with the objective of optimizing the location of nodes that use renewable energy such that the nonrenewable power consumption savings are maximized. The new LP model is subject to the same constraints in Section 12.3.1, except that Constraint 12.6 is replaced with Constraint 12.12 and a new constraint is added (Constraint 12.13). We consider time as a variable in this model (t is added to all the variables, where t is the time point of time set T). The model is given as follows:

Objective maximize [92]

$$\sum_{t \in T} \sum_{i \in N} \left(PR^s \left(Q_{it}^s + \sum_{p \in P} \delta_{ipt} Ws_{pt} \right) + \sum_{\theta \in E} PT^s ws_{\theta t} \delta_{i\theta t} + PMD^s DMs_{it} + PO_i^s y_{it} \right). \tag{12.11}$$

subject to the following Equations 12.2, 12.3, 12.4, 12.5, 12.7, 12.8, 12.9, 12.10 (every variable in these equations has had the time variable t augmented) [92]:

$$PR^s \left(Q_{it}^s + \sum_{p \in P} \delta_{ipt} Ws_{pt} \right) + \sum_{\theta \in E} PT^s ws_{\theta t} \delta_{i\theta t}$$

$$+ PMD^s DMs_{it} + PO_i^s y_{it} \le S_{it} \varepsilon_i \quad \forall i \in N, \quad t \in T, \tag{12.12}$$

$$\sum_{i \in N} \varepsilon_i = Ns. \tag{12.13}$$

where

$\varepsilon_i = 1$ if node i uses renewable energy; otherwise, $\varepsilon_i = 0$

Ns is the total number of nodes with access to renewable energy

Constraint (12.12) ensures that the renewable power consumption of each node at any time does not exceed the maximum output power available to it. In practice, batteries can be used to store the energy generated by solar cells; therefore, the solar energy availability constraint as a function of the time of day is relaxed. However, we do not include the use of energy storage elements in the current formulations. Constraint (12.13) gives the total number of nodes that use renewable energy Ns, which is set in advance.

The optimization results under different levels of the maximum renewable energy output power (20–80 kW), assuming $Ns = 5$ are as follows: nodes 4, 5, 6, 7, and 9.

In Figure 12.8, we verify the optimization results by evaluating the total nonrenewable power consumption under the REO-hop heuristic where we assume that only a single node uses renewable energy. We evaluate the performance under different values of the maximum solar energy per node. The REO-hop heuristic results match the optimization results where the reduction in the total nonrenewable power consumption is lower when the nodes in the center of the network (4, 5, 6, 7, and 9) use renewable energy.

Figure 12.9 shows the delay and power consumption of the REO-hop heuristic compared to the non-bypass without solar energy case. We consider a maximum renewable power of 60 kW per node when nodes at the core {4, 5, 6, 7, 9} or nodes at the edge {1, 2, 10, 11, 14} are selected to use renewable energy. The former node set results in a higher reduction in the nonrenewable power consumption compared to the latter node set. Also selecting the node set {4, 5, 6, 7, 9} results in a lower average propagation delay compared to the node set {1, 2, 10, 11, 14}.

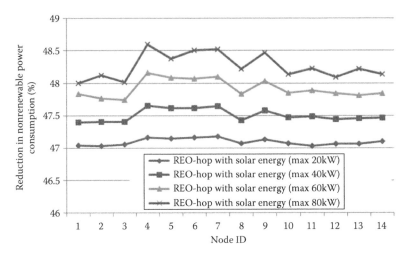

FIGURE 12.8 Reduction in the total nonrenewable power consumption (compared to non-bypass without solar energy) in 24 h period with different nodes using renewable energy.

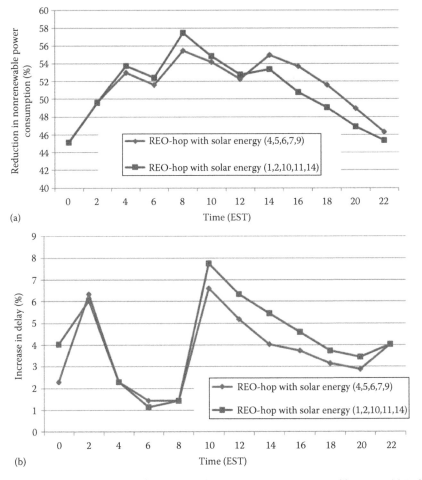

FIGURE 12.9 Performance under two different node selection scenarios using renewable energy. (a) Reduction in the nonrenewable power consumption. (b) Increase in the average propagation delay.

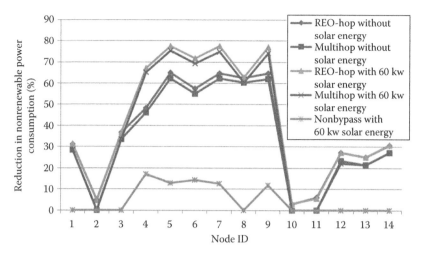

FIGURE 12.10 Reduction in the nonrenewable power consumption (compared to nonbypass without solar energy) of the nodes in a 24 h period under different heuristics.

12.3.3.3 Node Nonrenewable Power Consumption

In this section, we show the reduction in the nonrenewable power consumption experienced by each node individually under different heuristics. Considering a maximum solar energy output power of 60 kW, Figure 12.10 shows the reduction in the nonrenewable power consumption under the scenario when some nodes use renewable energy (nodes 4, 5, 6, 7, and 9) compared to the scenario when no renewable energy sources are used. Under the non-bypass heuristic, the nonrenewable power consumption of nodes varies significantly between nodes at the center and nodes at the edge of the network as nodes at the center consume more energy as more traffic flows are routed through them. The multihop-bypass and REO-hop heuristics have significantly reduced the nonrenewable power consumption and its variance. However, nodes at the center of network still have slightly more nonrenewable power consumption than the nodes at the edge of the network. As expected, the REO-hop heuristic results in further reductions compared to the multihop-bypass heuristic. The nonrenewable power consumption of nodes at the center of the network, where renewable sources are deployed, has significantly decreased when renewable energy sources are introduced to the network.

12.3.3.4 Nonrenewable Power Consumption under Adaptive Link Rate with the REO-Hop Heuristic

In this section, we investigate the impact of ALR on the nonrenewable power consumption. Different energy profiles are proposed to define of the dependency between equipment energy consumption and traffic load [87]. We consider the following:

1. "On–off" energy profile [87].
2. "Linear" energy profile: Here the energy consumption depends linearly on the traffic load, for example, in switch architectures such as batcher, crossbar, and fully connected [87,100].
3. "Log10" energy profile: This profile is employed in equipment that uses hibernation techniques such as the low-power idle technique for Ethernet [99,100]. In this approach, data are sent as fast as possible to allow the equipment to quickly return to the low-power idle state.
4. "Log100" energy profile: This profile is considered as a middle function between the "on–off" and the "Log10" profiles.
5. "Cubic" energy profile: Typical in equipment that uses dynamic voltage scaling (DVS) and dynamic frequency scaling (DFS) [87].

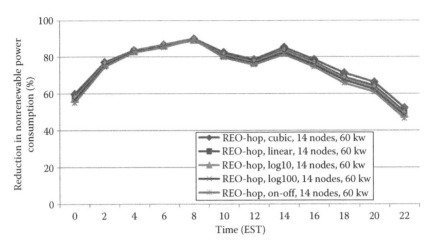

FIGURE 12.11 Reduction in the nonrenewable power consumption (compared to nonbypass without solar energy) of the REO-hop heuristic under different energy profiles.

Figure 12.11 shows the reduction in the nonrenewable power consumption of the REO-hop heuristic compared to the peak in Figure 12.4 (nonbypass without renewable energy) under different energy profiles for router ports and transponders (the two most power consuming components in the node) with 80 kW solar power at all the nodes. The nonrenewable power consumption is subject to the energy profile. The "cubic" profile results in the highest reduction in nonrenewable power consumption. Compared to the "on–off" profile, the "cubic" profile has reduced the nonrenewable power consumption by up to 9% between 12:00 and 20:00. This relatively small reduction in power consumption associated with ALR is commensurate with the fact that the ALR profiles will only make a difference to partially loaded wavelengths. Note that the power consumption continues to decrease beyond 6 pm due to the availability of solar power at all the nodes. A maximum reduction of 97% is achieved. The average savings are approximately 78% and vary slightly between the five profiles. Note that these savings are mostly due to architecture design (photonic switching instead of electronic routing) and powering down unused router ports and transponders as the renewable energy is low here and has limited effect.

12.4 Data Centers and Virtualization

Data centers are even more of a contributor to energy bills, and by extension carbon emission, than switching and communications lines. As well as housing many power-hungry server CPUs, data centers have moving parts—in other words disks. Of course, data centers also contain internal interconnects, and these are amenable to many of the power-reduction strategies and techniques as the external network itself (Figures 12.12 and 12.13). Modern data centers frequently make use of commodity hardware for the servers, since this is low cost and modular. This has some benefit in terms of energy savings since the design of a typical machine for the small office or home installation is often driven by operational as well as capital cost constraints, and so may tend to become more energy efficient over time.

The granularity of a single commodity server is often mismatched to the distribution of server demands for the many co-located customers in a large data center. To pool resources more efficiently, while still retaining partitioning or isolation of processing, I/O and storage demands, the typical modern data center makes heavy use of virtualization.

FIGURE 12.12 A machine room with typical wiring.

FIGURE 12.13 Entropy taking over.

12.4.1 Virtualization and Distributed Computing

There are very general lessons to be learned from distributed computing, which can be applied across a range of energy problems, as proposed in Andy Hopper's vision in [101]. More specifically, we have had the capability to determine where to place a job in a distributed system that is planet wide for many years [102]. Decision procedures for admission and placement of jobs in a system can and do leverage a past work with resource management in networks [103] with real-world workloads [104,105]. Global workload mobility allows everything from a disk drive [106] to a data center [107] to power down when electricity demand is high by allowing providers to maintain service levels by moving jobs to alternative locations where demand is low. Intermittent renewable energy can also be absorbed by this mechanism. Computing will be making use of energy that would otherwise be lost through overgeneration or use of inefficient generation techniques. Some researchers have considered a dedicated network for connecting a globally distributed set of data centers, so as to provide transparent workload migration [108] although energy conservation during migration will require high-bandwidth, low-latency connections.

12.4.2 An Agile, Power-Aware API

Provided with measurement and modeling of power utilization, information provided through a suitable application programming interface (API) can permit management and power-waste mitigation. The needs for hardware support with APIs and the functionality in protocols to exploit such hardware support

FIGURE 12.14   A rack with a top-of-rack switch.

for energy management and mitigating the possible negative impact on protocol and service performance has not been examined, and we propose to address these problems. There is also a need to permit agility in the network components; currently, there is a highly rigorous determination of the roles for each network subsystem (from host and server through NIC, switch, and router) (Figure 12.14). Work in this area aims to identify and evaluate a flexible interface that permits such extensions as end-point deflection routing (to support migration), network function agility (to accommodate both optimal performance and workload offloading—as determined by power needs and architectural constraints), and ideas of network and communications subsystems—such as NICs and switch subsystems that permit the powering down to sleep of entire systems while providing the lowest level of network response and a fast restart on demand—in the style of Somniloquy [65]. One can envisage such APIs providing the necessary flexibility to allow roles to be moved from host network stack and application through to being entirely offloaded into core switch components.

12.4.3 Agile Virtualization

It is estimated that servers and associated cooling overheads comprise around 1.2% of global energy demand [109]. SLAs typically require high levels of availability and so service providers also require highly reliable energy supplies. Some large-scale content distribution network (CDN) providers such as YouTube employ a high-speed, low-latency network to transparently replicate content so that most users perceive low-latency, high-performance downloads from nearby servers. Others such as Akamai already use energy signals such as price to route requests in their overlays [110]. One can envisage the same networks being used to move workloads around the world [111]. This network would allow the use of computing as a virtual battery ameliorating local changes in power availability by modulating energy consumption in sympathy with local grid demand. This workload mobility will allow continuous service provision, despite variations in power supply, which will allow us to locate computing installations near to remote renewable energy generation sites. This has the potential to improve the viability of remote renewable energy generation sites by allowing one to consume energy on site rather than incurring electricity transmission overheads [101,112–114]. Grid demand varies significantly over short timescales according to many different processes and so the National Grid is continually working to balance generation to demand. This is made possible through reserve generation capacity, which is provided either by operating large scale generators at less than full capacity or through the provision of standing reserves, which can be activated at short notice. Both of these approaches are less efficient than a large scale plant operating at full capacity. Our vision of global workload mobility will allow data centers to power down when demand is high by allowing providers to maintain service levels by moving jobs to alternative locations where demand is low. Intermittent renewable energy can also be absorbed by this mechanism. At the lowest level, we consider the physical link layer of the network. In particular, one expects future interconnects to be focused on optical networks supporting low-latency, low-power switching [115]. However, it takes 3 years of continuous operation to equal the manufacturing energy embodied in computing hardware [116], so powering down machines to release energy to the general grid is increasing the relative size of this cost. Relative impacts of provisioning fiber optic networks [117] or power transmission systems must also be considered.

12.4.4 Workload Suitability

Noninteractive or batch-computation jobs are becoming an increasingly significant proportion of computing workload [118]. Examples of this type of workload include indexing web pages and optical character recognition of books or scientific simulations. These types of workload provide flexibility to our system because they can have weaker constraints on acceptable levels of downtime. For example, one might choose to prioritize the transmission of interactive workloads when shutting down a data center or even to simply pause noninteractive jobs and wait for power to become available again. Interactive workloads are more challenging because they must continue to be available despite changes in power supply. Very highly responsive workloads might prove unsuitable to this approach because the downtime (however minimal) incurred by migration might not be tolerable. However, there is a large class of workloads for which small, occasional periods of downtime will be largely unnoticeable. Particular examples might include web, e-mail, or Domain Name System (DNS) services, this can benefit from the significant existing Internet-characterization experience [119–121].

12.4.4.1 Virtualization and Migration

The recent resurgence in virtualization provides the ideal software platform for mobile workloads [110]. Virtualization allows us to run and migrate existing workloads without requiring extensive engineering. There are a number of parameters to virtual machine migration, which will affect the performance of our network. The majority of a workload can often be migrated in the background while it continues to run on the host machine. This gives two quantities for describing workload movement. Migration time is the total amount of time to fully move the job from one host to another and downtime is the total amount of time for which the job was unavailable. Different workloads (with different patterns of memory use) require different amounts of migration and downtime. Another consideration is the location of any storage used by the workload. Conventionally, migration takes place in a local network using a storage area network (SAN). This means that any fixed storage is automatically available to the migrated workload. Suitable guarantees from the network might permit wide-area migration using a SAN. For example, one might imagine a scenario in which the computing workload is moved out of a data center, which then continues to operate but with less power demands providing storage capacity to these workloads. Alternatively, strategies exist for mirroring or live replication of remote storage, which can ensure that a workload has an up-to-date image to work from on arrival [122].

12.4.5 Dynamic Computation Placement Architecture

At present, data centers are provisioned with worst-case power and cooling capacity in order to enable them to meet user SLAs. However, this provisioning comes at the cost of wasted energy both as stored electrical energy and energy expended in the cooling subsystem. It is our intention to develop an architecture where computation can be moved in accordance with energy policy constraints, for example, where the cost of energy is cheapest or where there are reserves that would otherwise be wasted, with the goal of controlling the total power consumed by computation. Virtualization provides the benefit of abstraction—all the components that form the computation (the operating system, binaries, support libraries, etc.) are packaged together. This enables us to treat the entire software stack as a single unit for management purposes. Specifically, the migration feature present on most major virtualization platforms to build a dynamic computation placement architecture. Migration works by physically moving a computation unit from one host to another by moving all state and data information between the hosts. Current migration architectures have been designed for low-latency, low-bandwidth single-hop LAN setups where migrations are carried out occasionally within a small pool of well-known hosts. However, in our architecture all these parameters change. This architecture will require migrations be executed within pools of frequently changing hosts separated by large physical distances connected by high-bandwidth, high-latency networks. Moreover, the underlying network characteristics (both at the routing and physical layers) are likely to deviate substantially from current designs.

This suggests some of the following challenges to meet the goal of a dynamic computation placement architecture work:

1. Characterizing computation and workload: As a first step into realizing this architecture, we must characterize the computation and data footprints of a number of representative workloads for the purposes of determining which computation models are suitable for dynamic computation placement. One could start by leveraging the extensive network monitoring and measurement experience in our team [119,120,123,124].

2. Predicting migration: As a follow-on to the previous step, a model will be needed to predict the migration time for any computation in the system thereby creating a theoretical foundation on which we can build the system. Past work on migration–location–identification [102,125] will assist this phase significantly.

3. Network characterization: We expect the underlying energy-efficient network developed in Section 12.2, will inform us through its models of both network needs and load and of network topology, in combination with the opportunities for changes in the underlying physical network system, which we anticipate will contain significantly differing properties compared to current network designs. This will require a complete analysis and characterization of the network for the purposes of exploiting it for the purpose of system construction, possibly refining recent contributions in techniques for topology modeling and comparison [126].

4. Energy generation patterns: As we create ever more dynamic patterns of demand, it will be necessary to track the evolution of the generation of power, geographically and temporally. As renewable sources come online, these will be more decentralized and more variable; this might present a challenge for power grid engineers, but for the Internet and the web, it may represent an excellent opportunity for co-evolution, with small and very large data centers being co-located at all new tidal, wind, and solar sites to take up any local overproduction as and when needed. This is more a civil engineering challenge than one for computer science and electrical and power engineering.

5. Recent rapid evolution of smartphone operating systems such as the iPhone and Android Unix based operating system, together with energy-aware application design, has led the way. However, we expect to see the same family of techniques applied in operating system (OS) refinements for data centers and for the home and small office, with an emphasis on enabling efficient and timely migration within the architecture and by changing their design. We anticipate extensions of recent work on disk-energy needs [106], energy-aware whole-data center storage management, such as in Sierra [127], data center migration [102,107], and the vision of a Green ICT [101].

12.5 Advances in Photonic Physical Layer

The network concepts described in Sections 12.1 through 12.4 require an energy-efficient physical transport layer. It has been shown [128] that optical point-to-point data transmission over medium and extended distances has the potential to be extremely energy efficient, with optimal choice of the optical transmission technology, as is currently implemented between Ethernet switches and IP routers (e.g., a 40 channel 40 Gbps 1000 km transmission system requires 1.1 nJ/bit energy consumption). There is, however, a significant potential for further energy efficiency improvement by the use of optical technologies within IP routers, as there exists the inherent advantage that optical transmission systems and switches can effectively be bit-rate agnostic, having the potential to switch very high data rate signals at the packet and flow level, rather than at the bit level as in the purely electronically switched IP routers of today. In contrast, energy efficiency presents a real challenge to electronic switching, with current IP routers consuming 20 nJ/bit [129], and a 1 Tbps aggregate IP router consuming 20 kW—not only an energy consumption issue, but also a thermal management one.

With optical technologies enabling both energy-efficient transport and routing, energy-efficient networks can be devised in optimal form not only to minimize energy consumption but also to reduce related carbon emissions, for example, by enabling the ready use of low carbon energy sources.

We therefore concentrate on a discussion of optical switching technologies for high energy efficiency, and describe how optical switch–based technologies are starting to achieve the scalability and performance required for implementation within the next generation of switching and routing equipment. In describing such switches, it is important to note developments in transport technologies as any usable switch must be able to operate on the transmission signals. This section therefore first considers the development of energy-efficient optical communication links.

From the early 1990s, optical transmission systems typically connected telephone exchanges. These telecom systems used temperature-controlled semiconductor lasers in single wavelength low data rate circuits. In addition to the electro-optical conversion energy consumption, the photonic transmitters consumed a greater amount of power in the thermoelectric coolers required to maintain the laser operating temperature than in the light generation and transmission itself. They were also limited to transmission distances of less than 100 km before the signal needed to be regenerated electrically, as there were then no effective optical amplifiers in existence.

The invention of the EDFA-enabled [128,130] efficient optical amplification opening the way to wavelength division multiplexed transmission of many wavelengths over extended distances in an optical fiber, this improving long distance telecommunications. This technology provided a significant boost to IP communications between routers, but was not sufficiently low cost to allow the extension of high data rate communications from the router or switch to the desktop.

Gigabit Ethernet [131] and 10 Gb Ethernet [132] standards drew attention to the growing need for efficient and low-cost optical transmission systems over short distances between Ethernet switches and in certain cases from the switch to the user, requiring lasers to operate at high modulation rates at high ambient temperatures, without the need for a thermoelectric cooler. Extensive work on semiconductor laser design enabled operation at 10 Gbps direct modulation at 100°C at 1.3 μm [133–135], providing further reduced energy consumption. As a result, optical links can now be constructed with energy consumptions of 130 pJ/bit at rates of 10 Gbps and operating lengths of up to 10 km at ambient temperatures of 85°C [136]. With the advent of these energy-efficient optical links at certain wavelengths, it is interesting to develop optical switch technologies that have the potential for energy efficiency.

Broadly, optical switch systems can be classified into two types: those whose switching time is longer than a typical packet length for circuit switching [thermo-optic and micro electromechanical switches (MEMS)] and those switches that can switch in several nanoseconds or fewer, making them suitable for packet switching (lithium niobate and semiconductor).

MEMS are capable of being used in large array optical cross-connects with low insertion losses and low power consumption but with relatively slow switching speeds [137,138] (e.g., a 128 × 128 way cross-connect with 2.6 dB average insertion loss has a 57 ms switching time [139])

Thermo-optic switches are usually based on Mach–Zehnder modulators with a resistive heating element next to one arm, which is heated to produce an optical phase change and so switch the output of the modulator from one output port to the other.

Examples of this form of switch include silica (e.g., a 2 × 2 port Mach–Zehnder interferometer based switch, in which there is an electric heater in each of the arms. In order to switch the output from the interferometer the 45 mW heating power is initiated in one arm, causing an optical phase change and output switch in a response time of 3 ms [140]) and polymer-based switching elements (e.g., a 2 × 2 port polymer waveguide based Mach–Zehnder modulator, with a heater on one arm. Here, to initiate the switch, the heater is turned on, consuming 4 mW, and causing the output to switch from one port to the other in 160 μs [141]).

Switches that can operate on the packet timescale are generally of greater interest for applications within IP routers or Ethernet switches. To this end there are several approaches, such as the use of high-speed modulators, silicon photonic, and semiconductor optical amplifier (SOA)-based optical switches.

It is possible to create fast optical switches based on lithium niobate Mach–Zehnder modulators—these components offer fast switching speeds, but at the expense of high operating powers and significant losses [142,143]. For example, an 8 × 8 port switch was demonstrated but with 15 dB insertion loss [144].

Another potentially attractive technology for nanosecond scale optical switches is that of SOAs. These have received considerable interest in recent years as they have the potential of combining inherent optical gain, fast switching speed, potential for uncooled operation, and the capability of integrating large numbers of components, in order to make functionally compact switch subsystems.

Integrated 2 × 2 port crosspoint switches have been reported as early as 1990 [145], with designs incorporating low-loss passive waveguides and active SOA switching elements reported in 1993 [146,147]. A schematic of designs typical of early 2 × 2 port switches is depicted in Figure 12.15, with passive optical waveguides, splitters, and mirrors and SOAs as switching elements.

Clearly increased port count is of great importance for an optical switch, so considerable research effort has been expended in the design and realization of integrated optical switches. While SOAs have many advantages as optical switching elements, they can introduce excess amplified spontaneous emission noise and also can impart optical patterning distortion to the data signals if they are not designed and operated in the correct way. Two designs for higher port count switches are presented, one a semi-integrated 8 × 8 port switch subsystem and the other an integrated 16 × 16 port switch fabric, together with a route to a switch, which can operate uncooled at very high data capacities.

An 8 × 8 port switch has recently been demonstrated by Kai et al. of Fujitsu [148]. In this case, scheduled multiwavelength packets (10λ × 10 Gbps) are split in a 1 × 8 splitter, each output of which is fed to an input of an integrated array of 8 SOAs, which are combined and amplified by a further SOA within the integrated module, shown schematically in Figure 12.16a. Thus, any of the eight inputs can be directed to any of the eight outputs. This design uses integration with each of the modules being packaged and controlled, but fiber splitters connect the modules as can be seen in Figure 12.16b contained in a 19″ rack mounting enclosure.

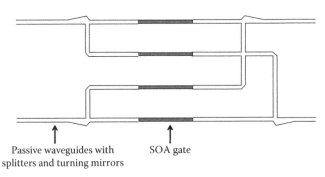

Passive waveguides with   SOA gate
splitters and turning mirrors

FIGURE 12.15   Schematic of designs typical of early 2 × 2 port switches, with passive optical waveguides, splitters and mirrors, and SOAs as switching elements.

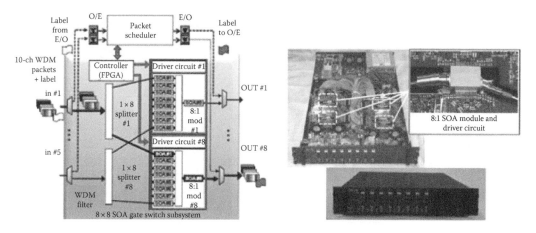

FIGURE 12.16 (a) A schematic of the 8 × 8 switch. (b) The completed switching subsystem from [148]. (From Kai, Y. et al., A compact and lossless 8 × 8 SOA gate switch subsystem for WDM optical packet interconnections, in *ECOC 2008*, pp. 1–2, September 21–25, 2008.)

TABLE 12.4 SOA Optical Switch Architecture

Switch Architecture	SOA Gates	Cascading Stages	Maximum Number of SOA Gates in Each Stage	Splitting Loss per Stage (dB)
Tree	256	1	256	24
Benes	224	7	32	6
Clos	192	3	64	12

Switches with port-counts of greater than 8 are desirable for use within Ethernet switches and IP routers, but the choices of architecture become key as larger port counts are considered. As an example of this, Table 12.4 lists, the number of SOAs, the number of cascaded switching components and the splitting losses between each switching stage which are required for 16 port Tree, Benes and Clos architecture based switch fabrics. It is desirable to limit both the total number of SOAs and the number of cascaded SOAs in any design, to limit the buildup of noise and distortion. This must however be balanced against the split-and-combine loss introduced by the number of cascaded splitting stages within any switch design. The choice of switch fabric design is thus an optimization of these competing factors. It can be seen that for the three architectures considered, the Clos is the best compromise between number of required SOA gates, cascaded SOAs and splitting losses.

Wang et al. [149] of Cambridge University have produced an integrated 16 × 16 port optical switch. Schematically the switch is made from 12 "4 × 4" port switching elements, each containing an input shuffle network, 16 gating SOAs, and an output shuffle network. These 4 × 4 switching elements are then combined in a Clos architecture, with two additional shuffle networks between them as shown in Figure 12.17.

The switch has been realized on indium phosphide and contains 1114 functional components, including 192 gating SOAs, being the highest component count active photonic integrated circuit yet constructed and is extremely compact, with dimensions of 6.3 × 6.5 mm. The device, shown in Figure 12.18 operates error free, with a power penalty of between 1.8 and 5 dB. This early demonstration of a 16 × 16 port integrated switch optical switch is fabricated from all active material, necessitating all of the shuffle networks (most of the switch) to be driven electrically.

It should be emphasized that this first design run of the 16 × 16 port integrated optical switch fabric is an all active construction, with all of the chip surface being pumped electrically. Improved fabrication

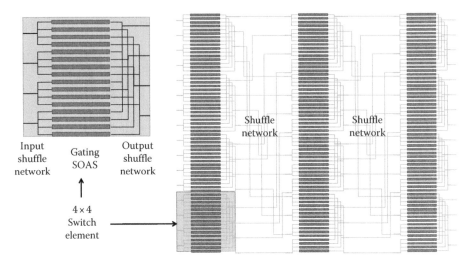

FIGURE 12.17 Schematic of the 16 × 16 port integrated switch in [149]. (From Wang, H. et al., Demonstration of a lossless monolithic 16 × 16 QW SOA switch, in *ECOC '09 Post-Deadline*, pp. 1–2, September 20–24, 2009.)

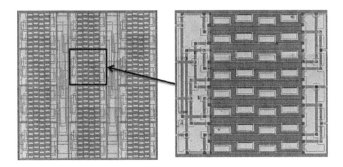

FIGURE 12.18 The fabricated 16 × 16 port switch, from [149], showing the whole device, and inset, a 4 × 4 port switching element, with optical splitters, 90° bends, and the 16 horizontal gating SOA switches.

with passive waveguides and active SOAs will not only reduce the electrical power consumption by more than half, but will also significantly improve the optical performance of the switch fabric, as the current noise source from the electrically pumped shuffle waveguides will be removed. It is calculated that the active passive version of this switch will require no more than 2 pJ/bit of energy in operation in its 16 × 16 port with 10 wavelengths at 10 Gbps per path operating regime. Although this is by no means all that is required for a router, the power consumption is far less than the 10 s of nJ/bit consumed by today's IP routers.

The integrated switch described earlier is rearrangeably nonblocking, requiring the switch to be reconfigured for all inputs simultaneously. This can easily be achieved in a time-slotted protocol, where all packets are of the same length, with centralized scheduling, with buffering being performed at the edges of the network [150]. These issues are inherent to most optical switch designs, as optical buffering is unavailable.

Bergman et al. [151,152] have produced a time-slotted optical switch based on SOAs, which reads WDM packet headers on the fly and performs deflection routing in a "data vortex." This solution is hard to integrate as it requires both electrical communication from neighboring nodes and latency between nodes to allow for on the fly processing.

Improvements in energy efficiency and functionality with SOA-based optical switch fabrics are anticipated within the next few years. One such key advance is the adoption of processing methodologies similar to those which have been learned in complementary metal-oxide semiconductor (CMOS) foundries, with the use of standardized "building blocks" [153]. This will enable the design of larger switch fabric designs including extensive active/passive material integration, with the SOA gates made from active material with high optical gain at high temperatures and the interconnects and splitters made from low-loss passive materials. This will dramatically reduce the energy consumption and cooling requirements of integrated SOA switch fabrics, making cooler-less operation possible, because of the low density of active switching elements on the largely passive switch fabric. In addition, the addition of on-chip power monitors and added functionality like integrated wavelength conversion [154] will greatly reduce the requirement for addition external parts.

The switches described here are constructed from the mature quantum well epitaxial technology, but work has been reported using quantum dots as the active medium for optical amplifiers, with their improved thermal and saturation performance [155]. An integrated 2×2 port switch, operating at 1.3 μm has been demonstrated to have stable gain in excess of 10 dB from room temperature to over 70°C [156]. This offers the possibility of large scale integrated optical switches without the need for thermoelectric coolers.

Additionally, recent advances in the field of silicon photonics have enabled optical sources, such as the hybrid silicon laser [157] to be integrated with modulators [158] and detectors to produce high-performance low-cost optical links for computing applications [159]. Further work based on ring resonators [160] has enabled initial demonstrations [161] of small port count integrated switch fabrics, with the possibility of moderate port count energy-efficient switch fabrics in the future.

It is thus envisaged that cooler-less optical switch fabrics will scale to larger than 32 or 64 ports, operating at energy efficiencies of 1 pJ/bit or better, being good candidates for use within tomorrow's routers and switches.

12.6 Conclusions

This chapter has presented and discussed different aspects of intelligent energy-aware networks. In Section 12.2, we discussed the techniques proposed to reduce the energy consumption of communication networks at the equipment and network levels. In Section 12.3, we have investigated the use of renewable energy in a hybrid-power IP-over-WDM network to reduce the nonrenewable energy consumption and consequently, CO_2 emissions. We have developed an LP optimization model and a novel heuristic, REO-hop, to optimize the use of renewable energy in the hybrid-power IP-over-WDM architecture. The results show that compared to the multihop-bypass, the REO-hop heuristic has reduced the nonrenewable energy consumption by 47%–52% while maintaining QoS. Another LP model has been developed to optimize the selection of nodes deploying renewable energy. The results show that compared to nodes at the edge, selecting nodes at the center of the network to deploy renewable energy results in higher reductions in the total nonrenewable energy consumption. Also we have investigated ALR where load-dependent energy consumption is assumed. The results show that the "cubic" energy profile gives the highest reduction in the nonrenewable energy consumption. Compared to the case where all nodes are statically dimensioned for the maximum traffic in terms of IP ports and optical layer, the REO-hop routing heuristic in the NSFNET network with the optimal node selection, each node has access to 80 kW renewable power each, total energy savings of 97% peak and 78% average are achieved. In Section 12.4, we have discussed using virtualization technologies for workload migration in data centers to minimize energy consumption. In Section 12.5, we have considered how photonic technology has the potential to provide energy-efficient integrated switch fabrics, reducing the energy consumption of Ethernet switches and IP routers.

References

1. Climate change: Valuing emissions, Department of Environment Food and Rural Affairs (DEFRA), U.K., 2008, http://www.defra.gov.uk/environment/
2. U.K. Parliament Select Committee on Science and Technology second report, Chapter 3, The economics of energy efficiency, 2005, http://www.publications.parliament.uk/pa/ld200506/ldselect/ldsctech/21/2106.htm
3. K.J. Christensena, C. Gunaratnea, B. Nordmanb, and A.D. George, The next frontier for communications networks: Power management, *Computer Communications*, 27(18), 2004, 1758–1770.
4. C. Bianco, F. Cucchietti, and G. Griffa, Energy consumption trends in the next generation access networks—A telco perspective. In *Proceedings of the 29th International Telecommunications Energy Conference, 2007 (INTELEC 2007)*, Rome, Italy, September 2007, pp. 737–742.
5. M. Etoh, T. Ohya, and Y. Nakayama, Energy consumption issues on mobile network systems. In *Proceedings of the International Symposium on Applications and the Internet (SAINT'08)*, IEEE, Turku, Finland, July 28, 2008–August 1, 2008, pp. 365–368.
6. Telecom Italia web site, The environment, http://www.telecomitalia.com/tit/en.html
7. BT Press, BT announces major wind power plans, October 2007, http://www.btplc.com/news/articles/showarticle.cfm?articleid=%7Bdd615e9c-71ad-4daa-951a-55651baae5bb%7D
8. ITWales, Green evangelist to call for big changes in computer use to aid environment, *ITWales Conference*, November 2007, http://www.itwales.com/997539.htm
9. An Inefficient Truth, Global action plan, London, U.K., December 2007, http://www.globalactionplan.org.uk/
10. K. Kawamoto, J.G. Koomey, B. Nordman, R.E. Brown, M.A. Piette, M. Ting, and A.K. Meier, Electricity used by office equipment and network equipment in the US, *Energy—The International Journal*, 27(3), 2002, 255–269.
11. J. Koomey, *Rebuttal to Testimony on 'Kyoto and the Internet: The Energy Implications of the Digital Economy*, Berkeley, CA: Lawrence Berkeley National Laboratory. LBNL-46509, 2000, http://enduse.lbl.gov/
12. M. Yamada, T. Yazaki, N. Matsuyama, and T. Hayashi, Power efficient approach and performance control for routers, *Workshop on Green Communications at IEEE ICC'09*, Dresden, Germany, June 14–18, 2009, pp. 1–5.
13. SMART 2020 Report, Enabling the low carbon economy in the information age, 2008. http://www.theclimategroup.org
14. T.E.H. El-Gorashi and J.M.H. Elmirghani, Differentiated resilience with dynamic traffic grooming for WDM mesh networks. In *Proceedings 11th International Conference on Transparent Optical Networks (ICTON'09)*, São Miguel, Portugal, June 28, 2009–July 2, 2009.
15. T.E.H. El-Gorashi and J.M.H. Elmirghani, Differentiated resilience for anycast flows in MPLS networks, ICTON 2009, Island of São Miguel, Azores, Portugal, June 28–July 2, 2009, pp. 1–5.
16. S. Nanda and T. Chiueh, A survey of virtualization technologies, Department of Computer Science, SUNY at Stony Brook, Tech. Rep., 2005.
17. R. Katz, Tech titans building boom, *IEEE Spectrum*, 46(2), 2009, 40–54.
18. X. Fan, W.-D. Weber, and L.A. Barroso, Power provisioning for a warehouse-sized computer. In *Proceedings of the 34th Annual International Symposium on Computer Architecture*, ser. ISCA'07, ACM, New York, 2007, pp. 13–23.
19. A.P. Bianzino, C. Chaudet, D. Rossi, and J.-L. Rougier, A survey of green networking research, Tech. Rep. arXiv:1010.3880, October 2010.
20. R. Bolla, R. Bruschi, F. Davoli, and F. Cucchietti, Energy efficiency in the future Internet: A survey of existing approaches and trends in energy-aware fixed network infrastructures, *IEEE Communications Surveys and Tutorials*, 13(2), 2010, 223–244.

21. M. Gupta and S. Singh, Greening of the internet. In *Proceedings of the 2003 Conference on Applications, Technologies, Architectures, and Protocols for Computer Communications*, ser. SIGCOMM'03, ACM, New York, 2003, pp. 19–26.

22. S. Nedevschi, L. Popa, G. Iannaccone, S. Ratnasamy, and D. Wetherall, Reducing network energy consumption via sleeping and rate-adaptation. In *Proceedings of the 5th USENIX Symposium on Networked Systems Design and Implementation*, ser. NSDI'08, USENIX Association, Berkeley, CA, 2008, pp. 323–336.

23. M. Gupta, S. Grover, and S. Singh, A feasibility study for power management in LAN switches. In *Proceedings of the 12th IEEE International Conference on Network Protocols*, IEEE Computer Society, Washington, DC, 2004, pp. 361–371.

24. M. Gupta and S. Singh, Dynamic Ethernet link shutdown for energy conservation on Ethernet links. In *IEEE International Conference on Communications, 2007, ICC'07*, Glasgow, Scotland, UK, June 24–28, 2007, pp. 6156–6161.

25. H. Tamura, Y. Yahiro, Y. Fukuda, K. Kawahara, and Y. Oie, Performance analysis of energy saving scheme with extra active period for LAN switches. In *IEEE GLOBECOM*, Washington, DC, November 26–30, 2007, pp. 198–203.

26. M. Rodríguez-Perez, S. Herreria-Alonso, M. Fernandez-Veiga, and C. Lopez-Garcia, Improved opportunistic sleeping algorithms for LAN switches. In *Proceedings of the 28th IEEE Conference on Global Telecommunications*, ser. GLOBECOM'09, IEEE Press, Piscataway, NJ, 2009, pp. 35–40.

27. P. Gunaratne and K. Christensen, Predictive power management method for network devices, *Electronics Letters*, 41(13), 2005, 775–777.

28. I. Kamitsos, L. Andrew, H. Kim, and M. Chiang, Optimal sleep patterns for serving delay-tolerant jobs. In *Proceedings of the 1st International Conference on Energy-Efficient Computing and Networking*, ser. e-Energy'10, ACM, New York, 2010, pp. 31–40.

29. M. Gupta and S. Singh, Using low-power modes for energy conservation in Ethernet LANs. In *Proceedings of the 26th IEEE International Conference on Computer Communications. IEEE INFOCOM 2007*, Anchorage, Alaska, USA, May 6–12, 2007, pp. 2451–2455.

30. G. Ananthanarayanan and R.H. Katz, Greening the switch. In *Proceedings of the 2008 Conference on Power Aware Computing and Systems, ser. Hot-Power'08*, USENIX Association, Berkeley, CA, 2008, pp. 7–7.

31. A. Bashar, G. Parr, S. McClean, B. Scotney, M. Subramanian, S. Chaudhari, and T. Gonsalves, Employing Bayesian belief networks for energy efficient network management. In *National Conference on Communications (NCC)*, Chennai, India, January 29–31, 2010, pp. 1–5.

32. T. Smith, R.S. Tucker, K. Hinton, and A.V. Tran, Implications of sleep mode on activation and ranging protocols in PONs. In *21st Annual Meeting of the IEEE Lasers and Electro-Optics Society (LEOS)*, Acapulco, Mexico, November 9–13, 2008, pp. 604–605.

33. R. Kubo, J.-I. Kani, Y. Fujimoto, N. Yoshimoto, and K. Kumozaki, Sleep and adaptive link rate control for power saving in 10G-EPON systems. In *Proceedings of the 28th IEEE Conference on Global Telecommunications*, ser. GLOBECOM'09, IEEE Press, Piscataway, NJ, 2009, pp. 1573–1578.

34. Y. Yan, S.-W. Wong, L. Valcarenghi, S.-H. Yen, D. Campelo, S. Yamashita, L. Kazovsky, and L. Dittmann, Energy management mechanism for Ethernet passive optical networks (EPONs). In *IEEE International Conference on Communications (ICC)*, Capetown, South Africa, May 23–27, 2010, pp. 1–5.

35. B. Nordman and K. Christensen, Reducing the energy consumption of network devices. In *Tutorial for the July 2005 IEEE 802 LAN/MAN Standards Committee Plenary Session*, ser. IEEE, San Francisco, California, USA, July 18, 2005.

36. 802.3az Energy Efficient Ethernet, meeting materials, IEEE, Tech. Rep., January 2008. Available online at http://www.ieee802.org/3/az/index.html

37. P. Reviriego, J. Maestro, J. Herna andndez, and D. Larrabeiti, Burst transmission for energy-efficient Ethernet, *IEEE Internet Computing*, 14(4), 2010, 50–57.

38. L.A. Barroso and U. Holzle, The case for energy-proportional computing, *Computer*, 40, December 2007, 33–37.
39. C. Gunaratne, K. Christensen, and B. Nordman, Managing energy consumption costs in desktop PCs and LAN switches with proxying, split TCP connections, and scaling of link speed, *International Journal on Network Management*, 15, September 2005, 297–310.
40. C. Gunaratne, K. Christensen, B. Nordman, and S. Suen, Reducing the energy consumption of Ethernet with adaptive link rate (ALR), *IEEE Transactions on Computing*, 57, April 2008, 448–461.
41. C. Gunaratne, K. Christensen, and S. Suen, NGL02-2: Ethernet adaptive link rate (ALR): Analysis of a buffer threshold policy. In *Proceedings of the IEEE Global Telecommunications Conference, 2006. GLOBECOM'06*, San Francisco, California, USA, November 27–December 1, 2006, pp. 1–6.
42. P. Mahadevan, P. Sharma, S. Banerjee, and P. Ranganathan, Energy aware network operations. In *Proceedings of the 28th IEEE International Conference on Computer Communications Workshops*, ser. INFOCOM'09, IEEE Press, Piscataway, NJ, 2009, pp. 25–30.
43. C. Gunaratne and K. Christensen, Ethernet adaptive link rate: System design and performance evaluation. In *Proceedings of the 31st IEEE Conference on Local Computer Networks*, Tampa, Florida, November 14–16, 2006, pp. 28–35.
44. B. Zhang, K. Sabhanatarajan, A. Gordon-Ross, and A. George, Real-time performance analysis of adaptive link rate. In *33rd IEEE Conference on Local Computer Networks 2008*, LCN 2008, Montreal, Quebec, Canada, October 14–17, 2008, pp. 282–288.
45. M. Weiser, B. Welch, A. Demers, and S. Shenker, Scheduling for reduced CPU energy. In *Proceedings of the 1st USENIX Conference on Operating Systems Design and Implementation*, ser. OSDI'94, USENIX Association, Berkeley, CA, 1994.
46. B. Zhai, D. Blaauw, D. Sylvester, and K. Flautner, Theoretical and practical limits of dynamic voltage scaling. In *Proceedings of the 41st Annual Design Automation Conference*, ser. DAC'04, ACM, New York, 2004, pp. 868–873.
47. E. Le Sueur and G. Heiser, Dynamic voltage and frequency scaling: The laws of diminishing returns. In *Proceedings of the 2010 Workshop on Power Aware Computing and Systems (HotPower'10)*, USENIX Association, Berkeley, CA, USA, 2010.
48. Y. Chen, T.X. Wang, and R.H. Katz. Energy efficient Ethernet encodings. In *Proceedings of the 33rd IEEE Conference on Local Computer Networks*, Montreal, Quebec, Canada, IEEE, Piscataway, NJ, October 14–17, 2008, pp. 122–129.
49. D.E. Taylor, Survey and taxonomy of packet classification techniques, *ACM Computing Surveys*, 37, September 2005, 238–275.
50. W. Jiang and V.K. Prasanna, Reducing dynamic power dissipation in pipelined forwarding engines. In *Proceedings of the 2009 IEEE International Conference on Computer Design*, ser. ICCD'09, IEEE Press, Piscataway, NJ, 2009, pp. 144–149.
51. K. Zheng, C. Hu, H. Lu, and B. Liu, A TCAM-based distributed parallel IP lookup scheme and performance analysis, *IEEE/ACM Transactions on Networking*, 14, August 2006, 863–875.
52. F. Zane, G. Narlikar, and A. Basu, Coolcams: Power-efficient TCAMs for forwarding engines. In *Proceedings of the 22nd Annual Joint Conference of the IEEE Computer and Communications, INFOCOM'03*, IEEE, Piscataway, NJ, 2003.
53. S. Kaxiras and G. Keramidas, IPStash: A set-associative memory approach for efficient IP-lookup. In *Proceedings of the 24th Annual Joint Conference of the IEEE Computer and Communications Societies, INFOCOM'05*, vol. 2, Miami, Florida, March 13–17, 2005, pp. 992–1001.
54. W. Eatherton, G. Varghese, and Z. Dittia, Tree bitmap: Hardware/software IP lookups with incremental updates, *SIGCOMM Computer Communication Reviews*, 34, April 2004, 97–122.
55. S. Sahni and K.S. Kim, Efficient construction of multibit tries for IP lookup, *IEEE/ACM Transactions on Networking*, 11(4), 2003, 650–662.

56. W. Jiang and V. Prasanna, Architecture-aware data structure optimization for green IP lookup. In *Proceedings of the International Conference on High Performance Switching and Routing (HPSR)*, Dallas, Texas, June 14–16, 2010, pp. 113–118.

57. F. Chung, R. Graham, and G. Varghese, Parallelism versus memory allocation in pipelined router forwarding engines. In *Proceedings of the 16th Annual ACM Symposium on Parallelism in Algorithms and Architectures*, ser. SPAA'04, ACM, New York, 2004, pp. 103–111.

58. W. Jiang, Q. Wang, and V. Prasanna, Beyond TCAMs: An SRAM-based parallel multi-pipeline architecture for terabit IP lookup. In *Proceedings of the 27th IEEE Conference on Computer Communications, INFOCOM'08*, Phoenix, Arisona, April 13–18, 2008, pp. 1786–1794.

59. S. Kumar, M. Becchi, P. Crowley, and J. Turner, CAMP: Fast and efficient IP lookup architecture. In *Proceedings of the 2006 ACM/IEEE Symposium on Architecture for Networking and Communications Systems*, ser. ANCS'06, ACM, New York, 2006, pp. 51–60.

60. F. Baboescu, D.M. Tullsen, G. Rosu, and S. Singh, A tree based router search engine architecture with single port memories. In *Proceedings of the 32nd Annual International Symposium on Computer Architecture*, ser. ISCA'05, IEEE Computer Society, Washington, DC, 2005, pp. 123–133.

61. L. Peng, W. Lu, and L. Duan, Power efficient IP lookup with supernode caching. In *Proceedings of the IEEE Global Telecommunications Conference*, Washington, DC, November 26–30, 2007, pp. 215–219.

62. M. Allman, K. Christensen, B. Nordman, and V. Paxson, Enabling an energy efficient future internet through selectively connected end systems. In *ACM SIGCOMM HotNets*, ACM, Atlanta, Georgia, November 14–15, 2007, pp. 1–7.

63. AMD, Magic packet technology application in hardware and software, 2003. http://support.amd.com/us/Embedded_Tech-Docs/20381.pdf

64. K. Sabhanatarajan and A. Gordon-Ross, A resource efficient content inspection system for next generation Smart NICs. In *Proceedings of the IEEE International Conference on Computer Design*, Lake Tahoe, California, October 12–15, 2008, pp. 156–163.

65. Y. Agarwal, S. Hodges, R. Chandra, J. Scott, P. Bahl, and R. Gupta, Somniloquy: Augmenting network interfaces to reduce PC energy usage. In *Proceedings of the 6th USENIX Symposium on Networked Systems Design and Implementation*, USENIX Association, Berkeley, CA, 2009, pp. 365–380.

66. M. Jimeno and K. Christensen, A prototype power management proxy for gnutella peer-to-peer file sharing. In *Proceedings of the 32nd IEEE Conference on Local Computer Networks*, IEEE Computer Society, Washington, DC, 2007, pp. 210–212.

67. J. Reich, M. Goraczko, A. Kansal, and J. Padhye, Sleepless in Seattle no longer. In *Proceedings of the 2010 USENIX Conference on USENIX Annual Technical Conference*, ser. USENIXATC'10, USENIX Association, Berkeley, CA, 2010, pp. 17–17.

68. S. Nedevschi, J. Chandrashekar, J. Liu, B. Nordman, S. Ratnasamy, and N. Taft, Skilled in the art of being idle: Reducing energy waste in networked systems. In *Proceedings of the 6th USENIX Symposium on Networked Systems Design and Implementation*, USENIX Association, Berkeley, CA, 2009, pp. 381–394.

69. Y. Agarwal, S. Savage, and R. Gupta, SleepServer: A software-only approach for reducing the energy consumption of PCs within enterprise environments. In *Proceedings of the 2010 USENIX Conference on USENIX Annual Technical Conference*, ser. USENIXATC'10, USENIX Association, Berkeley, CA, 2010, pp. 22–22.

70. J. Chabarek, J. Sommers, P. Barford, C. Estan, D. Tsiang, and S. Wright, Power awareness in network design and routing. In *Proceedings of the 27th IEEE Conference on Computer Communications (INFOCOM)*, Phoenix, AZ, IEEE, Piscataway, NJ, April 13–18, 2008, pp. 457–465.

71. R. Huelsermann, M. Gunkel, C. Meusburger, and D.A. Schupke, Cost modeling and evaluation of capital expenditures in optical multilayer networks, *Journal of Optical Networking*, 7(9), 2008, 814–833.

72. G. Shen and R.S. Tucker, Energy-minimized design for IP over WDM networks, *Journal of Optical Communications and Networking*, 1(1), 2009, 176–186.

73. F. Idzikowski, S. Orlowski, C. Raack, H. Woesner, and A. Wolisz, Dynamic routing at different layers in IP-over-WDM networks—Maximizing energy savings, Zuse Institute Berlin, TKN, TU Berlin, Germany, Tech. Rep., 2010.

74. B. Sanso and H. Mellah, On reliability, performance and internet power consumption. In *Proceedings of the 7th International Workshop on Design of Reliable Communication Networks*, Washington, DC, IEEE, Piscataway, NJ, October 25–28, 2009, pp. 259–264.

75. E. Palkopoulou, D. Schupke, and T. Bauschert, Energy efficiency and CAPEX minimization for backbone network planning: Is there a tradeoff? In *Proceedings of the 3rd IEEE International Symposium on Advanced Networks and Telecommunication Systems (ANTS)*, IEEE, Piscataway, NJ, 2009, pp. 1–3.

76. W. Shen, Y. Tsukishima, K. Yamada, and M. Jinno, Power-efficient multilayer networking: Design and evaluation, In *Proceedings of the 14th Conference on Optical Network Design and Modeling (ONDM)*, IEEE, Piscataway, NJ, 2010, pp. 1–6.

77. I. Cerutti, L. Valcarenghi, and P. Castoldi, Designing power-efficient WDM ring networks. In *Networks for Grid Applications*, Springer, Berlin, Heidelberg, vol. 25, 2010, pp. 101–108.

78. X. Xiao, A. Hannan, B. Bailey, and L. Ni, Traffic engineering with MPLS in the Internet, *IEEE Network*, 14(2), 2002, 28–33.

79. F. Giroire, D. Mazauric, J. Moulierac, and B. Onfroy, Minimizing routing energy consumption: From theoretical to practical results. In *IEEE/ACM International Conference on Green Computing and Communications (Green-Com)*, Hangzhou, China, December 18–20, 2010, pp. 252–259.

80. W. Fisher, M. Suchara, and J. Rexford, Greening backbone networks: Reducing energy consumption by shutting off cables in bundled links. In *Proceedings of the First ACM SIGCOMM Workshop on Green Networking*, New Delhi, India, ACM, New York, 2010, pp. 29–34.

81. L. Chiaraviglio, M. Mellia, and F. Neri, Reducing power consumption in backbone networks. In *Proceedings of the 2009 IEEE International Conference on Communications*, Dresden, Germany, IEEE, Piscataway, NJ, June 14–18, 2009, pp. 2298–2303.

82. L. Chiaraviglio, M. Mellia, and F. Neri, Energy-aware backbone networks: A case study. In *First International Workshop on Green Communications*, in Conjunction with *IEEE International Conference on Communications*, Dresden, Germany, June 14–18, 2009, pp. 1–5.

83. M. Zhang, C. Yi, B. Liu, and B. Zhang, GreenTE: Power-aware traffic engineering. In *Proceedings of the 18th IEEE International Conference on Network Protocols*, Kyoto, Japan, October 5–8, 2010, pp. 21–30.

84. P. Mahadevan, P. Sharma, S. Banerjee, and P. Ranganathan, Energy aware network operations. In *INFOCOM IEEE Conference on Computer Communications Workshops*, 2009, IEEE, Piscataway, NJ, 2009, pp. 1–6.

85. L. Chiaraviglio and I. Matta, GreenCoop: Cooperative green routing with energy-efficient servers. In *Proceedings of the 1st International Conference on Energy-Efficient Computing and Networking*, Passau, Germany, ACM, New York, 2010, pp. 191–194.

86. A.P. Bianzino, C. Chaudet, F. Larroca, D. Rossi, and J.-L. Rougier, Energy-aware routing: A reality check. In *3rd International Workshop on Green Communications (GreenComm3)*, in conjunction with *IEEE GLOBECOM 2010*, IEEE, Piscataway, NJ, Miami, Florida, December 6–10, 2010.

87. J. Restrepo, C.G. Gruber, and C. Machuca, Energy profile aware routing. In *First International Workshop on Green Communications, in Conjunction with IEEE International Conference on Communications*, Dresden, Germany, June 14–18, 2009, IEEE, Piscataway, NJ, pp. 1–5.

88. N. Vasic and D. Kostic, Energy-aware traffic engineering. In *Proceedings of the 1st International Conference on Energy-Efficient Computing and Networking*, Passau, Germany, ACM, New York, April 13, 2010, pp. 169–178.

89. A. Cianfrani, V. Eramo, M. Listanti, M. Marazza, and E. Vittorini, An energy saving routing algorithm for a green OSPF protocol. In *INFOCOM IEEE Conference on Computer Communications Workshops*, 2010, IEEE, Piscataway, NJ, 2010, pp. 1–5.

90. K.-H. Ho and C.-C. Cheung, Green distributed routing protocol for sleep coordination in wired core networks. In *Proceedings of the 6th International Conference on Networked Computing (INC)*, Gyeongju, Korea, IEEE, Piscataway, NJ, May 11–13, 2010, pp. 1–6.

91. S. Orlowski, R. Wessaly, M. Pioro, and A. Tomaszewski, SNDlib 1.0—Survivable network design library, *Networks*, 55, 2010, 276–286.

92. X. Dong, T. El-Gorashi, and J.M.H. Elmirghani, IP over WDM networks employing renewable energy sources, *Journal of Lightwave Technology*, 29(1), 2011, 3–14.

93. H. Wang, H. Yang, and H. Wu, A fine model for evaluating output performance of crystalline silicon solar modules, Photovoltaic Energy Conversion. In *IEEE 4th World Conference*, vol. 2, Waikoloa, HI, May 2006, pp. 2189–2192.

94. Y. Chen and C. Chou, Traffic modeling of a sub-network by using ARIMA, Info-tech and Info-net. In *Proceedings, ICII-2001*, vol. 2, Beijing, China, October 29–November 1, 2001, pp. 730–735.

95. J. Zhao, A. Wang, P.P. Altermatt, S.R. Wenham, and M.A. Green, 24% Efficient silicon solar cells, photovoltaic energy conversion, twenty fourth. In *IEEE Photovoltaic Specialists Conference*, vol. 2, Waikoloa, HI, December 5–9, 1994, pp. 1477–1480.

96. Cisco CRS-1 specification data sheet, http://www.cisco.com

97. Glimmerglass System-600 data sheet, http://www.glimmerglass.com

98. Cisco ONS 15454 data sheet, http://www.cisco.com

99. T.T. Ye, Analysis of power consumption on switch fabrics in network routers. In *Proceedings of the 39th Design Automation Conference*, New Orleans, LA, June 10–14, 2002, pp. 524–529.

100. R. Hays, Active/idle toggling with low-power idle. In presentation at *IEEE802.3az Task Force Group Meeting*, January 2008.

101. A. Hopper and A. Rice, Computing for the future of the planet, *Philosophical Transactions of the Royal Society A: Mathematical, Physical and Engineering Sciences*, 366(1881), 2008, 3685–3697.

102. D. Spence, J. Crowcroft, S. Hand, and T. Harris, Location based placement of whole distributed systems. In *Proceedings of the 2005 ACM Conference on Emerging Network Experiment and Technology*, Toulouse, France, ACM, New York, pp. 124–134.

103. A. Moore, An implementation-based comparison of measurement-based admission control algorithms, *Journal of High Speed Networks*, 13(2), 2004, 87–102.

104. T. Strayer, M. Allman, G. Armitage, S. Bellovin, S. Jin, and A. Moore, IMRG workshop on application classification and identification report, *ACM SIGCOMM Computer Communication Review*, 38(3), 2008, 87–90.

105. H. Jiang, A.W. Moore, Z. Ge, S. Jin, and J. Wang, Lightweight application classification for network management. In *Proceedings of the 2007 SIGCOMM Workshop on Internet Network Management*, Kyoto, Japan, ACM, New York, 2007, pp. 299–304.

106. A. Hylick, R. Sohan, A. Rice, and B. Jones, An analysis of hard drive energy consumption. In *Proceedings of the IEEE International Symposium on Modeling, Analysis and Simulation of Computers and Telecommunication Systems*, IEEE, Piscataway, NJ, 2008, pp. 103–112.

107. A. Rice, S. Akoush, and A. Hopper, Failure is an option, Microsoft Research Technical Report MSR-TR-2008-61, 2008.

108. B. Scheuermann, W. Hu, and J. Crowcroft, Near-optimal co-ordinated coding in wireless multihop networks. In *Proceedings of the 2007 ACM CoNEXT Conference*, ACM, New York, 2007, pp. 9:1–9:12.

109. J.G. Koomey, Estimating total power consumption by servers in the U.S. and the world, Lawrence Berkeley National Laboratory, Tech. Rep., 2007.

110. A. Qureshi, R. Weber, H. Balakrishnan, J. Guttag, and B. Maggs, Cutting the electric bill for internet-scale systems, *SIGCOMM Computer Communication Review*, 39, 2009, pp. 123–134.

111. B. Cully, G. Lefebvre, D. Meyer, M. Feeley, N. Hutchinson, and A. Warfield, Remus: High availability via asynchronous virtual machine replication. In *Proceedings of the 5th USENIX Symposium on Networked Systems Design and Implementation, ser. NSDI'08*, USENIX Association, Berkeley, CA, 2008, pp. 161–174.

112. D.J.C. MacKay, *Sustainable Energy—Without the Hot Air*, UIT, Cambridge, UK, 2008.

113. U.K. Renewable Energy News and Data Centers, 2010. Available online at http://www.bbc.co.uk/news/uk-england-kent-11395972

114. R. Miller, Data centers heat offices, greenhouses, pools, 2010, http://www.datacenterknowledge.com/archives/2010/02/03/data-centersheat-offices-greenhouses-pools/

115. E. Rodriguez-Colina, L. James, R. Penty, I. White, K. Williams, and A. Moore, TCP sending rate control at terabits per second. In *Proceedings of High-Speed Networking Workshop: The Terabits Challenge, INFOCOM*, IEEE Press, Piscataway, NJ, 2006, pp. 1–5.

116. E. Williams, Energy intensity of computer manufacturing: Hybrid assessment combining process and economic input-output methods, *Environmental Science & Technology*, 38(22), 2004, 6166–6174.

117. E. Wright, A. Azapagic, G. Stevens, W. Mellor, and R. Clift, Improving recyclability by design: A case study of fibre optic cable, *Resources, Conservation and Recycling*, 44(1), 2005, 37–50.

118. D.G. Murray and S. Hand, Scripting the cloud with skywriting. In *Proceedings of the 2nd USENIX Conference on Hot Topics in Cloud Computing, ser. HotCloud'10*, USENIX Association, Berkeley, CA, 2010, pp. 1–7.

119. A.W. Moore and K. Papagiannaki, Toward the accurate identification of network applications. In *Passive and Active Network Measurement*, ser. Lecture Notes in Computer Science, C. Dovrolis, Ed. Springer, Berlin/Heidelberg, vol. 3431, 2005, pp. 41–54.

120. R. Mondragon, A. Moore, J. Pitts, and J. Schormans, Analysis, simulation and measurement in large-scale packet networks, *IET Communications, Special Issue on Simulation, Analysis and Measurement of Broadband Network Traffic*, 3(6), 2009, 887–905.

121. M. Canini, W. Li, M. Zadnik, and A.W. Moore, Experience with high-speed automated application-identification for network-management. In *Proceedings of the 5th ACM/IEEE Symposium on Architectures for Networking and Communications Systems*, Princeton, NJ, ACM, New York, 2009, pp. 209–218.

122. R. Bradford, E. Kotsovinos, A. Feldmann, and H. Schioberg, Live wide-area migration of virtual machines including local persistent state. In *Proceedings of the 3rd International Conference on Virtual Execution Environments*, San Diego, California, ACM, New York, 2007, pp. 169–179.

123. R. Clegg, M. Withall, A. Moore, I. Phillips, D. Parish, M. Rio, R. Landa, H. Haddadi, K. Kyriakopoulos, J. Aug'e et al., Challenges in the capture and dissemination of measurements from high-speed networks, *IET Communications, Special Issue on Simulation, Analysis and Measurement of Broadband Network Traffic*, 3(6), 2009, 957–966.

124. A. Moore, J. Hall, C. Kreibich, E. Harris, and I. Pratt, Architecture of a network monitor. In *Proceedings of the Fourth Passive & Active Measurement Workshop 2003*, San Diego, CA, April 6–8, 2003.

125. S. Akoush, R. Sohan, A. Rice, A.W. Moore, and A. Hopper, Predicting the performance of virtual machine migration. In *International Symposium on Modeling, Analysis, and Simulation of Computer Systems*, Los Alamitos, CA, issue 1526–7539, 2010, pp. 37–46.

126. D. Fay, H. Haddadi, A. Thomason, A.W. Moore, R. Mortier, A. Jamakovic, S. Uhlig, and M. Rio, Weighted spectral distribution for internet topology analysis: Theory and applications, *IEEE/ACM Transactions on Networking*, 18, 2010, 164–176.

127. E. Thereska, A. Donnelly, and D. Narayanan, Sierra: A power-proportional, distributed storage system, Microsoft Research Technical Report MSR-TR-2009-153, November 2009.

128. R.S. Tucker, Green optical communications—Part I: Energy limitations in transport, *IEEE Journal of Selected Topics in Quantum Electronics*, 17, 2011, 245–260.

129. R.S. Tucker, Green optical communications—Part II: Energy limitations in networks, *IEEE Journal of Selected Topics in Quantum Electronics*, 17, 2011, 261–274.

130. E. Desurvire, J.R. Simpson, and P.C. Becker, High-gain erbium-doped traveling-wave fiber amplifier, *Optics Letter*, 12, 1987, 888–890.

131. IEEE, 802.3z-1998—IEEE media access control parameters, physical layers, repeater and management parameters for 1,000 Mb/s operation, 1998.

132. IEEE, 802.3ae-2002—IEEE standard for information technology—Local & metropolitan area networks—Part 3: Carrier sense multiple access with collision detection (CSMA/CD) access method and physical layer specifications—Media access control (MAC) parameters, physical layer, and management parameters for 10 Gb/s Operation, 2002.

133. A.B. Massara, K.A. Williams, I.H. White, R.V. Penty, A. Galbraith, P. Crump, and P. Harper, Ridge waveguide InGaAsP lasers with uncooled l0 Gbit/s operation at 70°C, *Electronics Letters*, 35, 1999, 1646–1647.

134. J.K. White, C. Blaauw, P. Firth, and P. Aukland, 85°C investigation of uncooled 10-Gb/s directly modulated InGaAsP RWG GC-DFB lasers, *IEEE Photonics Technology Letters*, 13, 2001, 773–775.

135. R. Paoletti, M. Agresti, G. Burns, G. Berry, D. Bertone, P. Charles, P. Crump, A. Davies, R.Y. Fang, R. Ghin, P. Gotta, M. Holm, C. Kompocholis, G. Magnetti, J. Massa, G. Meneghini, G. Rossi, P. Ryder, A. Taylor, P. Valenti, M. Meliga, 100°C, 10 Gb/s directly modulated InGaAsP DFB lasers for uncooled Ethernet applications, In *27th European Conference on Optical Communication, 2001 (ECOC'01)*, Amsterdam, the Netherlands, vol. 6, 2001, pp. 84–85.

136. http://www.avagotech.com/docs/5989-4527EN

137. A.C. O'Donnell and N.J. Parsons, 1×16 lithium niobate optical switch matrix with integral TTL compatible drive electronics, *Electronics Letters*, 27, 1991, 2367–2368.

138. A. Kazama, Y. Itou, M. Horino, K. Fukuda, M. Kanamaru, T. Akashi, T. Ishikawa, T. Harada, and R. Okada, Low-insertion-loss compact 3D-MEMS optical matrix switch with 18 input/output ports. In *Solid-State Sensors, Actuators and Microsystems, 2005.* Digest of Technical Papers. *The 13th International Conference on Transducers '05*, Seoul, South Korea, vol. 1, 2005, pp. 972–975.

139. M. Mizukami, J. Yamaguchi, N. Nemoto, Y. Kawajiri, H. Hirata, S. Uchiyama, M. Makihara, T. Sakata, N. Shimoyama, H. Ishii, and F. Shimokawa, 128×128 3D-MEMS optical switch module with simultaneous optical paths connection for optical cross-connect systems. In *International Conference on Photonics in Switching, 2009 (PS'09)*, Pisa, Italy, 2009, pp. 1–2.

140. S. Sohma, T. Goh, H. Okazaki, M. Okuno, and A. Sugita, Low switching power silica-based super high delta thermo-optic switch with heat insulating grooves, *Electronics Letters*, 38, 2002, 127–128.

141. N. Xie, T. Hashimoto, and K. Utaka, Ultimate-low-power-consumption, polarization-independent, and high-speed polymer Mach-Zehnder thermo-optic switch. In *Conference on Optical Fiber Communication, 2009 (OFC 2009)*, San Diego, CA, 2009, pp. 1–3.

142. W. Payne III, Design of lithium niobate based photonic switching systems, *IEEE Communications Magazine*, 25, 1987, 37–41.

143. P.J. Duthie and M.J. Wale, 16×16 single chip optical switch array in lithium niobate, *Electronics Letters*, 27, 1991, 1265–1266.

144. P. Granestrand, B. Lagerstrom, P. Svensson, H. Olofsson, J.E. Falk, and B. Stoltz, Pigtailed tree-structured 8×8 LiNbO$_3$ switch matrix with 112 digital optical switches, *IEEE Photonics Technology Letters*, 6, 1994, 71–73.

145. I.H. White, J.J.S. Watts, J.E. Carroll, C.J. Armistead, D.J. Moole, and J.A. Champelovier, InGaAsP 400×200 mu m active crosspoint switch operating at 1.5 mu m using novel reflective Y-coupler components, *Electronics Letters*, 26, 1990, 617–618.

146. G. Sherlock, J.D. Burton, P.J. Fiddyment, P.C. Sully, A.E. Kelly, and M.J. Robertson, Integrated 2×2 optical switch with gain, *Electronics Letters*, 30, 1994, 137–138.

147. J.D. Burton, P.J. Fiddyment, M.J. Robertson, and P. Sully, Monolithic InGaAsP-InP laser amplifier gate switch matrix, *IEEE Journal of Quantum Electronics*, 29, 1993, 2023–2027.

148. Y. Kai, K. Sone, S. Yoshida, Y. Aoki, G. Nakagawa, and S. Kinoshita, A compact and lossless 8 × 8 SOA gate switch subsystem for WDM optical packet interconnections, In *Optical Communication, 2008,* ECOC 2008. 34th European Conference, Brussels, Belgium, 2008, pp. 117–18.

149. H. Wang, A. Wonfor, K.A Williams, R.V. Penty, and I.H. White, Demonstration of a lossless monolithic 16 × 16 QW SOA switch. In *Optical Communication, 2009,* ECOC '09. 35th European Conference on, Post-Deadline, Vienna, Austria, 2009, pp. 1–2.

150. M. Glick, M. Dales, D. McAuley, L. Tao, K. Williams, R. Penty, and I. White, SWIFT: A testbed with optically switched data paths for computing applications. In *Proceedings of 2005 7th International Conference on Transparent Optical Networks,* Barcelona, Spain, vol. 2, 2005, pp. 29–32.

151. Q. Yang, K. Bergman, G.D. Hughes, F.G. Johnson, WDM packet routing for high-capacity data networks, *Journal of Lightwave Technology,* 19, 2001, 1420–1426.

152. A. Shacham, B.A. Small, O. Liboiron-Ladouceur, and K. Bergman, A fully implemented 12 × 12 data vortex optical packet switching interconnection network, *Journal of Lightwave Technology,* 23, 2005, 3066–3075.

153. http://europic.jeppix.eu/

154. H. Wang, A. Wonfor, S. Liu, R.V. Penty, and I.H. White, Novel 4 × 4 port integrated SOA switch and wavelength converter. In *International Conference on Photonics in Switching, 2009 (PS '09),* Pisa, Italy, 2009, pp. 1–2.

155. H. Wang, E.T. Aw, M. Xia, M.G. Thompson, R.V. Penty, I.H. White, and A.R. Kovsh, Temperature independent optical amplification in uncooled quantum dot optical amplifiers. In *Conference on Optical Fiber Communication/National Fiber Optic Engineers Conference, 2008 (OFC/NFOEC 2008),* San Diego, CA, 2008, pp. 1–3.

156. E.T. Aw, H. Wang, M.G. Thompson, A. Wonfor, R.V. Penty, I.H. White, and A.R. Kovsh, Uncooled 2 × 2 quantum dot semiconductor optical amplifier based switch. In *Conference on Lasers and Electro-Optics, 2008/2008 Conference on Quantum Electronics and Laser Science (CLEO/QELS 2008),* San Jose, CA, 2008, pp. 1–2.

157. J.E. Bowers, A.W. Fang, P. Hyundai, R. Jones, O. Cohen, and M.J. Paniccia, Hybrid silicon evanescent photonic integrated circuit technology. In *Conference on Lasers and Electro-Optics, 2007 (CLEO 2007),* Baltimore, MD, 2007, pp. 1–2.

158. D.J. Thomson, F.Y. Gardes, G.T. Reed, F. Milesi, and J.M. Fedeli, High speed silicon optical modulator with self aligned fabrication process, *Optics Express,* 18, 2010, 19064–19069.

159. A. Alduino, L. Liao, R. Jones, M. Morse, B. Kim, W. Lo, J. Basak, B. Koch, H. Liu, H. Rong, M. Sysak, C. Krause, R. Saba, D. Lazar, L. Horwitz, R. Bar, S. Litski, A. Liu, K. Sullivan, O. Dosunmu, N. Na, T.Yin, F. Haubensack, I. Hsieh, J. Heck, R. Beatty, H. Park, J. Bovington, S. Lee, H. Nguyen, H. Au, K. Nguyen, P. Merani, M. Hakami, and M. Paniccia, A 4 × 12.5 Gb/s CWDM Si photonics link using integrated hybrid silicon lasers. In *Lasers and Electro-Optics (CLEO), 2011* Conference on, Baltimore, Maryland, 2011, pp. 1–2.

160. M. Lipson, Silicon photonics: The optical spice rack, *Electronics Letters,* 45, 2009, 576–578.

161. B.G. Lee, A. Biberman, J. Chan, and K. Bergman, High-performance modulators and switches for silicon photonic networks-on-chip, *IEEE Journal of Selected Topics in Quantum Electronics,* 16, 2010, 6–22.

13

Green TCAM-Based Internet Routers

Tania Mishra
University of Florida

Sartaj Sahni
University of Florida

13.1 Introduction

Internet routers are devices that connect several packet switched networks to allow communication and resource sharing among a large group of users. A router implements the packet forwarding and routing functions of network layer which is the third layer of the seven-layer OSI model of computer networking. A router has several input and output ports through which it receives and sends packets. A packet arriving at an input port of a router is transferred to an appropriate output port and from here to its next hop on the path to its destination. A router maintains a list of rules in a forwarding table that is used to determine the next hops for packets during packet forwarding. A router updates its forwarding table continuously in response to changes in the Internet. All the routers in a network communicate with each other using routing protocols to remain updated of the changes and to select the best paths to reachable destination addresses.

As Internet packets get from source to destination they go through a number of routers. At each router, a forwarding engine uses the destination address of the packet and a set of rules to determine the next hop for the packet. A forwarding table often contains hundreds of thousands of rules. A packet forwarding rule (P, H) comprises a prefix P and a next hop H. A packet with destination address d is forwarded to H where H is the next hop associated with the rule that has the longest prefix matching d. Figure 13.1 shows a small forwarding table with six prefixes. The prefix associated with rule R5 is 100 (the * at the end indicates a sequence of don't care bits) and the associated next hop is H5. Rule R5 matches all destination addresses that begin with 100. The length of the prefix 100 associated with R5 is 3. A destination address that begins with 100 is matched by rules R1, R3, and R5. Among these rules,

283

	Prefixes	Next Hop
R1	*	H1
R2	00*	H2
R3	10*	H3
R4	11*	H4
R5	100*	H5
R6	111*	H6

FIGURE 13.1 An example six-prefix forwarding table.

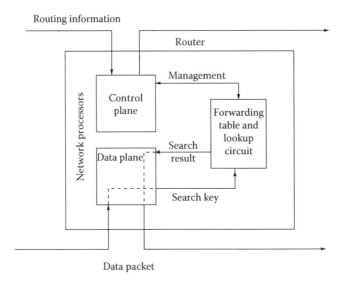

FIGURE 13.2 Control and data planes in routers.

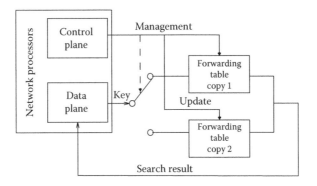

FIGURE 13.3 Router architecture with batch updating policy.

R5 is the one with the longest prefix. So, H5 is the next hop for packets with a destination address that begins with 100.

Figure 13.2 shows a block diagram of the internals of a router. The part of a router that deals with packet forwarding is called the data plane and the part that communicates with other routers and selects the best routes to different destinations and keeps the forwarding table updated is called the control plane. The control plane updates the forwarding table either in a batch or incrementally. When the updates are done in a batch, two copies of the forwarding table are maintained as shown in Figure 13.3. While one copy is used for data plane lookups, the other copy is made up-to-date by the control plane. When the updating

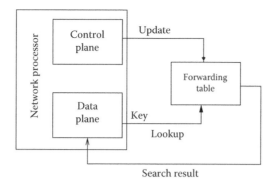

FIGURE 13.4 Router architecture with incremental updating policy.

is complete, the data plane switches to the freshly updated copy for lookups, and the outdated copy is updated by the control plane. Thus batch updates require twice the memory for storing two copies of the forwarding table and introduce a latency between arrival of an update request and the time the update is incorporated. In contrast, when the updates are applied incrementally, lookups and updates are done on the same table as shown in Figure 13.4.

With the rapid growth of the Internet, the number of routers being used is increasing dramatically. While routers used at homes consume about 5–10 W, edge routers consume about 4 kW of power and the core routers about 10 kW of power per rack [1]. It was measured that in Japan, ICT equipment in 2006 consumed about 45 TWh or 4% of the electricity generated and 1% of the total energy consumption of the country [2]. Approximately, 25% of the ICT equipment energy is consumed by routers. The packet forwarding engine of a router needs a lot of energy to perform high-speed lookups and packet switching. For example, to keep up to a line rate of 40 Gbps, a packet forwarding engine must perform 125 million searches per second, assuming the minimum IPv4 packet size of 40 bytes. It was found that about 62% of the router power consumption happens in the packet forwarding engine. With the growing trend of network usage, the total energy consumption in routers is only going to get worse. For example, with the current growth rate of Internet traffic at 40% per year in Japan, it is projected that by the year 2022, the energy consumption will exceed 10,000 TWh which was the total energy produced in Japan in 2005 [3]. Thus, unless we can make routers far more energy efficient, routers may soon consume most of the produced energy. This makes the design of low power packet forwarding engines essential.

In this chapter, we focus on reduction of energy consumption in a packet forwarding engine that uses a special type of memory called TCAM which is the acronym for ternary content addressable memory. A TCAM is different from a conventional memory in that, each bit may be set to one of the three states namely, 0, 1, and x (don't care). This makes it particularly convenient to store the destination prefixes along with their trailing sequence of 'x's. A TCAM has an associated SRAM which is used to store the next hops. A TCAM often has a priority encoder to choose the best match among multiple matches. TCAMs are attractive for use in edge and core routers, because like an associative memory, TCAMs enable a parallel search across all the rules and complete a table lookup in one clock cycle. Even though TCAMs support high-speed lookups, they are power hungry. So, a lot of research has been done to reduce the power consumption both in the hardware and software domains [4–12]. Pure hardware approaches for power reduction are presented in [9–12]. The software approaches are characterized by removing redundancies in the prefixes of the forwarding table and compacting them using different techniques, so that the number of prefixes to be stored in the TCAM decreases [7,8,13]. Using an indexed TCAM organization, the authors in [4,5,14] could obtain significant savings in power consumption. This chapter presents some of these latter techniques.

13.2 Simple TCAM (STCAM)

This is the basic scheme in which a single TCAM with its associated SRAM and a priority encoder is used to store the forwarding table. The prefixes are stored in decreasing order of length and the corresponding next hops are stored in the associated SRAM. In case of multiple matches, the priority encoder chooses the first match (which can be done in one clock cycle), and uses the address of the matched entry to index into the SRAM and obtain the next hop. Thus, the entire lookup completes in one TCAM clock cycle returning the next hop corresponding to the longest matching prefix.

Figure 13.5 shows the prefix assignment to a STCAM for the forwarding table in Figure 13.1. The prefixes are stored in the TCAM in decreasing order of prefix length and the next hops are stored in corresponding words of an SRAM. Now suppose the router receives a packet whose destination address begins with 111. Assuming the TCAM and SRAM words are indexed beginning at 0, a TCAM lookup returns the TCAM index 1 for the longest matching prefix. This index is used to access the SRAM and H6 is returned as the next hop for the packet.

13.3 Indexed TCAMs

In order to reduce power consumption during TCAM search, TCAM hardware supports a feature that searches only a portion of the TCAM. This reduces power consumption which is proportional to the size of the TCAM that needs to be searched. Zane et al. [4] proposed a two-level architecture in which the first-level TCAM is called the index TCAM (ITCAM) and the second-level TCAM is the data TCAM (DTCAM). The ITCAM is looked up using the destination address and the best match is used to activate a particular DTCAM segment of prefixes. The DTCAM segment is searched next for the longest matching prefix. To fill in the ITCAM and DTCAM segments with prefixes, Zane et al. [4] proposed two methods, namely, subtree split and postorder split. Both these methods split a 1-bit trie representation of prefixes repeatedly till all the prefixes are entered into the DTCAM. In subtree split, variable sized DTCAM segments are created, with a maximum bounding size of b entries, where $b > 1$. The split is done in such a way that all the DTCAM segments, except one, contain between $\lceil b/2 \rceil$ and b entries. There is a single segment that contains between 1 and b prefixes. The 1-bit trie is traversed in a postorder fashion and while visiting a node v if it is found that the subtree rooted at v contains at least $\lceil b/2 \rceil$ prefixes and the subtree rooted at its parent (if any) contains more than b prefixes, then the subtree is splitted (or carved) at node v. The prefixes in the carved out subtree are inserted into a DTCAM segment in decreasing order of length. The prefix represented by the path from the root of the trie to node v is inserted in the ITCAM, and the corresponding ISRAM entry points to the start and end addresses of the DTCAM segment created for storing prefixes in subtree rooted at v. ITCAM prefixes are entered in the same order as they are generated upon postorder traversal of the trie to fill up DTCAM segments. For a forwarding table with n prefixes,

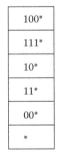

100*	H5
111*	H6
10*	H3
11*	H4
00*	H2
*	H1

FIGURE 13.5 Rules of Figure 13.1 stored in a STCAM organization.

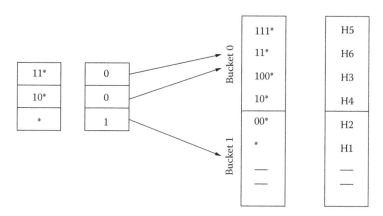

FIGURE 13.6 Six rules of Figure 13.1 stored in an indexed TCAM organization.

the number of ITCAM entries is at most $\lceil 2n/b \rceil$ and each bucket has at most $b + 1$ prefixes (including the covering prefix*) of TCAM entries activated during a search, the power of each lookup is bounded as $P \leq 2n/b + b + 1$ which is minimized when $b = \sqrt{2n}$. The minimum power required is $\sqrt[2]{2n} + 1$ compared to n required for a STCAM. Lu and Sahni [5] observed that the subtree split algorithm is suboptimal producing, in the worst case, about twice the optimal number of DTCAM segments and correspondingly twice the optimal number of ITCAM entries. They improved upon the subtree splitting techniques of [4] by following a greedy postorder traversal policy.

The postorder split algorithm of Zane et al. [4] generates buckets with a fixed number of prefixes. Each bucket additionally includes up to W covering prefixes, where W is the length of the longest prefix in the forwarding table. All buckets, except one, contains exactly b forwarding table prefixes, whereas the remaining bucket has fewer than b forwarding-table prefixes. This remaining bucket can be padded with empty TCAM entries to make it the same size as b. All the buckets could contain a variable number of covering prefixes up to a maximum of W covering prefixes. The algorithm for postorder split is different than subtree split in that now prefixes from several subtrees can be used to fill up a DTCAM segment of size b. As a result there may be more prefixes in the ITCAM, where multiple ITCAM prefixes point to the same DTCAM bucket. Note also that a bucket may contain up to 1 covering prefix for each subtree packed into it. Figure 13.6 shows the ITCAM, ISRAM, DTCAM, and DSRAM configurations for the six-prefix example of Figure 13.1. Lu and Sahni [5] have proposed a greedy heuristic for creating fixed size buckets which improves upon the number of prefixes to be entered into the ITCAM. For this greedy heuristic, the worst case number of ITCAM prefixes is $\lceil \log_2(b) \rceil \times \lceil n/b \rceil$.

13.4 TCAMs with Wide SRAMs

Lu and Sahni [5] proposed the use of wide SRAMs in association with a TCAM. It was observed that the number of next hops for a router is usually quite small (same as the number of output ports), and so it requires a few bits to encode these next hops. In that case, the remaining space in an SRAM word remains unutilized. Moreover, if a QDRII SRAM is used in association with a TCAM, then, 72 bits of SRAM memory could be accessed (in case of dual burst) or even 144 bits (quad burst) could be accessed simultaneously. To better utilize the SRAM space, the authors proposed to store a subtree of the 1-bit trie of prefixes in a forwarding table, which in turn reduces the number of TCAM entries and hence the power required for a lookup. Even though there is a minor increase in the lookup time due to additional

---

* The covering prefix for v is the longest-length forwarding table prefix that matches all destination addresses of the form $P*$, where P is the prefix defined by the path from the root of the 1-bit trie to v.

Match start position	Suffix count	Next hop of S0	Len (S1)	S2	Next hop of S1	...	Len (Sk)	Sk	Next hop of Sk	Unused

FIGURE 13.7 Suffix node of [5] with a 5-bit match start position field and an optimized representation of the first suffix.

processing needed to retrieve the nexthop from a wide SRAM word, it is still possible to complete a lookup in one clock cycle [8]. To store a subtree in an SRAM word, Lu and Sahni proposed a suffix node format which was later updated by Mishra and Sahni [8] and is included in Figure 13.7. A suffix node must fit in an SRAM word.

To fill up a suffix node, a subtree of the 1-bit trie T is carved and the size of the subtree is determined by the size of an SRAM word. If N is the root of the subtree being carved then $Q(N)$ is the prefix defined by the path from the root of the trie to the node N. Prefix $Q(N)$ is entered in the TCAM. Let $P_1, \ldots, P_k$ be the prefixes in the subtree rooted at N, excluding the prefix stored at N, if any. Then the data collected to create a suffix node are:

1. Length $|Q(N)|$ of the prefix stored in TCAM.
2. Suffix count $k+1$, due to prefixes $P_1, \ldots, P_k$ and either the prefix stored at node N or the covering prefix of node N.
3. Default next hop, which is the next hop corresponding to prefix (if any) stored at node N. Otherwise, the default next hop is set to the next hop of the covering prefix for node N. This field is marked as "next hop of S_0" in Figure 13.7.
4. For each prefix P_i, the suffix S_i of P_i consisting of bits from position $|Q(N)| + 1$ to $|P_i|$.
5. For each prefix P_i, the length of the suffix S_i, $|S_i| = |P_i| - |Q(N)|$.
6. For each prefix P_i, its next hop.

Let u be the number of bits allocated to the prefix length, suffix count and default next hop fields (items 1, 2, and 3 of the above list) of a suffix node and let v be the sum of the number of bits allocated to a suffix length and next hop fields (items 5 and 6 of the above list). Let $len(S_i)$ be the length of the suffix S_i. The space needed by the suffix node fields for $S_1 \ldots S_k$ is $u + kv + \sum_{i=1}^{k} len(S_i)$ bits. Thus it is required that $u + kv + \sum_{i=1}^{k} len(S_i)$ be less than or equal to the bandwidth (or word size) of the SRAM.

Example 13.1

Consider the six-prefix forwarding table of Figure 13.1. Suppose that a suffix node is 32 bits long and the SRAM word size is also 32 bits. Suppose 5 bits are used for length of prefix matched which allows a prefix in TCAM to be 31 bits in length, 2 bits for the suffix count field (this allows up to four suffixes in a node as the count must be more than 0), 2 bits for the suffix length field (permitting suffixes of length up to 3), and 10 bits for a next hop (permitting up to 1024 different next hops). With this bit allocation, a suffix node may store up to 2 next hops, including the default next hop. Figure 13.8 shows an example of a carving of the 1-bit trie for six-prefix example and the TCAM and SRAM bit assignments. This carving has the property that no subtree needs a covering prefix and each subtree may be stored in a suffix node using the stated format.

To search for a prefix, the TCAM is first looked up for the longest $Q(N)$ matching destination address d. Next, the SRAM word corresponding to the TCAM match is searched for the longest matching suffix. While matching the suffixes to d, the first $|Q(N)|$ bits of d are ignored and the matching is done starting from the $|Q(N)| + 1$ bit of d. The nexthop corresponding to the longest matching suffix is returned from the SRAM word search. In case there is no suffix that matches d, the default next hop is returned from the SRAM word.

If the average number of prefixes packed into a suffix node is a_1, then the TCAM size is approximately n/a_1, where n is the total number of forwarding table prefixes. So, the power needed for a lookup in

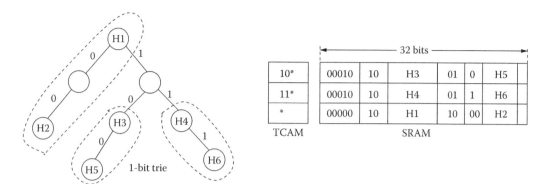

FIGURE 13.8 Suffix node example.

a forwarding table using a TCAM with a wide SRAM is about $1/a_1$ that required when the STCAM organization of Figure 13.1 is used.

For the carving heuristic, suppose u and v are as above and let w be the size of a suffix node. Also, for any node x in the 1-bit trie, let $ST(x)$ be the subtree rooted at x. Let $ST(x).numP$ be the number of prefixes in $ST(x)$ and let $ST(x).numB$ be the number of bits needed to store the suffix lengths, suffixes and next hops for these prefixes of $ST(x)$. For each node x the quantities $ST(x).numP$ and $ST(x).numB$ are computed by a postorder traversal of the trie. When x is null, $ST(x).numP = ST(x).numB = 0$. When x is not null, let l and r be its two children (either or both may be null). The following recurrence is obtained for $ST(x).numB$.

$$ST(x).numB = ST(l).numB + ST(l).numP + ST(r).numB + ST(r).numP \qquad (13.1)$$

Since each prefix in $ST(l)$ and $ST(r)$ has a suffix that is 1 bit longer, the extra suffix bits are accounted for in x by adding $ST(l).numP + ST(r).numP$. In addition to the extra suffix bits, x needs to also store the suffix length and nexthop fields which carry over from its left and right children. So, in total $ST(l).numB + ST(l).numP + ST(r).numB + ST(r).numP$ bits are needed to store the suffix length fields, suffixes and nexthops (Figure 13.9).

The size, $ST(x).size$, of the suffix node needed by $ST(x)$ is given by

$$ST(x).size = ST(x).numB + u \qquad (13.2)$$

The correctness of Equation 13.2 follows from the observation that in either case, u additional bits are needed for the prefix length, suffix count and the default next hop.

Example 13.2

Lets take the trie for the six-prefix forwarding table in Figure 13.1 and compute **ST(x).numP**, **ST(x).numB** and **ST(x).size**. To keep things simple, for **ST(x).numB** we consider the bits to store nexthop and suffix, ignoring the bits for the suffix length field. Similarly, for **u** in Equation 13.2, we count only the number of bits required to store the default nexthop. Figure 13.10 shows the triplets (**ST(x).numP**, **ST(x).numB**, **ST(x).size**) for each node. For example, if the node at 00 storing nexthop H2 is carved, then the corresponding SRAM word will just store the default nexthop H2. Hence, **ST(x).size** = 10, assuming we use 10 bits to store a next hop. Similarly, if the node at 1 is carved, the corresponding SRAM word will store four suffixes, which are: 0, 1, 00, 11. These suffix bits, along with the their nexthops, will occupy 46 bits of SRAM space. In this case, the default hop **H1** comes from the covering prefix and **ST(x).size** = 46 + 10 = 56.

Algorithm visit(x)

```
{
    if (ST(x).size < w) return;
    if (ST(x).size == w) {split(x); return;}
    // ST(x).size > w
    if (x has only 1 non-null child y) {split(y); return;}
    // x has 2 non-null children y and z
    if (ST(y).numB ≤ ST(z).numB) {
        split(z);
        recompute ST(x).size;
        if (ST(x).size < w) return;
        if (ST(x).size == w) split(x);
        else split(y);
    }
    else // ST(y).numB > ST(z).numB
    // this is symmetric to the case ST(y).numB ≤ ST(z).numB
}
```

FIGURE 13.9   Visit function for subtree carving heuristic.

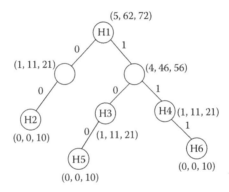

FIGURE 13.10 $(ST(x).numP, ST(x).numB, ST(x).size)$ computed for each node.

The carving heuristic performs a postorder traversal of the 1-bit trie T using the visit algorithm of Figure 13.9. Whenever a subtree is split from the 1-bit trie, the prefixes in that subtree are put into a suffix node and a TCAM entry for this suffix node generated. The complexity of the visit algorithm (including the time to recompute $ST(x).size$) is $O(1)$. So, the overall complexity of the tree carving heuristic is $O(nW)$, where n is the number of prefixes in the forwarding table and W is the length of the longest prefix.

It is possible to obtain further reduction in TCAM power by using the indexed TCAM schemes of Section 13.3 on TCAMs with wide SRAMs. The details of the method of adding an index TCAM to a TCAM with wide SRAM, can be found in [5]. Due to the necessity for storing covering prefixes in an indexed TCAM as well as in a wide SRAM, incremental update algorithms tend to be complex. For example, when a new prefix is added, it must be checked if the new prefix should replace any existing covering prefix. Similarly when a prefix is deleted from the forwarding table, it may have to be deleted from multiple locations, triggering fresh computations for covering prefixes. On the other hand, batch updating algorithms work perfectly for indexed TCAMs as well as those with wide SRAMs.

13.5 DUO

DUO is a dual TCAM architecture for faster prefix lookup and efficient storage. DUO uses advanced memory management schemes for inserting and deleting prefixes to and from the TCAM. The different versions of DUO are: (1) DUOS which is dual TCAM with simple SRAM, where both the TCAMs have a simple associated SRAM used for storing next hops. (2) DUOW which is dual TCAM with wide SRAM, where one or both the TCAMs have wide associated SRAMs that are used to store suffixes as well as next hops. (3) IDUOW which is indexed dual TCAM with wide SRAM, where either or both TCAMs have an associated index TCAM. All DUO versions support efficient incremental updates that do not degrade lookup speed.

13.5.1 DUOS

DUOS is a simple dual TCAM architecture, in which two TCAMs with associated SRAMs are used to store the prefixes in a forwarding table, as shown in Figure 13.11. The TCAMs are labeled as ITCAM (interior TCAM) and LTCAM (leaf TCAM). A prefix in a forwarding table rule is stored in either ITCAM or LTCAM. DUOS uses a reasonably efficient data structure, such as a binary trie, to store the prefixes in the control plane. This is needed for the initial prefix assignment to the TCAMs as well as for prefix assignment during incremental updates. A quick evaluation of a binary trie stored in a 100 ns DRAM shows that it permits about 300K IPv4 lookups, inserts, and deletes per second. This performance is adequate for the anticipated tens of thousands of control plane operations. The prefixes stored in the leaf nodes of the trie are entered in the LTCAM, and the remaining prefixes are entered in the ITCAM. Since

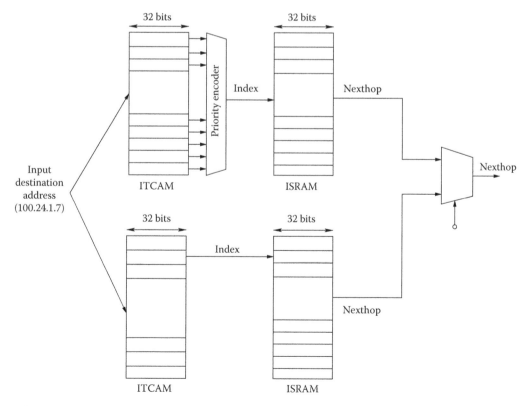

FIGURE 13.11   Dual TCAM with simple SRAM.

the LTCAM stores prefixes found in the leaf nodes of the trie, the prefixes in the LTCAM are disjoint and so at most one may match any given destination address. Consequently, the LTCAM prefixes, even though of varying length, may be stored in any order. Further, the LTCAM does not require a priority encoder and, as a result, the latency of an LTCAM search is up to 50% less than that of a search in a TCAM with a priority encoder [14]. A data plane lookup is performed by doing a search for the packet's destination address in both ITCAM and LTCAM. The ITCAM search yields the next hop associated with the longest matching non-leaf prefix while the LTCAM search yields the next hop associated with at most one leaf prefix that matches the destination address. Additional logic shown in Figure 13.11 returns the next hop (if any) from the LTCAM search; the next hop from the ITCAM search is returned only if the LTCAM search found no match. Note that since the LTCAM has no priority encoder, its search completes sooner than that in the ITCAM. The combining logic of Figure 13.11 can take advantage of this observation and abort the ITCAM search whenever the LTCAM search is successful, thereby reducing average lookup time. The correctness of the lookup is readily established. It was found from a number of forwarding tables downloaded from [15] and [16] that over 90% of the prefixes are stored in the LTCAM and less than 10% are stored in the ITCAM. Consequently, the LTCAM services most of the lookup requests.

Figure 13.12 shows the binary trie stored in the control plane for our six-prefix forwarding table of Figure 13.1 together with the content of the two TCAMs and the two SRAMs of DUOS. Each node of the control plane trie has fields such as *prefix*, *slot*, *nexthop*, and *length* which aid in storing the prefix (if any) on that node, the slot in ITCAM or LTCAM containing the prefix, nexthop and the length of the prefix, respectively. Functions for basic operations on the control plane trie (hereinafter simply referred to as trie) are assumed (see Figure 13.13).

As the control plane will modify the ITCAM, LTCAM, ISRAM, and LSRAM while the data plane performs lookups, the TCAMs need to be dual ported. Specifically, the following assumptions are made:

1. Each TCAM has two ports, which can be used to simultaneously access the TCAM from the control plane and the data plane.
2. Each TCAM entry/slot is tagged with a valid bit, that is set to 1 if the content for the entry is valid, and to 0 otherwise. A TCAM lookup engages only those slots whose valid bit is 1. The TCAM slots engaged in a lookup are determined at the start of a lookup to be those slots whose valid bits are 1 at that time. Changing a valid bit from 1 to 0 during a data plane lookup does not disengage that slot from the ongoing lookup. Similarly, changing a valid bit from 0 to 1 during a data plane lookup does not engage that slot until the next lookup.

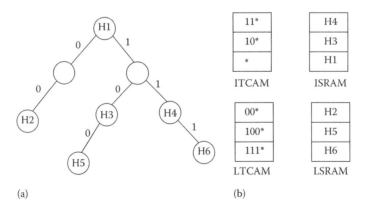

FIGURE 13.12 DUOS for the six-prefix forwarding table of Figure 13.1. Note that prefixes in ITCAM are stored in length order, whereas those in LTCAM are stored arbitrarily. (a) Trie and (b) DUOS.

```
Function: Trie.insert
(a, b) = Trie.insert(prefix, length, nextHop);
This function inserts a prefix given its length and next hop into the control-plane binary
trie. It returns the trie node a which stores the new prefix and a's nearest ancestor node
b that contains a prefix.
Function: Trie.delete
(a, b) = Trie.delete(prefix, length);
This function deletes a prefix from the control plane trie and returns the trie node a
that used to store the prefix just deleted and a's nearest ancestor node b that contains a
prefix.
Function: Trie.change
a = Trie.change(prefix, length, newHop);
This function changes the next hop associated with a prefix and returns the trie node a
that contains the prefix.
```

FIGURE 13.13   Table of control-plane trie functions.

```
DUOS:
insert
delete
change
      ITCAM (with simple SRAM):
      insert
      delete
      change
            getSlot
            freeSlot
                  movesFromAbove
                  movesFromBelow
                  getFromAbove
                  getFromBelow
LTCAM (with simple SRAM)
insert
delete
change
```

FIGURE 13.14   Table of functions used for incremental update.

The availability of the function *waitWriteValidate* which writes to a TCAM slot and sets the valid bit to 1, is assumed. In case the TCAM slot being written to is the subject of ongoing data plane lookup, the write is delayed till this lookup completes. During the write, the TCAM slot being written to is excluded from data plane lookups.* This is equivalent to the requirement that "After a rule is matched, resetting the valid bit has no effect on the action return process" [17], and to setting the valid entry to "hit" [18]. Similarly, the availability of the function *invalidateWaitWrite* is assumed, which sets the valid bit of a TCAM slot to 0 and then writes an address to the associated SRAM word in such a way that the outcome of the ongoing lookup is unaffected.

Note that *waitWriteValidate* may, at times, write the prefix and nexthop information in the TCAM and associated SRAM slot and validate it, without any wait. This happens, for example, when the writing is to be done to a TCAM slot that is not the subject of the ongoing data plane lookup. The wait component of the function *waitWriteValidate* is said to be null in this case.

Figure 13.14 lists the various update algorithms we define later in this section for DUOS and its associated ITCAM and LTCAM. The indentation represents the hierarchy of function calls. A function

---

* A possible mechanism to accomplish this exclusion is to set the valid bit to 0 before commencing the write and to change this bit to 1 when the write completes.

at one level of indentation calls one or more functions below it at the next level of indentation or at the same level of indentation.

13.5.2 DUOS Incremental Update Algorithms

13.5.2.1 Insert

Figure 13.15 gives the algorithm to insert a new prefix p of length l and nexthop h. For simplicity, it is assumed that p is, in fact new (i.e., p is not already in the rule table). First, p is inserted into the trie using the trie insertion algorithm, which returns nodes m and n, where m is the trie node storing p and n is the nearest ancestor (if any) of m that has a prefix. When m is a leaf of the trie, there is a possibility that the insertion of p transformed a prefix that was previously a leaf prefix into a non-leaf prefix. If so, this prefix is moved from the LTCAM to the ITCAM. Regardless, p is inserted into the LTCAM. When m is not a leaf, p is inserted into the ITCAM. Figure 13.16 illustrates the insertion of a new rule (000*, H7) to the six-prefix forwarding table of Figure 13.1.

13.5.2.2 Delete

Figure 13.17 gives the algorithm to delete the prefix p from DUOS. For simplicity, it is assumed that p is, in fact, present in the rule table and so may be deleted. First, p is deleted from the trie. The trie deletion function returns nodes m and n, where m is the trie node where p was stored and n is the nearest ancestor

---

Algorithm: insert (p, l, h)
(m, n) = Trie.insert(p, l, h)
if m is a leaf then begin
    if n exists and $n{\rightarrow}prefix$ was a leaf prefix then
        $slot$ = ITCAM.insert$(n{\rightarrow}prefix, n{\rightarrow}nexthop, n{\rightarrow}length)$;
                // $n{\rightarrow}prefix$ is no longer a leaf
        LTCAM.delete$(n{\rightarrow}slot)$;
        $n{\rightarrow}slot = slot$;
    endif
    $m{\rightarrow}slot$ = LTCAM.insert(p, h, l);
else $m{\rightarrow}slot$ = ITCAM.insert(p, h, l);
endif

FIGURE 13.15 Algorithm to insert into DUOS.

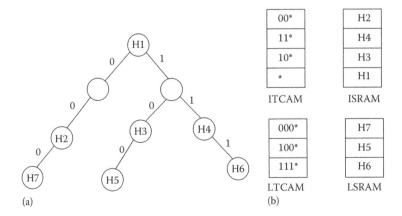

FIGURE 13.16 Insert new rule {000*, H7} to the six-prefix table. (a) Updated trie and (b) updated DUOS.

Algorithm: delete (*p*, *l*)

    (*m*, *n*) = Trie.delete(*p*, *l*)

    If *m* is a leaf then

        LTCAM.delete(*m*→*slot*)

        If *n* exists and *n* is now a leaf then

            slot = LTCAM.insert(*n*→*prefix*, *n*→*nexthop*, *n*→*length*)

            ITCAM.delete(*n*→*slot*, *n*→*length*) // since *n* is now a leaf prefix

            n→*slot* = slot;

        endif

    else

        ITCAM.delete(*m*→*slot*, *m*→*length*)

    endif

FIGURE 13.17   Algorithm to delete from DUOS.

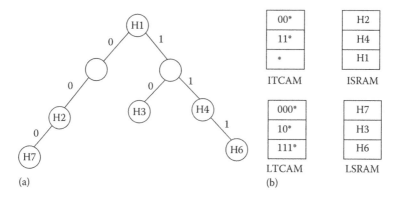

(a)   (b)

FIGURE 13.18   Delete rule {100*, H5} from the prefixes in Figure 13.16. (a) Updated trie and (b) updated DUOS.

(if any) of *m* that has a prefix. If *m* was a leaf, then *p* is to be deleted from the LTCAM. In this case, the prefix (if any) in *n* may become a leaf prefix. If so, the prefix in *n* is to be moved from the ITCAM to the LTCAM. When *m* is not a leaf, *p* is deleted from the ITCAM. Figure 13.18 illustrates the delete procedure for rule {100*, H5} from the seven-prefix forwarding table of Figure 13.16.

13.5.2.3 Change

To change the nexthop of an existing prefix to *newH*, first, the nexthop of the prefix in the trie is changed and the node *m* that contains *p* is returned. Then, depending on whether *m* is a leaf or non-leaf, the change function is invoked for the corresponding TCAM. Figure 13.19 gives the algorithm.

Algorithm: change (*p*, *length*, *newH*)

    m = Trie.change(*p*, *l*, *newH*)

    If *m* is a leaf then

        m→*slot* = LTCAM.change(*p*, *m*→*slot*, *newH*);

    else

        m→*slot* = ITCAM.change(*p*, *m*→*slot*, *newH*, *length*);

FIGURE 13.19   Algorithm to change a next hop in DUOS.

13.5.3 ITCAM Algorithms

The prefixes in the ITCAM are stored in a manner that supports the determination of the longest matching prefix (i.e., in any topological order that conforms to the precedence constraints defined by the binary trie–$p1$ must come before $p2$ whenever $p1$ is a descendent of $p2$ [19]). Decreasing order of length is a commonly used ordering. The function *getSlot*(*length*) returns an ITCAM slot such that insertion of the new prefix into this slot satisfies the ordering constraint in use provided the new prefix has the specified length; the function *freeSlot*(*slot*, *length*) frees a slot previously occupied by a prefix of the specified length and makes this slot available for reuse later. These functions, which are described in Section 13.5.5, are used in our ITCAM insert, delete, and change algorithms presented in Figure 13.20.

Notice that following the first step of the change algorithm, the prefix whose next hop is being changed is in two valid slots of the ITCAM–*oldSlot* and *slot*. This duplication does not affect correctness of data plane lookups as whichever one is matched by the ITCAM, the returned next hop that is valid either before or after the change operation. On the other hand, if an attempt was made to change the next hop in *ISRAM*[*oldSlot*] directly, an ongoing lookup may return a garbled next hop. Similarly, if an entry is first deleted and then inserted, lookups that take place between the delete and the insert may return a next hop that doesn't correspond to the forwarding table state either before or after the change. If a *waitWriteValidate* is used to change *ISRAM*[*oldSlot*] to nexthop, *oldSlot* becomes unavailable for data plane lookups during the write operation and inconsistent results are returned in case the prefix in TCAM[*oldSlot*] is the longest matching prefix.

13.5.4 LTCAM Algorithms

The prefixes in the LTCAM are disjoint and so may be stored in any order. The unused (or free) slots of the LTCAM/LSRAM are linked together into a chain using the words of the LSRAM to build this chain. AV is used to store the index of the first LSRAM word on the chain. So, the free slots are AV, *LSRAM*[*AV*], *LSRAM*[*LSRAM*[*AV*]], and so on. The last free slot on the AV chain has $LSRAM[last] = -1$. The LTCAM algorithms to insert, delete, and change are given in Figure 13.21. These algorithms are self-explanatory.

Example 13.3

Figure 13.22 shows a small LTCAM with five entries. Two out of the five entries store prefixes, whereas the other three are empty. So, **AV**, which is stored at a memory location of the control plane, is set to 1. LSRAM[1] is set to 3, and LSRAM[3] is set to 4. Since the word at the address 4 is the last free word, LSRAM[4] is set to −1. Now, as prefixes are inserted and deleted, **AV** and the values stored in the free LSRAM words are changed accordingly.

Algorithm: insert(prefix, nexthop, length)
    slot = getSlot(length);
    ITCAM.*waitWriteValidate*(slot, prefix, nexthop);
    return slot;

Algorithm: delete(slot, length)
    freeSlot(slot, length);

Algorithm: change(prefix, oldSlot, nexthop, length)
    slot = insert(prefix, nexthop, length);
    delete(oldSlot, length);
    return slot;

FIGURE 13.20 ITCAM algorithms.

Algorithm: insert(prefix, nexthop, length)
    if $(AV == -1)$ throw NoSlotException;
    slot = AV;
    AV = LSRAM[slot];
    LTCAM.*waitWriteValidate*(slot, prefix, nexthop);
    return slot;

Algorithm: delete(slot)
    LTCAM.*invalidateWaitWrite*(slot, AV); // AV is stored in LSRAM[slot]
                             // after waiting for an ongoing lookup to complete
    AV = slot;

Algorithm: change(prefix, oldSlot, nexthop, length)
    slot = insert(prefix, nexthop, length);
    delete (oldSlot);
    return slot;

FIGURE 13.21 LTCAM algorithms.

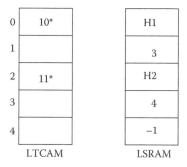

FIGURE 13.22 LTCAM-LSRAM layout showing free space management.

13.5.5 ITCAM Memory Management

The memory management scheme DLFS_PLO (Distributed and Linked Free Space with Prefix Length Ordering Constraint) offers the best algorithm known to us in terms of the number of prefix moves required when a prefix is to be inserted into or deleted from a TCAM [6]. Note that prefix moves are required since the ITCAM stores prefixes that could either enclose or be enclosed by other prefixes. Hence, to ensure that the first prefix matched in the TCAM is also the longest matching prefix, prefix insertions and deletions must be done carefully.

The description of memory management scheme DLFS_PLO includes an implementation of the *getSlot* and *freeSlot* functions used in Section 13.5.3 to get and free ITCAM slots. The implementations employ the function *move* (Figure 13.23) that moves the content of an in-use ITCAM slot to a free ITCAM slot in such a way as to maintain data plane lookup consistency. DLFS_PLO maintains the invariant that an ITCAM slot has its valid bit set to 0 iff that slot wasn't matched by the ongoing data plane lookup (if any); that is, iff the slot isn't involved in the ongoing data plane lookup.

Lookup consistent implementations of *getSlot* and *freeSlot* employ the following variables. W is the maximum prefix length (32 for IPv4), $top[i]$ is the slot where the first prefix of length i is stored and $bot[i]$ is the slot where the last prefix of length i is stored, $0 \le i \le W$ (i.e., these variables define the start and

Algorithm: *move (src, dest)*
    ITCAM.*waitWriteValidate*(dest, ITCAM[src], ISRAM[src]);

FIGURE 13.23 Move from ITCAM[*src*] to ITCAM[*dest*].

end of block i). Note that $top[i] \leq bot[i]$ for a nonempty block i and $top[i] > bot[i]$ for an empty block. Let, $top[0] = bot[0] = N + 1$ and $top[W + 1] = bot[W + 1] = -1$. For an empty ITCAM, $top[i] = N + 1$ for $1 \leq i \leq W$; $bot[i] = -1$ for $1 \leq i \leq W$. The forward links, called *next*[], of the doubly linked list are maintained using the ISRAM words corresponding to the free ITCAM slots with $AV[i]$ recording the first slot on the list for the ith block. The backward links, called *prev*[], are maintained in these ISRAM words in case an ISRAM word is large enough to accommodate two links and in the control plane memory otherwise. All variables, including the array $AV[]$, are stored in the control plane memory.

The *getSlot* algorithm (Figure 13.24) first attempts to make available a slot from the doubly linked list for the desired block. When this list is empty, the algorithm provides a free slot from either block boundary when there is a free slot on the block boundary. Otherwise, it moves a free slot from the nearest block boundary that has a free slot. This algorithm utilizes several supporting algorithms that are given in Figures 13.25 and 13.26. The algorithm *movesFromAbove* (*movesFromBelow*) returns the number of prefix moves that are required to get the nearest free slot from above (below) the block where it is needed and *getFromAbove* and *getFromBelow*, respectively, get the nearest free slot above or below the block where the free slot is needed.

The algorithm to free a slot (Figure 13.27) adds a freed slot inside the block to the doubly linked list of free slots. Again, correctness and consistency are established easily.

The scheme is shown in Figure 13.28a, where the ITCAM slots are indexed 0 through N. The prefixes are stored in decreasing order of length in the TCAM, which ensures that the longest matching prefix is returned as the first matching prefix. Prefixes of the same length appear one after another in a group in the TCAM and this is referred to as a prefix block. The free slots are distributed between the boundaries of adjacent prefix blocks. At the time the ITCAM is initialized, the available free slots are distributed in proportion to the number of prefixes in a block with the caveat that an empty block gets one free slot at its boundary. In addition to the free space at the boundary regions, a doubly linked list of free slots is maintained within each block. Free slots within a block are created as a result of prefix deletions from within the block.

Algorithm: getSlot(*len*)
```
    aP=0; bP=0; aC=0; bC=0;
    if (AV[len] == -1)
    // AV[len] stores the first free space in block of length len
        ma = movesFromAbove(len, &aP, &aC);
        mb = movesFromBelow(len, &bP, &bC);
        if (ma < mb)
            d = getFromAbove(len, aP, aC);
            if (top[len] > bot[len]) bot[len] = d;
            top[len] = d;
        else
            if (mb == W + 1) throw NoSpaceException; // no space
            d = getFromBelow(len, bP, bC);
            if (top[len] > bot[len]) top[len] = d;
            bot[len] = d;
        endif
    else
        d = AV[len];
        AV[len] = next[d];
    endif
    return d;
```

FIGURE 13.24 DLFS_PLO algorithm to get a free slot to insert a prefix whose length is *len*.

Algorithm: movesFromAbove(*len, *pos, *cur*)

    moves=0;

    // find max p $\leq$*len* such that block p is not empty

    for (p=*len*; top[p] > bot[p]; p−−);

    // find min c > *len* with space just below it

    for (c=*len*+1; c$\leq$*W*+1; c++)

        if (top[c] $\leq$ bot[c]) // not empty

            if (bot[c]+1 < top[p] || !valid[bot[c]])

                *cur = c; *pos = p;

                return moves;

            endif

            moves++;

            if (AV[c] >= 0) *pos = c; *cur = c; return moves; endif

            p = c;

        endif

    return $W + 1$;

Algorithm: movesFromBelow(*len, *pos, *cur*)

    moves=0;

    // find min p >= *len* such that block p is not empty

    for (p=*len*; top[p] > bot[p]; p++);

    // find min c > *len* with space just below it

    for (c=*len*-1; c>=0; c−−)

        if (top[c] $\leq$ bot[c]) // not empty

            if (top[c]-1 > bot[p] || !valid[top[c]])

                *pos = p; *cur = c;

                return moves;

            endif

            moves++;

            if (AV[c] >= 0) *pos = c; *cur = c; return moves; endif

            p = c;

        endif

    return $W + 1$;

FIGURE 13.25 Algorithms to compute the number of moves used by the algorithm of Figure 13.24.

The number of moves required by an update sequence is dependent on the number of free TCAM slots available. An experiment in [6] shows that even with 99% prefix occupancy and 1% free space, the total number of prefix moves using DLFS_PLO is at most 0.7% of the total number of prefix inserts and deletes.

13.5.6 DUOW

DUOW is a dual TCAM architecture in which a wide SRAM (say, 144-bit words or larger) is used with either or both the TCAMs. The method of adding a wide LSRAM to the LTCAM is described here. The technique of adding a wide ISRAM is almost identical to that used in Section 13.4. A wide LSRAM word is used to store a subtree of the binary trie of a forwarding table. Since the LTCAM stores prefixes in the leaf nodes of a binary trie, carving is done on a leaf trie storing the leaf prefixes only and not on the original binary trie storing all prefixes. When a subtree of the leaf trie is stored in an LSRAM word, that subtree is removed from (or carved out of) the leaf trie before another subtree is identified for carving. Let N be the root of the subtree being carved and let $Q(N)$ be the prefix defined by the path from the root of the trie to N. $Q(N)$ is stored in the LTCAM, and $|P_i| - |Q(N)|$ suffix bits, of each prefix P_i in the carved subtree rooted at N, are stored in the LSRAM word. Note that each suffix stored in the LSRAM word is a suffix

Algorithm: getFromAbove(*len, p, c*)
  if (top[*p*] > bot[*c*]+1) *d* = top[*p*]-1; *c* = *p*;
  else
    if (!valid[bot[*c*]])
      d = bot[*c*]−−;
      if (*d* == AV[*c*]) AV[*c*] = next[AV[*c*]];
      else
        next[prev[*d*]] = next[*d*];
        if (next[*d*] != -1) prev[next[*d*]] = prev[*d*];
      endif
    else
      d = AV[*c*];
      AV[*c*] = next[AV[*c*]];
      move(bot[*c*], *d*);
      d = bot[*c*]−−;
    endif
    c − −;
  endif
  for (; *c* > *len*; *c* − −)
    if (top[*c*] ≤ bot[*c*])
      move(bot[*c*]−−, −−top[*c*]);
      d = bot[*c*]+1;
    endif
  return *d*;

Algorithm: getFromBelow(*len, p, c*)
  if (top[*c*]-1 > bot[*p*]) *d* = bot[*p*]+1; *c* = *p*;
  else
    if (!valid[top[*c*]])
      d = top[*c*]++;
      if (d == AV[*c*]) AV[*c*] = next[AV[*c*]];
      else
        next[prev[*d*]] = next[*d*];
        if (next[*d*] != -1) prev[next[*d*]] = prev[*d*];
      endif
    else
      d = AV[*c*];
      AV[*c*] = next[AV[*c*]];
      move(top[*c*],*d*);
      d = top[*c*]++;
    endif
    c++;
  endif
  for (; *c* < *len*; *c*++)
    if (top[*c*] ≤ bot[*c*])
      move(top[*c*]++, ++bot[*c*]);
      d = top[*c*]-1;
    endif
  return *d*;

FIGURE 13.26 Algorithms to get a slot used by the algorithm of Figure 13.24.

of a leaf prefix that begins with $Q(N)$. By repeating this carving process, all leaf prefixes are allocated to the LTCAM and LSRAM. To obtain the mapping of leaf prefixes to the LTCAM and LSRAM, the carving algorithm must ensure that the $Q(N)$s stored in the LTCAM are disjoint. Since the carving algorithm in Section 13.4 does not ensure disjointedness, a new carving algorithm is needed. As an example, consider the binary trie of Figure 13.29a, which has been carved using a carving algorithm that ensures that each

Algorithm: freeSlot(*d*, *len*)
    if (top[*len*] == *d*) ITCAM[top[*len*]++].valid = 0;
    else if (bot[*len*] == *d*) ITCAM[bot[*len*]−−].valid = 0;
    else
        ITCAM.*invalidateWaitWrite*(*d*, AV[*len*]);
                // AV[*len*] is stored in ISRAM[*d*].
        if (AV[*len*] != -1) prev[AV[*len*]] = *d*;
        AV[*len*] = *d*;
    endif

FIGURE 13.27 DLFS_PLO algorithm to free a slot.

carved subtree has at most two leaf prefixes. The LTCAM will need to store $Q(N1)$, $Q(N2)$, and $Q(N3)$. Even though the prefixes in the binary trie are disjoint, the $Q(N)$s in the LTCAM are not disjoint (e.g., $Q(N1)$ is a descendant of $Q(N2)$ and so $Q(N2)$ matches all IP addresses matched by $Q(N1)$). To retain much of the simplicity of the LTCAM management scheme of DUOS it is necessary to carve the leaf trie in such a way that all $Q(N)$s in the LTCAM are disjoint.

In this case, carving is done via a postorder traversal of the binary trie, using the visit algorithm of Figure 13.30 to do the carving. In this algorithm, w is the number of bits in an LSRAM word and $x \rightarrow$ size is the number of bits needed to store (1) the suffix bits corresponding to prefixes in the subtrie rooted at x, (2) the length of each suffix, (3) the next hop for each suffix, (4) the number of suffixes in the word, and (5) the length of $Q(x)$, which is the corresponding prefix stored in the LTCAM. Algorithm *splitNode*(*q*) does the actual carving of the subtree rooted at node q, which involves setting child pointers appropriately for the parent of the carved node, as well as adjusting the number of prefixes and the SRAM bits needed for the rest of the trie in the absence of the subtree being carved. The basic idea in our carving algorithm is to forbid carving at two nodes that have an ancestor–descendent relationship. This ensures that the $Q(N)$s are disjoint. Figure 13.29b shows the subtrees carved by our algorithm. As can be seen, $Q(N1)$, $Q(N2)$, $Q(N3)$ are disjoint. Although our carving algorithm generally results in more $Q(N)$s than when the carving algorithm of [5] is used, our carving algorithm allows us to retain the flexibility to store the $Q(N)$s in any order in the LTCAM as the $Q(N)$s are independent.

The LTCAM algorithms to insert, delete, change, and necessary support algorithms are given in [6]. Figure 13.31 shows a possible assignment of the six-prefix example in Figure 13.1. The suffix node format need not store the default next hop, since in case of a "no match" situation the ITCAM provides the next hop. The intermediate prefixes for rules R1, R3, and R4 are stored in the ITCAM, while the leaf prefixes for rules R2, R5, and R6 are stored in the LTCAM using a wide LSRAM. The suffix nodes begin with the prefix length field of 2 bits in this example followed by the suffix count field of 2 bits. Next comes the (length, suffix, nexthop) triplet for each prefix encoded in the suffix node, the number of allocated bits being (2, 4, and 6 bits) respectively for the three fields in the triplet.

13.5.7 IDUOW

It is evident that by adding an index TCAM in conjunction with an LTCAM that uses a wide LSRAM (i.e., an index for the LTCAM of DUOW), there was further reduction in power consumption. Adding an index to the ITCAM of DUOW follows easily from [5]. When the LTCAM is indexed, we have two TCAMs replacing the LTCAM—a data TCAM referred to as DLTCAM and an index TCAM referred to as ILTCAM. The associated SRAMs are DLSRAM and ILSRAM.

Two index TCAM strategies based on the 1-12Wc and M-12Wb schemes of [5] were explored in [6]. The 1-12Wc scheme is best for power whereas the M-12Wb scheme is the best overall scheme consuming

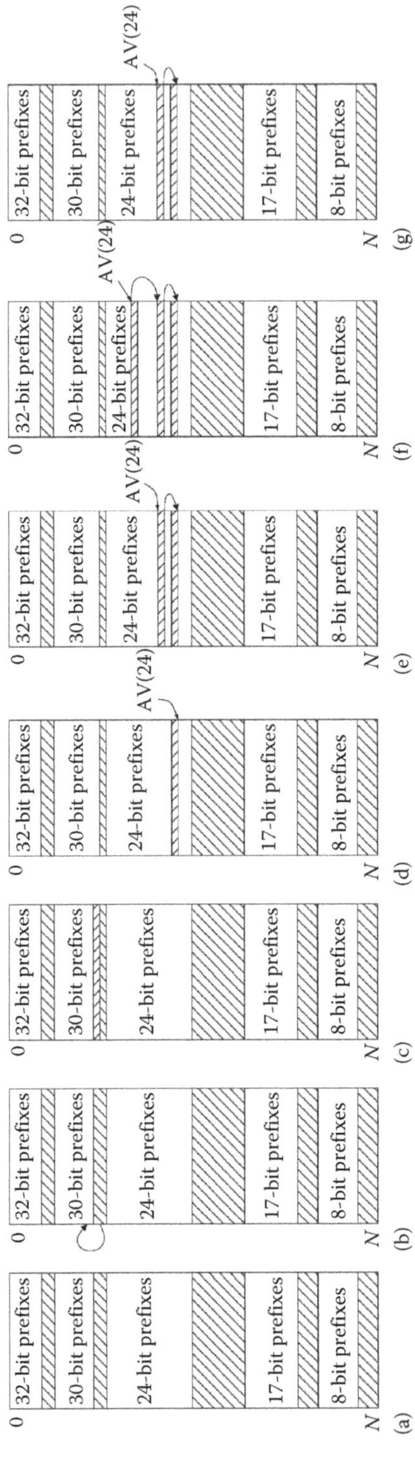

FIGURE 13.28 ITCAM layout for DLFS_PLO, with moves for insert and delete. The curved arrows on the right show the forward links in the list of free spaces. (a) Initial arrangement, (b) insert p/30, (c) free space available, (d) delete p/24, (e) delete p2/24, (f) delete p3/24, and (g) insert p/24.

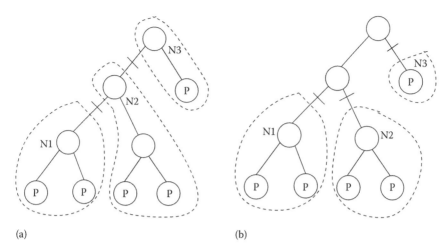

FIGURE 13.29 Carving using the method of [5] and our method. (a) Lu carving [4] and (b) our carving.

Algorithm: visit_postorder(x)

```
if (!x) return 0;
isSplit = visit_postorder(y);
isSplit | = visit_postorder(z);
if (isSplit ||x→size > w) then
    splitNode(y);
    splitNode(z); // where y and z are children of x
    return 1;
else if (x→size == w) then
    splitNode(x);
    return 1;
endif
return 0;
```

FIGURE 13.30 Algorithm to carve a leaf trie to obtain disjoint $Q(N)$s.

ITCAM:

11*
00*
*

ISRAM:

H4
H3
H1

LTCAM:

0*
1*

LSRAM:

01	01	01	0	H2			
01	10	10	00	H5	10	11	H6

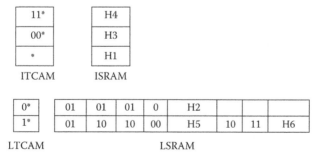

FIGURE 13.31 Assignment of prefixes of Figure 13.1 to DUOW.

least TCAM space and low power for lookups [5]. Both 1-12Wc and M-12Wb organize the DLTCAM into fixed size buckets that are indexed using the ILTCAM and ILSRAM, which also is a wide SRAM that stores suffixes and associated information.

The memory management scheme specifies how a new bucket is assigned when a prefix is to be added to a bucket that is already full. Similarly, as a prefix is deleted from a bucket, the memory management scheme explores the possibility of merging the contents of the bucket with another bucket and updating the DLTCAM and ILTCAM contents accordingly.

The algorithms for carving and prefix assignment for the 1-12Wc and M-12Wb prefix layout schemes as well as the memory management algorithms for both schemes can be found in [6].

13.6 Experimental Results

The performance evaluation of the different TCAM schemes are presented in [6] and are summarized here for 20 IPv4 routing tables and update sequences downloaded from [15,16]. Figure 13.32 gives the characteristics of these datasets. The update sequences for the first 20 routing tables were captured from files storing update announcements from 12 a.m. on February 1, 2009 for the stated number of hours. The columns labeled #RawInserts, #RawDeletes, and #RawChanges, respectively, give the number of insert, delete, and change next hop requests in the update sequences. Using consistent updates, a next hop change request is implemented (see Figure 13.20 for example) as an insert (of the prefix with the new next hop) followed by a delete (of the prefix with the old nexthop). Therefore, all results henceforth are in terms of the effective inserts and deletes. Note that the number of effective inserts (#Inserts) and deletes (#Deletes) is given by the following equations.

$$\#Inserts = \#RawInserts + \#RawChanges;\tag{13.3}$$

$$\#Deletes = \#RawDeletes + \#RawChanges;\tag{13.4}$$

Figures 13.33 and 13.34 show a comparison of the TCAM and SRAM power consumption among the TCAM schemes that support incremental updates. The IDUOW data assume that the LTCAM is indexed using an ILTCAM but the ITCAM is not indexed. The STCAM in Section 13.2 can be updated incrementally by using DLFS_PLO. The power numbers were estimated using CACTI [20] for SRAMs and "tcam-power" [21] for TCAMs. Additionally, the following settings were used: process technology of 70 nm, operating voltage of 1.12 V, TCAM operation frequency of 360 MHz [22]. For the IDUOW schemes, the bucket size for DLTCAM was set to 512 entries.

DataSet	#Prefixes	Hours	#RawInserts	#RawDeletes	#RawChanges
rrc00	294,098	75.7	39,553	40,051	368,013
rrc01	276,795	75.2	41,692	41,988	492,315
rrc03	283,754	42.7	27,702	27,914	292,454
rrc04	288,610	17	16,086	15,977	193,392
rrc05	280,041	103	20,276	18,285	439,647
rrc06	278,744	235	157,549	157,547	289,272
rrc07	275,097	0.417	247	218	179,835
rrc10	278,898	105	21,620	22,473	326,720
rrc11	277,166	80.2	58,115	58,378	290,621
rrc12	278,499	62.3	33,196	33,572	410,464
rrc13	284,986	57.8	23,920	23,713	284,710
rrc14	276,170	83.6	56,598	56,810	203,955
rrc15	284,047	134	95,790	93,750	183,131
rrc16	282,660	672	3,338	937	8,896
rv2	294,127	56.5	13,882	15,552	679,100
rv4	275,737	95	69,627	69,754	526,302
rv.eqix	275,736	70.3	51,104	51,066	253,693
rv.isc	281,095	68.2	44,286	44,444	292,323
rv.linx	278,196	49.1	23,137	23,413	384,344
rv.wide	283,569	174	101,821	103,862	372,035

FIGURE 13.32 Datasets used in the experiments.

DataSet	STCAM	DUOS	DUOW	IDUOW-(1-12Wc)	IDUOW-(M-12Wb)
rrc00	24,246.29	24,246.43	7,880.15	2,303.91	2,308.04
rrc01	22,820.04	22,820.18	7,375.72	2,090.03	2,093.65
rrc03	23,393.68	23,394.24	7,600.31	2,199.94	2,203.34
rrc04	23,794.03	23,794.59	7,623.57	2,119.8	2,124.49
rrc05	23,087.36	23,087.92	7,478.02	2,138.96	2,143.34
rrc06	22,980.46	22,981.02	7,438.55	2,124.73	2,130.96
rrc07	22,679.9	22,680.46	7,318.64	2,061.66	2,066.36
rrc10	22,993.33	22,993.47	7,422.78	2,093.83	2,097.4
rrc11	22,850.44	22,850.57	7,383.31	2,090.03	2,094.57
rrc12	22,960.24	22,960.8	7,422.87	2,099.1	2,102.59
rrc13	23,495.32	23,495.46	7,641.93	2,216.5	2,221.2
rrc14	22,768.28	22,768.84	7,343.06	2,065.21	2,068.7
rrc15	23,417.77	23,418.33	7,599.08	2,205.51	2,211.34
rrc16	23,303.6	23,303.74	7,539.56	2,155.93	2,160.39
rv2	24,248.63	24,249.2	7,847.89	2,263.41	2,268.03
rv4	22,732.87	22,733.01	7,336.02	2,069.9	2,075.65
rv.eqix	22,732.79	22,732.93	7,328.82	2,060.2	2,065.3
rv.isc	23,174.63	23,174.76	7,513.12	2,145.59	2,149.32
rv.linx	22,935.34	22,935.48	7,422.61	2,113.56	2,119.55
rv.wide	23,378.31	23,378.86	7,611.94	2,224.35	2,230.26

FIGURE 13.33   TCAM power consumption in mW.

DataSet	STCAM	DUOS	DUOW	IDUOW-(1-12Wc)	IDUOW-(M-12Wb)
rrc00	1412.07	1427.18	1639.49	170.98	173.78
rrc01	1329.46	1343.17	1549.85	157.49	159.91
rrc03	1362.26	1376.69	1585.82	164.29	166.6
rrc04	1385.34	1399.77	1608.5	159.77	162.9
rrc05	1345.26	1359.51	1566.66	160.58	163.53
rrc06	1339.18	1352.6	1558.92	159.59	163.76
rrc07	1322.17	1335.27	1539.72	155.66	158.83
rrc10	1339.18	1353.05	1561.08	157.76	160.18
rrc11	1331.88	1345.6	1551.24	157.49	160.55
rrc12	1337.96	1352.17	1560.03	157.98	160.34
rrc13	1368.34	1382.72	1593.89	165.5	168.67
rrc14	1327.03	1340.29	1545.43	155.83	158.19
rrc15	1364.69	1378.24	1584.76	164.67	168.58
rrc16	1357.4	1371.62	1579.02	161.69	164.7
rv2	1412.07	1426.95	1638.21	168.43	171.55
rv4	1324.6	1338.19	1542.99	156.16	160.01
rv.eqix	1324.6	1338.75	1542.33	155.5	158.94
rv.isc	1350.11	1364.87	1574.2	160.97	163.5
rv.linx	1336.75	1350.73	1556.88	159.03	163.05
rv.wide	1362.26	1376.97	1584.54	165.94	169.9

FIGURE 13.34   SRAM power consumption in mW.

The IDUOW schemes reduce TCAM power by an order of magnitude relative to the STCAM scheme and DUOS. For both the IDUOW schemes, the ILTCAM and DLTCAM together consume less than 70 mW; the ITCAM consumes the remaining approximately 2 W. When the ILTCAM is also indexed using its own index TCAM, the total power is less than 100 mW, which is a 200-fold reduction relative to STCAM and DUOS.

Incremental updates are efficiently performed in DUO. The DLFS_PLO memory management scheme used in DUO requires very few prefix moves, as can be seen in Figure 13.35 which gives the total and

Dataset	Total Moves		Average Moves		Standard Deviation	
	Insert	Delete	Insert	Delete	Insert	Delete
rrc00	401	0	0.000984	0	0.064	0
rrc01	0	0	0	0	0	0
rrc03	4630	0	0.0145	0	0.311	0
rrc04	2	0	0.00001	0	0.003	0
rrc05	541	0	0.00118	0	0.077	0
rrc06	8	0	0.00002	0	0.004	0
rrc07	0	0	0	0	0	0
rrc10	671	0	0.00193	0	0.069	0
rrc11	245	0	0.0007	0	0.0377	0
rrc12	4659	0	0.0105	0	0.27	0
rrc13	989	0	0.0032	0	0.114	0
rrc14	4	0	0.000015	0	0.005	0
rrc15	2769	0	0.0099	0	0.212	0
rrc16	0	0	0	0	0	0
rv2	14	0	0.00002	0	0.005	0
rv4	141	0	0.000236	0	0.034	0
rv.eqix	33	0	0.000108	0	0.0107	0
rv.isc	12	0	0.000036	0	0.006	0
rv.linx	0	0	0	0	0	0
rv.wide	1	0	0.000002	0	0.0015	0

FIGURE 13.35 Statistics on the number of prefix moves.

Dataset	DUOS	DUOW	IDUOW (1-12Wc)	IDUOW (M-12Wb)
rrc00	831,630	894,978	904,692	923,390
rrc01	1,083,395	1,151,102	1,160,645	1,179,951
rrc03	655,262	703,419	711,292	726,559
rrc04	425,587	453,943	458,190	477,705
rrc05	924,947	958,053	961,356	987,024
rrc06	950,924	1,194,841	1,224,466	1,254,236
rrc07	360,149	360,588	360,596	362,235
rrc10	703,412	746,524	749,215	769,919
rrc11	715,786	809,353	819,181	844,235
rrc12	908,971	964,395	975,039	993,097
rrc13	630,400	660,925	666,808	691,469
rrc14	536,465	632,699	640,566	658,301
rrc15	585,012	720,082	735,449	767,875
rrc16	22,919	26,107	28,480	52,885
rv2	1,397,932	1,419,396	1,424,525	1,443,112
rv4	1,225,347	1,352,059	1,368,462	1,400,013
rv.eqix	624,700	705,100	712,971	742,635
rv.isc	695,496	767,463	778,565	797,057
rv.linx	826,755	861,534	868,559	897,055
rv.wide	988,721	1,145,004	1,163,932	1,195,740

FIGURE 13.36 Total number of TCAM *waitWrite* operations.

average number of prefix moves as well as the standard deviation from the average, while applying the inserts and deletes on the STCAM. The average number of prefixes moved is obtained by dividing the total moves by the total number of inserts and deletes performed on the forwarding table.

The total number of *waitWrites* for the inserts and deletes are presented in the Figure 13.36. As noted in [6], the maximum number of *waitWrites* for the IDUOW scheme using M-12Wb layout is quite large, but it is easily reducible by fixing the size of a DLTCAM bucket to a smaller number.

The worst case number of writes needed for an insert or delete in DUOS is 34, in DUOW is 38, and in IDUOW 38+*bucketSize*.

13.7 Conclusion

The techniques of indexing a TCAM and the use of wide SRAMs were explored as mechanisms for reducing the power consumed by TCAM-based routers. It was found that the techniques when applied together reduced power consumption by more than two orders of magnitude. Using the dual TCAM architecture, DUO, for forwarding tables it was possible to have the benefits of low power consumption along with efficient incremental updates and faster TCAM lookup. The absence of a priority encoder in the LTCAM system of DUO made overall lookup faster by about 50% compared to a conventional TCAM, when the nexthop was found in the LTCAM. It was observed that over 90% prefixes are stored in the leaf TCAM system. Therefore, the probability of having a hit in the LTCAM system during lookup is very high. The lookup consistent algorithms for incremental updates allowed data plane lookup to proceed along with the control plane update operations.

Acknowledgment

This work was supported, in part, by the National Science Foundation, under grant 0829916 and the US Air Force under grant FA8750-10-1-0236.

References

1. R. S. Tucker, J. Baliga, R. Ayre, K. Hinton, and W. V. Sorin, Energy consumption in IP networks, *34th European Conference and Exhibition on Optical Communication*, Brussels, Belgium, September 2008.
2. A. M. Lyons, D. T. Neilson, and T. R. Salamon, Energy efficient strategies for high density telecom applications, *Workshop on Wireless Communications and Networks*, February 2008.
3. S. Sekiguchi, Grid around Asia, *Grid Asia*, 2009.
4. F. Zane, G. Narlikar, and A. Basu, CoolCAMs: Power-efficient TCAMs for forwarding engines, *Proceedings of INFOCOM*, San Francisco, CA, 2003.
5. W. Lu and S. Sahni, Low power TCAMs for very large forwarding tables, *Proceedings of INFOCOM*, Phoenix, AZ, 2008.
6. T. Mishra and S. Sahni, DUO-dual TCAM architecture for routing tables with incremental update. http://www.cise.ufl.edu/~sahni/papers/duo.pdf
7. H. Liu, Routing table compaction in ternary-CAM, *IEEE Micro*, 22(3), 58–64, 2002.
8. T. Mishra and S. Sahni, PETCAM—A power efficient TCAM architecture for forwarding tables, *IEEE Transactions on Computers*, July 2009, 224–229, 2010.
9. C. A. Zukowski and S. Wang, Use of selective precharge for low-power content-addressable memories, *IEEE International Symposium on Circuits and Systems*, Hong Kong, China, 1997.
10. N. Mohan and M. Sachdev, Low power dual matchline ternary content addressable memory, *IEEE International Symposium on Circuits and Systems*, Vancouver, Canada, 2004.
11. H. Miyatake, M. Tanaka, and Y. Mori, A design for high-speed low-power CMOS fully parallel content addressable memory macros, *IEEE Journal of Solid State Circuits*, 36(6), June 2001, 956–968.
12. C.-S. Lin, J.-C. Chang, and B.-D. Liu, A low-power pre-computation based fully parallel content addressable memory, *IEEE Journal of Solid State Circuits*, 38(4), April 2003, 654–662.
13. R. Draves, C. King, S. Venkatachary, and B. Zill, Constructing optimal IP routing tables, *Proceedings of INFOCOM*, New York, 1999, pp. 88–97.
14. M. Akhbarizadeh and M. Nourani, Efficient prefix cache for network processors, *IEEE Symposium on High Performance Interconnects*, Stanford, CA, August 2004.

15. RIPE Network Coordination Centre, RIS Raw Data, http://www.ripe.net/projects/ris/rawdata.html, 2008.

16. University of Oregon Route Views Project, http://www.routeviews.org, 2009.

17. Z. Wang, H. Che, M. Kumar, and S. K. Das, CoPTUA: Consistent policy table update algorithm for TCAM without locking, *IEEE Transactions on Computers*, 53(12), December 2004, 1602–1614.

18. G. Wang and N. Tzeng, TCAM-based forwarding engine with minimum independent prefix set (MIPS) for fast updating, *IEEE International Conference of Communications*, 1, June 2006, 103–109.

19. D. Shah and P. Gupta, Fast updating algorithms on TCAMs, *IEEE Micro*, 21(1), January-February 2001, 36–47.

20. N. Muralimanohar, R. Balasubramonian, and N. P. Jouppi, Optimizing NUCA organizations and wiring alternatives for large caches with CACTI 6.0, *Proceedings of the 40th International Symposium on Microarchitecture*, Chicago, IL, December 2007, pp. 3–14.

21. B. Agrawal and T. Sherwood, Ternary CAM power and delay model: Extensions and uses, *IEEE Transactions on Very Large Scale Integration (VLSI) Systems*, 16(5), May 2008, 554–564.

22. Renesas R8A20410BG 20Mb Quad search full ternary CAM. http://www.renesas.com/products/memory/TCAM/tcam_root.jsp, January 2010.

Algorithms

14

Algorithmic Aspects of Energy-Efficient Computing

Marek Chrobak
*University of California,
Riverside*

14.1 Introduction

The main objective of power management systems is to reduce the energy consumed by electrical devices while maintaining a satisfactory level of performance. Research on effective power management methods has intensified in recent years, driven by rising costs and detrimental environmental effects of energy production.

Information technology is estimated to consume about 10% of the overall energy production in the United States, and this number is expected to continue to grow. In addition to contributing to global energy needs, information technology faces its own specific energy limitations. Energy costs in server farms are now comparable to the cost of hardware. In microprocessor systems, heat dissipation is a major factor limiting improvements in performance. In laptop computers, cellphones, and other portable devices, reduced power consumption translates directly into extended battery life.

It is thus not surprising that energy consumption is now emerging as a dominant performance measure in computer systems, rivaling the running time. And just like with the run-time efficiency, where enormous advances have been accomplished by a combination of hardware, software, and algorithmic techniques, the design of energy-efficient computer systems will ultimately require the development of fundamental models, algorithmic tools, and principles that can be used to guide practical solutions.

Up to date, most of algorithms-related research on energy-efficient computing has been focused on task scheduling. The inspiration for these studies came from experimental research showing that significant energy savings can be accomplished by careful rescheduling of energy-consuming tasks and adjusting the system's parameters like the CPU frequency and the power level. The two paradigms that emerged from this work are referred to as *speed scaling* and *power-down scheduling*.

The term speed scaling (or frequency/voltage scaling) refers to the technique that reduces the energy consumption by lowering the processor's frequency. Lower frequency makes it possible to reduce voltage, and thus the overall energy required to complete the given workload will be reduced as well. Similar methods can be applied to other energy-consuming devices, for example to disk drives. The general goal of speed scaling algorithms is to schedule the given tasks and determine the optimum speed to execute them, while meeting the specified performance constraints. Algorithmic research on speed scaling has been initiated by Yao et al. [54], who developed a formal model for task scheduling with the objective function representing the total energy consumption. Algorithms for speed scaling are discussed in Section 14.2.

The power-down approach is very general, as it applies to almost any power consuming device, and its basic principle is to simply turn off the system during idle times. Since turning the system back on requires an additional energy overhead, powering down is beneficial only if the idle time is sufficiently long. Irani et al. [34,36] were the first to provide a formal setting for studying algorithmic aspects of power-down optimization in systems with multiple power levels. In their setting, the input is a sequence of idle and busy intervals of unknown length, and the objective is to design online algorithms with low competitive ratios with respect to the energy objective function. A more general model would be to consider a collection of energy-consuming tasks whose execution intervals are not fixed. Then one can attempt to group these tasks into a small number of contiguous blocks, allowing the system to be powered down in between these blocks. The constraints that need to be taken into account include task release times and deadlines, as well as the overhead associated with restarting the system after power-downs. This gives rise to the *power-down scheduling* problem discussed in Section 14.3.

In addition to speed scaling and power-down scheduling, we will briefly discuss a number of other topics related to energy-efficient computing. One of the main challenges facing microprocessor system designers is the CPU thermal management. High temperatures affect the chip's lifespan and reliability and cooling systems used to control the temperature increase further the energy consumption. In Section 14.4, we will describe past work on algorithmic issues in thermal management.

A significant fraction of energy consumption in modern memory systems can be attributed to internal and external memories. Other than power-down optimization discussed earlier, little algorithmic work has been done on optimizing energy required for data storage and access. The flash-memory technology offers potentially dramatic reduction in memory energy consumption, although its wide-scale deployment is currently hindered by high cost and limited lifespan: each block of flash memory "wears out" gradually with each write, and after a certain number of writes (specific to a given technology) the memory becomes unusable. This lifespan can be improved with appropriate wear-leveling strategies that we discuss in Section 14.5.

This survey is not meant to be exhaustive, but rather to highlight a few choice topics and research trends related to algorithms and energy minimization. Readers interested in learning more about this area are recommended to review the surveys by Irani and Pruhs [33], by Albers [1], or by Wierman et al. [52] (in this volume) that cover some topics not addressed in this chapter. Also, recent journal issues and conference proceedings have many papers related to speed scaling, power-down scheduling, and other algorithmic aspects of sustainable computing.

14.2 Speed Scaling

The power consumption of a CPU is known to be proportional to its speed (or frequency) s and the square of the voltage V, that is $P \sim sV^2$. Also, the minimum voltage required for reliable computation increases with speed. Although the exact nature of this dependency is not well understood, experimental studies show that the power consumption, expressed as a function $P(s)$ of speed, is roughly proportional to s^α, where $1 < \alpha \le 3$. Independently of the exact formula, $P(s)$ is a convex function of s; thus one can reduce the overall energy consumption by running the system at a lower speed. For example, for $P(s) = s^2$, if we decrease the speed by half, the processing time will double, but the power consumption will be reduced to $\frac{1}{4}$th of its original value, so the overall energy consumption will be cut by half. This observation is the basis of the speed-scaling method which, in essence, is to run the processor at a reduced speed whenever performance requirements allow that.

Yao et al. [54] proposed a formal model for studying speed scaling. In their model, we have a processor whose speed can vary arbitrarily over time. When the processor runs at speed s, its power consumption, that is the energy consumed per unit of time, is $P(s)$, where the function $P(\cdot)$ is assumed to be convex. An instance consists of a set of n jobs, with each job J_j specified by its release time r_j, deadline d_j, and processing time p_j. A schedule specifies, for each time t, what job is executed at time t and what is the processor's speed $s(t)$. The objective is to compute a preemptive schedule that minimizes the overall energy consumption $E = \int_0^\infty P(s(t))dt$.

In [54], the authors studied both offline and online algorithms for minimizing energy, and we discuss these two versions separately hereafter.

14.2.1 Offline Algorithms for Speed Scaling

As shown in [54], a speed-scaling schedule that minimizes the overall energy consumption can be computed in polynomial time. Their algorithm is in essence a greedy algorithm that at each phase identifies the most work-intensive time interval and then schedules the jobs contained in this interval at minimum possible speed. To formalize this idea, define the *intensity* of an interval $[x, y]$ to be the sum of the processing times of all jobs J_j wholly contained in $[x, y]$, that is those J_j that satisfy $x \le r_j < d_j \le y$. An interval $[x, y]$ with maximum intensity is called *critical*. If $[x, y]$ is critical then we can assume that $x = r_i$ and $y = d_j$, for some jobs J_i, J_j, for otherwise we could increase x or decrease y without decreasing the intensity of $[x, y]$.

Algorithm YDS: The algorithm proceeds in phases. In each phase, we find a critical interval $[r_a, d_b]$, with intensity, say, μ. In the computed schedule S^* the processor runs at speed μ between r_a and d_b and all jobs wholly contained in $[r_a, d_b]$ are scheduled using the earliest deadline policy. Then, we modify the instance by removing the jobs wholly contained in $[r_a, d_b]$ and "shrinking" this interval to a point. Specifically, each release time or deadline $\tau > r_a$ is changed to $r_a + \max(\tau - d_b, 0)$.

Note that after shrinking the interval $[r_a, d_b]$ the intensity values of other intervals may change, so they need to be recalculated. Figure 14.1 shows an example of an execution of Algorithm YDS. It consists of seven jobs, specified in the format $J_j = (r_j, d_j, p_j)$: $J_1 = (0, 11, 3)$, $J_2 = (1, 6, 4)$, $J_3 = (2, 5, 2)$, $J_4 = (2, 6, 2)$, $J_5 = (3, 12, 2)$, $J_6 = (8, 10, 1)$, and $J_7 = (8, 10, 2)$. The height of the rectangles in the schedule represents the processor speed. In the first phase the critical interval is $[1, 6]$, with intensity $(p_2 + p_3 + p_4)/(d_4 - r_2) = 8/5$. In the second phase the critical interval is $[8, 10]$ with intensity $3/2$. (In the algorithm, the endpoints of this interval will shift left due to shrinking of the previous interval $[1, 6]$.) The final critical interval is $[0, 12]$ with intensity 1.

The criticality of the interval $[r_a, d_b]$ chosen at each step and the earliest deadline policy used within this interval imply that the schedule S^* will never be idle in $[r_a, d_b]$ and that all jobs contained in $[r_a, d_b]$ will meet their deadlines. Thus S^* is feasible. The basic intuition behind the optimality of S^* is that, due to the convexity of $P(s)$, the speed used by an optimal schedule should be as level as possible over the whole range. The intensities of critical intervals decrease from phase to phase. (This is less obvious than it may seem at first, because the lengths of some intervals could decrease due to shrinking.) If we included in $[r_a, d_b]$ a portion of some job J_j that is not wholly contained in this interval, then a small piece of J_j could instead be executed in the next larger critical interval that contains J_j. This would decrease the

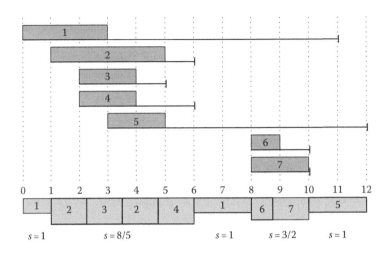

FIGURE 14.1　An example of a schedule produced by Algorithm YDS.

speed in $[r_a, d_b]$ and increase some speed outside this interval, thus flattening the overall speed profile and reducing the energy consumption.

A fully rigorous proof of optimality of Algorithm YDS was presented by Bansal et al. [8], who observed that it can be derived from the KKT criterion for the optimality of convex programs. To express the speed scaling problem as a convex program, let $\tau_1, \ldots, \tau_{2n}$ be the set of all release times and deadlines listed in nondecreasing order. Denote by $p_{j,h}$ the amount of job j processed in interval $[\tau_h, \tau_{h+1}]$. In an optimal schedule the processor will run at a constant speed in $[\tau_h, \tau_{h+1}]$; in fact, this speed will be exactly equal $\sum p_{j,h}/(\tau_{h+1} - \tau_h)$, where the sum is over all jobs j that are feasible in $[\tau_h, \tau_{h+1}]$, that is $r_j \leq \tau_h < d_j$. Thus the minimum total energy can be represented by the following program:

$$\text{minimize} \quad E$$

$$p_j \leq \sum_h p_{j,h} \quad \forall j$$

$$\sum_h (\tau_{h+1} - \tau_h)^{1-\alpha} (\sum_j p_{j,h})^\alpha \leq E$$

$$p_{j,h} \geq 0 \qquad \forall j, h$$

$$p_{j,h} = 0 \qquad \text{if } j \text{ not feasible in } [\tau_h, \tau_{h+1}]$$

It can be shown that this program is convex and, as explained in [8], the schedule produced by Algorithm YDS satisfies the KKT optimality conditions (see, for example, [18]).

What is the running time of Algorithm YDS? The intensity values can be computed in time $O(n^2)$: Divide the computation into stages, where in stage i we compute the intensity of all intervals $[r_i, d_j]$, by processing the d_j in increasing order while walking through the list of all jobs a for which $r_a \geq r_i$ (sorted in order of increasing deadlines) and incrementally computing the workloads of the intervals $[r_i, d_j]$. Then a naïve implementation of Algorithm YDS would update the instance after each phase, recompute all intensity values, and choose the maximum. Since there are $O(n)$ phases, the overall running time is $O(n^3)$. (See also [48].) As shown in [47], this running time can be improved to $O(n^2 \log n)$.

In the model in [54] the processor's speed can be adjusted to any value s—an assumption that is convenient for theoretical analysis, but also unrealistic, as in today's CPU technology only a finite (and small) number of speed values is available. Following some earlier work in [45], Li and Yao [48] studied the discrete version of the problem where the speed of the processor must be chosen from a given set $\{s_1, s_2, \ldots, s_\ell\}$ and showed that, rather surprisingly, this discrete variant can also be solved efficiently, in time $O(\ell n \log n)$. (To ensure feasibility, in this discrete–speed model we require that s_ℓ is large enough so that all jobs can be completed if executed at this speed.)

14.2.2 Online Algorithms for Speed Scaling

Yao et al. [54] also studied the online version of minimum energy scheduling with speed scaling, where the jobs arrive over time. Each job j arrives at its release time r_j, and at this time its processing time p_j and deadline d_j are announced. An online algorithm constructs the schedule in the online fashion, that is, at each time it chooses a pending job to process and the processor's speed.

Naturally, with only partial information about the input, an online algorithm cannot compute an optimal solution. A standard approach to measure the accuracy of online algorithms is to use competitive analysis. An online algorithm for a minimization problem is called *c-competitive* if it computes a solution whose value is at most c times the optimum value. The general objective of research in this area is to design algorithms with small competitive ratios and to establish lower bounds on the competitive ratio.

To illustrate the dilemma facing an online algorithm, imagine an instance where a job $J_1 = (0, 2, 2)$ is released at time 0. (We specify jobs in the format $J_j = (r_j, d_j, p_j)$.) If no other jobs are released, the optimal schedule would execute job J_1 at speed 1. But if we do execute J_1 at speed 1, it may happen that later, say at time $t = 1$, another job $J_2 = (1, 2, 2)$ could be released, and then we are forced to increase the speed to 3 in the interval $[1, 2]$. The optimum solution for this instance with two jobs would be to execute job J_1 at speed 2 in the interval $[0, 1]$ and job J_2 at speed 2 in the interval $[1, 2]$.

Two online algorithms were proposed in [54]. The first algorithm, Algorithm AR (average rate), allocates to each job the speed required to process this job at constant rate across its feasible interval, independently of the remaining jobs. The overall speed is the sum of the speeds allocated to the individual jobs.

Algorithm AR: For any job J_j let $\delta_j = (d_j - r_j)/p_j$ be the *density* of J_j. For each job J_j and time t define $\delta_j(t) = \delta_j$ for $r_j \leq t < d_j$ and 0 otherwise. Thus $\delta_j(t)$ represents the speed at which J_j should be executed if it were the only job in the instance. At any time t, the algorithm runs at speed $\sum_j \delta_j(t)$, using the earliest deadline policy to select the job to process.

Yao et al. [54] focus on the case when the power consumption function is $P(s) = s^\alpha$, for some constant $\alpha > 1$. This assumption is in fact quite common in the literature, and it is believed that in the current CMOS technology the power function has this form for some $1 < \alpha \leq 3$. They show that for $\alpha = 2$ the competitive ratio of Algorithm AR is between 4 and 8, and they write (without including a complete proof) that this result extends to any α, giving the competitive ratio between α^α and $2^{\alpha-1}\alpha^\alpha$.

Yao et al. also introduced another algorithm, Algorithm OA, although no analysis of its competitive ratio was given.

Algorithm OA: This algorithm runs at the speed that is optimal for the remaining portions of the pending jobs. More precisely, at each release time r_j, the algorithm computes the optimal schedule for the instances consisting of the unfinished portions of all jobs released up to time r_j, and uses the speed from this optimal schedule until the next job is released.

A comprehensive analysis of these algorithms was given by Bansal et al. [8]. They showed that for the power function $P(s) = s^\alpha$ the competitive ratio of both algorithms, AR and OA, is at least α^α. The example used in the lower bound proof consists of jobs $J_j = (r_j, d_j, p_j) = (j - 1, n, (1/(n - j))^{1/\alpha})$, for $j = 1, \ldots, n$. In the optimal schedule produced by Algorithm YDS, job J_j would be executed at speed

$(1/(n-j))^{1/\alpha}$, equal to its processing time, so that it will complete by time $r_j + 1 = j$. The total energy usage will be then $\sum_{j=0}^{n-1}(((1/(n-j))^{1/\alpha})^{\alpha} = H_n$, the nth harmonic number. On this instance both online algorithms, AR and OA, will produce the same schedules whose total energy, as shown in [8], is $\alpha^{\alpha}(H_n - \Theta(1))$. Thus the ratio indeed approaches α^{α} for $n \to \infty$.

In fact, Bansal et al. [8] also establish a matching upper bound for Algorithm OA, thus proving that its exact competitive ratio is α^{α}. As it turns out, the competitive ratio of Algorithm AR is worse, not better than $((2 - \delta_{\alpha})\alpha)^{\alpha}/2$, where $\delta_{\alpha} \to 0$ with $\alpha \to \infty$ [5]. (Note that this matches the upper bound for $\alpha \to \infty$.)

Both online algorithms from [54], AR and OA, are "conservative," in the following sense: if one job J_j is still pending and no new jobs arrive, job J_j will be processed with constant speed until its deadline. As illustrated in the example discussed earlier, it would make sense for an online algorithm to anticipate future job arrivals and process the currently pending jobs a little faster than necessary, to reduce the combined workload when other jobs are released in the future, if any. This idea leads to the algorithm called BKP proposed in [8]. At each time t this algorithm determines an estimate of the speed of Algorithm YDS and processes the pending jobs e times faster (where $e \approx 2.718$ is the Euler number).

Algorithm BKP: For $u < t \le v$, define $w(t, u, v)$ to be the sum of all processing times p_j over all jobs J_j with $u \le r_j \le t < d_j \le v$. In other words, this is the workload of the interval $[u, v]$ that is already known by time t. Let also

$$\kappa(t) = \max_{t' > t} \frac{w(t, et - (e-1)t', t')}{e(t' - t)}.$$

At time t, the algorithm uses speed $e\kappa(t)$, always executing the earliest deadline pending job.

Bansal et al. [8] prove that the competitive ratio of Algorithm BKP is at most $2(\alpha e/(e-1))^{\alpha}$, which is better than the competitive ratio of the other two algorithms for $\alpha \ge 4.21$.

A similar idea to speed up the computation in anticipation of future jobs was developed by Bansal et al. [7]. Their algorithm, called qOA, can be thought of as a fast-forward version of Algorithm OA: it processes jobs at the speed equal q times the minimum speed required to meet the deadlines, where $q \ge 1$ is a parameter that can be tuned to minimize the resulting competitive ratio.

Algorithm qOA: At any time t compute the optimal schedule for the remaining portions of the pending jobs. If this schedule uses speed s at time t, run the processor at speed qs, executing the earliest deadline pending job.

A thorough analysis of Algorithm qOA given in [7] shows that for $q = 2 - 1/\alpha$ its competitive ratio is at most $\frac{1}{2}4^{\alpha}/\sqrt{e\alpha}$. It is not known whether this choice of q is optimal. However, no other choice of q can significantly improve this bound, since, as shown in [7], for any q the competitive ratio of Algorithm qOA is no better than $\frac{1}{4\alpha}4^{\alpha}(1 - 2/\alpha)^{\alpha/2}$. Nevertheless, for α's between 1.7 and 10.6, for which the power function $P(s) = s^{\alpha}$ most accurately approximates the experimental dependence between s and $P(s)$, this algorithm's competitive ratios are best among the algorithms we discussed earlier. With a refined analysis, Bansal et al. [7] derive better estimates for the competitive ratios of Algorithm qOA for $\alpha = 2, 3$, respectively, 2.4 and 6.7.

It is not known what is the optimal competitive ratio for the power function $P(s) = s^{\alpha}$. As noted in [8], by parametrizing the two-job instance described earlier in this section and optimizing the parameters, a lower bound of $\frac{1}{2}(\frac{4}{3})^{\alpha}$ on the competitive ratio can be shown. Bansal et al. [7] show a better lower bound of $e^{\alpha-1}/\alpha$.

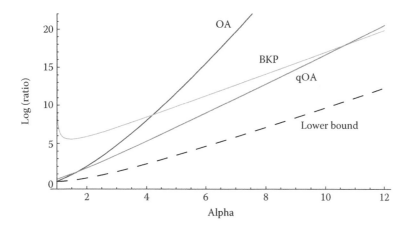

FIGURE 14.2 Comparisons of the competitive ratios of algorithms OA, BKP, and qOA (with $q = 2 - 1/\alpha$), and the lower bound of $e^{\alpha-1}/\alpha$.

The known bounds (as of January 2011) on the competitive ratios are summarized in Table 14.1 (courtesy [7]). The value of the general upper bound is equal to the minimum of the upper bounds for the listed algorithms, and it is achieved by different algorithms depending on the range of α. The table also shows the values of different ratios for $\alpha = 2$. Figure 14.2 shows the competitive ratios of algorithms OA, BKP, and qOA (with $q = 2 - 1/\alpha$), as functions of α. The lowest graph is the lower bound of $e^{\alpha-1}/\alpha$. The y-axis is the base-2 logarithm of the competitive ratios.

14.2.3 Other Scheduling Models with Speed Scaling

Our discussion of speed scaling has been limited to minimizing energy consumption for a collection of jobs with hard deadlines on a single processor. For multiple processors, the problem becomes $\mathbb{NP}$-hard, even in several special cases, with additional assumptions on processing times, release times or deadlines. For example, it is quite easy to show, by a reduction from the 3Partition problem, that the multiprocessor case is strongly $\mathbb{NP}$-hard for instances where all jobs are released at time 0 and have the same deadline d. Albers et al. [3] prove two additional hardness results: $\mathbb{NP}$-hardness for unit jobs on two processors, and strong $\mathbb{NP}$-hardness for unit jobs and arbitrary number of processors. In the same paper, some constant-ratio approximation algorithms were developed, offline and online, without any restrictions on release times, deadlines, or processing times.

Meeting hard deadlines is just one of many performance criteria studied in the job scheduling area. One can introduce the speed-scaling feature into other scheduling models, giving rise to bicriteria optimization problems that can be studied by either assuming a constraint on the energy budget or

TABLE 14.1 Known Bounds on the Competitive Ratios for Minimum-Energy Speed-Scaling Scheduling under Deadline Constraints

Algorithm	General α Upper	General α Lower	$\alpha = 2$ Upper	$\alpha = 2$ Lower
General		$e^{\alpha-1}/\alpha$ [7]		1.3
AR	$2^{\alpha-1}\alpha^{\alpha}$ [5,54]	$(2 - \delta_{\alpha})^{\alpha-1}\alpha^{\alpha}$ [5]	8	4
OA	α^{α} [8]	α^{α} [8]	4	4
BKP	$2(e\alpha/(\alpha - 1))^{\alpha}$ [8]		59.1	
qOA	$\frac{1}{2}4^{\alpha}/\sqrt{e\alpha}$ [7]	$\frac{1}{4\alpha}4^{\alpha}(1 - 2/\alpha)^{\alpha/2}$ [7]	2.4	

by minimizing objective functions that are linear combinations of total energy consumption and other performance measures. Objective functions that have been studied in the existing literature include flow time [2,9,51], makespan [19], throughput [6] (for bounded-speed processors), and various extensions to the multiprocessor case [32,46,50]. Among those different models, the $E + F$ measure, the sum of energy and flow time, has been studied most extensively, and is discussed in more detail in a companion article by Wierman et al. [52] in this volume.

14.3 Power-Down Scheduling

The essence of the *power-down method* is to simply suspend the system during idle times, or gaps. To achieve any energy savings, the idle time must be long enough to compensate for the energy overhead L required to transition the system back into the active state. Thus, normalizing energy values so that the system uses one unit of energy per unit of time, the system should be powered down if and only if the gap is longer than L. This L is sometimes called the *break-even* time. Note that, we have made two assumptions here: one, that the system does not use any energy when idle, and two, that the time required for reactivation is negligible.

The break-even principle mentioned earlier can be applied directly to systems where the schedule of energy-consuming tasks is fixed and known in advance. Often in practice, however, task execution times are not fixed and energy savings can be accomplished by grouping tasks into a small number of contiguous blocks, thus creating long gaps where the system can be powered down. Efficiently computing a schedule with minimum energy consumption is not trivial, even if the tasks are fully specified off line, because of the constraints imposed on the feasible schedules. The online case introduces additional challenges, due to incomplete information. For example, when no tasks are pending, how long should the system wait before transitioning to the power-down state? The trade-offs that need to be considered become increasingly subtle in more involved scenarios: with multiple power states, multiple components, or additional constraints.

14.3.1 Offline Power-Down Scheduling Algorithms

We now consider the offline case of power-down scheduling. Formally, we are given a set of jobs $J_1, J_2, \ldots, J_n$, with each job J_j specified by its release time r_j, deadline d_j, and processing time p_j. These jobs need to be executed, with preemptions, on a single processor. When the processor is on, it uses one unit of energy per unit of time. (The time is assumed to be discrete.) The energy required to wake up the processor is denoted by L. The objective is to compute a schedule that minimizes the overall energy consumption, or to report that no feasible schedule exists. Feasibility can be verified efficiently by scheduling all jobs preemptively according to the earliest deadline principle; thus we can as well assume that the input instance is feasible.

For feasible instances, the energy spent on task processing is equal to the total processing time of all jobs. This quantity is independent of the schedule and can be omitted in the objective function. For a gap of length g, the energy usage associated with this gap is $\min(g, L)$. Therefore, we can define our objective function E as the sum of all terms $\min(g, L)$, ranging over all gaps in the schedule. Note that for $L = 1$ our objective function is equal to the number of gaps.

Figure 14.3 shows an example of an instance with seven jobs and its schedule. Specified in the format $J_j = (r_j, d_j, p_j)$, these jobs are $J_1 = (0, 3, 2)$, $J_2 = (4, 6, 1)$, $J_3 = (5, 9, 2)$, $J_4 = (12, 14, 1)$, $J_5 = (15, 17, 1)$, $J_6 = (2, 18, 4)$, and $J_7 = (18, 20, 1)$. In the figure, each job J_j is represented by its index j. For $L = 2$, the shown schedule is optimal and has energy cost $E = 3$: the cost of the gap $[8, 13]$ is $L = 2$ and the gap $[17, 18]$ costs 1. (The infinite leading and trailing idle intervals do not count toward the energy value.)

To streamline the discussion, in the following, we will adopt the standard notation for scheduling problems, with E standing for the objective function. In this notation, our power-down scheduling

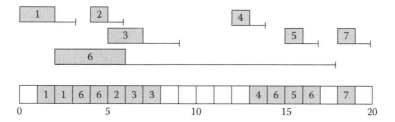

FIGURE 14.3 An example of a power-down schedule.

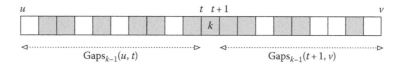

FIGURE 14.4 An illustration of the partitioning formula (14.1).

problem will be denoted $1|r_j; p_j; \text{pmtn}|E$, where we will additionally specify that $L = 1$ to indicate that we minimize the number of gaps or $p_j = 1$ when only unit jobs are considered.

The question whether the aforementioned power-down scheduling problem, $1|r_j; p_j; \text{pmtn}|E$, can be solved in polynomial time was posed by Irani and Pruhs [33]. Baptiste [10] reported the first breakthrough by providing a polynomial-time algorithm for the case of unit jobs and gap minimization, that is $1|r_j; p_j = 1; L = 1|E$. Based on his work, Chrobak et al. [11] developed a polynomial–time algorithm for the general case and a faster algorithm for unit jobs. The running times of the algorithms in [11] are $O(n^5)$ for $1|r_j; p_j; \text{pmtn}|E$ and $O(n^4)$ for $1|r_j; p_j = 1|E$.

All these algorithms use a dynamic programming approach whose main principle was developed by Baptiste [10]. We will now sketch the idea behind these algorithms, focusing on the case of unit jobs and gap minimization, that is $1|r_j; p_j = 1; L = 1|E$. We order the jobs $J_1, J_2, \ldots, J_n$, according to their deadlines. Denote by $\mathsf{Gaps}_k(u, v)$ the minimum number of gaps in the interval $[u, v]$ for the subinstance consisting of all jobs among $J_1, J_2, \ldots, J_k$ that are released between u and v (we also count the gap between u and the first job and the gap between the last job and v, if any.) Baptiste [10] showed that

$$\mathsf{Gaps}_k(u, v) = \mathsf{Gaps}_{k-1}(u, t) + \mathsf{Gaps}_{k-1}(t + 1, v), \tag{14.1}$$

for some time t. The minimum number of gaps is obtained from (14.1) by choosing $k = n$, u equal to the earliest release time and v to the latest deadline. We refer to the recurrence (14.1) as the *partitioning method*. The proof of (14.1) is actually quite simple (see Figure 14.4). Roughly: for any optimal schedule S of the subinstance considered in (14.1), we can reschedule all jobs in the earliest deadline order using only the busy time slots of S. This does not change the cost, so the new schedule is optimal as well. Let t be the time when job J_k is scheduled in S. At time t, J_k is the earliest deadline pending job, so the jobs released before time t are all completed before time t. This implies that the sets of jobs among $J_1, J_2, \ldots, J_{k-1}$ released before and after time t form disjoint subinstances, and (14.1) follows.

The recurrence (14.1) leads naturally to a dynamic programming algorithm, but its running time is not polynomial because it iterates over all candidate values of u, v, and t. To fix this, we can do a *domain reduction*, namely, to show that only polynomially many choices of u, v, t are needed. This is quite simple: by shifting busy blocks in an optimal schedule left or right, one can show that only numbers of the form $r_j + a$, $d_j + a$ need to be considered, where $-n \le a \le n$. This restricts the domain size of all three parameters u, v, t to $O(n^2)$. Since the range of k is $O(n)$, the resulting running time is $O(n^7)$, as in [10]. The domains of u and t were reduced to $O(n)$ in [11], which immediately improves the running time to $O(n^5)$.

The domain of v still has size $O(n^2)$. But $g = \mathsf{Gaps}_k(u, v)$ takes at most n values, so if we could somehow reverse the roles of v and g, by using g as an argument and v as the value of the dynamic programming function, we might obtain a faster algorithm. Indeed, this is the essence of the "inversion method" that leads to the $O(n^4)$-time algorithm in [11]. Specifically, let $V_k(u, g)$ be the maximum v for which there is a schedule that has at most g gaps and schedules the jobs among $J_1, J_2, \ldots, J_k$ that are released in $[u, v]$. The recurrence for $V_k(u, g)$ in [11] is rather involved, but it can be evaluated in time $O(n^4)$ due to the fact that each parameter k, u, and g has only $O(n)$ values. The desired values of $\mathsf{Gaps}_k(u, v)$ can be computed from the values $V_k(u, g)$ using binary search.*

The $O(n^5)$-time algorithm for arbitrary processing times in [11] uses similar ideas but is substantially more involved. The extension to minimizing energy (that is, to arbitrary L) is, however, quite simple; the basic idea is that the case for arbitrary L can be reduced in time $O(n^2 \log n)$ to the case $L = 1$.

14.3.1.1 Multiprocessor Scheduling

Demaine et al. [27] extended the partitioning method (14.1) to the m-processor case, obtaining a polynomial-time algorithm for unit jobs and $L = 1$. The difficulty here is that it is no longer possible to divide a schedule neatly using some cutoff time t (the time when job k is scheduled): among the jobs executed at time t it may be necessary to include some in the "left" instance and other in the "right" instance. But these jobs can be reordered so that the left-instance jobs occupy processors $1, 2, \ldots, \ell - 1$ while the right-instance jobs occupy processors $\ell + 1, \ldots, m$, with job k assigned to processor ℓ. Combined with the domain-reduction from (14.1), this yields $O(m^4 n^5)$ subinstances and running time $O(m^5 n^7)$. (This can be improved to $O(m^5 n^5)$ by applying the domain reduction from [11].)

14.3.1.2 Open Problems

The algorithms discussed earlier are not fully satisfactory: they are complicated, difficult to implement, and slow. Faster and simpler algorithms would be of considerable interest. Substantial improvements of the running times, to $O(n^3)$ or less, seem very unlikely. Therefore, it would be natural to ask whether there are fast, say $O(n\mathrm{polylog}(n))$-time algorithms that compute approximate solutions. The only result in this direction up to date is the three-approximation algorithm for unit jobs and $L = 1$ in [28], although this algorithm is also quite slow.

Another direction would be to study extensions of the model, for example, by allowing multiple power levels. In the multiprocessor case, the complexity of the problem for arbitrary L or arbitrary processing times remains open. It would also be quite natural to charge some cost for job migrations, or to incorporate into the model the feature where an idle interval common to all processors produces energy savings larger than the sum of savings for all processors.

The power-down scheduling model mentioned earlier has one undesirable feature: it is not well defined when the instance is not feasible. Thus it would be quite interesting to investigate its budgeted version, where the objective is to maximize throughput without exceeding a given energy budget B. A related approach was considered in [27] for the much more general multiple-interval case, where an $O(\sqrt{n})$-approximation algorithm was given for scheduling with a given number of gaps.

Another option is to consider the model combining power-down with speed scaling [33], whose offline complexity remains an open problem. A 2-competitive online algorithm for this problem (with two power states) appeared in [35].

14.3.2 Online Algorithms for Systems with Multiple Power States

The online variant of the power-down scheduling problem described earlier is poorly understood and, to the author's knowledge, its general version has not been yet addressed in the literature. One

---

* An implementation of this algorithm is available at http://www.lix.polytechnique.fr/~durr/MinBlocks/.

issue is that in the model of [10,11,33] the algorithm is required to schedule all jobs, which essentially forces the algorithm to follow the earliest deadline strategy; otherwise an adaptive adversary can force some job to miss its deadline. Thus to make the problem meaningful, some alternative formulation should be considered, for example, the budgeted version proposed in the previous section.

The existing work on this subject, pioneered by Irani et al. [34,36] (see also [4]) focuses on systems with multiple power levels, under the assumption that the schedule of all jobs is already given. The objective is to determine appropriate power states at all times.

Specifically, we consider a system with $m + 1$ power states $s_0, s_1, \ldots, s_m$ with decreasing power consumption rates $P_0 \geq P_1 \geq \cdots \geq P_m$. State s_0 represents the active state, namely, the only state where the processor can execute a job, while s_m is the power-off state. Assume, for now, that transitions from higher-to lower-power states do not require any additional energy and that a transition from a state s_i to the active state s_0 consumes energy L_i. The input specifies a set of intervals where the processor is idle. At the beginning of an idle interval the system is in state s_0. During the idle interval, as the time progresses, the algorithm is allowed to transition from its current state to a lower-power state, possibly several times. At the end of the interval, we must transition back to the active state s_0. The objective is to determine the sequence of states and the transition times so that the overall energy consumption is minimized. (We now assume that the time is continuous.) Clearly, each idle interval can be considered separately, so we can as well assume that we have just one interval, say $[0, z]$.

Let us analyze first the offline case, when the value of z is known in advance. It is easy to see that the optimal solution will change the state to some s_i at time 0 and stay in state s_i until time z. The cost of this strategy is $L_i + zP_i$. So we need to choose the state s_i that minimizes $L_i + zP_i$. This algorithm has a simple geometric interpretation. Define the *energy trajectory* of state s_i to be the linear function $L_i + tP_i$, where t stands for time. The lower envelope of the energy trajectories is $\min_i(L_i + tP_i)$. For a given z, the optimal solution is to choose the state s_i whose trajectory is in the lower envelope at $t = z$.

Consider, for example, the system illustrated in Figure 14.5, with four states: active state s_0, standby state s_1, sleep state s_2, and power-off state s_3, where $L_1 = 6, L_2 = 12, L_3 = 17, P_0 = 3, P_1 = 1, P_2 = 0.25$, and $P_3 = 0$. Suppose that $z = 10$. The energy trajectory with minimum value at z is $L_2 + tP_2$, and its value at $t = 10$ is 14.5. So the optimal state is state s_2.

We now turn to online algorithms. Here, "online" means that the value of z is not known in advance. An online algorithm starts in state s_0 and, as time passes, it can transition to lower-power states. Thus an online algorithm can be uniquely defined by its transition times t_i and the state $s(t_i)$ entered at each time t_i. This process is followed until the end z of the idle interval is reached. Recall that an online algorithm is said to be c-competitive if its cost, for any z, is at most c times the optimal cost for z.

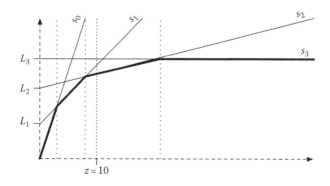

FIGURE 14.5 An example of power-down optimization with four power levels.

For two states and $P_1 = 0$, the problem is equivalent to the well-known ski rental problem and it is not hard to show that its optimal competitive ratio is 2. Without loss of generality, we can assume that $P_0 = 1$. (Otherwise, we can divide all parameters by P_0.) An online algorithm can transition from s_0 to s_1 at time $t_1 = L_1$, that is when its energy cost $P_0 t_1 = t_1$ is equal to the transition cost L_1. The cost of this algorithm is z for $z < L_1$ and $2L_1$ for $z \geq L_1$, while the optimum cost is $\min(z, L_1)$. So the competitive ratio of this algorithm is at most 2. On the other hand, no better ratio is possible. Indeed, if an online algorithm never transitions to state s_1, its competitive ratio is unbounded. On the other hand, if it changes its state to s_1 at time t' then consider $z = t' + \epsilon$, for some $\epsilon > 0$. For $\epsilon \to 0$, the algorithm's cost is $t' + L_1$, which is at least twice as large as the optimum cost of $\min(t', L_1)$.

The first contribution in [34] is an online algorithm that achieves the competitive ratio of 2 for any number of power states. Their Algorithm LEA is quite intuitive: it follows the lower envelope of the power trajectories, that is it uses state s_i at time t if and only if the energy trajectory for s_i is optimal for $z = t$. In the example from Figure 14.5, the lower envelope is indicated with thick line segments. The algorithm uses state s_0 until time 3, at which time it switches to state s_1, at time 8 it switches to s_2, and at time 20 it switches to s_3.

To analyze this algorithm, divide its cost into two parts: the energy *dissipated* during the idle interval and the *state-transition energy*, that is the energy required to transition to state s_0 at time z. We claim first that at each time t the dissipated energy is equal to the (total) optimal energy for the interval of length t. Indeed, at the beginning, for $t = 0$, both quantities are 0, and, by the definition of the algorithm, as the current time t increases, the dissipated energy grows at exactly the same rate as the optimum value for t. At time z the algorithm will be in the same state s_i as the optimum solution, so its transition cost L_i is of course at most the optimum value $L_i + zP_i$. Overall, we obtain that the algorithm's cost for z is at most twice the optimum, and we can conclude that this algorithm is 2-competitive.

In practice, the input data is not generated by a prescient adversary, and it is often the case that online algorithms used in practice compute solutions significantly below performance bounds derived with the competitive analysis. In order to establish bounds that more accurately reflect real-life scenarios, Irani et al. [34] refine their model by assuming that the z values are generated according to some unknown probability distribution. Extending the earlier work for two power states in [41] (which used a different, but equivalent setting), Irani et al. [34] show that in this model it is possible to achieve the competitive ratio of $e/(e - 1) \approx 1.58$. In fact, their algorithm outperforms some previously studied algorithms for this problem.

In the discussion so far, we have assumed that all power-down transitions incur no cost. One can consider the general model where any transition from state s_i to state s_j has some cost L_{ij} associated with it. If both the power-down transitions and power-up transitions are additive, that is $L_{ij} + L_{jk} = L_{ik}$ and $L_{kj} + L_{ji} = L_{ki}$ for $i < j < k$, then we can "charge" the power-down transition costs to the corresponding power-up transitions, and the analysis mentioned earlier still applies.

However, in realistic systems the transition costs are not additive. One assumption that we can make, of course, is that the costs satisfy the triangle inequality, in the sense that $L_{ij} + L_{jk} \geq L_{ik}$, for otherwise we could replace the transition $s_i \to s_k$ by two immediate transitions $s_i \to s_j \to s_k$. For this general case, Irani et al. [34] prove that the competitive ratio of $3 + 2\sqrt{2}$ can be achieved. It is not known what is the best competitive ratio for this problem; in [34] only a lower bound of 2.45 is given, with a better lower bound of 3.618 obtained by Damaschke [26].

The upper bounds mentioned earlier apply to arbitrary energy dissipation rates and state-transition energies, but the optimal competitive ratio for a given system depends on the specific choice of these parameters, and it could be much smaller than $3 + 2\sqrt{2}$. (In fact, even for two states the ratio is less than 2 if the power consumption rate in the idle state is not 0.) To address this issue, Irani et al. [34] also provide an efficient algorithm that, from the description of the system, computes an online strategy with a nearly optimal competitive ratio. More specifically, for any $\epsilon > 0$, if the system has an R-competitive online algorithm then their algorithm will compute a strategy with competitive ratio at most $R + \epsilon$.

14.4 Thermal Management

The term "thermal management" refers to methods for controlling the temperature of a processor. Detrimental effects of high temperatures are plenty, including reduced reliability and lifespan of a CPU's components. The issues of thermal and power management are, naturally, closely related. High temperatures are associated with power leakage, while, on the other hand, cooling systems required for temperature control contribute to increased energy consumption.

In addition to increasingly sophisticated hardware solutions, recent years have seen the emergence of "soft" thermal management tools that employ various frequency scaling, scheduling, or load balancing methods. A development of formal models and sophisticated algorithms can improve our fundamental understanding of these approaches, its potential and limitations, and lead to improved performance.

Modern CPUs are equipped with a hardware *dynamic thermal management* system (DTM) that controls the temperature by scaling down the frequency or idling the CPU when the temperature approaches the thermal threshold. Algorithmic approaches to temperature control fall roughly into two categories: architecture-level solutions and OS-level solutions. The first category includes DTM strategies—in essence, when and how to reduce the CPU frequency, which is fundamentally equivalent to speed scaling. We discuss the existing work on speed-scaling algorithms for temperature control in the following section. The other category involves approaches based on job scheduling or load balancing. The theoretical work in this category is discussed in Section 14.4.2.

14.4.1 Speed Scaling for Temperature Control

Bansal et al. [8] initiated the theoretical study of speed scaling as a tool for CPUs temperature control. The CPU's temperature $T(t)$ fluctuates over time. These variations are caused by two opposite factors: the heat generated by the electric current that increases the temperature and the process of heat dissipation into the environment (accelerated by the cooling system) that decreases it. The temperature function can be described by the differential equation

$$T'(t) = A \cdot PS(t) - B \cdot T(t), \tag{14.2}$$

where $PS(t)$ is the power supplied at time t and A, B are positive constants. The constant B represents the cooling rate of the system.

The model in [8] is an adaptation of their model (introduced in [54]) for energy minimization, and it was described in detail in Section 14.2. Recall that in that model, we had a collection of jobs J_j, each with a release time r_j, deadline d_j, and processing time p_j. It is assumed that the power supplied at time t is given as a function $P(s)$ of the processor's speed, that is $PS(t) = P(s(t))$. This time, instead of minimizing energy consumption, the objective is to compute a feasible schedule that minimizes the maximum temperature.

The offline case of temperature minimization seems difficult, yet Bansal et al. [8] show that the solution can be characterized by a convex program, and that, in spite of nonlinearity, it can be solved in polynomial time using the ellipsoid method.

In the online case, as it turns out, none of the algorithms AR nor OA is constant competitive with respect to the maximum temperature. However, Bansal et al. [8] prove that their new algorithm, Algorithm BKP, discussed earlier in Section 14.2, achieves a competitive ratio of at most $2^{\alpha+1}e^{\alpha}(6(\alpha/(\alpha-1))^{\alpha}+1)$, independently of the cooling factor B.

14.4.2 Thermal-Aware Scheduling

We focus now on solutions that do not require access to DTM. Instead, these methods use job scheduling or load balancing to spread heat contributions of tasks over time or space (machines). Most of the previous

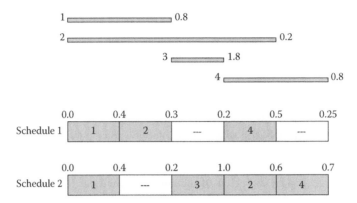

FIGURE 14.6 An example of temperature-aware scheduling, with $T^{max} = 1$ and $C = 2$.

work in this direction was concerned with multi-core systems, where one can move tasks between the processors to minimize the maximum temperature. However, even in single-core systems one can exploit variations in heat contributions among different tasks to reduce the processor's temperature through appropriate task scheduling [13,30,37,44,53,56]. One example is the Linux scheduler developed in [53,56]. That scheduler exploits variations between thermal characteristics of different tasks. Typically, running CPU-intensive tasks, like numerical computations, compression/decompression, etc., increases CPU's temperature faster than tasks requiring mainly data manipulation in memory. Heat contributions of different tasks can be predicted with high accuracy [53]. By reordering the tasks to spread these heat contributions evenly over time, it is possible to reduce the number of DTM interventions, which in turn results in improved performance. Although not fully formalized in [53], the setting used in that paper can be thought of as a budgeted optimization problem, where the objective is to maximize performance with a given maximum temperature budget.

A similar temperature-aware scheduling problem was studied by Chrobak et al. [23], and is defined as follows: We have a collection of unit-length jobs where job J_j has release time r_j, deadline d_j, and heat contribution h_j. If the processor executes job J_j at some time step, then its temperature changes from t to $(t + h_j)/C$, where $C > 1$ is the cooling factor. (This formula is a discrete approximation to the differential equation (14.2), with $C \approx 1/B$.) The objective is to maximize throughput without exceeding the given temperature threshold T^{max}. Note that the objective function is different from the one in [53], where all tasks needed to be completed and the objective was to minimize the makespan.

Figure 14.6 shows an example of an instance with $T^{max} = 1$, $C = 2$ and four jobs specified in the format $J_j = (r_j, d_j, h_j)$: $J_1 = (0, 2, 0.8)$, $J_2 = (0, 4, 0.2)$, $J_3 = (2, 3, 1.8)$, and $J_4 = (3, 5, 0.8)$. Each job J_j is identified by its index j. Two schedules are shown, with the current temperature at each step shown right above the schedule. In the first schedule, at time 2 the temperature is 0.3, so job J_3 cannot be executed, because otherwise the temperature after its execution would be $(0.3 + 1.8)/2 = 1.05$, exceeding the thermal threshold $T^{max} = 1$. Thus this schedule will complete only three jobs. In the second schedule, job J_2 is postponed, decreasing the temperature in step 3 to 0.2, so that job J_3 can be executed.

In [23] the authors focus on the case $C = 2$ and show that the offline version of the problem is $\mathbb{NP}$-hard, even if all jobs are released at time 0. They also give an online 2-competitive algorithm and prove a matching lower bound.

In subsequent work, Fellows et al. [29] proved that the problem is fixed-parameter tractable. The online case was studied by Birks and Fung [17] who considered arbitrary cooling factors and multiple processors.

Many open questions remain. Can the competitive ratio of 2 in [23] be improved with randomization? Instead of simply forbidding thermal threshold violations, as in [23], or idling, a more realistic approach would be to use the frequency-scaling model common in today's CPUs, as discussed earlier in this survey. The next step would be to extend this work to the multiprocessor case. Another variant would be to consider CPU components with different thermal thresholds, to model modern chips equipped with multiple temperature sensors, allowing for more fine-grained temperature control. Finally, one can, of course, consider arbitrary-length jobs, possibly with weights, other objective functions, etc., giving rise to a plethora of intriguing and never yet studied scheduling problems.

14.5 Algorithms for Flash Memories

Energy consumption of flash memories is significantly lower than that of RAM or plated disks, thus they are now commonplace in a variety of portable devices. The drawbacks of flash memory include high cost and limited "write-endurance": a sector of a flash drive can be written to only a limited number of times. As the longevity of a flash drive is determined by the *maximum* wear of any sector, various *wear-leveling* techniques are applied in practice to extend the lifetime of flash drives.

Ben-Aroya and Toledo [14,15] pioneered the study of competitive online algorithms for this problem. In their model, a flash drive consists of n disjoint *erasure units*, each unit having k sectors, where each sector can store one block of data. Typically, a flash drive stores some fixed number $m < kn$ of memory blocks. This internal organization of data is hidden behind the drive interface that maintains a mapping from the m data blocks to sectors. The m sectors that store the current data blocks are called *occupied*. The remaining sectors are either *clean* or *dirty*. To change the status of a sector from dirty to clean, the whole unit containing this sector needs to be erased. This erasure changes the status of all sectors in this block, and all data in this unit is lost. Thus the occupied sectors in this unit, if any, need to be moved to clean sectors outside this unit before the erasure. Dirty sectors are unoccupied sectors that have not been cleaned since the last write operation to them. Summarizing, the state transitions are

$$\text{clean} \xrightarrow{\text{write}} \text{occupied} \xrightarrow{\text{un-map}} \text{dirty} \xrightarrow{\text{unit erase}} \text{clean}$$

The essence of wear-leveling is to maintain the block-to-sector mapping in such a way that the maximum number of unit erasures is minimized. In addition to allocating sectors to the accessed blocks, a wear-leveling algorithm may perform some remapping that moves current data blocks from one sector to another. This remapping also contributes to increased wear.

In the formal setting in [14,15], we assume that each unit can be erased at most H times. For any sequence of write requests, the objective is to maximize the number of writes before this limit on the number erasures is reached.

To illustrate where the challenge is, let us focus on the case $k = 1$, when sectors are identical to erasure units. Suppose first that $m = n$, which means that the drive is always full. For any deterministic online algorithm, the adversary can create a request sequence which at each step will write to the block stored in sector 1. Since we do not have any "spare" sectors, no matter how we rearrange the mapping, Sector 1 will need to be erased before each write. So we can only guarantee a total of H writes. For arbitrary $m \le n$, an analogous argument shows that at most $(n - m + 1)H$ writes can be guaranteed with a deterministic algorithm. This bound can be achieved with a simple greedy algorithm that always writes to the least used empty sector. As shown in [14], the optimal competitive ratio (for $k = 1$) is between $\frac{n-m+1}{n+(n-m)/H}$ and $\frac{n-m+1}{n-n/H}$. The difference between the two bounds is the result of using different lower and upper estimates on the optimal solution. Note that when the drive is nearly full, that is when $n - m$ is small, the competitive ratio is close to zero; in other words, most of the drive's potential "write capacity" is wasted.

In order to achieve better performance, we need to hide from the adversary the block-to-sector mapping, which, naturally, begs for randomization. Ben-Aroya and Toledo [14,15] analyze randomized algorithms and show that, with some reasonable assumptions, competitive ratios close to 1 can be achieved using randomization.

Somewhat surprisingly, the case when k is large is qualitatively different from the case $k = 1$. To understand the difference, consider requests sequences where at each step we write to a block chosen uniformly at random. Such sequences are easy to serve in the $k = 1$ case, as they uniformly spread the wear of different sectors. For large k, random sequences spread wear evenly among different units. As a result, when the system needs to erase a unit, with high probability each unit will contain a significant number of sectors that are either clean or occupied, forcing the algorithm to perform redundant erasures or remapping. Ben-Aroya and Toledo [14] give some performance bounds for arbitrary k, although they are not as crisp as those for $k = 1$.

Rapidly growing importance of the flash-memory technology justifies further study of wear-leveling algorithms. For example, the offline complexity of minimizing wear-leveling is not fully resolved in [14,15], and the bounds on the competitive ratios, although close, are not yet tight. More importantly though, some flash-memory technologies allow the system to function even after some segments have been corrupted, by employing appropriate *bad block management* policies. For such systems, minimizing the maximum wear is not necessarily the best strategy, especially that it often requires additional erasures.

14.6 Final Comments

In this survey, we highlighted some research directions on algorithmic aspects of energy minimization, focusing on issues related to microprocessor power management: speed scaling, power-down strategies, thermal management, and flash-memory leveling.

Among the topics we did not address, it is worth to mention research on minimizing energy consumption in wireless networks. Today's radio devices, cell phones, laptops, sensors, etc., are often battery operated and minimizing energy consumed by radio transmissions increases the battery performance. The general goals in this area are to design low-energy strategies for network primitives, including broadcasting, data aggregation, all-to-all communication, leader election, and others; see, for example, [12,20,22,25]. This is a dynamic and fast growing area of research and it would take a separate survey to describe the models and algorithms from the existing literature.

Finally, a number of authors attempted to formalize the notion of "energy complexity of algorithms," in order to improve our understanding of the relation between the structure of an algorithm and the energy required for its execution. The need for progress in this direction has been expressed by many researchers, for example, in [38–40,49,55], and emphasized in the report from the NSF Workshop on the Science of Power Management [21]. These efforts, so far, have been generally inconclusive.* For example, some researchers explored circuit-based approaches, including the ET^{α}-measure [16], to study the trade-off between energy and time (see also [31,49]). Others focus on parallel algorithms and study the relation between the energy consumption and the number of cores [42,43]. Approaches for measuring energy complexity of algorithms have been studied in [24,57].

Acknowledgments

This work was partially supported by research grant CCF-0729071 from the National Science Foundation.

References

1. S. Albers. Energy-efficient algorithms. *Communications of the ACM*, 53:86–96, 2010.
2. S. Albers and H. Fujiwara. Energy-efficient algorithms for flow time minimization. *ACM Transactions on Algorithms*, 3:49, 2007.

---

* Indeed, one indication of the state of the art of this area is the striking abundance of papers with the word "toward" in the title.

3. S. Albers, F. Müller, and S. Schmelzer. Speed scaling on parallel processors. In *Proceedings of the 19th Annual ACM Symposium on Parallel Algorithms and Architectures (SPAA'07)*, San Diego, CA, pp. 289–298, 2007.

4. J. Augustine, S. Irani, and C. Swamy. Optimal power-down strategies. In *Proceedings of the 45th Symposium Foundations of Computer Science (FOCS'04)*, Rome, Italy, pp. 530–539, 2004.

5. N. Bansal, D. P. Bunde, H.-L. Chan, and K. Pruhs. Average rate speed scaling. In *Proceedings of the 8th Latin American Conference on Theoretical Informatics (LATIN'08)*, Búzios, Brazil, pp. 240–251, 2008.

6. N. Bansal, H.-L. Chan, T.-W. Lam, and L.-K. Lee. Scheduling for speed bounded processors. In *Proceedings of the 35th International Colloquium on Automata, Languages and Programming (ICALP'08)*, Reykjavik, Iceland, pp. 409–420, 2008.

7. N. Bansal, H.-L. Chan, K. Pruhs, and D. Katz. Improved bounds for speed scaling in devices obeying the cube-root rule. In *Proceedings of the 36th International Colloquium on Automata, Languages and Programming (ICALP'09)*, Rhodes, Greece, pp. 144–155, 2009.

8. N. Bansal, T. Kimbrel, and K. Pruhs. Speed scaling to manage energy and temperature. *Journal of the ACM*, 54(1):1–39, 2007.

9. N. Bansal, K. Pruhs, and C. Stein. Speed scaling for weighted flow time. In *Proceedings of the 18th Annual ACM-SIAM Symposium on Discrete Algorithms (SODA'07)*, New Orleans, LA, pp. 805–813, 2007.

10. P. Baptiste. Scheduling unit tasks to minimize the number of idle periods: A polynomial time algorithm for offline dynamic power management. In *Proceedings of the 17th Annual ACM-SIAM Symposium on Discrete Algorithms (SODA'06)*, Miami, FL, pp. 364–367, 2006.

11. P. Baptiste, M. Chrobak, and C. Dürr. Polynomial time algorithms for minimum energy scheduling. In *Proceedings of the 15th European Symposium on Algorithms (ESA'07)*, Eilat, Israel, pp. 136–150, 2007.

12. L. Becchetti, A. Marchetti-Spaccamela, A. Vitaletti, P. Korteweg, M. Skutella, and L. Stougie. Latency-constrained aggregation in sensor networks. *ACM Transactions on Algorithms*, 6:13:1–13:20, 2009.

13. F. Bellosa, A. Weissel, M. Waitz, and S. Kellner. Event-driven energy accounting for dynamic thermal management. In *Proceedings of the Workshop on Compilers and Operating Systems for Low Power (COLP'03)*, New Orleans, LA, pp. 1–10, 2003.

14. A. Ben-Aroya and S. Toledo. Competitive analysis of flash-memory algorithms. *ACM Transactions on Algorithms*, 7(2):23, 2011.

15. A. Ben-Aroya and S. Toledo. Competitive analysis of flash-memory algorithms. In *Proceedings of the 14th European Symposium on Algorithms (ESA'06)*, Jurich, Switzerland, pp. 100–111, 2006.

16. B. D. Bingham and M. R. Greenstreet. Computation with energy-time trade-offs: Models, algorithms and lower-bounds. In *IEEE International Symposium on Parallel and Distributed Processing with Applications*, Sydney, New South Wales, Australia, pp. 143–152, 2008.

17. M. Birks and S. Fung. Temperature aware online scheduling with a low cooling factor. In *Proceedings of the 7th Annual Conference on Theory and Applications of Models of Computation (TAMC'10)*, vol. 6108, Prague, Czech Republic, pp. 105–116, 2010.

18. S. Boyd and L. Vandenberghe. *Convex Optimization*. Cambridge University Press, Cambridge, U.K., 2004.

19. D. P. Bunde. Power-aware scheduling for makespan and flow. In *Proceedings of the 18th Annual ACM Symposium on Parallelism in Algorithms and Architectures (SPAA'06)*, Cambridge, MA, pp. 190–196, 2006.

20. T. Calamoneri, A. E. F. Clementi, M. D. Ianni, M. Lauria, A. Monti, and R. Silvestri. Minimum-energy broadcast and disk cover in grid wireless networks. *Theoretical Computer Science*, 399:38–53, 2008.

21. K. W. Cameron, K. Pruhs, S. Irani, P. Ranganathan, and D. Brooks. Report of the science of power management workshop. http://scipm.cs.vt.edu/SciPM-ReportToNSF-Web.pdf, 2009.

22. I. Caragiannis, M. Flammini, and L. Moscardelli. An exponential improvement on the MST heuristic for minimum energy broadcasting in ad hoc wireless networks. In *Proceedings of the 34th International Colloquium on Automata, Languages and Programming (ICALP'07)*, Wroclaw, Poland, pp. 447–458, 2007.

23. M. Chrobak, C. Dürr, M. Hurand, and J. Robert. Algorithms for temperature-aware task scheduling in microprocessor systems. In *Proceedings of the 4th International Conference on Algorithmic Aspects in Information and Management (AAIM'08)*, Shanghai, China, pp. 120–130, 2008.

24. A. Cisternino, P. Ferragina, and D. Morelli. Information processing at work: On energy-aware algorithm design. In *Proceedings of the International Conference on Green Computing*, San Diego, CA, pp. 407–415, 2010.

25. A. E. F. Clementi, P. Crescenzi, P. Penna, G. Rossi, and P. Vocca. On the complexity of computing minimum energy consumption broadcast subgraphs. In *Proceedings of the 18th Annual Symposium on Theoretical Aspects of Computer Science (STACS'01)*, Dresden, Germany, pp. 121–131, 2001.

26. P. Damaschke. Nearly optimal strategies for special cases of on-line capital investment. *Theoretical Computer Science*, 302(1–3):35–44, 2003.

27. E. D. Demaine, M. Ghodsi, M. T. Hajiaghayi, A. S. Sayedi-Roshkhar, and M. Zadimoghaddam. Scheduling to minimize gaps and power consumption. In *Proceedings of the 19th ACM Symposium on Parallelism in Algorithms and Architectures (SPAA'07)*, San Diego, CA, pp. 46–54, 2007.

28. U. Feige, S. Khanna, S. Naor, and M. T. Hajiaghayi. Gap minimization. Unpublished manuscript, 2008.

29. M. R. Fellows, S. Gaspers, and F. A. Rosamond. Parameterizing by the number of numbers. arXiv:1007.2021v2 [cs.DS] October 31, 2010.

30. M. Gomaa, M. D. Powell, and T. N. Vijaykumar. Heat-and-run: Leveraging SMT and CMP to manage power density through the operating system. *SIGPLAN Notices*, 39(11):260–270, 2004.

31. R. Gonzalez and M. Horowitz. Energy dissipation in general purpose microprocessors. *IEEE Journal on Solid-State Circuits*, 31:1277–1284, 1996.

32. G. Greiner, T. Nonner, and A. Souza. The bell is ringing in speed-scaled multiprocessor scheduling. In *Proceedings of the 21st Annual Symposium on Parallelism in Algorithms and Architectures (SPAA'09)*, Calgary, Alberta, Canada, pp. 11–18, 2009.

33. S. Irani and K. R. Pruhs. Algorithmic problems in power management. *SIGACT News*, 36(2):63–76, 2005.

34. S. Irani, S. Shukla, and R. Gupta. Competitive analysis of dynamic power management strategies for systems with multiple power savings states. In *Proceedings of the Conference on Design, Automation and Test in Europe (DATE)*, Paris, France, pp. 117–123, 2002.

35. S. Irani, S. Shukla, and R. Gupta. Algorithms for power savings. In *Proceedings of the 14th Annual ACM-SIAM Symposium on Discrete Algorithms (SODA'03)*, Baltimore, MD, pp. 37–46, 2003.

36. S. Irani, S. Shukla, and R. Gupta. Online strategies for dynamic power management in systems with multiple power-saving states. *Transactions on Embedded Computing Systems*, 2(3):325–346, 2003.

37. M. Martonosi and J. Donald. Techniques for multicore thermal management: Classification and new exploration. In *Proceedings of the 33rd International Symposium on Computer Architecture*, Boston, MA, pp. 78–88, 2006.

38. R. Jain, D. Molnar, and Z. Ramzan. Towards a model of energy complexity for algorithms. In *IEEE Wireless Communications and Networking Conference*, New Orleans, LA, pp. 1884–1890, 2005.

39. R. Jain, D. Molnar, and Z. Ramzan. Towards understanding algorithmic factors affecting energy consumption: Switching complexity, randomness, and preliminary experiments. In *Proceedings of the 2005 Joint Workshop on Foundations of Mobile Computing*, Cologne, Germany, pp. 70–79, 2005.

40. K. Kant. Toward a science of power management. *IEEE Computer*, 42:99–101, 2009.

41. A. Karlin, M. Manasse, L. McGeoch, and S. Owicki. Competitive randomized algorithms for nonuniform problems. *Algorithmica*, 11:542–571, 1994.

42. V. A. Korthikanti and G. Agha. On the energy complexity of parallel algorithms. Technical report, Department of Computer Science, University of Illinois at Urbana-Champaign, 2010.

43. V. A. Korthikanti and G. Agha. Towards optimizing energy costs of algorithms for shared memory architectures. In *Proceedings of the 22nd Annual ACM Symposium on Parallel Algorithms and Architectures*, Thira, Santorini, Greece, pp. 157–165, 2010.

44. A. Kumar, L. Shang, L.-S. Peh, and N. K. Jha. HybDTM: A coordinated hardware-software approach for dynamic thermal management. In *Proceedings of the 43rd Annual Conference on Design Automation (DAC'06)*, San Francisco, CA, pp. 548–553, 2006.

45. W.-C. Kwon and T. Kim. Optimal voltage allocation techniques for dynamically variable voltage processors. In *Proceedings of the 40th Annual Design Automation Conference (DAC'03)*, Anaheim, CA, pp. 125–130, 2003.

46. T.-W. Lam, L.-K. Lee, I. K. K. To, and P. W. H. Wong. Energy efficient deadline scheduling in two processor systems. In *Proceedings of the 18th International Conference on Algorithms and Computation (ISAAC'07)*, Sendai, Japan, pp. 476–487, 2007.

47. M. Li, A. C. Yao, and F. F. Yao. Discrete and continuous min-energy schedules for variable voltage processors. *Proceedings of the National Academy of Sciences*, 103:3983–3987, 2006.

48. M. Li and F. F. Yao. An efficient algorithm for computing optimal discrete voltage schedules. *SIAM Journal on Computing*, 35(3):658–671, 2005.

49. A. J. Martin. Towards an energy complexity of computation. *Information Processing Letters*, 77: 181–187, 2001.

50. K. Pruhs, R. V. Stee, and P. Uthaisombut. Speed scaling of tasks with precedence constraints. *Theory of Computing Systems*, 43(1):67–80, 2008.

51. K. Pruhs, P. Uthaisombut, and G. Woeginger. Getting the best response for your erg. *ACM Transactions on Algorithms*, 4:38:1–38:17, 2008.

52. A. Wierman, L. L. H. Andrew, and M. Lin. Speed scaling: An algorithmic perspective. In *Handbook on Energy-Aware and Green Computing*. CRC Press, Boca Raton, FL, 2011.

53. J. Yang, X. Zhou, M. Chrobak, and Y. Zhang. Dynamic thermal management through task scheduling. In *Proceedings of IEEE International Symposium on Performance Analysis of Systems and Software*, Austin, TX, pp. 191–201, 2008.

54. F. Yao, A. Demers, and S. Shenker. A scheduling model for reduced CPU energy. In *Proceedings of the 36th Annual IEEE Symposium on Foundations of Computer Science (FOCS'95)*, Milwaukee, WI, pp. 374–382, 1995.

55. T. Yokoyama, G. Zeng, H. Tomiyama, and H. Takada. Analyzing and optimizing energy efficiency of algorithms on DVS systems: A first step towards algorithmic energy minimization. In *Proceedings of Asia and South Pacific Design Automation Conference*, Pacific Yokohama, Yokohama, Japan, pp. 727–732, 2009.

56. X. Zhou, J. Yang, M. Chrobak, and Y. Zhang. Performance-aware thermal management via task scheduling. *ACM Transactions on Architecture and Code Optimization*, 7:1–31, 2010.

57. K. Zotos, A. Litke, E. Chatzigeorgiou, S. Nikolaidis, and G. Stephanides. Energy complexity of software in embedded systems. In *IASTED International Conference on Automation, Control and Applications*, Novosibirsk, Russia, pp. 146–150, 2005.

15

Algorithms and Analysis of Energy-Efficient Scheduling of Parallel Tasks

Keqin Li
*State University of New York
at New Paltz*

15.1 Introduction

15.1.1 Motivation

For six decades, the concept of the performance of a computer has been equivalent to the computing speed measured by floating-point operations per second (FLOPS). The peak speed of high-performance supercomputers has increased at an exponential speed. At the same time, the peak power requirements also increase at the same rate [12]. To achieve higher computing performance per processor, microprocessor manufacturers have doubled the power density at an exponential speed over decades, which will soon reach that of a nuclear reactor [40]. The emphasis on speed has led to the emergence of supercomputers that consume tremendous amounts of electrical power and produce so much heat that excessive cooling facilities must be constructed to ensure proper operation. Furthermore, the adoption of speed as the ultimate performance metric has caused other metrics such as reliability, availability, and usability to be largely ignored. Consequently, there has been an extraordinary increase in the total cost of ownership of a supercomputer.

Such increased energy consumption causes severe economic, ecological, and technical problems. A large-scale multiprocessor computing system consumes millions of dollars of electricity and natural resources every year, equivalent to the amount of energy used by tens of thousands of U.S. households [13]. A large data center such as Google can consume as much electricity as a city. Furthermore, the cooling bill for heat dissipation can be as high as 70% of the cost mentioned previously [11]. A recent report reveals that the global information technology industry generates as much greenhouse gas as the world's airlines, about 2% of global carbon dioxide (CO_2) emissions. Despite sophisticated cooling facilities constructed to ensure proper operation, the reliability and availability (mean time between failures and interrupts) of large-scale multiprocessor computing systems is measured in hours, and the main source of outage is hardware (CPU, memory, storage, and third-party hardware) failure caused by excessive heat. It is conceivable that a supercomputing system with 10^5 processors would spend most of its time checkpointing and restarting [16]. Furthermore, the hourly cost of downtime can be as high as millions of US dollars. The cost of ownership can easily exceed the initial acquisition cost.

In recent years, there has been rapidly growing interest and importance in developing high-performance and energy-efficient computing systems (see [1,5,39,40] for comprehensive surveys). Low power consumption and high system reliability, availability, and usability are main concerns of modern high-performance computing system development. In addition to the traditional performance measure using FLOPS, the Green500 list uses FLOPS per watt to rank the most energy-efficient supercomputers in the world, so that the awareness of other performance metrics such as performance per watt, energy efficiency, system reliability, and total cost of ownership can be raised [12]. The philosophy is that high-performance supercomputers should only simulate and predict climate and weather, but should not change or create them.

It can be found from the Green500 list that all the current supercomputing systems which can achieve at least 400 MFLOPS/W are clusters of low-power processors, aiming to achieve high performance/power and performance/space. For instance, the Dawning Nebulae, currently the world's second fastest computer which achieves peak performance of 2.984 PFLOPS, is also the fourth most energy-efficient supercomputer in the world with an operational rate of 492.64 MFLOPS/W. Intel's tera-scale research project has developed the world's first programmable processor that delivers supercomputer-like performance from a single 80-core chip which uses less electricity than most of today's home appliances and achieves over 16.29 GFLOPS/W.

One effective and widely adopted approach to reducing energy consumption in computing systems is the method of using a mechanism called *dynamic voltage scaling* (equivalently, dynamic frequency scaling, dynamic speed scaling, dynamic power scaling). Many modern components allow voltage regulation to be controlled through software, for example, the BIOS or applications such as PowerStrip. It is usually possible to control the voltages supplied to the CPUs, main memories, local buses, and expansion cards. Processor power consumption is proportional to frequency and the square of supply voltage. A power-aware algorithm can change supply voltage and frequency at appropriate times to optimize a combined consideration of performance and energy consumption. There are many existing technologies and commercial processors that support dynamic voltage (frequency, speed, power) scaling. SpeedStep is a series of dynamic frequency scaling technologies built into some Intel microprocessors that allow the clock speed of a processor to be dynamically changed by software. LongHaul is a technology developed by VIA technologies which supports dynamic frequency scaling and dynamic voltage scaling. By executing specialized operating system instructions, a processor driver can exercise fine control on the bus-to-core frequency ratio and core voltage according to how much load is put on the processor. LongRun and LongRun2 are power management technologies introduced by Transmeta. LongRun2 has been licensed to Fujitsu, NEC, Sony, Toshiba, and NVIDIA.

15.1.2 Related Research

Dynamic power management at the operating system level refers to supply voltage and clock frequency adjustment schemes implemented while tasks are running. These energy conservation techniques explore the opportunities for tuning the energy delay trade-off [38]. Power-aware task scheduling on processors with variable voltages and speeds has been extensively studied since mid 1990s. In a pioneering paper [41], the authors first proposed the approach to energy saving by using fine grain control of CPU speed by an operating system scheduler. The main idea is to monitor CPU idle time and to reduce energy consumption by reducing clock speed and idle time to a minimum. In a subsequent work [43], the authors analyzed off-line and online algorithms for scheduling tasks with arrival times and deadlines on a uniprocessor computer with minimum energy consumption. These researches have been extended in [3,7,21,28–30,44] and inspired substantial further investigation, much of which focus on real-time applications, namely, adjusting the supply voltage and clock frequency to minimize CPU energy consumption while still meeting the deadlines for task execution. In [2,17,18,20,23,31,32,34,36,37,42,46–49] and many other related works, the authors addressed the problem of scheduling independent or precedence-constrained tasks on uniprocessor or multiprocessor computers where the actual execution time of a task may be less than the estimated worst-case execution time. The main issue is energy reduction by slack time reclamation.

There are two considerations in dealing with the energy delay trade-off. On one hand, in high-performance computing systems, power-aware design techniques and algorithms attempt to maximize performance under certain energy consumption constraints. On the other hand, low-power and energy-efficient design techniques and algorithms aim to minimize energy consumption while still meeting certain performance goals. In [4], the author studied the problems of minimizing the expected execution time given a hard energy budget and minimizing the expected energy expenditure given a hard execution deadline for a single task with randomized execution requirement. In [6], the author considered scheduling jobs with equal requirements on multi-processors. In [9], the authors studied the relationship among parallelization, performance, and energy consumption, and the problem of minimizing energy-delay product. In [19,22], the authors attempted joint minimization of energy consumption and task execution time. In [35], the authors investigated the problem of system value maximization subject to both time and energy constraints.

In [25,27], we addressed energy- and time-constrained power allocation and task scheduling on multiprocessor computers with dynamically variable voltage and frequency and speed and power as combinatorial optimization problems. In particular, we defined the problem of minimizing schedule length with energy consumption constraint and the problem of minimizing energy consumption with schedule length constraint on multiprocessor computers [25]. The first problem has applications in general multiprocessor and multi-core processor computing systems where energy consumption is an important concern and in mobile computers where energy conservation is a main concern. The second problem has applications in real-time multiprocessing systems and environments such as parallel signal processing, automated target recognition, and real-time MPEG encoding, where timing constraint is a major requirement. Our scheduling problems are defined such that the energy-delay product is optimized by fixing one factor and minimizing the other. The investigation in [25,27] is for sequential tasks. A sequential task is executed on one processor. The motivation of this chapter is to study energy-efficient scheduling of parallel tasks.

15.1.3 Our Contributions

In this chapter, we investigate energy-efficient scheduling of parallel tasks on multiprocessor computers with dynamically variable voltage and speed. As in traditional scheduling theory, our problems are defined as combinatorial optimization problems. In particular, we define the problem of minimizing schedule length with energy consumption constraint and the problem of minimizing energy consumption with schedule length constraint for parallel tasks on multiprocessor computers. It is noticed that power-aware scheduling of parallel tasks has not been well studied before. Most existing studies are on scheduling sequential tasks which require one processor to execute, while a parallel task requires multiple processors to execute. It is clear that our scheduling problems introduce new challenges to energy-aware computing.

Our investigation in this chapter, together with our recent study in [26], make some initial attempt to energy-efficient scheduling of parallel tasks on multiprocessor computers with dynamic voltage and speed.

It is found that our scheduling problems contain three nontrivial sub-problems, namely, system partitioning, task scheduling, and power supplying. These subproblems are described as follows:

- *System partitioning*: Since each parallel task requests for multiple processors, a multiprocessor computer should be partitioned into clusters of processors to be assigned to the tasks.
- *Task scheduling*: Parallel tasks are scheduled together with system partitioning, and it is NP-hard even scheduling sequential tasks without system partitioning.
- *Power supplying*: Tasks should be supplied with appropriate powers and execution speeds such that the schedule length is minimized by consuming given amount of energy or the energy consumed is minimized without missing a given deadline.

Each subproblem should be solved efficiently, so that heuristic algorithms with overall good performance can be developed.

We consider two types of algorithms to solve our energy-aware parallel task scheduling problems, depending on the order of solving the subproblems.

- *Pre-Power-Determination algorithms*: In this type of algorithms, power allocation and execution speed determination are performed before tasks are scheduled. When tasks are scheduled, their execution times are available.
- *Post-Power-Determination algorithms*: In this type of algorithms, power allocation and execution speed determination are performed after tasks are scheduled. When a system is partitioned and tasks are scheduled, their execution times are not available, and tasks are scheduled based on their execution requirements.

The decomposition of our optimization problems into three subproblems makes design and analysis of heuristic algorithms tractable. We will develop a number of pre-power-determination algorithms and post-power-determination algorithms for energy-aware scheduling of parallel tasks. Furthermore, we will show that our heuristic algorithms are able to produce solutions very close to optimum.

15.2 Background Information

15.2.1 Power Consumption Model

Power dissipation and circuit delay in digital CMOS circuits can be accurately modeled by simple equations, even for complex microprocessor circuits. CMOS circuits have dynamic, static, and short-circuit power dissipation; however, the dominant component in a well-designed circuit is dynamic power consumption p (i.e., the switching component of power), which is approximately $p = aCV^2f$, where a is an activity factor, C is the loading capacitance, V is the supply voltage, and f is the clock frequency [8]. Since $s \propto f$, where s is the processor speed, and $f \propto V^\gamma$ with $0 < \gamma \leq 1$ [45], which implies that $V \propto f^{1/\gamma}$, we know that power consumption is $p \propto f^\alpha$ and $p \propto s^\alpha$, where $\alpha = 1 + 2/\gamma \geq 3$. It is clear from $f \propto V^\gamma$ and $s \propto V^\gamma$ that linear change in supply voltage results in up to linear change in clock frequency and processor speed. It is also clear from $p \propto V^{\gamma+2}$ and $p \propto f^\alpha$ and $p \propto s^\alpha$ that linear change in supply voltage results in at least quadratic change in power supply, and that linear change in clock frequency and processor speed results in at least cubic change in power supply.

Assume that we are given n independent parallel tasks to be executed on m identical processors. Task i requires π_i processors to execute, where $1 \leq i \leq n$, and any π_i of the m processors can be allocated to task i. We call π_i the *size* of task i. It is possible that in executing task i, the π_i processors may have different execution requirements (i.e., the numbers of CPU cycles or the numbers of instructions executed on the processors). Let r_i represent the maximum execution requirement on the π_i processors executing task i. We use p_i to represent the power supplied to execute task i. For ease of discussion, we will assume that p_i is simply s_i^α, where $s_i = p_i^{1/\alpha}$ is the execution speed of task i. The execution time of task i is $t_i = r_i/s_i = r_i/p_i^{1/\alpha}$. Note that all the π_i processors allocated to task i have the same speed s_i for duration t_i, although some of the π_i processors may be idle for some time. The energy consumed to execute task i is $e_i = \pi_i p_i t_i = \pi_i r_i p_i^{1-1/\alpha} = \pi_i r_i s_i^{\alpha-1} = w_i s_i^{\alpha-1}$, where $w_i = \pi_i r_i$ is the amount of work to be performed for task i.

We would like to mention a number of important observations.

- $f_i \propto V_i^\phi$ and $s_i \propto V_i^\phi$: Linear change in supply voltage results in up to linear change in clock frequency and processor speed.
- $p_i \propto V_i^{\phi+2}$ and $p_i \propto f_i^\alpha$ and $p_i \propto s_i^\alpha$: Linear change in supply voltage results in at least quadratic change in power supply, and linear change in clock frequency and processor speed results in at least cubic change in power supply.
- $s_i/p_i \propto V_i^{-2}$ and $s_i/p_i \propto s_i^{-(\alpha-1)}$: The processor energy performance, measured by speed per Watt, is at least quadratically proportional to the supply voltage and speed reduction.
- $w_i/e_i \propto V_i^{-2}$ and $w_i/e_i \propto s_i^{-(\alpha-1)}$: The processor energy performance, measured by work per Joule, is at least quadratically proportional to the supply voltage and speed reduction.
- $e_i \propto p_i^{1-1/\alpha} \propto V_i^{(\phi+2)(1-1/\alpha)} = V_i^2$: Linear change in supply voltage results in quadratic change in energy consumption.
- $e_i = w_i s_i^{\alpha-1}$: Linear change in processor speed results in at least quadratic change in energy consumption.
- $e_i = w_i p_i^{1-1/\alpha}$: Energy consumption reduces at a sublinear speed as power supply reduces.
- $e_i t_i^{\alpha-1} = \pi_i r_i^\alpha$ and $p_i t_i^\alpha = r_i^\alpha$: For a given task, there exist energy delay and power delay trade-offs. (Later, we will extend such trade-off to a set of parallel tasks, i.e., the energy delay trade-off theorem.)

15.2.2 Definitions

Problem 15.1 *(Minimizing schedule length with energy consumption constraint)*
Input: A set of n parallel tasks with task sizes $\pi_1, \pi_2, \ldots, \pi_n$ and task execution requirements $r_1, r_2, \ldots, r_n$, a multiprocessor computer with m identical processors, and energy constraint E.
Output: Power supplies $p_1, p_2, \ldots, p_n$ to the n tasks and a non-preemptive schedule of the n parallel tasks on the m processors such that the schedule length is minimized and the total energy consumed does not exceed E.

Problem 15.2 *(Minimizing energy consumption with schedule length constraint)*
Input: A set of n parallel tasks with task sizes $\pi_1, \pi_2, \ldots, \pi_n$ and task execution requirements $r_1, r_2, \ldots, r_n$, a multiprocessor computer with m identical processors, and time constraint T.
Output: Power supplies $p_1, p_2, \ldots, p_n$ to the n tasks and a non-preemptive schedule of the n parallel tasks on the m processors such that the total energy consumed is minimized and the schedule length does not exceed T.

In the previous description, the energy-aware parallel task scheduling problems on multiprocessor computers with energy and time constraints addressed in this chapter are defined as optimization problems.

When all the π_i's are identical, the scheduling problems discussed earlier are equivalent to scheduling sequential tasks discussed in [25,27]. Since both scheduling problems are NP-hard in the strong sense for all rational $\alpha > 1$ in scheduling sequential tasks [14,25], our problems for scheduling parallel tasks are also NP-hard in the strong sense for all rational $\alpha > 1$. Hence, we will develop fast polynomial time heuristic algorithms to solve these problems.

Let T_A denote the length of the schedule produced by algorithm A, and T_{OPT} denote the length of an optimal schedule. Similarly, let E_A denote the total amount of energy consumed by algorithm A, and E_{OPT} denote the minimum amount of energy consumed by an optimal schedule. The following performance measures are used to analyze and evaluate the performance of our heuristic power allocation and parallel task scheduling algorithms.

Definition 15.1   The *performance ratio* of an algorithm A that solves the problem of minimizing schedule length with energy consumption constraint is defined as $\beta_A = T_A/T_{\text{OPT}}$. If $\beta_A \leq B$, we call B a *performance bound* of algorithm A.

Definition 15.2   The *performance ratio* of an algorithm A that solves the problem of minimizing energy consumption with schedule length constraint is defined as $\gamma_A = E_A/E_{\text{OPT}}$. If $\gamma_A \leq C$, we call C a *performance bound* of algorithm A.

When parallel tasks have random sizes and/or random execution requirements, T_A, T_{OPT}, β_A, B, E_A, E_{OPT}, γ_A, and C are all random variables. Let $\bar{x}$ be the expectation of a random variable x.

Definition 15.3   If $\beta_A \leq B$, then $\bar{\beta}_A \leq \bar{B}$, where $\bar{B}$ is an *average-case performance bound* of algorithm A.

Definition 15.4   If $\gamma_A \leq C$, then $\bar{\gamma}_A \leq \bar{C}$, where $\bar{C}$ is an *average-case performance bound* of algorithm A.

15.2.3 Lower Bounds

The performance of our heuristic algorithms will be compared with optimal solutions analytically. Unfortunately, it is infeasible to compute optimal solutions in reasonable amount of time. To make such comparison possible, we have derived lower bounds for the optimal solutions in Theorems 15.1 and 15.2 [26]. The significance of these lower bounds is that they can be used to evaluate the performance of heuristic algorithms when they are compared with optimal solutions.

Let $W = w_1 + w_2 + \cdots + w_n = \pi_1 r_1 + \pi_2 r_2 + \cdots + \pi_n r_n$ denote the total amount of work to be performed for the n parallel tasks. The following theorem gives a lower bound for the optimal schedule length T_{OPT} for the problem of minimizing schedule length with energy consumption constraint.

Theorem 15.1   *For the problem of minimizing schedule length with energy consumption constraint in scheduling parallel tasks, we have the following lower bound*

$$T_{\text{OPT}} \geq \left(\frac{m}{E} \left(\frac{W}{m} \right)^{\alpha} \right)^{1/(\alpha-1)}$$

for the optimal schedule length.

The following theorem gives a lower bound for the minimum energy consumption E_{OPT} for the problem of minimizing energy consumption with schedule length constraint.

Theorem 15.2   *For the problem of minimizing energy consumption with schedule length constraint in scheduling parallel tasks, we have the following lower bound*

$$E_{\text{OPT}} \geq m \left(\frac{W}{m} \right)^{\alpha} \frac{1}{T^{\alpha-1}}$$

for the minimum energy consumption.

The lower bounds in Theorems 15.1 and 15.2 essentially state the following important theorem.

$ET^{\alpha-1}$ **Lower Bound Theorem (Energy-Delay Trade-off Theorem)** *For any execution of a set of parallel tasks with total amount of work W on m processors with schedule length T and energy consumption E, we must have the following trade-off*

$$ET^{\alpha-1} \geq m \left(\frac{W}{m} \right)^{\alpha},$$

by using any scheduling algorithm.

Therefore, our scheduling problems are defined such that the energy delay product is optimized by fixing one factor and minimizing the other.

It is noticed that when $\alpha = 3$, we have $ET^{\alpha-1} = ET^2$, which was proposed as a measure of the energy efficiency of a computation [33].

Notice that the lower bounds in Theorems 15.1 and 15.2 and the energy delay trade-off theorem are applicable to various parallel task models (independent or precedence constrained, static or dynamic tasks), various processor models (regular homogeneous processors with continuous or discrete voltage/frequency/speed/power levels, bounded or unbounded voltage/frequency/speed/power levels, with/without overheads for voltage/frequency/speed/power adjustment and idle processors), and all scheduling models (preemptive or non-preemptive, online or off-line, clairvoyant or non-clairvoyant scheduling). (See [27] for description of these models.)

15.3 Pre-Power-Determination Algorithms

15.3.1 Overview

In pre-power-determination algorithms, we first determine power supplies to the n parallel tasks and then partition the system and schedule the tasks. We use $A_1 - A_2$ to represent a pre-power-determination algorithm, where A_1 is a strategy for power supplying and A_2 is an algorithm for system partitioning and task scheduling. Algorithm $A_1 - A_2$ works in the following way. First, algorithm A_1 is used to allocate powers to the n tasks and to solve the subproblem of power supplying. Second, algorithm A_2 is used to partition the system with m processors and to schedule the n tasks whose execution times are known based on the power allocation, and to solve the subproblems of system partitioning and task scheduling simultaneously.

In this chapter, we consider three strategies (i.e., algorithm A_1) for power allocation.

- *Equal-time (ET)*: The power supplies $p_1, p_2, \ldots, p_n$ are determined such that all the n parallel tasks have the same execution time, i.e., $t_1 = t_2 = \cdots = t_n$.
- *Equal-energy (EE)*: The power supplies $p_1, p_2, \ldots, p_n$ are determined such that all the n parallel tasks consume the same amount of energy, i.e., $e_1 = e_2 = \cdots = e_n$.
- *Equal-speed (ES)*: The power supplies $p_1, p_2, \ldots, p_n$ are determined such that all the n parallel tasks have the same execution speed, i.e., $s_1 = s_2 = \cdots = s_n$.

In all cases, the execution times of all the n parallel tasks are available before the tasks are scheduled.

We consider the following four methods (i.e., algorithm A_2) for scheduling parallel tasks.

- SIMPLE: The n parallel tasks are put into a list, which is divided into groups, where each group contains a sublist of consecutive tasks. In the beginning and whenever a task is completed, the next group of tasks are scheduled for execution. Each group includes as many tasks as possible for simultaneous execution.

- SIMPLE*: This method works in the same way as SIMPLE, except that tasks are arranged in a nonincreasing order of sizes, i.e., $\pi_1 \geq \pi_2 \geq \cdots \geq \pi_n$, before scheduling.
- GREEDY: In the beginning and whenever a task is completed, each remaining unscheduled task is examined for possible execution, i.e., whether there are enough available processors for the task.
- GREEDY*: This method works in the same way as GREEDY, except that tasks are arranged in a nonincreasing order of sizes, i.e., $\pi_1 \geq \pi_2 \geq \cdots \geq \pi_n$, before scheduling.

Algorithms SIMPLE and GREEDY were considered in [24] for non-clairvoyant scheduling of parallel tasks, since these algorithms do not require the information of task execution times.

To summarize, we have developed 12 pre-power-determination algorithms for power allocation and parallel task scheduling, namely, ET-A, EE-A, and ES-A, where $A \in \{\text{SIMPLE}, \text{SIMPLE}^*, \text{GREEDY}, \text{GREEDY}^*\}$.

For n parallel tasks with sizes $\pi_1, \pi_2, \ldots, \pi_n$ and execution times $t_1, t_2, \ldots, t_n$, we use $A(\pi_1, t_1, \pi_2, t_2, \ldots, \pi_n, t_n)$ to denote the length of the schedule produced by algorithm A.

15.3.2 Analysis of Equal-Time Algorithms

15.3.2.1 Minimizing Schedule Length

To solve the problem of minimizing schedule length with energy consumption constraint E by using an equal-time algorithm ET-A, we have

$$t_1 = t_2 = \cdots = t_n = t,$$

where t is the identical task execution time. Since

$$t_i = \frac{r_i}{s_i} = t,$$

we get

$$s_i = \frac{r_i}{t},$$

and

$$p_i = s_i^\alpha = \left(\frac{r_i}{t}\right)^\alpha,$$

and

$$e_i = \pi_i r_i p_i^{1-1/\alpha} = \frac{\pi_i r_i^\alpha}{t^{\alpha-1}},$$

for all $1 \leq i \leq n$. Since

$$e_1 + e_2 + \cdots + e_n = E,$$

that is,

$$\frac{\pi_1 r_1^\alpha + \pi_2 r_2^\alpha + \cdots + \pi_n r_n^\alpha}{t^{\alpha-1}} = E,$$

we obtain

$$t = \left(\frac{\pi_1 r_1^\alpha + \pi_2 r_2^\alpha + \cdots + \pi_n r_n^\alpha}{E}\right)^{1/(\alpha-1)}.$$

Consequently, the length of the schedule generated by algorithm ET-*A* is

$$T_{\text{ET-}A} = A(\pi_1, t, \pi_2, t, \ldots, \pi_n, t).$$

By Theorem 15.1, the performance ratio of algorithm ET-*A* is

$$\beta_{\text{ET-}A} = \frac{T_{\text{ET-}A}}{T_{\text{OPT}}} \leq \frac{A(\pi_1, t, \pi_2, t, \ldots, \pi_n, t)}{((m/E)(W/m)^\alpha)^{1/(\alpha-1)}}.$$

We observe that when all tasks have the same execution time t, the problem of scheduling parallel tasks with sizes $\pi_1, \pi_2, \ldots, \pi_n$ such that the total execution time is minimized is equivalent to the classic bin packing problem, i.e., packing n items of sizes $\pi_1, \pi_2, \ldots, \pi_n$ into bins of size m, such that the number of bins used is minimized. Let $b_1, b_2, b_3, \ldots$ be the sequence of bins used. The following four methods are well known bin packing algorithms [10].

- *Next fit (NF):* Assume that b_i is the current bin being packed. b_i is packed as many items as possible, until there is π_j which cannot be packed into b_i. Then, a new bin b_{i+1} is opened to pack π_j.
- *Next fit decreasing (NFD):* This algorithm is the same as *NF*, except that the sizes are sorted in a nonincreasing order before packing, i.e., $\pi_1 \geq \pi_2 \geq \cdots \geq \pi_n$.
- *First fit (FF):* Let $b_1, b_2, \ldots, b_i$ be the bins ever used. To pack an item π_j, each bin of $b_1, b_2, \ldots, b_i$ is examined in this order to see whether π_j can be packed. π_j is packed into the first bin which can accommodate π_j. If no such bin is found, a new bin b_{i+1} is opened to pack π_j.
- *First fit decreasing (FFD):* This algorithm is the same as *FF*, except that the sizes are sorted in a nonincreasing order before packing, i.e., $\pi_1 \geq \pi_2 \geq \cdots \geq \pi_n$.

Each parallel task scheduling algorithm A is equivalent to a bin packing algorithm A'. It is clear that for $A =$ SIMPLE, SIMPLE*, GREEDY, GREEDY*, we have $A' =$ NF, NFD, FF, FFD, respectively.

Let $A'(\pi_1, \pi_2, \ldots, \pi_n)$ denote the number of bins used by algorithm A'. Then, we have

$$\text{SIMPLE}(\pi_1, t, \pi_2, t, \ldots, \pi_n, t) = tNF(\pi_1, \pi_2, \ldots, \pi_n),$$

$$\text{SIMPLE}^*(\pi_1, t, \pi_2, t, \ldots, \pi_n, t) = tNFD(\pi_1, \pi_2, \ldots, \pi_n),$$

$$\text{GREEDY}(\pi_1, t, \pi_2, t, \ldots, \pi_n, t) = tFF(\pi_1, \pi_2, \ldots, \pi_n),$$

$$\text{GREEDY}^*(\pi_1, t, \pi_2, t, \ldots, \pi_n, t) = tFFD(\pi_1, \pi_2, \ldots, \pi_n),$$

and

$$T_{\text{ET-}A} = tA'(\pi_1, \pi_2, \ldots, \pi_n).$$

By Theorem 15.1, the performance ratio of algorithm ET-*A* is

$$\beta_{\text{ET-}A} = \frac{T_{\text{ET-}A}}{T_{\text{OPT}}}$$

$$\leq \frac{tA'(\pi_1, \pi_2, \ldots, \pi_n)}{((m/E)(W/m)^\alpha)^{1/(\alpha-1)}}$$

$$= \left(\frac{\pi_1 r_1^\alpha + \pi_2 r_2^\alpha + \cdots + \pi_n r_n^\alpha}{E}\right)^{1/(\alpha-1)} \frac{A'(\pi_1, \pi_2, \ldots, \pi_n)}{((m/E)(W/m)^\alpha)^{1/(\alpha-1)}}$$

$$= \left(\frac{m(\pi_1 r_1^\alpha + \pi_2 r_2^\alpha + \cdots + \pi_n r_n^\alpha)^{1/(\alpha-1)}}{W^{\alpha/(\alpha-1)}}\right) A'(\pi_1, \pi_2, \ldots, \pi_n).$$

This discussion can be summarized in the following theorem.

Theorem 15.3 *By using an equal-time algorithm ET-A to solve the problem of minimizing schedule length with energy consumption constraint, the schedule length is*

$$T_{ET\text{-}A} = A(\pi_1, t, \pi_2, t, \ldots, \pi_n, t),$$

where

$$t = \left(\frac{\pi_1 r_1^\alpha + \pi_2 r_2^\alpha + \cdots + \pi_n r_n^\alpha}{E} \right)^{1/(\alpha-1)},$$

or,

$$T_{ET\text{-}A} = \left(\frac{\pi_1 r_1^\alpha + \pi_2 r_2^\alpha + \cdots + \pi_n r_n^\alpha}{E} \right)^{1/(\alpha-1)} A'(\pi_1, \pi_2, \ldots, \pi_n).$$

The performance ratio is $\beta_{ET\text{-}A} \leq B_{ET\text{-}A}$, where the performance bound is

$$B_{ET\text{-}A} = \frac{A(\pi_1, t, \pi_2, t, \ldots, \pi_n, t)}{((m/E)(W/m)^\alpha)^{1/(\alpha-1)}},$$

or,

$$B_{ET\text{-}A} = \left(\frac{m \left(\pi_1 r_1^\alpha + \pi_2 r_2^\alpha + \cdots + \pi_n r_n^\alpha \right)^{1/(\alpha-1)}}{(\pi_1 r_1 + \pi_2 r_2 + \cdots + \pi_n r_n)^{\alpha/(\alpha-1)}} \right) A'(\pi_1, \pi_2, \ldots, \pi_n).$$

For the purpose of average-case performance analysis, throughout this chapter, we make the following assumptions of task sizes and task execution requirements.

- $\pi_1, \pi_2, \ldots, \pi_n$ are independent and identically distributed (i.i.d.) discrete random variables with a common probability distribution in the range $[1 \ldots m]$.
- $r_1, r_2, \ldots, r_n$ are i.i.d. continuous random variables.
- The probability distribution of the π_i's and the probability distribution of the r_i's are independent of each other.

These assumptions make the probabilistic analysis of SIMPLE feasible.

Let us consider the following packing model. Assume that there is a bag of capacity m and there are n objects of sizes $\pi_1, \pi_2, \ldots, \pi_n$. These objects are to be packed into the bag as many as possible, i.e., we need to find i such that

$$\pi_1 + \pi_2 + \cdots + \pi_i \leq m,$$

but

$$\pi_1 + \pi_2 + \cdots + \pi_i + \pi_{i+1} > m.$$

If $\pi_1, \pi_2, \ldots, \pi_n$ are i.i.d. random variables, the total size of the objects packed into the bag, i.e., $\pi_1 + \pi_2 + \cdots + \pi_i$, is also a random variable. Let $P(n, m)$ denote the mean of this random variable, and define $P(m) = \lim_{n \to \infty} P(n, m)$. In fact, if $\pi_1, \pi_2, \ldots, \pi_n$ are discrete integer random variables, we have $P_m = P(n, m)$ for all $n \geq m$, since the bag can accommodate at most m objects. It was shown in [24] that P_m can be calculated easily by using a recurrence relation for any probability distribution of the π_i's.

Furthermore, it was known that by the previously mentioned assumptions of task sizes and task execution requirements, for algorithm SIMPLE, we have

$$\mathbb{E}(\text{SIMPLE}(\pi_1, t_1, \pi_2, t_2, \ldots, \pi_n, t_n)) \approx \frac{1}{P_m} \mathbb{E}(\pi_1 t_1 + \pi_2 t_2 + \cdots + \pi_n t_n),$$

when n is large [24]. (Notation: $\mathbb{E}(x)$ and $\bar{x}$ represent the expectation of a random variable x.) Based on the previous result, we obtain

$$\mathbb{E}(\text{SIMPLE}(\pi_1, t, \pi_2, t, \ldots, \pi_n, t)) \approx \frac{1}{P_m} \mathbb{E}(\pi_1 t + \pi_2 t + \cdots + \pi_n t)$$

$$\approx \frac{1}{P_m} \mathbb{E}(t) \mathbb{E}(\pi_1 + \pi_2 + \cdots + \pi_n),$$

which implies the following approximation of the average-case performance bound $\bar{B}_{\text{ET-SIMPLE}}$

$$\bar{B}_{\text{ET-SIMPLE}} \approx \frac{\mathbb{E}(\text{SIMPLE}(\pi_1, t, \pi_2, t, \ldots, \pi_n, t))}{\mathbb{E}(((m/E)(W/m)^\alpha)^{1/(\alpha-1)})}$$

$$\approx \frac{1}{P_m} \cdot \frac{\mathbb{E}(t) \mathbb{E}(\pi_1 + \pi_2 + \cdots + \pi_n)}{\mathbb{E}(((m/E)(W/m)^\alpha)^{1/(\alpha-1)})}$$

$$\approx \frac{m}{P_m} \cdot \frac{\mathbb{E}\left(\left(\pi_1 r_1^\alpha + \pi_2 r_2^\alpha + \cdots + \pi_n r_n^\alpha\right)^{1/(\alpha-1)}\right) \mathbb{E}(\pi_1 + \pi_2 + \cdots + \pi_n)}{\mathbb{E}((\pi_1 r_1 + \pi_2 r_2 + \cdots + \pi_n r_n)^{\alpha/(\alpha-1)})}.$$

We will provide numerical and simulation data to demonstrate the accuracy of the previous approximation.

15.3.2.2 Minimizing Energy Consumption

To solve the problem of minimizing energy consumption with schedule length constraint T by using an equal-time algorithm ET-A, we need to provide enough energy $E_{\text{ET-}A}$, so that the deadline T is met, i.e., $T_{\text{ET-}A} = T$,

$$\left(\frac{\pi_1 r_1^\alpha + \pi_2 r_2^\alpha + \cdots + \pi_n r_n^\alpha}{E_{\text{ET-}A}}\right)^{1/(\alpha-1)} A'(\pi_1, \pi_2, \ldots, \pi_n) = T.$$

This equation implies that the amount of energy $E_{\text{ET-}A}$ consumed by algorithm ET-A is

$$E_{\text{ET-}A} = \left(\frac{A'(\pi_1, \pi_2, \ldots, \pi_n)}{T}\right)^{\alpha-1} \left(\pi_1 r_1^\alpha + \pi_2 r_2^\alpha + \cdots + \pi_n r_n^\alpha\right).$$

By theorem 15.2, the performance ratio of algorithm ET-A is

$$\gamma_{\text{ET-}A} = \frac{E_{\text{ET-}A}}{E_{\text{OPT}}}$$

$$\leq \frac{(A'(\pi_1, \pi_2, \ldots, \pi_n)/T)^{\alpha-1} \left(\pi_1 r_1^\alpha + \pi_2 r_2^\alpha + \cdots + \pi_n r_n^\alpha\right)}{(m/T^{\alpha-1})(W/m)^\alpha}$$

$$= \frac{(mA'(\pi_1, \pi_2, \ldots, \pi_n))^{\alpha-1} \left(\pi_1 r_1^\alpha + \pi_2 r_2^\alpha + \cdots + \pi_n r_n^\alpha\right)}{W^\alpha}.$$

This discussion basically proves the following theorem.

Theorem 15.4 *By using an equal-time algorithm ET-A to solve the problem of minimizing energy consumption with schedule length constraint, the energy consumed is*

$$E_{\text{ET-A}} = \left(\frac{A'(\pi_1, \pi_2, \ldots, \pi_n)}{T} \right)^{\alpha-1} \left(\pi_1 r_1^\alpha + \pi_2 r_2^\alpha + \cdots + \pi_n r_n^\alpha \right).$$

The performance ratio is $\gamma_{\text{ET-A}} \le C_{\text{ET-A}}$, where the performance bound is

$$C_{\text{ET-A}} = \left(\frac{m^{\alpha-1} \left(\pi_1 r_1^\alpha + \pi_2 r_2^\alpha + \cdots + \pi_n r_n^\alpha \right)}{(\pi_1 r_1 + \pi_2 r_2 + \cdots + \pi_n r_n)^\alpha} \right) \left(A'(\pi_1, \pi_2, \ldots, \pi_n) \right)^{\alpha-1}.$$

Since $C_{\text{ET-A}} = B_{\text{ET-A}}^{\alpha-1}$, we obtain the following approximation of the average-case performance-bound $\overline{C}_{\text{ET-SIMPLE}}$, i.e.,

$$\overline{C}_{\text{ET-SIMPLE}} = \mathbb{E}(B_{\text{ET-SIMPLE}}^{\alpha-1}) \approx \overline{B}_{\text{ET-SIMPLE}}^{\alpha-1}$$

$$\approx \left(\frac{m}{P_m} \cdot \frac{\mathbb{E}\left(\left(\pi_1 r_1^\alpha + \pi_2 r_2^\alpha + \cdots + \pi_n r_n^\alpha \right)^{1/(\alpha-1)} \right) \mathbb{E}(\pi_1 + \pi_2 + \cdots + \pi_n)}{\mathbb{E}((\pi_1 r_1 + \pi_2 r_2 + \cdots + \pi_n r_n)^{\alpha/(\alpha-1)})} \right)^{\alpha-1}.$$

15.3.3 Analysis of Equal-Energy Algorithms

15.3.3.1 Minimizing Schedule Length

To solve the problem of minimizing schedule length with energy consumption constraint E by using an equal-energy algorithm EE-A, we have

$$e_1 = e_2 = \cdots = e_n = \frac{E}{n}.$$

Hence, we get

$$e_i = \pi_i r_i p_i^{1-1/\alpha} = w_i p_i^{1-1/\alpha} = \frac{E}{n},$$

which gives

$$p_i = \left(\frac{E}{nw_i} \right)^{\alpha/(\alpha-1)},$$

and

$$s_i = p_i^{1/\alpha} = \left(\frac{E}{nw_i} \right)^{1/(\alpha-1)},$$

and

$$t_i = \frac{r_i}{s_i} = r_i w_i^{1/(\alpha-1)} \left(\frac{n}{E} \right)^{1/(\alpha-1)},$$

for all $1 \le i \le n$.

It is noticed that for all $\phi \ge 0$, we have

$$A(\pi_1, \phi t_1, \pi_2, \phi t_2, \ldots, \pi_n, \phi t_n) = \phi A(\pi_1, t_1, \pi_2, t_2, \ldots, \pi_n, t_n).$$

In other words, the schedule length is changed by a factor of ϕ if all the task execution times are changed by a factor of ϕ. This observation implies that the length of the schedule produced by algorithm EE-A is

$$
\begin{aligned}
T_{\text{EE-}A} &= A(\pi_1, t_1, \pi_2, t_2, \ldots, \pi_n, t_n) \\
&= A\left(\pi_1, r_1 w_1^{1/(\alpha-1)}, \pi_2, r_2 w_2^{1/(\alpha-1)}, \ldots, \pi_n, r_n w_n^{1/(\alpha-1)}\right) \left(\frac{n}{E}\right)^{1/(\alpha-1)}.
\end{aligned}
$$

By theorem 15.1, the performance ratio of algorithm EE-A is

$$
\begin{aligned}
\beta_{\text{EE-}A} &= \frac{T_{\text{EE-}A}}{T_{\text{OPT}}} \\
&\leq \frac{A\left(\pi_1, r_1 w_1^{1/(\alpha-1)}, \pi_2, r_2 w_2^{1/(\alpha-1)}, \ldots, \pi_n, r_n w_n^{1/(\alpha-1)}\right) (n/E)^{1/(\alpha-1)}}{((m/E)(W/m)^\alpha)^{1/(\alpha-1)}} \\
&= \frac{mn^{1/(\alpha-1)}A\left(\pi_1, r_1 w_1^{1/(\alpha-1)}, \pi_2, r_2 w_2^{1/(\alpha-1)}, \ldots, \pi_n, r_n w_n^{1/(\alpha-1)}\right)}{W^{\alpha/(\alpha-1)}} \\
&= \frac{mn^{1/(\alpha-1)}A\left(\pi_1, r_1 w_1^{1/(\alpha-1)}, \pi_2, r_2 w_2^{1/(\alpha-1)}, \ldots, \pi_n, r_n w_n^{1/(\alpha-1)}\right)}{(\pi_1 r_1 + \pi_2 r_2 + \cdots + \pi_n r_n)^{\alpha/(\alpha-1)}}.
\end{aligned}
$$

This discussion leads to the following theorem.

Theorem 15.5 *By using an equal-energy algorithm EE-A to solve the problem of minimizing schedule length with energy consumption constraint, the schedule length is*

$$
T_{\text{EE-}A} = A\left(\pi_1, r_1 w_1^{1/(\alpha-1)}, \pi_2, r_2 w_2^{1/(\alpha-1)}, \ldots, \pi_n, r_n w_n^{1/(\alpha-1)}\right) \left(\frac{n}{E}\right)^{1/(\alpha-1)}.
$$

The performance ratio is $\beta_{\text{EE-}A} \leq B_{\text{EE-}A}$, where the performance bound is

$$
B_{\text{EE-}A} = \frac{mn^{1/(\alpha-1)}A\left(\pi_1, r_1 w_1^{1/(\alpha-1)}, \pi_2, r_2 w_2^{1/(\alpha-1)}, \ldots, \pi_n, r_n w_n^{1/(\alpha-1)}\right)}{(\pi_1 r_1 + \pi_2 r_2 + \cdots + \pi_n r_n)^{\alpha/(\alpha-1)}}.
$$

For algorithm SIMPLE, we have

$$
\begin{aligned}
&\mathbb{E}\left(\text{SIMPLE}\left(\pi_1, r_1 w_1^{1/(\alpha-1)}, \pi_2, r_2 w_2^{1/(\alpha-1)}, \ldots, \pi_n, r_n w_n^{1/(\alpha-1)}\right)\right) \\
&\approx \frac{1}{P_m}\mathbb{E}\left(w_1^{\alpha/(\alpha-1)} + w_2^{\alpha/(\alpha-1)} + \cdots + w_n^{\alpha/(\alpha-1)}\right).
\end{aligned}
$$

Hence, we get

$$
\begin{aligned}
\overline{B}_{\text{EE-SIMPLE}} &\approx \frac{mn^{1/(\alpha-1)}\mathbb{E}\left(\text{SIMPLE}\left(\pi_1, r_1 w_1^{1/(\alpha-1)}, \pi_2, r_2 w_2^{1/(\alpha-1)}, \ldots, \pi_n, r_n w_n^{1/(\alpha-1)}\right)\right)}{\mathbb{E}((\pi_1 r_1 + \pi_2 r_2 + \cdots + \pi_n r_n)^{\alpha/(\alpha-1)})} \\
&\approx \frac{m}{P_m} \cdot \frac{n^{1/(\alpha-1)}\mathbb{E}\left(w_1^{\alpha/(\alpha-1)} + w_2^{\alpha/(\alpha-1)} + \cdots + w_n^{\alpha/(\alpha-1)}\right)}{\mathbb{E}((w_1 + w_2 + \cdots + w_n)^{\alpha/(\alpha-1)})},
\end{aligned}
$$

for large n.

15.3.3.2 Minimizing Energy Consumption

To solve the problem of minimizing energy consumption with schedule length constraint T by using an equal-energy algorithm EE-A, we need to provide enough energy $E_{\text{EE-}A}$, so that the deadline T is met, i.e., $T_{\text{EE-}A} = T$,

$$
A\left(\pi_1, r_1 w_1^{1/(\alpha-1)}, \pi_2, r_2 w_2^{1/(\alpha-1)}, \ldots, \pi_n, r_n w_n^{1/(\alpha-1)}\right)\left(\frac{n}{E_{\text{EE-}A}}\right)^{1/(\alpha-1)} = T.
$$

The last equation implies that the amount of energy $E_{\text{EE-}A}$ consumed by algorithm EE-A is

$$
E_{\text{EE-}A} = \frac{n}{T^{\alpha-1}}\left(A\left(\pi_1, r_1 w_1^{1/(\alpha-1)}, \pi_2, r_2 w_2^{1/(\alpha-1)}, \ldots, \pi_n, r_n w_n^{1/(\alpha-1)}\right)\right)^{\alpha-1}.
$$

By theorem 15.2, the performance ratio of algorithm EE-A is

$$
\begin{aligned}
\gamma_{\text{EE-}A} &= \frac{E_{\text{EE-}A}}{E_{\text{OPT}}} \\[2mm]
&\leq \frac{(n/T^{\alpha-1})\left(A\left(\pi_1, r_1 w_1^{1/(\alpha-1)}, \pi_2, r_2 w_2^{1/(\alpha-1)}, \ldots, \pi_n, r_n w_n^{1/(\alpha-1)}\right)\right)^{\alpha-1}}{(m/T^{\alpha-1})(W/m)^{\alpha}} \\[2mm]
&= \frac{nm^{\alpha-1}\left(A\left(\pi_1, r_1 w_1^{1/(\alpha-1)}, \pi_2, r_2 w_2^{1/(\alpha-1)}, \ldots, \pi_n, r_n w_n^{1/(\alpha-1)}\right)\right)^{\alpha-1}}{W^{\alpha}} \\[2mm]
&= \frac{nm^{\alpha-1}\left(A\left(\pi_1, r_1 w_1^{1/(\alpha-1)}, \pi_2, r_2 w_2^{1/(\alpha-1)}, \ldots, \pi_n, r_n w_n^{1/(\alpha-1)}\right)\right)^{\alpha-1}}{(\pi_1 r_1 + \pi_2 r_2 + \cdots + \pi_n r_n)^{\alpha}}.
\end{aligned}
$$

This discussion essentially proves the following theorem.

Theorem 15.6 *By using an equal-energy algorithm EE-A to solve the problem of minimizing energy consumption with schedule length constraint, the energy consumed is*

$$
E_{\text{EE-}A} = \frac{n}{T^{\alpha-1}}\left(A\left(\pi_1, r_1 w_1^{1/(\alpha-1)}, \pi_2, r_2 w_2^{1/(\alpha-1)}, \ldots, \pi_n, r_n w_n^{1/(\alpha-1)}\right)\right)^{\alpha-1}.
$$

The performance ratio is $\gamma_{\text{EE-}A} \leq C_{\text{EE-}A}$, where the performance bound is

$$
C_{\text{EE-}A} = \frac{nm^{\alpha-1}\left(A\left(\pi_1, r_1 w_1^{1/(\alpha-1)}, \pi_2, r_2 w_2^{1/(\alpha-1)}, \ldots, \pi_n, r_n w_n^{1/(\alpha-1)}\right)\right)^{\alpha-1}}{(\pi_1 r_1 + \pi_2 r_2 + \cdots + \pi_n r_n)^{\alpha}}.
$$

Since $C_{\text{EE-}A} = B_{\text{EE-}A}^{\alpha-1}$, we obtain the following approximation of the average-case performance-bound $\overline{C}_{\text{EE-SIMPLE}}$, i.e.,

$$
\begin{aligned}
\overline{C}_{\text{EE-SIMPLE}} &= \mathbb{E}(B_{\text{EE-SIMPLE}}^{\alpha-1}) \approx \overline{B}_{\text{EE-SIMPLE}}^{\alpha-1} \\[2mm]
&\approx \left(\frac{m}{P_m} \cdot \frac{n^{1/(\alpha-1)}\mathbb{E}\left(w_1^{\alpha/(\alpha-1)} + w_2^{\alpha/(\alpha-1)} + \cdots + w_n^{\alpha/(\alpha-1)}\right)}{\mathbb{E}((w_1 + w_2 + \cdots + w_n)^{\alpha/(\alpha-1)})}\right)^{\alpha-1}.
\end{aligned}
$$

15.3.4 Analysis of Equal-Speed Algorithms

15.3.4.1 Minimizing Schedule Length

To solve the problem of minimizing schedule length with energy consumption constraint E by using an equal-speed algorithm ES-A, we have

$$s_1 = s_2 = \cdots = s_n = s,$$

where s is the identical task execution speed. Hence, we get

$$e_i = \pi_i r_i p_i^{1-1/\alpha} = w_i p_i^{1-1/\alpha} = w_i s_i^{\alpha-1} = w_i s^{\alpha-1},$$

for all $1 \le i \le n$. Since

$$\sum_{i=1}^{n} e_i = s^{\alpha-1} \sum_{i=1}^{n} w_i = W s^{\alpha-1} = E,$$

we obtain

$$s = \left(\frac{E}{W}\right)^{1/(\alpha-1)},$$

and

$$t_i = \frac{r_i}{s} = r_i \left(\frac{W}{E}\right)^{1/(\alpha-1)},$$

for all $1 \le i \le n$. Thus, we get the length of the schedule generated by algorithm ES-A as

$$T_{\text{ES-}A} = A(\pi_1, r_1, \pi_2, r_2, \ldots, \pi_n, r_n) \left(\frac{W}{E}\right)^{1/(\alpha-1)}.$$

By theorem 15.1, the performance ratio of algorithm ES-A is

$$\begin{aligned}
\beta_{\text{ES-}A} &= \frac{T_{\text{ES-}A}}{T_{\text{OPT}}} \\
&\le \frac{A(\pi_1, r_1, \pi_2, r_2, \ldots, \pi_n, r_n)(W/E)^{1/(\alpha-1)}}{((m/E)(W/m)^\alpha)^{1/(\alpha-1)}} \\
&= \frac{mA(\pi_1, r_1, \pi_2, r_2, \ldots, \pi_n, r_n)}{\pi_1 r_1 + \pi_2 r_2 + \cdots + \pi_n r_n}.
\end{aligned}$$

This discussion is summarized in the following theorem.

Theorem 15.7 *By using an equal-speed algorithm* ES-A *to solve the problem of minimizing schedule length with energy consumption constraint, the schedule length is*

$$T_{\text{ES-}A} = A(\pi_1, r_1, \pi_2, r_2, \ldots, \pi_n, r_n) \left(\frac{W}{E}\right)^{1/(\alpha-1)}.$$

The performance ratio is $\beta_{\text{ES-}A} \le B_{\text{ES-}A}$, *where the performance bound is*

$$B_{\text{ES-}A} = \frac{mA(\pi_1, r_1, \pi_2, r_2, \ldots, \pi_n, r_n)}{\pi_1 r_1 + \pi_2 r_2 + \cdots + \pi_n r_n}.$$

For algorithm SIMPLE, we have

$$\overline{B}_{\text{ES-SIMPLE}} \approx \frac{m\mathbb{E}(\text{SIMPLE}(\pi_1, r_1, \pi_2, r_2, \ldots, \pi_n, r_n))}{\mathbb{E}(\pi_1 r_1 + \pi_2 r_2 + \cdots + \pi_n r_n)} \approx \frac{m}{P_m},$$

for large n.

15.3.4.2 Minimizing Energy Consumption

To solve the problem of minimizing energy consumption with schedule length constraint T by using an equal-speed algorithm ES-A, we need to provide enough energy $E_{\text{ES-}A}$, so that the deadline T is met, i.e., $T_{\text{ES-}A} = T$,

$$A(\pi_1, r_1, \pi_2, r_2, \ldots, \pi_n, r_n) \left(\frac{W}{E_{\text{ES-}A}}\right)^{1/(\alpha-1)} = T.$$

This equation implies that the amount of energy $E_{\text{ES-}A}$ consumed by algorithm ES-A is

$$E_{\text{ES-}A} = \left(\frac{A(\pi_1, r_1, \pi_2, r_2, \ldots, \pi_n, r_n)}{T}\right)^{\alpha-1} W.$$

By theorem 15.2, the performance ratio of algorithm ES-A is

$$\begin{aligned}
\gamma_{\text{ES-}A} &= \frac{E_{\text{ES-}A}}{E_{\text{OPT}}} \\
&\leq \frac{(A(\pi_1, r_1, \pi_2, r_2, \ldots, \pi_n, r_n)/T)^{\alpha-1} W}{(m/T^{\alpha-1})(W/m)^\alpha} \\
&= \frac{(A(\pi_1, r_1, \pi_2, r_2, \ldots, \pi_n, r_n))^{\alpha-1}}{(W/m)^{\alpha-1}} \\
&= \left(\frac{mA(\pi_1, r_1, \pi_2, r_2, \ldots, \pi_n, r_n)}{\pi_1 r_1 + \pi_2 r_2 + \cdots + \pi_n r_n}\right)^{\alpha-1}.
\end{aligned}$$

This discussion gives rise to the following theorem.

Theorem 15.8 *By using an equal-speed algorithm ES-A to solve the problem of minimizing energy consumption with schedule length constraint, the energy consumed is*

$$E_{\text{ES-}A} = \left(\frac{A(\pi_1, r_1, \pi_2, r_2, \ldots, \pi_n, r_n)}{T}\right)^{\alpha-1} W.$$

The performance ratio is $\gamma_{\text{ES-}A} \leq C_{\text{ES-}A}$, where the performance bound is

$$C_{\text{ES-}A} = \left(\frac{mA(\pi_1, r_1, \pi_2, r_2, \ldots, \pi_n, r_n)}{\pi_1 r_1 + \pi_2 r_2 + \cdots + \pi_n r_n}\right)^{\alpha-1}.$$

Since $C_{\text{ES-}A} = B_{\text{ES-}A}^{\alpha-1}$, we obtain the following approximation of the average-case performance-bound $\overline{C}_{\text{ES-SIMPLE}}$, i.e.,

$$\overline{C}_{\text{ES-SIMPLE}} = \mathbb{E}(B_{\text{ES-SIMPLE}}^{\alpha-1}) \approx \overline{B}_{\text{ES-SIMPLE}}^{\alpha-1} \approx \left(\frac{m}{P_m}\right)^{\alpha-1}.$$

15.3.5 Performance Data

In this section, we demonstrate numerical and simulation data for the average-case performance bounds derived for pre-power-determination algorithms in Theorems 15.3 through 15.8.

Assume that there are $n = 1000$ parallel tasks to be scheduled on a parallel computing system with $m = 128$ processors. The parameter α is set as 3. Although these parameters are for demonstration, they do not affect the observations and conclusion to be made.

The task execution requirements $r_1, r_2, \ldots, r_n$ are treated as i.i.d. continuous random variables uniformly distributed in $[0, 1)$. The task sizes $\pi_1, \pi_2, \ldots, \pi_n$ are i.i.d. discrete random variables in $[1 \ldots m]$.

We consider three types of probability distributions of task sizes with about the same expected task size $\bar{\pi}$. Let a_b be the probability that $\pi_i = b$, where $b \geq 1$.

- Uniform distributions in the range $[1 \ldots u]$, i.e., $a_b = 1/u$ for all $1 \leq b \leq u$, where u is chosen such that $(u + 1)/2 = \bar{\pi}$, i.e., $u = 2\bar{\pi} - 1$.
- Binomial distributions in the range $[1 \ldots m]$, i.e.,

$$a_b = \frac{\binom{m}{b} p^b (1 - p)^{m-b}}{1 - (1 - p)^m},$$

for all $1 \leq b \leq m$, where p is chosen such that $mp = \bar{\pi}$, i.e., $p = \bar{\pi}/m$. However, the actual expectation of task sizes is

$$\frac{\bar{\pi}}{1 - (1 - p)^m} = \frac{\bar{\pi}}{1 - (1 - \bar{\pi}/m)^m},$$

which is slightly greater than $\bar{\pi}$, especially when $\bar{\pi}$ is small.

- Geometric distributions in the range $[1 \ldots m]$, i.e.,

$$a_b = \frac{q(1 - q)^{b-1}}{1 - (1 - q)^m},$$

for all $1 \leq b \leq m$, where q is chosen such that $1/q = \bar{\pi}$, i.e., $q = 1/\bar{\pi}$. However, the actual expectation of task sizes is

$$\frac{1/q - (1/q + m)(1 - q)^m}{1 - (1 - q)^m} = \frac{\bar{\pi} - (\bar{\pi} + m)(1 - 1/\bar{\pi})^m}{1 - (1 - 1/\bar{\pi})^m},$$

which is less than $\bar{\pi}$, especially when $\bar{\pi}$ is large.

In Table 15.1, we show the average-case performance-bound $\overline{B}_{\text{ET-SIMPLE}}$. For each combination of the expected task size $\bar{\pi} = 10, 20, 30, 40, 50, 60$ and the three probability distributions of task sizes, we

TABLE 15.1 Simulation Data for $\overline{B}_{\text{ET-SIMPLE}}$

Average Task Size	Uniform Distribution	Binomial Distribution	Geometric Distribution
10	1.4925696	1.4799331	1.5272638
20	1.5727972	1.5363483	1.6507309
30	1.6701122	1.5875832	1.7432123
40	1.7975673	1.6175625	1.7904198
50	1.9119195	1.8022895	1.8189020
60	1.9069588	1.6583340	1.8336815

show $\overline{B}_{\text{ET-SIMPLE}}$ obtained by random sampling as follows. We generate 500 sets of n parallel tasks, find $B_{\text{ET-SIMPLE}}$ by using theorem 15.3, and report the average of the 500 values of $B_{\text{ET-SIMPLE}}$, which is the experimental value of $\overline{B}_{\text{ET-SIMPLE}}$. In Table 15.2, we show our approximation of the average-case performance-bound $\overline{B}_{\text{ET-SIMPLE}}$. These data are obtained by using the same method of Table 15.1. In Table 15.3, we show the average-case performance-bound $\overline{B}_{\text{ET-GREEDY}}$.

In Table 15.4, we show the average-case performance-bound $\overline{C}_{\text{ET-SIMPLE}}$. For each combination of the expected task size $\bar{\pi} = 10, 20, 30, 40, 50, 60$ and the three probability distributions of task sizes, we show $\overline{C}_{\text{ET-SIMPLE}}$ obtained by random sampling as follows. We generate 500 sets of n parallel tasks, find $C_{\text{ET-SIMPLE}}$ by using theorem 15.4, and report the average of the 500 values of $C_{\text{ET-SIMPLE}}$, which is the experimental value of $\overline{C}_{\text{ET-SIMPLE}}$. In Table 15.5, we show our approximation of the average-case performance-bound $\overline{C}_{\text{ET-SIMPLE}}$. These data are obtained by using the same method of Table 15.4. In Table 15.6, we show the average-case performance-bound $\overline{C}_{\text{ET-GREEDY}}$.

TABLE 15.2 Simulation Data for the Approximation of $\overline{B}_{\text{ET-SIMPLE}}$

Average Task Size	Uniform Distribution	Binomial Distribution	Geometric Distribution
10	1.4838727	1.4712699	1.5175525
20	1.5673887	1.5321343	1.6497503
30	1.6657088	1.5823904	1.7527788
40	1.7812809	1.6123531	1.8235302
50	1.9030034	1.8004366	1.8632664
60	1.9571228	1.6383452	1.8903620

TABLE 15.3 Simulation Data for $\overline{B}_{\text{ET-GREEDY}}$

Average Task Size	Uniform Distribution	Binomial Distribution	Geometric Distribution
10	1.4310459	1.4299945	1.4431398
20	1.4385093	1.4304722	1.4728624
30	1.4752343	1.4447528	1.5030419
40	1.4796543	1.4884609	1.5269353
50	1.6343792	1.7111951	1.5403827
60	1.6434003	1.5080826	1.5556964

TABLE 15.4 Simulation Data for $\overline{C}_{\text{ET-SIMPLE}}$

Average Task Size	Uniform Distribution	Binomial Distribution	Geometric Distribution
10	2.2253142	2.1877799	2.3346273
20	2.4741347	2.3623308	2.7328413
30	2.7900444	2.5184688	3.0317472
40	3.2155913	2.6227083	3.1991694
50	3.6596691	3.2529483	3.3139123
60	3.6380009	2.7527856	3.3663785

TABLE 15.5 Simulation Data for the Approximation of $\overline{C}_{\text{ET-SIMPLE}}$

Average Task Size	Uniform Distribution	Binomial Distribution	Geometric Distribution
10	2.2013749	2.1650992	2.3096532
20	2.4662319	2.3443439	2.7233384
30	2.7792308	2.5079857	3.0749527
40	3.1699565	2.5930330	3.3145158
50	3.6138148	3.2443710	3.4685996
60	3.8208861	2.6877475	3.5659465

In Tables 15.7 through 15.12, we duplicate the same work for algorithms EE-SIMPLE and EE-GREEDY. In Tables 15.13 through 15.18, we repeat the same work for algorithms ES-SIMPLE and ES-GREEDY. Notice that the data in Tables 15.14 and 15.17 are obtained by numerical calculation.

We would like to mention that the 99% confidence interval of all the data in the same table is no more than ±0.8%.

TABLE 15.6   Simulation Data for $\overline{C}_{\text{ET-GREEDY}}$

Average Task Size	Uniform Distribution	Binomial Distribution	Geometric Distribution
10	2.0460829	2.0405601	2.0883940
20	2.0758287	2.0537063	2.1842573
30	2.1838240	2.0850937	2.2613991
40	2.1899949	2.2190162	2.3271776
50	2.6855169	2.9297280	2.3798546
60	2.7006062	2.2793767	2.4128476

TABLE 15.7   Simulation Data for $\overline{B}_{\text{EE-SIMPLE}}$

Average Task Size	Uniform Distribution	Binomial Distribution	Geometric Distribution
10	1.3463462	1.2348196	1.6112312
20	1.4298048	1.2560974	1.7861327
30	1.5369151	1.2928216	1.8688076
40	1.7015489	1.3220516	1.8833816
50	1.8260972	1.4423315	1.8774630
60	1.7745509	1.3775116	1.8612660

TABLE 15.8   Simulation Data for the Approximation of $\overline{B}_{\text{EE-SIMPLE}}$

Average Task Size	Uniform Distribution	Binomial Distribution	Geometric Distribution
10	1.3264845	1.2174092	1.5773889
20	1.4117628	1.2450593	1.7214958
30	1.5017826	1.2792598	1.7990156
40	1.6086828	1.2981559	1.8269234
50	1.7165236	1.4474411	1.8372114
60	1.7664283	1.3150274	1.8392359

TABLE 15.9   Simulation Data for $\overline{B}_{\text{EE-GREEDY}}$

Average Task Size	Uniform Distribution	Binomial Distribution	Geometric Distribution
10	1.2918285	1.1932594	1.5244995
20	1.3051615	1.1673758	1.5757269
30	1.3561090	1.1688848	1.5902360
40	1.3677546	1.1954258	1.5857617
50	1.5496023	1.3769236	1.5722843
60	1.5302645	1.2082060	1.5644404

TABLE 15.10   Simulation Data for $\overline{C}_{\text{EE-SIMPLE}}$

Average Task Size	Uniform Distribution	Binomial Distribution	Geometric Distribution
10	1.8114682	1.5255576	2.5931999
20	2.0482069	1.5777715	3.1880987
30	2.3612833	1.6696594	3.5059091
40	2.9001191	1.7473951	3.5492390
50	3.3359420	2.0831180	3.5155132
60	3.1424992	1.9012417	3.4714851

TABLE 15.11 Simulation Data for the Approximation of $\overline{C}_{EE\text{-}SIMPLE}$

Average Task Size	Uniform Distribution	Binomial Distribution	Geometric Distribution
10	1.7574598	1.4814359	2.4787496
20	1.9903160	1.5500319	2.9644179
30	2.2544732	1.6362618	3.2282039
40	2.5863755	1.6841439	3.3426538
50	2.9486812	2.0959486	3.3784349
60	3.1174110	1.7281041	3.3864316

TABLE 15.12 Simulation Data for $\overline{C}_{EE\text{-}GREEDY}$

Average Task Size	Uniform Distribution	Binomial Distribution	Geometric Distribution
10	1.6706632	1.4231051	2.3331851
20	1.7052260	1.3624873	2.5014480
30	1.8417139	1.3675744	2.5252715
40	1.8759147	1.4285966	2.4987712
50	2.3967202	1.8971794	2.4801984
60	2.3425787	1.4592520	2.4528297

TABLE 15.13 Simulation Data for $\overline{B}_{ES\text{-}SIMPLE}$

Average Task Size	Uniform Distribution	Binomial Distribution	Geometric Distribution
10	1.0612181	1.0519759	1.0904810
20	1.1183020	1.0897683	1.1905421
30	1.1916076	1.1270139	1.2693422
40	1.2985623	1.1529977	1.3120221
50	1.3915928	1.2759972	1.3375258
60	1.3963486	1.1952081	1.3525453

TABLE 15.14 Numerical Data for the Approximation of $\overline{B}_{ES\text{-}SIMPLE}$

Average Task Size	Uniform Distribution	Binomial Distribution	Geometric Distribution
10	1.0491804	1.0403200	1.0756198
20	1.1099183	1.0836783	1.1677707
30	1.1784512	1.1202275	1.2417018
40	1.2610587	1.1400834	1.2889239
50	1.3448804	1.2739033	1.3179703
60	1.3831981	1.1585732	1.3363460

TABLE 15.15 Simulation Data for $\overline{B}_{ES\text{-}GREEDY}$

Average Task Size	Uniform Distribution	Binomial Distribution	Geometric Distribution
10	1.0164086	1.0145500	1.0246841
20	1.0200092	1.0138670	1.0470670
30	1.0472299	1.0226195	1.0700968
40	1.0506965	1.0535358	1.0857458
50	1.1677252	1.2109840	1.0971333
60	1.1684183	1.0682941	1.1036731

We have the following observations from our simulations.

- The performance of the three power allocation algorithms are ranked as ET, EE, ES, from the worst to the best.
- The task scheduling algorithm GREEDY performs noticeably better than algorithm SIMPLE.
- All our approximate performance bounds are very accurate.

TABLE 15.16 Simulation Data for $\overline{C}_{\text{ES-SIMPLE}}$

Average Task Size	Uniform Distribution	Binomial Distribution	Geometric Distribution
10	1.1262599	1.1070876	1.1897538
20	1.2499551	1.1877677	1.4151224
30	1.4202129	1.2702833	1.6086410
40	1.6843206	1.3293841	1.7232975
50	1.9345899	1.6277295	1.7879323
60	1.9475815	1.4293379	1.8290383

TABLE 15.17 Numerical Data for the Approximation of $\overline{C}_{\text{ES-SIMPLE}}$

Average Task Size	Uniform Distribution	Binomial Distribution	Geometric Distribution
10	1.1007795	1.0822657	1.1569579
20	1.2319186	1.1743586	1.3636884
30	1.3887472	1.2549096	1.5418234
40	1.5902689	1.2997902	1.6613248
50	1.8087032	1.6228296	1.7370457
60	1.9132369	1.3422920	1.7858207

TABLE 15.18 Simulation Data for $\overline{C}_{\text{ES-GREEDY}}$

Average Task Size	Uniform Distribution	Binomial Distribution	Geometric Distribution
10	1.0340945	1.0288788	1.0494240
20	1.0405856	1.0282521	1.0972321
30	1.0974053	1.0459606	1.1451099
40	1.1046457	1.1095928	1.1824923
50	1.3555731	1.4671864	1.2022531
60	1.3672342	1.1413527	1.2230523

Due to space limitation, we do not provide the performance data for algorithms SIMPLE* and GREEDY*. Our main finding is that the performance of SIMPLE* is slightly better than that of SIMPLE, while the performance of GREEDY* is about the same as that of GREEDY.

15.4 Post-Power-Determination Algorithms

15.4.1 Overview

In post-power-determination algorithms, we first partition the system and schedule the tasks, and then determine power supplies to the n parallel tasks. Unfortunately, if we apply algorithm SIMPLE or GREEDY to generate a schedule of the tasks, it is not clear at all how power allocation can be performed, since power supplies also determine (and change) task execution speeds and times, and such change affects the schedule already produced. What we need to make sure is that power supplies do not change the schedule. Therefore, new system partitioning and task scheduling methods are required.

Our strategies for solving the three subproblems are described as follows.

- *System partitioning*: We use the harmonic system partitioning and processor allocation scheme, which divides a multiprocessor computer into clusters of equal sizes and schedules tasks of similar sizes together to increase processor utilization.
- *Task scheduling*: Our approach is to divide a list of tasks into sublists such that each sublist contains tasks of similar sizes which are scheduled on clusters of equal sizes. Scheduling such parallel tasks on clusters is no more difficult than scheduling sequential tasks and can be performed by list scheduling algorithms.

- *Power supplying*: We adopt a three-level energy/time/power allocation scheme for a given schedule, namely, optimal energy/time allocation among sublists of tasks (theorems 15.11 and 15.14), optimal energy allocation among groups of tasks in the same sublist (theorems 15.10 and 15.13), and optimal power supplies to tasks in the same group (theorems 15.9 and 15.12).

The post-power-determination algorithms and their analysis presented in this section are developed in [26].

15.4.2 System Partitioning

Our post-power-determination algorithms are called H_c-A, where "H_c" stands for the harmonic system partitioning scheme with parameter c to be presented later, and A is a list scheduling algorithm to be presented in the next section.

To schedule a list of n parallel tasks, algorithm H_c-A divides the list into c sublists according to task sizes (i.e., numbers of processors requested by tasks), where $c \geq 1$ is a positive integer constant. For $1 \leq j \leq c - 1$, we define sublist j to be the sublist of tasks with

$$\frac{m}{j+1} < \pi_i \leq \frac{m}{j},$$

i.e., sublist j contains all tasks whose sizes are in the interval $I_j = (m/(j+1), m/j]$. We define sublist c to be the sublist of tasks with $0 < \pi_i \leq m/c$, i.e., sublist c contains all tasks whose sizes are in the interval $I_c = (0, m/c]$. The partition of $(0, m]$ into intervals $I_1, I_2, \ldots, I_j, \ldots, I_c$ is called *the harmonic system partitioning scheme* whose idea is to schedule tasks of similar sizes together. The similarity is defined by the intervals $I_1, I_2, \ldots, I_j, \ldots, I_c$. For tasks in sublist j, processor utilization is higher than $j/(j+1)$, where $1 \leq j \leq c - 1$. As j increases, the similarity among tasks in sublist j increases and processor utilization also increases. Hence, the harmonic system partitioning scheme is very good at handling small tasks.

Algorithm H_c-A produces schedules of the sublists sequentially and separately. To schedule tasks in sublist j, where $1 \leq j \leq c$, the m processors are partitioned into j *clusters* and each cluster contains m/j processors. Each cluster of processors is treated as one unit to be allocated to one task in sublist j. This is basically the harmonic system partitioning and processor allocation scheme. Therefore, scheduling parallel tasks in sublist j on the j clusters where each task i has processor requirement π_i and execution requirement r_i is equivalent to scheduling a list of sequential tasks on j processors where each task i has execution requirement r_i. It is clear that scheduling of the list of sequential tasks on j processors can be accomplished by using algorithm A, where A is a list scheduling algorithm.

15.4.3 Task Scheduling

When a multiprocessor computer with m processors is partitioned into $j \geq 1$ clusters, scheduling tasks in sublist j is essentially dividing sublist j into j groups of tasks, such that each group of tasks are executed on one cluster. Such a partition of sublist j into j groups is essentially a schedule of the tasks in sublist j on m processors with j clusters. Once a partition (i.e., a schedule) is determined, we can use the methods in the next section to find power supplies.

We propose to use the list scheduling algorithm and its variations to solve the task scheduling problem. Tasks in sublist j are scheduled on j clusters by using the classic list scheduling algorithm [15] and by ignoring the issue of power supplies. In other words, the task execution times are simply the task execution requirements $r_1, r_2, \ldots, r_n$, and tasks are assigned to the j clusters (i.e., groups) by using the list scheduling algorithm, which works as follows to schedule a list of tasks $1, 2, 3 \ldots$.

- *List scheduling* (*LS*): Initially, task k is scheduled on cluster (or group) k, where $1 \leq k \leq j$, and tasks $1, 2, \ldots, j$ are removed from the list. Upon the completion of a task k, the first unscheduled task in the list, i.e., task $j + 1$, is removed from the list and scheduled to be executed on cluster k. This process repeats until all tasks in the list are finished.

Algorithm LS has many variations, depending on the strategy used in the initial ordering of the tasks. We mention several of them here.

- *Largest requirement first (LRF)*: This algorithm is the same as the LS algorithm, except that the tasks are arranged such that $r_1 \geq r_2 \geq \cdots \geq r_n$.
- *Smallest requirement first (SRF)*: This algorithm is the same as the LS algorithm, except that the tasks are arranged such that $r_1 \leq r_2 \leq \cdots \leq r_n$.
- *Largest size first (LSF)*: This algorithm is the same as the LS algorithm, except that the tasks are arranged such that $\pi_1 \geq \pi_2 \geq \cdots \geq \pi_n$.
- *Smallest size first (SSF)*: This algorithm is the same as the LS algorithm, except that the tasks are arranged such that $\pi_1 \leq \pi_2 \leq \cdots \leq \pi_n$.
- *Largest task first (LTF)*: This algorithm is the same as the LS algorithm, except that the tasks are arranged such that $\pi_1^{1/\alpha} r_1 \geq \pi_2^{1/\alpha} r_2 \geq \cdots \geq \pi_n^{1/\alpha} r_n$.
- *Smallest task first (STF)*: This algorithm is the same as the LS algorithm, except that the tasks are arranged such that $\pi_1^{1/\alpha} r_1 \leq \pi_2^{1/\alpha} r_2 \leq \cdots \leq \pi_n^{1/\alpha} r_n$.

We call algorithm LS and its variations simply as list scheduling algorithms.

15.4.4 Power Supplying

Once the n parallel tasks are divided into c sublists and tasks in sublist j are further partitioned into j groups, power supplies which minimize the schedule length within energy consumption constraint or the energy consumption within schedule length constraint can be determined. We adopt a three-level energy/time/power allocation scheme for a given schedule, namely,

- Optimal power supplies to tasks in the same group (theorems 15.9 and 15.12)
- Optimal energy allocation among groups of tasks in the same sublist (theorems 15.10 and 15.13)
- Optimal energy/time allocation among sublists of tasks (theorems 15.11 and 15.14)

15.4.4.1 Minimizing Schedule Length

We first consider optimal power supplies to tasks in the same group. Notice that tasks in the same group are executed sequentially. In fact, we consider a more general case, i.e., n parallel tasks with sizes $\pi_1, \pi_2, \ldots, \pi_n$ and execution requirements $r_1, r_2, \ldots, r_n$ to be executed sequentially one by one. Let

$$M = \pi_1^{1/\alpha} r_1 + \pi_2^{1/\alpha} r_2 + \cdots + \pi_n^{1/\alpha} r_n.$$

The following result gives the optimal power supplies when the n parallel tasks are scheduled sequentially.

(*Note*: Due to space limitation, the proofs of all theorems in this section are omitted, and the interested reader is referred to [26].)

Theorem 15.9 *When n parallel tasks are scheduled sequentially, the schedule length is minimized when task i is supplied with power $p_i = (E/M)^{\alpha/(\alpha-1)}/\pi_i$, where $1 \leq i \leq n$. The optimal schedule length is $T = M^{\alpha/(\alpha-1)}/E^{1/(\alpha-1)}$.*

Now, we consider optimal energy allocation among groups of tasks in the same sublist. Again, we discuss group-level energy allocation in a more general case, i.e., scheduling n parallel tasks on m processors, where $\pi_i \leq m/j$ for all $1 \leq i \leq n$ with $j \geq 1$. In this case, the m processors can be partitioned into j clusters, such that each cluster contains m/j processors. Each cluster of processors are treated as one unit to be allocated to one task. Assume that the set of n tasks is partitioned into j groups, such that all the tasks in group k are executed on cluster k, where $1 \leq k \leq j$. Let M_k denote the total $\pi_i^{1/\alpha} r_i$ of the tasks in group k. For a given partition of the n tasks into j groups, we are seeking power supplies that minimize the schedule length. Let E_k be the energy consumed by all the tasks in group k. The following result characterizes the optimal power supplies.

Theorem 15.10 *For a given partition $M_1, M_2, \ldots, M_j$ of n parallel tasks into j groups on a multiprocessor computer partitioned into j clusters, the schedule length is minimized when task i in group k is supplied with power $p_i = (E_k/M_k)^{\alpha/(\alpha-1)}/\pi_i$, where*

$$E_k = \left(\frac{M_k^\alpha}{M_1^\alpha + M_2^\alpha + \cdots + M_j^\alpha} \right) E,$$

for all $1 \le k \le j$. The optimal schedule length is

$$T = \left(\frac{M_1^\alpha + M_2^\alpha + \cdots + M_j^\alpha}{E} \right)^{1/(\alpha-1)},$$

for the previously mentioned power supplies.

To use algorithm H_c-A to solve the problem of minimizing schedule length with energy consumption constraint E, we need to allocate the available energy E to the c sublists. We use $E_1, E_2, \ldots, E_c$ to represent an energy allocation to the c sublists, where sublist j consumes energy E_j, and $E_1 + E_2 + \cdots + E_c = E$. By using any of the list scheduling algorithms to schedule tasks in sublist j, we get a partition of the tasks in sublist j into j groups. Let R_j be the total execution requirement of tasks in sublist j, and $R_{j,k}$ be the total execution requirement of tasks in group k, and $M_{j,k}$ be the total $\pi_i^{1/\alpha} r_i$ of tasks in group k, where $1 \le k \le j$. Theorem 15.11 provides optimal energy allocation to the c sublists for minimizing schedule length with energy consumption constraint in scheduling parallel tasks by using scheduling algorithms H_c-A, where A is a list scheduling algorithm. The theorem also gives the performance bound when algorithms H_c-A are used to solve the problem of minimizing schedule length with energy consumption constraint.

Theorem 15.11 *For a given partition $M_{j,1}, M_{j,2}, \ldots, M_{j,j}$ of the tasks in sublist j into j groups produced by a list scheduling algorithm A, where $1 \le j \le c$, and an energy allocation $E_1, E_2, \ldots, E_c$ to the c sublists, the length of the schedule produced by algorithm H_c-A is*

$$T_{H_c\text{-}A} = \sum_{j=1}^c \left(\frac{M_{j,1}^\alpha + M_{j,2}^\alpha + \cdots + M_{j,j}^\alpha}{E_j} \right)^{1/(\alpha-1)}.$$

The energy allocation $E_1, E_2, \ldots, E_c$ which minimizes $T_{H_c\text{-}A}$ is

$$E_j = \left(\frac{N_j^{1/\alpha}}{N_1^{1/\alpha} + N_2^{1/\alpha} + \cdots + N_c^{1/\alpha}} \right) E,$$

where $N_j = M_{j,1}^\alpha + M_{j,2}^\alpha + \cdots + M_{j,j}^\alpha$, for all $1 \le j \le c$, and the minimized schedule length is

$$T_{H_c\text{-}A} = \frac{\left(N_1^{1/\alpha} + N_2^{1/\alpha} + \cdots + N_c^{1/\alpha} \right)^{\alpha/(\alpha-1)}}{E^{1/(\alpha-1)}},$$

by using the previous energy allocation. The performance ratio is $\beta_{H_c\text{-}A} \leq B_{H_c\text{-}A}$, *where*

$$B_{H_c\text{-}A} = \left(\frac{\left(N_1^{1/\alpha} + N_2^{1/\alpha} + \cdots + N_c^{1/\alpha} \right)^{\alpha}}{m(W/m)^{\alpha}} \right)^{1/(\alpha-1)}.$$

Furthermore, we have

$$B_{H_c\text{-}A} \approx \left(\left(\sum_{j=1}^{c} \frac{R_j}{j} \left(\frac{2j+1}{2j+2} \right)^{1/\alpha} \right) / \left(\frac{W}{m} \right) \right)^{\alpha/(\alpha-1)}.$$

Theorems 15.10 and 15.11 give the power supply to the task i in group k of sublist j as

$$\frac{1}{\pi_i} \left(\frac{E_{j,k}}{M_{j,k}} \right)^{\alpha/(\alpha-1)} = \frac{1}{\pi_i}$$

$$\left(\left(\frac{M_{j,k}^{\alpha}}{M_{j,1}^{\alpha} + M_{j,2}^{\alpha} + \cdots + M_{j,j}^{\alpha}} \right) \left(\frac{N_j^{1/\alpha}}{N_1^{1/\alpha} + N_2^{1/\alpha} + \cdots + N_c^{1/\alpha}} \right) \frac{E}{M_{j,k}} \right)^{\alpha/(\alpha-1)},$$

for all $1 \leq j \leq c$ and $1 \leq k \leq j$.

15.4.4.2 Minimizing Energy Consumption

The following result gives the optimal power supplies when n parallel tasks are scheduled sequentially.

Theorem 15.12 *When n parallel tasks are scheduled sequentially, the total energy consumption is minimized when task i is supplied with power $p_i = (M/T)^{\alpha}/\pi_i$, where $1 \leq i \leq n$. The minimum energy consumption is $E = M^{\alpha}/T^{\alpha-1}$.*

The following result gives the optimal power supplies that minimize energy consumption for a given partition of n tasks into j groups on a multiprocessor computer.

Theorem 15.13 *For a given partition $M_1, M_2, \ldots, M_j$ of n parallel tasks into j groups on a multiprocessor computer partitioned into j clusters, the total energy consumption is minimized when task i in group k is executed with power $p_i = (M_k/T)^{\alpha}/\pi_i$, where $1 \leq k \leq j$. The minimum energy consumption is*

$$E = \frac{M_1^{\alpha} + M_2^{\alpha} + \cdots + M_j^{\alpha}}{T^{\alpha-1}}$$

for the previously mentioned power supplies.

To use algorithm $H_c\text{-}A$ to solve the problem of minimizing energy consumption with schedule length constraint T, we need to allocate the time T to the c sublists. We use $T_1, T_2, \ldots, T_c$ to represent a time allocation to the c sublists, where tasks in sublist j are executed within deadline T_j, and $T_1 + T_2 + \cdots + T_c = T$. Theorem 15.14 provides optimal time allocation to the c sublists for minimizing energy consumption with schedule length constraint in scheduling parallel tasks by using scheduling algorithms $H_c\text{-}A$, where A is a list scheduling algorithm. The theorem also gives the performance bound when algorithms $H_c\text{-}A$ is used to solve the problem of minimizing energy consumption with schedule length constraint.

Theorem 15.14 *For a given partition $M_{j,1}, M_{j,2}, \ldots, M_{j,j}$ of the tasks in sublist j into j groups produced by a list scheduling algorithm A, where $1 \leq j \leq c$, and a time allocation $T_1, T_2, \ldots, T_c$ to the c sublists, the amount of energy consumed by algorithm H_c-A is*

$$E_{H_c\text{-}A} = \sum_{j=1}^{c} \left(\frac{M_{j,1}^\alpha + M_{j,2}^\alpha + \cdots + M_{j,j}^\alpha}{T_j^{\alpha-1}} \right).$$

The time allocation $T_1, T_2, \ldots, T_c$ which minimizes $E_{H_c\text{-}A}$ is

$$T_j = \left(\frac{N_j^{1/\alpha}}{N_1^{1/\alpha} + N_2^{1/\alpha} + \cdots + N_c^{1/\alpha}} \right) T,$$

where $N_j = M_{j,1}^\alpha + M_{j,2}^\alpha + \cdots + M_{j,j}^\alpha$, for all $1 \leq j \leq c$, and the minimized energy consumption is

$$E_{H_c\text{-}A} = \frac{\left(N_1^{1/\alpha} + N_2^{1/\alpha} + \cdots + N_c^{1/\alpha} \right)^\alpha}{T^{\alpha-1}},$$

by using the previous time allocation. The performance ratio is $\gamma_{H_c\text{-}A} \leq C_{H_c\text{-}A}$, where

$$C_{H_c\text{-}A} = \frac{\left(N_1^{1/\alpha} + N_2^{1/\alpha} + \cdots + N_c^{1/\alpha} \right)^\alpha}{m(W/m)^\alpha}.$$

Furthermore, we have

$$C_{H_c\text{-}A} \approx \left(\left(\sum_{j=1}^{c} \frac{R_j}{j} \left(\frac{2j+1}{2j+2} \right)^{1/\alpha} \right) \Big/ \left(\frac{W}{m} \right) \right)^\alpha.$$

Theorems 15.13 and 15.14 give the power supply to task i in group k of sublist j as

$$\frac{1}{\pi_i} \left(\frac{M_{j,k}}{T_j} \right)^\alpha = \frac{1}{\pi_i} \left(\frac{M_{j,k} \left(N_1^{1/\alpha} + N_2^{1/\alpha} + \cdots + N_c^{1/\alpha} \right)}{N_j^{1/\alpha} T} \right)^\alpha,$$

for all $1 \leq j \leq c$ and $1 \leq k \leq j$.

15.4.5 Performance Data

In this section, we demonstrate numerical and simulation data for the average-case performance bounds derived for pre-power-determination algorithms in theorems 15.11 and 15.14.

All parameters are set in the same way as pre-power-determination algorithms.

In Table 15.19, we show the average-case performance-bound $\overline{B}_{H_c\text{-}LS}$. For each combination of the expected task size $\bar{\pi} = 10, 20, 30, 40, 50, 60$ and the three probability distributions of task sizes, we show $\overline{B}_{H_c\text{-}LS}$ obtained by random sampling as follows. We generate 200 sets of n parallel tasks, find $B_{H_c\text{-}LS}$ by using theorem 15.11, and report the average of the 200 values of $B_{H_c\text{-}LS}$, which is the experimental value of $\overline{B}_{H_c\text{-}LS}$. In Table 15.20, we show our approximation of the average-case performance-bound $\overline{B}_{H_c\text{-}LS}$. These data are obtained by using the same method of Table 15.19.

In Table 15.21, we show the average-case performance-bound $\overline{C}_{H_c\text{-}LS}$. For each combination of the expected task size $\bar{\pi} = 10, 20, 30, 40, 50, 60$ and the three probability distributions of task sizes, we show $\overline{C}_{H_c\text{-}LS}$ obtained by random sampling as follows. We generate 200 sets of n parallel tasks, find $C_{H_c\text{-}LS}$ by

TABLE 15.19   Simulation Data for $\overline{B}_{H_c\text{-LS}}$

Average Task Size	Uniform Distribution	Binomial Distribution	Geometric Distribution
10	1.1310152	1.0710001	1.2195685
20	1.1374080	1.0998380	1.2254590
30	1.1921506	1.1376307	1.2639101
40	1.3722257	1.2113868	1.2818346
50	1.3957714	1.2486290	1.2907563
60	1.3266285	1.2804629	1.2935460

TABLE 15.20   Simulation Data for the Approximation of $\overline{B}_{H_c\text{-LS}}$

Average Task Size	Uniform Distribution	Binomial Distribution	Geometric Distribution
10	1.1850743	1.0637377	1.3192655
20	1.1495195	1.0823137	1.2730873
30	1.2030697	1.1397329	1.3026445
40	1.4479539	1.2382998	1.3166660
50	1.4619769	1.2776203	1.3207317
60	1.3441788	1.3188035	1.3190466

TABLE 15.21   Simulation Data for $\overline{C}_{H_c\text{-LS}}$

Average Task Size	Uniform Distribution	Binomial Distribution	Geometric Distribution
10	1.2775988	1.1481701	1.4868035
20	1.2942669	1.2084384	1.5021708
30	1.4215784	1.2940830	1.5991783
40	1.8829298	1.4667344	1.6447990
50	1.9496348	1.5596231	1.6669842
60	1.7650820	1.6392893	1.6759810

TABLE 15.22   Simulation Data for the Approximation of $\overline{C}_{H_c\text{-LS}}$

Average Task Size	Uniform Distribution	Binomial Distribution	Geometric Distribution
10	1.4023173	1.1313445	1.7373603
20	1.3197428	1.1711625	1.6161178
30	1.4463168	1.2997302	1.7008003
40	2.0986717	1.5306295	1.7312586
50	2.1356210	1.6311186	1.7435018
60	1.8069105	1.7392907	1.7393141

using theorem 15.14, and report the average of the 200 values of $C_{H_c\text{-LS}}$, which is the experimental value of $\overline{C}_{H_c\text{-LS}}$. In Table 15.22, we show our approximation of the average-case performance-bound $\overline{C}_{H_c\text{-LS}}$. These data are obtained by using the same method of Table 15.21.

We would like to mention that the 99% confidence interval of all the data in the same table is no more than $\pm 0.6\%$.

We have the following observations from our simulations.

- $\overline{B}_{H_c\text{-LS}}$ is less than 1.4 and $\overline{C}_{H_c\text{-LS}}$ is less than 1.95. Since $\overline{B}_{H_c\text{-LS}}$ and $\overline{C}_{H_c\text{-LS}}$ only give upper bonds for the average-case performance ratios, the performance of our heuristic algorithms is even better and our heuristic algorithms are able to produce solutions very close to optimum.
- The performance of algorithm $H_c\text{-}A$ for A other than LS (i.e., LRF, SRF, LSF, SSF, LTF, STF) is very close (within $\pm 1\%$) to the performance of algorithm $H_c\text{-LS}$. Since these data do not provide further insight, they are not shown here.
- Our approximate performance bounds are very accurate.

It is interesting to compare the performance of pre-power-determination algorithms and post-power-determination algorithms. We find that algorithm H_c-LS performs better than ET-SIMPLE, ET-GREEDY, EE-SIMPLE, EE-GREEDY, and is comparable with ES-SIMPLE; however, it performs worse than ES-GREEDY. Although post-power-determination provides an optimal power allocation for a given schedule, the performance of H_c-LS is still worse than ES-GREEDY due to the inefficiency of the harmonic system partition scheme.

15.5 Summary

We have developed a number of pre-power-determination algorithms and post-power-determination algorithms for energy-aware scheduling of parallel tasks. Our best heuristic algorithms (ES-GREEDY and ES-GREEDY*) are able to produce solutions very close to optimum.

It is of great interest to extend our algorithms to other task models, processor models, and scheduling models described in [27].

Acknowledgment

The material presented in this chapter is based in part upon the author's work in [26].

References

1. S. Albers, Energy-efficient algorithms, *Communications of the ACM*, 53(5), 86–96, 2010.
2. H. Aydin, R. Melhem, D. Mossé, and P. Mejía-Alvarez, Power-aware scheduling for periodic real-time tasks, *IEEE Transactions on Computers*, 53(5), 584–600, 2004.
3. N. Bansal, T. Kimbrel, and K. Pruhs, Dynamic speed scaling to manage energy and temperature, *Proceedings of the 45th IEEE Symposium on Foundation of Computer Science*, Rome, Italy, pp. 520–529, 2004.
4. J. A. Barnett, Dynamic task-level voltage scheduling optimizations, *IEEE Transactions on Computers*, 54(5), 508–520, 2005.
5. L. Benini, A. Bogliolo, and G. De Micheli, A survey of design techniques for system-level dynamic power management, *IEEE Transactions on Very Large Scale Integration (VLSI) Systems*, 8(3), 299–316, 2000.
6. D. P. Bunde, Power-aware scheduling for makespan and flow, *Proceedings of the 18th ACM Symposium on Parallelism in Algorithms and Architectures*, Cambridge, MA, pp. 190–196, 2006.
7. H.-L. Chan, W.-T. Chan, T.-W. Lam, L.-K. Lee, K.-S. Mak, and P. W. H. Wong, Energy efficient online deadline scheduling, *Proceedings of the 18th ACM-SIAM Symposium on Discrete Algorithms*, New Orleans, LA, pp. 795–804, 2007.
8. A. P. Chandrakasan, S. Sheng, and R. W. Brodersen, Low-power CMOS digital design, *IEEE Journal on Solid-State Circuits*, 27(4), 473–484, 1992.
9. S. Cho and R. G. Melhem, On the interplay of parallelization, program performance, and energy consumption, *IEEE Transactions on Parallel and Distributed Systems*, 21(3), 342–353, 2010.
10. E. G. Coffman, Jr., M. R. Garey, and D. S. Johnson, Approximation algorithms for bin packing: A survey, in *Approximation Algorithms for NP-Hard Problems*, D. S. Hochbaum, ed., PWS Publishing Company, Boston, MA, 1997.
11. W.-C. Feng, The importance of being low power in high performance computing, *Cyber Infrastructure Watch Quarterly*, 1(3), Los Alamos National Laboratory, August 2005.
12. W.-C. Feng and K. W. Cameron, The Green500 list: Encouraging sustainable supercomputing, *Computer*, 40(12), 50–55, 2007.

13. A. Gara, et al., Overview of the Blue Gene/L system architecture, *IBM Journal of Research and Development*, 49(2/3), 195–212, 2005.

14. M. R. Garey and D. S. Johnson, *Computers and Intractability—A Guide to the Theory of NP-Completeness*, W. H. Freeman, New York, 1979.

15. R. L. Graham, Bounds on multiprocessing timing anomalies, *SIAM Journal of Applied Mathematics*, 2, 416–429, 1969.

16. S. L. Graham, M. Snir, and C. A. Patterson, eds., *Getting up to Speed: The Future of Supercomputing*, Committee on the Future of Supercomputing, National Research Council, National Academies Press, Washington, DC, 2005.

17. I. Hong, D. Kirovski, G. Qu, M. Potkonjak, and M. B. Srivastava, Power optimization of variable-voltage core-based systems, *IEEE Transactions on Computer-Aided Design of Integrated Circuits and Systems*, 18(12), 1702–1714, 1999.

18. C. Im, S. Ha, and H. Kim, Dynamic voltage scheduling with buffers in low-power multimedia applications, *ACM Transactions on Embedded Computing Systems*, 3(4), 686–705, 2004.

19. S. U. Khan and I. Ahmad, A cooperative game theoretical technique for joint optimization of energy consumption and response time in computational grids, *IEEE Transactions on Parallel and Distributed Systems*, 20(3), 346–360, 2009.

20. C. M. Krishna and Y.-H. Lee, Voltage-clock-scaling adaptive scheduling techniques for low power in hard real-time systems, *IEEE Transactions on Computers*, 52(12), 1586–1593, 2003.

21. W.-C. Kwon and T. Kim, Optimal voltage allocation techniques for dynamically variable voltage processors, *ACM Transactions on Embedded Computing Systems*, 4(1), 211–230, 2005.

22. Y. C. Lee and A. Y. Zomaya, Energy conscious scheduling for distributed computing systems under different operating conditions, *IEEE Transactions on Parallel and Distributed Systems*, 22(8), 1374–1381, 2011.

23. Y.-H. Lee and C. M. Krishna, Voltage-clock scaling for low energy consumption in fixed-priority real-time systems, *Real-Time Systems*, 24(3), 303–317, 2003.

24. K. Li, An average-case analysis of online non-clairvoyant scheduling of independent parallel tasks, *Journal of Parallel and Distributed Computing*, 66(5), 617–625, 2006.

25. K. Li, Performance analysis of power-aware task scheduling algorithms on multiprocessor computers with dynamic voltage and speed, *IEEE Transactions on Parallel and Distributed Systems*, 19(11), 1484–1497, 2008.

26. K. Li, Energy efficient scheduling of parallel tasks on multiprocessor computers, *Journal of Supercomputing*, DOI: 10.1007/S11227-010-0416-0, published online First on March 12, 2010. http://www.springerlink.com/content/v5242235k151h5v3/

27. K. Li, Power allocation and task scheduling on multiprocessor computers with energy and time constraints, *Energy Aware Distributed Computing Systems*, Y.-C. Lee and A. Zomaya, eds., Wiley Series on Parallel and Distributed Computing, 2011.

28. M. Li, B. J. Liu, and F. F. Yao, Min-energy voltage allocation for tree-structured tasks, *Journal of Combinatorial Optimization*, 11, 305–319, 2006.

29. M. Li, A. C. Yao, and F. F. Yao, Discrete and continuous min-energy schedules for variable voltage processors, *Proceedings of the National Academy of Sciences USA*, 103(11), pp. 3983–3987, 2006.

30. M. Li and F. F. Yao, An efficient algorithm for computing optimal discrete voltage schedules, *SIAM Journal on Computing*, 35(3), 658–671, 2006.

31. J. R. Lorch and A. J. Smith, PACE: A new approach to dynamic voltage scaling, *IEEE Transactions on Computers*, 53(7), 856–869, 2004.

32. R. N. Mahapatra and W. Zhao, An energy-efficient slack distribution technique for multimode distributed real-time embedded systems, *IEEE Transactions on Parallel and Distributed Systems*, 16(7), 650–662, 2005.

33. A. J. Martin, Towards an energy complexity of computation, *Information Processing Letters*, 77, 181–187, 2001.

34. G. Quan and X. S. Hu, Energy efficient DVS schedule for fixed-priority real-time systems, *ACM Transactions on Embedded Computing Systems*, 6(4), Article no. 29, 2007.
35. C. Rusu, R. Melhem, and D. Mossé, Maximizing the system value while satisfying time and energy constraints, *Proceedings of the 23rd IEEE Real-Time Systems Symposium*, Austin, TX, pp. 256–265, 2002.
36. D. Shin and J. Kim, Power-aware scheduling of conditional task graphs in real-time multiprocessor systems, *Proceedings of the International Symposium on Low Power Electronics and Design*, Fukuoka, Japan, pp. 408–413, 2003.
37. D. Shin, J. Kim, and S. Lee, Intra-task voltage scheduling for low-energy hard real-time applications, *IEEE Design & Test of Computers*, 18(2), 20–30, 2001.
38. M. R. Stan and K. Skadron, Guest editors' introduction: Power-aware computing, *IEEE Computer*, 36(12) 35–38, 2003.
39. O. S. Unsal and I. Koren, System-level power-aware design techniques in real-time systems, *Proceedings of the IEEE*, 91(7), 1055–1069, 2003.
40. V. Venkatachalam and M. Franz, Power reduction techniques for microprocessor systems, *ACM Computing Surveys*, 37(3), 195–237, 2005.
41. M. Weiser, B. Welch, A. Demers, and S. Shenker, Scheduling for reduced CPU energy, *Proceedings of the First USENIX Symposium on Operating Systems Design and Implementation*, San Diego, CA, pp. 13–23, 1994.
42. P. Yang, C. Wong, P. Marchal, F. Catthoor, D. Desmet, D. Verkest, and R. Lauwereins, Energy-aware runtime scheduling for embedded-multiprocessor SOCs, *IEEE Design & Test of Computers*, 18(5), 46–58, 2001.
43. F. Yao, A. Demers, and S. Shenker, A scheduling model for reduced CPU energy, *Proceedings of the 36th IEEE Symposium on Foundations of Computer Science*, Washington DC, pp. 374–382, 1995.
44. H.-S. Yun and J. Kim, On energy-optimal voltage scheduling for fixed-priority hard real-time systems, *ACM Transactions on Embedded Computing Systems*, 2(3), 393–430, 2003.
45. B. Zhai, D. Blaauw, D. Sylvester, and K. Flautner, Theoretical and practical limits of dynamic voltage scaling, *Proceedings of the 41st Design Automation Conference*, San Diego, CA, pp. 868–873, 2004.
46. X. Zhong and C.-Z. Xu, Energy-aware modeling and scheduling for dynamic voltage scaling with statistical real-time guarantee, *IEEE Transactions on Computers*, 56(3), 358–372, 2007.
47. D. Zhu, R. Melhem, and B. R. Childers, Scheduling with dynamic voltage/speed adjustment using slack reclamation in multiprocessor real-time systems, *IEEE Transactions on Parallel and Distributed Systems*, 14(7), 686–700, 2003.
48. D. Zhu, D. Mossé, and R. Melhem, Power-aware scheduling for AND/OR graphs in real-time systems, *IEEE Transactions on Parallel and Distributed Systems*, 15(9), 849–864, 2004.
49. J. Zhuo and C. Chakrabarti, Energy-efficient dynamic task scheduling algorithms for DVS systems, *ACM Transactions on Embedded Computing Systems*, 7(2), Article no. 17, 2008.

16

Power Saving by Task-Level Dynamic Voltage Scaling: A Theoretical Aspect

Taewhan Kim
Seoul National University

16.1 Introduction

During the last two decades, there have been enormous efforts in minimizing the energy consumption of CMOS circuit systems. Dynamic voltage scaling (DVS), involving dynamic adjustments of the supply voltage and the corresponding operating clock frequency, has emerged as one of the most effective energy minimization techniques. The one-to-one correspondence between the supply voltage and the clock frequency in CMOS circuits imposes an inherent constraint on DVS techniques to ensure that voltage adjustments do not violate the target system's timing (or deadline) constraint.

Many previous works have focused on hard real-time systems with multiple tasks. Their primary concern is to assign a proper operating voltage to each task while satisfying the task's timing constraint. In these techniques, determination of the voltage is carried out on a task-by-task basis and the voltage assigned to the task is unchanged during the entire execution of the task, which is referred to as *intertask voltage scheduling*. Yao et al. [35] proposed an optimal intertask voltage scheduling algorithm for independent tasks in which each task is characterized by its arrival time, deadline, and required CPU cycles. (In fact, many researchers have used Yao's algorithm as a basis of their DVS algorithms. More on Yao's algorithm is given in Section 16.3.) The proposed scheduling technique computes the speed of execution at any given time (and thus automatically determines each task's starting and ending times) so that the total energy consumption is minimized. Although they formulated the problem without the constraint that each task should be executed using a single operating voltage, by the convexity of the power function each task is given only one "middle" voltage that is proved to be optimal. This is because of the underlying assumption that the speed of any specific time is *constant*, which is not true since the required number of processor cycles, on which the calculation of the speed depends, may vary depending on the behavior of the task (for example, encountering conditional branch during execution). That means, the proposed intertask voltage scheduling technique is optimal only if each task execution follows the worst-case execution path. Leaving the same assumption untouched, many works in the literature have tried to formulate new intertask voltage scheduling problems considering other issues. Some instances of such issues include tasks with dependency relations [1,7,24,33,37], discretely variable voltage processors [7,16], multi-processor environments [1,7,24,33,37], and voltage transition overheads [1].

It has been reported [30] that the voltage transition overheads are not trivial in terms of energy and delay, and should be taken into account in DVS as well. In addition, DVS requires additional circuitry of DC–DC converter to adjust the supply voltage dynamically. Consequently, the energy efficiency of the DC–DC converter in voltage scaling also needs to be addressed [5]. Furthermore, the focus by DVS researchers has been moved from minimizing switching (dynamic) energy to minimizing both dynamic and leakage energy by maximally exploiting sleep state of the system. The fundamental difference between the DVS systems with and without sleep state will be described in Section 16.5.

On the other hand, a number of studies in other direction (e.g., [25,27,29]) have added a new dimension to the voltage scheduling problem, by considering energy-saving opportunities within the task boundaries. In their approaches, the operating voltage of the task is dynamically adjusted according to the execution behavior to accurately reflect the changes in the required number of cycles to finish the task before the deadline, which is referred to as *intratask voltage scheduling*. Shin et al. [29] proposed a remaining worst-case path-based algorithm which achieves the best granularity by executing the basic blocks with possibly different operating voltages. To obtain tight operating points that lead to a minimum energy consumption, the algorithm updates the remaining path length as soon as the execution deviates from the previous remaining worst-case path. More recently, a profile-based optimal intratask voltage scheduling technique was introduced in [25,27]. It shows the best energy savings by incorporating the task's execution profile into the calculation of the operating voltages. This algorithm is proven to be optimal in that when the task is executed repeatedly or periodically, it achieves a minimum average energy consumption. A recent work [26] attempts to solve the intertask and intratask DVS problems simultaneously so as not to miss the energy-saving opportunity at each granularity of intertask and intratask DVS.

On the other side, as the power density on a processor increases very rapidly, managing or controlling thermal profiles becomes another hot issue in DVS. Thus, in addition to minimizing energy consumption by DVS, the peak temperature caused by the task execution should be controlled via the execution scheduling of tasks and DVS to be applied to the task. The problems addressed are divided into a number of cases: DVS for minimizing total execution time under peak temperature constraint [36]; task scheduling and DVS for minimizing peak temperature under deadline constraint [10]; distributing idle times between the executions of tasks with DVS for minimizing dynamic and leakage energy including the temperature-induced leakage under deadline constraint [2]; thermal-constrained energy optimization using DVS and energy-constrained thermal optimization using DVS under deadline constraint in multiprocessor systems [18].

In this chapter, we survey and describe, in a theoretical aspect, state-of-the-art techniques of DVS problems, which include (1) intertask DVS problem, (2) intratask DVS problem, (3) integrated intertask and intratask DVS problems, (4) DVS problem with sleep state, (5) DC–DC converter-aware DVS problem, (6) transition-aware DVS problem, and (7) thermal-aware DVS problem. (The preliminary version of this work can be found in [14,15].) The scope of this survey is confined to single processor DVS techniques. The objective to be achieved by DVS for all topics except (4) and (7) is to minimize the amount of dynamic energy consumption whereas the objective for topics (4) and (7) is to minimize the total amount of dynamic and leakage energy.

16.2 Intratask DVS

The amount of energy dissipation for the execution of a task is

$$E \propto V_{DD}^{2} \times N_{tot} \tag{16.1}$$

where N_{tot} is the total number of instruction cycles executed for a task. Thus, the *intratask voltage scheduling* problem is to assign a proper voltage to each basic block of the task so that the energy consumption in Equation 16.1 is minimized.

The relationship between clock frequency and voltage in CMOS circuits is

$$\frac{f_{CLK} \propto (V_{DD} - V_T)^{\alpha}}{V_{DD}} \tag{16.2}$$

where

V_T is the threshold voltage

α is the velocity saturation index

If the value of V_T is small enough, the expression is reduced to $f_{CLK} \propto V_{DD}^{\alpha-1}$.

Since the clock frequency determines the voltage, the scheduling problem can be stated as:

(Intratask DVS Problem) *The intratask voltage scheduling problem for a task's CFG (control flow graph) is to determine the clock frequency for each node (i.e., basic block) of the CFG so that the total energy consumed by the task is minimized while satisfying the timing constraint of the task.*

The key problem to solve is to determine what clock frequency should be set to the entry point of each basic block so that the overall energy consumption of the task is minimized. Existing intratask DVS techniques can be classified according to the way of determining the lowest clock frequency to be set at the entry point of each basic block.

1. (*RWCEP-based DVS*): This technique [29] uses the lowest clock frequency by which the remaining WCEP (worst case execution path) can be completed within the deadline of the task.
2. (*RACEP-based DVS*): This technique [28] uses the lowest clock frequency by which the remaining ACEP (average case execution path) can be completed within the deadline of the task.
3. (*ROCEP-based DVS*): This technique [27] uses the lowest clock frequency by which the remaining OCEP (energy-optimal case execution path) can be completed within the deadline of the task.

For example, consider a simple hard real-time task with deadline of 100 ms and three basic blocks. Its control flow graph is shown in Figure 16.1a where the number inside each node indicates the number of execution cycles of the block and the number assigned to each arc indicates the probability that the control flow follows the edge. If DVS technique is not used, the speed for the task should be set tightly to (length of the critical path)/(remaining time to deadline) = $[(2 + 8)10^7$ cycles$]/(100 \cdot 10^{-3}$s$) = 1$ GHz (see Figure 16.1a). If DVS technique follows the worst-case execution path, the speed of block_2 will be set to (length of the (RWCEP from block_2)/(remaining time to deadline) $= [1 \cdot 10^7$ cycles$]/(80 \cdot 10^{-3}$ s$) = 125$ MHz) (see Figure 16.1b). On the other hand, if DVS technique follows the average case execution path, the speed is set to (length of the (RACEP from block_0)/(remaining time to deadline) $= [(2+1) \cdot 10^7$ cycles$]/(100 \cdot 10^{-3}$ s$) = 300$ MHz) (see Figure 16.1c). Finally, if DVS follows energy-optimal execution path, the speed to be set in block_0 is calculated based on the probabilities of its succeeding basic blocks. Here, the speed is 573 MHz (see Figure 16.1d). In summary, we can see that the ROCEP-based DVS technique outperforms the others because the clock

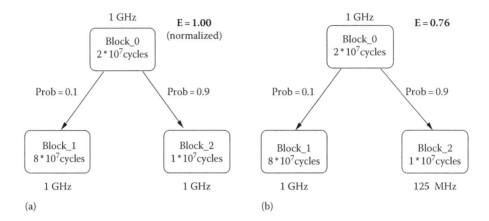

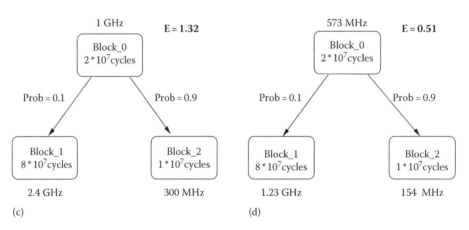

FIGURE 16.1 Example showing the calculations of clock frequency in basic block_0. (a) No DVS, (b) RWCEP-based DVS, (c) RACEP-based DVS, and (d) ROCEP-based DVS. (Taken from Seo, J. et al., *IEEE Trans. Comput. Aided Des. Integrated Circ. Syst.*, 25(1), 47, 2006.)

speed used at each basic block always leads to the total energy consumption which is optimal on the average.

The work in [27] contains detailed procedure of energy-optimal speed calculation of basic blocks. Here, we give a summary of the speed calculation. A task τ is represented with its CFG $G_\tau = (V, A)$, where V is the set of basic blocks in the task and A is the set of directed edges which impose precedence relations between basic blocks. (For example, see Figure 16.2a.) The set of immediate successor basic blocks of any $b_i \in V$ is denoted by $succ(b_i)$. Each basic block b_i is annotated with its non-zero number of execution cycles n_i and each arc (b_i, b_j) is given a probability p_j that the execution follows the arc.

Given a task's CFG and its execution profile that offers the probabilities, we execute each basic block b_i at a speed of $\delta_i/(remaining\ time)$ and adjust the supply voltage accordingly, where δ_i is defined as:

$$\delta_i = \begin{cases} n_i, & \text{if } succ(b_i) = \emptyset \\ n_i + \sqrt[3]{\sum_{\forall b_j \in succ(b_i)} p_j \delta_j^3}, & \text{otherwise} \end{cases} \tag{16.3}$$

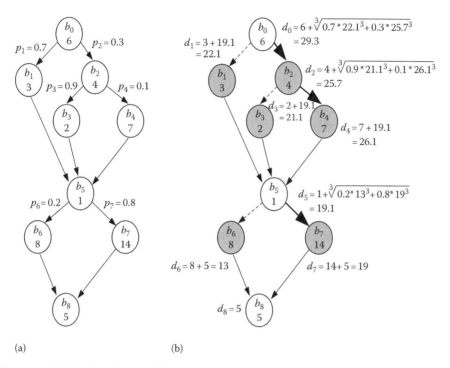

FIGURE 16.2 (a) CFG of a task τ_{simple}; (b) calculation of δ values. (Taken from Seo, J. et al., *IEEE Trans. Comput. Aided Des. Integrated Circ. Syst.*, 25(1), 47, 2006.)

The corresponding average energy consumption is proved to be optimal and expressed as [27]:

$$E_{intra} = C \cdot \left(\frac{\delta_0}{(relative)\ deadline} \right)^2 \cdot \delta_0 \tag{16.4}$$

where

    C is a system-dependent constant

    δ_0 is the δ value of the top basic block b_0 and called *energy-optimal path length* of the task

One interesting interpretation of Equation 16.4 is that it can be considered as the energy consumed in the execution of δ_0 cycles at a speed of $\delta_0/deadline$. Note that $\delta_0/deadline$ is the initial speed of the task.

For example, consider a real-time task τ_{simple} shown in Figure 16.2a. Figure 16.2b shows the procedure of calculating each δ_i value according to Equation 16.3. Once we have performed the calculation, the operating speed of basic block b_i is simply obtained by dividing δ_i by the remaining time to deadline. For example, suppose that *deadline* = 10 (unit time) and the execution follows the path $(b_0, b_2, b_3, b_5, b_6, b_8)$. Then, the corresponding speed/ending time changes as follows:

Speed: 2.93 > 3.24 > 3.14 > 3.14 > 2.26 > 2.26

Ending time: 2.05 > 3.29 > 3.92 > 4.24 > 7.78 > 10

In Figure 16.2b, the thick, dotted, and regular arrows, respectively, indicate the increase, decrease, and no change in the processor speed, and the basic blocks with changed operating speed are marked with gray color.

16.3 Intertask DVS

Most of DVS works belong to the intertask DVS. The intertask voltage scheduling problem can be stated as:

(Intertask DVS Problem) *Given an instance of tasks with deadlines and voltages, find a feasible task-level schedule and task-level voltage allocation to tasks that minimizes the total energy consumption while satisfying the deadline constraints of tasks.*

Table 16.1 summarizes the current status of the energy-optimal works for the various constrained cases of intertask DVS problem.

The most cited and fundamental DVS algorithm is given by Yao et al. [35]. The algorithm solves the intertask DVS problem optimally under the assumption of continuously variable voltage. For any set of n tasks, the proposed voltage scheduling leads to an $O(n \log_2 n)$ time algorithm. The algorithm is based on the notion of a *critical interval* for an input task set, which is an interval in which a group of tasks must be scheduled at maximum, constant speed in any optimal schedule for the task set. The algorithm proceeds by identifying such a critical interval for the task set, scheduling those tasks whose arrival times and deadlines are within the critical interval, then constructing a subproblem for the remaining tasks and solving it recursively. Even though the Yao's algorithm is not realistic due to the use of continuously variable voltage, it gives a starting basis for many algorithms of several variant DVS problems such as DVS problem with discretely variable voltage and DVS problem with sleep state.

Note that in addition to the voltage, the amount of energy consumption is affected by the switched capacitance of the task. The value of capacitance is determined according to the execution characteristics of the task. If the task requires hardware components with high switched capacitance, such as multipliers, for execution, the capacitance value will be large, and vice versa. Consequently, to reduce the total energy consumed by tasks, it is desirable to execute the tasks with low switched capacitance using high supply voltages while the tasks with high switched capacitance using low supply voltages. The unsolved case is that of continuous voltages with nonuniform capacitance of tasks. The work by Kwon and Kim [16] for uniform capacitance case actually makes use of the optimal work by Ishihara and Yasuura [9] and Yao et al. [35]. The work by Kwon and Kim [16] for nonuniform capacitance case uses a linear programming (LP) formulation. Here, we review the algorithm of Kwon and Kim [16] for uniform capacitance case.

The procedure starts from the results of the possibly invalid voltage allocation with the feasible task schedule obtained from Yao et al.'s algorithm [35], and transforms it into that of valid voltage allocation with a feasible schedule. More precisely, it retains the schedule of tasks during transformation, but changes the voltages so that they are all valid. Then, the question is what and how the valid supply voltages are selected and used. A valid voltage for each (scheduled) task is determined by performing the following three steps: (Step 1: *Merge time intervals*) all the scheduled time intervals that were allotted to execute the

TABLE 16.1 Polynomial-Time Optimality of Intra- and Intertask DVS Techniques

Voltage	Tasks		Optimal?/Ref.
Cont. voltage	Single task		Yes/trivial
	Multiple tasks	Uniform cap.	Yes/Yao et al. [35]
		Nonuniform cap.	Unknown
Disc. voltage	Single task		Yes/Ishihara and Yasuura [9]
	Multiple tasks	Uniform cap.	Yes/Kwon and Kim [16]
		Nonuniform cap.	Yes/Kwon and Kim [16]

Source: Taken from Kim, T., Application driven low-power techniques using dynamic voltage scaling, in: *Proceedings of the IEEE International Conference on Embedded and Real-Time Computing Systems and Application*, Sydney, Australia, pp. 199–206, 2006.

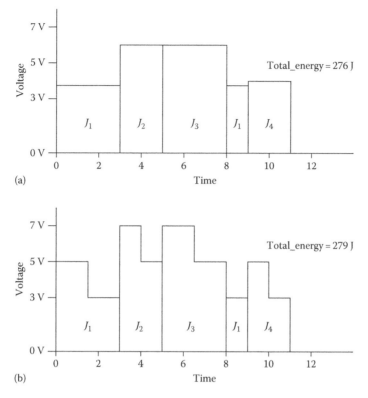

FIGURE 16.3 An example illustrating the transformation of continuously variable voltage allocation into discontinuously variable voltage allocation. (a) A continuously variable voltage allocation for tasks; (b) a discontinuously variable voltage allocation derived from (a). (Taken from Kwon, W.C. and Kim, T., *ACM Trans. Embedded Comput. Syst.*, 4(1), 211, 2005.)

task are merged into one; (Step 2: *Voltage reallocation*) the invalid supply voltage is replaced with a set of valid voltages; (Step 3: *Split time interval*) the merged time interval is then split into the original time intervals.

For example, suppose that we have three voltages 7.0, 5.0, and 3.0 V available for use and their corresponding clock speeds are 70, 50, and 30 MHZ, respectively. Then, for each scheduled task with the ideal voltage in Figure 16.3a, the three steps are applied. Figure 16.4 shows the results of three steps for task J_1. Initially, J_1 is scheduled to be executed in two time intervals [0,3] and [8,9] with the voltage being 3.75 V, as shown in Figure 16.4a. (According to the results in [35] each task always uses the same voltage.) Consequently, in Step 1 the time intervals are merged into [0,4] as shown in Figure 16.4b. The supply voltage in Step 2 is then updated. To do this, we make use of Ishihara and Yasuura's results [9]: For a given ideal (optimal) voltage for a task, the valid (optimal) voltage allocation is to use the two immediately neighboring valid voltages to the ideal voltage. (For details on how to find the time point at which the clock speed changes, see [9].) Figure 16.4c shows the result of voltage reallocation where two voltages 3.0 and 5.0 V are used because the ideal voltage (=3.75 V) is in between 3.0 and 5.0 V, and no other valid voltages are in the interval. Finally, in Step 3 we restore the time intervals while preserving the voltage reallocation obtained in Step 2, as shown in Figure 16.4d. By repeating these three steps for J_2, J_3, and J_4 in Figure 16.3a, a voltage allocation for all tasks with a feasible schedule is obtained, as shown in Figure 16.3b. Note that because we used only a number of discrete voltages, the energy consumption, which is 279 J, increases from that in Figure 16.3a, which is 276 J. However, according to [16], the amount of the increase is minimal.

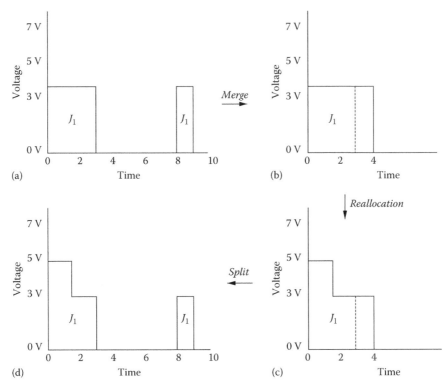

FIGURE 16.4   The three steps of voltage allocation procedure for task J_1 in Figure 16.3. (a) An initial schedule in Figure 16.3a; (b) the result after merging time intervals; (c) the result after voltage reallocation; (d) the result after splitting the time interval. (Taken from Kwon, W.C. and Kim, T., *ACM Trans. Embedded Comput. Syst.*, 4(1), 211, 2005.)

16.4 Integration of Intratask DVS into Intertask DVS

The combined problem of the inter and intratask DVS problems can be described as:

(Combined DVS Problem) *Given an instance of tasks with deadlines and voltages, find a feasible task-level schedule and task-level voltage allocation to tasks that minimizes the total energy consumption while satisfying the deadline constraints of tasks.*

There are two optimal results in the literature related to the combined DVS problem: (i) an optimal intertask DVS scheme [35] that determines the operating voltage of each task assuming the worst-case execution path and (ii) an optimal intratask DVS scheme [27] that determines the operating voltage of each basic block in a single task. From the analysis of the procedures of (i) and (ii), the work in [14] found that an energy-optimal integration of (i) and (ii) is possible with a slight modification of the procedures. The proposed integrated DVS approach is a two-step method:

1. Statically determine energy-optimal starting and ending times (s_i and e_i) for each task τ_i.
2. Execute τ_i within $[s_i, e_i]$ while varying the processor speed according to the voltage scales obtained by an existing optimal intratask DVS scheme.

The key concern is to develop a new intertask scheduling algorithm that finds starting and ending times (s_i and e_i) for each task, which leads to a minimum value of total energy consumption when an optimal intratask scheme is applied to the tasks. For details on the optimal algorithm, readers can be referred to the work in [14].

16.5 DVS with Sleep State

The DVS approaches mentioned so far assume that the system has no sleep state. In this environment, it is always the best for the task scheduler in DVS to execute the task as slowly as possible under the tasks timing constraint because due to the convexity of power function, the slower the speed is, the lesser the energy consumption is. By contrast, if a system has no job to run (i.e., idle), it can be put into a low-power sleep state. Even though the system consumes less power in a sleep state, a fixed amount of energy is needed to make a transition from the sleep state to an active state in which tasks can be executed. Thus, the decision on whether the system is put into a sleep state or not should be made according to the length of idle period. There have been many techniques that address DVS problem with sleep state (e.g., [4,8,11–13,17]). A common objective of the pervious off-line algorithms on DVS with sleep state is to schedule tasks so that the number of idle intervals is as small as possible while the length of each idle period is as long as possible. There are many intertask DVS algorithms with sleep state. All the existing algorithms on DVS with sleep state have assumed a fixed number of execution cycles of each task and not considered the varying execution cycles. (There is no work which addresses the effect of intratask DVS on intertask DVS with sleep state.)

Given a set of tasks representing an application, let us consider a *periodic task model*. The periodic task model consists of N (periodic) tasks, $T = \{\tau_1, \tau_2, \ldots, \tau_N\}$ in which each task $\tau_i \in T$ consists of a set of basic blocks $\{b_{i,1}, b_{i,2}, \ldots\}$.

Each task $\tau_i \in T$ is associated with the following parameters: $n_{i,j}$ which is the required number of CPU cycles to complete execution of basic block $b_{i,j}$, a_i which is the arrival or release time, d_i which is the deadline ($e_i \geq s_i$), and $s(t)$ which is the clock speed (or voltage level) at time t, $a_i \geq t \geq d_i$. Note that $n_{i,j}$, a_i, and d_i are given for each task τ_i, while $s(t)$ should be determined by DVS in a way to minimize the resulting energy consumed by completing the task.

The system can always be in one of the following three states:

- *Active*, which is the state where the system is on and running a task
- *On-idle* state, which is the state where the system is on, but not running a task
- *Sleep*, which is the state where the system is in a very low power mode or the system power is off

For the DVS problem with no sleep state, the system is in one of active or on-idle state, while for dynamic voltage scaling problem with sleep state, the system is in one of the three states.

The total power consumed in the system can be decomposed into a number of factors summarized as follows:

- P_{on}: Power required to retain the system in *on* state
- $P_{AC}(s)$: Power function of speed s, which is the power spent in running a task in *active* state
- P_{DC}: Power leakage when system is either in *on* or *idle* state

Note that $P_{AC}(s)$ is the power contributed by the dynamic and leakage powers. $P_{AC}(s)$ is a monotonically increasing function of s. P_{on} is the power contributed mainly by the PLL circuitry, which drives up to 200 mA current and the I/O subsystem with a supply voltage higher than the processor core voltage and peak current of 400 mA during I/O. Though the current is comparatively lower when there is no I/O, the power consumption adds to a significant portion of the idle power consumption. The power consumption of these components will scale with technology and architectural improvements, and a conservative value of P_{on} is about 0.1 W. Thus, the total power consumption P is given as,

$$P = P_{on} + P_{AC} + P_{DC}. \tag{16.5}$$

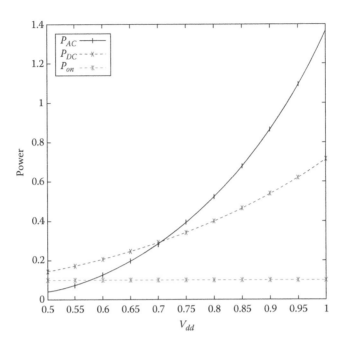

FIGURE 16.5   Power consumption of 70 nm technology for Crusoe processor: P_{DC} is the leakage power, P_{AC} is the dynamic power , and P_{on} is the intrinsic (I/O) power consumption in *on* state. (Taken from Jejurikar, R. et al., Leakage aware dynamic voltage scaling for real-time embedded systems, in: *Proceedings of ACM/IEEE Design Automation Conference*, Anaheim, CA, pp. 275–280, 2004.)

The change in the values of P_{on}, P_{AC}, and P_{DC} as the supplied voltage V_{dd} varies is shown in Figure 16.5 [13] where the linear dependence of leakage power consumption on voltage and the quadratic dependence of dynamic power on voltage are seen in the figure. Though the total power consumption decreases as V_{dd} is scaled, it does not imply energy savings. Figure 16.6 [13] shows the change in the values of energy $E_{on}(= P_{on} \times \Delta T)$, $E_{AC}(= P_{AC} \times \Delta T)$, and $E_{DC}(= P_{DC} \times \Delta T)$ as the supplied voltage V_{dd} varies where ΔT is one clock cycle time period. The dynamic energy per cycle, E_{AC}, is given in Equation 16.6. The leakage energy per cycle, E_{DC}, is given by

$$E_{DC} = f^{-1} \cdot L_g \cdot (I_{subn} \cdot V_{dd} + |V_{bs}| \cdot I_j) \tag{16.6}$$

where
   f^{-1} is the delay per cycle
   L_g is the number of devices in the circuit

The decrease in the supply voltage increases the leakage and I/O energy consumptions E_{DC} and E_{on}, which can surpass the gains of DVS. We see that total energy E_{total} in Figure 16.6 decreases as V_{dd} is scaled up to 0.7 V, beyond which leakage energy consumption dominates. The total energy per cycle to be minimized is

$$E = E_{on} + E_{AC} + E_{DC}. \tag{16.7}$$

The total energy consumption increases with further slowdown, and it is not energy efficient to slowdown beyond $V_{dd} = 0.7$ V. Executing at $V_{dd} = 0.7$ V and shutting down the system is more energy efficient than executing at lower voltage levels. The operating point that minimizes the energy consumption per cycle is called the *critical speed*. From the figure, it is seen that the critical speed of operation is

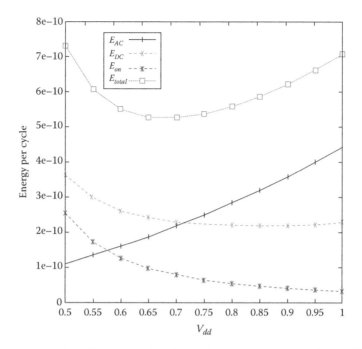

FIGURE 16.6   Energy per cycle for 70 nm technology for the Crusoe processor: E_{AC} is the switching energy, E_{DC} is the leakage energy, and E_{on} is the intrinsic energy to keep the processor *on*. (Taken from Jejurikar, R. et al., Leakage aware dynamic voltage scaling for real-time embedded systems, in: *Proceedings of ACM/IEEE Design Automation Conference*, Anaheim, CA, pp. 275–280, 2004.)

$V_{dd} = 0.7$ V. The critical speed will be computed by evaluating the gradient of the energy function with respect to V_{dd}.

Another term to be added in the total energy consumption is the energy (E_{wake}) required to wake up the system from *sleep* to *on* state. Usually it is assumed E_{wake} is constant and the transition time from *on* to *sleep* state and the time from *sleep* to *on* state are ignored in DVS. Some previous work does consider the transition time from *on* to *sleep* state and vice versa [3]. The work by Ramanathan et al. [22] empirically shows that an algorithm which does not incorporate the state transition latency performs very well even when the transition latency is taken into account.

The conventional algorithms of intertask DVS with sleep state basically utilize the following two facts: (fact 1) if a system can be in *sleep* state during the time to wait for the next task to be performed, the power can further be saved if the waiting time is long enough to compensate for the power consumed in waking up the system to *on* state; (fact 2) since the total energy curve follows that in Figure 16.6 when system has sleep state, the energy consumption is minimal if the system's clock speed is set to the critical speed (i.e., at 0.7 V) in Figure 16.6 where the power spent on waking up is ignored. Left-To-Right is one notable algorithm, which shows that it always produces a DVS schedule whose total energy consumption is never more than a factor of three of optimal. For details on the algorithm, readers can refer to the work in [8].

16.6 DC–DC Converter-Aware DVS

Almost all modern digital systems are supplied with power through DC–DC converters as high-performance CMOS devices are optimized to specific supply voltage ranges. DC–DC converters are generally classified into two types: linear voltage regulators and switching voltage regulators, according

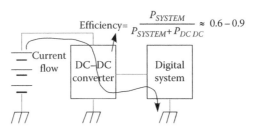

$$\text{Efficiency} = \frac{P_{SYSTEM}}{P_{SYSTEM} + P_{DC\,DC}} \approx 0.6 - 0.9$$

FIGURE 16.7 A typical current flow path from battery: a non-trivial power loss in DC–DC converter, resulting in short battery life despite low-power dissipation in the digital system. (Taken from Choi, Y. et al., *IEEE Trans. Comput. Aided Des. Integrated Circ. Syst.*, 26(8), 1367, 2007.)

to the circuit implementation. However, the power dissipation in both types of voltage conversion is unavoidable, and directly affects the lifetime of battery in the whole system. Figure 16.7 shows the path of current flow through DC–DC converter from battery. It is reported that there is always a non-trivial power loss in the converter, the amount of which is 10%–40% of the total energy consumed in the system.

The divergence of digital devices and technology innovation make it hard to use a single supply voltage to all devices or even to an individual device. Since all supply voltages to the devices are generally received from a single battery source, the voltage regulators (DC–DC converters) control the supply voltage for each device, as indicated in Figure 16.8, which shows a simplified power supply network for battery-operated embedded systems.

The primary role of a DC–DC converter is to generate a regulated power source. Unlike passive components, logic devices do not draw constant current from a power supply. The power supply current rapidly changes according to the changes of its internal states. An unregulated power supply may induce IR drop corresponding to the load current, whereas a regulated power supply keeps the same output voltage regardless of the load current variation. Note that IR drop is caused by non-zero resistance of power supply. Thus, even though we use a single power supply voltage, a DC–DC converter for voltage regulation is required.

Even though DC–DC converters are significant in saving energy, most of the proposed DVS schemes except the work in [5] have not taken into account the output voltage scaling of a DC–DC converter and thus the load current variation. Finally, the output voltage and load current variations due to DVS will cause efficiency variation of the DC–DC converter. In the following, we review the DC–DC converter-aware power management technique proposed by [5], which uses, as DC–DC converter, switching regulator whose block diagram and current flow are shown in Figure 16.9. A switching regulator is a circuit that uses an inductor, a transformer, or a capacitor as an energy-storage element to transfer energy from a power source to a system. The amount of power dissipated by voltage conversion in a switching

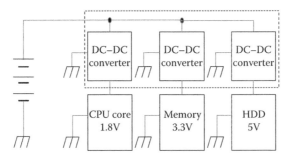

FIGURE 16.8 DC–DC converters generate different supply voltages to CPU, memory, and HDD devices from a single battery. (Taken from Choi, Y. et al., *IEEE Trans. Comput. Aided Des. Integrated Circ. Syst.*, 26(8), 1367, 2007.)

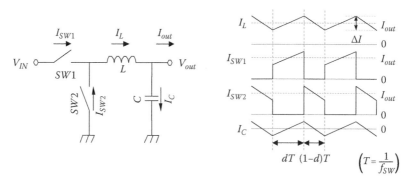

FIGURE 16.9 Simplified block diagram of buck converter and current flow at each component. (Taken from Choi, Y. et al., *IEEE Trans. Comput. Aided Des. Integrated Circ. Syst.*, 26(8), 1367, 2007.)

regulator is relatively low, mainly due to the use of low-resistance switches and energy storage elements, while the amount of power dissipated in a linear regulator, as opposed to the switching regulator, is rather high, mainly from the fact that the power efficiency of a linear regulator is upper-bounded by the value of output voltage divided by the input voltage. In addition, switching regulators can step up (i.e., boost), step down (i.e., buck), and invert input voltage with a simple modification of the converter topology, unlike linear regulators. A switching regulator contains a circuit, positioned on the path between the external power supply and the energy-storage element to control switches.

The work in [5] models the power consumption of DC–DC converter in terms of two variables, load current I and the parameter W, which controls a trade-off between load-dependent power consumption and load-independent power consumption such as gate width of two switches.

$$P_{DC}(I, W) = \left(\frac{c_1}{W} + c_2\right) I^2 + c_3 W + c_4, \quad \text{when } I \neq 0,$$
$$P_{DC}(I, W) = 0, \quad \text{otherwise} \tag{16.8}$$

where $c_1, \ldots, c_4$ are constants. If $I = 0$, we can consider $P_{DC}(I, W)$ as zero because many DC–DC converters enter the shutdown state with very little power loss when there is no load current.

Figure 16.10 shows the relation between the converter's energy efficiency and the converter configuration parameter W. We can see that the energy efficiency of a DC–DC converter increases as W becomes large at heavy load, or W becomes small at light load.

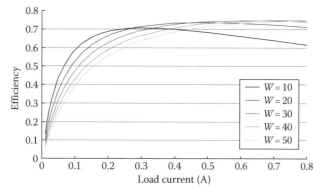

FIGURE 16.10 The energy efficiency of DC–DC converter with a set of different values of parameter W. (Taken from Choi, Y. et al., *IEEE Trans. Comput. Aided Des. Integrated Circ. Syst.*, 26(8), 1367, 2007.)

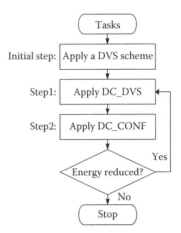

FIGURE 16.11   DC–DC converter-aware DVS scheme proposed by [5]. (Taken from Choi, Y. et al., *IEEE Trans. Comput. Aided Des. Integrated Circ. Syst.*, 26(8), 1367, 2007.)

Solving the problem of determining a configuration of a DC–DC converter and a DVS result that lead to a minimal energy consumption simultaneously would be a quite complex problem. To make the problem tractable to solve in a systematic way, the work in [5] proposed a simple framework, called DC-lp, of converter-aware energy minimization algorithm. DC-lp combines two core techniques: DC_DVS and DC_CONF; the objective of DC_DVS is to refine the DVS result by considering the energy efficiency of the DC–DC converter to be used, while the objective of DC_CONF to refine the configuration of the DC–DC converter (i.e., determining an optimal value of parameter W) according to the update of the DVS result. Figure 16.11 shows the flow of the integrated algorithm. Initially, it is given a DVS result A for input tasks, and a converter configuration B. Then, the two steps in Figure 16.11 are performed iteratively until there is no further reduction in total energy consumption: (step 1) DC_DVS is applied to A by using B to produce a new DVS result A'; (step 2) DC_CONF is applied to A' to produce a new configuration B'. Refer to [5] for details on DC_DVS and DC_CONF. It is reported that using the converter-aware DVS scheme saved energy by 16.0%–22.1% on average, which otherwise was dissipated in the DVS schemes with no consideration of DC–DC converter efficiency variation.

16.7 Transition-Aware DVS

During the execution of tasks, three types of transition overhead are encountered for each voltage transition: transition cycle, transition interval, and transition energy.

A *transition cycle* (denoted as Δn) is the number of instruction cycles of the transition code itself. We assume that the energy for executing the transition instruction varies depending on the voltages in transition.

A *transition interval* (denoted as Δt) is the time taken during the voltage transition, which is not constant. Note that in the current commercial designs, the phase-locked loop that is used to set the clock frequency requires a fixed amount of time to lock on a new frequency. This locking time is known to be independent of the source and target frequencies, and is typically much smaller than the time it takes for the voltage to change [6]. Therefore, it is desirable to assume that the transition interval Δt is not a fixed value. A reasonable assumption for the variable voltage processor is that the transition interval is proportional to the difference between the starting and ending transition voltages [32].

A *transition energy* (denoted as ΔE) is the amount of energy consumed during the transition interval by the systems. The value of ΔE may vary depending on the starting and ending voltage levels. The DVS problem with the additional consideration of transition energy is even more difficult to solve because the generalized model of the transition energy for various systems/processors is hard to obtain and even a simplified version of the problem with the assumption that the transition energy is a constant looks a non-trivial task to solve. Let $\Delta E_{(v_i \to v_j)}$ denote the energy dissipated by the transition of voltage from v_i to v_j. We assume that the processor has a set $\mathcal{V}$, called *voltage set*, of voltages available to use, i.e., $\mathcal{V} = \{v_1, v_2, \ldots, v_M\}$ ($v_1 < v_2 < \cdots < v_M$). In addition, we assume that the values of $\Delta E_{(v_i \to v_j)}$, $v_i, v_j \in \mathcal{V}$ have already been given, and the values are stored in a table, called *transition-energy table*. Furthermore, we reasonably assume that $\Delta E_{(v_i \to v_j)} < \Delta E_{(v_i \to v_{j'})}$ if $|v_i - v_j| < |v_i - v_{j'}|$.

(Transition Energy-Aware DVS Problem) *Given a fixed schedule of tasks, and voltage set $\mathcal{V}$ and the associated transition-energy table, assign the voltages to the tasks so that the total energy consumption for the execution of tasks, together with the energy consumed by the voltage transitions, is minimized.*

Unfortunately, most of the DVS methods never take into account the minimization of the amount of energy consumed during the voltage transition. Mochocki et al. [20] and Shin and Kim [30] addressed the DVS problem with the consideration of transition overheads and discrete voltages. The work in [20] put the primary emphasis on the consideration of transition intervals by modifying the optimal scheduling algorithm in [35] together with a simple treatment on both the transition energy and discrete voltages. On the other hand, the work in [30] attempted the limitation of the work in [20] in that it tries to solve the problem optimally for a constrained case. We review the work by [30] here. The key contribution of the work is the network formulation of the problem. Here, we introduce the procedure of single task case only. It was naturally extended to the cases of multiple tasks.

Suppose an instance of TE-VA has a task τ_1 with the schedule interval $[t_{start}, t_{end}]$ and R cycles to be executed. If the overheads of voltage transition are not taken into account, an optimal result can be obtained by using the voltage assignment technique in [9], i.e., the optimal voltage assignment is to use the two voltages in $\mathcal{V}$ that are the immediate neighbors to the (ideal) voltage corresponding to the lowest possible clock speed which results in an execution of task τ_1, exactly starting at time t_{start} and ending at time t_{end}; the ideal voltage can be obtained according to the voltage and delay (clock speed) relation [9].

Lemma 16.1 *([30]) For an instance of the transition-aware DVS problem with a single task, an energy-optimal voltage assignment uses at most two voltages for the execution of task. (The proof described below becomes the foundation of the proposed network formulation.)*

Proof 16.1 Suppose there is an optimal voltage assignment $\mathcal{V}A$ which uses more than two voltages for the execution of the task during the scheduled interval of $[t_{start}, t_{end}]$. Let $v_1, \ldots, v_K$ ($K > 2$) be the sequence of voltages applied to the task, starting from t_{start} to t_{end}, by the optimal voltage assignment, and let $[t_{start} + \Delta t, t_2 - \Delta t/2]$, $[t_2 + \Delta t/2, t_3 - \Delta t/2]$, $\ldots$, $[\Delta t/2 + t_r, t_{end} - \Delta t]$ be the corresponding execution intervals. (See Figure 16.12a for an example.) Note that the length of the actual execution interval, not including the transition interval, is $T_L = t_{end} - t_{start} - (r + 1)\Delta t$. Since the voltages before and after the execution of the task are $0\,V$, transition interval Δt is required from $0\,V$ to v_1 at the beginning (i.e., $[t_{start} + \Delta t, t_2 - \Delta t/2]$) and from v_r to $0\,V$ at the end (i.e., $[\Delta t/2 + t_r, t_{end} - \Delta t]$) Also, a transition interval is required between the two consecutive execution intervals (e.g., Δt around time t_i in $[t_{i-1} + \Delta t/2, t_i - \Delta t/2]$ and $[t_i + \Delta t/2, t_{i+1} - \Delta t/2]$).

Now, consider another voltage assignment $\mathcal{V}A'$ which uses two voltages for the actual execution length of $T_L' = t_{end} - t_{start} - 3\Delta t$ (see Figure 16.12b). Let v_1' and v_2' be the two voltages which lead to a minimum total energy consumption except the transition energy. (Note that the two voltages can be obtained by applying the technique in [9] to the task of R cycles with an execution interval of length T_L.) Then, we want to compare the amount of energy consumptions excluding the transition energy used in $\mathcal{V}A$ and $\mathcal{V}A'$, and compare the amounts of transition energy used in $\mathcal{V}A$ and $\mathcal{V}A'$.

Obviously, we can see that $max\{v_1, \ldots, v_r\} \geq max\{v_1', v_2'\}$ since the execution interval for $\mathcal{V}A$ is shorter than that for $\mathcal{V}A'$ due to more transition intervals in $\mathcal{V}A$ than that of $\mathcal{V}A'$. This means that the transition

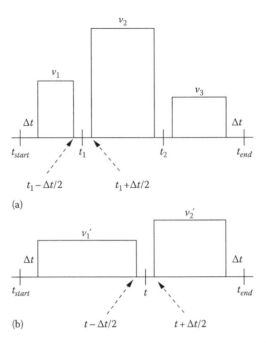

FIGURE 16.12 Transition intervals and actual execution intervals for an example of voltage assignment using more than two voltages and voltage assignment using two voltages. (a) An example of more than two voltage assignment for a task execution. (b) An example of two voltage assignment for the same task in (a) (v_1' and v_2' are adjacent voltages). (Taken from Shin, J. and Kim, T., *IEEE Trans. Integrated Circ. Syst.*, 53(9), 956, 2006.)

energy from $0\,V$ at the start to eventually $max\{v_1, \ldots, v_r\}$ is greater than that from $0\,V$ to $max\{v_1', v_2'\}$ by the assumption $\Delta E_{(v_i \to v_j)} < \Delta E_{(v_i \to v_{j'})}$ if $|v_i - v_j| < |v_i - v_{j'}|$. Thus, the total transition energy for $\mathcal{V}A$ is greater than that for $\mathcal{V}A'$. On the other hand, since from the fact that v_1' and v_2' are the voltages of optimal voltage assignment with time length of T_L', and $T_L' < T_L$, the total energy consumption without transition energy for $\mathcal{V}A'$ is less than that for $\mathcal{V}A$. Thus, the assumption that $\mathcal{V}A$ is optimal is false.   □

From Lemma 16.1, the solution space for a single task can be confined to the solutions which only use at most two voltages in $\mathcal{V}$. The remaining issue is to give, for a given execution interval for the task, a technique of finding the two (optimal) voltages and their execution intervals. According to [30]: A network $N(V, A)$ is constructed for a task with R instruction cycles, execution interval $[t_{start}, t_{end}]$, voltage set $\mathcal{V}$, and transition energy table where V is the set of nodes and A is the set of arcs between two nodes. Note that Lemma 16.1 tells there is always an optimal voltage assignment which uses two voltages only. (See the upper figure in Figure 16.13.) If we know the two voltages used, then the two actual execution intervals can be computed accordingly using the speed and voltage relation [9,16]. For each of the two execution intervals, $|\mathcal{V}|$ nodes are arranged vertically, each node representing a unique voltage in $\mathcal{V}$. Then two additional nodes are included in V. One is *start-node*, placing at the front of $N(V, A)$ and the other is *end-node* at the end. (See the lower figure in Figure 16.13.) We insert arcs from the nodes in the first column to those in the second column, from the start node to the nodes in the first column, and from the nodes in the second column to the end node. (See the lower figure in Figure 16.13.) We then assign weight to each arc. The weight of an arc from the start node to a node labeled as v_i in the first column indicates $\Delta E_{(0V \to v_i)}$, and the weight of an arc from a node labeled as v_i to the end node indicates $\Delta E_{(v_i \to 0v)}$. The weight of an arc from a node labeled as v_i to a node labeled as v_j represents the total sum of the (minimal)

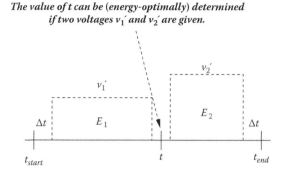

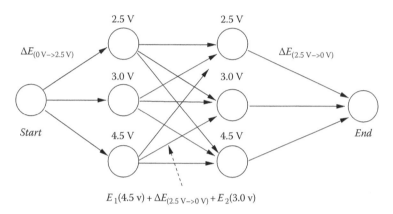

FIGURE 16.13 The construction of network for modeling an instance of the TE-VA problem with a single task. (Taken from Shin, J. and Kim, T., *IEEE Trans. Integrated Circ. Syst.*, 53(9), 956, 2006.)

energy consumed by the execution of the task (i.e., $E1(v_1) + E2(v_2)$ in Figure 16.13) and the transition energy $\Delta E_{v_i \to v_j}$.[*]

Then, from the network $N(V, A)$, a shortest path can be found from the start node to the end node. The cost of the shortest path is exactly equivalent to the total amount of energy consumption including the transition energy for the execution of the task. The procedure [30] performs two steps: (step 1) construct network $N(V, A)$ for the task; (step 2) find a shortest path (SP) on $N(V, A)$.

16.8 Temperature-Aware DVS

As the power density on a chip increases rapidly, controlling or minimizing the increase in temperature is one of the most important design factors to be considered in the course of task execution, because it increases circuit delay, in particular spatially unbalanced delays on a chip, causing system function failure. Furthermore, the leakage power increases exponentially as the temperature increases.

(Processor Thermal Model) Most of the temperature-aware DVS works (e.g., [2,10,18,36]) use, as the thermal model, the lumped RC model similar to that in [31] to capture the heat transfer phenomena as shown in Figure 16.14. (For the additional models, see the references [19,21,23,34].) In the figure, T, C, and R represent the processor's die temperature, the thermal capacitance of the die, and the thermal resistance, respectively. T_{amb} indicates the ambient temperature and $P(t)$ the power consumption of the

---

[*] In case of $v_i = v_j$, there will be no voltage transition and $\Delta E_{(v_i \to v_j)} = 0$.

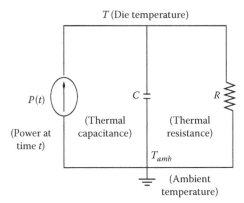

FIGURE 16.14 Processor heat transfer model. (Taken from Bao, M. et al., Temperature-aware idle time distribution for energy optimization with dynamic voltage scaling, in: *Proceedings of ACM/IEEE Design Automation and Test in Europe*, Dresden, Germany, pp. 21–26, 2010.)

processor at current time t. For the given values of T_{amb}, C, and R, the relation between die temperature T and power consumption $P(t)$ can be expressed as

$$R \cdot C \cdot \frac{dT}{dt} + T - R \cdot P = T_{amb}. \tag{16.9}$$

In the following, we introduce two fundamental problems on the temperature-aware voltage scaling and their solutions.

(Temperature-Constrained DVS Problem for Performance Optimization) *Given an execution schedule of tasks and peak temperature constraint T_{max}, determine the voltages to be applied to tasks, if needed, inserting sleep time intervals between task executions so that the completion time of all tasks is minimized while satisfying the peak temperature constraint.*

One notable work on the temperature-constrained DVS problem is done by [36] in which the NP-hardness of the problem is shown when a set of discrete voltages is assumed to be used. The optimal formulation is followed by introducing two types of variables defined as follows:

1. Suppose there are n tasks $\tau_1, \tau_2, \ldots, \tau_n$ that should be executed in that order. Each task can be assigned to a voltage among m discrete voltages $v_1, \ldots, v_m$. Let $x_{i,j}$ ($1 \leq i \leq n, 1 \leq j \leq m$) denote a variable having a value of 0 or 1 such that $x_{i,j} = 1$ if τ_i is executed with v_j, and $x_{i,j} = 0$ if τ_i is executed with a voltage other than v_j. Thus, when $exe\_t(\tau_i, v_j)$ denotes the time spent on executing task τ_i on v_j, the total execution time spent on executing all tasks is $\sum_{i=1,\ldots,n} \sum_{j=1,\ldots,m} x_{i,j} \cdot exe\_t(\tau_i, v_j)$.

2. In addition, there are total of $n+1$ idle intervals, one for each of two consecutive task executions, one before τ_1, and one after τ_n. The time length of each idle interval will be upper-bounded by t_{max_sleep}: the value of t_{max_sleep} is the time that is minimally required to cool down T_{max} to T_{amb}. We discretize $[0, t_{max,leep}]$, so that q sub-intervals of equal length are formed, i.e., $[t_0, t_1, t_2, \cdot, t_i, \ldots, t_q]$ where $t_i = i \cdot (t_{max,leep}/q), i = 0, \ldots, q$. ($q$ is a user-defined value.) Now, let variable $y_{i,j}$ ($1 \leq i \leq n$, $1 \leq j \leq q$) be such that $y_{i,j} = 1$ if the time spent on idle state right after τ_i is t_j. Then, the total idle time spent during the execution of tasks is $\sum_{i=1,\ldots,n+1} \sum_{j=0,\ldots,q} y_{i,j} \cdot t_j$.

Thus, the formulation is expressed as

$$minimize \ D = \sum_{i=1}^{n} \sum_{j=1}^{m} x_{i,j} \cdot exe\_t(\tau_i, v_j) + \sum_{i=1}^{n} \sum_{j=0}^{q} y_{i,j} \cdot t_j \tag{16.10}$$

such that

$$\sum_{j=1}^{m} x_{i,j} = 1, \sum_{j=0}^{q} y_{i,j} = 1, \quad \forall i \in \{1, \ldots, n\}; \tag{16.11}$$

$$R \cdot C \cdot \frac{dT}{dt} + T - R \cdot P = T_{amb}; \tag{16.12}$$

$$T \le T_{max}; \quad x_{i,j} = \{0, 1\}; \quad y_{i,j} = \{0, 1\}; \tag{16.13}$$

$$T(t) \le T_{max}, 0 \le t \le D; \quad T(t = 0) = T(t = D) \tag{16.14}$$

in which the last constraint ensures that the temperature at the beginning of the execution of tasks must be the same as that at the time when the execution of all tasks is completed to support the periodic execution of task set.

The problem can also be optimally formulated using the dynamic programming (DP) approach that runs in pseudo-polynomial time [36]. Let $T_1(\tau_i, D)$ denote the minimum temperature at D which is the time when τ_i just finishes the execution, and let $T_2(\tau_i, D)$ denote the minimum temperature at D which is the time just when τ_{i+1} starts execution. Then, the optimal D^* is determined by the smallest value of D such that $T_2(\tau_n, D) \le T_{max}$. The recurrence relation can be given as

$$T_1(\tau_i, D) = min_{j=1,\ldots,m}\{T = T_2(\tau_{i-1}, D - exe\_t(\tau_i, v_j))$$
$$+ \Delta T(exe\_t(\tau_i, v_j))|T \le T_{max}\};$$

$$T_2(\tau_i, D) = min_{j=0,\ldots,q}\{T = T_1(\tau_i, D - t_j) + \Delta T(t_j, v_{sleep})|T \le T_{max}\}$$

where v_{sleep} is the voltage at idle state. (The derivation of termination cases can be easily obtained.)

The time complexity of the DP algorithm is polynomial in terms of n, q, and D_{UB}, where D_{UB} is a large number that surely exceeds D^*. The work in [36] also proposes a fully polynomial time approximation scheme (FPTAS).

(Time-Constrained DVS Problem for Peak Temperature Minimization) *Given a set of tasks and time constraint D_{max}, determine the schedule of tasks and voltages to be applied to tasks so that the peak temperature is minimized while satisfying the time constraint.*

The work in [10] showed that the problem is NP-hard even when the voltage to each task is given, and proposed a greedy heuristic which attempts to schedule such that a cold task is put right after a hot task and a hot task right after a cold task. The heuristic is further exploited repeatedly to gradually find the voltages to tasks that minimize the peak temperature.

For a given task sequence, the voltage assignment under D_{max} constraint for minimizing peak temperature can be formulated in pseudo-polynomial time [10], which is similar to the DP algorithm shown earlier.

Unfortunately, the work does not address the allocation of idle states between task executions. The work in [2] proposes a solution to the idle allocation problem in which the objective is to minimize the total power including the temperature-induced leakage power while meeting D_{max} constraint. For example, Figure 16.15 [2] shows two different idle allocations for a sequence of five tasks $\tau_1, \ldots, \tau_5$ with predefined DVS. Clearly, Figure 16.15b will lead to less power consumption than that in Figure 16.15a. Figure 16.16 [2] shows how the temperature changes as the voltages and idle times are allocated in a sequence like that in Figure 16.15b.

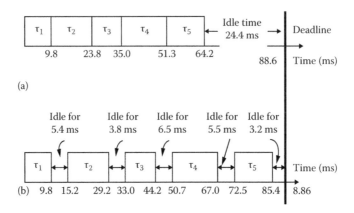

(a)

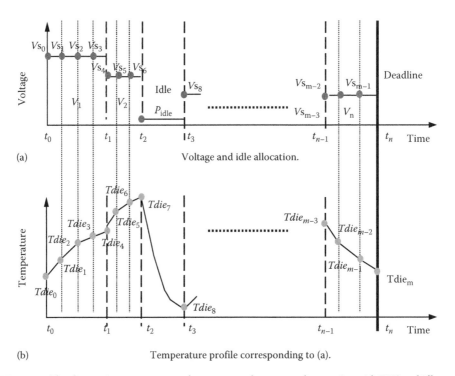

FIGURE 16.15 Two different idle time allocations; distributing idle times, as shown in (b), between task execution can reduce the temperature-induced leakage power. (a) Idle times are allocated to the remaining time after the completion of all task executions, causing high leakage power. (b) A better idle time allocation, which results in low temperature, causing less leakage power than that in (a). (Taken from Bao, M. et al., Temperature-aware idle time distribution for energy optimization with dynamic voltage scaling, in: *Proceedings of ACM/IEEE Design Automation and Test in Europe*, Dresden, Germany, pp. 21–26, 2010.)

FIGURE 16.16 The changes in temperature as the processor alternates task execution with DVS and idle state, like that in Figure 16.15b; task executions heat up the die while idle states cool down the die. (a) Voltage and idle allocation. (b) Temperature profile corresponding to (a). (Taken from Bao, M. et al., Temperature-aware idle time distribution for energy optimization with dynamic voltage scaling, in: *Proceedings of ACM/IEEE Design Automation and Test in Europe*, Dresden, Germany, pp. 21–26, 2010.)

16.9 Conclusions

In this chapter, we described the current status of the research works on DVS in view of the theoretical optimality of energy minimization. The DVS problems we covered in this chapter were intertask DVS problem, intratask DVS problem, integrated intertask and intratask DVS problem, DVS problem with sleep state, DC–DC converter-aware DVS problem, transition-aware DVS problem, and thermal-aware DVS problem. It should be mentioned that in addition to the techniques described in this chapter, there are many effective DVS techniques which are targeted under other various DVS constraints or environment such as fixed priority scheduling, jitter constraint, soft deadline, multiprocessor systems, etc.

Acknowledgment

This work was supported by Basic Science Research Program through National Research Foundation (NRF) grant funded by the Korea Ministry of Education, Science and Technology (No. 2010-0028711), and also supported by the MKE (The Ministry of Knowledge Economy), Korea, under the ITRC (Information Technology Research Center) support program supervised by the NIPA (National IT Industry Promotion Agency) (NIPA-2011-C1090-1100-0010).

References

1. A. Andrei, M. Schmitz, P. Eles, Z. Peng, and B. M. Al-Hashimi. Overhead-conscious voltage selection for dynamic and leakage energy reduction of time-constrained systems. In *Proceedings of the IEEE Design Automation and Test in Europe*, Paris, France, pp. 518–523, 2004.
2. M. Bao, A. Andrei, P. Eles, and Z. Peng. Temperature-aware idle time distribution for energy optimization with dynamic voltage scaling. In *Proceedings of the ACM/IEEE Design Automation and Test in Europe*, Dresden, Germany, pp. 21–26, 2010.
3. L. Benini, A. Bogliolo, and G. De Micheli. A survey of design techniques for system-level dynamic power management. *IEEE Transactions on Very Large Scale Integration Systems*, 8(3):299–316, 2000.
4. J.-J. Chen and T.-W. Kuo. Procrastination determination for periodic real-time tasks in leakage aware dynamic voltage scaling systems. In *Proceedings of the ACM/IEEE International Conference on Computer-Aided Design*, San Jose, CA, pp. 290–294, 2007.
5. Y. Choi, N. Chang, and T. Kim. Dc-Dc converter-aware power management for low-power embedded systems. *IEEE Transactions on Computer-Aided Design of Integrated Circuits and Systems*, 26(8): 1367–1381, 2007.
6. Analog Device, Inc. Fundamentals of phase locked loops. http://www.analog.com/library/analogDialogue/archives/30-3/single-chip.html, 2011.
7. B. Gorji-Ara, P. Chou, N. Bagherzadeh, M. Reshadi, and D. Jensen. Fast and efficient voltage scheduling by evolutionary slack distribution. In *Proceedings of the IEEE Asia-South Pacific Design Automation Conference*, Yobohama, Japan, pp. 659–662, 2004.
8. S. Irani, S. Shukla, and R. Gupta. Algorithm for power saving. *ACM Transactions on Algorithms*, 3(4):41:1–41:23, 2007.
9. T. Ishihara and H. Yasuura. Voltage scheduling problem for dynamically variable voltage processors. In *Proceedings of the ACM International Symposium on Low Power Electronic Design*, Monterey, CA, pp. 197–202, 1998.
10. R. Jayaseelan and T. Mitra. Temperature aware task sequencing and voltage scaling. In *Proceedings of the ACM/IEEE International Conference on Computer-Aided Design*, San Jose, CA, pp. 618–623, 2008.

11. R. Jejurikar and R. K. Gupta. Procrastination scheduling in fixed priority real-time systems. In *Proceedings of the ACM Conference on Languages, Compilers, and Tools for Embedded Systems*, Ottawa, Canada, pp. 57–60, 2004.
12. R. Jejurikar and R. K. Gupta. Dynamic slack reclamation with procrastination scheduling in real-time embedded systems. In *Proceedings of the ACM/IEEE Design Automation Conference*, Dana Point, CA, pp. 111–116, 2005.
13. R. Jejurikar, C. Pereira, and R. K. Gupta. Leakage aware dynamic voltage scaling for real-time embedded systems. In *Proceedings of the ACM/IEEE Design Automation Conference*, Anaheim, CA, pp. 275–280, 2004.
14. T. Kim. Application driven low-power techniques using dynamic voltage scaling. In *Proceedings of the IEEE International Conference on Embedded and Real-Time Computing Systems and Application*, Sydney, Australia, pp. 199–206, 2006.
15. T. Kim. Task-level dynamic voltage scaling for embedded system design: Recent theoretical results. *Journal of Computing Science and Engineering*, 4(3):189–206, 2010.
16. W. C. Kwon and T. Kim. Optimal voltage allocation techniques for dynamically variable voltage processors. *ACM Transactions on Embedded Computing Systems*, 4(1):211–230, 2005.
17. Y.-H. Lee, K. P. Reddy, and C. M. Krishna. Scheduling techniques for reducing leakage power in hard real-time systems. In *Proceedings of the Euromicro Conference on Real-Time Systems*, Porto, Portugal, pp. 105–112, 2003.
18. Y. Liu, H. Yang, R. Dick, H. Wang, and L. Shang. Thermal vs energy optimization for DVFS-enabled processors in embedded systems. In *Proceedings of the IEEE International Symposium on Quality Electronic Designs*, San Jose, CA, pp. 204–209, 2007.
19. S. Martin, K. Flautner, T. Mudge, and D. Blaauw. Combined dynamic voltage scaling and adaptive body biasing for low power microprocessor under dynamic workloads. In *Proceedings of the ACM/IEEE International Conference on Computer Aided Design*, San Jose, CA, pp. 721–725, 2002.
20. B. Mochocki, X. S. Hu, and G. Quan. A realistic variable voltage scheduling model for real-time applications. In *Proceedings of the ACM/IEEE International Conference on Computer-Aided Design*, San Jose, CA, pp. 726–731, 2002.
21. P. Pillai and K. G. Shin. Real-time dynamic voltage scaling for low-power embedded operating systems. In *Proceedings of the ACM Symposium on Operating Systems Principles*, Banff, Canada, pp. 89–102, 2001.
22. D. Ramanathan, S. Irani, and R. K. Gupta. Latency effects of system level power management algorithms. In *Proceedings of the ACM/IEEE International Conference on Computer-Aided Design*, San Jose, CA, pp. 111–116, 2000.
23. R. Rao, S. Vrudhula, C. Chakrabarti, and N. Chang. An optimal analytical processor speed control with thermal constraint. In *Proceedings of the ACM/IEEE International Conference on Computer Aided Design*, San Jose, CA, pp. 292–297, 2006.
24. M. T. Schmitz, B. M. Al-Hashimi, and P. Eles. Energy-efficient mapping and scheduling for DVS enabled distributed embedded systems. In *Proceedings of the IEEE Design Automation and Test in Europe*, Paris, France, pp. 514–521, 2002.
25. J. Seo, T. Kim, and K. Chung. Profile-based optimal intra-task voltage scheduling for hard real-time applications. In *Proceedings of the ACM/IEEE Design Automation Conference*, San Diego, CA, pp. 87–92, 2004.
26. J. Seo, T. Kim, and N. D. Dutt. Optimal integration of intra and inter task dynamic voltage scaling for hard real-time applications. In *Proceedings of the ACM/IEEE International Conference on Computer-Aided Design*, San Jose, CA, pp. 450–455, 2005.
27. J. Seo, T. Kim, and J. Lee. Optimal intra-task dynamic voltage scaling and its practical extensions. *IEEE Transactions on Computer-Aided Design of Integrated Circuits and Systems*, 25(1):47–57, 2006.

28. D. Shin and J. Kim. A profile-based energy-efficient intra-task voltage scheduling algorithm for hard real-time applications. In *Proceedings of the ACM International Symposium on Low-Power Electronics and Design*, Huntington Beach, CA, pp. 271–274, 2001.

29. D. Shin, J. Kim, and S. Lee. Intra-task voltage scheduling for low-energy hard real-time applications. *IEEE Design and Test of Computers*, 18(2):20–30, 2001.

30. J. Shin and T. Kim. Technique for transition energy-aware dynamic voltage assignment. *IEEE Transactions on Integrated Circuits and Systems II*, 53(9):956–960, 2006.

31. K. Skadron, T. Abdelzaher, and M. R. Stan. Control-theoretic techniques and thermal remodeling for accurate and localized dynamic thermal management. In *Proceedings of the ACM International Symposium on High Performance Computer Architecture*, Boston, MA, pp. 17–28, 2002.

32. Texas-Instruments, Inc. High-speed, fully differential, continuously variable gain amplifier. http://focus.ti.com/analog/docs, April 2010.

33. G. Varatkar and R. Marculescu. Communication-aware task scheduling and voltage selection for total systems energy minimization. In *Proceedings of the ACM/IEEE International Conference on Computer-Aided Design*, San Jose, CA, pp. 510–515, 2003.

34. F. Xie, M. Martonosi, and S. Malik. Bounds on power saving using run-time dynamic voltage scaling: An exact algorithm and linear-time heuristic approximation. In *Proceedings of the ACM International Symposium on Low Power Electronic Design*, San Jose, CA, pp. 287–292, 2005.

35. F. Yao, A. Demers, and S. Shenker. A scheduling model for reduced CPU energy. In *Proceedings of the ACM Symposium on Foundations of Computer Science*, Santa Fe, NM, pp. 374–377, 1995.

36. S. Zhang and K. S. Chatha. Approximation algorithm for the temperature-aware scheduling problem. In *Proceedings of the ACM/IEEE International Conference on Computer-Aided Design*, San Jose, CA, pp. 281–288, 2007.

37. Y. Zhang, X. Hu, and D. Z. Chen. Task scheduling and voltage selection for energy minimization. In *Proceedings of the ACM/IEEE Design Automation Conference*, San Diego, CA, pp. 183–188, 2002.

17

Speed Scaling: An Algorithmic Perspective*

Adam Wierman
*California Institute of
Technology*

**Lachlan L.H.
Andrew**
*Swinburne University of
Technology*

Minghong Lin
*California Institute of
Technology*

17.1 Introduction

Computer systems must make a fundamental trade-off between performance and energy usage. The days of "faster is better" are gone—energy usage can no longer be ignored in designs, from chips to mobile devices to data centers. This has led speed scaling, a technique once only applied in microprocessors, to become an important technique at all levels of systems. At this point, speed scaling is quickly being adopted across systems from chips [32] to disks [51] and data centers [50] to wireless devices [35].

Speed scaling works by adapting the "speed" of the system to balance energy and performance. It can be highly sophisticated—adapting the speed at all times to the current state (*dynamic speed scaling*)—or very simple—running at a static speed chosen *a priori* to balance energy and performance, except when idle (*gated-static speed scaling*) as used in current disk drives [1].

Dynamic speed scaling leads to many challenging algorithmic problems. Fundamentally, when implementing speed scaling, an algorithm must make two decisions at each time: (1) a *scheduling policy* decides

* This work is based on an earlier work: Andrew, L.H., Lin, M., and Wierman, A., Optimality, fairness and robustness in speed scaling designs, in *ACM SIGMETRICS Performance Evaluation Review*, Vol. 38, No. 1, June 1, 2010. © ACM, 2010. http://doi.acm.org/10.1145/1811099.1811044

which job(s) to serve, and (2) a *speed scaler* decides how fast to run. Further, there is an unavoidable interaction between these decisions.

The growing adoption of speed scalers and the inherent complexity of their design has led to a significant and still growing analytic literature studying the design and performance of speed scaling algorithms. The analytic study of the speed scaling problem began with the seminal work of Yao et al. [59]. Since [59], three main performance objectives have emerged: (1) the total energy used in order to meet job deadlines, e.g., [10,46] (2) the average response time given an energy/power budget, e.g., [16,60], and (3) a linear combination of expected response time and energy usage per job [3,9].

The goal of this chapter *is not* to provide a complete survey of analytic work on speed scaling. For such purposes we recommend the reader consult the recent survey [2] in addition to the current chapter. Instead, the goal of the current chapter *is* to provide examples of the practical insights about fundamental issues in speed scaling that come from the analytic work. Consequently, to limit the scope of the chapter, we focus on only the third performance objective listed previously, a linear combination of response time and energy usage, which captures how much reduction in response time justifies using one extra joule and applies where there is a known monetary cost to extra delay (e.g., in many web applications). Note that (3) is related to (2) by duality. Further, we additionally simplify the setting considered in this chapter by studying only a simple, single resource model, which is described in detail in Section 17.2.

Through the simplifications described earlier, we avoid complexity in the model and can focus on the insights provided by the results we survey. More specifically, we focus on four fundamental issues in speed scaling design for which recent analytic results have been able to provide new, practical insights.

1. *What structure do (near-)optimal algorithms have? Can an online speed scaling algorithm be optimal?*
2. *How does speed scaling interact with scheduling?*
3. *How important is the sophistication of the speed scaler? What benefits come from using dynamic versus static speed scaling?*
4. *What are the drawbacks of speed scaling?*

To provide insight into these issues, we provide an overview of recent work from two analytic frameworks—worst-case analysis and stochastic analysis. For each question we provide a summary of insights from both worst-case and stochastic analysis as well as a discussion of interesting open questions that remain. To conclude the introduction we provide a brief, informal overview of the insights into issues 1–4 that follow from recent results in the analytic community. The remainder of the chapter will then provide a more formal description of the results and insights referred to briefly in the following text.

Issue 1: The analytic results we survey prove that *energy-proportional speed scaling provides near-optimal performance*. Specifically, by using s_n, the speed to run at given n jobs, such that $P(s_n) = n\beta$ (where $P(s)$ is the power needed to run at speed s and $1/\beta$ is the price of energy), it is possible to be too competitive for general P; that is, to incur a total cost at most twice the optimum obtainable by a system with full prior knowledge. This provides analytic justification for a common heuristic applied by system designers, e.g., [13]. However, note that this algorithm does not match the offline optimal performance. In fact, no "natural" speed scaling algorithm (see Definition 17.1) can be better than 2-competitive; hence no online energy-proportional speed scaler matches the offline optimal. Thus, optimizing for energy and delay is, in some sense, fundamentally harder than optimizing for delay alone.*

Issue 2: The analytic results we survey related to issue 2 uncover two useful insights. First, at least among the policies analyzed so far, *speed scaling can be decoupled from scheduling* in that energy-proportional speed scaling provides nearly optimal performance for three common scheduling policies. Second, scheduling is not as important once energy is considered. Specifically, scheduling policies that are not constantly competitive for delay alone become $O(1)$-competitive for the linear combination of energy

---

* A simple, greedy policy (Shortest Remaining Processing Time) optimizes mean delay in a single server system of fixed rate [49].

and response time once speed scaling is considered. These insights allow designers to deal individually with two seemingly coupled design decisions.

Issue 3: The analytic results we survey highlight that *increased sophistication provides minimal performance improvements in speed scaling designs*. More specifically, the optimal gated-static speed scaling performs nearly as well as the optimal dynamic speed scaler. However, increased sophistication does provide improved robustness. These insights have the practical implication that it may be better to design "optimally robust" speeds instead of "optimal" speeds, since robustness is the main benefit of dynamic scaling.

Issue 4: The analytic results we survey show that, unfortunately, *dynamic speed scaling can magnify unfairness*. This is perhaps surprising at first, but follows intuitively from the fact that dynamic speed scaling selects faster speeds when there are more jobs in the system: If some property of a job is correlated with the occupancy while it is in service, then dynamic speed scaling gives an unfairly high service rate to jobs with that property. In contrast to the conclusion for issue 2, these results highlight that designers should be wary about the interaction between the scheduler and speed scaler when considering fairness.

17.2 Model and Notation

To highlight the insights that follow from analytic results, we focus on a simple, single server model and ignore many practical issues such as deadlines and the costs of switching speeds. It is important to emphasize, however, that there is a broad literature of analytic work that has begun to study the impact of such issues, and pointers to this work are given.

Throughout this chapter we consider a single server with two control decisions: speed scaling and scheduling. Further, we take as the performance objective the minimization of a linear combination of the expected time between the arrival of a job and its completion (called response time, sojourn time, or flow time), denoted by T, and energy usage per job, $\mathcal{E}$:

$$z = \mathbb{E}[T] + \mathbb{E}[\mathcal{E}]/\beta. \tag{17.1}$$

Using Little's law, this may be more conveniently expressed as

$$\lambda z = \mathbb{E}[N] + \mathbb{E}[P]/\beta \tag{17.2}$$

where
    N is the number of jobs in the system
    $P = \lambda \mathcal{E}$ is the power expended

Note that though we take a linear combination of energy and response time as our performance metric, there is a wide variety of other metrics that have been studied in the speed scaling literature. The first to be studied in an algorithmic context was minimizing the total energy consumption given deadlines for each job [11,44,59]. In this setting, the energy reduction incurs no performance penalty, since it is assumed that there is no benefit to finishing jobs before the deadline. Another objective is to minimize the makespan (time to finish all jobs) given a constraint on the total energy [17,44]. This is called the "laptop problem," since the constrained energy consumption matches the fixed capacity of a portable device's battery. Circuit designers are often interested in the product of power and delay [20], which represents the energy per operation. Many other formulations have been considered, such as minimizing delay given a trickle-charged energy storage [8], and cases in which jobs need not be completed but can provide revenue dependent on their delay [45].

Further, though we focus only on a single server, there is recent work that studies speed scaling in multiserver environments as well in the context of jobs with deadlines [4,27], the laptop problem [44], or with the preceding objective [27,37,38,53].

17.2.1 Modeling Energy Usage

To model the energy usage, first let $n(t)$ denote the number of jobs in the system at time t and $s(t)$ denote the speed at which the system is running at time t. Further, define $P(s)$ as the power needed to run at speed s. Then, the energy used until time t is $\mathcal{E}(t) = \int_0^t P(s(\tau))d\tau$.

In many applications $P(s) \approx ks^\alpha$ with $\alpha \in (1,3)$. For example, for dynamic power in CMOS chips $\alpha \approx 1.8$ [57]. However, wireless communication has an exponential P [22] or even unbounded power at finite rate [28]. Some of the results we discuss assume a polynomial form, and simplify specifically for $\alpha = 2$. Others hold for general, even nonconvex and discontinuous, power functions. Several results are limited to *regular* power functions [9], which are differentiable on $[0,\infty)$, strictly convex, nonnegative, and have $P(0) = 0$.

Note that implicit in the previous model is the assumption that all jobs have the same $P(s)$. In practice, different classes of jobs may have different mixes of requirements for CPU, disk, IO, etc. This leads to each class of jobs having a different $P(s)$ function characterizing its power-speed trade-off curve.

17.2.2 Modeling Speed Scaling and Scheduling

A speed scaling algorithm $\mathcal{A} = (\pi, \Sigma)$ consists of a scheduling discipline π that defines the order in which jobs are processed, and a speed scaler Σ that defines the speed as a function of system state, in terms of the power function, P. Throughout, we consider the speed to depend only on the number of jobs in the system, i.e., s_n is the speed when the occupancy is n.*

We consider online preempt-resume schedulers, which are not aware of a job j until it arrives at time $r(j)$, at which point they learn its size, x_j, and which may interrupt serving a job and later restart it from the point it was interrupted without overhead. We focus on shortest remaining processing time (SRPT), which preemptively serves the job with the least remaining work, and processor sharing (PS), which serves all jobs in the system at equal rate. SRPT is important due to its optimality properties: in a classical, constant speed system, SRPT minimizes the expected response time [49]. PS is important because it is a simple model of scheduling that is commonly applied for, e.g., operating systems and web servers. A third policy that we discuss occasionally is first come first served (FCFS), which serves jobs in order of arrival.

The speed scaling rules, s_n, we consider can be *gated-static*, which runs at a constant speed while the system is nonempty and sleeps while the system is empty, i.e., $s_n = s_{gs}1_{n\neq 0}$, or more generally *dynamic* $s_n = g(n)$ for some function $g : \mathbb{N} \cup \{0\} \to [0,\infty)$. To be explicit, we occasionally write s_n^π as the speed under policy π when the occupancy is n. The queue is single-server in the sense that the full speed s_n can be devoted to a single job.

17.2.3 Analytic Framework

Analytic research studying speed scaling has used two distinct approaches: Worst-case analysis and stochastic analysis. We discuss results from both settings in this paper, and so we briefly introduce the settings and notation for each in the following.

---

* This suits objective (17.1); e.g., it is optimal for an isolated batch arrival, and the optimal s is constant between arrival/departures. For other objectives, it is better to base the speed on the unfinished work instead [12].

17.2.3.1 Notation for the Worst-Case Model

In the worst-case model we consider finite, arbitrary (maybe adversarial) deterministic instances of ν arriving jobs. Let $\mathcal{E}(I)$ be the total energy used to complete instance I, and T_j be the response time of job j, the completion time minus the release time $r(j)$. The analog of (17.1) is to replace the ensemble average by the sample average, giving the cost of an instance I under a given algorithm $\mathcal{A}$ as

$$z^{\mathcal{A}}(I) = \frac{1}{\nu}\left(\sum_j T_j + \frac{1}{\beta}\mathcal{E}(I)\right). \tag{17.3}$$

In this model, we compare the cost of speed scaling algorithms to the cost of the optimal offline algorithm, OPT. In particular, we study the competitive ratio, defined as

$$CR = \sup_I z^{\mathcal{A}}(I)/z^O(I),$$

where $z^O(I)$ is the optimal cost achievable on I. A scheme is "*c*-competitive" if its competitive ratio is at most c.

17.2.3.2 Notation for the Stochastic Model

In the stochastic model, we consider an M/GI/1 (or sometimes GI/GI/1) queue with arrival rate λ. Let X denote a random job size with cumulative distribution function (c.d.f.) $F(x)$, complementary c.d.f. (c.c.d.f.) $\bar{F}(x) = 1 - F(x)$, and continuous probability density function (p.d.f.) $f(x) = dF(x)/dx$. Let $\rho = \lambda\mathbb{E}[X] \in [0, \infty)$ denote the load of arriving jobs. Note that ρ is not the utilization of the system and that many dynamic speed scaling algorithms are stable for all ρ. When the power function is $P(s) = s^{\alpha}$, it is natural to use a scaled load, $\gamma := \rho/\beta^{1/\alpha}$ (see [57]).

Denote the response time of a job of size x by $T(x)$. We consider the performance metric (17.1) where the expectations are averages per job. In this model the goal is to optimize this cost for a specific workload, ρ. Let z^O be the average cost of the optimal offline algorithm, and define the competitive ratio in the M/GI/1 model as

$$CR = \sup_{F, \lambda} z^{\mathcal{A}}/z^O.$$

17.3 Optimal Speed Scaling

The most natural starting point for discussing speed scaling algorithms is to understand

What structure do (near-)optimal algorithms have? Can a speed scaling algorithm be optimal?

In this section, we discuss the stream of research that has sought to answer these questions.

To begin, recall that a speed scaling algorithm depends on two choices, the scheduler and the speed scaler. However, when determining the optimal speed scaling algorithm, the optimality properties of SRPT in the static speed setting [49] make it a natural choice for the scheduler. In fact, it is easy to argue that any speed scaling algorithm which does not use SRPT can be improved by switching the scheduler to SRPT and leaving the speeds $s(t)$ unchanged [9].

Thus, in this section we focus on speed scaling algorithms that use SRPT scheduling and seek to understand the structure of the (near-)optimal speeds. The study of optimal speed scaling algorithms has primarily been performed in the worst-case analytic framework. The reason for this is simple: SRPT is difficult to analyze in the stochastic model even in the case when the speed is fixed. So, when dynamic, controllable speeds are considered the analysis is, to this point, seemingly intractable.

However, in the worst-case framework considerable progress has been made. The focus of this research has primarily been on one particularly promising algorithm: $(SRPT, P^{-1}(n))$, which uses policy $\pi = SRPT$ and speed scaler Σ which sets $s_n = P^{-1}(n)$, and there has been a stream of upper bounds on its competitive ratio for objective (17.1): for unit-size jobs in [3,12] and for general jobs with $P(s) = s^{\alpha}$ in [7,39]. A major breakthrough was made in [9], which shows the 3-competitiveness of $(SRPT, P^{-1}(n+1))$ for general P. Following this breakthrough, [5] was able to use similar techniques to reduce the competitive ratio slightly and prove a matching lower bound. In particular, [5] establishes the following:

Theorem 17.1   *For any regular power function P, $(SRPT, P^{-1}(n\beta))$ has a competitive ratio of exactly 2.*

Thus, $(SRPT, P^{-1}(n))$ is not optimal, in that an offline algorithm could do better. However, it is highly robust—guaranteeing to provide cost within a factor of two of optimal.

We will not dwell on the analytic approaches in this chapter; however, it is useful to give some feel for the proof techniques for the major results. In this case, the proof uses amortized local competitiveness arguments with the potential function

$$\Phi(t) = \int_0^\infty \sum_{i=1}^{n[q;t]} \frac{1+\eta}{\beta} P'\left(P^{-1}(i\beta)\right) dq, \tag{17.4}$$

where $n[q;t] = \max(0, n^{\mathcal{A}}[q;t] - n^O[q;t])$ with $n^{\mathcal{A}}[q;t]$ and $n^O[q;t]$ the number of unfinished jobs at time t with remaining size at least q under $(SRPT, P^{-1}(n\beta))$ (algorithm $\mathcal{A}$) and the optimal (offline) algorithm O, respectively. Specifically, it first shows that

$$z^{\mathcal{A}}(t) + \frac{d\Phi}{dt} \le (1+\eta)z^O(t), \tag{17.5}$$

where $z^{\mathcal{A}}(t)$ and $z^O(t)$ are the instantaneous costs of $\mathcal{A}$ and O respectively, given by $n(t) + P(s(t))/\beta$. It then shows that, despite discontinuities, integrating this with $\eta = 1$ gives the preceding theorem.

The following more general form can be proven analogously.

Corollary 17.2   *Let $\eta \ge 1$ and Φ be given by (17.4). Any scheme $\mathcal{A} = (SRPT, s_n)$ with $s_n \in [P^{-1}(n\beta), P^{-1}(\eta n\beta)]$ is $(1+\eta)$-competitive.*

The previous results motivate the question:

Are there other speed scaling algorithms with lower competitive ratios?

A recent result [5] suggests that the answer is "No." In particular, no "natural" speed scaling algorithm can be better than 2-competitive.

Definition 17.1   *A speed scaling algorithm $\mathcal{A}$ is defined to be **natural** if it runs at speed s_n when it has n unfinished jobs, and for convex P, one of the following holds:*

   (a) The scheduler is work conserving and works on a single job between arrival/departure events
   (b) $g(s) + P(s)/\beta$ is convex, for some g with $g(s_n) = n$
   (c) The speeds s_n satisfy $P(s_n) = \omega(n)$
   (d) The speeds s_n satisfy $P(s_n) = o(n)$

Theorem 17.3   *For any $\varepsilon > 0$ there is a regular power function P_ε such that any natural algorithm $\mathcal{A}$ on P_ε has competitive ratio larger than $2 - \varepsilon$.*

Note that natural algorithms include SRPT scheduling for any s_n, energy-proportional speeds for all schedulers and any algorithm with speeds consistently faster or slower than energy proportional.

We conjecture that Theorem 17.3 holds for all speed scaling algorithms, which would imply that the competitive ratio of (SRPT, $P^{-1}(n\beta)$) is minimal.

Interestingly, the proof considers only two classes of workloads: one consisting of an isolated burst of jobs and the other consisting of a long train of uniformly spaced jobs. Any "natural" scheduler which processes the long train fast enough to avoid excessive delay costs must process the burst so fast that it incurs excessive energy costs; thus optimizing over these two specific workloads is already impossible for a "natural" speed scaler.

17.3.1 Open Questions

The results in the preceding text highlight that energy-proportional speed scaling ($P(s_n) = n\beta$) is nearly optimal, which provides analytic justification of a common design heuristic, e.g., [13]. Further, they highlight that optimizing for the combination of energy and delay is fundamentally harder than optimizing for delay alone (where SRPT is optimal, i.e., 1-competitive). However, they also leave a number of interesting and important questions unanswered. Three such questions that are particularly interesting are the following.

First, the fact that Theorem 17.3 holds only for "natural" algorithms is bothersome and likely only a technical restriction. It would be very nice to remove this restriction and provide a true lower bound on the achievable competitive ratio.

Additionally, the fact that (SRPT, $P^{-1}(n)$) is exactly 2-competitive for every power function highlights that it may be possible to outperform this algorithm for specific power functions. In particular, for the realistic case of $P(s) = s^\alpha$ with $\alpha \in (1, 3)$ it is possible to derive other speed scalers besides $P^{-1}(n)$ that have competitive ratios smaller than 2. In the extreme case that P is concave, it is optimal to run at the maximum possible speed whenever there is work in the system and be idle otherwise. It would be interesting to understand in general what the lower bound on the achievable competitive ratio for a fixed $P(s)$ is, and what speed scaler can achieve it.

Finally, a third open problem is to understand the optimal speeds for SRPT scheduling in the stochastic model, i.e., in an M/GI/1 queue. This problem is completely open, and seemingly difficult, but would provide interesting new insight.

These three open questions all relate to the simple model which is the focus of this chapter; however, the issues discussed in this section are clearly also interesting (and open) when the model is made more complex. For example, by allowing class-dependent $P(s)$ functions.

17.4 Interaction between Speed Scaling and Scheduling

In the prior section, we focused on optimal speed scaling algorithms and thus limited our attention to SRPT. However, in practice, SRPT may not be appropriate, for a variety of reasons. In many situations, other scheduling policies are desirable (or required) and in particular, policies such as PS are used. Once a different scheduling policy is considered it is not clear how the optimal speeds should be adjusted. In particular, there is a clear coupling between the speed scaling and the scheduling. So, the focus of this section is to understand

How does speed scaling interact with scheduling?

To shed light on this question we focus primarily on PS in this section, due to its importance in application such as operating systems and web servers. However, the insights we describe certainly hold more generally.

There is work relating to the interaction between scheduling and speed scaling from both the worst-case and stochastic frameworks, so we will discuss each briefly. Interestingly, the two distinct sets of results highlight similar insights. In particular they highlight that (1) scheduling and speed scaling can be largely decoupled with minimal performance loss and (2) scheduling is much less important when considering speed scaling than it is in standard models with constant speeds.

17.4.1 Worst-Case Analysis

In contrast to the long stream of research focusing on SRPT, until recently there was no worst-case analysis of speed scaling under PS. The first result relating to PS is from [5], which shows that (PS, $P^{-1}(n\beta)$) is $O(1)$-competitive for $P(s) = s^\alpha$ with fixed α. In particular

Theorem 17.4 *If $P(s) = s^\alpha$ then (PS, $P^{-1}(n\beta)$) is $\max(4\alpha - 2, 2(2 - 1/\alpha)^\alpha)$-competitive.*

As with SRPT, this admits a more general corollary

Corollary 17.5 *For s_n given in Corollary 17.2, the scheme (PS, s_n) is $O(1)$-competitive in the size of the instance.*

Notice that the competitive ratio in Theorem 17.4 reduces to $(4\alpha - 2)$-competitive for the α typical in CPUs, i.e., $\alpha \in (1, 3]$.

The proof of Theorem 17.4 in [5] builds on the analysis in [18] of LAPS, another blind policy. (LAPS, $P^{-1}(n\beta)$) is also $O(1)$-competitive in this case, and actually has better asymptotic performance for large α. However, as $\alpha \in (1, 3]$ in most computer subsystems, the performance for small α is more important than asymptotics in α, and PS outperforms LAPS in that case. In both the case of PS and LAPS the proof uses the potential function

$$\Phi = (1 + \eta)(2\alpha - 1)(H/\beta)^{1/\alpha} \sum_{i=1}^{n^{\mathcal{A}}(t)} i^{1-1/\alpha} \max(0, q^{\mathcal{A}}(j_i; t) - q^O(j_i; t)) \qquad (17.6)$$

where

$q^\pi(j; t)$ is the remaining work on job j at time t under scheme $\pi \in \{\mathcal{A}, O\}$

$\{j_i\}_{i=1}^{n^{\mathcal{A}}(t)}$ is an ordering of the jobs in increasing order of release time: $r(j_1) \leq r(j_2) \leq \cdots \leq r(j_{n^{\mathcal{A}}(t)})$.

This result has been generalized in [19] to show that weighted PS is again $O(1)$-competitive in the case when jobs have differing weights. If the processor has a maximum speed, then the result still holds under a resource augmentation assumption.

There are a number of remarks we would like to highlight about Theorem 17.4. First, notice that this scheme uses the same energy-proportional speeds $s_n = P^{-1}(n\beta)$ that achieve the optimal competitive ratio under SRPT. Further, the same speeds are also used for LAPS in [18]. This suggests that the two design decisions of speed scaling and scheduling are not as closely coupled as may be first thought.

A second remark about Theorem 17.4 is that, although this competitive ratio of around 10 is larger than may be hoped, it is very much smaller than the competitive ratio of PS on a fixed-speed processor. On a fixed-speed processor, PS has an unbounded competitive ratio: PS is $\Omega(v^{1/3})$-competitive for instances with v jobs [43]. This is because PS is "blind" to the job sizes, which can cause it to make a poor decision early on, from which it can never recover. With unbounded speeds and dynamic speed scaling, the processor can always run a little faster to recover quickly, and thus continue with a modest additional cost. Note however, that difference in cost across scheduling policies grows if there is a bounded maximum speed.

To explain the limited impact of scheduling, one can view speed scaling as a physical realization of the analytical technique of resource augmentation [31]. The latter shows that many algorithms with high competitive ratios could achieve almost the same cost as the optimal algorithm, if they were given the advantage of a slightly faster processor.

Finally, a third remark about Theorem 17.4 is that it is much weaker than Theorem 17.1 for SRPT: Not only is the competitive ratio larger, but the result is only valid for power functions of the form $P(s) = s^\alpha$, and even depends strongly on the exponent α. This is not simply because the result is incomplete, but because the problem is fundamentally less well approximable. In fact, all policies blind to job sizes have unbounded competitive ratio as $\alpha \to \infty$ [18], which shows that the form of power function must unavoidably enter any competitive ratio result.

17.4.2 Stochastic Analysis

In contrast to the limited amount of research studying PS in the worst-case framework, there has been a substantial amount of work applicable to PS in the stochastic framework, due to its analytic tractability. In particular, in the M/GI/1 model, [15,23,26,55]* study speed scaling in the context of generic "operating costs."

This work formulates the determination of the optimal speeds as a dynamic programming (DP) problem and provides numerical techniques for determining the optimal speeds. Further, it has been proven that the optimal speeds are monotonic in the queue length [52] and that the optimal speeds can be bounded as follows [57]. Recall that $\gamma = \rho/\beta^{1/\alpha}$.

Proposition 17.6 *Consider an M/GI/1 PS queue with controllable service rates s_n. Let $P(s) = s^\alpha$. The optimal dynamic speeds are concave and satisfy the dynamic program given in [57]. For $\alpha = 2$ and any $n \geq 2\gamma$, they satisfy*

$$\gamma + \sqrt{n - 2\gamma} \leq \frac{s_n^{DP}}{\sqrt{\beta}} \leq \gamma + \sqrt{n} + \min\left(\frac{\gamma}{2n}, \gamma^{1/3}\right). \tag{17.7}$$

For general $\alpha > 1$, they satisfy

$$\frac{s_n^{DP}}{\beta^{1/\alpha}} \leq \left(\frac{n}{\alpha - 1}\right)^{1/\alpha} + \gamma\frac{1}{\alpha - 1} + O\left(n^{-1/\alpha}\right) \tag{17.8}$$

$$\frac{s_n^{DP}}{\beta^{1/\alpha}} \geq \left(\frac{n}{\alpha - 1}\right)^{1/\alpha}. \tag{17.9}$$

Interestingly, the bounds in Proposition 17.6 are tight for large n and have a form similar to the form of the worst-case speeds for SRPT and PS in Theorems 17.1 and 17.4.

The asymptotic form of these results also matches the form of the optimal speed when there are n jobs in the system and no more jobs will arrive before the queue is emptied, as occurs when the instance is a single batch arrival. For both SRPT and PS, this batch speed is given by the solution s_n^* to

$$\beta n = s_n^* P'\left(s_n^*\right) - P\left(s_n^*\right). \tag{17.10}$$

This is not surprising, since when the backlog is large, the arrival rate will be negligible compared with the optimal speed. This insight is further supported by looking at the fluid model, e.g., [21], where the differences between scheduling policies are negligible.

As we have discussed previously, the analysis of the optimal dynamic speeds given SRPT scheduling is seemingly intractable. However, taking inspiration from the fact that the speeds given by the worst-case analysis of SRPT and PS match, combined with the fact that (17.10) is the optimal batch speed for both

---

* Since M/GI/1 PS with controllable service rates is symmetric [33], it has the same occupancy distribution and mean delay as M/M/1 FCFS studied in the references.

SRPT and PS and also matches the asymptotic form of the stochastic results for PS in Proposition 17.6, a natural proposal for the dynamic speeds under SRPT is to use the optimal PS speeds.

It turns out that this heuristic works quite well. In particular the simulation experiments in Figure 17.1 compare the performance of this dynamic programming speed scaling algorithm ("DP") both to the form $P^{-1}(n\beta)$ motivated by the worst-case results ("INV") and to the following lower bound, which was proven by [57] and holds for all scheduling policies.

Proposition 17.7   *In a GI/GI/1 queue with $P(s) = s^{\alpha}$,*

$$z^O \geq \frac{1}{\lambda} \max \left(\gamma^{\alpha}, \gamma \alpha (\alpha - 1)^{(1/\alpha)-1} \right).$$

The previous was stated in [57] for M/GI/1 PS, but the same proof holds more generally.

Simulation experiments also allow us to compare (1) the performance of the worst-case schemes for SRPT and PS with the stochastic schemes and (2) the performance of SRPT and PS in the speed scaling model. In these experiments, the optimal speeds for PS in the stochastic model are found using the numerical algorithm described in [26,57], and then these speeds are also used for SRPT.

Figure 17.2 shows that the optimal speeds from the DP have a similar form to the speeds $P^{-1}(n\beta)$ appearing in the worst-case results, differing by γ for high queue occupancies. Figure 17.1 shows how the total cost (17.1) depends on the choice of speeds and scheduler. At low loads, all schemes are indistinguishable. At higher loads, PS-INV degrades significantly, but SRPT-INV maintains good performance.

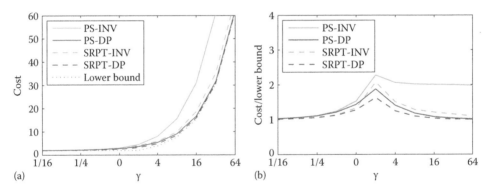

FIGURE 17.1   Comparison of SRPT and PS under both $s_n = P^{-1}(n\beta)$ and speeds optimized for an M/GI/1 PS system, using Pareto(2.2) job sizes and $P(s) = s^2$.

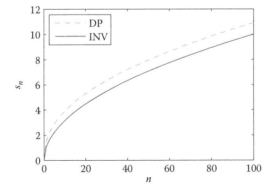

FIGURE 17.2   Comparison of $s_n = P^{-1}(n\beta)$ with speeds "DP" optimized for an M/GI/1 system with $\gamma = 1$ and $P(s) = s^2$.

Note though that if $P(s) = s^\alpha$ for $\alpha > 3$, SRPT-INV degrades significantly too. In contrast, the DP-based schemes benefit significantly from having the slightly higher speeds chosen to optimize (17.1) rather than minimize the competitive ratio. Finally, SRPT-DP performs nearly optimally, which justifies the heuristic of using the optimal speeds for PS in the case of SRPT. However, PS-DP performs nearly as well as SRPT-DP. Together, these observations suggest that it is important to optimize the speed scaler, but not necessarily the scheduler.

17.4.3 Open Questions

The results in this section have highlighted a number of important insights regarding the decoupling of speed scaling and scheduling. In particular, it seems that energy-proportional speed scaling ($s_n = P^{-1}(n)$) performs well for a variety of scheduling policies and it seems that performance differences between scheduling policies shrink dramatically when speed scaling is considered.

However, to this point these insights are derived by comparing a small number of common policies. We conjecture that these insights will apply much more broadly; however, it is an open question to understand exactly how broadly they apply. More specifically, it would be interesting to characterize the class of scheduling policies for which $s_n = P^{-1}(n)$ guarantees $O(1)$-competitiveness. This would jointly highlight a decoupling between scheduling and speed scaling and show that scheduling has much less impact on performance. The earlier question proposes to study the decoupling of scheduling and speed scaling in the worst-case framework. In the stochastic framework however, the parallel question can be asked: for what class of policies will $s_n = P^{-1}(n)$ guarantee $O(1)$-competitiveness in the M/GI/1 model?

17.5 Impact of Speed Scaler Sophistication

Up until this point we have focused on dynamic speed scaling, which allows the speed to change as a function of the number of jobs in the system. Dynamic speed scaling can perform nearly optimally; however its complexity may be prohibitive. In contrast, *gated-static* speed scaling, where $s_n = s_{gs} 1_{n \neq 0}$ for some constant speed s_{gs}, requires minimal hardware to support; e.g., a CMOS chip may have a constant clock speed but AND it with the gating signal to set the speed to 0.

In this section our goal is to understand

How important is the sophistication of the speed scaler? What benefits come from using dynamic versus static speed scaling?

Note that we can immediately see that gated-static speed scaling can be arbitrarily bad in the worst case since jobs can arrive faster than s_{gs}. Thus, we study gated-static speed scaling only in the stochastic model, where the constant speed s_{gs} can depend on the load.

We focus our discussion on SRPT and PS, though the insights we highlight do not depend on the workings of these particular policies and should hold more generally.

Overall, there are three main insights that have emerged from the literature regarding the aforementioned questions. First, and perhaps surprisingly, the simplest policy (gated-static) provides nearly the same expected cost as the most sophisticated policy (optimal dynamic scaling). Second, the performance of gated-static under PS and SRPT are not too different, thus scheduling is less important to optimize than in systems in which the speed is fixed in advance. This aligns with the observations discussed in the previous section for the case of dynamic speed scaling. Third, though dynamic speed scaling does not provide significantly improved cost, it does provide significantly improved robustness (e.g., to time-varying and/or uncertain workloads).

17.5.1 Optimal Gated-Static Speeds

To begin our discussion, we must first discuss the optimal speeds to use for gated-static designs under both PS and SRPT. In the case of PS, this turns out to be straightforward; however in the case of SRPT it is not.

Before stating the results, note that since the power cost is constant at $P(s_{gs})$ whenever the server is running, the optimal speed is

$$s_{gs} = \arg\min_s \beta \mathbb{E}[T] + \frac{1}{\lambda} P(s) \Pr(N \neq 0). \tag{17.11}$$

In the second term $\Pr(N \neq 0) = \rho/s$, and so multiplying by λ and setting the derivative to 0 gives that the optimal gated-static speed satisfies

$$\beta \frac{d\mathbb{E}[N]}{ds} + r \frac{P^*(s)}{s} = 0, \tag{17.12}$$

where $r = \rho/s$ is the utilization and $P^*(s) \equiv sP'(s) - P(s)$. Note that if P is convex, then P^* is increasing and if P'' is bounded away from 0 then P^* is unbounded.

Under PS, $\mathbb{E}[N] = \rho/(s - \rho)$, and so $d\mathbb{E}[N]/ds = \mathbb{E}[N]/(\rho - s)$. By (17.12), the optimal speeds satisfy [57]

$$\beta \mathbb{E}[N] = (1 - r)rP^*(s). \tag{17.13}$$

For SRPT, things are not as easy. For $s = 1$, we have [36]

$$\mathbb{E}[T] = \int\limits_{x=0}^{\infty} \int\limits_{t=0}^{x} \frac{dt}{1 - \lambda \int_0^t \tau \, dF(\tau)} + \frac{\lambda \int_0^x \tau^2 \, dF(\tau) + x^2 \bar{F}(x)}{2(1 - \lambda \int_0^x \tau \, dF(\tau))^2} \, dF(x)$$

The complexity of this equation rules out calculating the speeds analytically. So, instead [5] derives simpler forms for $\mathbb{E}[N]$ that are exact in asymptotically heavy or light traffic. Note that using a heavy-traffic approximation is natural for this setting because, if energy is important (β small), then the speed will be provisioned so that the server is in "high" load.

We state the heavy-traffic results for distributions whose c.c.d.f. $\bar{F}$ has lower and upper Matuszewska indices [14] of m and M. Intuitively, this means that $C_1 x^m \lesssim \bar{F}(x) \lesssim C_2 x^M$ as $x \to \infty$ for some C_1, C_2, so the Matuszewska index can be thought of as a "moment index." Further, let $G(x) = \int_0^x tf(t) \, dt/\mathbb{E}[X]$ be the fraction of work coming from jobs of size at most x, and define $h(r) = (G^{-1})'(r)/G^{-1}(r)$. The following was proven in [41].

Proposition 17.8　*For an M/GI/1 under SRPT with speed 1, $\mathbb{E}[N] = \Theta(H(\rho))$ as $\rho \to 1$, where*

$$H(\rho) = \begin{cases} E[X^2]/((1 - \rho)G^{-1}(\rho)) & \text{if } M < -2 \\ E[X]\log(1/(1 - \rho)) & \text{if } m > -2. \end{cases} \tag{17.14}$$

For the case of unit speed, this suggests the heavy traffic approximation

$$\mathbb{E}[N] \approx CH(\rho) \tag{17.15}$$

where the constant C depends on the distribution of job sizes. This is validated numerically in Figure 17.3.

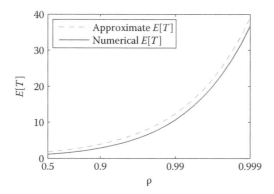

FIGURE 17.3   Validation of the heavy-traffic approximation (17.15) by simulation using Pareto(3) job sizes with $E[X] = 1$.

Theorem 17.9   *Under the assumption that (17.15) holds with equality*

(i) *If $M < -2$, then for the optimal gated-static speed,*

$$\beta \mathbb{E}[N] \left(\frac{2-r}{1-r} - rh(r) \right) = rP^*(s). \tag{17.16a}$$

(ii) *If $m > -2$, then for the optimal gated-static speed,*

$$\beta \mathbb{E}[N] \left(\frac{1}{(1-r)\log(1/(1-r))} \right) = P^*(s). \tag{17.16b}$$

Moreover, if $P^*(s)$ is unbounded as $s \to \infty$ and $-2 \notin [m, M]$ then as $\rho \to \infty$, (17.16) induces the heavy-traffic regime, $\rho/s \to 1$.

Although the heavy-traffic results hold for large ρ, for small ρ the performance of the optimal GS speed for PS gives better performance. To combine these, [5] proposes using the speed

$$s_{gs}^{SRPT} = \min \left(s_{gs}^{PS}, s_{gs}^{SRPT(HT)} \right), \tag{17.17}$$

where s_{gs}^{PS} satisfies (17.13), and $s_{gs}^{SRPT(HT)}$ is given by (17.16) with $\mathbb{E}[N]$ estimated by (17.15).

17.5.2 Dynamic versus Gated-Static: Cost Comparison

Given the optimal gated-static speeds derived in the previous section, we can now contrast the cost of gated-static, the simplest scheme, with that of dynamic speed scaling, the most sophisticated, under both SRPT and PS.

As Figure 17.4 shows, the performance of a well-tuned gated-static system is almost indistinguishable from that of the optimal dynamic speeds. Moreover, there is little difference between the cost under PS-GATED and SRPT-GATED, again highlighting that scheduling becomes less important when speed can be scaled than in traditional models.

Additionally, [57] has recently provided analytic support for the empirical observation that gated-static schemes are near optimal. Specifically, in [57] it was proven that PS-GATED is within a factor of 2 of PS-DP when $P(s) = s^2$. With the results of Section 17.3, this implies the following.

Corollary 17.10   *If $P(s) = s^2$ then the optimal PS and SRPT gated-static designs are $O(1)$-competitive in an M/GI/1 queue with load ρ.*

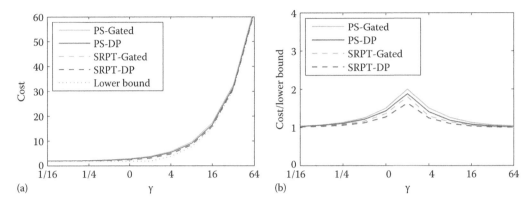

FIGURE 17.4 Comparison of PS and SRPT with gated-static speeds (17.13) and (17.17), versus the dynamic speeds optimal for an M/GI/1 PS. Job sizes are distributed as Pareto(2.2) and $P(s) = s^2$.

17.5.3 Dynamic versus Gated-Static: Robustness Comparison

The previous section shows that near-optimal performance is obtained by running at a static speed when not idle. *Why then do CPUs have multiple speeds?* The reason is that the optimal gated-static design depends intimately on the load ρ. This cannot be exactly known in advance, especially since workloads vary with time. So, an important property of a speed scaling design is *robustness* to uncertainty in the workload (ρ and F) and to model inaccuracies.

Figure 17.5 shows that the performance of gated-static degrades dramatically when ρ is mispredicted. If the load is lower than expected, excess energy will be used; if the system has static speed s and $\rho \geq s$ then the cost is unbounded. In contrast, Figure 17.5 also shows that dynamic speed scaling (SRPT-DP) is significantly more robust to misprediction of the workload.

Recently [5], proved the robustness of SRPT-DP and PS-DP by providing worst-case guarantees for SRPT-DP and PS-DP showing that they are both constantly competitive. These results are nearly unique in that they provide worst-case guarantees for stochastic control policies. In the following, let s_n^{DP} denote the speeds used for SRPT-DP and PS-DP.

Corollary 17.11 *Consider $P(s) = s^\alpha$ with $\alpha \in (1, 2]$ and algorithm $\mathcal{A}$ which chooses speeds s_n^{DP} optimal for PS in an M/GI/1 queue with load ρ. If $\mathcal{A}$ uses either PS or SRPT, then $\mathcal{A}$ is $O(1)$-competitive in the worst case.*

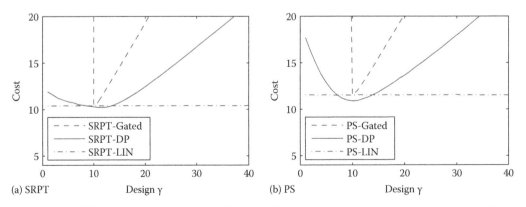

FIGURE 17.5 Effect of misestimating γ under PS and SRPT: cost when $\gamma = 10$, but s_n are optimal for a different "design γ." Pareto(2.2) job sizes; $P(s) = s^2$.

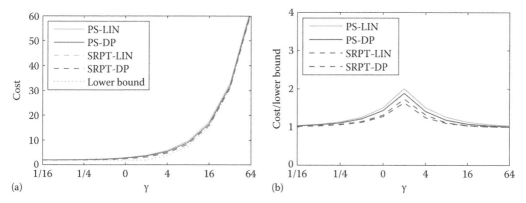

FIGURE 17.6 Comparison of PS and SRPT with linear speeds, $s_n = n\sqrt{\beta}$, and with dynamic speeds optimal for PS. Job sizes are Pareto(2.2) and $P(s) = s^2$.

The proof of the result follows from combining (17.8) and (17.9) with Corollaries 17.2 and 17.5.

Notice that for $\alpha = 2$, Proposition 17.6 implies $s_n^{DP} \leq (2\gamma + 1)P^{-1}(n\beta)$, whence (SRPT, s_n^{DP}) is $(4\gamma^2 + 4\gamma + 2)$-competitive.

Finally, though Corollary 17.11 shows that s_n^{DP} designed for a given ρ leads to a "robust" speed scaler, note that the cost still degrades significantly when ρ is mispredicted badly (see Figure 17.5). To address this issue, we consider a second, weaker form of robustness: robustness only to misestimation of ρ, not to the underlying stochastic model. In particular, in this model the arrivals are known to be Poisson, but ρ is unknown. In this setting, it was shown in [57] that using "linear" speeds, $s_n = n\sqrt{\beta}$, gives near-optimal performance when $P(s) = s^2$ and PS scheduling is used. Specifically, this scheme ("LIN") outperforms the other load-independent scaling, $s_n = P^{-1}(n\beta)$, as shown in Figure 17.5. Further, the decoupling of scheduling and speed scaling suggested in Section 17.4 motivates using LIN also for SRPT. Figure 17.6 also shows that SRPT-LIN is again nearly optimal when the exact load is known. However, it is important to note that LIN is not robust to model inaccuracies: it is not $O(1)$-competitive for non-Poisson workloads.

17.5.4 Open Questions

This section has highlighted that the increased sophistication of dynamic speed scaling as compared with gated-static speed scaling does not provide much reduction in cost. However, it does provide at least one important benefit—improved robustness. This message of "sophistication improves robustness" is supported by results that combine worst-case and stochastic analysis to provide worst-cast guarantees for optimal stochastic control.

One key implication of these insights is that dynamic speed scalers should be designed with robustness as the goal, not cost minimization (since cost can be minimized without dynamic speeds). Thus, an open question is how to formulate and solve a stochastic control problem for optimizing robustness of speed scalers.

A second set of open problem relates to Corollary 17.11, which provides worst-case guarantees for the optimal stochastic control policy. This is the first result of its kind in this area, and thus the guarantees it provides are likely loose and should be able to be improved, either via a tighter analysis or by finding a new algorithm with a better tradeoff between optimality in the stochastic model and robustness in the worst-case model.

Finally, and perhaps most importantly, throughout this chapter we are ignoring the costs (both in terms of energy and delay) of switching the speed of the server and we are allowing arbitrary speeds to be chosen. In practice these simplifications are significant, and certainly should affect the design of the speed scaler. In particular, they are important to consider when comparing dynamic and gated static speed scaling. Some analytic work has begun to take these issues into consideration [24], however there are many unanswered questions that remain.

17.6 Drawbacks of Speed Scaling

We have seen that speed scaling has many benefits; however, it is also important to consider potential drawbacks of speed scaling.

Are there unintended drawbacks to speed scaling?

In this section we discuss one unintended, undesirable consequence of speed scaling—magnifying unfairness.

Fairness is important in many applications, and as a result there is a large literature of work studying the fairness of scheduling policies. However, the interaction of fairness and energy efficiency has received little attention so far, e.g., [5,54]. In the domain of speed scaling, [5] was the first to identify that speed scaling can have unintended negative consequences with respect to fairness. Intuitively, speed scaling can create unfairness as follows: If there is some job type that is always served when the queue length is long/short it will receive better/worse performance than it would have in a system with a static speed. To see that this magnifies unfairness, note that the scheduler has greatest flexibility in job selection when the queue is long, and so jobs served at that time are likely to be those that already get better service.

In fact, it is possible to prove that the intuition mentioned previously is correct and that the service rate differential can lead to unfairness in a rigorous sense under SRPT and nonpreemptive policies such as FCFS. However, under PS, speed scaling does not lead to unfairness because of the inherent fairness of the policy.

17.6.1 Defining Fairness

The fairness of scheduling policies has been studied extensively, leading to a variety of fairness measures, e.g., [6,48,58], and the analysis of nearly all common scheduling policies, e.g., [34,47,58]. See also the survey [56].

We compare fairness not between individual jobs, but between classes of jobs, consisting of all jobs of a given size. Since this chapter focuses on delay, we compare $\mathbb{E}[T(x)]$ across x, where $\mathbb{E}[T(x)]$ denotes the expected response time for a job of size x. Fairness when $s = 1$ has been previously defined as [56]

Definition 17.2   *A policy π is fair if for all x*

$$\frac{\mathbb{E}[T^{\pi}(x)]}{x} \leq \frac{\mathbb{E}[T^{PS}(x)]}{x}.$$

This metric is motivated by the fact that (1) PS is intuitively fair since it shares the server evenly among all jobs at all times; (2) for $s = 1$, the slowdown ("stretch") of PS is constant, i.e., $\mathbb{E}[T(x)]/x = 1/(1 - \rho)$; (3) $\mathbb{E}[T(x)] = \Theta(x)$ [29], so normalizing by x when comparing the performance of different job sizes is appropriate. Additional support is provided by the fact that $\min_{\pi} \max_{x} \mathbb{E}[T^{\pi}(x)]/x = 1/(1 - \rho)$ [58].

By this definition, the class of large jobs is treated fairly under all work-conserving policies, i.e., $\lim_{x \to \infty} \mathbb{E}[T(x)]/x \leq 1/(1 - \rho)$ [29]—even policies such as SRPT that seem biased against large jobs. In contrast, all non-preemptive policies, e.g., FCFS have been shown to be unfair to small jobs [58].

The foregoing applies only when $s = 1$. However, the following proposition shows that PS still maintains a constant slowdown for arbitrary speeds, and so Definition 17.2 is still a natural notion of fairness.

Proposition 17.12 *Consider an M/GI/1 queue with a symmetric scheduling discipline, e.g., PS with controllable service rates. Then,* $\mathbb{E}[T(x)] = x\,(\mathbb{E}[T]/\mathbb{E}[X])$.

17.6.2 Speed Scaling Magnifies Unfairness

Using the earlier criterion for fairness, we can now illustrate that speed scaling creates/magnifies unfairness under SRPT and nonpreemptive policies such as FCFS.

17.6.2.1 SRPT

We first show that SRPT treats the largest jobs unfairly in a speed scaling system. Recall that the largest jobs are always treated fairly in the case of a static speed ($s = 1$).

Let $\bar{s}^\pi$ be the time average speed under policy π, and let $\pi + 1$ denote running policy π on a system with a permanent customer in addition to the stochastic load (e.g., $\bar{s}^{PS+1}$).

Theorem 17.13 *Consider a GI/GI/1 queue with controllable service rates and unbounded inter-arrival times. Let $s_n^{SRPT} \leq s_n^{PS}$ be weakly monotone increasing and satisfy $\bar{s}^{PS+1} > \rho$ and $\bar{s}^{SRPT+1} > \rho$. Then*

$$\lim_{x\to\infty} \frac{T^{PS}(x)}{x} <_{a.s.} \lim_{x\to\infty} \frac{T^{SRPT}(x)}{x}.$$

The intuition behind Theorem 17.13 is the following. An infinitely sized job under SRPT will receive almost all of its service while the system is empty of smaller jobs. Thus it receives service during the idle periods of the rest of the system. Further, if $s_n^{SRPT} \leq s_n^{PS}$ then the busy periods will be longer under SRPT and so the slowdown of the largest job will be strictly greater under SRPT. This intuition also provides an outline of the proof.

Figure 17.7 shows that, under SRPT, large jobs suffer a significant increase in slowdown as compared to PS, although only 10% of the jobs are worse off. Since this setting has a moderate load, SRPT with static speeds would be fair to all job sizes. Figure 17.7a shows 90% confidence intervals.

Theorem 17.13 proves that SRPT cannot use dynamic speeds and provide fairness to large jobs; however, by using gated-static speed scaling SRPT can provide fairness, e.g., [58]. Further, as Figure 17.4 illustrates, gated-static speed scaling provides nearly optimal cost. So, it is possible to be fair and near-optimal using SRPT scheduling but, to be fair, robustness must be sacrificed.

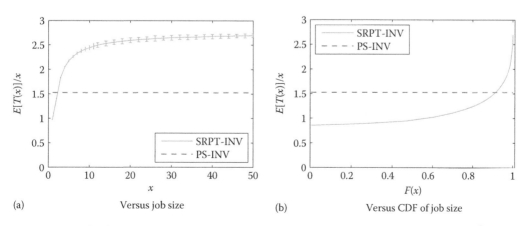

(a)   Versus job size   (b)   Versus CDF of job size

FIGURE 17.7 Slowdown of large jobs under PS and SRPT under Pareto(2.2) job sizes, $\gamma = 1$, $s_n = P^{-1}(n)$, and $P(s) = s^2$. Note the fairness of PS.

17.6.2.2 Nonpreemptive Policies

The magnification of unfairness by speed scaling also occurs for all nonpreemptive policies.

For a fixed speed, all nonpreemptive policies are unfair to small jobs [58] since the response time must include at least the residual of the job size distribution if the server is busy

$$\mathbb{E}[T(x)]/x \geq 1 + \rho \mathbb{E}[X^2]/(2\mathbb{E}[X]x),$$

which grows unboundedly as $x \to 0$. However, if we condition on the arrival of a job to an empty system (i.e., the work in system at arrival is $W = 0$), then nonpreemptive policies are "fair," in the sense that the slowdown is constant: $T(x|W = 0)/x = 1$. Speed scaling magnifies unfairness under nonpreemptive policies in the sense that it allows $T(x|W = 0)/x$ to differ dramatically across job sizes.

Proposition 17.14   *Consider a nonpreemptive GI/GI/1 speed scaling queue with mean inter-arrival time* $1/\lambda$ *and speeds* s_n *monotonically approaching* $s_\infty \in (0, \infty]$ *as* $n \to \infty$. *Then, with probability 1*

$$\lim_{x \to 0} \frac{T(x|W = 0)}{x} = \frac{1}{s_1} \quad and \quad \lim_{x \to \infty} \frac{T(x|W = 0)}{x} = \frac{1}{s_\infty}.$$

The intuition for this result is that small jobs receive their whole service while alone in the system, whereas large jobs have a large queue build up behind them, and therefore get served at a faster speed. Thus, the service rates of large jobs exceeds that of small jobs, magnifying the unfairness of nonpreemptive policies.

17.6.3 Open Questions

This section has highlighted that dynamic speed scaling has an important drawback—it magnifies unfairness. However, the results summarized here represent only a first step toward understanding this phenomenon. In particular, they only highlight the existence of the phenomenon and say little else.

There are many remaining questions that should be answered. For example, what is the magnitude of the unfairness created? How does this depend on the workload? the speed scaler? and the scheduler? Further, the results summarized previously only focus on the largest or smallest jobs. Is it possible to characterize the fairness experienced by intermediate sized jobs?

Additionally, note that the discussion of fairness in this section has focused on one particular measure of fairness. There are many other fairness measures that have been proposed and studied in the classic, static speed model. It would be interesting to see whether the results summarized in this section are corroborated by results for other fairness metrics or not.

Finally, note that this section has focused on only one particular drawback of dynamic speed scaling—unfairness. There are certainly other possible drawbacks, such as increased switching costs (both in terms of added delay and energy), which warrant study. The particular case of switching costs becomes very important when considering scaling the "speed" of a data center by turning servers on and off. This has been the focus of some work, e.g., [25,30,40,42]; however, there is much yet to be understood.

17.7 Concluding Remarks

In this chapter we have reviewed recent results on algorithms for speed scaling. Our focus has not been on providing a complete overview of the literature, which is large and growing quickly. Rather, our goal has been to illustrate the insight provided by recent analytic results for four fundamental issues in speed scaler design.

We have highlighted that significant progress has been made in understanding speed scaling design and, specifically, in providing insight into these four issues. However, we have also illustrated throughout that there are many remaining open questions that provide interesting directions for future work.

To end, it is interesting to tie together the four issues we have discussed throughout the chapter to highlight one final important direction for future work. The results we have surveyed demonstrate that there is a tension in the design of speed scaling algorithms between three desirable objectives: near-optimal performance, robustness, and fairness. As we have seen, SRPT with dynamic speed scaling is robust and nearly optimal, but is unfair. However, SRPT can be fair and still nearly optimal if gated-static speed scaling is used, but this is not robust. On the other hand, dynamic speed scaling with PS can be fair and robust but, in the worst case, pays a significant performance penalty compared to using SRPT. Thus, among the policies considered in this chapter, *it is possible to be any two of near-optimal, fair, and robust—but not all three*. So, the question remains for future research: is it possible for a speed scaling algorithm to be near-optimal, robust, and fair or alternatively, is there a fundamental limitation that prevents this?

References

1. WD green power: A new benchmark in HDD acoustics & power. Online. ⟨http://www.silentpcreview. com/article786-page2.html⟩.
2. S. Albers. Energy-efficient algorithms. *Communications of the ACM*, 53(5):86–96, 2010.
3. S. Albers and H. Fujiwara. Energy-efficient algorithms for flow time minimization. In *Lecture Notes in Computer Science (STACS)*, Vol. 3884, Marseille, France, pp. 621–633, 2006.
4. S. Albers, F. Müller, and S. Schmelzer. Speed scaling on parallel processors. In *Proceedings of the 19th Annual ACM Symposium on Parallel Algorithms and Architectures (SPAA)*, San Diego, CA, pp. 289–298, 2007.
5. L. H. Andrew, M. Lin, and A. Wierman. Optimality, fairness, and robustness in speed scaling designs. In *Proceedings of ACM Sigmetrics*, New York, pp. 37–48, 2010.
6. B. Avi-Itzhak, H. Levy, and D. Raz. A resource allocation fairness measure: Properties and bounds. *Queueing Systems Theory and Applications*, 56(2):65–71, 2007.
7. N. Bansal, H.-L. Chan, T.-W. Lam, and L.-K. Lee. Scheduling for speed bounded processors. In *Proceedings of the 35th International Colloquium on Automata, Languages and Programming*, Reykjavik, Iceland, pp. 409–420, 2008.
8. N. Bansal, H.-L. Chan, and K. Pruhs. Speed scaling with a solar cell. *Theoretical Computer Science*, 410(45):4580–4587, 2009.
9. N. Bansal, H.-L. Chan, and K. Pruhs. Speed scaling with an arbitrary power function. In *Proceedings of 12th Annual ACM-SIAM Symposium on Discrete Algorithms SODA*, New York, pp. 693–701, 2009.
10. N. Bansal, H.-L. Chan, K. Pruhs, and D. Katz. Improved bounds for speed scaling in devices obeying the cube-root rule. In *Automata, Languages and Programming*, Rhodes, Greece, pp. 144–155, 2009.
11. N. Bansal, T. Kimbrel, and K. Pruhs. Speed scaling to manage energy and temperature. *Journal of the ACM*, 54(1):1–39, March 2007.
12. N. Bansal, K. Pruhs, and C. Stein. Speed scaling for weighted flow times. In *Proceedings of the 18th Annual ACM-SIAM Symposium on Discrete Algorithms SODA*, New Orleans, LA, pp. 805–813, 2007.
13. L. A. Barroso and U. Hölzle. The case for energy-proportional computing. *Computer*, 40(12):33–37, 2007.
14. N. H. Bingham, C. M Goldie, and J. L. Teugels. *Regular Variation*. Cambridge University Press, Cambridge, U.K., 1987.
15. J. R. Bradley. Optimal control of a dual service rate M/M/1 production-inventory model. *European Journal of Operations Research*, 161(3):812–837, 2005.
16. D. P. Bunde. Power-aware scheduling for makespan and flow. In *Proceedings of the 18th Annual ACM Symposium Parallel Algorithms and Architectures*, Cambridge, MA, pp. 190–196, 2006.
17. D. P. Bunde. Power-aware scheduling for makespan and flow. *Journal of Scheduling*, 12(5):489–500, 2009.

18. H.-L. Chan, J. Edmonds, T.-W. Lam, L.-K. Lee, A. Marchetti-Spaccamela, and K. Pruhs. Non-clairvoyant speed scaling for flow and energy. In *Proceedings of the 26th International Symposium on Theoretical Aspects of Computer Science STACS*, Freiburg, Germany, pp. 255–264, 2009.

19. S.-H. Chan, T.-W. Lam, L.-K. Lee, H.-F. Ting, and P. Zhang. Non-clairvoyant scheduling for weighted flow time and energy on speed bounded processors. In *Proceedings of Computing: The Australasian Theory Symposium (CATS)*, Brisbane, Australia, pp. 3–10, 2010.

20. A. P. Chandrakasan, S. Sheng, and R. W. Brodersen. Low-power CMOS digital design. *IEEE Journal of Solid-State Circuits*, 27(4):473–484, April 1992.

21. W. Chen, D. Huang, A. Kulkarni, J. Unnikrishnan, Q. Zhu, P. G. Mehta, S. Meyn, and A. Wierman. Approximate dynamic programming using fluid and diffusion approximations with applications to power management. In *Proceedings of the 48th IEEE Conference on Decision and Control of CDC*, Shanghai, China, pp. 3575–3580, 2009.

22. T. M. Cover and J. A. Thomas. *Elements of Information Theory*. Wiley, New York, 1991.

23. T. B. Crabill. Optimal control of a service facility with variable exponential service times and constant arrival rate. *Management Science*, 18(9):560–566, 1972.

24. F. Dabiri, A. Vahdatpour, M. Potkonjak, and M. Sarrafzadeh. Energy minimization for real-time systems with non-convex and discrete operation modes. In *Design, Automation and Test in Europe (DATE)*, Nice, France, pp. 1416–1421, 2009.

25. A. Gandhi, V. Gupta, M. Harchol-Balter, and M. Kozuch. Optimality analysis of energy-performance trade-off for server farm management. In *Proceedings of IFIP Performance*, Namur, Belgium, pp. 1–23, 2010.

26. J. M. George and J. M. Harrison. Dynamic control of a queue with adjustable service rate. *Operations Research*, 49(5):720–731, 2001.

27. G. Greiner, T. Nonner, and A. Souza. The bell is ringing in speed-scaled multiprocessor scheduling. In *Proceedings of the 21st Annual ACM Symposium on Parallel Algorithms and Architectures (SPAA)*, Alberta, Canada, pp. 11–18, 2009.

28. S. V. Hanly. Congestion measures in DS-CDMA networks. *IEEE Transactions on Communications*, 47(3):426–437, March 1999.

29. M. Harchol-Balter, K. Sigman, and A. Wierman. Asymptotic convergence of scheduling policies with respect to slowdown. *Performance Evaluation*, 49(1–4):241–256, September 2002.

30. S. Irani, S. Shukla, and R. Gupta. Algorithms for power savings. *ACM Transactions on Algorithms*, 3(4), November 2007.

31. B. Kalyanasundaram and K. Pruhs. Speed is as powerful as clairvoyance. *Journal of the ACM (JACM)*, 47(4):617–643, July 2000.

32. S. Kaxiras and M. Martonosi. *Computer Architecture Techniques for Power-Efficiency*. Morgan and Claypool, Seattle, WA, 2008.

33. Frank P. Kelly. *Reversibility and Stochastic Networks*. Wiley, New York, 1979.

34. A. A. Kherani and R. Núñez Queija. TCP as an implementation of age-based scheduling: Fairness and performance. In *IEEE INFOCOM*, Barcelona, Catalunya, Spain, pp. 1–12, 2006.

35. S.-L. Kim, Z. Rosberg, and J. Zander. Combined power control and transmission rate selection in cellular networks. In *Proceedings of IEEE 50th Vehicular Technology Conference*, Amsterdam, the Netherlands, pp. 1653–1657, 1999.

36. L. Kleinrock. *Queueing Systems Volume II: Computer Applications*. Wiley Interscience, New York, 1976.

37. T.-W. Lam, L.-K. Lee, I. K. K. To, and P. W. H. Wong. Competitive non-migratory scheduling for flow time and energy. In *Proceedings of the 20th Annual ACM Symposium on Parallel Algorithms and Architectures (SPAA)*, Munich, Germany, pp. 256–264, 2008.

38. T.-W. Lam, L.-K. Lee, I. K. K. To, and P. W. H. Wong. Nonmigratory multiprocessor scheduling for response time and energy. *IEEE Transactions on Parallel and Distributed Systems*, 19(11):1527–1539, November 2008.

39. T.-W. Lam, L.-K. Lee, I. K. K. To, and P. W. H. Wong. Speed scaling functions for flow time scheduling based on active job count. In *Proceedings of the 16th Annual European Symposium on Algorithms*, Karlsruhe, Germany, pp. 647–659, 2008.

40. M. Lin, A. Wierman, L. L. H. Andrew, and E. Thereska. Dynamic right-sizing for power-proportional data centers. In *Proceedings of IEEE INFOCOM*, Shanghai, China, April 10–15, 2011.

41. M. Lin, A. Wierman, and B. Zwart. Heavy-traffic analysis of mean response time under shortest remaining processing time. *Performance Evaluation*, 68(10):955–966, 2011.

42. Z. Liu, M. Lin, A. Wierman, L. H. Andrew, and S. Low. Greening geographical load balancing. In *Proceedings of ACM SIGMETRICS*, San Jose, CA, pp. 233–244, 2011.

43. R. Motwani, S. Phillips, and E. Torng. Nonclairvoyant scheduling. *Theoretical Computer Science*, 130(1):17–47, 1994.

44. K. Pruhs, R. V. Stee, and P. Uthaisombut. Speed scaling of tasks with precedence constraints. *Theory of Computing Systems*, 43(1):67–80, 2008.

45. K. Pruhs and C. Stein. How to schedule when you have to buy your energy. In *Approximation, Randomization, and Combinatorial Optimization (LNCS)*, Barcelona, Spain, pp. 352–365, 2010.

46. K. Pruhs, P. Uthaisombut, and G. Woeginger. Getting the best response for your erg. In *Scandinavian Workshop on Algorithm Theory*, Humleback, Denmark, pp. 14–25, 2004.

47. I. A. Rai, G. Urvoy-Keller, and E. Biersack. Analysis of LAS scheduling for job size distributions with high variance. In *Proceedings of ACM Sigmetrics*, San Diego, CA, pp. 218–228, 2003.

48. W. Sandmann. A discrimination frequency based queueing fairness measure with regard to job seniority and service requirement. In *Proceedings of Euro NGI Conference on Next Generation Internet Networks*, Rome, Italy, pp. 106–113, 2005.

49. L. E. Schrage. A proof of the optimality of the shortest remaining processing time discipline. *Operations Research*, 16:678–690, 1968.

50. R. K. Sharma, C. E. Bash, C. D. Patel, R. J. Friedrich, and J. S. Chase. Balance of power: Dynamic thermal management for internet data centers. *IEEE Internet Computing*, 9(1):42–49, 2005.

51. S. W. Son and M. Kandemir. Energy-aware data prefetching for multi-speed disks. In *Proceedings of the Third ACM International Conference on Computing Frontiers*, Ischia, Italy, pp. 105–114, 2006.

52. S. Stidham and R. R. Weber. Monotonic and insensitive policies for control of queues. *Operations Research*, 87(4):611–625, 1989.

53. H. Sun, Y. Cao, and W.-J. Hsu. Non-clairvoyant speed scaling for batched parallel jobs on multiprocessors. In *ACM Conference on Computing Frontiers*, Ischia, Italy, pp. 99–108, 2009.

54. P. Tsiaflakis, Y. Yi, M. Chiang, and M. Moonen. Fair greening for DSL broadband access. In *GreenMetrics*, Seattle, WA, pp. 74–78, 2009.

55. R. Weber and S. Stidham. Optimal control of service rates in networks of queues. *Advances in Applied Probability*, 19:202–218, 1987.

56. A. Wierman. Fairness and classifications. *Performance Evaluation Review*, 34(4):4–12, 2007.

57. A. Wierman, L. L. H. Andrew, and A. Tang. Power-aware speed scaling in processor sharing systems. In *Proceedings of IEEE INFOCOM*, Rio de Janeiro, Brazil, pp. 2007–2015, 2009.

58. A. Wierman and M. Harchol-Balter. Classifying scheduling policies with respect to unfairness in an M/GI/1. In *Proceedings of ACM Sigmetrics*, San Diego, CA, pp. 238–249, 2003.

59. F. Yao, A. Demers, and S. Shenker. A scheduling model for reduced CPU energy. In *Proceedings of the 36th Annual IEEE Symposium on Foundations of Computer Science (FOCS)*, Milwaukee, WI, pp. 374–382, 1995.

60. S. Zhang and K. S. Catha. Approximation algorithm for the temperature-aware scheduling problem. In *Proceedings of the IEEE International Conference Computer Aided Design*, San Jose, CA, pp. 281–288, November 2007.

18

Processor Optimization for Energy Efficiency

Omid Azizi
Stanford University

Benjamin C. Lee
Duke University

The design of a microprocessor, like the design of any engineered system, is an optimization problem: the designer has various resources (many of which may be limited and/or costly), design constraints, and a design objective that needs to be maximized or minimized. Faced with this problem, the task of the designer is to find a solution that best achieves the design goals. In the past, the primary design constraint was chip area, and the optimization problem was mostly an evaluation of performance versus chip area. As semiconductor technologies progressed, however, power dissipation in chips gradually rose, reaching a point where hard power constraints became necessary. Today, these power constraints significantly limit the performance of the system. Thus, in the current era of design, design for energy efficiency is important for all classes of systems, from low-power embedded devices to high-performance servers.

In this chapter, we examine the general problem of designing and optimizing systems for energy efficiency. At the core of this optimization process is the development of models that predict performance and energy as a function of changing design parameters. With accurate models in hand, the architect can then explore the space—either with manual sweeps or with automated optimization routines—to find an efficient design implementation. Developing accurate models that can capture large design spaces and searching the resulting design space efficiently are both challenging problems, however, as the sheer size of the design space that needs to be explored grows combinatorially with the number of design parameters. This modeling and optimization problem is addressed in Section 18.1, where we show how data fitting and statistical inference methods can enable a designer to explore large spaces efficiently. In Sections 18.2 and 18.3, we then use these approaches to present various case studies.

18.1 Processor Modeling and Optimization

To achieve energy efficiency in a digital system, a designer needs to consider all the different design options and features and must then evaluate them for their benefits and costs. By applying a trade-off analysis, where the benefits of a design decision are weighed against its costs, the more efficient design options can be identified, and a designer can optimize the design. This process, however, requires models that indicate how design decisions affect the power and performance characteristics of the system.

Traditionally, these models are realized through the development of architectural simulators. Because of the flexibility and ease of understanding that simulators offer, simulation models can achieve high accuracies while allowing the architect to model complex systems that depend on the application input. Thus, today, architectural simulators are the primary evaluation tool of the processor architect.

While simulators are invaluable to architectural evaluations, they do have their drawbacks. In particular, simulations typically require long run times, taking hours to even days to produce a result. Unfortunately, for performing a large-scale design space exploration in which numerous design decisions are evaluated, this simulation time becomes prohibitive.

To resolve this issue, there are a number of approaches that have been developed. One effective approach is to reduce the simulation time of an individual simulation run; there are a variety of techniques that help achieve this, from faster simulation technologies to the use of representative code samples. However, even if the simulation time of a single design is reduced, performing a large-scale design space exploration is challenging because of a second factor: the sheer size of the space that needs to be explored, which grows combinatorially with each added parameter.

To help explore large design spaces effectively, one needs to find a way to limit the number of simulations that need to be run. One effective technique to achieve this goal is the use of statistical sampling and inference to create fitted models from a relatively small sample of design points [2,6–8]. The basic idea in this approach is that one does not necessarily have to simulate every single design point to get a good indication of how different parameters in the system behave and interact. One can learn a lot about the behavior of the system simply by simulating a relatively small number of designs with different configurations and observing the output. Essentially, these techniques take advantage of the fact that the performance surface is usually fairly smooth; they work by sampling the surface and then interpolating for missing data.

In these approaches, the design (i.e., the simulator) is treated as a black box with various configurable/-tunable design parameters and a user-provided, but fixed, application workload. By randomly setting the design parameters to create random designs, and then running simulations for a handful of the possible designs in the design space, one can obtain enough data to develop a fitted model of how the system behaves as a function of the design parameters for the given workload.*

While this chapter considers random sampling, more sophisticated strategies exist and may improve accuracy for specific scenarios by increasing sample coverage over the design space or emphasizing particularly interesting regions of the design space. Weighted or regional sampling might emphasize particular design points or design regions to increase their influence during model fitting. Adaptive sampling might estimate model error variances for each sampled designs. If samples with larger variances are more prone to inaccurate prediction, including more of these samples might improve accuracy. Adapting to already collected samples, new samples are chosen to be most different from older ones. Lastly, design space coverage can be guaranteed with structured sampling (e.g., Latin hypercube or space-filling techniques).

Once collected, these samples are used to fit models. The result of these fitted models is a predictive mathematical model that can be quickly evaluated to predict the performance in any part of the captured space. The number of samples required to create the models can be dramatically smaller than the number of designs in the space, with several hundred to a few thousand samples typically being sufficient to model complex processor systems with spaces that span billions of possible design configurations. Thus, these approaches have shown to be powerful tools for design space exploration. This approach, and how it contrasts to traditional simulation-based optimization, is shown in Figure 18.1.

---

* To model mixed application workloads, one can either develop a single model directly by using the mixed workload as the simulator input or can develop individual models for each of the different benchmarks from which an average performance can be computed later.

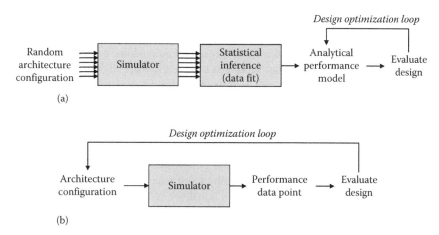

FIGURE 18.1 (a) Inference-based design optimization. (b) In traditional optimization, the simulator directly serves as the system model, so all design points need to be simulated to determine their performance. This results in a design optimization loop around the simulator. In inference-based methods, an analytical model is generated from design space samples (acquired from the same simulator). Typically, several hundred to a few thousand samples are sufficient to capture a large space of billions of design points. Using the generated mathematical model then results in a much more efficient exploration and optimization of the system.

This fit-based modeling offers several advantages. First, because it still relies on simulators to generate the initial samples, the approach offers a lot of flexibility; one can model any system so long as an appropriate performance simulator for the system exists (assuming an appropriate fit function is found). The designer simply uses the simulator to obtain design space samples from which he then creates fitted model. At the same time, because this approach generates mathematical models of the system, it can be very effective in quickly exploring the space and evaluating different design configurations. In this way, regression-based inference approaches can be viewed as a hybrid of direct analytical modeling and purely simulation-based methods, achieving the strengths of both methods.

18.1.1 Regression Models

To create regression models, suppose we generate simulation samples of random designs within a design space of interest. simulation. Each sampled design point is denoted by a vector of design parameters $x_1, \ldots, x_n$. In the context of processor design, for example, these parameters might include datapath organization and/or the cache hierarchy parameters. For each design point, simulators provide measures of design metrics, which we denote as y. For processor efficiency, the metrics of interest are performance and power. In the simplest regression model, we might express design metrics as a function of design parameters with regression coefficients β. In practice, however, transformations are often required of y and x_i to capture interesting performance and power surfaces. Following is a general regression form that often works well:

$$y = \beta_0 + \sum_{i=1}^{n} \beta_i x_i + \sum_{(j,k) \in S} \beta_{j,k} x_j x_k + \epsilon \tag{18.1}$$

In this regression, the first summation term accounts for each architectural parameter's independent influence on the response y. Such a term is included for each parameter of interest. The second summation term accounts for interactions between pairs of parameters (j and k), with the set S denoting parameter pairs in the model. For example, one important interaction might include the L1 cache size and the L2 cache latency; L2 access times become more important as L1 miss rates increase. Expressing interactions as products within the regression can capture this mutual dependence; $\delta y/\delta x_j$ is a function of x_k and vice versa. The choice of parameters and interaction terms requires some attention. If some important terms are not included, the model will not be able to capture the space; if too many superfluous terms are included, then finding the fit coefficients becomes harder and over-fitting the data may become an issue. A judicious choice of model parameters will produce a more compact model and reduce the cost of optimization.

As part of this fitting process, therefore, it is important to identify the right parameter interactions. One could use domain knowledge to manually specify expected interactions, but this requires that the system already be well understood. Alternatively, important interactions might be identified heuristically. To determine the interaction terms in an automated fashion, we can perform an exhaustive search. We first create a base fit with no interaction terms. We then initialize S to be empty and iterate over all possible pairs. For each visited pair, we attempt a new fit with that term. If the resulting fit improves the model by some threshold, the interaction terms are added to S.

The interaction terms we have discussed have been limited to parameter pairs, but one can also look for more complex interaction terms that involve three or more parameters. The number of terms to consider when looking at three parameters, however, is exponentially larger, making it more difficult to identify such terms. Nevertheless, one way to check whether the exclusion of any term (not only triplets, but any single parameter or otherwise) is adversely affecting the fit is to check for residual errors in the fit that are correlated to the parameters under consideration. For the processors we examine in the case studies later in this chapter, we limit our fits to interaction pairs, as it yields fairly good accuracy and seems to be sufficient for modeling the systems.

18.1.2 Splines and Posynomials

In applying a fit-based modeling approach, the challenge is to find a fitting function that can serve as a good model for the system by effectively fitting the design space samples. Several different functional forms are possible to achieve this goal; the different forms offer different trade-offs. Cubic splines are powerful fitting functions that are commonly used in regressions; they can fit complex surfaces and the resulting mathematical fit can be evaluated quickly [8]. An alternative approach is to use a class of functions referred to as posynomials; these functions are more restrictive, but result in smoother surfaces that enable the use of powerful convex optimization packages that can search the large design space to perform fast optimization [1]. Other approaches, such as the use of artificial neural networks [6,7], have also been proposed. In the rest of this section, we elaborate on the first two approaches.

18.1.2.1 Spline-Based Regression

To capture non-linear trends, regression models might rely on polynomial transformations for predictors suspected of having a non-linear correlation with the response. However, polynomials have undesirable peaks and valleys that are determined by the degree of the polynomial and are difficult to manipulate. Furthermore, a good fit in one region of the predictor's values may unduly impact the fit in another region of values. For these reasons, we consider splines a more effective technique for modeling non-linearity.

Spline functions are piecewise polynomials used in curve fitting [5]. The function is divided into intervals defining multiple different continuous polynomials with endpoints called knots. The number of knots can vary depending on the amount of available data for fitting the function, but more knots generally lead to better fits. Relatively simple linear splines may be inadequate for complex, highly curved relationships. Splines of higher order polynomials may offer better fits and cubic splines have been found particularly effective. Unlike linear splines, cubic splines may be made smooth at the knots by forcing the first and second derivatives of the function to agree at the knots. Equation 18.2 illustrates a cubic spline for parameter x using three knots k. Noting that $(u)_+ = max(u, 0)$, the knots effectively allow $\beta_4, \ldots, \beta_6$ to adjust coefficients depending on the value of x.

$$x_s = \beta_0 + \beta_1 x + \beta_2 x^2 + \beta_3 x^3$$
$$= +\beta_4 (x - k_1)_+^3 + \beta_5 (x - k_2)_+^3 + \beta_6 (x - k_3)_+^3 \tag{18.2}$$

The choice and position of knots are variable parameters when specifying non-linearity with splines. Placing knots at fixed quantiles of parameter's range of values is a good approach for most datasets as this approach ensures a sufficient number of points in each piece of the piecewise polynomial. As the number of knots increases, flexibility improves at the risk of over-fitting the data. In many cases, four knots offer an adequate fit of the model and is a good compromise between flexibility and loss of precision from over-fitting. We vary the number of knots to explore trade-offs between flexibility and fit. As flexibility increases, we observe rapidly diminishing marginal returns in fit from more than five knots. Thus, the benefits from a large number of knots may not justify the larger number of terms in the model.

Intuitively, the designer might desire a larger number of knots for parameters that strongly affect design metrics, such as performance and power. A lack of fit for these parameters will have a greater negative impact on accuracy and we assign more knots to such parameters. For example, the pipeline depth parameter might use four knots, given its broad impact on efficiency, while particular queue sizes might use three knots, given their narrower impact.

18.1.2.2 Posynomial-Based Regression

Cubic splines are not the only possible function to use to perform inference. An alternative is the use of posynomial functions, which are mathematical expressions consisting of the sum of any number of positive monomial terms, where monomials are the product of powers of variables. For example, $kx^a y^b z^c$ is a monomial in the variables x, y, and z (with k, a, b, and c as constants), while $k_1 x^a y^b z^c + k_2 x^d$ with $k_1 \geq 0$, $k_2 \geq 0$, is a posynomial in x, y, and z because it is the sum of two positive monomials.

Using posynomial functions offers the advantage that, as log-convex functions, they open the door to structured convex optimization algorithms that can search the space very efficiently. The potential drawback, on the other hand, is that posynomial functions are less general than other forms such as cubic splines; posynomials, for example, can have difficulty capturing spaces with hills and valleys. While it would be ineffective to attempt to capture a space with a more restricted form such as a posynomial if the space was intrinsically more complex, if the space does not require more complex mathematical forms, then it makes sense to use the simpler posynomial form: the simpler form makes over-fitting less of a concern, and, more importantly, the use of posynomials enables the application of some powerful mathematical techniques for performing optimization.

As it turns out, in examining the architectural performance space of digital systems like processors, many of the design knobs exhibit behavior that seems to be well-suited for posynomial modeling. Specifically, a large number of tunable design knobs have a smooth, monotonic profile that can typically be captured well by posynomials. For example, reducing a unit's latency or increasing the size of a queue, buffer, or memory structure typically only improves the number of cycles per instruction (*CPI*). There are numerous other parameters that exhibit this smooth, monotonic, and often diminishing returns profile, including cache sizes, the reorder buffer size, reservation station sizes, and instruction queues to name only a handful. Even in cases where a parameter exhibits more complex behavior that results in a peak or valley—for example, performance as a function cache block size—posynomials may still be able to capture the effect by using the sum of multiple different monomial terms (e.g., $1/x + x$ is a posynomial that results in a single valley) to model the more complex behavior.

In the following, we present a posynomial form that we have found effective in modeling processor architecture performance. As inputs to the model, we have the various tunable design knobs within the system (x_i). The other variables are fitting parameters, with only the constraint that a_i, b_i, and c all be positive. Thus, the models we generate predict *CPI* as a function of the latencies of units and sizes of structures like caches, buffers, and queues:

$$y = \sum_{i=1}^{n} a_i(x_i)^{d_i} + \sum_{(j,k) \in S} b_{jk}(x_j)^{e_{jk}}(x_k)^{g_{jk}} + c \qquad (18.3)$$

18.1.3 Validation

To validate and check the accuracy of our fits, we set aside a fraction of the simulation samples specifically for the purpose of checking our fits; these samples are not used in producing the fit. We can then compare how well the fitted model would predict the performance of a design configuration in the design space which it has not seen before. To measure error, we use the following metric:

$$Error = \frac{|predicted - actual|}{actual}$$

This metric is applied to each validation sample, and the median error achieved is reported. In this section, we describe the accuracy of posynomial-based regression, which is comparable to that of spline-based regression.

Generally, the number of samples required to generate a good fit depends on the size and complexity of the system. For example, we have found 200 samples often enough for simple in-order processors with 11 parameters, and 500 samples to be sufficient for a complex superscalar out-of-order processor with 18 parameters. The average of median errors over different benchmarks range from less than 1% to 6%, with more complex high-level architectures such as out-of-order processors tending to be harder to fit.

Figure 18.2 shows some sample fits, which are scatter plots of the actual performance of a design configuration versus the performance predicted by the posynomial model. Thus, the closer the points are to the diagonal, the better the fit is. Each plot represents the result of running a particular application on a given architecture. The figure shows results of varying qualities: a very accurate fit, a typical fit, and a worse fit. Even in the worst case, the fitting error is below 10%. The cumulative distribution functions (CDFs) of these three fits are also shown to give the reader a sense of the distribution of errors in each of these generated models.

To check the degree to which our posynomial fits may be restricting the quality of the models we generate, we compare our fits with other fits that use a more flexible form; this checks to see whether more freedom would produce a better fit. Along these lines, we have compared our fits to cubic spline fits and more relaxed versions of our posynomial functions with the positive constraints on coefficients removed. In some cases, we find that we could produce a fit that may be two or three percentage points more accurate by using the more flexible forms. In these cases, these results indicate that our particular choice of posynomial function is restricting the quality of the fit somewhat. Investigating these cases shows that the loss in accuracy can be attributed to smoothness issues. For example, we sometimes find that a parameter, though still monotonic, has a sharp change in curvature or slope. Our posynomial function often fits this data in a smoother way, resulting in some increased error at the point of sharp curvature change.

Despite these occasional fitting effects, architectural regression modeling produces median errors of less than 10%, which is sufficiently accurate for the large-scale optimization that we are interested in; a second optimization iteration around the identified area of interest can always be used to refine the optimization results to mitigate minor fitting effects. Finally, it should be noted that when optimizing over a suite of benchmarks, averaging makes the effects of these occasional fitting issues less significant.

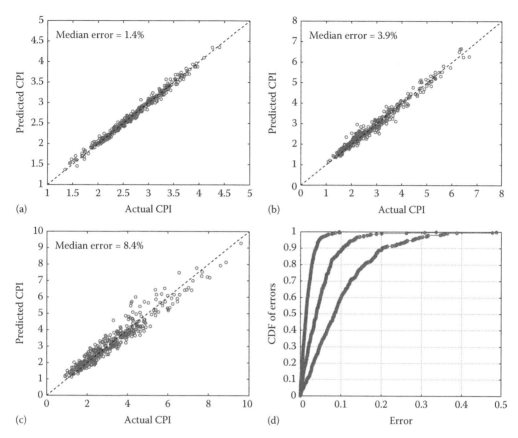

FIGURE 18.2 Validation of three architectural models generated through posynomial design space fitting. (a)–(c) compare model predictions to the results of the simulator. Each fit is for a particular application running on a given processor architecture. (a) is a very accurate fit generated for a single-issue in-order machine, (b) is a typical fit, in this case generated for a quad-issue out-of-order machine, and (c) is a worse fit, also for the quad-issue out-of-order machine (but using a different application). (d) shows the cumulative distribution of errors for these three models. Even in the worst case, the median error is less than 10% for a performance range that spans around 10×.

18.2 Case Study: Energy-Performance Trade-Offs and Marginal Costs

The previous section discussed modeling methodologies for characterizing and exploring large design spaces. This section uses these techniques to present a case study on the energy efficiency of general-purpose processors. This case study explores a number of design decisions in the processor design space, from the choice of high-level architecture to various lower-level microarchitectural knobs and circuit design trade-offs. The case study begins with an evaluation of energy-performance trade-offs of different design decisions, using posynomial fitted models and an automated convex optimization framework to find the optimal design at different performance targets. The case study then introduces the voltage knob into the design optimization to study how voltage scaling effects processor energy-efficiency.

TABLE 18.1 Microarchitectural Design Space Parameters

Parameter	Units	Min.	Max.
Branch pred. size	Entries	0	1024
BTB size	Entries	0	1024
I-cache (2-way) size	kB	2	32
D-cache (4-way) size	kB	4	64
Fetch latency	Cycles	1	3
Decode/reg file lat.	Cycles	1	3
Retire latency	Cycles	1	3
Integer ALU latency	Cycles	1	4
FP ALU latency	Cycles	3	12
L1 D-cache latency	Cycles	1	3
ROB size	Entries	4	32
IW (centralized) size	Entries	2	32
LSQ size	Entries	1	16
L2 cache latency	Cycles	8	64
DRAM latency	Cycles	50	200
Cycle time	ns	0	Unrestricted
Supply voltage	V	0.7	1.4

Source: B.C. Lee and D. Brooks. Illustrate design space studies with microarchitectural regression models. In *Proceedings of the 13th IEEE International Symposium on High Performance Computer Architecture (HPCA)*, Phoenix, AZ, pp. 340–351, 2007. Washington, DC: IEEE Computer Society.

18.2.1 Design Space

To study energy-performance trade-offs within the processor design space, we consider six different high-level processor architectures: single-issue, dual-issue, and quad-issue designs, each with both in-order and out-of-order execution. This covers a large range of the traditional architecture space, from a simple lower-energy, low-performance single-issue in-order processor to an aggressive higher-energy, high-performance quad-issue out-of-order processor.

For each of these high-level architectures, there are then various tunable microarchitectural parameters that trade off energy and performance. Table 18.1 lists the parameters for the design space this case study explores. This microarchitectural space consists of billions of possible design configurations for each high-level architecture. In addition to these microarchitectural parameters, various lower-level circuit trade-offs are also considered that further expand the size of the design space being explored. The inclusion of the circuit design space enables the best circuit to be used that meets the microarchitecture's needs, an important consideration when there are a variety of underlying circuit implementations—with different energy and delay characteristics—that can potentially be used.

For this study, we use a CMOS 90 nm technology. This technology's leakage current is relatively low, so for the results we present, we consider dynamic energy consumption only. We discuss how the inclusion of leakage currents would affect the results in Section 18.2.4.

We use a combination of synthesis tools and the SRAM modeling tool CACTI to characterize the energy characteristics at different operating delays of the major blocks in the processor: the ALUs, the caches, the reorder buffer, the instruction window (IW), etc. It should be noted that these energy models are approximate: First, while all major blocks are considered, there are often numerous smaller units and state registers that are present in commercial designs that we are not including. Second, the models are generated from synthesis tools which only have approximate wireload models. One could, with more data to draw on, improve these models; while the detailed results will change as the underlying models improve, we believe that the general trends and conclusions in our study will still hold true.

TABLE 18.2   Percent Errors of Architectural Performance(CPI) Models Generated for SPECint Benchmarks

	1-Issue In-Order	2-Issue In-Order	4-Issue In-Order	1-Issue ooo	2-Issue ooo	4-Issue ooo
Min	0.49	0.55	0.41	3.41	4.10	3.80
Max	6.18	9.25	8.25	6.56	7.74	8.99
Average	2.61	3.37	4.42	4.86	5.90	5.83

Source: O. Azizi, A. Mahesri, B.C. Lee, and S.J. Patel, and M. Horowitz. Energy-performance tradeoffs in processor architecture and circuit design: A marginal cost analysis. In *Proceedings of the 37th IEEE/ACM International Symposium on Computer Architecture (ISCA)*, Saint-Malo, France, pp. 26–36, 2010. New York: ACM.

Finally, as benchmark suites for this case study, we use SPEC CPU integer benchmarks. We simulate 500 randomly generated design configurations per benchmark, from which we generate our architectural models through statistical inference, using posynomial fitting functions. Table 18.2 shows the maximum, minimum, and average errors of these models when validated against a separate set of simulation samples set aside for validation purposes (i.e., not used for fitting). Averaged over all benchmarks, the generated models have errors of 6.0% or less when validated against a separate set of simulation samples.

18.2.2 Energy-Performance Trade-Off Space

Using the fitted posynomial models, we can perform an automated design space exploration using a convex optimization engine. Provided with a performance target, the optimization engine evaluates the performance benefit of changing a design parameter versus its energy cost. This optimization process essentially computes marginal costs—energy cost per unit performance offered—for all the potential design knobs. By selecting design features with lower costs, and exchanging higher cost options for lower cost alternatives, the optimization procedure automatically finds the optimal design.*

Using this optimization framework, one can obtain the energy-performance trade-offs for the entire system by sweeping the performance target while running optimizations. Applying this procedure to the six architectures mentioned earlier yields the energy-performance trade-offs shown in Figure 18.3. These Pareto-optimal curves show the entire range of trade-offs. As performance is pushed, each architecture uses more aggressive structures and circuits, causing the energy consumed per instruction to increase. Given an energy budget or performance target, a designer can use these curves to identify the most appropriate design.

It is common practice to use energy-delay metrics when optimizing for energy efficiency; these Pareto-optimal trade-off curves between performance and energy, however, are more general than metrics like ED or ED^2. In particular, ED^n metrics essentially set an exchange ratio between energy and performance, with higher powers of n favoring more performance. Optimizing for ED^n with a particular value of n corresponds to a particular point on the Pareto-optimal curve. Since one generally wants to design for a specific performance target or energy budget, representing the results as a trade-off curve between energy per operation and performance provides a more complete picture of the design space to designers.

For the results of this case study, the overall trade-off space spans approximately 6.5× in performance—from about 300 MIPS (million instructions per second) to 1950 MIPS—and 4.25× in energy—from about 80 to 340 pJ/op. The various architectures contribute different segments to the overall energy-efficient frontier. As one might expect, the single-issue in-order architecture is appropriate for very low energy

---

* Since the problem was formed with smooth posynomial functions that are convex, the problem is ensured to have only a single global optimum; this simplifies the search algorithm, since there are no local optima that need to be avoided.

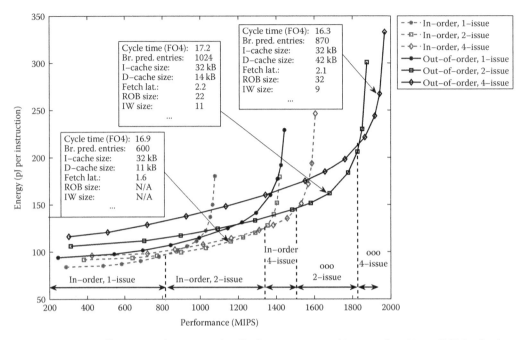

FIGURE 18.3 Overall energy-performance trade-offs of our six macro-architectures for a 90 nm CMOS technology, produced by jointly optimizing microarchitectural and circuit parameters. As the performance is pushed, the optimal choice of macro-architecture changes to progressively more aggressive machines. A sampling of design parameters for the three of the designs are also shown.

design points, while the quad-issue out-of-order is only appropriate at very high performance points. In between these two extremes, we find that the dual-issue in-order and out-of-order processors are efficient for large parts of the design space. Thus, when starting from a basic single-issue in-order design, the order in which high-level architectural features should be considered is, first, superscalar issue, and then—if more performance is still needed—out-of-order processing. From the perspective of the marginal energy cost per unit performance, the move to a superscalar design is cheaper than investing in out-of-order processing. The quad-issue in-order design is only efficient for a small performance range, not being as energy-efficient as the dual-issue in-order design at lower energy points, and being outmatched at high performance points by the dual-issue out-of-order design. Finally, the single-issue out-of-order design is never efficient and does not contribute to the overall efficient frontier. This architecture represents a design that is out of balance. Being able to issue only a single instruction becomes a bottleneck to the out-of-order processor, resulting in wasted effort.

We can also examine how the various underlying parameters are changing throughout the design space. A few selected design parameters are shown for three of the designs in Figure 18.3. Not surprisingly, as we push for more performance, the frequency and structure sizes generally increase, while latencies generally decrease. Parameters such as latencies show fractional values which would need to be snapped to discrete values, although techniques such as time borrowing and register retiming can also be used to work with the results. Throughout the optimization process, the optimization is evaluating the marginal costs of each of the parameters to find the optimal design; the optimization tries to equalize the marginal costs of all parameters to achieve efficiency, since unequal marginal costs imply that a more expensive parameter can be "sold" to save energy, with a "cheaper" parameter being used to reclaim the lost performance. In this fashion, the optimization tunes each parameter until each parameter is tuned and in balance for the efficient operation of the whole system.

While the general trends produced by the optimization are mostly expected, there are a few cases, however, in which the general trends are broken. First, both the I-cache and D-cache tend to stay away from small sizes, even when targeting lower-performance points. Although smaller caches result in less expensive individual accesses, a larger cache potentially saves energy by reducing the number of misses that incur a more expensive access to higher level caches. Thus, for lower power design points, an optimization process determines that the marginal savings it can achieve by reducing misses outweighs the access cost of the larger caches, ultimately finding the right balance and settling on the chosen values. These results show the importance of caches in energy-efficient designs as a way to both save energy and increase performance.

Second, we note that the IW is relatively small compared to the maximum available IW of size 32. This effect is mostly a performance issue. While a larger IW improves the architectural performance and increases *CPI*, a larger IW also increases the complexity (and delay) of the instruction dispatch logic. Since the dispatch logic in a traditional out-of-order machine must execute every cycle, the delay of the dispatch circuitry can adversely affect the clock frequency. The optimization process realizes this trade-off and finds the right balance between architectural performance through a higher CPI and pipeline performance through a higher cycle time. We see a similar effect in the branch predictor size, another structure that needs to execute once per cycle.

18.2.3 Voltage Optimization and Marginal Costs

It is well-known that an important consideration in energy-efficient design is the choice of operating voltage. One needs only to scale voltage by a few tenths of a volt to see significant increases in both performance and energy consumption. Thus, it becomes important to optimize the design along with the supply voltage.

Figure 18.4 shows the energy-delay trade-offs of the supply voltage parameter. This data shows that, by itself, voltage tuning from 0.7 to 1.4 V provides a range of about 3× in performance and 4× in energy. More importantly, the profile of the energy-performance curve is relatively shallow throughout this entire range. This means that the marginal cost of increasing performance through voltage scaling does not change much as we continue to increase the voltage parameter. At low voltages, the marginal cost is at about 0.80% in energy for 1% in performance; at the high end, this marginal cost reaches 2.3% in energy for 1% in performance.

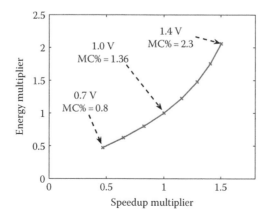

FIGURE 18.4 Effect of voltage on delay and energy. Percentage marginal costs (MC%) is the percentage energy cost required to increase performance by 1%. For a wide range of energy and performance, marginal costs change only modestly, making voltage a powerful knob for energy-efficiency.

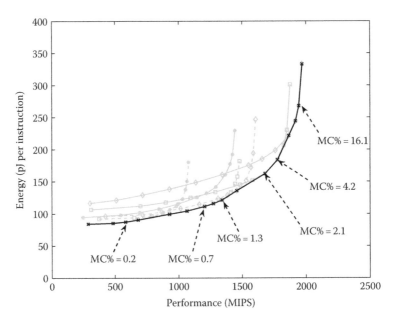

FIGURE 18.5 Marginal costs of the joint architecture and circuit design space. In light are the energy-performance curves of our six architectures. In dark is the overall, composite energy-performance frontier. Data is normalized to 0.9 V to allow for comparison to the voltage marginal costs. Marginal costs in the architecture/circuit space vary considerably more than voltage marginal costs. For most practical design objectives, the optimal architecture should be selected from the narrow band of designs with a marginal cost of 0.80%–2.3% to match voltage marginal costs.

We can contrast this marginal cost profile against the marginal cost profile of achieving performance through circuits and architecture, shown in Figure 18.5. This space shows a much larger range of marginal costs. At the low performance points, the marginal costs are very cheap, while at the high performance points, the marginal costs are very expensive.

We recall that to optimize a design, the marginal costs of all parameters should be equal. If this were not the case, then an arbitrage opportunity would exist, and the more expensive parameters could be exchanged for cheaper parameters: selling the expensive parameter would cause some performance loss, but this performance could be recovered at a lower cost through the cheaper parameter. Comparing the marginal costs of voltage versus architectural parameters, this suggests that, unless we are trying to achieve the very extremes of performance or low power (to the point that the voltage knob is constrained by its maximum or minimum voltage, respectively), the optimal set of designs should lie roughly in the range of marginal costs from 0.80% to 2.3% in order to match the marginal costs of voltage scaling.* This results in a narrow band of architectural and circuit designs being optimal when the voltage scaling parameter is available.

---

* Strictly speaking, this is an approximate statement as it requires that all components in the system scale uniformly with voltage. While this is mostly the case, in reality, components like L2 caches use low-swing bitlines that change their scaling behavior. This means that one cannot simply look at the marginal cost profile of the whole system when considering voltage scaling; one must, rather, separate out the L2 cache component and treat it differently. This effect applies to our case study as well, since we assume L2 caches and main memory have a fixed access cost and physical latency (i.e., we specifically exclude these components from scaling with voltage because they are outside the design space). Nevertheless, the rule of matching marginal costs still applies to a high degree and is useful as a rule of thumb. It is particularly so in our study because the activity factors of the L2 cache and main memory are not high, and the energy spent in these components is not a dominating factor. Regardless, in the results we present next in Figure 18.6, these effects are all modeled correctly, as the optimizer is aware of which components scale with voltage, and it therefore evaluates the correct set of energy-performance trade-offs to find the energy-efficient frontier.

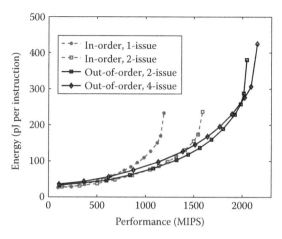

FIGURE 18.6 Energy-performance trade-offs for the processor design space with voltage scaling. The dual-issue out-of-order design now dominates an even larger part of the design space; the dual-issue in-order design is optimal at low energy points. The quad-issue in-order design and the single-issue out-of-order design are not shown to simplify the plot; they are never efficient.

Figure 18.6 shows the optimization results when the supply voltage parameter is included in the design space. Confirming the marginal cost analysis, we see that a smaller set of architectures cover a larger part of the energy-efficient frontier. The dual-issue out-of-order processor is energy-efficient for a large part of the design space. At low performance targets, the dual-issue in-order processor takes over, although the dual-issue out-of-order processor is still not overly inefficient. Only at the very extremes, when the voltage knob becomes capped, do the single-issue in-order and quad-issue out-of-order designs play a role, and these represent designs with very low and very high marginal costs, respectively.

This result suggests that a small number of properly tuned designs can cover most of the overall energy-performance frontier at near optimal efficiencies simply by voltage and frequency scaling. In fact, for the results presented here, we have been able to show that one dual-issue in-order processor and one dual-issue out-of-order processor (each with fixed microarchitectural and circuit parameters) can be used to cover a large part of the energy-performance space with only inefficiencies of 3% or less. Of course, this result requires that we start with the right two designs in the architecture/circuit space sweet spot. Thus, with two carefully selected designs and voltage scaling, we can operate at near-optimal energy efficiencies over a broad performance range.

18.2.4 Discussion

In design for efficiency, a designer needs to consider the cost-performance trade-offs of all available design options, using this information to make the best decisions. This process needs to be done in a disciplined, systematic way, with the designer always adhering to the principles of minimizing marginal costs and ensuring that marginal costs of all design options match.

The results in this section have shown the potential pitfalls of ignoring these principles. One needs to be careful to neither over- nor under-design any aspect of the system; all components need to be in balance, with all design decisions having matching marginal costs. Without applying this principle, inefficiencies arise, with a cheaper source of performance being left untapped. The importance of this principle was particularly clear when considering voltage scaling versus architectural/circuit design decisions as a means of achieving performance. The results of the last section showed that it makes little sense to over-aggressively design the system when the marginal costs of voltage—an equally powerful design parameter—changes relatively slowly over its range. Since the marginal costs of voltage vary between 0.8% and 2.3% energy for 1% in performance (over the range of 0.7–1.4 V), this implies that a potential rule of thumb should be that the marginal cost of any design decision should fall within this range to be acceptable, with lower or higher marginal costs only being considered if the voltage parameter becomes constrained, and the designer has no other feasible options.

As a result of the steady marginal cost profile of voltage versus the rapidly changing marginal costs of architectural and circuit design, we found that the optimal architecture/circuit design was limited to a small sweet spot; most other designs fell outside the 0.8%–2.3% marginal cost range. Some design features landed on a very cheap part of the trade-off curve, meaning they should virtually always be used, while many other design options came at very expensive rates, meaning they should be avoided. In between these two extremes, the set of design knobs did not vary much. This suggested that with a few fixed designs from within this sweet spot, voltage scaling could be a very effective means of achieving different design objectives, a hypothesis that our results seemed to confirm.

One should be careful not to misinterpret this result as indicating that the architecture and circuit design are irrelevant. In fact, the conclusion one should draw is quite the opposite. There are many ways to build an inefficient design, and the designer has to make a concerted effort to find a design within this sweet spot. If the initial system is inefficient or lies outside this sweet spot, voltage scaling cannot make up for the initial inefficiency. It becomes critical to tune the design to include the right set of features with the right marginal costs.

The results presented earlier were optimized only for dynamic power, but it is important to consider how these results would change with higher leakage future technologies. There are at least a couple ways that the results would be affected by leakage. First, as leakage is highly correlated to area, one would expect that structures with larger area would be penalized somewhat during the optimization; this is especially true if the structures are less frequently used (i.e., have a lower activity factor), because then the amount of leakage energy per instruction rises, requiring a higher increase in performance per instruction to make the structure attractive from an energy-performance perspective. Second, in cases where the chip can power down to a low-power idle mode, optimizing with a high leakage technology can actually favor more aggressive, higher performance features. This is because leakage is a rate of energy consumption that gets multiplied by the execution time; with leakage considerations, the whole system would like to "run" to the finish line of a task, and then power down, reducing the leakage contribution which would otherwise be incurred over a longer time period. We have confirmed these results with simple experiments for our optimizations. Including leakage causes the trade-off curves rise due to the additional leakage energy, but in a skewed way—the design points at the left-hand side (lower performance), rise more as a percentage of their original power. This has two consequences: first, the low-energy tails of the trade-off curves get cut off—at some point it does not make sense to run any slower because the small savings in dynamic energy per operation are offset by the increase in leakage energy—and second, because the trade-off curves rise in a skewed fashion, the more aggressive architectures become somewhat more important from an energy-efficiency perspective and end up covering a larger area of the energy-efficient frontier. Thus, we saw that the dual-issue out-of-order design was optimal over an even larger percentage of performance targets.

While this case study has presented some results of optimally tuned processors within the design space we have considered, one should be looking into the results to glean insights on how the design could be modified to achieve further energy efficiency. For example, initially, each of our high-level architectures were configured to fetch instructions according to the width of the machine (e.g., one word for the single-issue machines, etc.). This led to wider machines being more energy efficient than single issue machines even at low performance. Clearly, because of high instruction locality, it makes sense to have all architectures fetch multiple instructions at a time to amortize the cost of going to cache. This example stresses that one should never blindly optimize across a modeled design space, but should rather examine and interpret the results to find new design directions that might improve energy efficiency further.

Finally, in addition to exploring new architectures, the designer may desire to explore different optimization objectives. The results presented in this chapter have focused on performance-energy trade-offs, and not considered area (die cost) or what happens for threaded or data parallel applications. One can still apply the regression-based methodology to handle these cases, but the optimization metrics and/or the application benchmarks need to change to account for the different design objective. For example, in the case of multi-core designs for highly parallel workloads, one needs to change the performance objective because the number of cores that can fit on a die is critical to performance, and one must consider both the performance and area of the cores. As another example, optimizing for different applications can yield significantly different optimization results, as the needs of the system change. The next case study examines the problem of identifying and optimizing for different application classes.

18.3 Case Study: Heterogeneity and Clustering

An understanding of performance trade-offs and marginal costs reveals design optima, such as the most energy-efficient architecture for a given performance target. Homogeneous processor designs are designed as a single compromise between a variety of per-application optima to accommodate diverse application demands. Such compromise trades efficiency for generality.

In contrast, heterogeneous processor design mitigates efficiency penalties. Two questions arise in heterogeneous design: what degree of heterogeneity is most effective and what do those heterogeneous designs look like? By combining fitted performance and energy models with clustering heuristics, one can answer these questions [9]. Models, constructed with splines or posynomials, identify an optimally efficient architecture for each application of interest. By partitioning the set of per-application optima into subsets of similar designs, cluster heuristics capture the degree of heterogeneity (i.e., the optimal number of clusters) and the nature of those heterogeneous designs (i.e., the location of clusters within the design space).

18.3.1 Clustering Methodology

K-means clustering is a heuristic that partitions a set into K subsets according to some clustering criterion. If we desire objects within a cluster to be very similar and any two objects from different clusters to be very different, we need to define a similarity metric. Clustering provides a mechanism to understand design heterogeneity.

First, we characterize the design space. In this case study, we construct spline-based regression models for the comprehensive space of Table 18.3. Optimizing these models, we identify efficient designs according to the desired metric. In this instance, we optimize bips^3/w, which integrates performance and energy aspects of efficiency into a single metric.* As shown in Table 18.4, these efficient designs occupy very different parts of the space and encompass a diverse set of design parameter values. For example, pipeline depth ranges from 15 to 30 FO4 delays per stage, superscalar width ranges from 2 to 8 instructions decoded per cycle, and L2 caches range from 0.25 to 4 MB. Each application's execution characteristics are reflected in its optimal architecture. For example, compute-intensive gzip has the smallest L2 cache while memory-intensive mcf has the largest.

K-means clustering these nine efficient architectures identifies heterogeneous design compromises. The heuristic for K clusters consists of the following:

1. Define K centroids, one for each cluster, and place randomly at initial locations in architectural design space.
2. Assign application-specific optimum to cluster with closest centroid.

---

* bips^3/w is a voltage invariant power-performance metric derived from the cubic relationship between power and voltage. This metric is inverse ED^2 as described in Section 18.2.2.

TABLE 18.3 Clustering Design Space with Range $i::j::k$ Denoting Values from i to k in Steps of j

	Set	Parameters	Measure	Range	$|S_i|$
S_1	Depth	Depth	FO4	9::3::36	10
S_2	Width	Width	Decode b/w	2,4,8	3
		L/S queue	Entries	15::15::45	
		Store queue	Entries	14::14::42	
		Functional units	Count	1,2,4	
S_3	Physical	General purpose	Count	40::10::130	10
	registers	Floating-point	Count	40::8::112	
		Special purpose	Count	42::6::96	
S_4	Reservation	Branch	Entries	6::1::15	10
	stations	Fixed-point	Entries	10::2::28	
		Floating-point	Entries	5::1::14	
S_5	I-L1 cache	i-L1 cache size	kB	16::2×::256	5
S_6	D-L1 cache	d-L1 cache size	kB	8::2×::128	5
S_7	L2 cache	L2 cache size	MB	0.25::2×::4	5

TABLE 18.4 Efficiency ($bips^3/w$) Maximizing Architectures per Application

	Depth (FO4)	Width	Reg.	Resv.	I–$ (kB)	D–$ (kB)	L2–$ (MB)	Delay Model	Err (%)	Power Model	Err (%)
ammp	27	8	130	12	32	128	2	1.0	0.2	35.9	−3.9
applu	27	8	130	15	16	8	0.25	0.8	−0.8	39.6	0.1
equake	27	8	130	15	64	8	0.25	1.2	−0.8	41.5	−3.0
gcc	15	2	70	9	16	8	1	1.2	5.2	44.1	−6.0
gzip	15	2	70	6	16	8	0.25	0.8	8.8	24.2	0.0
jbb	15	8	80	12	16	128	1	0.6	−4.7	80.9	1.6
mcf	30	2	70	6	256	8	4	3.5	2.4	12.9	−3.0
mesa	15	8	80	13	256	32	0.25	0.4	5.2	86.9	−7.1
twolf	27	8	130	15	128	128	2	1.1	−1.2	34.5	−3.6

3. When all application-specific optima have been assigned, re-compute placement of K centroids such that their distances to application-specific optima within respective clusters are minimized.

4. Since centroids may have moved in step 3, assignment of applications to clusters may change. Thus, steps 2 and 3 are repeated until centroid placement is stable.

In steps 2 and 3, similarity is measured using a normalized and weighted Euclidean distance. For each design parameter, we normalize its values by subtracting its mean and dividing by its standard deviation. We then weight parameters by their correlation with the objective metric (e.g., $bips^3/w$), thereby giving parameters with a greater impact on efficiency a bigger role in the distance calculation. If correlation coefficients $\rho_i > \rho_j$, an increase in parameter i will change the distance more than the same increase in parameter j. If two architectures are represented by vectors $\vec{a}, \vec{b}$ of parameter values, then the distance between them is determined by normalizing and weighting the values in $\vec{a}, \vec{b}$ and computing the Euclidean distance.

For example, pipeline depth values range from 12 to 30 FO4 in increments of 3 with a mean of 21 and standard deviation of 6.48. The normalized depth values range from −1.39 to 1.39 with mean 0 and standard deviation of 1.0. We then take the samples collected to fit the regression model and compute the correlation between depth and $bips^3/w$ to obtain depth's weighting factor.

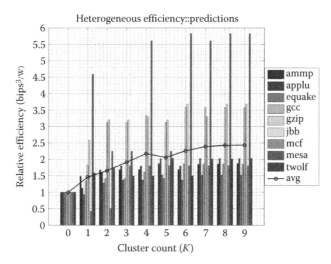

FIGURE 18.7   Predicted efficiency gains where cluster 0 is general-purpose architecture, cluster 1 is a homogeneous design targeting applications of interest, and cluster 9 are heterogeneous designs targeting each of nine applications (Table 18.4).

18.3.2 Heterogeneity Efficiency

Each K-means cluster comprises a set of similar architectures. The cluster centroid is the design compromise identified for those architectures. In this framework, the number of clusters is a measure of the degree of heterogeneity. As illustrated in Figure 18.7, exploring values of K will reveal the point of diminishing marginal returns. Efficiency is presented relative to cluster count zero, which we take to represent a general-purpose processor (e.g., IBM Power4).

Cluster count one identifies a single, homogeneous design for nine applications of interest. Such specialization improves efficiency by $1.5\times$ with some applications benefiting disproportionately and at the expense of others. For three designs, all applications benefit from heterogeneity with efficiency improving by $1.9\times$ on average. For this particular set of applications, we observe diminishing marginal returns beyond the four designs of Table 18.5. Representing a bound on heterogeneous efficiency, a cluster count that equals the number of applications represents complete specialization in which no compromises are made (e.g., nine applications lead to nine designs).

Figure 18.8 compares simulator reported heterogeneity gains against those of our regression models. The point of diminishing marginal returns is predicted to within 10% and the overall trends are consistent across both analyses.

After determining the number of clusters, which determines the degree of heterogeneity, we examine the centroids to examine the compromise architecture. Table 18.5 uses a $K = 4$ clustering to identify compromise architectures and their power-delay characteristics when executing their associated applications. This analysis illustrates the ability of regression models to identify optima and compromises occupying diverse parts of the design space. For example, the four compromise architectures capture all combinations of pipeline depths and widths. Cluster 1 suggests a pipeline with many short stages and multiple execution units for jbb and mesa. For a memory-intensive application, like mcf, cluster 4 suggests a shallow, narrow pipeline and larger caches to mitigate the impact of memory stalls in the datapath (Table 18.6).

Figure 18.9 plots heterogeneous designs in delay and power coordinates with applications executing on their corresponding designs. Architectures with deep pipelines operating at high clock frequencies and wide issue sustained by multiple functional units occupy the upper left quadrant of the space. Designs

TABLE 18.5 $K = 4$ Architectural Compromises

Cluster	Depth	Width	Reg.	Resv.	I–$ (KB)	D–$ (KB)	L2–$ (MB)	Avg. Delay Model	Avg. Power Model
1	15	8	80	12	64	64	0.5	2.26	82.17
2	27	8	130	14	32	32	0.5	1.05	32.53
3	15	2	70	8	16	8	0.5	0.93	37.55
4	30	2	70	6	256	8	4	0.29	12.91

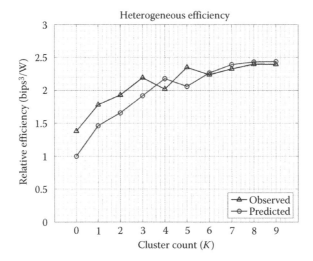

FIGURE 18.8   Average bips^3/w average from regression models and detailed simulation where x-axis interpreted as in Figure 18.7.

TABLE 18.6 $K = 4$ Cluster to Application Mapping

Cluster	Applications
1	jbb, mesa
2	ammp, applu, equake, twolf
3	gcc, gzip
4	mcf

with the fewest hardware resources occupy the lower right quadrant. Moderate designs occupy the center. The figure illustrates compromise architectures (circles) executing their clusters' respective application. Each compromise is located close to its cluster's application-specific designs (radial points), indicating that design compromises incur modest performance and power penalties. Although we cluster in a high-dimensional microarchitectural space (e.g., Table 18.3), the strong relationship between an architecture and its delay and power characteristics means we also observe clustering in the two dimensional (2D) delay-power space. Thus, the efficiency of an application suite executing on four heterogeneous designs is similar to the efficiency when executing on nine application-specific architectures. Thus, applications might achieve near ideal bips^3/w efficiency at much lower cost.

Figure 18.9 also suggests new opportunities for application similarity analysis based on hardware resource requirements. For example, ammp, applu, equake, and twolf may be similar applications since they are most efficient when running on similar pipeline designs and cache hierarchies. While other approaches might reduce the number of applications required in a benchmark suite for microarchitectural simulation [3], hardware-based clustering has direction implications for hardware design. For example,

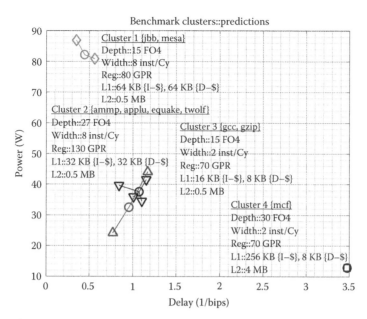

FIGURE 18.9 Performance and power for application-specific optima of Table 18.4 (radial points) and resulting compromises of Table 18.5 (circles).

accelerators might be designed for every application or kernel of interest. However, resource constraints necessitate compromises and the penalties from such compromises may be minimized by designing an accelerator to meet the needs of multiple similar kernels.

18.3.3 Discussion

Heterogeneity specializes hardware for particular applications and deploys a mix of hardware for a mix of application needs. Specialization improves efficiency as overheads from supporting general, arbitrary computation are reduced. For example, application-specific integrated circuits (ASICs) are orders of magnitude more energy efficient than general-purpose architectures. Seeking a more general approach to efficiency, we might reason about a spectrum of specialization and heterogeneity scenarios.

In a conservative scenario, heterogeneity might rely on differentiated, commodity hardware. By assembling a system from off-the-shelf architectures already serving disparate markets, the costs of heterogeneity might be mitigated. For example, datacenters might deploy a mix of low-power, mobile-class processors with high-performance, server-class ones. This mix might simultaneously provide mobile efficiency and quality-of-service guarantees for heavy-weight tasks requiring more substantial capability.

In a moderate scenario, heterogeneity might rely on components designed explicitly for that goal. This is the scenario we evaluate in our case study. By differentiating architectures to target specific application clusters, we improve efficiency by reducing the efficiency penalties incurred by design compromises. While efficiency favors more heterogeneity, economics favors less heterogeneity so that each design targets larger markets. Clustering heuristics help designers navigate this trade-off.

Finally, in an extreme scenario, designers might pursue ASICS, sacrificing generality for efficiency [4]. These efficiencies are possible as overheads in control and communications are reduced. For example, efficiency might arise from mechanisms that extract instruction- and data-level parallelism with VLIW and SIMD. Fusing common operation sub-graphs to create larger instructions similarly reduces overheads. These "general-purpose," modestly programmable customizations can enhance efficiency by an order of magnitude. More substantial customizations can further improve efficiency by another order of magnitude with highly specific functional units that effect hundreds of operations per invocation.

Thus, in the pursuit of heterogeneity, we require methodologies to systematically extract an order of magnitude in efficiency from large, complex design spaces. We require application domain knowledge to extract another order of magnitude.

References

1. O. Azizi, A. Mahesri, B.C. Lee, S.J. Patel, and M. Horowitz. Energy-performance tradeoffs in processor architecture and circuit design: A marginal cost analysis. In *Proceedings of the 37th IEEE/ACM International Symposium on Computer Architecture (ISCA)*, Saint-Malo, France, pp. 26–36, 2010. New York: ACM.
2. C. Dubach, T. Jones, and M. O'Boyle. Microarchitectural design space exploration using an architecture-centric approach. In *Proceedings of the 40th IEEE/ACM International Symposium on Microarchitecture (MICRO)*, Chicago, IL, pp. 262–271, 2007. Washington, DC: IEEE Computer Society.
3. L. Eeckhout, H. Vandierendonck, and K. De Bosschere. Quantifying the impact of input data sets on program behavior and its applications. *Journal of Instruction-Level Parallelism*, 5:1–33, 2003.
4. R. Hameed, W. Qadeer, M. Wachs, O. Azizi, A. Solomatnikov, B.C. Lee, S. Richardson, C. Kozyrakis, and M. Horowitz. Understanding sources of inefficiency in general-purpose chips. In *Proceedings of the 37th IEEE/ACM International Symposium on Computer Architecture (ISCA)*, Saint-Malo, France, pp. 37–47, 2010. New York: ACM.
5. F. Harrell. *Regression Modeling Strategies*, 2001. New York: Springer-Verlag.
6. E. Ipek, S.A. McKee, R. Caruana, B.R. de Supinski, and M. Schulz. Efficiently exploring architectural design spaces via predictive modeling. In *Proceedings of the 12th ACM International Conference on Architectural Support for Programming Languages and Operating Systems (ASPLOS)*, San Jose, CA, pp. 195–206, 2006. New York: ACM.
7. P.J. Joseph, K. Vaswani, and M.J. Thazhuthaveetil. A predictive performance model for superscalar processors. In *Proceedings of the 39th IEEE/ACM International Symposium on Microarchitecture (MICRO)*, Orlando, FL, pp. 161–170, 2006. Washington, DC: IEEE Computer Society.
8. B.C. Lee and D. Brooks. Accurate and efficient regression modeling for microarchitectural performance and power prediction. In *Proceedings of the 12th ACM International Conference on Architectural Support for Programming Languages and Operating Systems (ASPLOS)*, San Jose, CA, pp. 185–194, 2006. New York: ACM.
9. B.C. Lee and D. Brooks. Illustrate design space studies with microarchitectural regression models. In *Proceedings of the 13th IEEE International Symposium on High Performance Computer Architecture (HPCA)*, Phoenix, AZ, pp. 340–351, 2007. Washington, DC: IEEE Computer Society.

19

Energy-Aware SIMD Algorithm Design on GPU and Multicore Architectures

Da Qi Ren
The University of Tokyo

Reiji Suda
The University of Tokyo

19.1 Introduction

GPU and multicore hybrid platforms deliver excellent performance on massively parallel processing and vector computing. It has become one of the most popular processing element (PE) to construct a modern parallel computer. Energy saving is an important issue that influences the design development of high performance computing (HPC) because large scale scientific computing may lead to an enormous energy predicament. The power efficiency of a multiprocessing system is dependent on not only the electrical features of hardware components but also the high level algorithms and programming paradigms. Enhancing the utilization of each individual PE to reach its best computation capability and power efficiency is important for optimizing overall system power performance. In this chapter, a power model based on measurement and experimental evaluation of SIMD computations on GPU and multicore architectures has been introduced. Three primary energy aware CUDA algorithm design methods have been investigated and illustrated, including: building a processing element with single CPU core and parallel GPUs; splitting GPU workloads to CPU components; removing GPU computing overheard by executing small tasks with CPU functions. The improvements of these approaches on computation time and power consumption have been validated by examining the CUDA programs executing on real systems.

General-purpose computing on graphics processing units (GPGPU) has become a serious challenger for large-scale scientific computing solutions due to their multicore architecture and suitability for high performance vector processing. A modern computing application may use thousands of GPU and multicore processing elements [1], and hundreds of hours of continuous execution. A GPU computing program starts with C/CUDA code, compiler, and an assembler that translates it into a compact sequence of instructions to create object code. A linker puts the objects together into an executable, which can then be run on GPU and multicore PEs. The power consumption of the C/CUDA program is eventually originated from the micro-architecture of each of the hardware components including CPU, GPU, memory, and PCI bus. The binary executable operations on hardware are controlled by software algorithms and high-level design strategies that significantly affect the performance and energy expense of the computing.

The power consumption of an HPC program has been modeled at different levels of abstractions, such as simulating the circuit at the transistor or switch level [2] and modeling the middle-grained components such as multipliers and registers [3]. At lower-level models, the available physical information allows acquiring accurate power estimates; however, higher-level models depend on the abstraction of hardware system descriptions using more indirect and approximate design parameters. There are many research efforts on CPU based power models, such as the method illustrated in [4] uses a coordinated measurement approach combining overall power measurement with performance - counter - based, per-unit power estimation. A GPU-based power model can be created in a similar way to CPU-based models. An integrated power model is proposed in [5] for GPUs to predict execution times and calculate dynamic power events. Power efficiency is a system issue that is related to each level of an HPC system. CUDA on CPU-GPU platform is becoming a major choice of HPC in various applications; however, much less research has been carried out to evaluate the power performance for the processing element of CPU-GPU with integrated parallel programming paradigms [6,7].

This chapter's focus is on handling the critical design constraints in the level of software that runs beyond a hardware system composed of huge numbers of power-hungry parallel components in order to optimize the program design and thus achieve the best power performance. In this chapter, detailed power measurements have been performed to each major power-consuming component in a CUDA PE; therefore, power factors and identities of the system can be abstracted and captured accordingly. Based on these, hardware power parameters are imported to software study, energy consumption is estimated and optimized by analyzing the CPU-GPU architecture and program design. One of the advantages is that it allows obtaining design characteristic values at the early programming stage, thus benefiting programmers by providing necessary environment information for choosing the best power-efficient design alternative [17].

19.2 Energy-Aware SIMD Algorithm Optimization Framework on GPU-Multicore Architecture

An algorithm level energy-aware program design framework has been purposed and illustrated in Figure 19.1. On top of hardware, a power model is built using hierarchical modules representing the electrical characters abstracted from each hardware component inside the GPU and multicore unit. The links between the modules describe functional or algorithmic behaviors of each component under the control of a CUDA program. The hardware components involved in a SIMD computing include CPU core, GPU, and main memory; their power features can be obtained from measurements and analysis.

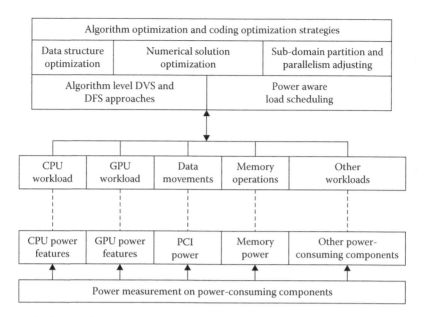

Algorithm optimization and coding optimization strategies		
Data structure optimization	Numerical solution optimization	Sub-domain partition and parallelism adjusting
Algorithm level DVS and DFS approaches		Power aware load scheduling

CPU workload	GPU workload	Data movements	Memory operations	Other workloads

CPU power features	GPU power features	PCI power	Memory power	Other power-consuming components

Power measurement on power-consuming components

FIGURE 19.1   The framework on energy-aware algorithm design optimization. (© 2010 IEEE.)

Knowing the power features of each such individual CUDA processing element, an overall power model can be built up for the entire multiprocessing platform [8].

The computation capability of a CUDA PE is physically determined by hardware microarchitecture, compiler technique, programming language, and the character of applications performed on it. In SIMD computing, a compiler will generate an identical instruction sequence for continual execution. Power parameters are used by algorithm design approaches; algorithm optimization and coding optimization strategies manage workload to assign them to different CPU and GPU components or component combinations based on the different computation capability and the corresponding power-consuming feature.

Other power factors, from top to bottom, include program parallel overheads such as kernel launch time and synchronization time; the instruction structures; hardware temperatures; frequency scaling; chip microarchitectures; memory bandwidth and bus bandwidth; interconnection topology, etc.

The performance improvement procedure usually consists of several incremental iterations. The result of each iteration will be checked with original design objectives, then according to the improvement satisfaction to decide the necessity of further refinement until the required power performance is reached. An important issue that needs to be considered is the trade-off in each optimization alternative between the energy consumption and the computing performance. These are significant in the early program design.

19.3 Power Model for SIMD Computing on CUDA Processing Element

A CUDA kernel is launched by a host CPU, a certain numbers of GPU blocks and threads are defined in the kernel to instruct the operations of the kernel on the target GPU hardware, as shown in Figure 19.2a. The case studies in this chapter use NVIDIA GeForce 8800 GTS 512 GPU with 16 MPs (multiprocessors), each MP has 8 SPs (streaming processors) to execute multiple threads [9]. CUDA manual introduces one warp including a group of 32 parallel scalar threads executing one common instruction at a time;

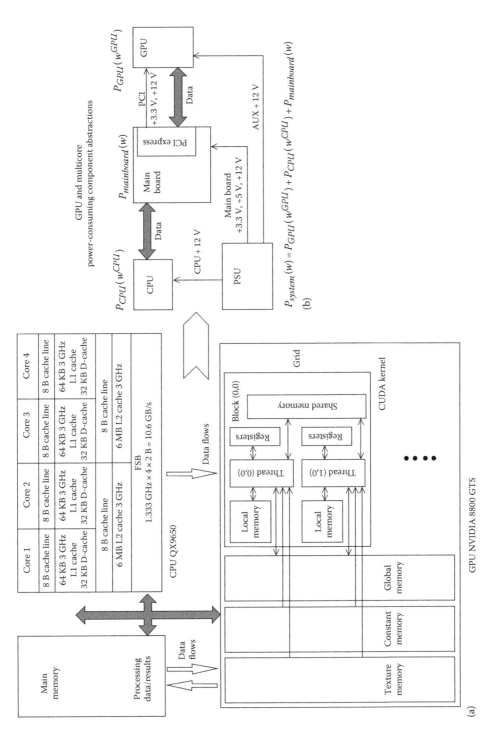

FIGURE 19.2 (a) Data flows of CUDA kernel execution and (b) the energy-consuming component abstraction on a GPU and multicore architecture.

this collection of functional units is recognized as a core. A sample CUDA kernel processing flow is shown in Figure 19.2a: the CPU host firstly controls copying data from main memory to GPU memory; next CPU launches CUDA kernel and starts running the kernel on GPU; GPU executes threads in parallel by each core; finally the results are copied from GPU memory to the main memory [9].

19.3.1 Power Model

Three major components involved in the CUDA kernel's execution are multicore, GPU, and main board, as shown in Figure 19.2b. The main memory and PCI buses dominate power consumption on the main board [8]. The power consumption of the CUDA kernel can therefore be modeled as

$$P_{system}(w) = P_{GPU}(w^{GPU}) + P_{CPU}(w^{CPU}) + P_{mainboard}(w) \tag{19.1}$$

where P_{system}, P_{CPU}, P_{GPU}, and $P_{mainboard}$ represent the power of the overall system, GPU, CPU, and main board, respectively. w, w^{GPU}, w^{CPU} represent the total workload, GPU and CPU workload, respectively.

Power efficiency of a GPU and multicore PE is dependent on a certain number of software and hardware factors as listed in Table 19.1. CPU power efficiency is determined by the number of CPU cores and threads that are involved in the CUDA computing, the operations of instruction flows, the frequency setup of CPU cores. GPU power efficiency lies on the GPU microarchitecture, number of GPU cores used by a CUDA kernel, number of CUDA kernels performed at a time, GPU frequency, PCI bus speed, and the CUDA kernel programming pattern. Main board power efficiency depends on the amount and speed of data transmission between CPU and GPU, memory operations, and the bandwidth and electrical features of PCI bus. User data transfer rates depend on higher layer applications and the way the data is moved.

The software and hardware environment in this work is introduced in Table 19.2. Intel Core 2 Quad Extreme QX9650 CPU is packaged in LGA775 socket on the main board and its power is connected from PSU directly. A GPU card is plugged in a PCI-Express slot on main board. It is mainly powered from PCI-Express pins, and additional power comes directly from PSU. The CPU input current and voltage at LGA775 socket is measured in an approximate way. GPU power is measured on an auxiliary power line with a clamp probe, also on the PCI bus by approximation from main board power inputs. National Instruments USB-6216 BNC data acquisition, Fluke i30s/i310s current probes, and Yokogawa 700925 voltage probe are used. The room was air conditioned in 23°C. [10–12]. The LabView 8.5 is used as oscilloscopes and analyzers for the results data analysis. By testing the power responses of each component involved in a sample matrix multiplication, the real-time voltage and current are recorded from measurement readings. The product of voltage and current is the instant power at each sampling point during measurement. [10–12]. All power sampling points are plotted together into a power chart

TABLE 19.1 Power-Consuming Components and Data Flows in CPU-GPU Architecture

Major Components	Power Performance Factors
$P_{GPU}(w^{GPU})$	1. GPU microarchitecture 2. Number of GPU cores employed 3. Number of CUDA kernels performed 4. GPU frequency 5. PCI bus speed 6. CUDA kernel programming pattern
$P_{CPU}(w^{CPU})$	1. Number of CPU cores and threads involved in computing 2. CPU frequency
$P_{mainboard}(w)$	1. The PCI bandwidth for data transmission 2. The amount of data transferred between CPU and GPU 3. The electrical feature of PCI bus

TABLE 19.2 List of the Major Components and Their Power Features in the Working Environment

Power-Related Components	Model/Power Specification
CUDA driver	Version 185.18
Compiler	CUDA2.2/gcc 4.3.2
OS	Ubuntu 8.10
GPU	NVIDIA GeForce 8800
Power supply unit	Seasonic SS-700HM
Main board	GIGABYTE GA-58-UD3R
CPU	QX9650 3 GHz / LGA775
Main memory	3 SDRAM PC3-8500 8 GB
CPU FAN	Intel E29477 / 9.0 W

Source: D. Q. Ren and R. Suda, Power model of large-scale matrix multiplication on multi-core CPUs and GPUs platform, in *The 8th International Conference on Parallel Processing and Applied Mathematics (PPAM 2009)*, pp. 421–428, Wroclaw, Poland, September 13–16, 2009. D. Q. Ren and R. Suda, Investigation on the power efficiency of multi-core and GPU processing element in large scale SIMD computation with CUDA, in *The 2010 International Green Computing Conference*, pp. 309–316, Chicago, IL, August 15–18, 2010. © 2010 IEEE.

which shows the power usage of each component during the program execution. The area between power curve and time axis is an approximation of energy cost of the program execution [10]. We have introduced the detailed power modeling approach in [8,10].

19.3.2 Power Parameters

Flops per *watt* is a measure of power efficiency. It shows the rate of computation that can be provided by a processing element for every watt of power consumed [8]. The following formulas illustrate the relation of Flops/watt with the corresponding computation time, power, and energy consumption:

$$Time(T) = \frac{Workload(w)}{Gflops(s)} = \frac{W}{s} \tag{19.2}$$

$$Power(P) = \frac{Gflops(s)}{Gflops/watt(s/w)} = \frac{s}{f} \tag{19.3}$$

$$Energy(E) = Time(T) \cdot Power(P) = \frac{Workload(W)}{Gflops/watt(s/w)} = \frac{w}{f} \tag{19.4}$$

Where
    w represents the workload
    T is computing time
    E is energy consumed by the computation
    s stands for computing speed, i.e., the number of *Gflops*
    f represents the power measure of *Gflops/watt* [13,14]

When a SIMD computation is performed in a streaming way on a CUDA PE, the PE's power consumption is originated by the same instruction flow executing repeatedly on the identical structure. If the PE's frequency and temperature is invariant, its power can be approximated as a constant [8,10]. This feature can help to determine the energy consumption function and power efficiency for a specific algorithm. Note that the processor's temperature scaling is not discussed in here even though we understand it will contribute errors, because modern CMOS technique has many solutions to control the processor's

TABLE 19.3 Summarization of the Energy-Consuming Features

PE Components	Peak Gflops	Mflops/W	CPU Frequency (GHz)	GPU NV/Memory Frequency (MHz)
Intel QX9650 CPU GF 8800 GTS/512	50.1	73.6	3	650/972

temperature to be within a tolerable range. A software solution to this problem is finding an average of the measurements when power shifts with the temperature during a computation (Table 19.3) [17].

19.3.3 Conditions and Limitations

The model has been built based on the following assumptions: the bandwidth of PCI bus between CPU and GPU is fixed and data move along the memory hierarchy at a consistent speed; system resources are used by the CUDA program exclusively; CPU and GPU's temperature stays in a tolerable range. The accuracy of the power model is limited by problem size. Because CUDA kernel overheads and underfoot may change at each time when the CPU launches the CUDA kernel, the power phase leg to the code execution will cause calculation errors especially when a problem size is small [8].

19.4 Enhance the Power Efficiency of SIMD CUDA Programs

The CPU and GPU have different computation capabilities and different power features. A parallel algorithm on the top of a CUDA PE instructs the use of CPU, GPU and the data flowing between them. An ideal program utilizes each hardware component to reach the best parallel speedup and overall power efficiency. A basic principle considered is to minimize the number of employed CPU, GPU and other power-hungry components in order to reduce the total power consumption and enhance the usage of each employed component in order to achieve the best computing performance.

19.4.1 Reduce the Number of Computing Components

If a program can be optimized to use fewer power-consuming components without delaying the total execution time, the overall energy consumption will be decreased and the PE's power efficiency will be promoted.

When a CUDA kernel is running at a high working frequency, operations such as *CudaMemCpy* in the CUDA kernel require the host CPU to respond correctly to the GPU, the CPU core will then be blocked in the loop. Until the computing is finished, the CPU can be released to respond to other calls. This minimizes the responding latency of CPU to GPU, but will take all the CPU resources.

In the case of one CPU core working with one GPU device as shown in Figure 19.3a, saving one entire CPU core is not practical. Because a CPU core's frequency is generally higher than the need to respond a GPU's polling, if the GPU polling frequency can be estimated, the CPU core frequency can be scaled down to a proper value that is exactly enough to respond the GPU polling (CUDA kernel calls). The CPU power can therefore be saved. This approach is recommended when hardware parameters are known and the CUDA program is granted system authorization to change CPU frequencies [8].

When one CPU works with multi-GPUs, as shown in Figure 19.3b, CPU efficiency can be improved by making good use of CPU clocks. In detail, let one of the CPU cores run in high frequency to support two or more GPU devices; thus at least one core can be saved comparing with the one core on one GPU solution. The CPU core will be used in multithreads that each thread occupies one segment of the CPU clocks controlling one GPU device. The CPU core's frequency is required to be configured higher than the total frequencies for polling all GPUs [16].

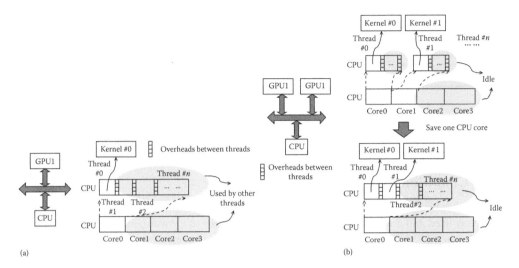

FIGURE 19.3 The usage of CPU and GPU energy-consuming components in a GPU and multicore PE: (a) single CPU thread and single CUDA kernel on a one-CPU-one-GPU platform; (b) single CPU with multiple threads for multiple CUDA kernels on a one-CPU-two-GPU platform, one CPU core is saved by multithreading method. (Ren, D. Q. and Suda, R. Investigation on the power efficiency on multi-core and GPU processing element in large scale SIMD Computation with CUDA, in *The 2010 International Green Computing Conference*, pp. 309–316, Chicago, IL, August 15–18, 2010; *Intel® 64 and IA-32 Architectures Software Developer's Manual*, Volume 1: Basic Architecture, Order Number: 253665-028US, September 2008.)

19.4.2 Enhancing the Usage of CPU Component

An improvement will be done by letting a CPU inside a CUDA and multicore PE to share some workloads originally belonging to GPU(s). The purpose is to increase the usage of CPU resources in order to obtain computing speedup on the program, and thereby enhance the PE's power efficiency. Multithreads on the CPU core(s) will be created and a small portion of the GPU jobs will be relocated to the CPU cores and threads.

19.4.3 Parallel Algorithm Design

Dense matrix multiplication in an optimal way is a fundamental operation in solving many HPC problems requiring large numbers of elements for high-accuracy solutions. Faster matrix multiplication will give more efficient algorithms for many standard problems. The performance improvements of design optimization methods introduced in Sections 19.4.1 and 19.4.2 have been validated by using the following matrix multiplication examples. For given matrices $A \in R^{n \times n}$ and $B \in R^{n \times n}$, there are a variety of ways to implement the calculation of $C = AB$ in parallel. One typical method is to partition matrix A firstly, i.e., let $A = (A_0^T, A_1^T \ldots A_{n-1}^T)^T$, then distribute the computation of $C_0 = A_0 B$, $C_1 = A_1 B$ until $C_{n-1} = A_{n-1} B$ to n different parallel processors in the multiprocessing system.

The above parallelization method is adopted and implemented on parallel GPU and multicore platforms: let the matrix A being split into n submatrices on main memory; next initializing n CUDA kernels to compute $C_0 = A_0 B$, $C_1 = A_1 B$ until $C_{n-1} = A_{n-1} B$ in parallel asynchronously on n different GPUs. The results $C_0, C_1, \ldots C_{n-1}$ are finally copied back to the main memory for building up the final result C. Three C/CUDA programs are implemented to calculate the matrices multiplication $C = AB$ on a multiprocessing platform including on Intel QX9650 CPU and two GF 8800/512 GTS GPUs.

Algorithm 19.1

$C = AB$ is computed by one CPU and one GPU with a CUDA kernel directly. The matrices A and B are generated on main memory, they are future loaded to GPU shared memory in 2D blocks during CUDA kernel execution.

Algorithm 19.2

In algorithm 19.2, the multiplication $C = AB$ is computed using two GPUs in parallel. Matrices A and B are firstly located and generated on main memory. Two multiple threads are created and initialed on a host CPU core; each thread is in charge of launching and managing one CUDA kernel on each of the two GPUs, respectively. Global synchronization marks are set to control and synchronize the parallel processes among all CPU cores and GPUs. [16]. Matrix A is split into $A = (A_0^T, A_1^T)^T$, computations of $C_0 = A_0B$ and $C_1 = A_1B$ are distributed to GPU_0 and GPU_1, respectively. When each CUDA kernel completes its execution, the results C_0 and C_1 are copied to the main memory. The diagram of algorithm 19.2 is described in Figure 19.4.

The multiplication $C = AB$ is computed in parallel with two GPUs. Let $A = (A_1^T, A_2^T)^T$, distribute the computation of $C_1 = A_1B$ and $C_2 = A_2B$ to GPU1 and GPU2, respectively. Matrices A and B are firstly located and generated on main memory; a global synchronization mark is set up to synchronize the parallel computation between two GPUs. Multithreads are created and initialed by the host CPU that ensures two CUDA kernels are computing $C_1 = A_1B$ and $C_2 = A_2B$ in parallel synchronously. Finally the results C_1 and C_2 are copied to the main memory (Figure 19.5) [16].

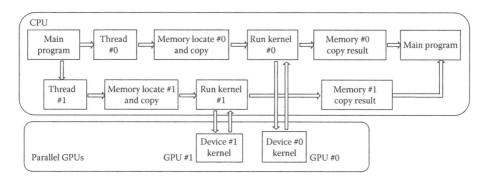

FIGURE 19.4   The data flow of CUDA kernel using Algorithm 19.2.

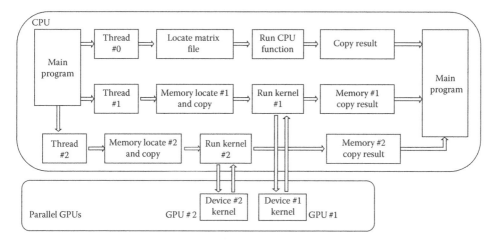

FIGURE 19.5   The data flow of CUDA kernel using Algorithm 19.3.

Algorithm 19.3

Three multiple threads on one CPU core are created and initialized in algorithm 19.3: thread #0 operates on the CPU core; thread #1 and thread #2 control CUDA kernel#1 and kernel#2 on GPU_1 and GPU_2, respectively. The matrix A is split into $A = (A_0^T, A_1^T, A_2^T)^T$. Workload $C_0 = A_0 B$ is assigned to the CPU core; workloads $C_1 = A_1 B$ and $C_2 = A_2 B$ are assigned to GPU_1 and GPU_2, respectively. The diagram of algorithm 19.3 is described in Figure 19.5.

19.4.4 Implementation and Results

We define matrix size $A \in R^{3200 \times 3200}$, $B \in R^{3200 \times 3200}$. In Algorithm 19.2, the submatrices A_1 and A_2 are the same size that half split matrix A. In Algorithm 19.3, A_1 and A_2 have the same size, each of them takes 1596 rows from matrix A, the rest eight rows as submatrix A_0 is to be assigned to the host CPU. We show the result measurement power and energy charts of CPU, GPU, AUX, and main board during the executions of all three different algorithms in Figure 19.6b through d, respectively.

In Figure 19.6b, when the CPU works on Algorithm 19.2, the average execution power is around 161.31 W, which is higher than the average power of 148.12 W when it works on Algorithm 19.1. When CPU works on Algorithm 19.3 with sharing workload, the average power is around 184.63 W. Algorithm 19.3 takes 0.85 s and Algorithm 19.2 takes 0.82 s to complete the work, both of them are faster than 1.37 s by Algorithm 19.1.

The total CPU energy consumption from Algorithms 19.1 to 19.3 is 0.055, 0.047, and 0.054 Wh, respectively. The cases of GPU and main board are similar with CPU, as shown in Figure 19.6c and d.

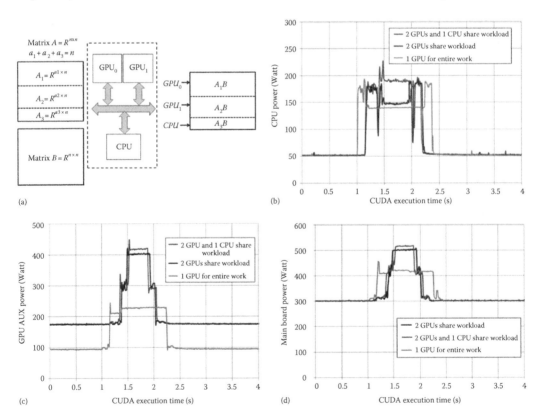

FIGURE 19.6 (a) The description of parallel workload sharing and scheduling for matrix multiplication among the CPU and GPU components (b)–(d) Power measure results for Algorithms 19.1, 19.2, and 19.3: (a) CPU +12 V power; (c) GPU AUX +12 V power; (d) the power measured from main board +3.3, +5, and +12 V. (© 2010 IEEE.)

Parallel GPUs consume more power than a single GPU at AUX + 12 V power input; however, they speed up the execution time by 0.55 s (Algorithm 19.2) and 0.52 s (Algorithm 19.3). The energy saving is 0.004 and 0.002 Wh, respectively. On the mother board, parallel GPUs contribute time speedup with 0.55 s (Algorithm 19.2) and 0.52 s (Algorithm 19.3) and energy saving with 0.0042 Wh (Algorithm 19.2) and 0.0031 Wh (Algorithm 19.3) comparing with Algorithm 19.1.

The multithreading CPU and two GPU combinations can achieve 39.7% (Algorithm 19.2) and 37.2% (Algorithm 19.3) speedup in kernel execution time, 21.4% (Algorithm 19.2) and 20.1% (Algorithm 19.3) speedup in CPU execution time compared with the power consumed by one CPU and one GPU combination (Algorithm 19.1). For energy consumptions, even with one more GPU component the Algorithm 19.2 and 19.3 have decreased 6.17% and 2.32% of the overall energy consumption compared with Algorithm 19.1, respectively, as shown in Figure 19.6 [17].

19.5 Workload Scheduling for GPU and Multicore

During the execution of a C/CUDA program, a multicore CPU initalizes the main program first then executes serial parts of the program. Parallel SIMD parts of the program are executed by CUDA kernels on GPU. The purpose of splitting GPU workload to multicore is to improve the utilization of CPU; reduce the GPU workload in order to reduce the overall execution time and energy consumption. Because multicore and GPU have different program patterns, computational capabilities and power features, the workload assignment on multicore has to be restricted within a predefined threshold to avoid the potential delay that the multicore may cause to the overall performance.

19.5.1 Workload Sharing between GPU and Multicore

Let W_{CPU} and W_{GPU} represent the workload assigned to CPU and GPU; T_{CPU} and T_{GPU} represent the execution time of CPU and GPU; S_{CPU} and S_{GPU} represent the execution speed of the CPU and GPU, respectively. Their relations satisfy equations (19.5) and (19.6), respectively.

$$T_{CPU} = \frac{W_{CPU}}{S_{CPU}} \tag{19.5}$$

$$T_{GPU} = \frac{W_{GPU}}{S_{GPU}} \tag{19.6}$$

The computing time of the overall PE is determined by the longest one among the CPU and GPU elements that work in parallel, i.e.

$$T = \max(T_{CPU}, T_{GPU}) \tag{19.7}$$

The energy consumptions of CPU and GPU are shown in equation 19.8 and 19.9, respectively:

$$E_{CPU} = T \cdot P_{CPU} \tag{19.8}$$

$$E_{GPU} = T_{GPU} \cdot P_{GPU} \tag{19.9}$$

Note that in (19.8), if a CPU finishes its assignment earlier than its co-working GPU, it will still work on responding to CUDA kernel's polling from the GPU until the kernel finishes, thus the CPU execution time is equal to the overall computation time of the PE. Thereby, a PE's overall energy consumption is defined as the summation of the energy consumptions from each of the CPU and GPU elements:

$$E = E_{CPU} + E_{GPU} \tag{19.10}$$

The aforementioned formulas express a CUDA PE constructed by one-CPU-one-GPU. One-CPU-more-GPU cases can be analyzed in a similar way.

19.5.2 Performance Analysis

In Figure 19.7a through d, the legends marked with (*) describe the measurement results on each component performing the algorithm with multicore CPU sharing workload; the legends with no mark describe the measurement results of the same problem without multicore CPU sharing workload.

Figure 19.7a and b show that when CPU workload is small (about 0%–0.7% of the total workload), the CPU execution time is less than GPU, the overall execution time will become shorter compared with GPU only. Computation performance can be improved up to 2% at the point where CPU time and GPU time are equal, as shown in Figure 19.7b.

Figure 19.7c and d show the relations between PE's energy consumption and CPU/GPU workload. The maximum power saving can be predicted by the formula

$$E_{\min} = (E_{CPU} + E_{GPU})_{\min} = (T \cdot P_{CPU} + T_{GPU} \cdot P_{GPU})_{\min} \qquad (19.11)$$

An optimized minimum energy value can be obtained from Figure 19.7d when CPU workload share is around 0.83%, the maximum energy saving can reach around 1.3%.

When CPU workload shares are greater than the proper period area marked in the figures, CPU workload will decrease the overall computation performance and power efficiency of the PE. We have validated the previous analysis by testing CPU workloads on different CUDA PEs as shown in Table 19.4. The real measurements agree with the workload scheduling analysis.

19.5.3 Conditions and Limitations

A CUDA kernel takes most of the PCI bandwidth for transferring data between CPU and GPU, and also occupies a large portion of the main memory interface for read/write operations. The workload assigned to CPU needs to be restricted by the following conditions: (1) there is no collision between the CPU workload and CUDA kernel workload on the use of PCI bandwidth; and (2) there is no collision on the main memory operation between CPU workload and CUDA kernel workload. Any interference from extra utilization of the resource of CPU or GPU may decrease the computation performance and power efficiency of the entire PE [17].

19.6 Optimization to Reduce Computation Overhead for Small Workloads

While benefiting large scale massive parallel computing, a C/CUDA program has to process an overhead that is used for kernel initialization, memory copy, and kernel launch before each time it starts to run the kernel. When the size of a problem is sufficiently big, CUDA overhead acts as a trade-off of the GPU performance enhancement; however, when problem size is small, the time and energy cost of the overhead becomes so significant that it will decrease the program's performance.

In order to minimize the impacts of CUDA kernel overheads on small size problems, a method is introduced in this section to remove the overheads by using a hardware selection approach that can place and perform the computations of small problems on multicore CPU instead of GPU.

A parallel C/OMP/MPI function is embedded in a C/CUDA program that is able to compute matrix multiplications by only CPU cores. A switch threshold k is predefined in the program that ensures $E_{CPU}(k) \leq E_{PE}(k)$, as described in function 19.12. $E_{CPU}(k)$ and $E_{PE}(k)$ represent the energy consumed by multicore only and by GPU and multicore CUDA PE, on a problem of size k, respectively. $T_{CPU}(k)$

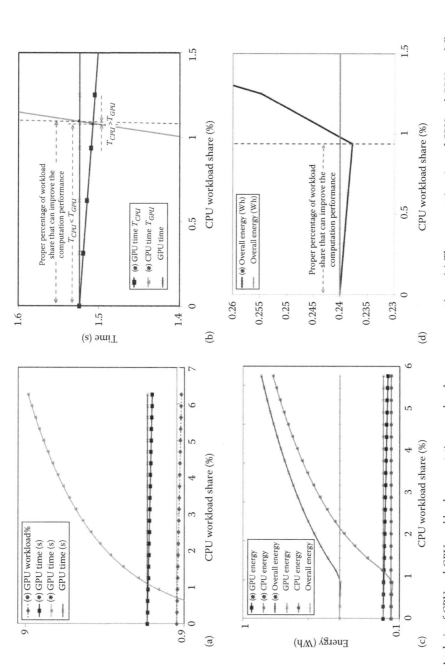

FIGURE 19.7 Analysis of CPU and GPU workload, computation speed and energy consumptions: (a) The computation time of CPU and GPU on different percentage of workload sharing; (b) the proper workload sharing period and the best speedup; (c) the power chart of CPU and GPU on different percentage of workload sharing; (d) the proper workload sharing period and the best power performance. (© 2010 IEEE.)

TABLE 19.4   The Measurement Results and Energy-Consuming Features from Different Algorithms on the GPU and Multicore Platform

PE Components	Peak Gflops	Mflops/W	CPU Frequency (GHz)	GPU NV/Memory Frequency (MHz)	Total Time (s)	Total Energy (Wh)
Intel QX9650 CPU GF 8800 GTS/512	50.1	73.6	3	650/972	1.37	0.259
Intel QX9650 + 2 GF 8800 GTS/512	80	75.1	3	650/972	0.82	0.243
QX9650 (CPU share 1%) GF 8800 GTS/512	53	77.9	3	650/972	1.23	0.232
QX9650 (CPU share 1%) 2 GF 8800 GTS/512	77	71.3	3	650/972	0.85	0.253

Source: © 2010 IEEE.

and $T_{PE}(k)$ represent the time consumed by multicore and by CUDA PE, respectively. P_{CPU} and P_{PE} represent the power of multicore and CUDA PE during the computation, respectively.

$$T_{PE}(k) = T_{GPUoverhead}(k) + T_{CUDAkernel}(k)$$

$$E_{CPU}(k) = P_{CPU} \times T_{CPU}(k)$$

$$E_{PE}(k) = P_{PE} \times T_{PE}(k) \tag{19.12}$$

Therefore when the sizes of input matrices are less than k, main program will call the C/OMP/MPI function to perform the computing on multicore rather than on the entire CUDA PE. For a general SIMD application, threshold k is determined by the character of the specific problem; the numerical solution and computing algorithm applied; and the target GPU and multicore platform where the program is executed. The value of k is obtained by experimental power measurements.

Fig. 19.8a and b show the performance and power efficiency of a 4 core CPU (QX9650) in computing multiplications of small matrices sized from 100 to 500. The computation times and energy consumptions are less than those needed to finish the same tasks on CUDA PEs.

We examine the performance of the preceding design by computing square matrices sized from 100 to 500 on 4 core CPU (QX9650), one-CPU-one-GPU(8800GTS) CUDA PE, and one-CPU-two-GPU(8800GTS) CUDA PE, respectively. Figure 19.8c and d show the performance comparisons between the C function on 1–4 CPU cores and the CUDA kernel on one-GPU and two-GPU PEs. It shows that in average 4 cores CPU can save 71.2% of the power consumed by a two-GPU CUDA PE and 49.4% of the power consumed by one-GPU CUDA PE. Comparing it with using one-GPU and two-GPU PEs, the computation time that has been saved is 45.3% and 25.1% on average by using 4 CPU cores, respectively.

19.7 Conclusion

Power efficiency is a crucial design issue because a large-scale scientific computation may require a huge amount of power consumption. Program design strategies will significantly affect the overall energy expense. Targeting on improving the power efficiency of an individual multicore and GPU processing element, we investigate software methodologies to optimize the power utilization through algorithm design and programming technique. Based on power parameters captured from real measurements on CPU–GPU architectures, modeling and evaluation of the power efficiency for CUDA PEs is illustrated. Following the approach, we provide a load sharing method to manage the workload distribution among each CPU and GPU component inside a PE in order to optimize the overall power consumption by tuning the size of workload assignment. We have also introduced a hardware component selection method to

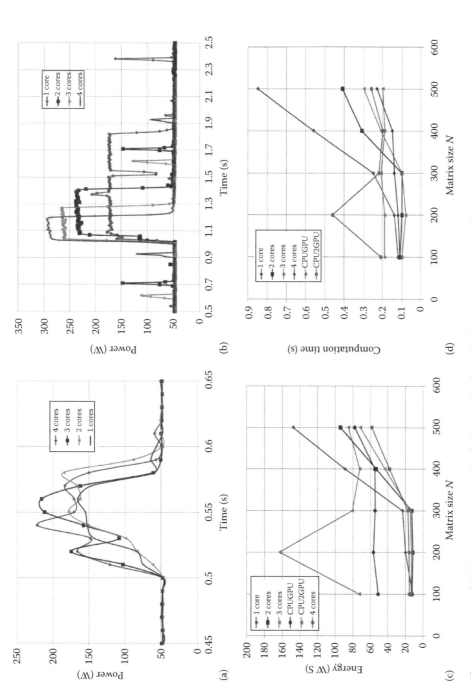

FIGURE 19.8 The measurement of CUDA computation overhead when workload is mall: (a) matrix size $n = 100$; (b) matrix size $= 500$; (c) energy consumption comparison of 1–4 cores, one-GPU PE and two GPU PE; (d) computation time comparison of 1–4 cores, one-GPU PE and two-GPU PE. (© 2010 IEEE.)

remove the computation overhead of CUDA PE when workloads are small. The performance improvement of our design on computation speed and power consumption has been validated by measuring the programs running on real systems with previous methods applied.

References

1. Nvidia, NVIDIA Tesla GPUs power world's fastest supercomputer, NVIDIA Press, press-room.nvidia.com, Oct 27, 2010.
2. S. Ravi, A. Raghunathan, and S. T. Chakradhar, Efficient RTL power estimation for large designs, in: *Proceedings of the 16th International Conference on VLSI Design*, pp. 431–439, New Delhi, India, January 2003.
3. B. Arts, N. Eng, M. J. M. Heijligers et al., Statistical power estimation of behavioral descriptions, in: *Proceedings of the 13th International Workshop on Integrated Circuit and System Design, Power and Timing Modeling, Optimization and Simulation (PATMOS'03)*, pp. 197–207, Springer, Torino, Italy, September 2003.
4. C. Isci and M. Martonosi, Runtime power monitoring in high-end processors: Methodology and empirical data, in: *Proceedings of the 36th Annual International Symposium on Microarchitecture, MICRO*, pp. 93–104, San Diego, CA, December 3–5, 2003.
5. S. Hong and H. Kim, An integrated GPU power and performance model, in: *The 37th International Symposium on Computer Architecture, ISCA*, pp. 280–289, Saint-Malo, France, June 19–23, 2010.
6. J. Li, J. Martinez, and M. Huang, The thrifty barrier: Energy-aware synchronization in shared-memory multiprocessors, in: *The 10th International Conference on High-Performance Computer Architecture (HPCA)*, pp. 14–23, Madrid, Spain, February 14–18, 2004.
7. B. Rountree, D. Lowenthal, B. Supinski, and M. Schulz, Adagio: Making DVS practical for complex HPC applications, in: *Proceedings of The 23rd International Conference on Supercomputing (ICS)*, pp. 460–469, New York, June 8–12, 2009.
8. D. Q. Ren, Algorithm level power efficiency optimization for CPU–GPU processing element in data intensive SIMD/SPMD computing, *Journal of Parallel and Distributed Computing*, 71(2):245–253, February 2011.
9. Nvidia, *CUDA 2.0 Programming Guide*, Santa Clara, CA, 2008.
10. D. Q. Ren and R. Suda, Power model of large-scale matrix multiplication on multi-core CPUs and GPUs platform, in: *The 8th International Conference on Parallel Processing and Applied Mathematics (PPAM 2009)*, pp. 421–428, Wroclaw, Poland, September 13–16, 2009.
11. Intel Corp., Intel Core i7 processor integration overview (LGA1366-land package), www.intel.com/support/processors/corei7/sb/cs-030866.htm, Mar 29, 2011.
12. Intel Corporation, *ATX12V Power Supply Design Guide*, Version 2.2, Santa Clara, CA, March 2005.
13. M. Rabaey, *Digital Integrated Circuits*, Prentice Hall, Upper Saddle River, NJ, 1996.
14. E. Grochowski, M. Annavaram, Energy per instruction trends in Intel microprocessors, http://support.intel.co.jp/pressroom/kits/core2duo/pdf/epi-trends-final2.pdf, 2006.
15. *Intel® 64 and IA-32 Architectures Software Developer's Manual*, Volume 1: Basic Architecture, Order Number: 253665-028US, September 2008.
16. D. Q. Ren and R. Suda, Power efficient large matrices multiplication by load scheduling on multi-core and GPU platform with CUDA, in: *Proceedings of the 12th IEEE International Conference on Computational Science and Engineering (CSE'09)*, pp. 424–429, Vancouver, Canada, August 29–31, 2009.
17. D. Q. Ren and R. Suda, Investigation on the power efficiency of multi-core and GPU processing element in large scale SIMD computation with CUDA, in: *The 2010 International Green Computing Conference*, pp. 309–316, Chicago, IL, August 15–18, 2010.

Memetic Algorithms for Energy-Aware Computation and Communications Optimization in Computing Clusters

Johnatan E. Pecero
University of Luxembourg

Bernabé Dorronsoro
University of Luxembourg

Mateusz Guzek
University of Luxembourg

Pascal Bouvry
University of Luxembourg

20.1 Introduction

Recently, many researchers are working on the design of new schedulers for the minimization of the energy required to compute the assigned tasks, in addition to the minimization of the time when the latest task is finished (called the *makespan* time). Some examples of such techniques are the use of dynamic voltage and frequency scaling (or DVFS) to reduce the processors' frequency for avoiding idle times [18,19,29], to maximize simultaneously the savings of different providers in the grid, instead of the total power saving [33], the use of dynamic power management (DPM) to reduce the processors' consumption [5], or maximizing the energy saved in communications [12], between others.

Indeed, this is a major issue for saving energy in computational clusters, grids, and clouds, and there are results in the literature reporting high levels of saved energy just by using these techniques [12,18,19,29].

Additionally, reducing the heating of processors consequently alleviates the cooling needs at the same time.

However, there is a lack in the literature in the sense that all the existing energy-aware scheduling techniques are focused on the energy saved at the computation level, while there are also important energy consumption requirements at the communications level. And this is specially important in highly distributed systems as computational clusters, grids, or cloud systems are.

Therefore, we focus in this chapter on the design of new schedulers that are minimizing not only the makespan, but also the energy required to execute all the tasks both at the computation and communication levels.

The contributions of this chapter are threefold. First, we propose two new memetic algorithms and apply them to solve the problem of scheduling tasks with dependencies with the goals of minimizing makespan while reducing the energy consumption both in the communications and computations required. One of these algorithms is an adaptation of the genetic acyclic clustering algorithm (mGACA in short) [28] to the considered problem, while the other one is a new cellular grouping memetic algorithm. To the best of our knowledge, this is the first time in which cellular populations are used in grouping genetic algorithms (GGA) [11]. Second, we adapt the earliest task first (or ETF) list algorithm to solve our problem [13]. Third, we compare the performance of all the three algorithms on a wide benchmark of problem instances.

The structure of this chapter is detailed next. Section 20.2 describes the system, application, energy, and scheduling models used in this chapter. A brief description about the DVFS technique is given in Section 20.3. Section 20.4 describes in detail the proposed approach. It introduces the new memetic algorithm and the adapted state-of-the-art scheduling algorithms, then the voltage scaling algorithm is provided. The results of our comparative evaluation study are presented in Section 20.5. Finally, Section 20.6 concludes the chapter.

20.2 Preliminaries

In this section, we describe the system, application, energy, and scheduling models used in this work.

20.2.1 System Model

Although many studies in scheduling have been focusing on heterogeneous systems, scheduling in homogeneous systems are still of concern because of its wide use and relative simplicity of modeling. In this chapter, the target execution support is a cluster computing system composed of a collection of a set M of m homogeneous processors/machines. The processors have the same processing speed or provide the same processing performance in terms of MIPS (million instruction per second). Each processor $m_j \in M$ is DVFS-enabled; that is, it can be operated on a set of supply voltages V. Hence, different relative speed performances are associated to different clock frequencies. Furthermore, the DFVS is identical for all the processors, that is, all the processors have the same profile in terms of consumption, computational capabilities, etc., and are interchangeable. Moreover, two processors running at a same speed execute the same amount of execution units, and all the processors have the same minimal and maximal speed. The processors can operate at different speeds at the same time, and they are referred to as independent [21,26,35]. Formally, the sets of voltages and speed are defined as follows:

$$V = \bigcup_{1 \le k \le K} \{v_k\}, \tag{20.1}$$

$$RS = \bigcup_{1 \le k \le K} \{RS_k\}, \tag{20.2}$$

where,

v_k is the kth processor operating voltage

rs_{ops_k} is the kth processor speed

$v_{min} = v_1 \leq v_2 \leq \ldots \leq v_K = v_{max}$

$rs_{ops_{min}} = rs_1 \leq rs_2 \leq \ldots \leq rs_K = rs_{ops_{max}}$

$1 \leq k \leq K$, K is the total number of processor operating points

In this work, we assume that the processors operate under the same set of pairs of operating voltage and its corresponding speed (v_k, rs_{ops_k}).

Processors consume energy while idling; that is, when a processor is idling it is assumed that the lowest voltage is supplied. Since clock frequency transition overheads take negligible amount of time (e.g., 10–150 µs), they are not considered in our study, such as in many researches [14,15,19,25]. For the sake of simplicity, and without any loss of generality, we assume that all the processing elements are fully connected through network resource. That is, every processor has a direct link to any other one. The network is considered to be nonblocking; that is, the communication of two processors does not block the connection of any other two processors in the network. Additionally, we assume that the links have the same data transfer speed Bw—the transmission rate of a link connecting two processors measured by megabits per second (Mbps), and the same energy consumption rate ce measured in terms of J/Mb. It is also assumed that a message can be transmitted from one processor to another while a task is executed on the recipient processor, which is possible in many systems. Finally, communications between tasks executed onto the same processor are neglected.

20.2.2 Application Model

Scientific applications represent an important class of programs that take advantage of high-performance computing systems like computer clusters. Scientists and engineers are building more and more complex applications to manage and process large data sets, and execute scientific experiments on these computing systems. In this chapter, as it is common, we model scientific applications by a parallel program. The parallel program is represented by a directed acyclic graph (DAG). The DAG, as a graph theoretic representation of a program, is a precedence-constrained-directed and acyclic graph $G = (T, E)$, where T is a finite set of n nodes (vertices) and E is the set of directed edges between the tasks that maintain a partial order among them. The node $n_j \in T$ is associated with one task t_j of the modeled program and each edge $e_{ij} \in E$ (with $t_i, t_j \in T$) is a precedence constraint between tasks and is associated with the communication from task t_i to task t_j. A task with no predecessors is called an entry task, whereas an exit task is one that does not have any successors. Among the predecessors of a task t_i, the predecessor which completes the communication at the latest time is called the most influential parent (MIP) of the task denoted as $MIP(t_i)$. Every task t_j is associated with a value, denoted as p_j, that represents its computation time. Each edge e_{ij} has an associated parameter c_{ij}, measured in Mb, which is the volume of transferred data, as well as a data transfer time or inter-processor communication cost defined as: $Tc_{ij} = \frac{c_{ij}}{Bw}$. However, the communication cost is only required when two tasks with precedence relations are allocated on different processors, as described in the system model. In other case, the communication cost is neglected. The longest path of the graph is called the critical path (CP).

Let $\prec$ be the partial order of the tasks in G, the partial order $t_i \prec t_j$ models precedence constraints. That is, if there is an edge $e_{ij} \in E$ then task t_j cannot start its execution before task t_i completes. Hence, the result of task t_i must be available before task t_j starts its execution. We can define a clustering of the tasks in graph G for this application model as:

Definition 20.1 (clustering) *A clustering $R = \{V_i, \prec_i\}_i$ of G is a mapping of the graph onto groups of tasks associated to a total linear order extension of the original partial order $\prec$.*

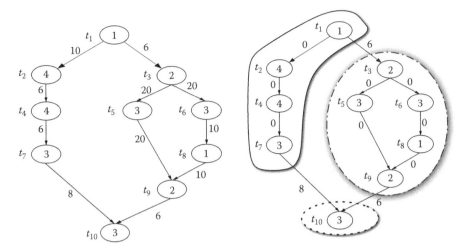

FIGURE 20.1   A simple DAG and a possible clustering of it.

TABLE 20.1   Computation Cost p_i
at Maximum Voltage and Task
Priorities (t-Level and b-Level)

Task	p_i	t-Level(t_i)	b-Level(t_i)
1	1	0	64
2	4	11	34
3	2	7	57
4	4	21	24
5	3	29	34
6	3	29	35
7	3	31	14
8	1	42	22
9	2	53	11
10	3	61	3

Figure 20.1 depicts a simple precedence task graph and a clustering of the graph. The details of the task graph are given in Table 20.1. The top level (t-level) and bottom level (b-level) values are two frequently used task prioritization methods. The t-level of a task t_j is defined as the summation of the computation and communication costs along the longest path of the node from the entry task in the task graph to t_j (excluding t_j). The t-level(t_j) is computed recursively by traversing the DAG downwards, starting from the entry task. In contrast, the b-level of a task t_i is computed by adding the computation and communication costs along the longest path of the task from the exit task in the task graph (including the task t_i). The b-level(t_i) is computed recursively by traversing the DAG upwards, starting from the exit task.

The applications can be classified based on the communication to computation ratio CCR, which is a measure that indicates whether a workflow application is communication intensive or computation intensive. For a given application, it is computed by the average communication cost divided by the average computation cost on a target system. A high CCR means that the application is considered as communication intensive.

20.2.3 Energy Model

The energy model used in this chapter is derived from the power consumption model in digital complementary metal-oxide semiconductor (CMOS) logic circuitry. The power consumption of a CMOS-based microprocessor is defined to be the summation of capacitive power, which is dissipated whenever active computations are carried out, short-circuit, and leakage power (static power dissipation). The capacitive power (P_c) (dynamic power dissipation) is the most significant factor of the power consumption. It is directly related to frequency and supply voltage, and is defined as [6,7,18,31]

$$P_c = AC_{eff} V^2 f,\tag{20.3}$$

where

A is the number of switches per clock cycle
C_{eff} denotes the effective charged capacitance
V is the supply voltage
f denotes the operational frequency

The relationship between circuit delay (T_d) and the supply voltage is approximated by (Equation 20.4)

$$T_d \propto \frac{C_L V}{(V - V_{th})^\alpha},\tag{20.4}$$

where

C_L is the load capacitance
V_{th} is the threshold voltage
α is the velocity saturation index that varies between one and two ($\alpha = 2$)

Because the clock frequency is proportional to the inverse of the circuit delay, the reduction of the supply voltage results in reduction of the clock frequency. It would not be beneficial to reduce the CPU frequency without also reducing the supply voltage, because in this case the energy per operation would be constant.

Equation 20.3 indicates that the supply voltage is the dominant factor; therefore, its reduction would be most influential to lower power consumption. The energy consumption in the execution of any application used in this chapter is defined as

$$E_c = \sum_{i=1}^{n} AC_{ef} V_i^2 f p_i^* = \sum_{i=1}^{n} KV_i^2 p_i^*,\tag{20.5}$$

where V_i is the supply voltage of the processor on which task t_i is executed, and p_i^* is the computation cost of task t_i (the amount of time taken for t_i's execution at speed rs_i) on the scheduled processor. On the other hand, the energy consumption during idle time is defined as

$$E_i = \sum_{j=1}^{m} \sum_{idle_{jk} \in IDLES_j} \alpha V_{min_j}^2 I_{jk},\tag{20.6}$$

where $IDLES_j$ is the set of idling slots on processor m_j, V_{min_j} is the lowest supply voltage on m_j, and I_{jk} is the amount of idling time for $idle_{jk}$. Then the total energy consumption is defined as

$$E_t = E_c + E_i.\tag{20.7}$$

The energy for communications in this work is computed based on a model derived from findings reported in the literature [38]. The communication energy model adopted in this work is defined as

$$E_{comm} = ce \sum_{i=1}^{n} \sum_{j=1, j \neq i}^{n} c_{ij},\tag{20.8}$$

where the term $\sum_{i=1}^{n} \sum_{j=1, j \neq i}^{n} c_{ij}$ is the total of the active and effective communication or volume of data transmission for the current schedule.

20.2.4 Scheduling Model

Let us consider an application composed of n tasks to be processed on a set M of m (almost) homogeneous processors. *Scheduling* is defined as a pair of functions, namely, $\pi : V \to M$ and $\sigma : V \to \mathbb{N}^{+}$, where $\pi(j)$ provides the processor on which task j is executed and $\sigma(j)$ represents the starting time of task j. The feasibility of the schedule means the respect of the precedence constraints which are guaranteed by the following relations:

1. $\forall (i, j) \in E$, $\sigma(j) \geq \sigma(i) + p_i$ if i and j are allocated on the same processor ($\pi(i) = \pi(j)$) and, $\sigma(j) \geq \sigma(i) + p_i + c_{ij}$ otherwise ($\pi(i) \neq \pi(j)$). The task j cannot be executed before task i has been finished.
2. For any pair of tasks $\{i, j\}$, if $\pi(i) = \pi(j)$, then $\sigma(i) + p_i \leq \sigma(j)$ or $\sigma(j) + p_j \leq \sigma(i)$. One processor may not execute more than one task at a time.

Let (σ, π) be a feasible schedule for G. The *completion time* of each task $j \in V$ is defined by $C_j \equiv \sigma(j) + p_j$. The *schedule length* or *makespan* of (σ, π) is the maximum completion time (denoted by C_{max}) of a task of G. Formally, $C_{max} \equiv max_{j \in V} C_j$. The objective is to minimize C_{max}. This problem is known to be NP-hard in its simplest version (without communications, see Ullman [36]). Even when the number of processors is greater than or equal to the number of tasks (i.e., unbounded number of processors), the problem still falls in the class of NP-hard problems [30]. Let us mention that tasks of the applications in our study are not associated with deadlines as in real-time systems.

20.2.4.1 Scheduling with Communications

Scheduling with consideration of communication delays is an active research area, primarily because of the various possible technical preconditions under which communication takes place. The transmission technique used in the network, the chosen communication protocol, and the performance of network components are examples of factors that influence the speed and the performance of the network. Knowledge of these communication properties is of course essential in the attempt to process the tasks in some optimal manner. In addition to the first, in actual real parallel and distributed systems communication among processors takes considerable time; specially in distributed systems, where communication costs tend to be high with respect to processor speed. Moreover, the energy consumption is important because of the idle time where processors are waiting for data without performing work.

The inclusion of communication costs has made the scheduling problem much more difficult to solve. There is a trade-off between parallelism and communication which determines the optimal grain size of the task graph. Exposing maximum parallelism in the program, by decomposing tasks into as fine a granularity as possible, comes at the cost of increased inter-processor communication, which counteracts the potential benefits of the extra parallelism. This is true not only in the decomposition stage but also in the scheduling stage, where parallelism is exploited by spreading the computation over many processors, while communication is minimized by restricting this spread.

Different approaches have been designed to deal with the problem of scheduling with communication costs. The most widely used methods of the existing heuristics are so far an extension of the *list scheduling algorithms, task clustering*, and *genetic algorithms* [16,17,32]. List scheduling is a class of scheduling heuristics in which tasks are assigned with priorities and placed in a list ordered in decreasing magnitude of priority. First, list scheduling sorts the task of the DAG to be scheduled according to a priority scheme, while respecting the precedence constraints of the tasks; that is, the resulting task list is in topological order after each task of the list is successively scheduled to a processor chosen for the task. Usually, the chosen is the one that allows the earliest start time of the task.

The task clustering algorithms separate the scheduling of a task graph into two steps. The idea is detailed next: in the first step, a heuristic is used to gather tasks into clusters, and thus to reduce the unbalance between communication and execution times. This clustering operation is made assuming unbounded number of processors. In the second step, the resulting schedule is fold onto restricted number of processors. The basic rule of task clustering is that all the tasks of a given cluster will be allocated into the same processor.

A genetic algorithm (GA) is a guided random search method which is based on the principles of evolution and natural genetics. It combines the exploitation of past results with the exploration of new areas of the search space. By using survival of the fittest techniques, a structured randomized information exchange and adaptation, genetic algorithm can mimic some of the innovative flair of human search. Genetic algorithm is randomized but it is not a set of simple random walks. It exploits historical information efficiently to speculate on new search points with expected improvement. Since they operate on more than one solution at once, GAs are typically good at both the exploration and exploitation of the search space.

In this chapter, we propose a new scheduling algorithm that combines task clustering with a class of genetic algorithms for optimizing the communication costs and the makespan. The proposed scheduling algorithm will be further described later on in the chapter.

20.3 Dynamic Voltage and Frequency Scaling

Modern processors incorporate the capability of dynamically changing its working voltage and frequency without stopping or pausing the execution of any instruction. This mechanism, called dynamic voltage and frequency scaling (DVFS), was initially included in laptops and other mobile devices, and nowadays we can find this feature in desktop computers and servers too. DVFS scaling is one of the most effective energy reduction mechanisms for CMOS circuitry. The mechanism allows system software to increase or decrease the supply voltage and the operating frequency of a CMOS circuit at run time.

We may consider lowering the voltage and frequency of a processor to reduce the heat generated and the power required for its functioning, consequently saving energy. However, it is obvious that the processor is working at lower frequency, and therefore longer times will be required to execute a given job.

This technique is appropriate to reduce processors' idle times when scheduling a set of tasks with dependencies among them, since the amount of energy consumed by the processor in idle state is important as when it is working at full capacity. It often occurs that in a given schedule some processor must wait for some time between the execution of one assigned task and the next one due to some data dependency. This happens when the next task needs some data from some other task that is being executed in another processor, and therefore the processor is required to wait until the data is received before starting its assigned task. Therefore, it is possible to reduce the working voltage and frequency of the processor to delay the finishing time of a task in order to avoid the processor to be in idle state while waiting for the data to run the next task.

20.4 Proposed Approach

The energy consumption and task scheduling in multiprocessor systems can be addressed in two ways: simultaneously and independently [29]. In the simultaneous mode, the scheduling is computed in a global cost function including performance and energy saving to satisfy both time and energy constraints at the same time. In this mode because these constraints have opposite relation with each other, the result of such algorithms is typically a trade-off between saving energy and less computing time. Different solutions produce trade-offs between the two objectives, which means there is no single optimal solution. In this case, the problem is modeled as a multi-constrained multi-objective optimization problem and the objective becomes finding Pareto optimal schedules (i.e., non-dominated schedules). In other words, a solution is Pareto optimal if it is not possible to improve a given objective without deteriorating at least another objective. This set of solutions represents the compromise solutions between the different conflicting objectives. In literature there are two main ways to tackle such problems: (a) optimize objectives concurrently, and (b) optimize one objective first, then make that as a constraint the second objective [15]. Moreover, different methods could be used to optimize objectives concurrently and they are classified according to the strategy to evaluate the quality of the solution, for example, scalar approaches, criterion-based, dominance-based, and indicator-based approaches [8,9,34].

In the independent mode, which is considered a two-phase optimization technique, a *best-effort* scheduling algorithm is combined with a *slack reclamation technique* (see Algorithm 20.1). The slack reclamation technique is used as a second step to minimize the energy consumption of tasks in a schedule generated by the best-effort scheduling algorithm. In this mode, energy and time constraints are independent and new scheduling algorithms or the existing scheduling algorithms in the literature are adapted to become energy efficient. The principle of slack reclamation techniques is simply to exploit the slack times during software execution to accommodate the negative impacts on performance when applying DVFS. The challenge is to make appropriate decisions on processor speeds to guarantee the timing requirements while also considering the timing requirements of all tasks in a system.

The solution provided in this chapter follows the independent mode approach. For the first step of the provided algorithms we look for solutions with near-optimal makespan, and for that a best-effort scheduling algorithm is used. We propose three different best-effort scheduling algorithms: a new scheduling algorithm and extend two state-of-the-art scheduling algorithms. The new scheduling algorithm is based on an evolutionary technique. More specifically, it is a cellular memetic genetic algorithm. The state-of-the-art scheduling algorithms are an evolutionary-based scheduling algorithm [28] and the well-known ETF which is a list-based scheduling algorithm [13]. For the second step, a voltage scaling algorithm is used to reduce the energy consumption by scaling down the processor voltages to a proper level, thus extending the execution time of jobs without increasing the makespan computed by the best-effort scheduling algorithms. The voltage scaling algorithm exploits the slack time in the schedule. Intuitively, the slack time represents the amount of idle time a task can be delayed to start or complete its execution without incurring an increase in performance degradation. This slack time may occur mainly due to precedence constraints or communication delays.

Algorithm 20.1 Performance-Energy Independent Optimization Technique

---
1: Makespan optimization via best-effort scheduling algorithms
2: Slack reclamation technique via voltage scaling algorithm

The following sections describe the best-effort scheduling algorithms. First, the new evolutionary memetic scheduling algorithm is introduced. Next, the two state-of-the-art scheduling algorithms are described. Then, the voltage scaling algorithm is detailed.

20.4.1 Cellular Memetic-Based Scheduling Algorithm

We now describe the proposed scheduling algorithm. It intrinsically optimizes the energy communication dissipation by properly allocating and scheduling precedence-constrained applications on the processing elements, reducing application completion time and inter-processor communications at the same time. Before describing the new proposed algorithm in detail, we provide a brief description of cellular GAs.

20.4.1.1 Cellular GAs

Cellular GAs [1,23,37] are structured population algorithms with a high explorative capacity. The individuals composing their population are (usually) arranged in a two-dimensional (2D) toroidal mesh. This mesh is also called grid. Only neighboring individuals (i.e., the closest ones measured in Manhattan distance) are allowed to interact during the breeding loop. This way, we introduce a form of isolation in the population that depends on the distance between individuals. Hence, the genetic information of a given individual spreads slowly through the population (since neighborhoods overlap). The genetic information of an individual will need a high number of generations to reach distant individuals, thus avoiding premature convergence of the population. By structuring the population in this way, we achieve a good exploration/exploitation trade-off on the search space. This improves the capacity of the algorithm to solve complex problems [1,2].

A canonical CGA follows the pseudo-code of Algorithm 20.2. In this basic CGA, the population is usually structured in a regular grid of d dimensions ($d = 1, 2, 3$), with a neighborhood defined on it. Each individual in the grid is iteratively evolved (line 2). A generation is the evolution of all individuals of the population. Individuals may only interact with individuals belonging to their neighborhood (line 3), so parents are chosen among the neighbors (line 4) with a given criterion. Recombination and mutation operators are applied to the individuals in lines 5 and 6, with probabilities p_comb and p_mut, respectively. Afterward, the algorithm computes the fitness value of the new offspring individual (or individuals) (line 7), and replaces the current individual (or individuals) in the population (line 8), according to a given replacement policy. This loop is repeated until a termination condition is met (line 1); for example, the total elapsed processing time or a number of generations.

The CGA described by Algorithm 20.2 is called asynchronous, because the population is updated with next-generation individuals immediately after their creation. These new individuals can interact with those belonging to their parent's generation. Alternatively, we can place all the offspring individuals into an auxiliary population, and then replace all the individuals of the population with those from the auxiliary population, at once. This last version is referred to as the synchronous CGA model. As it was studied in [1,4], the asynchronous CGAs converge the population faster than the synchronous CGAs. For more details about CGAs we invite the reader to see Ref. [1].

20.4.1.2 Cellular Memetic GA

Memetic algorithms are search algorithms in which some knowledge of the problem is used in one or more operators. The objective is to improve the behavior of the original algorithm. Not only local search, but also restart, structured, and intensive search is commonly considered in MAs [1,3,27]. The proposed cellular memetic (cMGA) based scheduling algorithm combines task clustering with a cellular GA. That is, we have designed the genetic operators (e.g., recombination, mutation) considering groups of tasks instead of applying them directly on the tasks. The rational is that in task clustering the clusters are the

---

Algorithm 20.2 Pseudo-Code for a Canonical CGA (Asynchronous)

```
 1: while ! StopCondition() do
 2:     for all ind in population do
 3:         neigh ← get_neighborhood(ind);
 4:         parents ← select(neigh);
 5:         offspring ← recombine(p_comb, parents);
 6:         mutate(p_mut, offspring);
 7:         evaluate(offspring);
 8:         replace(ind, offspring);
 9:     end for
10: end while
```

significant building blocks, that is, the smallest piece of a solution, which can convey information on the expected quality of the solution they are part of. A salient feature of the proposed algorithm is that it takes advantage of the good exploration/exploitation balance performed by the cellular genetic algorithm on the search space, improved with the effectiveness of task clustering for reducing inter-processor communication.

20.4.1.3 Algorithm Design

The cMGA algorithm follows the principle of the GGA. GGA is a genetic algorithm heavily modified to suit the structure of grouping problems [11]. Those are the problems where the aim is to find a good partition of a set, or to group together the members of the set, but in most of these problems, not all possible groupings are allowed. GGA manipulates groups rather than individual objects; similarly cMGA manipulates clusters rather than tasks.

In the proposed algorithm, the population is structured in a bi-dimensional grid of individuals and the neighborhood is a L5 neighborhood shape [1]. As we have already mentioned, the aim of the cellular memetic algorithm is to gather the tasks on clusters (one cluster for one physical processor) and schedule them efficiently onto m processors.

String Representation: Solutions are represented as a string of integers of length n, where n is the number of tasks in the given DAG. The value in the i-th position in the string represents the cluster to which task i is scheduled. The maximum number of clusters is bounded to the number of processors in the system. Due to the specification of our problem, namely large communication delays, it is important to avoid too fine-grained scheduling. We have enriched this representation with a group part (C_i), encoding the clusters on one gene for one cluster basis. The group part is a string of integers, a random permutation of clusters used in the scheduling. The length of the group part representation is variable. The genetic operators are applied to the group part of the chromosomes; the standard task part of the chromosomes identify which tasks actually form which cluster.

Figure 20.2 depicts a string representation for the clustering of Figure 20.1. The clustering is composed of three clusters. This representation is enriched by the cluster part. For this example, cluster C_1 is composed of tasks $\{t_1, t_2, t_4, t_7\}$, cluster C_2 is composed of tasks $\{t_3, t_5, t_6, t_8, t_9\}$, and cluster C_3 is composed of task $\{t_{10}\}$.

Normalization: In this type of group-number encoding, there are multiple possible representations of the same solutions: as all processors in the system are identical, then if all tasks are rescheduled from processor j to processor k, and all tasks from processor k are rescheduled to processor j, the resulting clustering will be equivalent (i.e., redundant). However, the difference in labeling hides this similarity, so that the redundancy prevents cMGA from finding similarities among the solutions and increases the difficulty of finding good "building blocks." To avoid that, we adopt the *normalization* procedure used in [40]. This procedure takes place after applying diversification operators. It reorders numbers of the clusters such that cluster number one is assigned to the first task and every task assigned to its same cluster in the original solution. Then, we proceed to assign in the same way consecutive numbers to identify the

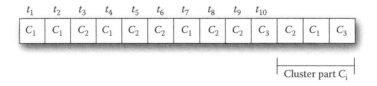

FIGURE 20.2   A string representation of the clustering depicted in Figure 20.1.

other clusters we find in the solution starting from task two to the last one. Such procedure creates finally bijective map, which maps clusters from the existing schedule to the new numbering scheme.

Initial Population: The first population of solutions is initialized randomly.

Fitness Function: The fitness of each offspring S_i is a measurement of performance of the design variables as defined by the objective function and the constraints of the problem. First, makespan SL_{S_i} is calculated for each individual of the population. Hence, the fitness function is (Equation 20.9).

$$f_{S_i} = \frac{1}{SL_{S_i}} \tag{20.9}$$

To obtain the makespan for a particular clustering (i.e., offspring), each task is scheduled in decreasing order priority. That is, to schedule a task into the first slot available after the specified earliest start time on the assigned processor. The earliest start time $est(t_i)$ for task t_i is calculated using Equation 20.10:

$$est(t_i) = MAX\left[\sigma(t_i) + p_i + Tc_{ij}, (i,j) \in E\right] , \tag{20.10}$$

where $\sigma(t_i)$ is the start execution time of task t_i. Let us recall that $Tc_{ij} = 0$ if tasks t_i and t_j are placed on the same processor. Once all tasks have been scheduled, the algorithm uses Equation 20.11 to determine the makespan of the scheduled DAG.

$$SL_{S_i} \equiv MAX\left[\sigma(t_i) + p_i\right] \tag{20.11}$$

To reduce the computation required, it only checks the sink tasks, since by definition they should finish after any of their predecessors.

Stopping Condition: cMGA stops the execution when it reaches the maximum number of iterations.

Selection: The chosen selection operator is tournament best of two.

Recombination: The most important operator for finding new clusterings is recombination or crossover. It is based on the GGA crossover operator [11]. This operator uses the group part of the parents to change the structure of the clustering. When recombination occurs, it randomly chooses one crossover point in each group part of both parent solutions. Then, considering creation of the first new solution, it copies the assignments of tasks from the clusters of the second parent whose identifiers are on the right side of crossover point. The rest of the tasks assignments are copied from the first parent. To create the other new solution, analogical procedure takes place, with switched roles of parents. Because of described behavior, the resulting solutions may have less clusters than the initial ones.

The pseudo-code of recombination is shown in Algorithm 20.3. The *get_clusters_amount(S_2)* function (line 1) obtains the number of clusters of parent S_2. Parent S_1 is copied to the new offspring S_1' in line 2. *random(2, ca)* (line 3) generates a random crossing point to delimit the crossing section in the group part of individual S_2. The **for** loops are used to copy the contents (i.e., alleles) of the crossing section of S_2 to the offspring S_1' (line 4–11). The *get_cluster(S_2, i)* function (line 5) gets the allele on the gene i from the group part of S_2. *get_task(S_2, j)* and *set_task(S_1', j, cn)* finally copy the alleles from S_2 to S_1'.

Mutation: First, an individual is randomly chosen. Next, the mutation operator is applied to the selected chromosome with a certain probability *p_mut*. The mutation operator randomly changes one task in the scheduling to another different cluster. The new cluster is chosen from the range of all possible clusters. It means that when not all clusters are used, mutation of one task may produce a new cluster. Therefore, it has the opposite influence on the number of used clusters than the recombination operator.

Algorithm 20.3 Pseudo-Code for Recombination Operator

```
1:  ca ← get_clusters_amount(S₂);
2:  S'₁ ← S₁;
3:  r ← random(2, ca);
4:  for all i = r to ca do
5:      cn ← get_cluster(S₂, i);
6:      for all j = 1 to n do
7:          if get_task(S₂, j) == cn then
8:              set_task(S'₁, j, cn)
9:          end if
10:     end for
11: end for
```

20.4.2 Best-Effort Algorithm Based on a Memetic Genetic Algorithm

We describe now a memetic genetic algorithm which is an adaptation of the mGACA algorithm [28] to the considered problem. We call the modified algorithm energy-efficient memetic algorithm for clustering and scheduling (EMACS). The main idea used by the algorithm is based on the decomposition of the task graph on structured properties. The principle is to assign tasks to locations into *convex clusters* [10,20,22], meaning that for any path whose extremities are in the same cluster, all intermediate tasks are also in that cluster. The algorithm partitions the graph into convex sets of tasks with the property that the final clustering remains acyclic (i.e., a DAG). The algorithm also belongs to the family of the GGA problems. One important characteristic of the algorithm is that the genetic operators could produce individuals with a number of clusters greater than the number of physical processors. Thus, a merging clustering algorithm is applied after the reproduction process until the number of clusters has been reduced to the actual number of processors. The genetic operators and the merging cluster algorithm have been designed to produce only legal solutions not violating convexity constraints. The main modifications are the local heuristic used after the recombination operator to repair clusters violating the convexity constraint and the selection operator. mGACA uses the proportionate selection scheme based on roulette wheel principle. EMACS implements tournament selection best of two. Before presenting the algorithm, we provide some basic definitions.

Definition 20.2 (Convex cluster) *A group of tasks $T \in V$ is convex if and only if all pairs of tasks $(t_x, t_z) \in T$, all the intermediate tasks t_y on any path between t_x and t_z belongs to T.*

Figure 20.3 depicts a sample of a convex clustering. On the contrary, the clustering in the right of Figure 20.3 is non-convex since the path from task t_1 to t_{10} through tasks t_2, t_4, and t_7 is not contained within the cluster. The clustering depicted in Figure 20.1 is a convex clustering too.

For each clustering $R = \{C_i, \prec_i\}_i$, we can build a corresponding *cluster-graph* denoted by $G_R^{cluster}$ and defined as follows: the nodes of $G_R^{cluster}$ are the clusters of R. There exists an edge between two distinct cluster nodes C_i and C_j ($i \neq j$), if and only if there exist two tasks $t_x \in C_i$ and $t_y \in C_j$ such that $(t_x, t_y) \in E$. Moreover, the graph is weighted by *max Tc_{xy}* on each edge, and each cluster node C_i is weighted by $v(C_i) = \sum_{k \in C_i} p_k$.

Definition 20.3 (Acyclic clustering) *A clustering R is acyclic if and only if all the clusters are convex and $G_R^{cluster}$ remains a DAG.*

Algorithm 20.4 shows the pseudo-code of the memetic genetic algorithm. The algorithm starts by initializing a population of individuals (line 1). Each individual of the population is evaluated (line 2); then based on the fitness of each individual, individual solutions are selected from the population (line 4). Recombination and mutation operators are applied to the individuals in lines 5 and 9, with probabilities

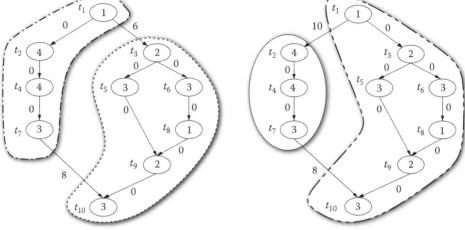

FIGURE 20.3   Left, the clustering of the graph is convex. In the right the clustering is nonconvex.

p_comb and *p_mut*, respectively. A local heuristic (line 7) is implemented after the recombination operator to repair some individuals complaining of the convexity constraint. If the number of clusters (nC_i) in the individual is greater than the number of processors (m), then some clusters are merged to the number of physical processors (line 11). Individuals are normalized (line 13) to avoid the redundancy problem. Afterward, EMACS evaluates the new offsprings, and replaces the current individuals in the population (line 15). This loop is repeated until a termination condition is met (line 3).

20.4.2.1 Algorithm Design

The algorithm uses some functions and parameters utilized by cMGA. For example, the same string representation, fitness function, normalization, selection, and stopping condition.

Initial Population: The algorithm uses a *bottom-level clustering algorithm* (BCA) to create the initial population. BCA clusters the DAG according to the *bottom level* (blevel) value of every task. The BCA algorithm works as follows: the algorithm starts by randomly selecting the number of clusters *nCl* to be generated (between 1 and *m*) for any individual. After that, the clusters' grain size *nt_Cl* is calculated. Next, compute the *blevel* for each task and sort the tasks in decreasing order according to their *blevel* value (tie-breaking is done randomly). Afterwards, assign the sorted tasks in a list. For each cluster T_i, assign the *nt_Cl* tasks in the top of the list and remove them from that. This process is repeated until the maximal number of clusters (nCL) has been reached. The authors in [28] showed that this algorithm always produces legal individuals.

Recombination: The recombination operator follows the pseudo-code of Algorithm 20.3; however, after the recombination operator, a local heuristic is used to repair some altered clusters, *"if necessary"*. We have developed a simple local heuristic as follows: We remove the task of the altered cluster violating the convex constraint and allocate it and the set of its predecessors into a new cluster. We only select the predecessors belonging to the altered cluster.

Mutation: The mutation operator works by changing a task's cluster to a new cluster. First, an individual is randomly chosen. Next, the mutation operator is applied to the selected chromosome with a certain probability p_m. Then, the mutation operator selects a cluster T_i randomly (from the cluster part of the individual). After that, the mutation operator selects a task t_i from the top or the bottom of the selected cluster T_i, creates a new cluster, and puts the task t_i in the new cluster. We recall that the tasks in the clusters are sorted by topological order. Thus, if the mutation operator selects any task from the top or

---

Algorithm 20.4 Pseudo-Code for the EMACS Algorithm

 1: *Initializethepopulation*;
 2: *Evaluationofpopulation*;
 3: **while** ! *StopCondition*() **do**
 4:    *parents* $\leftarrow$ *select*(*ind*);
 5:    *offspring* $\leftarrow$ *recombine*(*p_comb*, *parents*);
 6:    **if** *clustering is not convex* **then**
 7:      Repair the clustering with a local heuristic;
 8:    **end if**
 9:    *mutate*(*p_mut*, *offspring*);
10:    **if** $nC_i \geq m$ **then**
11:      Merge clusters;
12:    **end if**
13:    *normalize the ind*;
14:    *evaluate*(*offspring*);
15:    *replace*(*ind*, *offspring*);
16: **end while**

---

the bottom of any cluster and assigns it in a new cluster, it does not violate the convex cluster constraint and the resulting clustering is acyclic.

Merging: A merging algorithm is used because genetic operators (i.e., recombination and mutation) can generate individuals with a number of clusters greater than the available physical processors. The algorithm merges existing clusters in a pairwise fashion until the number of clusters is reduced to m. The merging clusters algorithm works over the cluster-graph $G_R^{cluster}$. Remember that the clusters represent the task nodes in $G_R^{cluster}$. This procedure initially sorts the elementary clusters by $v(C_i)$ to allow the smallest cluster node to be considered for merging at each step. For each cluster calculates its blevel. After that, chooses the smallest cluster. If the chosen cluster will be merged with one of its immediate predecessors, then select the predecessor with the smallest blevel (ties are broken by selecting the immediate predecessor which communicates the most) and merge them in one cluster. In other case, if the chosen task will be merged with one of its immediate successors, then select the immediate successor with the greatest blevel value and merge them in one cluster. Repeat this process until the number of cluster nodes is less than or equal to the number of processors m. Let us remark that if the chosen task is an entry task, then the algorithm merges it with any of its immediate successors. In other case, if it is an exit task then the algorithm chooses a task among its immediate predecessors and merges them.

20.4.3 Best-Effort Algorithm Based on List Scheduling Heuristic

The earliest task first (ETF) algorithm was developed for scheduling tasks on multiprocessor architectures with an arbitrary topology. It is a list scheduling algorithm based on a dynamic task priority scheme. The core of the ETF heuristic is a simple greedy strategy: schedule a *ready* task to an available processor, in which it can start as early as possible; a task is ready when all its immediate predecessors have already been scheduled. The interesting feature of the algorithm is that it makes a scheduling decision based on the delay associated with message transfer between any pair of nodes in the network and the amount of data to be transferred between any pair of tasks t_i and t_j such that t_i is an immediate predecessor of t_j. At each scheduling step (triggered by a processor becoming available), the ETF algorithm aims at finding the task t_i in the set of ready tasks which has the earliest start time (min [EST]), and assigns it to the available processor yielding this start time. Ties are broken by considering the statically computed task priority (e.g., tlevel, blevel, number of successors, etc.). In this chapter, we use the blevel. The ETF algorithm provided in this chapter is described in Algorithm 20.5. We called this algorithm modified earliest task first (METF). The algorithm starts by allocating each task with the blevel priority (line 1) and sorting them by descending order of priority (line 2). Then, the algorithm selects a ready task (line 4) from the top of the list. At each scheduling step, the algorithm computes the earliest available time on each processor (lines 5–7). This is the main modification that we propose. In ETF the ready task is assigned to the processor that becomes available first and arranges the communication delays. In METF, the task is assigned to the processor on which the task starts as earliest as possible based on communication delays even if this processor is not available in the current time (i.e., it is executing another task). That is, the algorithm selects the available processor among all the processors not only in the available set. Then, the algorithm assigns the task to the selected processor (line 8) and calculates its start and finish time (line 9). Afterward, the list of ready tasks is updated (line 10) and the priority of tasks is again calculated (line 11). This process is repeated until the list of tasks is empty (line 3).

Algorithm 20.5 Pseudo-Code for METF

1: Calculate the priority of each task according to the *b-level*;
2: Sort the tasks in a ready tasks list by decreasing order of priorities;
3: **while** *there are unscheduled tasks in the list* **do**
4:     Remove the first task, t_i, from the list;
5:     **for** *each processor m_j in M* **do**
6:       Compute earliest starting time (t_i, m_j) by considering communication delays;
7:     **end for**
8:     Assign t_i on the processor m_j which has the earliest available time;
9:     Calculate the starting and finish time of t_i on m_j;
10:    Update the list of ready tasks;
11:    Compute priority of ready tasks and sort the list;
12: **end while**

20.4.4 Voltage Scaling Algorithm

We use a slack reclamation algorithm after the best-effort scheduling algorithm to decrease still the energy consumption of the schedule. The slack reclamation is based on a voltage scaling algorithm that exploits the time a task can be delayed to start or complete its execution without degrading the performance of the schedule (i.e., the makespan) or without breaking any constraint. This allowable delay time or slack may occur due to precedence constraints. Intuitively, the allowable delay time represents the idle time that can be used to increase the execution of the tasks with the advantage of reducing the energy consumption. We only apply the voltage scaling for jobs on noncritical path. We first identify the sets of critical and noncritical tasks; for that we compute the slack of each task and the tasks with slack equal to zero are critical tasks, and noncritical otherwise. For a given schedule, we compute the slack as follows:

The earliest start time (EST) of, and the earliest finish time (EFT) of, a task t_i on a processor m_j are defined as follows (Equations 20.12 and 20.13, respectively):

$$EST(t_i, m_j) = \begin{cases} 0 & \text{if } t_i = t_{entry} \\ EFT(MIP(t_i), m_k) + Tc_{MIP(t_i),i} & \text{otherwise} \end{cases} \quad (20.12)$$

$$EFT(t_i, m_j) = EST(t_i, m_j) + p_i, \quad (20.13)$$

where m_k is the processor on which $MIP(t_i)$ is scheduled.

The earliest start time for a task is equal to the largest of all the earliest finish times for all its immediate predecessors since the task cannot be started until all other preceding tasks have finished. We make a forward pass through the whole network determining all the earliest start and finish times for all tasks. The earliest finish time for the last task will be how long the schedule will take to complete.

We now perform some backward steps through the network to determine the critical tasks.

The latest start (LST) and finish times (LFT) of a task t_i on a resource m_j are defined as

$$LST(t_i, m_j) = LFT(t_i, m_j) - p_i, \tag{20.14}$$

$$LFT(t_i, m_j) = \left\{ \begin{array}{ll} EFT(t_i) & \text{if } t_i = t_{exit} \\ min\{min_{t_k \in succ(t_i)}(LST(t_k, m_l) - Tc_{ik}), & \\ LST(t_{next}, m_j)\} & \text{otherwise} \end{array} \right\} \tag{20.15}$$

where $succ(t_i)$ is the set of a successor tasks of task t_i, m_l is the computing resource executing task t_k, and $LST(t_{next}, m_j)$ is the actual latest start time of the next task scheduled after t_i on the same computing resource m_j. For a given task, its LST/LFT may differ from EST/EFT if the minimum EST in all its successor tasks is later than the time it finishes its communication (data transfer); this may occur due to precedence constraints. Note that LST and LFT of a task have a cumulative effect on those of its parent tasks. Moving backward, the latest finish time for a task is the smallest of the latest start times for all tasks that immediately follow that task. The latest finish time is the earliest start time for the immediately following tasks.

Once we have completed the forward and backward steps through the schedule, we can determine how much slack is in each task. Slack is determined as follows:

$$Slack(t_i) = LFT(t_i) - EFT(t_i). \tag{20.16}$$

Now, let us assume that task t_i is a noncritical task (i.e., Slack $(t_i) > 0$) and is executed on processor m_j. Then the execution time of t_i can be extended to Slack (t_i) without modifying the finish time of its predecessors and the start time of its successors (i.e., without violating precedence constraints). The operating relative speed (rs_{ops}) of processor m_j to execute a task t_i for the Slack (t_i) is determined as

$$rs_{ops}(m_j) = \frac{p_i}{p_i + Slack(t_i)}. \tag{20.17}$$

Once we have determined the operating relative speed, we identify the corresponding operating voltage from the pair $(v_k(m_j), rs_{opt}(m_j))$. Algorithm 20.6 shows how to scale down noncritical tasks. First, the algorithm calculates the slack for each task (line 1–3). Then, the set of noncritical tasks is calculated (line 4). For each noncritical task, the algorithm extends the execution time to the slack time if possible and determines the operating relative speed, scales down the operating voltage (lines 5–8) to a proper

---

Algorithm 20.6 Pseudo-Code for the Voltage Scaling Algorithm (VSA)

1: **for** *each task t_i* **do**
2:    Calculate Slack(t_i) as Equation 20.16;
3: **end for**
4: Identify the set of non-critical task;
5: **for** *each non-critical task on m_j* **do**
6:    Determine the operating relative speed $rs_{ops}(m_j)$ as Equation 20.17;
7:    Determine the operating voltage;
8: **end for**
9: **for** *each processor m_j* **do**
10:    **for** *each idle time due to communications on m_j* **do**
11:       Scale down the operating voltage to lowest;
12:    **end for**
13: **end for**

---

value. Afterward, for each idle time due to transfer data in a processor, the algorithm scales down the operating voltage to the lowest (lines 9–13).

20.5 Experiments

In this section, we compare the proposed solutions. The performance comparison is made in terms of solution quality of the makespan and the energy consumption during the schedule. We compute the energy of the schedule due to computation and the inter-processor communications. Furthermore, we compare the number of resources used for each processor to compute the schedule. Although the algorithms assume m number of processors, let us remark that the way each algorithm spends resources is different and the optimization of resources could lead to gain in the energy. In some works, it is doing by task and resource consolidation.

Kwok and Ahmad [16] proposed a set of task graphs instances to benchmark scheduling algorithms that take into account communication delays on homogeneous processors. The benchmark contains some trace graphs which are actual instances extracted from parallel compiler logs. They are shown in Table 20.2, together with their number of tasks. These task graphs can be characterized by the size of the input data matrix because the number of nodes and edges in the task graph depends on the size of this matrix. For example, the number of tasks for LU decomposition graph is equal to $(N^2 + N - 2)/2$ where N is the size of the matrix. This section presents the results obtained using each of the three scheduling algorithms on those graphs. However, we have only used the structure of the graphs and we have varied the processing time and communication delays to different CCRs according to the studied problem. For each graph we have varied the CCR ratio to the following values (0.1, 0.5, 1) for computation-intensive applications and CCRs to (3,5,10) for communication-intensive applications.

We have simulated the three algorithms on 64 processors. Table 20.3 lists the voltage-speed of these processors [18].

The following parameters were used by the memetic algorithms. For cMGA, a bidimensional grid of 100 individuals (10×10) has been chosen for structuring the population. As we have already mentioned, the neighborhood used is composed of L5 individuals or NEWS neighborhood (the considered individuals plus those located at its north, east, west, south). The number of generations is 100, recombination probability of 0.85, and mutation probability of 0.005. The same parameters were adopted by EMACS for the generation limit, population size, and recombination and mutation probabilities.

20.5.1 Results

Tables 20.4 through 20.11 show the results that were obtained with cMGA and EMACS algorithms. Specifically, the tables show the average values obtained by the algorithms after 30 independent runs

TABLE 20.2 Problem Instances Studied

Instance	Size	Instance	Size
	324		484
	464	laplace	625
cholesky	629	mva	784
	819		961
	1034		1156
	348		299
	493		434
gauss	663	lu	594
	858		779
	1078		989

TABLE 20.3   Voltage-Relative Speed Pairs for the
DVFS-Enabled Processors

Voltage Level	Nominal Voltage	Relative Speed (%)
0	1.5	100
1	1.4	90
2	1.3	80
3	1.2	70
4	1.1	60
5	1.0	50
6	0.9	40

together with the standard deviation. These algorithms have been compared with METF. The results for METF algorithm are shown in Appendix 20.A. The gray background of the cells emphasizes that the corresponding result is of better quality than the one reported by METF. Additionally, we compared the results of cMGA and EMACS in every case, and the best results are emphasized in bold font. Underlined values are statistically significant in the comparison of the two memetic algorithms according to the Wilcoxon [31,39] test with 95% confidence.

Tables 20.4 and 20.5 show the results of the algorithms in terms of the makespan of the solutions found. We can see that the two proposed algorithms clearly outperform METF for most problems. There are only a few exceptions for laplace and mva problems with the smallest CCR values. Comparing the results of the two memetic algorithms, we see that EMACS is outperforming cMGA for most problems, with only seven exceptions.

The algorithms are compared in Tables 20.6 and 20.7 by means of the number of resources (i.e., processors) used in their solutions. Although cMGA is finding solutions with higher makespan values (as shown in Tables 20.4 and 20.5), it requires less processors than EMACS in most cases. The exceptions are one-third of the problems with CCR = 5 and all the problems with the highest CCR ratio studied.

Regarding the energy used by the different solutions in the computation of the jobs, we can see in Tables 20.8 and 20.9 that the results provided by cMGA and EMACS are clearly outperforming METF. There are only 20 exceptions out of the 360 studied comparisons on the different problem instances. Comparing the two genetic algorithms, we can see that cMGA generally outperforms EMACS for the three smallest CCR values, while the latter is better for the other three cases, in general.

Finally, we compare the algorithms in Tables 20.10 and 20.11 in terms of the energy used in the communications. In this case, all the solutions reported by the two MAs outperform those of METF, with no exceptions. Additionally, EMACS is reporting solutions requiring less energy for communications than those of cMGA for every problem, with no exceptions in this case neither.

From the commented results, we can conclude that both cMGA and EMACS clearly outperform METF algorithm for all the studied problems and CCRs by means of the four considered performance metrics, namely, makespan, number of processors used, and energy used in both computations and communications. When comparing cMGA versus EMACS, we conclude that EMACS is the best algorithm attending to the makespan. If we consider the cost of the solutions, then cMGA might be a better option for low CCR ratios. However, for systems with expensive communications, EMACS is definitely the best of the studied algorithms.

20.6 Conclusion and Future Work

We present in this work two new energy-aware memetic algorithms (cMGA and EMACS) for the problem of scheduling tasks with communications. The aim of the algorithms is to minimize the makespan of the solution, but minimizing the energy used both in the computations and communications. Additionally, cMGA is aiming at reducing the number of resources used too.

TABLE 20.4 Average Makespan Obtained for Computation-Intensive Applications

	Problem/CCR	0.1	0.5	1
cMGA	cholesky324	296.26 1.21	350.04 3.53	420.32 6.02
	cholesky464	362.71 1.26	440.69 3.15	538.00 5.33
	cholesky629	414.47 1.63	504.99 4.87	622.16 7.74
	cholesky819	495.25 1.95	595.39 5.64	732.26 8.63
	cholesky1034	563.13 1.88	687.71 4.61	844.41 6.84
	gauss348	320.20 1.29	369.59 3.48	440.27 5.78
	gauss493	360.54 1.23	399.52 3.29	476.13 5.32
	gauss663	413.21 1.17	499.34 4.38	611.07 6.34
	gauss858	562.30 1.58	685.63 5.61	847.28 11.53
	gauss1078	**574.11** 2.32	697.22 5.98	864.94 8.44
	laplace484	347.68 1.84	440.18 2.05	562.93 3.33
	laplace625	423.00 3.53	535.06 2.75	648.26 3.38
	laplace784	469.11 4.08	592.08 2.43	765.32 3.74
	laplace961	548.14 4.17	670.09 3.57	**852.02** 3.92
	laplace1156	599.67 4.41	742.13 4.15	951.88 4.44
	lu299	305.02 1.46	354.50 4.35	427.57 5.23
	lu434	350.52 1.23	417.61 4.28	504.70 4.66
	lu594	390.48 1.32	471.96 3.90	576.00 5.82
	lu779	494.66 1.79	603.83 4.28	742.90 6.53
	lu989	543.53 2.04	**667.03** 4.97	823.70 9.32
	mva484	371.63 2.04	467.61 3.20	600.42 3.22
	mva625	417.44 2.15	527.48 2.93	666.63 2.72
	mva784	483.35 3.38	604.90 2.80	**783.31** 4.07
	mva961	550.31 4.42	687.24 4.31	880.09 4.01
	mva1156	608.82 4.99	**754.59** 5.03	**963.36** 5.40
EMACS	cholesky324	**290.92** 1.92	**336.23** 5.69	**396.59** 5.86
	cholesky464	**357.37** 2.19	**425.88** 6.01	**501.12** 7.53
	cholesky629	**408.26** 2.57	**479.99** 6.96	**568.35** 7.55
	cholesky819	**487.97** 2.66	**572.57** 7.96	**679.03** 11.35
	cholesky1034	**555.40** 2.57	**661.37** 7.77	**840.90** 10.88
	gauss348	**318.42** 1.38	**362.11** 5.15	**421.70** 7.35
	gauss493	**355.18** 1.64	**384.63** 4.51	**446.52** 7.28
	gauss663	**404.87** 1.96	**480.00** 5.64	**572.19** 9.20
	gauss858	**555.72** 2.82	**656.33** 9.51	**775.30** 13.12
	gauss1078	601.71 5.78	715.49 7.41	**856.41** 14.47
	laplace484	**340.12** 1.51	**428.03** 3.59	**545.28** 4.79
	laplace625	**411.74** 3.52	**521.68** 3.89	**631.41** 5.56
	laplace784	**457.86** 3.60	**580.08** 3.54	**744.03** 7.42
	laplace961	**533.49** 4.05	**652.98** 5.58	906.74 10.91
	laplace1156	**585.19** 5.51	**728.73** 5.09	**930.92** 5.37
	lu299	**299.91** 2.11	**340.10** 5.40	**394.63** 5.18
	lu434	**344.50** 1.93	**398.18** 6.04	**471.65** 6.84
	lu594	**383.07** 2.64	**453.33** 5.49	**536.95** 8.86
	lu779	**488.88** 2.25	**578.97** 6.15	**685.39** 10.54
	lu989	**535.38** 2.71	678.49 6.75	**766.48** 12.03
	mva484	**363.24** 1.59	**450.97** 2.46	**578.35** 5.94
	mva625	**410.94** 1.86	**514.13** 3.40	**644.61** 6.07
	mva784	**469.12** 3.62	**590.15** 3.52	804.12 10.56
	mva961	**538.50** 3.81	**671.46** 3.35	**852.94** 7.76
	mva1156	**595.68** 4.29	872.18 16.78	1035.99 16.55

TABLE 20.5 Average Makespan Obtained for Communication-Intensive Applications

	Problem/CCR	3	5	10
cMGA	cholesky324	755.37 13.13	1095.95 28.78	1951.60 70.42
	cholesky464	974.36 15.59	1431.82 36.91	2556.10 94.03
	cholesky629	1123.18 18.76	1649.14 38.61	2927.50 138.61
	cholesky819	1335.23 27.99	1973.06 40.44	3501.65 117.38
	cholesky1034	1535.05 23.28	2241.84 30.59	4063.99 119.00
	gauss348	779.80 10.99	1126.49 38.79	2024.69 76.62
	gauss493	837.06 11.88	1220.66 17.82	2200.46 78.08
	gauss663	1098.92 16.71	1594.23 40.03	2834.19 140.88
	gauss858	1508.83 39.09	2200.15 74.42	3522.29 164.68
	gauss1078	1598.81 24.27	2342.73 34.26	4206.90 179.50
	laplace484	1080.56 9.09	1708.15 22.42	3094.91 84.20
	laplace625	1274.13 8.59	1905.42 15.42	3468.77 100.94
	laplace784	1484.39 10.88	2219.62 16.41	4039.30 153.91
	laplace961	1635.41 6.45	2450.19 15.59	4449.31 96.62
	laplace1156	1757.31 7.85	2626.38 12.89	4790.80 68.53
	lu299	756.83 14.24	1101.91 30.86	1983.74 53.15
	lu434	903.62 17.20	1325.78 30.75	2364.40 109.87
	lu594	1041.89 19.43	1527.73 34.70	2743.36 129.50
	lu779	1346.70 26.05	1970.82 52.81	3472.86 191.57
	lu989	1496.02 28.86	2182.13 55.04	3913.92 161.68
	mva484	1165.31 10.79	1715.85 23.68	3076.02 95.60
	mva625	1275.90 7.91	1880.32 21.43	3351.02 143.26
	mva784	1520.47 10.62	2246.81 24.73	4040.10 106.52
	mva961	1689.89 11.60	2511.22 20.13	4506.12 153.72
	mva1156	1853.62 7.79	2769.86 15.46	4947.37 147.37
EMACS	cholesky324	632.67 12.92	781.45 21.05	1025.71 24.81
	cholesky464	803.98 13.46	1010.91 18.63	1322.85 20.67
	cholesky629	906.90 13.86	1154.50 12.91	1511.31 22.18
	cholesky819	1099.11 16.01	1400.39 21.05	1874.17 27.46
	cholesky1034	1299.76 18.05	1610.26 21.64	2143.95 23.38
	gauss348	652.50 13.28	800.10 15.61	1016.71 18.04
	gauss493	736.07 16.83	923.00 16.41	1227.76 20.66
	gauss663	918.89 14.33	1165.44 18.21	1521.71 19.93
	gauss858	1194.14 19.10	1494.55 17.05	1831.29 22.30
	gauss1078	601.71 5.78	715.49 7.41	856.41 14.47
	laplace484	953.55 17.34	1251.67 37.38	1686.28 71.24
	laplace625	1099.93 17.10	1408.21 23.51	1901.90 50.20
	laplace784	1303.64 20.17	1692.25 30.23	2271.26 58.70
	laplace961	1485.42 18.75	1910.41 48.60	2629.15 71.44
	laplace1156	1581.30 29.56	2082.15 51.27	2860.45 84.44
	lu299	610.47 12.27	760.67 18.66	989.65 19.73
	lu434	764.28 17.12	969.11 17.95	1276.34 15.71
	lu594	846.82 17.87	1072.78 15.98	1425.10 20.79
	lu779	1083.99 21.76	1370.57 18.72	1831.08 26.45
	lu989	1187.77 17.43	1531.68 18.92	2042.60 20.16
	mva484	1000.29 20.19	1267.35 30.76	1674.99 47.04
	mva625	1133.01 17.19	1422.70 40.15	1925.08 44.84
	mva784	1313.45 20.96	1701.62 47.43	2312.00 66.56
	mva961	1500.99 26.02	1931.30 47.47	2670.60 66.62
	mva1156	1661.16 25.99	2157.08 37.69	2991.36 81.24

TABLE 20.6 Average Number of Non-Used Processors from a Maximum of 64 Allowed Ones When Scheduling Computation-Intensive Applications

	Problem/CCR	0.1		0.5		1	
cMGA	cholesky324	7.50	3.04	9.73	3.08	9.40	3.07
	cholesky464	3.63	2.17	4.60	2.50	6.53	2.98
	cholesky629	2.13	1.83	3.37	2.24	3.17	1.72
	cholesky819	0.67	0.80	2.60	2.09	2.30	1.68
	cholesky1034	1.03	1.38	1.20	1.24	1.50	1.07
	gauss348	9.37	3.37	9.53	2.66	9.87	2.83
	gauss493	4.87	2.70	4.90	2.62	4.97	3.25
	gauss663	1.50	1.46	2.37	1.85	2.27	1.89
	gauss858	1.57	1.72	2.67	2.22	2.27	2.02
	gauss1078	0.57	0.82	0.83	1.15	1.63	1.77
	laplace484	0.60	1.00	1.03	1.13	2.33	1.84
	laplace625	0.57	0.90	0.40	0.62	1.60	1.45
	laplace784	0.13	0.35	0.37	0.76	0.47	0.78
	laplace961	0.17	0.38	0.10	0.31	0.50	0.73
	laplace1156	0.13	0.35	0.17	0.38	0.17	0.46
	lu299	10.37	3.05	11.33	3.03	11.00	3.17
	lu434	3.70	2.74	5.27	2.83	6.37	2.88
	lu594	2.73	2.23	3.93	2.02	4.57	2.58
	lu779	1.60	1.35	1.63	1.43	2.87	2.01
	lu989	1.03	1.10	1.43	1.14	1.63	1.61
	mva484	0.80	0.81	1.40	1.54	2.00	1.91
	mva625	0.63	0.96	0.53	0.78	1.33	1.52
	mva784	0.27	0.52	0.27	0.52	0.67	0.99
	mva961	0.07	0.25	0.40	0.81	0.60	1.10
	mva1156	0.00	0.00	0.17	0.53	0.10	0.31
EMACS	cholesky324	0.57	0.73	0.83	1.09	0.47	0.63
	cholesky464	0.10	0.31	0.07	0.25	0.10	0.31
	cholesky629	0.00	0.00	0.00	0.00	0.00	0.00
	cholesky819	0.00	0.00	0.00	0.00	0.00	0.00
	cholesky1034	0.00	0.00	0.00	0.00	0.07	0.37
	gauss348	0.27	0.45	0.23	0.43	0.50	0.82
	gauss493	0.10	0.31	0.07	0.25	0.00	0.00
	gauss663	0.00	0.00	0.00	0.00	0.00	0.00
	gauss858	0.00	0.00	0.00	0.00	0.00	0.00
	gauss1078	0.00	0.00	0.03	0.18	0.13	0.35
	laplace484	0.00	0.00	0.03	0.18	0.03	0.18
	laplace625	0.00	0.00	0.00	0.00	0.00	0.00
	laplace784	0.00	0.00	0.00	0.00	0.00	0.00
	laplace961	0.00	0.00	0.00	0.00	0.03	0.18
	laplace1156	0.00	0.00	0.00	0.00	0.00	0.00
	lu299	0.77	0.90	1.03	0.96	0.93	0.94
	lu434	0.07	0.25	0.03	0.18	0.10	0.31
	lu594	0.03	0.18	0.00	0.00	0.00	0.00
	lu779	0.00	0.00	0.00	0.00	0.00	0.00
	lu989	0.00	0.00	0.03	0.18	0.00	0.00
	mva484	0.00	0.00	0.03	0.18	0.03	0.18
	mva625	0.00	0.00	0.03	0.18	0.03	0.18
	mva784	0.00	0.00	0.00	0.00	0.27	0.45
	mva961	0.00	0.00	0.00	0.00	0.00	0.00
	mva1156	0.00	0.00	0.00	0.00	0.00	0.00

TABLE 20.7　Average Number of Non-Used Processors from a Maximum of 64 Allowed Ones When Scheduling Communication-Intensive Applications

	Problem/CCR	3	5	10
cMGA	cholesky324	12.03 (3.40)	13.80 (3.95)	13.30 (3.32)
	cholesky464	7.17 (2.73)	8.27 (3.69)	8.40 (2.92)
	cholesky629	4.07 (2.35)	6.73 (2.99)	4.13 (1.94)
	cholesky819	3.17 (1.62)	2.33 (1.63)	3.10 (1.86)
	cholesky1034	1.60 (1.45)	1.60 (1.28)	1.27 (1.20)
	gauss348	10.53 (3.27)	10.83 (3.57)	11.77 (3.33)
	gauss493	6.40 (2.14)	6.03 (2.03)	6.20 (2.76)
	gauss663	4.10 (2.44)	4.73 (2.41)	4.73 (2.49)
	gauss858	2.30 (1.74)	3.27 (1.84)	2.87 (1.72)
	gauss1078	1.87 (1.74)	1.90 (1.52)	1.70 (1.51)
	laplace484	5.80 (2.76)	6.67 (2.62)	7.00 (3.41)
	laplace625	3.53 (2.16)	4.53 (3.03)	5.50 (2.37)
	laplace784	2.53 (2.08)	3.13 (1.94)	3.43 (2.53)
	laplace961	1.00 (1.23)	1.43 (1.25)	2.77 (1.85)
	laplace1156	0.57 (0.73)	0.90 (1.12)	1.40 (1.43)
	lu299	15.13 (3.73)	13.40 (4.06)	14.77 (4.64)
	lu434	7.63 (3.03)	7.03 (3.07)	9.67 (3.60)
	lu594	5.00 (2.53)	6.27 (2.60)	5.33 (3.00)
	lu779	2.83 (2.67)	3.30 (2.09)	4.37 (2.20)
	lu989	1.83 (1.39)	2.37 (1.43)	2.10 (1.56)
	mva484	5.40 (3.28)	6.97 (3.36)	7.47 (3.70)
	mva625	3.50 (2.35)	5.77 (2.70)	5.27 (2.86)
	mva784	2.53 (2.05)	3.10 (1.88)	3.80 (1.67)
	mva961	1.10 (1.12)	1.53 (1.53)	2.40 (2.55)
	mva1156	0.53 (0.94)	0.83 (1.05)	1.77 (1.50)
EMACS	cholesky324	2.33 (1.45)	9.60 (3.72)	25.80 (5.82)
	cholesky464	1.03 (1.47)	5.87 (3.72)	22.77 (4.23)
	cholesky629	0.43 (1.04)	4.87 (3.43)	21.70 (5.98)
	cholesky819	0.07 (0.25)	3.23 (2.25)	28.73 (7.09)
	cholesky1034	3.13 (3.47)	3.57 (3.23)	16.33 (4.40)
	gauss348	1.57 (1.25)	7.17 (4.78)	21.57 (3.84)
	gauss493	0.83 (1.93)	4.63 (3.22)	19.83 (5.06)
	gauss663	0.23 (0.50)	4.67 (3.60)	18.60 (5.57)
	gauss858	0.07 (0.25)	2.57 (2.54)	15.80 (4.65)
	gauss1078	2.27 (2.55)	2.50 (2.52)	25.53 (5.15)
	laplace484	2.33 (4.01)	14.70 (6.34)	29.60 (4.51)
	laplace625	3.17 (5.29)	11.57 (7.32)	37.03 (5.18)
	laplace784	0.90 (2.28)	17.53 (4.15)	32.63 (5.18)
	laplace961	0.40 (0.89)	6.07 (4.61)	32.90 (6.18)
	laplace1156	0.23 (0.50)	11.20 (5.95)	28.43 (5.96)
	lu299	2.73 (2.07)	10.07 (5.13)	26.10 (6.62)
	lu434	1.07 (1.51)	7.37 (4.27)	39.93 (5.91)
	lu594	0.47 (0.90)	7.40 (4.32)	20.97 (5.46)
	lu779	0.23 (0.63)	4.17 (3.07)	29.47 (7.37)
	lu989	0.27 (1.11)	15.17 (5.67)	28.77 (7.33)
	mva484	3.27 (3.19)	10.40 (5.03)	28.37 (4.94)
	mva625	0.57 (1.14)	11.43 (9.22)	26.03 (3.59)
	mva784	0.63 (1.03)	8.60 (4.30)	25.93 (5.94)
	mva961	1.03 (2.65)	8.67 (4.80)	32.60 (6.09)
	mva1156	0.33 (0.92)	5.87 (4.34)	27.33 (6.16)

TABLE 20.8 Average Energy Used for the Computations on Computation-Intensive Applications

	Problem/CCR	0.1	0.5	1
cMGA	cholesky324	18375.09 893.73	20694.99 1089.73	24654.62 1252.37
	cholesky464	24170.15 743.14	28407.22 1130.62	33164.76 1629.68
	cholesky629	28696.16 706.80	33687.53 1100.80	40826.32 1093.53
	cholesky819	35205.86 438.12	40410.50 1187.00	48924.27 1220.35
	cholesky1034	40772.91 720.32	48401.20 872.18	57813.36 983.20
	gauss348	19020.39 1052.24	21673.96 951.64	25383.86 1320.66
	gauss493	23631.45 923.75	25945.13 1051.21	30417.49 1563.33
	gauss663	29067.59 614.11	33893.49 1002.24	40725.62 1174.55
	gauss858	39072.87 881.85	45987.46 1434.29	56036.92 1681.56
	gauss1078	41825.41 445.33	49325.46 820.20	59060.97 1637.93
	laplace484	24965.47 338.78	30617.33 510.61	37567.27 1092.50
	laplace625	30685.54 420.82	37859.32 336.90	44058.25 924.69
	laplace784	34634.09 283.84	42295.79 466.75	53177.47 589.41
	laplace961	40947.06 309.84	48642.57 236.61	59824.76 608.44
	laplace1156	45146.17 318.70	54211.55 325.60	67525.85 418.65
	lu299	17864.05 945.71	20230.86 1095.60	24212.53 1352.98
	lu434	23362.22 932.83	26780.27 1202.34	31359.42 1421.15
	lu594	26698.21 834.05	31093.35 1009.69	36939.70 1479.65
	lu779	34826.10 694.63	41520.42 885.77	49094.70 1501.64
	lu989	39185.19 549.89	46529.12 812.47	56016.31 1416.64
	mva484	26533.24 292.61	32313.58 707.37	40212.69 1188.44
	mva625	30198.44 396.02	37217.58 435.89	45429.13 971.27
	mva784	35644.77 328.47	43406.48 374.99	54357.34 898.77
	mva961	40965.65 286.28	49480.66 517.88	61511.15 973.29
	mva1156	45855.89 303.61	55068.44 495.50	68368.81 409.13
EMACS	cholesky324	19978.94 224.77	22811.95 486.22	26855.29 469.48
	cholesky464	24964.92 176.79	29338.69 381.24	34201.47 513.62
	cholesky629	28997.02 151.63	33632.76 411.88	39301.46 438.71
	cholesky819	34867.71 182.80	40336.32 469.49	47163.61 654.39
	cholesky1034	40657.64 137.60	47360.77 422.03	59687.74 769.03
	gauss348	21585.61 155.15	24452.44 345.45	28198.03 551.84
	gauss493	24811.60 150.78	26817.42 266.05	30878.58 431.64
	gauss663	28943.65 127.96	33679.79 339.36	39594.30 543.48
	gauss858	39227.03 192.71	45578.31 562.68	53194.90 795.55
	gauss1078	44730.50 388.62	51941.98 514.99	60857.66 839.31
	laplace484	24716.60 98.00	30302.24 228.98	37765.82 327.89
	laplace625	30243.05 216.61	37248.27 236.61	44049.12 353.02
	laplace784	34009.82 216.64	41796.08 218.93	52210.36 454.60
	laplace961	40151.53 249.42	47662.83 354.80	63623.11 688.17
	laplace1156	44344.88 332.49	53541.50 314.59	66391.55 337.90
	lu299	20288.92 318.53	22783.96 441.16	26364.92 527.61
	lu434	24133.88 146.49	27638.27 358.04	32340.40 413.43
	lu594	27167.36 126.58	31624.88 306.99	37005.12 503.01
	lu779	34928.54 115.29	40644.01 346.67	47446.09 645.32
	lu989	38906.37 185.89	48869.98 412.84	53627.07 691.29
	mva484	26310.10 102.31	31915.66 172.09	40013.23 349.79
	mva625	30058.15 114.43	36659.46 243.90	44924.43 401.53
	mva784	34910.90 225.15	42647.25 216.59	55940.57 827.02
	mva961	40296.37 245.80	48791.72 201.96	60321.53 497.77
	mva1156	45049.06 273.40	62515.50 1052.31	72994.93 1029.96

TABLE 20.9   Average Energy Used for the Computations on Communication-Intensive Applications

	Problem/CCR	3		5		10	
cMGA	cholesky324	40888.67	2582.83	56644.06	4905.99	100588.57	8257.57
	cholesky464	57467.09	3004.38	81836.55	6175.77	144137.11	10024.83
	cholesky629	70132.18	2763.52	97091.30	5052.15	178017.71	11608.85
	cholesky819	84614.06	2697.00	124857.10	4325.75	216425.36	9308.40
	cholesky1034	100451.59	2496.03	144416.78	3502.38	259259.54	9043.03
	gauss348	43189.01	2762.51	61318.23	4341.23	107244.92	9090.23
	gauss493	50453.88	2031.64	72932.75	2709.14	129276.53	7717.92
	gauss663	68599.69	2546.04	97258.59	4484.40	170694.21	11776.87
	gauss858	96574.71	3115.53	137043.80	5367.23	218601.32	11336.01
	gauss1078	103909.59	3219.43	149729.08	3939.07	266377.25	13997.31
	laplace484	65610.49	3204.38	100667.84	4824.17	179116.34	12070.71
	laplace625	80463.47	2798.81	116660.49	5948.22	206256.41	11387.87
	laplace784	95572.09	3347.68	139311.20	4147.40	249057.92	17202.13
	laplace961	108432.28	2105.52	158595.76	3032.71	277661.65	11402.84
	laplace1156	117797.45	1422.10	171899.88	3217.94	305948.20	8042.83
	lu299	38482.42	2865.72	57232.39	5016.74	99123.42	10034.62
	lu434	53087.45	2996.43	77574.61	4574.45	130568.62	11125.83
	lu594	63956.16	2809.94	90585.03	4170.91	163436.23	12925.18
	lu779	85651.83	3839.18	122842.91	4485.07	210179.63	13116.64
	lu989	97495.02	3256.64	138826.37	4498.25	246386.39	11850.28
	mva484	71060.48	3881.51	100595.58	6448.26	176706.53	14445.13
	mva625	80685.95	3080.26	112930.37	5490.89	200340.72	14969.16
	mva784	97915.83	3090.38	141189.82	4441.49	247587.20	10808.60
	mva961	111652.39	1975.74	162087.02	4146.28	283002.55	18595.37
	mva1156	124132.69	1853.98	181276.74	3167.17	314214.47	13225.26
EMACS	cholesky324	40809.22	1276.60	44473.88	3468.51	41240.79	6094.02
	cholesky464	53083.68	1444.20	61511.48	3961.92	57484.93	5691.13
	cholesky629	60977.97	1248.78	71880.38	4152.21	67794.04	9063.36
	cholesky819	74446.71	1083.60	89864.72	3284.67	71281.85	13412.27
	cholesky1034	85261.38	4561.04	103606.34	5320.60	108810.31	9330.16
	gauss348	42387.66	1246.10	47353.21	3677.27	45177.18	3723.76
	gauss493	48963.69	1618.17	57519.63	2993.83	57150.73	6236.09
	gauss663	61968.67	958.02	72926.95	4341.59	73108.95	8449.12
	gauss858	80814.57	1253.28	96942.94	3734.25	93586.11	8567.22
	gauss1078	89324.05	3651.08	108715.45	4037.14	93165.25	11721.86
	laplace484	61594.03	4355.49	64754.49	8677.44	61215.11	8835.46
	laplace625	70454.26	6207.16	77458.36	10822.67	55050.20	10075.54
	laplace784	86693.92	3260.48	83172.09	7098.58	75989.30	11962.97
	laplace961	99977.19	1939.68	116413.50	10529.11	87895.19	17375.73
	laplace1156	107273.34	2225.96	116610.95	13363.29	108740.01	18046.54
	lu299	39097.87	1639.27	42861.52	4147.77	39446.15	6465.72
	lu434	50487.73	1547.75	57526.98	4323.82	33617.27	7605.29
	lu594	56866.02	1230.18	64113.62	4564.49	64917.48	8098.38
	lu779	73242.70	1325.87	86668.86	4205.52	68323.12	13750.03
	lu989	81154.18	1298.03	80948.76	9043.29	78259.69	14845.30
	mva484	63663.83	3963.76	70889.96	7090.25	62814.37	8992.45
	mva625	75407.27	1838.88	78609.64	14219.96	76852.08	7631.81
	mva784	87824.25	1923.96	98974.25	8484.13	93102.28	15355.68
	mva961	100034.37	4461.63	112570.43	10499.31	89820.60	16957.76
	mva1156	112361.15	2861.91	132159.89	10290.49	116882.53	19583.86

TABLE 20.10 Average Energy Used for Communications on Computation-Intensive Applications

	Problem/CCR	0.1		0.5		1	
cMGA	cholesky324	318.73	4.29	956.04	15.89	1745.11	21.95
	cholesky464	461.16	3.27	1420.23	14.54	2600.47	28.29
	cholesky629	616.46	3.54	1839.68	19.31	3362.38	29.87
	cholesky819	817.01	6.45	2471.04	19.85	4516.60	36.71
	cholesky1034	1051.69	6.98	3255.04	20.07	5961.79	36.77
	gauss348	330.71	5.57	1060.39	15.12	1976.57	34.19
	gauss493	482.42	4.99	1401.24	17.75	2545.14	27.87
	gauss663	655.93	4.86	1944.35	20.94	3540.74	32.10
	gauss858	869.53	6.26	2754.84	30.77	5088.88	68.37
	gauss1078	1104.32	6.28	3334.32	22.01	6096.57	45.07
	laplace484	490.40	3.47	1501.18	15.58	2723.49	40.60
	laplace625	642.92	4.76	1959.73	16.71	3507.48	35.05
	laplace784	796.59	3.90	2406.96	18.86	4376.83	33.68
	laplace961	1001.44	4.17	3074.13	15.72	5636.38	38.66
	laplace1156	1184.30	5.32	3622.54	18.64	6645.31	38.34
	lu299	294.49	3.40	905.61	13.87	1660.44	36.57
	lu434	434.14	3.73	1297.88	13.96	2354.78	37.99
	lu594	574.71	5.95	1702.81	13.41	3090.87	43.35
	lu779	778.77	5.43	2395.28	24.48	4375.89	39.89
	lu989	988.46	6.37	3020.97	23.85	5550.38	48.79
	mva484	504.26	2.80	1539.19	10.79	2797.39	27.55
	mva625	638.36	3.20	1995.33	10.54	3510.03	26.63
	mva784	813.61	4.41	2538.74	14.30	4633.50	31.82
	mva961	989.80	3.41	3088.09	16.92	5667.38	37.45
	mva1156	1189.67	5.40	3735.47	17.95	6898.32	37.63
EMACS	cholesky324	**275.77**	10.92	**840.15**	27.10	**1515.64**	73.32
	cholesky464	**386.51**	11.43	**1200.74**	46.04	**2210.77**	88.21
	cholesky629	**490.83**	13.10	**1438.64**	278.90	**2752.00**	111.71
	cholesky819	**627.65**	19.26	**1949.80**	71.86	**3568.81**	197.86
	cholesky1034	**782.19**	31.03	**2385.86**	116.56	**4563.93**	208.94
	gauss348	**292.29**	11.80	**921.78**	54.46	**1636.42**	151.46
	gauss493	**401.75**	13.62	**1191.37**	42.27	**2077.41**	411.33
	gauss663	**526.90**	17.48	**1577.30**	73.69	**2859.30**	123.40
	gauss858	**669.69**	22.22	**2187.81**	75.73	**3978.21**	180.10
	gauss1078	**817.04**	32.46	**2526.59**	76.28	**4566.22**	220.31
	laplace484	**311.92**	108.77	**1093.87**	58.46	**2004.74**	155.10
	laplace625	**449.42**	27.11	**1309.35**	257.37	**2443.51**	126.00
	laplace784	**526.59**	30.15	**1538.61**	305.87	**2901.03**	128.41
	laplace961	**643.23**	36.49	**1945.79**	137.48	**3616.60**	178.76
	laplace1156	**734.16**	39.96	**2233.57**	96.41	**4193.55**	241.64
	lu299	**260.31**	6.65	**806.54**	31.60	**1464.60**	82.16
	lu434	**363.77**	8.14	**1111.60**	51.68	**2035.22**	81.18
	lu594	**467.21**	14.67	**1392.78**	55.68	**2558.30**	154.43
	lu779	**606.35**	19.76	**1866.73**	72.57	**3461.26**	159.87
	lu989	**753.12**	26.98	**2299.54**	84.14	**4219.90**	193.53
	mva484	**365.23**	31.88	**1109.63**	59.70	**2025.11**	176.77
	mva625	**440.68**	29.21	**1378.16**	88.71	**2289.80**	637.43
	mva784	**550.60**	31.60	**1687.01**	121.15	**3099.54**	210.24
	mva961	**623.13**	41.57	**1976.67**	130.95	**3660.40**	237.73
	mva1156	**742.52**	45.66	**2348.64**	115.52	**4344.01**	211.74

TABLE 20.11   Average Energy Used for Communications on
Communication-Intensive Applications

	Problem/CCR	3		5		10	
cMGA	cholesky324	4798.42	147.29	7785.92	312.01	15070.99	1254.64
	cholesky464	7238.74	167.92	11699.55	466.68	22692.73	1509.81
	cholesky629	9208.24	224.33	15018.17	435.63	28069.94	2967.83
	cholesky819	12632.49	216.98	20549.61	593.23	38965.81	2923.99
	cholesky1034	16571.00	308.53	26851.42	767.69	52128.63	3482.86
	gauss348	5482.57	130.73	8808.77	417.63	17366.44	1111.43
	gauss493	6996.00	134.72	11279.60	286.33	22173.64	1106.89
	gauss663	9684.02	171.04	15444.44	669.04	29259.36	3060.17
	gauss858	14067.02	409.39	22422.10	1226.27	39396.34	3482.68
	gauss1078	16813.68	263.12	27248.37	692.90	52229.68	3875.90
	laplace484	7495.13	133.99	13059.66	228.86	25375.91	1441.76
	laplace625	9695.99	115.64	15623.23	299.24	29989.08	2450.65
	laplace784	11951.98	135.83	19446.94	343.66	36483.90	3724.52
	laplace961	15695.91	173.54	25619.19	249.50	48414.03	3657.09
	laplace1156	18277.90	149.03	29804.81	351.93	57229.07	3091.90
	lu299	4600.09	125.95	7525.84	373.74	14728.02	708.95
	lu434	6450.34	156.13	10691.00	348.18	20032.41	2290.48
	lu594	8578.93	199.80	13833.01	345.34	26531.67	1934.66
	lu779	12255.90	223.05	19851.83	695.66	37730.23	4252.08
	lu989	15475.62	220.74	24720.71	956.96	48138.50	3540.02
	mva484	7605.61	107.88	12157.62	240.29	23030.04	1789.57
	mva625	9609.57	124.19	15323.17	419.32	28331.60	3208.39
	mva784	12682.01	128.89	20230.71	634.99	37161.28	3217.69
	mva961	15735.19	139.70	25379.26	479.63	46957.84	5174.90
	mva1156	19153.56	137.63	31128.37	289.92	55086.83	6187.00
EMACS	cholesky324	275.77	10.92	840.15	27.10	1515.64	73.32
	cholesky464	386.51	11.43	1200.74	46.04	2210.77	88.21
	cholesky629	490.83	13.10	1438.64	278.90	2752.00	111.71
	cholesky819	627.65	19.26	1949.80	71.86	3568.81	197.86
	cholesky1034	782.19	31.03	2385.86	116.56	4563.93	208.94
	gauss348	292.29	11.80	921.78	54.46	1636.42	151.46
	gauss493	401.75	13.62	1191.37	42.27	2077.41	411.33
	gauss663	526.90	17.48	1577.30	73.69	2859.30	123.40
	gauss858	669.69	22.22	2187.81	75.73	3978.21	180.10
	gauss1078	817.04	32.46	2526.59	76.28	4566.22	220.31
	laplace484	311.92	108.77	1093.87	58.46	2004.74	155.10
	laplace625	449.42	27.11	1309.35	257.37	2443.51	126.00
	laplace784	526.59	30.15	1538.61	305.87	2901.03	128.41
	laplace961	643.23	36.49	1945.79	137.48	3616.60	178.76
	laplace1156	734.16	39.96	2233.57	96.41	4193.55	241.64
	lu299	260.31	6.65	806.54	31.60	1464.60	82.16
	lu434	363.77	8.14	1111.60	51.68	2035.22	81.18
	lu594	467.21	14.67	1392.78	55.68	2558.30	154.43
	lu779	606.35	19.76	1866.73	72.57	3461.26	159.87
	lu989	753.12	26.98	2299.54	84.14	4219.90	193.53
	mva484	365.23	31.88	1109.63	59.70	2025.11	176.77
	mva625	440.68	29.21	1378.16	88.71	2289.80	637.43
	mva784	550.60	31.60	1687.01	121.15	3099.54	210.24
	mva961	623.13	41.57	1976.67	130.95	3660.40	237.73
	mva1156	742.52	45.66	2348.64	115.52	4344.01	211.74

The two new algorithms are compared with METF (a state-of-the-art list algorithm) for validation. Our first conclusion is that the two proposed approaches are clearly more appropriate for the considered problem than METF, since they outperform it in most cases. Comparing the two proposed techniques, cMGA seems more suitable in the cases where the number of available resources is limited (or they must be shared), or when the communication to computation ratio is low. However, when the makespan is a priority, the communications are expensive, or the application is communications intensive, then EMACS stands as the most appropriate choice, among the studied ones.

As future work, we plan to extend our study to allow the use of heterogeneous resources. Additionally, we are working on the parallelization of the algorithms (either for multi-core or GPU architectures). This will allow us to perform longer runs in reasonable time, and we expect it will help in greatly improving the results obtained.

Acknowledgments

The authors would like to acknowledge the support of Luxembourg FNR in the framework of GreenIT project (C09/IS/05) and the Luxembourg AFR for providing a research grant (PDR-08-010).

20.A Appendix

Tables 20.12 through 20.14 contain the results reported by METF for all the studied problems in terms of makespan and energy used both in computations and communications. The table for the number of non-used processors is not given because METF always uses all the available processors to build the solution.

TABLE 20.12 Average Makespan Obtained

Problem/CCR	0.1	0.5	1	3	5	10
cholesky324	311.09	417.46	550.99	1085.14	1625.65	2990.64
cholesky464	377.82	508.46	674.97	1347.21	2029.55	3743.82
cholesky629	432.18	585.92	779.99	1583.89	2405.02	4461.33
cholesky819	510.62	672.23	887.94	1784.90	2686.45	4944.84
cholesky1034	579.22	760.69	996.96	1967.62	2958.91	5454.42
gauss348	339.73	437.66	562.12	1087.75	1624.96	2968.80
gauss493	376.88	462.34	600.87	1185.16	1793.10	3324.60
gauss663	428.96	567.49	748.27	1505.19	2285.22	4248.28
gauss858	582.69	787.05	1045.13	2084.55	3130.95	5042.79
gauss1078	591.61	775.55	1021.44	2048.78	3101.10	5777.85
laplace484	343.03	447.49	593.00	1235.70	1985.03	3708.79
laplace625	403.49	532.49	679.29	1448.23	2224.57	4165.41
laplace784	453.20	594.02	787.98	1672.45	2567.20	4805.03
laplace961	525.56	646.91	866.80	1794.62	2738.06	5114.46
laplace1156	554.67	717.27	946.37	1934.78	2955.98	5516.77
lu299	323.67	432.19	569.37	1123.33	1677.30	3062.21
lu434	366.55	489.91	644.51	1285.54	1944.44	3614.22
lu594	406.26	534.28	699.81	1394.24	2110.59	3901.46
lu779	513.69	685.88	904.06	1792.77	2686.03	4934.52
lu989	560.38	746.33	982.38	2004.77	3013.76	5547.08
mva484	367.70	471.30	623.34	1310.15	2014.17	3779.94
mva625	415.85	536.03	695.24	1426.67	2185.96	4084.19
mva784	458.94	595.01	804.35	1683.59	2591.87	4869.75
mva961	533.48	680.48	900.39	1867.25	2848.02	5312.59
mva1156	580.27	738.27	970.81	1989.35	3034.01	5670.76

TABLE 20.13   Average Energy Used for the Computations

Problem/CCR	0.1	0.5	1	3	5	10
cholesky324	21366.11	27970.12	36455.18	70597.88	105200.45	192530.57
cholesky464	26342.88	34439.68	44901.90	87826.65	131425.56	241167.01
cholesky629	30716.79	40210.01	52465.32	103827.89	156356.16	287909.26
cholesky819	36842.98	46854.62	60504.41	117652.02	175293.92	319680.76
cholesky1034	42676.22	53816.01	68697.21	130564.40	193902.50	353329.09
gauss348	23111.66	29263.12	37175.70	70779.39	105159.78	191159.69
gauss493	26345.87	31639.90	40365.79	77682.70	116511.99	214598.18
gauss663	30707.91	39318.15	50749.41	98992.12	148790.57	274406.50
gauss858	41639.41	54349.88	70750.61	137120.91	203902.69	326147.69
gauss1078	43916.87	55380.16	70759.95	136184.80	203355.45	374550.54
laplace484	24926.66	31566.11	40816.87	81755.84	129725.64	239978.12
laplace625	29802.23	38020.13	47167.32	96118.82	145725.67	269840.89
laplace784	33837.51	42750.96	55137.60	111459.61	168637.14	311697.66
laplace961	39876.53	47443.10	61379.55	120321.54	180474.91	332475.52
laplace1156	42684.11	53033.72	67617.56	130254.77	195413.36	359095.62
lu299	21944.14	28704.81	37414.96	72822.76	108230.39	196835.93
lu434	25530.41	33163.48	42860.85	83718.71	125940.50	232774.33
lu594	28810.13	36697.79	47112.02	91483.26	137282.47	251963.42
lu779	36885.35	47622.13	61448.67	118112.73	175208.07	319054.02
lu989	41117.85	52633.12	67493.15	132807.80	197281.67	359378.57
mva484	26633.74	33237.86	42883.21	86587.71	131570.96	244500.76
mva625	30505.90	38153.36	48219.54	94858.23	143311.99	264669.17
mva784	34400.34	43050.31	56382.65	112308.91	170262.03	315828.53
mva961	40226.58	49561.47	63445.50	124944.88	187533.37	345137.38
mva1156	44390.59	54395.60	69215.31	133931.38	200462.20	369080.96

TABLE 20.14   Average Energy Used for the Communications

Problem/CCR	0.1	0.5	1	3	5	10
cholesky324	628.61	1930.44	3558.19	10155.88	16531.70	33036.00
cholesky464	909.65	2805.91	5184.43	14650.50	24238.82	47770.90
cholesky629	1211.32	3723.24	6893.33	19469.89	32028.80	63983.95
cholesky819	1573.97	4928.28	9109.74	25865.67	42630.50	84358.06
cholesky1034	2067.77	6427.50	11888.48	33672.44	55453.85	109952.52
gauss348	691.21	2179.93	4056.74	11572.60	19145.40	37962.70
gauss493	974.29	2897.28	5306.55	15032.41	24635.46	49212.63
gauss663	1323.90	4001.75	7404.25	20836.21	34169.05	67810.45
gauss858	1756.93	5534.69	10279.80	29281.19	48073.55	89325.27
gauss1078	2194.24	6822.13	12574.25	35658.08	58678.38	116385.58
laplace484	957.86	2927.17	5390.59	15245.66	26646.09	52871.66
laplace625	1251.68	3866.99	7019.08	19884.53	32603.09	64737.68
laplace784	1559.71	4738.11	8716.85	24597.01	40417.25	80379.52
laplace961	1948.54	6036.57	11188.46	31691.85	52259.58	103634.45
laplace1156	2320.06	7162.66	13230.26	36020.97	59376.06	116961.56
lu299	591.89	1834.25	3406.15	9683.00	15944.72	31802.77
lu434	860.80	2658.04	4925.30	13980.52	22999.94	45432.85
lu594	1126.36	3431.73	6304.35	17903.02	29488.43	58201.71
lu779	1545.14	4752.69	8790.35	25154.54	41092.07	81429.48
lu989	1935.79	6010.29	11091.52	31667.97	52249.76	103578.15
mva484	973.78	3071.59	5672.77	16133.32	26602.71	52518.98
mva625	1238.21	3845.19	6973.14	19862.17	32645.32	64774.48
mva784	1592.00	4936.21	9106.93	25811.74	42725.61	84579.98
mva961	1924.20	6047.98	11159.29	31622.97	52103.26	103663.68
mva1156	2316.29	7246.80	13387.71	38021.61	62683.32	124167.54

References

1. E. Alba and B. Dorronsoro. *Cellular Genetic Algorithms*. Operations Research/Computer Science Interfaces Series, Vol. 42. Springer-Verlag, Heidelberg, Germany, 2008.
2. E. Alba and M. Tomassini. Parallelism and evolutionary algorithms. *IEEE Transactions on Evolutionary Computation*, 6(5):443–462, 2002.
3. E. Alba, B. Dorronsoro, and H. Alfonso. Cellular memetic algorithms. *Journal of Computer Science and Technology*, 5(4):257–263, 2006.
4. E. Alba, B. Dorronsoro, M. Giacobini, and M. Tomassini. Decentralized cellular evolutionary algorithms. In *Handbook of Bioinspired Algorithms and Applications*, CRC Press, Boca Raton, FL, pp. 103–120, 2006.
5. L. Benini, A. Bogliolo, R. Bogliolo, and G.D. Micheli. A survey of design techniques for system-level dynamic power management. *IEEE Transactions on VLSI Systems*, 8:299–316, 2000.
6. T.D. Burd, T.A. Pering, A.J. Stratakos, and R.W. Brodersen. A dynamic voltage scaled microprocessor system. *IEEE Journal Solid-State Circuits*, 35(11):1571–1580, 2000.
7. A.P. Chandrakasan and R.W. Brodersen. *Low Power Digital CMOS Design*. Kluwer Academic Publisher, Boston, MA, 1995.
8. C.A. Coello, D.A. Van Veldhuizen, and G.B. Lamont. *Evolutionary Algorithms for Solving Multi-Objective Problems*. Genetic Algorithms and Academic Publisher, Evolutionary Computation. Kluwer Dordrecht, the Netherlands, 2002.
9. K. Deb. *Multi-Objective Optimization Using Evolutionary Algorithms*. John Wiley & Sons, New York, 2001.
10. E. Deelman, G. Singh, M.H. Su, J. Blythe, A. Gil, C. Kesselman, G. Mehta, K. Vahi, G.B. Berriman, J. Good, A. Laity, J.C. Jacob, and D.S. Katz. Pegasus: A framework for mapping complex scientific workflows onto distributed systems. *Scientific Programming Journal*, 13:219–237, 2005.
11. E. Falkenauer. *Genetic Algorithms and Grouping Problems*. John Wiley & Sons Ltd., London, 1999.
12. M. Guzek, J.E. Pecero, B. Dorronsoro, and P. Bouvry. A cellular genetic algorithm for scheduling applications and energy-aware communication optimization. In *Workshop on Optimization Issues in Energy Efficient Distributed Systems (OPTIM), Part of the International Conference on High Performance Computing & Simulation (HPCS)*, Caen, France, pp. 241–248, 2010.
13. J.J. Hwang, Y.-C. Chow, F.D. Angers, and C.-Y. Lee. Scheduling precedence graphs in systems with interprocessor communication times. *SIAM Journal on Computing*, 18(2):244–257, 1989.
14. S. Irani, S. Shukla, and R. Gupta. Online strategies for dynamic power management in system with multiple power-saving states. *ACM Transactions on Embedded Computing Systems*, 2(3):325–346, 2003.
15. S.U. Khan and I. Ahmad. A cooperative game theoretical technique for joint optimization of energy consumption and response time in computational grids. *IEEE Transactions on Parallel and Distributed Systems*, 20(3):346–360, 2009.
16. Y.K. Kwok and I. Ahmad. Benchmarking and comparison of the task graph scheduling algorithms. *Journal of Parallel and Distributed Computing*, 59(3):381–422, 1999.
17. Y.K. Kwok and I. Ahmad. Static scheduling algorithms for allocating directed task graphs to multiprocessors. *ACM Computing Surveys*, 31(4):406–471, 1999.
18. Y.C. Lee and A.Y. Zomaya. Minimizing energy consumption for precedence-constrained applications using dynamic voltage scaling. In *Ninth IEEE/ACM International Symposium on Cluster Computing and the Grid*, IEEE Computer Society, Washington, DC, pp. 92–99, 2009.
19. Y.C. Lee and A.Y. Zomaya. Energy efficient utilization of resources in cloud computing systems. *Journal of Supercomputing*, 1–13, DOI: 10.1007/s11227-010-0421-3, 2011.
20. R. Lepère and D. Trystram. A new clustering algorithm for large communication delays. In *Proceedings of the 16th International Parallel and Distributed Processing Symposium, IPDPS'02*, IEEE Computer Society, Fort Lauderdale, FL, pp. 68–73, April 2002.

21. G. Magklis, G. Semeraro, D.H. Albonesi, S.G. Dropsho, S. Dwarkadas, and M.L. Scott. Dynamic frequency and voltage scaling for a multiple-clock-domain microprocessor. *IEEE Micro*, 23:62–68, 2003.

22. A. Mahjoub, J.E. Pecero, and D. Trystram. Scheduling with uncertainties on new computing platforms. *Journal Computational Optimization and Applications*, 48(2):369–398, 2011, DOI: 10.1007/s10589-009-9311-0.

23. B. Manderick and P. Spiessens. Fine-grained parallel genetic algorithm. In J.D. Schaffer, ed., *Proceedings of the Third International Conference on Genetic Algorithms (ICGA)*, Morgan Kaufmann, San Francisco, pp. 428–433, 1989.

24. S.M. Martin, K. Flautner, T. Mudge, and D. Blaauw. Combined dynamic voltage scaling and adaptive body biasing for lower power microprocessors under dynamic workloads. In *Proceedings of the 2002 IEEE/ACM International Conference on Computer-Aided Design*, ICCAD'02, ACM, New York, pp. 721–725, 2002.

25. P. Mejia-Alvarez, E. Levner, and D. Mossé. Adaptive scheduling server for power-aware real-time tasks. *ACM Transactions on Embedded Computing Systems*, 2:284–306, 2004.

26. V. Nélis and J. Goossens. Mora: An energy-aware slack reclamation scheme for scheduling sporadic real-time tasks upon multiprocessor platforms. In *International Conference on Real-Time Computing Systems and Applications*, pp. 210–215, 2009.

27. Q.H. Nguyen, Y.-S. Ong, M.-H. Lim, and N. Krasnogor. Adaptive cellular memetic algorithms. *Evolutionary Computation*, 17(2):231–256, 2009.

28. J.E. Pecero, D. Trystram, and A.Y. Zomaya. A new genetic algorithm for scheduling for large communication delays. In *Euro-Par*, Delft, the Netherlands, pp. 241–252, 2009.

29. N.B. Rizvandi, J. Taheri, A.Y. Zomaya, and Y.C. Lee. Linear combinations of DVFS-enabled processor frequencies to modify the energy-aware scheduling algorithms. In *10th IEEE/ACM International Conference on Cluster, Cloud and Grid Computing*, IEEE, Melbourne, Victoria, Australia, pp. 388–397, 2010.

30. V. Sarkar. *Partitioning and Scheduling Parallel Programs for Execution on Multiprocessors*. MIT Press, Cambridge, MA, 1989.

31. D.J. Sheskin. *Handbook of Parametric and Nonparametric Statistical Procedures*. CRC Press, Boca Raton, FL, 2003.

32. O. Sinnen. *Task Scheduling for Parallel Systems*. Wiley-Interscience, Hoboken, NJ, 2007.

33. R. Subrata, A.Y. Zomaya, and B. Landfeldt. Cooperative power-aware scheduling in grid computing environments. *Journal of Parallel and Distributed Computing*, 70:84–91, 2010.

34. E.-G Talbi. *Metaheuristics: From Design to Implementation*. John Wiley & Sons, Hoboken, NJ, 2009.

35. E. Talpes and D. Marculescu. Toward a multiple clock/voltage island design style for power-aware processors. *IEEE Transactions on Very Large Scale Integration Systems*, 13:591–603, 2005.

36. J.D. Ullman. NP-complete scheduling problems. *Journal of Computer and System Sciences*, 10(3):384–393, 1975.

37. D. Whitley. Cellular genetic algorithms. In S. Forrest, ed., *Proceedings of the Fifth International Conference on Genetic Algorithms (ICGA)*, Morgan Kaufmann, Urbana-Champaign, IL, p. 658, 1993.

38. T. Xie, X. Qin, and M. Nijim. Solving energy-latency dilemma: Task allocation for parallel applications in heterogeneous embedded systems. In *Proceedings of the International Conference on Parallel Processing (ICPP '06)*, Washington, DC, pp. 12–22, 2006.

39. J.H. Zar. *Biostatistical Analysis*. Prentice Hall, Upper Saddle River, NJ, 1999.

40. A.Y. Zomaya and G. Chan. Efficient clustering for parallel tasks execution in distributed systems. *International Journal of Foundations of Computer Science*, 16(2):281–299, 2005.

V

Real-Time Systems

21

Online Job Scheduling Algorithms under Energy Constraints

Fei Li
George Mason University

21.1 Introduction

Along with the rapid advance of chip design technologies, the energy consumption and heat dissipation are also increasing at a similar pace [39]. Such a trend raises a few critical engineering problems. For example, an inefficient energy management for an embedded system may keep the processor run at its highest *clock frequency* (full *speed*), and thus deplete the battery power in a short time. Executing jobs at a high speed also results in more heat dispersion per unit time by the processor. If the temperature rises beyond some threshold, the processor's circuit timing will be affected which may result in an unpredictable behavior and cause the processor to function erratically or even halt. It has been widely accepted that energy consumption has become the bottleneck for improving a system's performance and has been one of the most significant factors to optimize.

For embedded systems and portable devices that mainly rely on batteries for energy, the benefit of completing a job is usually diluted by its energy expense. How to develop effective algorithms to economically manage energy expenditure and prolong battery life is a challenging problem. This research topic has attracted much attention recently in the algorithm community; see surveys [3,29,31] for a general introduction. In this chapter, we mainly talk about online algorithms at the operating system level to schedule jobs under energy constraints. The objective of such an energy-aware scheduling algorithm is to maximize some QoS measures by efficiently consuming limited available energy.

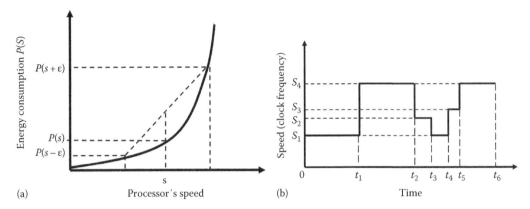

FIGURE 21.1   (a) Energy consumption is a convex function of the processor's speed. (b) A speed scheduler describes the time intervals and their corresponding processor's speeds.

21.1.1 Hardware Support for Energy Saving

In general, there are two hardware techniques supporting algorithms to reduce energy consumption: *dynamic voltage scaling* (DVS) and *dynamic power management* (DPM).

- DVS (also called *speed-scaling technique*) is a technique to dynamically set the processor's voltage and *clock frequency* (also called *speeds* or *frequency,* for short) to meet service performance requirements. We note that in scaling speeds, both the frequency and supply voltage of a processor are adjusted. Frequency/speed change always involves the corresponding voltage change [42]. Energy consumption is reduced due to the fact that completing a job at a lower speed uses less energy than at a higher speed. DVS has been implemented in modern processors—such examples include the SpeedStep technology of Intel processors [28] and the PowerNow! technology of AMD processors [2].

  Consider a processor in the DVS setting. The processor has variable clock frequencies. Under speed s, the processor consumes energy $P(s)$ per unit time and it takes time p_j/s to complete a job J_j with processing time p_j. In general, the function $P(\cdot)$ is convex (see Figure 21.1a): $P(\cdot)$ satisfies $P(0) = 0$ and $P(s + \epsilon) + P(s - \epsilon) \geq 2P(s), \forall\, 0 \leq \epsilon \leq s$.

  The total energy consumed by a schedule is the integral of energy consumption over time. For example, consider scheduling one job J_j with processing time $p_j = 10$. Assume $P(s) = s^3$. If we execute the job at speed $s = 1$, it takes time $t = 10$ to finish the job and the energy consumption is $P(s) \cdot t = 1^3 \cdot 10 = 10$. If we run the job at one-half speed $s' = 1/2$, it will take time $t' = 10/(2^{-1}) = 20$ to complete the job and the energy consumption is reduced to $P(s') \cdot t' = (1/2)^3 \cdot 20 = 2.5$.

  An algorithm in the DVS setting can be viewed as a *speed scheduler.* It is a piecewise continuous curve, specifying at which speed the processor runs jobs during which time interval. Figure 21.1b illustrates an example of a schedule. The scheduler employs 4 distinct speeds s_1, s_2, s_3, s_4 in 6 time intervals.

- DPM (also called *power-down strategy*) is an architecture-level design based on a function component called *clock gating* or *power gating.* The system has a *highest-power* active state and one or more *lower-power* sleep or standby states. Jobs can be processed only when the system is at its active state but not at any lower-power state. DPM is commonly used to cut a system's energy cost during the idle times, via eliminating or reducing energy consumption of one or more of its components by setting the components at the lower-power state (for instance, setting the C-states for Intel processors [28]). A constant amount of energy, called *transition energy,* is usually associated with transitions between different states. This additional energy cost makes that suspending the system is only beneficial when the idle time is long enough.

An algorithm in the DPM setting is to specify at which time the system is set at which state.

Both DVS and DPM have been employed either individually or jointly for reducing system energy consumption [7,30,31,34].

21.1.2 Online Computing and Competitive Analysis

Scheduling jobs arriving over time is essentially an online decision-making process. One common way of solving it is to design *online algorithms*. For online algorithms, the input is revealed one by one over time and the complete information is not known beforehand. A job's characteristics such as processing time, deadline, and value are known to the online algorithms only at the time when the job arrives. Once the decision is made, the cost and effect cannot be redone.

In this survey, we focus on online algorithmic solutions to job scheduling problems. In order to evaluate the worst-case performance of an online algorithm lacking knowledge of the future, we compare it against an *optimal offline algorithm* which is also called *adversary*. The adversary is assumed to be clairvoyant, that is, all input information is known a priori. We use *competitiveness* as the worst-case metric.

Definition 21.1 (Competitive Analysis.) *[14] Given a maximization (minimization) problem, an online algorithm is called c*-competitive *if its objective on* any *instance is at least* $1/c$ *(at most c) times of the objective of an optimal off-line algorithm on this finite instance.*

Using a maximization problem as an example, the *upper bound* of competitive ratio is one achieved by some algorithms. A competitive ratio smaller than the *lower bound* cannot be reached by any online algorithm. Competitive analysis has been one of the standard measures for analyzing online algorithms [14,35].

We underline that competitive analysis is certainly not the only way to evaluate the performance of an online scheduling algorithm with uncertainty of future input instance. For example, if we have a reasonable approximation of the input probability distribution, *average-case analysis* can be undertaken either analytically [19,40,41] or experimentally. However, when such information is unavailable or unreliable, and/or when *analytical worst-case performance guarantees* are sought, competitive analysis is of fundamental importance. Note that in competitive analysis, the input can be intentionally generated in an adversarial manner to worsen the online algorithm's competitive ratio. Thus, sometimes, competitive ratio provides the worst-case guarantees but they are pessimistic.

If an online algorithm is offered more information (rather than an assumed probability distribution) about the complete input instance, better competitive ratio may be achieved. These online algorithms are called *semi-online* [35] and it has been shown that even with only limited information revealed in advance, semi-online algorithms perform much better [26]. Another algorithm design framework is called *resource augmentation* [32], and it has been used to evaluate an online algorithm's performance in practical situations. In this framework, an online algorithm is given *additional* resources to compensate for the lack of knowledge about the future. An online algorithm is called β-*speed c*-competitive if this online algorithm is *c*-competitive when it is equipped with a processor whose speed is β (>1) times of the adversary's speed. Both semi-online algorithms and resource augmentation online algorithms will be discussed for some models in this article.

21.2 Maximizing Weighted Throughput

Weighted throughput is one of the most important QoS measures for computer systems.

Consider the problem of scheduling a set of jobs in an online manner in the one-processor setting. Each job J_j has four parameters, a *release time* r_j, a *processing time* p_j, a *deadline* d_j, and a *value (reward)* v_j.

A job's *slack time* is defined as the time period between its release time and its deadline. Jobs are allowed to be preempted without cost and they can resume at later times. Let $k_j = v_j/p_j$ denote the job J_j's *value density*. If all jobs have the same value density, we call this a *uniform-value setting*, otherwise, we call this a *non-uniform value setting*. The objective is to maximize *throughput* $\sum_j U_j$ or *weighted throughput* $\sum_j v_j U_j$ under some specified energy constraints. Here $U_j = 1$ if J_j is completed by its deadline and $U_j = 0$, otherwise. We note that in these settings, due to the energy hard constraint, an algorithm may not be able to complete all the released jobs.

In the following, we consider the problem of maximizing (weighted) throughput under the energy constraint of either a limited energy budget E_{budget} or a bounded highest clock frequency S_{max}.

21.2.1 Bounded Energy Budget

At first, we consider the simplest DPM setting with only two states: active and sleep. For simplicity, let the energy cost transitioning from the sleep (respectively, active) state to the active (respectively, sleep) state be E_{tran} (respectively, 0).

Theorem 21.1 *In a two-state DPM setting with $E_{tran} \neq 0$, there exists an instance such that no online algorithm has a constant competitive ratio.*

Proof. To prove Theorem 21.1, we will construct an input instance step by step, based on the online algorithm's decision, such that no matter what the decision an online algorithm makes, it cannot gain a constant fraction of what an optimal off-line algorithm gains.

Consider the following instance. We assume the system is at its sleep state initially. The system consumes energy δ (respectively, 0) per unit time when it is active (respectively, sleep). Each job J_j is with unit length $p_j = 1$ and takes energy δ to be finished. Let the energy budget $E_{budget} = 2E_{tran} + \delta$, where E_{tran} is an integer multiplier of δ. This setting implies that after the online algorithm powers on the system from the sleep state to the active state to execute one job, if the system goes back to the sleep state again, the online algorithm has no sufficient remaining energy to power on the system and execute jobs.

Define an *atom job* as a job with processing time 1 and slack time 1. Assume the first atom job J_0 is released at time 0. If the online algorithm skips executing J_0, we release the second atom job J_1 at time 1. We keep releasing jobs in such a manner: The ith atom job is released at an integer time i. If the online algorithm skips executing $(E_{tran} + \delta)/\delta = 1 + E_{tran}/\delta$ atom jobs, we do not release any more jobs. The adversary powers on the system at time 0 and executes all of these atom jobs. The competitive ratio of throughput is arbitrarily large since the online algorithm has a throughput of 0.

If the online algorithm executes the first atom job J_t at time t before the adversary releases $1 + E_{tran}/\delta$ atom jobs, then the adversary will stop releasing new jobs at time $t + 1$ till a later time $T = t + 1 + E_{tran}/\delta$. If the online algorithm remains active till T, then it will deplete all its energy without executing any more jobs and its throughput is only 1 job. Otherwise, if the online algorithm powers the system off at some time $t' < T$ after it finishes J_t, then no sufficient energy remains for it to power the system on again. Starting from time $t'' = \max\{t', t + 1 + E_{tran}/\delta\}$, the adversary releases $1 + E_{tran}/\delta$ atom jobs consecutively, one in each unit time. The adversary powers on the system at t'' and executes all these jobs released at time t'' and onward in a back-to-back manner. The competitive ratio of throughput is $1 + E_{tran}/\delta$, which can be arbitrarily large if δ is set ϵ.   □

According to Theorem 21.1, in the remaining part of this section, we will talk about maximizing weighted throughput in the two-state DPM setting with $E_{tran} = 0$ under a bounded energy budget E_{budget}.

Consider the case when there is no energy constraint (i.e., $E_{budget} = +\infty$). For underloaded systems in which there exists an algorithm to schedule all the jobs successfully by their deadlines using the given energy, preemptive EDF (earliest-deadline-first) achieves the best throughput [23]. For overloaded systems for which it is possible that no algorithm is capable of scheduling all the jobs by their deadlines, the off-line version of maximizing throughput is a NP-hard problem. For the online version, Baruah

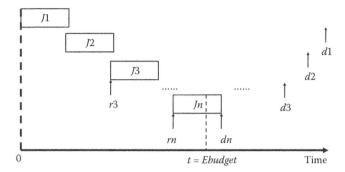

FIGURE 21.2   The worst-case instance for preemptive EDF.

et al. [13] proved that the lower bound of competitive ratio is 4 for uniform value setting and is $(1 + \sqrt{k})^2$ for nonuniform value setting, where k is the ratio of the largest value density and the smallest value density of the jobs in the input instance. Remarkably, Koren and Shasha provided an optimal online algorithm D^{over} which achieved these upper bounds [33].

Whereas the previous work looked at the well-studied online overloaded real-time scheduling settings [13,33] where there is sufficient energy but insufficient time, we study the impact of the energy constraint on the competitiveness of online real-time scheduling in settings where there is *sufficient time* but *insufficient energy* to complete all the workload. Note that the off-line version of maximizing the total system throughput can easily be shown to be *NP-hard* even for the simple case where all jobs have the same deadline, by slightly modifying the intractability proof given in [38]. For the online version, a natural idea to explore is to examine whether preemptive EDF remains its optimality.

Lemma 21.1   *[24,25] If the energy budget cannot satisfy scheduling all the jobs by their deadlines, then the preemptive EDF cannot guarantee a non-zero total value (hence, it cannot guarantee a non-zero competitive ratio).*

The proof of Lemma 21.1 depends on the following job release pattern: Later released jobs have earlier deadlines and they are released at the time just before the preemptive EDF finishes a job. Thus, the preempted EDF wastes its energy on unfinished jobs and at last all the budgeted energy is exhausted before any job is finished. See Figure 21.2 for a pictorial illustration.

Lemma 21.1 motivates us to design better competitive online algorithms than preemptive EDF for this model. Let us consider the uniform-value density setting first. In this setting, all jobs share the same value density. Without loss of generality, we set $v_j/p_j = 1$ and the system consumes energy $\delta = 1$ per unit time when it is active. Let $\mathcal{J}$ denote the set of jobs to be scheduled in the input instance. We define a constant p_{max}. p_{max} represents the largest job size in the *actual* input instance (i.e., $p_{max} = \max p_j$, $\forall j \in \mathcal{J}$). Hence, $p_{max} \leq E_{budget}$. The actual input sequence $\mathcal{J}$ is guaranteed to contain at least one job of size p_{max}. The following Theorem 21.2 presents the lower bound of competitive ratio for this problem.

Theorem 21.2   *[24,25] In the deadline-underloaded setting, no online algorithm can achieve a competitive ratio smaller than $E_{budget}/(E_{budget} - p_{max})$.*

The proof of Theorem 21.2 is slightly involved, mainly because the input instance created by the adversary must comply with the system model and settings. The idea of proving Theorem 21.2 is similar to the one used in the proof of Theorem 21.1. The adversary generates jobs with smaller values to deplete the online algorithm's energy such that the algorithm's remaining energy is not sufficient to execute a later released job which is associated with the largest value p_{max}.

Now, we design competitive online algorithms and provide upper bounds of competitive ratio for this model. The observation is as follows. Consider a time t. There are two kinds of jobs: Some jobs have been executed but have not been finished; the others have not been executed since they are released. For the jobs that have not been executed, it is naturally to apply the greedy approach to select a set of candidate jobs to execute using the remaining energy at time t, with the goal of maximizing throughput. The dilemma of scheduling jobs under a hard energy constraint lies at dealing with those jobs executed but unfinished. Should we invest more energy to complete a job that is being executed but has not been finished yet, or should we preempt this job and reserve energy for later released jobs with greater values? Lemma 21.1 tells us that we should complete jobs that have been started to avoid wasting energy on jobs that we will never finish. This intuition leads to the algorithm EC-EDF. EC-EDF effectively *commits* to *all* the admitted jobs, in the sense that it guarantees their timely completion without violating the system's energy budget limit. Further, all admitted jobs are scheduled according to the well-known preemptive EDF policy. Formally speaking, let $\mathcal{C}$ denote the set of committed jobs. Let E^r and p_j^r represent the remaining energy with the system and the remaining execution time of job J_j at time t, respectively. We assume that both E^r and p_j^r are properly updated at run-time.

- A job J_j with size p_j arriving at time t is admitted to the system *if and only if* (1) all jobs in $\{J_j\} \cup \mathcal{C}$ can be completed by their deadlines in the preemptive EDF manner, and (2) $E^r \geq p_j + \sum_{j \in \mathcal{C}} p_j^r$.
- All admitted jobs are scheduled according to the preemptive EDF policy.

It is obvious that EC-EDF guarantees the completion of *all* the admitted jobs before their deadlines, without violating the system's energy budget limit. Given that EC-EDF stops admitting new jobs for execution when it has cumulated a total energy of $E_{budget} - p_{max}$, we have

Theorem 21.3 *[24,25] EC-EDF has a competitive ratio of $E_{budget}/(E_{budget} - p_{max})$.*

Theorem 21.3 indicates that knowledge of the largest job size p_{max} in the actual input instance turns out to be very helpful. For the uniform-value density model where the value of a job is equal to its execution time, the largest-size job *in the input instance* is also the most valuable job in the set. Thus, if we know that a job with value $p_{max} \geq E_{budget}/2$ will definitely appear in the input instance, we shall wait for its arrival and execute it for a profit of p_{max} ($\geq E_{budget}/2$). If $p_{max} < E_{budget}/2$, we prefer to apply EC-EDF, collecting a total value of no less than $E_{budget} - p_{max}$. Thus, given the extra information p_{max}, we have a semi-online algorithm EC-EDF* with a competitive ratio of 2. A similar instance like the one in the proof of Theorem 21.2 shows that the lower bound of competitive ratio for semi-online algorithms knowing the largest-value job is 2. Thus, EC-EDF* is optimal.

The idea of EC-EDF can be extended to the *nonuniform value setting* in which jobs may have different value densities. The value of a job J_j with value density k_j is given by $v_j = k_j \cdot p_j$. Let k_{min} and k_{max} denote the smallest and largest value densities that can be associated with any job. Without loss of generality, we assume that $k_{min} = 1$, and thus the ratio of k_{max}/k_{min} is simply k_{max}. Different from the uniform-value density setting, the largest-size job is no longer guaranteed to be the most valuable job. A rather involved proof shows

Theorem 21.4 *[24,25] In the nonuniform value setting where $k_{max} > 1$, no online algorithm can achieve a competitive ratio smaller than $(k_{max})^{\frac{p_{max}}{E}}$.*

A similar proof technique like the one used for Theorem 21.3 shows that algorithm EC-EDF has a competitive ratio of $2k_{max}$. This leaves a big gap $[(k_{max})^{\frac{p_{max}}{E}}, 2k_{max}]$ between lower bound and upper bound of competitive ratio for this model.

21.2.2 Bounded Processor's Highest Speed

The problem of online maximizing throughput in the DVS setting subject to the total energy consumption bounded by a given amount has not been discussed in literature. The reason is obvious since an online algorithm cannot determine how much energy it needs to invest to complete a released job. For example, let us consider a job J_j with processing time p_j and deadline d_j at time t. An online algorithm cannot decide whether to execute J_j using some energy, say $e_j = P(p_j/(d_j - t))$, because it is possible that many jobs with much larger slack times are released later and the energy e_j can be used to complete those jobs with larger slack times to maximize the throughput. This observation leads to the fact that there is no online algorithm with a constant competitive ratio in maximizing throughput in the DVS setting when a total energy E_{budget} is enforced.

In the DVS setting, [8,16,17] studied the problem of *online maximizing throughput and minimizing energy* subject to the energy consumption per unit time bounded by some value. In this model, jobs arrive over time and they have processing time and deadlines. The DVS-capable processor's highest speed is no more than a predefined value S_{max}. [The energy consumed per unit time is then below $P(S_{max})$.] Note that if we ignore the competitiveness of energy usage, in order to maximize throughput subject to the jobs' deadline constraints, the processor should run at its full speed in executing jobs, and an online algorithm only needs to decide whether to run a released job or not. This model then becomes the one studied in [13]. The lower bound of competitive ratio 4 (in maximizing throughput for job overloaded systems) developed in [13] holds for the speed-bounded model in [8,16,17].

Recall that the energy consumption function $P(s(t))$ is a convex one of the speed $s(t)$ at time t. In our following discussion, we assume that $P(s(t)) = s(t)^\alpha$, where $\alpha \geq 2$ [15].

There are two hard constraints in designing online algorithms for the model [8,16,17]: *jobs' deadlines* and *processor's highest speed*.

If we relax the restriction of the processor's speed, all the released jobs can be finished by their deadlines, and thus the off-line version of this problem becomes the first theoretical min-energy job scheduling model studied by Yao et al. [42]. Due to the convexity of $P(s(t)) = s(t)^\alpha$, if the processor is to complete $s_1 + s_2$ units of jobs by working at speeds of s_1 and s_2 in two time units, the total energy consumption reaches its minimum at $s_1 = s_2$ for both time units. Thus, as long as the jobs can be scheduled by their deadlines, it is better to make the schedule as "flat" as possible to reduce the total energy consumption. The off-line optimal algorithm given by Yao et al. [42] is based on this insight and it naturally leads to an online min-energy algorithm called OA (optimal available) in [42]:

- At any time, upon each newly arriving job, OA recomputes the optimal schedule for all the jobs that have been released but have not been completed yet. Each job's remaining processing time is considered in OA's schedule and thus, the recomputation is regarded as a procedure for a set of "jobs" which have the same arrival time.

Bansal et al. [11] proved the following theorem on OA's competitiveness.

Theorem 21.5 *[11] Consider a DVS setting in which the processor can have arbitrary speed. The algorithm OA achieves a competitive ratio α^α in minimizing the total energy cost and this ratio is tight.*

If the processor's highest speed is bounded by S_{max}, an online algorithm may not finish all the jobs by their deadlines. Thus, an online algorithm needs to specify how to select jobs to execute (i.e., *job admission policy*) and at which speed to run the jobs (i.e., *speed scheduling policy*). The first idea might be targeting on the objective of maximizing throughput alone in a greedy manner subject to the processor's bounded speed. However, remember that not all the jobs are able to be completed successfully by their deadlines. We also need to specify which jobs to give up upon the system being overloaded. In this case, we admit a newly arriving job only if it has sufficiently large value, compared with those to-be-discarded ones. Chan et al. [16] proposed an online algorithm called FSA-OAT (full speed admission, optimal available with speed $\leq S_{max}$) based on this idea: Consider a newly arriving job J_j and all the admitted jobs in a set $\mathcal{C}$; its job admission policy (FSA) works as follows:

- If $\{J_j\} \cup \mathcal{C}$ can be scheduled by their deadlines when the processor runs at its highest speed S_{max} (we say $\{J_j\} \cup \mathcal{C}$ is *full-speed admissible*), then J_j is accepted and $\mathcal{C}$ is updated with $\{J_j\} \cup \mathcal{C}$.

- Otherwise (if some job in $\{J_j\} \cup \mathcal{C}$ cannot be scheduled successfully), then the job admission policy accepts J_j only if
    1. Its processing time p_j is at least two times larger than the first l earliest-deadline jobs in $\mathcal{C}$.
    2. J_j, along with $\mathcal{C}$ without these l jobs, is admissible.

  After accepting J_j, the l earlier-deadline jobs are discarded from $\mathcal{C}$.

In determining the speed of scheduling jobs in $\mathcal{C}$, FSA-OAT follows Yao et al. [42]'s online algorithm OA, running at speed $\min\{S_{\max}, s^{OA}(t)\}$, where $s^{OA}(t)$ is calculated by OA. Chan et al. proved

Theorem 21.6 *[16] FSA is 14-competitive on maximizing throughput, the total size of jobs completed by their deadlines.*

Theorem 21.7 *[16] FSA-OAT works according to the speed function $\min\{S_{\max}, s^{OA}(t)\}$, where $s^{OA}(t)$ is the speed of a simulated algorithm OA at time t. FSA-OAT is $(\alpha^{\alpha} + \alpha^2 4^{\alpha})$-competitive for energy against any off-line algorithm that maximizes the throughput.*

For the setting in which all the jobs may not have the same value density, the analysis of Theorems 21.6 and 21.7 can be adapted easily. We have

Theorem 21.8 *[17] FSA-OAT is $(12k_{\max}/k_{\min} + 2)$-competitive on weighted throughput and $(\alpha^{\alpha} + \alpha^2 (4k_{\max}/k_{\min})^{\alpha})$-competitive on energy, where $k_{\min} = \min\{v_j/p_j\}$ and $k_{\max} = \max\{v_j/p_j\}$.*

Recall that the lower bound of competitive ratio on throughput is 4 [13]. To achieve better competitive ratios, we consider the resource augmentation setting in which an online algorithm's speed is relaxed to $(1 + \epsilon)S_{\max}$ while the off-line algorithm keeps its original speed $S_{\max}$. Here, an online algorithm called FSA' works similar to FSA in admitting jobs:

- Upon each newly arriving job J_j, as long as all the admitted jobs can be completed successfully under the speed $(1 + \epsilon)S_{\max}$, FSA' admits J_j. Unlike FSA, FSA' does not expel any admitted jobs.

To analyze the competitiveness of throughput for FSA', we partition the job set into two sets: one set completed by both the optimal off-line algorithm and FSA', and one set completed only by the optimal off-line algorithm. Note that for each job J_i finished by the off-line algorithm but not by FSA', if any, the resources on throughput for FSA' from time r_i to time d_i is at least ϵ times of p_i. Based on this observation, Chan et al. proved

Theorem 21.9 *[17] FSA' is $(1+1/\epsilon)$-competitive on throughput with its maximum speed of $(1+1/\epsilon)S_{\max}$ (against any off-line algorithm with maximum speed $S_{\max}$).*

Remember that if there is no energy constraint, the well-known online algorithm D^{over} achieves an optimal competitive ratio of 4 in maximizing throughput [13]. Theorem 21.6 motivates us to consider the following question: Can we borrow the job admission policy from D^{over} to replace the greedy approach used by FSA-OAT for better competitiveness? The answer is positive. Combining the admission policy of D^{over} and the job scheduling policy of OA leads to a new algorithm called Slow-D [8]. The algorithm Slow-D is sensitive to the speed function of OA. All released jobs are stored in two queues named Q_{work} and Q_{wait}. Jobs are classified as t-urgent and t-slack, according to the relationship between the jobs' deadlines and the latest time when a simulated algorithm OA reaches a speed $\geq S_{\max}$.

- At time t, if a newly arriving job can be scheduled successfully along with t-urgent jobs in Q_{work} using speed $\leq S_{\max}$ or this job is t-slack, then this job is accepted in Q_{work}. Otherwise, this job is put in Q_{wait}. Note that a job J_j in Q_{wait} may be admitted at a later time, when J_j reaches its latest start time and its value is large enough to evict some jobs in Q_{work} to give room for J_j.
- At any time t, the algorithm processes the job in Q_{work} with the earliest deadline at speed $s(t) = \min\{s^{OA}(t), S_{\max}\}$.

Bansal et al. showed

Theorem 21.10 *[8] Slow-D is 4-competitive with respect to throughput and $(\alpha^\alpha + \alpha^2 4^\alpha)$-competitive with respect to energy consumption.*

Details of the proofs of Theorems 21.7 to 21.10 can be found in [8,16]. Notice that Slow-D has achieved the optimal competitive ratio 4 on throughput. The lower bound of competitive ratio for speed scaling algorithms in minimizing energy consumption is $e^{\alpha-1}/\alpha$ [10]. It is an open problem to shrink or close the gap $[e^{\alpha-1}/\alpha, \alpha^\alpha + \alpha^2 4^\alpha]$ of competitive ratio with respect to energy consumption while keeping 4-competitive with respect to throughput.

Recently, Han et al. [27] considered both DVS and DPM in one setting and extended the idea of Slow-D in admitting jobs in such a setting. In their model, the processor is DVS-equipped and it can enter the sleep state with a wake-up (from the sleep state to the active state) cost of E_{tran}. At the active state, the processor consumer energy $s^\alpha + \delta$ per unit time. Based on the idea and analysis of Slow-D in [8], Han et al. got an algorithm named Slow-D(SOA) and proved

Theorem 21.11 *[27] In the bounded speed model, Slow-D(SOA) is $(\alpha^\alpha + \alpha^2 4^\alpha + 2)$-competitive for energy and 4-competitive for throughput.*

21.3 Minimizing Weighted Flow Time and Energy

In many computational tasks, jobs may not have deadlines as their time constraints. In this case, a scheduling algorithm's performance is measured by how fast the system/processor responses to the requests (i.e., how long a job can be finished after it is requested).

Consider a job J_j. Its flow time f_j is defined as the difference between its completion time c_j and its release time $r_j, f_j = c_j - r_j$. For each job J_j, it is associated with a weight (value) v_j. The sum of weighted flow time is defined as $\sum v_j f_j$. In scheduling jobs without time constraints, flow time is one of the most significant QoS measures in computer science literature [35]. Note that minimizing flow time and minimizing energy consumption are two orthogonal objectives in the DVS setting: If we want to save energy, we have to employ a slower frequency, and thus a job's flow time is elongated. If we want to get a job done sooner, we have to run the processor at a higher speed and pay more energy.

In this section, we consider the problems of optimizing flow time and energy consumption in scheduling jobs either under a budget energy E_{budget} or in a setting where the processor's highest speed is bounded by S_{max}.

21.3.1 Bounded Energy Budget

Let us consider minimizing total flow time $\sum f_j$ subject to the total energy consumption bounded by an energy budget E_{budget}. If jobs have different release time, the problem of minimizing average flow time subject to the bounded energy consumption gets much more difficult, even for the off-line version (see [37]). Note that with a limited energy budget, an online algorithm may not complete all the jobs released. Without knowing future released jobs, it is hard for a policy to determine how much energy to invest for each newly released job. Bansal et al. [12] proved

Lemma 21.2 *[12] Assume that there is some fixed energy-bound E_{budget} and the objective is to minimize total flow time. Then there is no $O(1)$-competitive online speed scaling algorithm, even for jobs with unit size and unit weight.*

To prove Lemma 21.2, the authors constructed an input instance in which time is divided into *batches*. All the jobs in each batch arrive at some time after the online algorithm has processed all the jobs released earlier. They then showed that in each batch, tuning the number of jobs makes the energy used by the online algorithm to be within a constant factor of the total energy E_{budget}. If an online algorithm does not know how many batches of jobs are to be released, a nonconstant number of batches of jobs in the input instance ≈ makes the online algorithm not $O(1)$-competitive.

Based on the observation of Lemma 21.2, Albers and Fujiwara [4] studied an online version of the model whose objective is to minimize the *sum of the total energy cost E and w times of the total flow time*, where w is a constant. From an economic viewpoint, the motivation of this model is to assume that users want to reduce one unit of flow time at the cost of extra w units of energy. With some reasonable scaling, w can be assumed 1. Note that the algorithms developed for this model in [4] can be adapted to the setting of minimizing the total flow time subject to a fixed energy budget; however, the competitive ratios are not maintained any more.

In minimizing the sum of energy and total flow time, if preemption of jobs is not allowed, we have a negative result:

Lemma 21.3 *[4] In non-preemptive setting, the competitive ratio of any deterministic online algorithm is at least $n^{1-\frac{1}{\alpha}}$-competitive, if the job processing time can take arbitrary values.*

The proof of Lemma 21.3 is based on the following insight: If an online algorithm starts to execute a job which has a larger processing time, then the adversary releases many jobs with smaller processing time. If the online algorithm runs at a high speed to execute the larger job first, then its energy cost is larger, compared with the adversary which processes the smaller jobs first and then the larger one. If the online algorithm runs at a low speed to execute the larger job, then it introduces more flow time for the smaller jobs released later. Choosing the jobs' processing time and the online algorithm's speed carefully, the adversary can beat the online algorithm with a competitive ratio of at least $n^{1-\frac{1}{\alpha}}$.

Albers and Fujiwara [4] studied the online version of scheduling unit-size jobs in minimizing the sum of energy cost and total flow time. Consider Lemma 21.3 and the assumption of unit-size jobs. Intuitively, in each time unit, we prefer to have the incurred energy equal to the additional flow time accumulated from all the unfinished jobs in this time unit. Based on this idea, Albers and Fujiwara proposed to let the processor run at an energy equal to the number of jobs that have been released but not finished yet. The presented algorithm in [4] slightly modifies this idea and processes jobs in *phases* and *batches* manner:

- For the first group of n_1 jobs released at time 0, they are processed at speed $\sqrt[\alpha]{n_1/c}$, $\sqrt[\alpha]{(n_1-1)/c}, \ldots, \sqrt[\alpha]{1/c}$, respectively, where c is a constant depending on α. For any jobs released in the first phase (after time 0 but before the completion of the last job in the first batch), they will be executed in the second phase and the processor's speed are determined likewise.
- For each batch, the algorithm performs its speed scaling similar to what it does for the first group.

Theorem 21.12 *[4] In scheduling unit-size jobs, no preemption is necessary. There is an algorithm achieving a competitive ratio of $(1+\phi)\left(1+\phi^{\frac{\alpha}{2\alpha-1}}\right)\frac{\alpha^\alpha}{(\alpha-1)^{\alpha-1}} \min\left\{\frac{5\alpha-2}{2\alpha-1}, \frac{4}{2\alpha-1}+\frac{4}{\alpha-1}\right\}$, where $\phi = \frac{1+\sqrt{5}}{2} \approx 1.618$. This competitive ratio is bounded by $8.22e(1+\phi)^\alpha$.*

Bansal et al. [9] considered the preemption setting. They modified the algorithm in [4] slightly, choosing a speed of $\sqrt[\alpha]{l+1}$ whenever there are l *active* jobs (an active job is one that has been released but has not been completed yet). The SRPT (shortest remaining processing time) policy is used to schedule these active jobs in each phase. Both algorithms in the following theorems permit preemption over jobs.

Theorem 21.13 *[9] Consider the scheduling algorithm that uses SRPT for job selection and energy equal to one more than the number of unfinished jobs for speed scaling. In the general model, this scheduling algorithm is $(3+\epsilon)$-competitive for the objective of total flow plus energy on arbitrary-size unit-weight jobs.*

Theorem 21.14 *[9] Consider the scheduling algorithm that uses highest density first (HDF) for job selection and energy equal to the fractional weight of the unfinished jobs for speed scaling. In the general model, this* *scheduling algorithm is* $(2 + \epsilon)$*-competitive for the objective of fractional weighted flow plus energy on arbitrary-work arbitrary-weight jobs.*

What is interesting is that Andrew et al. [6] found that the algorithm running SRPT at an energy consumption equal to the *queue length* (the number of active jobs) is 2-competitive for a very wide class of energy-speed trade-off functions. Further, [6] shows that no online algorithm can attain a competitive ratio less than 2. This result closes the gap of upper bound and lower bound of competitive ratio for the problem of scheduling jobs in the DVS setting for the objective of minimizing a linear combination of energy consumption and flow time.

21.3.2 Bounded Processor's Highest Speed

The first online algorithm dealing with the sum of energy and weighted flow time was proposed in [12]. In the model studied in [12], the processor's speed is assumed infinite. In [8], a practical case is considered in which the processor's speed is bounded by S_{max} while the objective is to minimize the energy plus the total weighted flow time.

Define a job's *fractional weight* as the ratio of its remaining work to the original work. One important insight is: The larger a job's fractional work is, the more weighted flow time is reduced if the HDF (highest density first) policy is picked to schedule jobs. Bansal et al. [8] provided the following algorithm based on this observation:

- When the processor has a ceiling of running speed S_{max}, the algorithm runs the processor at the speed of $\min\{S_{max}, w_a(t)^{\alpha^{-1}}\}$, where $w_a(t)$ is the fractional weight of a job. Jobs are scheduled by HDF.

Theorem 21.15 *[8] For unit-size jobs, the algorithm runs the processor at speed* $\min\{S_{max}, w_a(t)^{\alpha^{-1}}\}$, *where* $w_a(t)$ *is the fractional weight of a job, and runs jobs using HDF is 4-competitive.*

Resource augmentation approach also helps in this setting and Bansal et al. showed

Theorem 21.16 *[8] For arbitrary-size jobs, the algorithm runs the processor at speed* $\min\{S_{max}, w_a(t)^{\alpha^{-1}}\}$, *where* $w_a(t)$ *is the fractional weight of a job, and runs jobs using HDF is* $(\max\{1 + 1/\epsilon, (1 + \epsilon)^{\alpha}\}(2 + o(1))\alpha/\ln \alpha)$*-competitive when the algorithm is equipped with a processor with speed* $(1 + \epsilon)S_{max}$.

For the problem of minimizing the total weighted flow and total energy consumption under the constraint that the processor's speed is bounded by S_{max}, there is a big gap of upper bounds and lower bounds of competitive ratios that we need to shrink or close.

21.4 Maximizing Net Profit

A newly emerging class of problems of efficiently managing limited energy comes from the monetary cost of energy point of view. We note that the benefit of completing a job is usually diluted by its energy expense, and thus, the *efficiency* of spending energy is very important. Consider the following motivating example. Assume all jobs will have the same reward upon being finished. An algorithm spending amount E of energy in finishing n jobs may not be considered more "efficient" than an algorithm spending amount $E/4$ of energy in finishing $3n/4$ jobs. Thus, we need to design schedules whose objective is more realistic: Maximize the *net profit*, defined as the difference between the aggregated income due to scheduling jobs and the energy cost that is associated with this schedule. As a result, some jobs may not even be considered for scheduling in this setting, due to the incorporation of energy term in the objective.

21.4.1 In the DPM Setting

Zhang and Li [43] studied energy management algorithms for scheduling jobs in a two-state DPM system. In this model, each job J_j has a release time r_j, a processing time p_j, a reward (value) v_j, and a deadline d_j. There is a wake-up energy cost $E_{tran} \neq 0$ when the system goes from the sleep state to the active state. Without loss of generality, the transition energy from the active state to the sleep state is assumed 0. The objective is to maximize net profit, which is defined as the difference between the total reward achieved by completing jobs by their deadlines and the total energy consumption accrued during this course.

For jobs sharing the same release time or the same deadline, they can be finished in a back-to-back manner. Thus, powering on the system once is sufficient. For this specific case, the schedule that maximizes the net profit is the same as the one that maximizes the total reward. If jobs have different release time or deadlines, the problem gets more difficult. In the first attempt, Zhang and Li studied the off-line version of this model and considered scheduling jobs with *agreeable deadlines*. Jobs are with agreeable deadlines if for any two jobs i and J_j, if $r_i < r_j$, then we have $d_i < d_j$.

Lemma 21.4 *[43] In an optimal schedule for maximizing net profit for agreeable-deadline instances, the selected jobs can be scheduled in the order of their release time and they are finished by their deadlines. Also, no job is preempted.*

Consider a case in which the transition energy cost E_{tran} equals the energy cost per unit time when the system is active. For this case, the machine can be regarded as never *spinning* (being active but not processing any jobs). At any time when the processor finishes a job and there are no pending jobs, the machine will shut down and will restart only when requested to process the next job in the given schedule. We can treat an optimal schedule which maximizes the net profit as multiple, say m, nonoverlapping "batches of jobs" B_1, B_2, ..., B_m. Each batch B_i consists of a few jobs consecutively executed in a back-to-back manner without any "gap" in between. All jobs in a given batch might be shifted as a whole part, but in the same relative order of execution in the optimal schedule, either to the left or to the right with respect to the constraints from the jobs' release time and deadlines. If a batch cannot move to the left or to the right further without generating a "gap," we define such a batch *fixed at its position*. Applying the algorithmic techniques of dynamic programming and divide-and-conquer, we have

Theorem 21.17 *[43] For the variant in which all jobs have agreeable deadline and $E_{tran} = \delta$, there is an algorithm with a running time of $O(n^2 \log n)$ in maximizing net profit, where n is the number of jobs and δ is the energy consumption per time unit when the system is at its active state.*

We note that without energy restriction, there are many papers on maximizing net profit subject to the constraint of jobs' deadlines. Taking energy into account, there exists many open problems in maximizing the net profit in the DPM setting for both off-line and online versions. In maximizing net profit in the online setting, it is possible that any online algorithm may achieve only a negative net profit while the off-line algorithm achieves a positive net profit, even when all jobs have the same value and same processing time. Semi-online algorithms or resource augmentation algorithms are expected to maximize net profit. For semi-online algorithms, the input instance's *loading factor*,* which is an important parameter in real-time system literature, has not been taken into account in online algorithm design and analysis. More work is expected in this area.

21.4.2 In the DVS Setting

In [18], the off-line version of maximizing net profit is considered in the DVS setting. The profit (benefit) is defined as the total reward $\sum v_j$ by completing jobs by their deadlines d_j [38]. There is one processor with a finite number of speeds. The energy budget is E_{budget}. The goal is to select a subset of jobs from

---

* The loading factor is defined as the ratio between the total amount of workload of jobs in an interval over this interval's length.

the input instance such that these jobs can be finished by their deadlines and the net profit is optimized. In [18], all the jobs are assumed to share the same deadline D. This assumption conveniently leads to an optimal algorithm to finish all the jobs with the minimum energy consumption $\sum p_j/D$, due to the convexity of the energy consumption function. With the total energy restriction E_{budget}, a greed-based approximation algorithm can be derived by simply removing the job(s) with the minimum value density, which is defined as v_j/p_j, till the remaining jobs can be finished successfully under the budget E_{budget}. This greedy algorithm is 2-approximation. Note that this problem is similar to packing items in a bag with a fixed capacity. A PTAS (polynomial-time approximation scheme) exists, using a dynamic programming approach over the geometrically rounded job values. However, in [18], no online solutions were given for such models.

The first online algorithm studying maximizing net profit in the DVS setting was provided in [36]. Instead of having a fixed value associated with each job, Pruhs and Stein defined a job's *income as a function of its completion time incurred*. This income is a nonnegative, nonincreasing function of the time when a job is finished. The authors in [36] considered the multiple identical processor case. A scheduler determines which job to run on which processor and at which speed to run the processors. Obviously, some jobs may not be executed in order to maximize income.

It is easy to see that the competitive ratio of online algorithms cannot be bounded by any function of the number of jobs. The underlying idea, similar to the one presented in the proof of Theorem 21.1, is to construct an input instance such that an online algorithm is forced to spend some energy and run a job. However, a job with a higher income (using a different income functions) will be released later and the off-line optimal algorithm will reserve its energy to run this later released job for a better income. We have

Theorem 21.18 *[36] The competitive ratio of any deterministic algorithm cannot be bounded by any function of n, even if jobs have unit work.*

Pruhs and Stein [36] considered resource augmentation approaches and they showed

Theorem 21.19 *[36] In maximizing income using multiple identical speed scaling processors, there is a $(1 + \epsilon)$-speed $O(1/\epsilon^3)$-competitive online algorithm.*

There are many open problems with the objective of maximizing net profit. As stated in [36], "it would be interesting to investigate other problems within this general framework of maximizing profit when you have to buy your energy."

21.5 Discussion

Despite plenty of excellent work on scheduling jobs to reduce energy and maximize some QoS measures has been done, there are still many interesting open problems that need to be investigated further. For example, how to manage temperature (peak temperature or average temperature) to minimize thermal dissipation is of high importance for scheduling jobs. Also, emerging multi-chips raises many combinatorial problems for energy management; see examples in [5]. At the same time, there are other energy management–related problems that have not been formulated successfully. In addition to the open problems that have been mentioned in the previous sections, we address two issues that have not received broad attention in the online algorithm community.

21.5.1 Interplay of DPM and DVS

It is not surprising to see that some existing algorithmic solutions of optimizing different objectives conflict with each other. The interference of various energy-aware techniques for different system components dilutes the overall performance improvement in saving energy. For example, DVS and DPM adversely

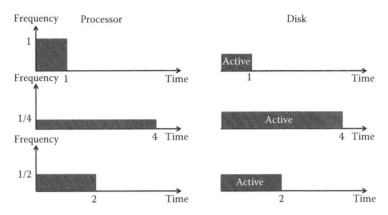

FIGURE 21.3   An example of system-level energy-aware computation.

affect each other at the system level, though both being effective energy management techniques. DPM is usually used for system devices while DPM is usually used for processors. For DVS, reducing the processor's speed to minimize energy consumption may keep the devices active longer, which negatively affects DPM's benefit in reducing energy cost.

Refer to Figure 21.3 as an example. Suppose that there is a job which requires 1 unit time to finish when the processor runs at its full speed. Let the processor and the disk consume energy $s(t)^3$ and 0.25 per unit time, respectively. When the job is running, the disk must be at its active state and the disk has no DVS functionality. If we run the processor at its full speed, it takes 1 unit time to finish the job. The disk is active during this period and the total energy consumption is thus $1^3 + 0.25 = 1.25$. In order to minimize its energy consumption, the processor reduces the clock frequency to a quarter of the full speed. In this situation, the job needs 4 time units to be completed. The total energy consumption for both processor and disk is $(1/4)^3 \cdot 4 + 0.25 \cdot 4 = 1.0625$. However, neither of these strategies achieves the optimal energy cost. A better solution is to run the processor at its half speed with a total energy cost of $(1/2)^3 \cdot 2 + 0.25 \cdot 2 = 0.75$. This simple example shows the need for balancing the energy cost of various system components in a global setting.

Effectively integrating DVS and DPM in a single framework poses several nontrivial challenges. A realistic formulation on the interplay of DPM and DVS has not been proposed in the algorithm research community.

21.5.2 Interplay of Energy-Aware Algorithms and Chip's Life-Time Reliability

Existing studies consider energy minimization through speed scaling without much attention to the impact of frequency changes. Such changes typically involve time and energy overhead. Moreover, recent studies indicate that the *lifetime reliability* of a CMOS circuit is directly related to the number and span of frequency changes. For example, in [20,22], it is reported that *(hardware) failures, such as cracks and fatigue failures, are created not by sustained high temperatures, but rather by the repeated heating and cooling of sections of the processor. This phenomenon is referred to as "thermal cycling."* Thermal cycling is caused by the large difference in thermal expansion coefficients of metallic and dielectric materials, and leads to cracks and other permanent failures. Using MTTF (mean-time-to-failure) to describe the expected processor's life, the following Coffin–Mansion formula [1] is used to characterize a processor's lifetime reliability.

$$\text{MTTF} \propto \frac{1}{C_o(\Delta T - \Delta T_o)^{-q}m}, \tag{21.1}$$

where C_o is a material-dependent constant, ΔT is the entire temperature cycle range of the device, ΔT_o is the portion of the temperature range in the elastic region, q is the Coffin–Mansion exponent, and m is the frequency (number of occurrences per unit time) of thermal cycles [1]. Formula 21.1 clearly indicates that an energy-aware speed scheduler which constantly and frequently changes the processor's speeds results in large m and ΔT. Thus, such a schedule may introduce a large temperature cycle range, and therefore hurts the processor's reliability badly. Simulations in [21,22] have confirmed that various speed-scaling energy-aware policies have different impacts on processor's reliability in terms of MTTF.

For now, no algorithms have been proposed with rigorous mathematical analysis to study the impacts of energy-aware job scheduling algorithms on a chip's lifetime reliability. The challenges are summarized as follows. (1) Minimizing energy consumption and maximizing a chip's lifetime reliability are two orthogonal objectives. If we want to save energy, we have to employ speed-scaling policies which are sensitive to processes' demands, and thus jeopardize a chip's lifetime reliability. If we want to maximize a chip's reliability, we prefer to run the processor at a constant speed, which consumes more energy. (2) Modeling the impact of energy-aware scheduling algorithms in a combinatorial optimization way is very difficult. The study of energy-aware scheduling algorithms and their impacts on chip's reliability is a critical and interesting problem, and calls for efficient solutions.

References

1. Failure mechanisms and models for semiconductor devices, JEDEC publication JEP122C. http://www.jedec.org (accessed January 4, 2011).
2. ACP. The truth about power consumption starts here. http://www.amd.com/us/Documents/43761C_ACP_WP_EE.pdf, 2009 (accessed January 4, 2011).
3. S. Albers. Energy-efficient algorithms. *Communications of the ACM (CACM)*, 53(5):86–96, 2010.
4. S. Albers and H. Fujiwara. Energy-efficient algorithms for flow time minimization. *ACM Transactions on Algorithms (TALG)*, 3(4):Article No. 49, 2007.
5. S. Albers, F. Muller, and S. Schmelzer. Speed scaling on parallel processors. In *Proceedings of the 19th Annual ACM Symposium on Parallel Algorithms and Architectures (SPAA)*, San Diego, CA, pp. 289–298, 2007.
6. L. Andrew, A. Wierman, and A. Tang. Optimal speed scaling under arbitrary power functions. *ACM SIGMETRICS Performance Evaluation Review*, 37(2):39–41, 2009.
7. J. Augustine, S. Irani, and C. Swamy. Optimal power-down strategies. *SIAM Journal on Computing (SICOMP)*, 27(5):1499–1516, 2008.
8. N. Bansal, H.-L. Chan, T.-W. Lam, and L.-K. Lee. Scheduling for speed bounded processors. In *Proceedings of the 35th International Colloquium on Automata, Languages and Programming (ICALP)*, Reykjavik, Iceland, pp. 409–420, 2008.
9. N. Bansal, H.-L. Chan, and K. Pruhs. Speed scaling with an arbitrary power function. In *Proceedings of the 20th Annual ACM-SIAM Symposium on Discrete Algorithms (SODA)*, New York, pp. 693–701, 2009.
10. N. Bansal, H.-L. Chan, K. Pruhs, and D. Katz. Improved bounds for speed scaling in devices obeying the cube-root rule. In *Proceedings of the 36th International Colloquium on Automata, Languages and Programming (ICALP)*, Rhodes, Greece, pp. 144–155, 2009.
11. N. Bansal, T. Kimbrel, and K. Pruhs. Speed scaling to manage energy and temperature. *Journal of the ACM (JACM)*, 54(1):Article No. 3, 2007.
12. N. Bansal, K. Pruhs, and C. Stein. Speed scaling for weighted flow time. In *Proceedings of the 18th Annual ACM-SIAM Symposium on Discrete Algorithms (SODA)*, New Orleans, LA, pp. 805–813, 2007.
13. S. Baruah, G. Koren, D. Mao, B. Mishra, A. Raghunathan, L. Rosier, D. Shasha, and F. Wang. On the competitiveness of on-line real-time task scheduling. *Real-Time Systems*, 4(2):125–144, 1992.

14. A. Borodin and R. El-Yaniv. *Online Computation and Competitive Analysis*. Cambridge University Press, Cambridge, U.K., 1998.

15. D. M. Brooks, P. Bose, S. E. Schuster, H. Jacobson, P. N. Kudva, A. Buyuktosunoglu, J.-D. Wellman, V. Zyuban, M. Gupta, and P. W. Cook. Power-aware microarchitecture: Design and modeling challenges for next-generation microprocessors. *IEEE Micro*, 20(6):26–44, 2000.

16. H.-L. Chan, W.-T. Chan, T.-W. Lam, L.-K. Lee, K.-S. Mak, and P. W. H. Wong. Energy efficient online deadline scheduling. In *Proceedings of the 18th Annual ACM-SIAM Symposium on Discrete Algorithms (SODA)*, New Orleans, LA, pp. 795–804, 2007.

17. H.-L. Chan, W.-T. Chan, T.-W. Lam, L.-K. Lee, K.-S. Mak, and P. W. H. Wong. Optimizing throughput and energy in online deadline scheduling. *ACM Transactions on Algorithms*, 6(1):Article 10, 2009.

18. J.-J. Chen, T.-W. Kuo, and C.-L. Yang. Profit-driven uniprocessor scheduling with energy and timing constraints. In *Proceedings of the 2004 ACM Symposium on Applied Computing (SAC)*, Nicosia, Cyprus, pp. 834–840, 2004.

19. T. H. Cormen, C. E. Leiserson, R. L. Rivest, and C. Stein. *Introduction to Algorithms*. MIT Press, Cambridge, MA, 3rd edn., 2009.

20. A. K. Coskun, T. S. Rosing, and K. Gross. Utilizing predictors for efficient thermal management in multiprocessor SoCs. *IEEE Transactions on Computer Aided Design (IEEECAD)*, 28(10):1503–1516, 2009.

21. A. K. Coskun, T. S. Rosing, Y. Leblebici, and G. D. Micheli. A simulation methodology for reliability analysis in multi-core socs. In *Proceedings of the 16th ACM Great Lakes Symposium on VLSI*, Philadelphia, PA, pp. 95–99, 2006.

22. A. K. Coskun, R. Strong, D. M. Tullsen, and T. S. Rosing. Evaluating the impact of job scheduling and power management on processor lifetime for chip multiprocessors. In *Proceedings of the 11th ACM International Joint Conference on Measurement and Modeling of Computer Systems (SIGMETRICS/Performance)*, Seattle, WA, pp. 169–180, 2009.

23. M. Dertouzos. Control robotics: The procedural control of physical processes. In *Proceedings of the IFIP Congress*, Stockholm, Sweden, pp. 807–813, 1974.

24. V. Devadas, F. Li, and H. Aydin. Competitive analysis of energy-constrained real-time scheduling. In *Proceedings of the 21st Euromicro Conference on Real-Time Systems (ECRTS)*, Dublin, Ireland, pp. 217–226, 2009.

25. V. Devadas, F. Li, and H. Aydin. Analysis of online real-time scheduling algorithms under hard energy constraint. *Real-Time Systems*, 46(1):88–120, 2010.

26. T. Ebenlendr and J. Sgall. Semi-online preemptive scheduling: One algorithm for all variants. In *Proceedings of the 26th International Symposium on Theoretical Aspects of Computer Science (STACS)*, Freiburg, Germany, pp. 349–360, 2009.

27. X. Han, T.-W. Lam, L.-K. Lee, I. K. K. To, and P. W. H. Wong. Deadline scheduling and power management for speed bounded processors. *Theoretical Computer Science*, 411(40–42):3587–3600, 2010.

28. Intel. Power and thermal management in the Intel core duo processor. *Intel Technology Review*, 10(2), 109–212, 2006.

29. S. Irani and K. R. Pruhs. Algorithmic problems in power management. *ACM SIGACT News*, 36(2): 63–76, 2005.

30. S. Irani, S. Shukla, and R. Gupta. Online strategies for dynamic power management in systems with multiple power-saving states. *ACM Transactions on Embedded Computing Systems (TECS)*, 2(3):325–346, 2003.

31. S. Irani, G. Singh, S. K. Shukla, and R. K. Gupta. An overview of the competitive and adversarial approaches to designing dynamic power management strategies. *IEEE Transactions on Very Large Scale Integration Systems (IEEE TVLSI)*, 13(2):1349–1361, 2005.

32. B. Kalyanasundaram and K. Pruhs. Speed is as powerful as clairvoyance. *Journal of the ACM*, 47:617–643, 2000.

33. G. Koren and D. Shasha. *d*-over: An optimal online scheduling algorithm for overloaded real-time systems. *SIAM Journal on Computing*, 24(2):318–339, 1995.

34. S. Nedevschi, L. Popa, G. Iannaccone, S. Ratnasamy, and D. Wetherall. Reducing network energy consumption via sleeping and rate-adaptation. In *Proceedings of the 5th USENIX Symposium on Networked Systems Design and Implementation (NSDI)*, San Francisco, CA, pp. 323–336, 2008.

35. K. Pruhs, J. Sgall, and E. Torng. Online scheduling. *Handbook of Scheduling, Algorithms, Models and Performance Analysis*, CRC Press, Boca Raton, FL, 2004.

36. K. Pruhs and C. Stein. How to schedule when you have to buy your energy. In *Proceedings of International Workshop on Approximation, Randomization, and Combinatorial Optimization. Algorithms and Techniques (APPROX 2010 and RANDOM 2010)*, Barcelona, Spain, pp. 352–365, 2010.

37. K. Pruhs, P. Uthaisombut, and G. Woeginger. Getting the best response for your erg. *ACM Transactions on Algorithms (TALG)*, 4(3):Article No. 38, 2008.

38. C. Rusu, R. Melhem, and D. Mosse. Maximizing rewards for real-time applications with energy constraints. *ACM Transactions of Embedded Computing Systems (TECS)*, 2(4):537–559, 2003.

39. H. D. Simon. Petascale computing in the U.S. *ACTS Workshop*, Berkeley, CA, pp. 1–72, 2006.

40. W. Szpankowski. Average case analysis of algorithms, *Algorithms and Theory of Computation Handbook*, vol. 1, 2nd edn., Chapter 11. Chapman & Hall/CRC, Boca Raton, FL, 2010.

41. J. S. Vitter and P. Flajolet. Average-case analysis of algorithms and data structures, *Handbook of Theoretical Computer Science*, vol. A: Algorithms and Complexity, Chapter 9. Elsevier, Amsterdam, the Netherlands, 1990.

42. F. Yao, A. Demers, and S. Shenker. A scheduling model for reduced CPU energy. In *Proceedings of the 36th Annual IEEE Symposium on Foundations of Computer Science (FOCS)*, Milwaukee, WI, pp. 374–382, 1995.

43. Z. Zhang and F. Li. Dynamic power management algorithms in maximizing net profit. In *Proceedings of the 44th Annual Conference on Information Sciences and Systems (CISS)*, Princeton, NJ, pp. 1–5, 2010.

22

Reliability-Aware Power Management for Real-Time Embedded Systems

Dakai Zhu
The University of Texas at San Antonio

Hakan Aydin
George Mason University

22.1 Introduction

The performance of modern computing systems has steadily increased in the past decade, thanks to the ever-increasing processing frequencies and integration levels. However, such performance improvements have resulted in drastic increases in the power consumption of computing systems and promoted energy to be a first-class system resource. Hence, *power-aware computing* has become an important research area and several hardware and software power management techniques have been proposed. The common strategy to reduce power consumption in a computing system is to operate system components at low-performance (thus, low-power) states, whenever possible.

As one of the popular and widely exploited power management techniques, *dynamic voltage and frequency scaling* (DVFS) reduces the processor energy consumption by scaling down the supply voltage and processing frequency simultaneously [46,48]. However, executing an application at lower processing frequencies (i.e., at lower speeds) normally increases the computation time. Consequently, for real-time embedded systems with stringent timing constraints, special provisions are needed to avoid deadline misses when DVFS is employed to save energy. For various real-time task models, a number of power management schemes have been proposed to minimize the energy consumption while meeting the deadlines [4,10,12,31,36,37].

Reliability has been a traditional requirement for computer systems. During the operation of a computing system both *permanent* and *transient faults* may occur due to, for instance, the effects of hardware defects, electromagnetic interferences, or cosmic ray radiations, and result in system *errors*. In general, fault tolerance techniques exploit *space* and *time redundancy* [33] to detect and possibly recover from system errors caused by various faults. It has been shown that transient faults occur much more frequently than permanent faults [9,23], especially with the continued scaling of CMOS technologies and reduced design margins for higher performance [19]. For transient faults, which will be the focus of this chapter, the *backward error recovery* is an effective fault tolerance technique. This technique restores the system state to a previous *safe state* and repeats the computation when an error has been detected [33].

Until recently, energy management through DVFS and fault tolerance through redundancy have been studied independently in the context of real-time systems. However, there is an interesting trade-off between system energy efficiency and reliability as both DVFS and backward recovery techniques are based on (and compete for) the active use of the available CPU time (also known as *slack*). Moreover, it has been recently shown that DVFS has a direct and adverse effect on the transient fault rates, especially for those induced by cosmic ray radiations [15,19,61], which further complicates the problem. Therefore, for safety-critical real-time embedded systems (such as satellite and surveillance systems) where both reliability and energy efficiency are important, *reliability-aware power management* (RAPM) becomes a necessity. A number of reliability-aware power management schemes have been recently developed by the research community to address the negative effects of DVFS on system reliability. These schemes, which are the main focus of this chapter, typically guarantee system reliability requirements by scheduling proper recovery tasks while still saving energy with the remaining system slack.

In this chapter, we first present system models and state our assumptions (Section 22.2). Then, the fundamental idea of reliability-aware power management is introduced in the context of a single real-time task (Section 22.3.1). For systems with multiple real-time tasks that share a common deadline, the reliability-aware power management framework with individual recovery tasks is presented (Section 22.3.2), followed by the scheme that adopts a shared recovery task (Section 22.3.3). For general periodic real-time tasks that have different deadlines, the task-level recovery technique is first introduced (Section 22.4.1); then the utilization-based RAPM schemes for the earliest deadline-first (EDF) algorithm and priority-monotonic RAPM schemes for the rate-monotonic scheduling (RMS) algorithm are discussed (in Sections 22.4.2 and 22.4.3, respectively). The dynamic RAPM schemes that exploit dynamic slack generated at runtime for better energy savings are also covered (Section 22.5). We further discuss the RAPM schemes for energy-constrained systems aiming at maximizing system reliability (Section 22.6). We provide an overview of other related studies and identify some open research problems (Section 22.7). At the end, a brief summary concludes this chapter (Section 22.8).

22.2 Background and System Models

In this chapter, we consider a single processor system with DVFS capability. We consider the problems of how to minimize energy consumption without sacrificing system reliability, and to maximize system reliability with a given energy budget, while guaranteeing the timing constraints of real-time tasks. To better characterize the settings and define the scope of the discussion, in what follows, we first present the task, system power, and fault models and state our assumptions.

22.2.1 Real-Time Tasks

In real-time systems, applications are generally modeled by a set of *tasks*, which arrive periodically and need to finish their executions before a certain *deadline* after their arrivals [29]. More specifically, we consider an application that consists of a set of n independent real-time tasks $\Gamma = \{T_1, \ldots, T_n\}$. The task T_i is characterized by a pair (c_i, p_i), where c_i denotes its worst-case execution time (WCET) and p_i represents its period (which coincides with its relative deadline). That is, the task T_i generates an infinite number of task instances (also called *jobs*), and the inter-arrival time of two consecutive jobs of T_i is p_i. Once a job of task T_i arrives at time t, it needs to execute (in the worst case) for c_i time units before its deadline $t + p_i$ (which is also the arrival time of the task T_i's next job).

Given that the system under consideration adopts a variable-frequency processor, we assume that the WCET c_i of the task T_i is obtained under the maximum processing frequency f_{max} of the processor. Note that a number of studies have indicated that the execution time of tasks may not always scale linearly due to memory accesses or I/O operations [6,40]. However, to simplify our discussions, in this chapter, the execution time of a task is assumed to scale *linearly* with the processing frequency. That is, at the scaled processing frequency f ($\leq f_{max}$), task T_i will take (in the worst case) $c_i \cdot f_{max}/f$ time units to complete its execution.

22.2.2 Dynamic Voltage and Frequency Scaling

In many execution scenarios, real-time tasks may complete their executions well before their deadlines and leave the processor idle. Such idle time can be exploited through the *dynamic power management (DPM)* technique by putting the system to low-power sleep states to save energy. However, a more preferable strategy is to execute tasks at low processing frequencies and complete them just in time before their deadlines. This is due to the fact that the dynamic power P_d normally dominates in processors. P_d is linearly related to the processing frequency f and quadratically related to the supply voltage V_{dd} (i.e., $P_d \approx f \cdot V_{dd}^2$) [7]. Moreover, the operating frequency for CMOS circuits is almost linearly related to the supply voltage. Hence, the dynamic power becomes essentially a convex function of the processor frequency when applying DVFS [46] where the supply voltage and frequency are adjusted simultaneously. In modern processors, such frequency changes can be performed quite efficiently (in a few cycles or microseconds [2,13]). Therefore, we assume that the overhead for DVFS is negligible (or such overhead can be incorporated into the WCETs of tasks). Note that we will use the term *frequency change* to stand for scaling both processing frequency and supply voltage in the rest of this chapter.

In addition to dynamic power, another component in system power consumption is leakage power, which becomes increasingly important due to scaled technology size and increased levels of integration [25]. Moreover, it is necessary to consider all power-consuming system components (such as memory [27]) in an effective power management framework. Several such system-level power models have been recently proposed [12,22,25]. In this chapter, we adopt a simple *system-level power model*, where the system power consumption $P(f)$ of a computing system at processing frequency f is given by [61]:

$$P(f) = P_s + \hbar(P_{ind} + P_d) = P_s + \hbar(P_{ind} + C_{ef}f^m) \tag{22.1}$$

Here, P_s denotes the *static power*, which includes the power to maintain basic circuits and to keep the clock running. It can be removed only by powering off the whole system. When there is computation in progress, the system is *active* and $\hbar = 1$. When the system is turned off or in power-saving sleep modes, $\hbar = 0$.

P_{ind} is the *frequency-independent active power*, which corresponds to the power that is independent of processing frequency. For simplicity, unless specified otherwise, P_{ind} is assumed to be a constant and is the same for all the tasks; it can be efficiently removed (typically, with acceptable overhead) by putting the system components (e.g., main memory) into sleep state(s) [27]. P_d is the *frequency-dependent active power*, which includes the processor's dynamic power and *any* power that depends on the system processing frequency f (and the corresponding supply voltage) [7,27]. The effective switching capacitance C_{ef} and the dynamic power exponent m (in general, $2 \leq m \leq 3$) are system-dependent constants [7]. Despite its simplicity, this power model captures the essential components of an effective system-wide energy management framework.

22.2.2.1 Energy-Efficient Frequency

Since *energy* is the integral of power over time, the energy consumed by a task executing at constant frequency f ($\leq f_{max}$) is $E(f) = P(f) \cdot t(f) = P(f) \cdot c \cdot f_{max}/f$. Here, $t(f) = c \cdot f_{max}/f$ denotes the execution time of the task at frequency f. From the power model in Equation 22.1, intuitively, executing the task at lower processing frequencies can result in less energy consumption due to the frequency-dependent active power P_d. However, at lower frequencies, the task needs more time to complete its execution, and thus consumes more energy due to the static and frequency-independent active power. Therefore, there exists a minimal *energy-efficient frequency* f_{ee} below which a task starts to consume more system energy [22,25]. Considering the prohibitive overhead of turning on and off a system (e.g., tens of seconds), we assume that the system is on for the operation interval considered and P_s is always consumed. By differentiating $E(f)$ with respect to f and setting it to 0, we can find that [61]:

$$f_{ee} = \sqrt[m]{\frac{P_{ind}}{C_{ef} \cdot (m-1)}} \cdot f_{max} \tag{22.2}$$

Consequently, for energy efficiency, the frequency f to execute any task should be limited to the range $f_{ee} \leq f \leq f_{max}$. Moreover, to simplify the discussion, we assume that *normalized* frequencies are used and that the maximum frequency is $f_{max} = 1$.

22.2.3 Transient Faults and Backward Recovery

Unlike *crash failures* that result from permanent faults, a *soft error* that follows a transient fault typically does not last for long and disappears when the computation is repeated. With the aim of tolerating transient faults, we exploit temporal redundancy (i.e., system slack) and employ *backward recovery* techniques to tolerate soft errors. More specifically, when soft errors due to transient faults are detected by, for example, *sanity* (or *consistency*) checks at the completion time of a task, a recovery task is dispatched in the form of re-execution [33]. The overhead for error detection is assumed to be incorporated into tasks' worst-case execution times.

Traditionally, transient faults have been modeled through a Poisson process with an average arrival rate of λ [49,51]. With the continued scaling of technology sizes and reduced design margins, it has been shown that DVFS has a direct and negative effect on the arrival rate λ due to the increased number of transient faults (especially the ones induced by cosmic ray radiations) at lower supply voltages [15,19]. For systems with a DVFS-enabled processor, the average rate of soft errors caused by transient faults at scaled processing frequency f (and the corresponding supply voltage) can be expressed as [61]:

$$\lambda(f) = \lambda_0 \cdot g(f) \tag{22.3}$$

where λ_0 is the average error rate corresponding to the maximum processing frequency f_{max}. That is, $g(f_{max}) = 1$. With reduced processing frequencies and supply voltages, the average error rate generally increases and $g(f) > 1$ for $f < f_{max}$. In other words, $g(f)$ is a nonincreasing function of the processor frequency f.

22.2.3.1 Exponential Fault Rate Model

The radiation-induced transient faults in semiconductor circuits have been known and well studied for decades [67]. When high-energy particles strike a sensitive region in semiconductor devices, a dense track of electron-hole pairs are deposited, which can accumulate and exceed the minimum charge (i.e., the *critical charge*) required to flip the value stored in a memory cell [21], or be collected by pn-junctions via drift and diffusion mechanisms to form a current pulse and cause a logic error [26]. However, it has been a great challenge to model such soft errors caused by transient faults considering the various factors, such as cosmic ray flux (i.e., number of particles per area), technology feature size, chip capacity, supply voltage, and operating frequency [39,41,68].

In general, when the supply voltage of a semiconductor device decreases, the critical charge becomes smaller, which can lead to exponentially increased transient fault rates [21,41]. Such effects have been observed on both processors [38] and memory subsystems [68]. Moreover, in addition to high-energy particles such as cosmic rays, which are more likely to cause transient faults in circuits with smaller critical charge, lower-energy particles do also exist, and in much larger quantities [67]. Therefore, considering the number of particles in the cosmic rays and the relationship between transient fault rate, critical charge, and supply voltage, an *exponential* rate model for soft errors caused by transient faults has been derived as [61]:

$$\lambda(f) = \lambda_0 \cdot g(f) = \lambda_0 10^{\frac{d(1-f)}{1-f_{ee}}} \tag{22.4}$$

In this expression, the exponent d (>0) is a constant which indicates the sensitivity of the rate of soft errors to DVFS. The maximum error rate at the minimum energy-efficient processing frequency f_{ee} (and corresponding supply voltage) is assumed to be $\lambda_{max} = \lambda_0 10^d$. For example, when $d = 3$, the average rate of soft errors at the minimum frequency can be 1000 times higher than that at the maximum frequency.

22.3 Reliability-Aware Power Management

The preceding discussion shows that when the processing frequency (and supply voltage) of a task is scaled down through DVFS to save energy, the probability of incurring soft errors due to transient faults during the scaled execution will drastically increase, which in turn leads to significantly reduced reliability. To compensate such reliability loss at lower supply voltage, special provisions are needed when DVFS is employed. Suppose that the *original reliability* of a system denotes the probability of successful operation without incurring any errors caused by transient faults for a given interval when there is no power management (i.e., when all tasks are executed at the maximum processing frequency f_{max}). With the goal of preserving system's original reliability in the presence of DVFS, we now present the details of the *reliability-aware power management* (RAPM) framework.

Specifically, we first introduce the fundamental idea of reliability-aware power management in the context of a single real-time task. The approach involves scheduling an additional recovery task to recuperate reliability loss induced by power management (DVFS). Next, for systems with multiple real-time tasks that share a common deadline/period, the RAPM scheme with individual recovery tasks is discussed, followed by the scheme with a single shared recovery task. The RAPM schemes for general periodic tasks with different deadlines will be presented in Section 22.4.

22.3.1 The Case with a Single Real-Time Task

For systems with a single real-time task T, let R^0 denote the *original* reliability of an instance of task T when it runs at the maximum frequency f_{max}. R^0 is the probability of successfully completing the execution of the task instance without incurring soft errors caused by transient faults. Since system reliability depends on the correct execution of every instance of the task, to achieve the objective of maintaining the system's original reliability, we can preserve the original reliability R^0 for each instance of task T.

The central idea behind RAPM is to schedule a recovery task using the available slack before exploiting the slack for DVFS to save energy [55,56]. As a concrete example, we consider a single task T that has

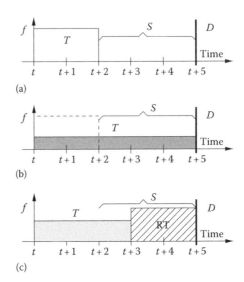

FIGURE 22.1 (a) Slack S is available at time t. (b) Ordinary power management (c) reliability-aware power management.

the worst-case execution time c and period p as 2 and 5, respectively. Suppose that an instance of task T arrives at time t, which needs to complete its execution by time $t + 5$. As shown in Figure 22.1a, we can see that there are $S = 3$ units of slack available.

Without paying special attention to system reliability, the *ordinary* power management would utilize all the available slack to scale down the processing frequency of the task instance through DVFS for maximum energy savings [4,46]. Therefore, the scaled frequency for the task instance under the ordinary power management will be $f = 2/(2 + 3) \cdot f_{max} = 0.4$ (recall that $f_{max} = 1$), as shown in Figure 22.1b. In the figures, the x-axis represents time, the y-axis represents processing frequency (i.e., cycles executed per time unit), and the area of the task box defines the workload (i.e., number of cycles) of the task instance. By focusing on active power and assuming that $P_{ind} = 0.1, C_{ef} = 1$, and $m = 3$ [7], from Equation 22.1, we can easily find that the ordinary power management can save 63% of the *active energy* when compared to that of *no power management* (NPM) where all tasks run at the maximum frequency.

However, as discussed earlier, with reduced processing frequency and supply voltage, the execution of task T is more susceptible to errors caused by transient faults [19,61]. Suppose that the exponent in the exponential fault rate model is $d = 2$ (see Equation 22.4). At the reduced frequency $f = 0.4$, the probability of incurring errors due to transient fault(s) during T's execution can be found as

$$\rho_1 = 1 - R_1 = 1 - e^{-\lambda_0 10^{\frac{d(1-f)}{1-f_{ee}}} \frac{c}{f}}$$

$$= 1 - e^{-\lambda_0 10^{\frac{d(1-f)}{1-f_{ee}}} \frac{c}{f}} = 1 - e^{-\lambda_0 c 10^{\frac{2(1-0.4)}{1-0.37}} \frac{1}{0.4}}$$

$$\approx 1 - (R^0)^{200} = 1 - (1 - \rho^0)^{200} \approx 200\rho^0 \qquad (22.5)$$

Here, the minimum energy-efficient frequency can be found as $f_{ee} = 0.37$. R^0 is the original reliability of task T, and $\rho^0 (= 1 - R^0)$ is the corresponding *probability of failure* of task T's execution. Since ρ^0 is in general a very small number (usually $<10^{-4}$), we can see that, executing the task instance of T at the scaled frequency $f = 0.4$ can lead to approximately 200 times higher probability of failure figures. Such increase in the probability of failure during the execution of individual tasks will degrade the overall system reliability, which is unbearable, especially for safety-critical systems, where the requirement for high levels of reliability is strict.

Instead of using all the available slack for DVFS to save energy, we can also reserve part of the available slack for temporal redundancy to re-execute the tasks incurring soft errors and thus increase system reliability [33]. Hence, the central idea of *reliability-aware power management* is to reserve part of the available slack and schedule a recovery task, which can recuperate the reliability loss due to energy management. Then, the remaining slack can be utilized by DVFS to scale down the execution of tasks for energy savings. For simplicity, we assume that the recovery task takes the form of *re-execution* and has the same size of the task to be recovered.

For the earlier example, 2 units of the available slack can be reserved for scheduling a recovery task RT, as shown in Figure 22.1c. The remaining 1 unit of slack can be utilized by DVFS to scale down the processing frequency of task T to $f = 2/(2 + 1) = 0.67$. Note that the recovery task RT will be invoked only if the scaled execution of task T is subject to a soft error caused by transient faults. Without considering the energy consumption of the recovery task, executing task T at the scaled frequency $f = 0.67$ could yield 26% energy savings when compared to that of no power management. Moreover, since the recovery task takes the form of re-execution of its primary task, it will be executed at the maximum frequency $f_{max} = 1$ to ensure that there is no deadline miss.

With the additional recovery task RT, the reliability R of task T will be the summation of the probability of the primary task T being executed correctly and the probability of incurring errors due to transient faults during task T's execution while the recovery task RT is executed correctly. With the assumption that the recovery task RT is essentially the re-execution of task T at the maximum frequency f_{max}, its probability of successful execution will be $R^0 = e^{-\lambda_0 c}$. Thus, we have

$$R = e^{-\lambda(f)S} + \left(1 - e^{-\lambda(f)S}\right) \cdot R^0 > R^0 \tag{22.6}$$

where $\lambda(f)$ is the average rate of soft errors due to transient faults at the reduced frequency f. From this equation, we can see that the resulting reliability for task T under RAPM is always better than its original reliability R^0.

Therefore, when the amount of available slack is larger than the worst-case execution time of a task, the RAPM scheme can reserve part of the slack to schedule a recovery task while using the remaining slack for DVFS to save energy. With the help of the recovery task, which is assumed to take the form of re-execution at the maximum frequency in case the scaled execution of the primary task fails, the RAPM scheme guarantees to preserve a real-time task's original reliability while still obtaining energy savings using the remaining slack, regardless of different error rate increases (i.e., different values of d) and the scaled processing frequency [55,56].

22.3.2 Multiple Tasks with a Common Deadline

When there are multiple real-time tasks, the reliability-aware power management problem gains a new dimension. Here, we need to allocate the available slack to multiple tasks, *possibly in different amounts*, to maximize the energy savings while preserving system reliability. We start our analysis with systems where all tasks have the same period, which have been denoted as *frame-based task systems* [35]. In such systems, the common period is considered as the *frame* of the application. Due to their periodicity, we consider only the tasks within one frame of such systems, where all tasks arrive at the beginning of a frame and need to complete their executions by the end of the frame (i.e., the common deadline).

Recall that the reliability of a system generally depends on the correct execution of all its tasks. Although it is possible to preserve the original reliability of a given system while sacrificing the reliability of some individual tasks, for simplicity, we focus on preserving the original reliability of each task to guarantee the system original reliability. From the previous discussions, we know that for any task whose execution is scaled down, the reliability-aware power management can schedule a recovery task to recuperate the reliability loss due to power management at the reduced supply voltage levels.

For ordinary power management that does not consider system reliability, the optimal scheme for obtaining the maximum energy savings is to allocate the available slack to all tasks proportionally and scale down their executions uniformly [3,46]. The optimality comes from the convex relationship between the energy consumption of tasks and their processing frequencies. However, when system reliability is considered, the proportional slack allocation scheme may not be the most energy-efficient approach, especially for cases where the amount of available slack is not enough to accommodate a separate recovery task for each and every task.

22.3.2.1 An Example

We explain the key idea of the reliability-aware power management for systems with multiple real tasks that share a common deadline, again, through a concrete example. Here, the system consists of four tasks, which have the same period of 7. If the worst-case execution time of each task is 1 time unit, 3 units of slack time exist within each frame, as shown in Figure 22.2a. It is not hard to see that the available slack is not sufficient to accommodate a separate recovery task for each of the four tasks, and not all tasks can be scaled down under RAPM for energy savings.

If we adopt a greedy approach and allocate all the available slack to the first task T_1, a recovery task RT_1 (which takes 1 unit of slack) can be scheduled and the remaining 2 units of slack can be utilized to scale down the processing frequency of task T_1 to $f = 1/3$, as shown in Figure 22.2b. Assuming the same parameters as in the power model in Section 22.3.1, simple calculation shows that about 22% energy savings can be obtained compared to that of no power management case. As explained earlier, the original reliability of task T_1 is preserved with the help of the recovery task RT_1. Moreover, as tasks T_2, T_3, and T_4 are executed at the maximum frequency f_{max}, their original reliabilities are preserved as well. Therefore, the overall system reliability is guaranteed to be no worse than the system original reliability.

However, the greedy strategy of allocating all available slack to one task may not be the most energy-efficient one and we can select more tasks for more energy savings. In fact, the available slack is not enough to slow down all four tasks. Moreover, if we select three tasks, all 3 units of available slack will be utilized for scheduling the recovery tasks, which leaves no slack for DVFS and no energy savings can be obtained. Instead, suppose that two tasks T_1 and T_2 are selected. After scheduling their recovery tasks

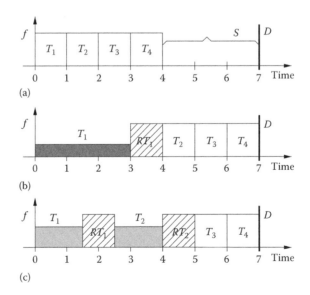

(a)

(b)

(c)

FIGURE 22.2 RAPM for multiple tasks with a common deadline. (a) 4 tasks with 3 units slack. (b) Task T_1 is selected for management. (c) Tasks T_1 and T_2 are selected for management.

RT_1 and RT_2 and using the remaining 1 unit of slack to scale down the processing frequency of tasks T_1 and T_2, we obtain the schedule shown in Figure 22.2c. Here, the scaled frequency for tasks T_1 and T_2 is $f_{12} = 2/3$ and the energy savings can be calculated as 28%—a significant improvement over the greedy approach. Observe that in this solution, the original system reliability is still preserved.

22.3.2.2 Optimal Task Selection

From this example, we can see that one of the key problems in the reliability-aware power management for frame-based real-time tasks is to select an *appropriate* subset of tasks to apply DVFS. The execution of the selected tasks will be scaled down after reserving part of the slack to schedule a separate recovery task for each of them. The remaining tasks are left intact and they run at the maximum processing frequency in order to preserve their original reliability.

Intuitively, for a system with a given amount of available slack S, when more tasks are selected, more slack needs to be reserved for the recovery tasks, which reduces the amount of slack for DVFS to scale down the selected tasks and thus reduce the energy savings. Similarly, if fewer tasks are selected, using the large amount of slack to scale down a few tasks may not be energy efficient either. Therefore, there should exist an optimal subset of selected tasks for which the energy savings are maximized. A natural question to ask is whether there exists a fast (i.e., polynomial-time) solution to the problem of reliability-aware energy management for multiple tasks. Unfortunately, the answer is negative, as we argue next.

Let us denote the total workload (computation requirement) of all tasks by $L = \sum_{i=1}^{n} c_i$. The available slack is $S = D - L$, where D is the common deadline (also the frame of the task set). Without loss of generality, assume that a subset of tasks are selected, where the total computation time of the selected tasks is X. We have $X \leq L$ and $X \leq S$. After reserving X units of slack for recovery tasks, the remaining $S - X$ units of slack could be used to scale down the processing frequency for the selected tasks. Considering the convex relationship between energy consumption and processing frequency, the optimal solution to get the maximum energy savings can be obtained by uniformly scaling down the execution of the selected tasks. Therefore, the amount of *fault-free* energy consumption (without considering the execution of recoveries that are needed only with a small probability) will be:

$$E_{total} = P_s \cdot D + S \left(P_{ind} + c_{ef} \cdot \left(\frac{X}{S} \right)^m \right)$$

$$+ (L - X) \left(P_{ind} + c_{ef} \cdot f_{max}^m \right) \tag{22.7}$$

where the first part represents the energy consumption due to the *static power*, which is always consumed during the operation. The second part is the energy consumption for the *selected tasks* and the third part is the energy consumption of the *unselected tasks*. Simple algebra shows that to minimize E_{total}, the total size of the selected tasks should be equal to

$$X_{opt} = S \cdot \left(\frac{P_{ind} + C_{ef}}{m \cdot C_{ef}} \right)^{\frac{1}{m-1}} \tag{22.8}$$

Hence, the value of X_{opt} can serve as a guideline for task selection to obtain the best energy savings. If $X_{opt} \geq L$, all tasks should be selected and scaled down uniformly for maximum energy savings. Otherwise (the case where $X_{opt} < L$), if there exist a subset of tasks such that the summation of their computation is *exactly* X_{opt}, selecting that subset would definitely be optimal.

However, finding a subset of tasks with exactly X_{opt} units of total computation time turns out to be NP-hard [57]. Note that, the reliability of any single selected task under RAPM is better than its original reliability with the help of its recovery task (Equation 22.6). To select as many tasks as possible, we can adopt an efficient task selection heuristic, namely the *smallest-task-first* (STF) scheme. That is, after obtaining X_{opt} for a given task set, we can find the largest value of k, such that the total computation of the first k smallest tasks is no more than X_{opt}. The drawback of this heuristic is that the difference between X_{opt} and the total computation of the selected tasks (denoted as *selection error*) can be large, which may result in less energy savings. As a second scheme, we can select tasks following the *largest-task-first* (LTF) heuristic. That is, the tasks can be selected by processing them in decreasing order of their size, as long as the selection error is larger than the current task. Hence, LTF can ensure that the selection error is bounded by the size of the smallest task. Interested readers are referred to [57] for further details.

The results of performance evaluation through extensive simulations, given in [55–57], show that the RAPM technique is able to preserve (even *improve*) the overall system reliability, while the ordinary (but reliability-ignorant) power management technique can result in reliability degradations of several orders of magnitude. Moreover, RAPM schemes can still obtain up to 40% energy savings where the gains tend to be higher at lower workloads. However, since some amount of CPU time needs to be reserved for recoveries, the energy savings achieved by RAPM schemes are generally 20%–30% lower compared to those of ordinary power management schemes.

22.3.3 Shared Recovery Technique

The probability of incurring an error due to transient faults during the execution of a task is rather low, even for scaled execution at low supply voltages. Hence, most of the recovery tasks will not be invoked at run-time. Moreover, the required slack for multiple separate recovery tasks reduces the prospects for energy savings with less available slack for DVFS. That is, the RAPM scheme with individual recovery tasks is, in some sense, *conservative*. In fact, when the execution of a scaled task completes successfully without incurring any error, its recovery task will not be activated and the corresponding slack time can immediately be made available for the *next* scaled task. Furthermore, to preserve the original reliability of a task, it is sufficient to ensure that, at the dispatch time of a task whose execution is scaled down, there is sufficient slack time for a recovery (to re-execute the task) at the maximum frequency.

The key idea of the *shared recovery*–based RAPM scheme is to reserve slack time for only one recovery block, which can be shared by all selected tasks at run-time [53]. Therefore, more slack can be left for DVFS to scale down the execution of the selected tasks and save more energy. Here, to ensure that the recovery block is large enough to re-execute any scaled task at the maximum frequency, its size needs to be the same as the largest selected task. As long as there is no error caused by transient faults during the execution of scaled tasks at run-time, the next scaled task can run at its pre-calculated low processing frequency. When an error is detected during the execution of a scaled task, the recovery will be triggered and the faulty task will be re-executed at the maximum frequency. After that, a *contingency* schedule is adopted in the sense that all remaining selected tasks are executed at the maximum frequency until the end of the current frame to preserve the system reliability.

The idea can be further illustrated with the following example. As shown in Figure 22.3a, there are two tasks T_1 and T_2 that share a common period/deadline of 7. The worst-case execution times of T_1 and T_2 are 2 and 1, respectively. The system potentially has access to $7 - 3 = 4$ units of slack when both tasks execute at the maximum frequency. If we consider the individual recovery–based RAPM scheme, it can be found that selecting only task T_1 to utilize all slack (while task T_2 runs at the maximum frequency) will yield maximum energy savings. With the same parameters for the system power model as in Section 22.3.1, the energy savings can be calculated as 33% with respect to no power management scheme.

For the shared recovery–based RAPM scheme, only one recovery of size 2 units (the maximum size of tasks T_1 and T_2) is scheduled, which can be shared by both tasks. The remaining 2 units of slack can be used to scale down the execution of both tasks T_1 and T_2 at the frequency of $f = 3/5$, as shown in Figure 22.3b. Compared to the no power management scheme, it can be found that 40% energy can be saved, which is better than that of the individual recovery–based RAPM scheme.

Figure 22.3c further shows the scenario where, after the detection of an error due to transient faults, the shared recovery block is utilized to re-execute task T_1 at the maximum frequency. From previous discussions, we know that the reliability of task T_1 is no worse than its original reliability. Moreover,

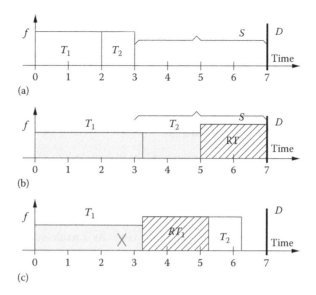

FIGURE 22.3 Two tasks managed by the shared recovery technique. (a) Two tasks with 4 units slack. (b) A shared recovery task is scheduled. (c) Task T_1 fails and recovers; T_2 runs at f_{max}.

by switching to the contingency schedule after recovering the faulty task T_1 and enforcing task T_2 run at the maximum frequency, the original reliability of task T_2 is also preserved. To conclude, the shared recovery–based RAPM scheme can still preserve the original reliability of a task system [53].

Our evaluation results show that the energy savings performance of the shared recovery–based RAPM scheme is excellent—in fact, very close (within 5% difference for most cases) to that of the ordinary power management. Perhaps unexpectedly, the results also indicate that the RAPM scheme with a shared recovery task can offer nontrivial gains on the reliability side over individual recovery–based RAPM schemes as well [53]. This comes from the fact that the shared recovery–based RAPM scheme provides a better protection for *all* single-fault scenarios that are typically much more likely than some multiple-fault scenarios that the individual recovery–based RAPM schemes provision for. Since optimizing the most common case is a well-known system design technique, overall reliability achieved by the shared-recovery technique tends to be better in most cases.

22.4 RAPM for Periodic Real-Time Tasks

For frame based real-time tasks that share a common period/deadline, the available slack within a frame is accessible to all tasks, which allows the use of the shared recovery–based RAPM scheme. However, for general periodic real-time tasks with different periods, the active task instances (or jobs) of different tasks can have different deadlines, preventing them from sharing a common recovery task. In addition, the available system slack may also have different expiration times [4], which makes it difficult (if not impossible) to develop the shared recovery–based RAPM scheme for general periodic real-time tasks. Hence, in this section, we focus on the individual recovery–based RAPM schemes, where each task instance whose execution is scaled down will have a separate recovery task.

For periodic real-time tasks, the workload in a system is normally represented by *system utilization*. The utilization of a periodic real-time task T_i is defined as $u_i = c_i/p_i$, where c_i is its worst-case execution time and p_i is its period. The system utilization of a task set Γ with n tasks is correspondingly defined as $U = \sum_{i=1}^{n} u_i$. For a task set to be schedulable on a uniprocessor system, it is necessary to have $U \leq 1$ [29].

For task sets with $U < 1$, the spare processor capacity is denoted as $1 - U$, which indicates the amount of available *static slack* in the system. In this section, we address the problem of exploiting spare processor capacity (i.e., static slack) to maximize energy savings while guaranteeing the original reliability of the system under consideration.

22.4.1 Task-Level RAPM Techniques

Theoretically, it is possible to select only a subset of jobs of each task to obtain the maximum energy savings. That is, the jobs of a given task can be handled differently, where the execution of only a subset of selected jobs are scaled down (with a separate recovery job scheduled before the deadline for each), while the remaining jobs can run at the maximum frequency. However, such a job-oriented approach requires the consideration of all jobs within the *hyper-period* (defined as the least common multiple, LCM, of all tasks' periods) of the task set, which may be arbitrarily large. As a result, such an approach cannot be generally considered as a computationally efficient technique.

This section focuses on a particular efficient *task-level* RAPM technique, where all jobs of the same task will get the same treatment [58,59,65]. In particular, the jobs of all the unselected tasks will run at the maximum frequency for reliability preservation. Moreover, all the jobs of a given selected task T_k will be scaled down to the same low processing frequency f_k, and each such job will need to have a recovery scheduled before its deadline to preserve its original reliability. To provide the recovery time needed by all the jobs of T_k, we can construct a *periodic recovery task* (PRT), with the same timing parameters (WCET and period) as those of task T_k. By incorporating such a periodic recovery task into the task set, we can ensure that there is a recovery job scheduled before the deadline of each job of T_k.

22.4.1.1 Periodic RAPM Problem

Let us denote by Φ ($\subseteq \Gamma$) the subset of the selected tasks. A periodic recovery task will be constructed for each task T_k ($\in \Phi$) and the scaled frequency for task T_k is f_k ($< f_{max}$). To preserve reliability, the remaining (unselected) tasks run at the maximum frequency f_{max}. Assume that the *augmented* task set, which incorporates the newly created recovery tasks and the scaled execution of the selected tasks, is schedulable under a given scheduling policy. Without considering the energy consumed by recovery tasks (which normally have a small probability of being executed), the *fault-free* energy consumption within the hyper-period (LCM) of the task set is

$$E(\Phi) = \sum_{T_i \in (\Gamma - \Phi)} \frac{LCM}{p_i} P(f_{max}) c_i + \sum_{T_k \in \Phi} \frac{LCM}{p_k} P(f_k) \frac{c_k f_{max}}{f_k} \qquad (22.9)$$

where the first part represents the energy consumed by the *unselected* tasks and the second part corresponds to the energy consumed by the *selected* tasks. Hence, the RAPM problem for periodic real-time tasks can be stated as follows: for a given scheduling policy, find the subset Φ of tasks and their corresponding scaled frequencies to minimize $E(\Phi)$ while preserving the system original reliability and meeting all deadlines.

Not surprisingly, finding the optimal subset Φ and the scaled frequencies to minimize $E(\Phi)$ is NP-hard [59]. In fact, if all tasks have the same period, the special case of the periodic RAPM problem becomes essentially the RAPM problem for multiple tasks sharing a common deadline, which is NP-hard as discussed in Section 22.3.2. However, for different uniprocessor scheduling policies, such as preemptive *earliest-deadline first* (EDF) and *rate-monotonic-scheduling* (RMS), there are different schedulability conditions that result in different task selection and frequency assignment strategies.

22.4.2 Utilization-Based RAPM for EDF Scheduling

Under preemptive EDF scheduling, a set of n periodic real-time tasks are schedulable on a uniprocessor system as long as the system utilization $U \leq 1$ [29]. The spare processor capacity is denoted as $sc = 1 - U$, which can be exploited for energy and reliability management. Suppose that the total utilization of the selected tasks in Φ is $X = \sum_{T_i \in \Phi} u_i < sc$. After incorporating the newly constructed periodic recovery tasks, the remaining spare processor capacity is $(sc - X)$, which can be used to scale down the execution of all jobs of the selected tasks to save energy. Again, due to the convex relation between power and processing frequency (see Equation 22.1), the optimal solution that minimizes system energy consumption will consist in uniformly scaling down all jobs of the selected tasks. The scaled processing frequency can be found as $f = X/(X + (sc - X)) = X/(sc)$. Note that, with the newly constructed recovery tasks and the scaled execution of the selected tasks, the system utilization of the augmented task set can be calculated as

$$U' = (U - X) + \frac{X}{f} + X = U + \frac{X}{X/sc} = 1 \tag{22.10}$$

That is, the augmented task set is still schedulable with EDF.

Based on system utilization, Equation 22.9 can be rewritten as

$$E(\Phi) = \text{LCM} \cdot P_s + \text{LCM}(U - X) \left(P_{ind} + c_{ef} \cdot f_{max}^m \right)$$

$$+ \text{LCM} \cdot sc \left(P_{ind} + c_{ef} \cdot \left(\frac{X}{sc} \right)^m \right) \tag{22.11}$$

where the first part denotes the energy consumption due to static power, the second part captures the active energy consumption of unselected tasks, and finally, the third part represents the active energy consumption of the selected tasks.

The formulation is similar to that of the case with multiple tasks sharing a common deadline (Section 22.3.2). We can find that when the total utilization of the selected tasks is $X_{opt} = sc \cdot \left((P_{ind} + C_{ef})/(m \cdot C_{ef}) \right)^{1/(m-1)}$, $E(\Phi)$ is minimized. Again, finding the optimal subset of tasks with total utilization equal to exactly X_{opt} turns out to be NP-hard. Following a similar reasoning, the efficient heuristics *largest-utilization-first* (LUF) and *smallest-utilization-first* (SUF) can be adopted when selecting tasks [58,59].

22.4.3 Priority-Monotonic RAPM for Rate Monotonic Scheduling

In rate monotonic scheduling (RMS), tasks are assigned static priorities and the ones with smaller periods have higher priorities. Several feasibility tests have been proposed for task sets under RMS with different levels of accuracy and complexity. A simple well-known feasibility test is based on the system utilization of a task set. A task set is feasible as long as its utilization $U \leq LLB(n) = n(2^{1/n} - 1)$, where n is the number of tasks in the task set and $LLB(n)$ denotes the *Liu-Layland bound* [29]. Following the same approach as that for EDF scheduling, similar utilization-based RAPM schemes can be developed using the Liu-Layland bound for the feasibility of task sets scheduled by RMS. Interested readers can refer to [65] for detailed discussions.

Here, we focus on the feasibility test with the exact *time demand analysis* (TDA) technique. The time demand function $wq_i(t)$ of task T_i is defined as [28]

$$wq_i(t) = c_i + \sum_{k=1}^{i-1} \left\lceil \frac{t}{p_k} \right\rceil c_k, \text{for } 0 < t \leq p_i \tag{22.12}$$

The task set is considered feasible if, for every task T_i in the task set under consideration, it is possible to find a time instant t such that $wq_i(t) \leq t \leq p_i$.

First, by incorporating the scaled tasks in Φ and their corresponding recovery tasks, the *modified* time demand function $mwq_i(t)$ for task T_i ($\in \Gamma$) can be defined as

$$mwq_i(t) = \sum_{k=1}^{i-1} \left\lceil \frac{t}{p_k} \right\rceil c_k + \sum_{T_k \in \Phi, 1 \leq k \leq i-1} \left\lceil \frac{t}{p_k} \right\rceil \frac{c_k \cdot f_{max}}{f_k}$$

$$+ \begin{cases} c_i & \text{if } T_i \notin \Phi; \\ c_i \cdot \left(1 + \frac{f_{max}}{f_i} \right) & \text{if } T_i \in \Phi. \end{cases} \tag{22.13}$$

This function incorporates the time demand from all (scaled and unscaled) tasks as well as the required recovery tasks. It is not difficult to see that the augmented task set is schedulable if, for every task $T_i \in \Gamma$, there is a time instant t such that $mwq_i(t) \leq t \leq p_i$.

Therefore, the periodic RAPM problem for RMS with TDA-based analysis can be formally expressed as: Find the subset Φ and the scaled frequencies for selected tasks so as to

$$minimize(E(\Phi))$$

subject to

$$\forall T_i \in \Gamma, \exists t, mwq_i(t) \leq t \leq p_i, \quad \text{where } 0 < t \leq p_i$$

Following the idea of *priority-monotonic* frequency assignment [36], we assume that the first x highest priority tasks are selected and $\Phi = \{T_1, \ldots, T_x\}$. That is, it is assumed that tasks are ordered by their priorities where T_1 has the highest priority and T_n has the lowest priority. Depending on how the scaled frequencies for the selected tasks are determined, we present two priority-monotonic RAPM schemes, which assume identical and different frequency settings for the selected tasks, respectively.

22.4.3.1 Single Frequency for Selected Tasks

Starting with the single frequency assignment for the selected tasks, the RAPM-TDA scheme resorts to the exact TDA test that considers all the time instants within a task's period to get a lower-scaled frequency and thus better energy savings. The scaled frequency $f_{RAPM-TDA}(x)$ for the selected x highest priority tasks is given by

$$f_{RAPM-TDA}(x) = \max \left\{ f_{ee}, \max_{T_i \in \Gamma} \{f_i(x)\} \right\} \tag{22.14}$$

$$f_i(x) = \min_{0 < t \leq p_i} \{f_i(x, t)\} \tag{22.15}$$

$$f_i(x, t) = \begin{cases} \dfrac{\sum_{k=1}^{i} \left\lceil \frac{t}{p_k} \right\rceil c_k}{\sum_{k=1}^{i} \left\lceil \frac{t}{p_k} \right\rceil c_k + (t - mwq_i(t))} f_{max} & \text{if } i \leq x; \\ \dfrac{\sum_{k=1}^{x} \left\lceil \frac{t}{p_k} \right\rceil c_k}{\sum_{k=1}^{x} \left\lceil \frac{t}{p_k} \right\rceil c_k + (t - mwq_i(t))} f_{max} & \text{if } i > x. \end{cases} \tag{22.16}$$

where $f_i(x)$ is the scaled frequency determined for task T_i and $mwq_i(t)$ is the modified time demand function as defined in Equation 22.13 with $f_k = f_{max}$.

If there does not exist a feasible time instant for any task T_i (i.e., $mwq_i(t) > t$ for $\forall t\ 0 < t \leq p_i$), it is not valid to select exactly x highest priority tasks for slow-down. Otherwise, by applying Equation 22.9, we can get the energy consumption when the x highest priority tasks are scaled down to the frequency of $f_{RAPM\text{-}TDA}(x)$. Searching through all the feasible task selections, RAPM-TDA can find out the optimal number of highest priority tasks x_{opt} and the corresponding scaled frequency $f_{RAPM\text{-}TDA}(x_{opt})$ that give the minimal energy consumption in pseudo-polynomial time. The complexity of RAPM-TDA can be easily found to be $O(n^3 r)$, where $r = p_n/p_1$ is the ratio of the largest period to the smallest period.

22.4.3.2 Multiple Frequencies for Selected Tasks

In RAPM-TDA, it is possible that the resulting single-scaled frequency is constrained by a high priority task T_k in the subset Φ. That is, T_k has more stringent timing constraints and requires a higher scaled frequency. For such cases, by exploiting the slack time from the high frequency assignment for high priority tasks, the new scheme (called RAPM-TDAM) iteratively recalculates and assigns a lower frequency for low priority tasks in the subset Φ, and thus saves more energy.

More specifically, for a given subset Φ with x highest priority tasks, if the single-scaled frequency obtained by RAPM-TDA is $f_{RAPM\text{-}TDA}(x) = f_k(x)$ and $k < x$, we can assign the frequency $f_{RAPM\text{-}TDA}(x)$ to the first k highest priority tasks. Then, we can recalculate the scaled frequency for the remaining tasks in the subset Φ. When the frequency assignment for the first k highest priority tasks is fixed, the modified work demand function and scaling factor for task T_i ($k < i \le n$) can be recalculated as

$$mwq_i(t) = \sum_{j=1}^{k} \left\lceil \frac{t}{p_j} \right\rceil \left(1 + \frac{f_{max}}{f_j}\right) c_j +$$

$$\begin{cases} \sum_{j=k+1}^{i} \left\lceil \frac{t}{p_j} \right\rceil 2 \cdot c_j & \text{if } i \le x; \\ \sum_{j=k+1}^{x} \left\lceil \frac{t}{p_j} \right\rceil 2 \cdot c_j + \sum_{j=x+1}^{i} \left\lceil \frac{t}{p_j} \right\rceil c_j & \text{if } i > x. \end{cases} \tag{22.17}$$

$$f_i(x, t) = \begin{cases} \dfrac{\sum_{j=k+1}^{i} \left\lceil \frac{t}{p_j} \right\rceil c_j}{\sum_{j=1}^{i} \left\lceil \frac{t}{p_j} \right\rceil c_j + (t - mwq_i(t))} f_{max} & \text{if } i \le x; \\[4mm] \dfrac{\sum_{j=k+1}^{x} \left\lceil \frac{t}{p_j} \right\rceil c_j}{\sum_{j=1}^{x} \left\lceil \frac{t}{p_j} \right\rceil c_j + (t - mwq_i(t))} f_{max} & \text{if } i > x. \end{cases} \tag{22.18}$$

After recalculating the scaled frequencies for tasks T_{k+1} to T_n, we can obtain a new maximum frequency $f^{new}(x)$. Suppose that $f^{new}(x) = f_q^{new}(x)$, where $k + 1 \le q \le n$. If $q \ge x$, the scaled frequency for the remaining tasks in the subset Φ will be $f^{new}(x)$. Otherwise, we can assign $f^{new}(x)$ as the scaled frequency for tasks T_{k+1} to T_q, and then repeat the earlier process until we complete the frequency assignment for all tasks in the subset Φ. Then, the energy consumption for the case of selecting x highest priority tasks can be calculated. Checking through all possible values of x, finally we could obtain the optimal value of x_{opt} and corresponding frequency settings that result in the minimum energy consumption. With an additional round to assign the possible different scaled frequencies for tasks in the subset, one can derive the complexity of the RAPM-TDAM scheme as $O(n^4 r)$, where, again, $r = p_n/p_1$ is the ratio of the largest period to the smallest period.

The performance of the task-level RAPM schemes for periodic real-time tasks follows a trend similar to those of the individual recovery–based RAPM schemes for frame-based tasks. First, system reliability can be preserved by all the RAPM schemes. Second, energy savings of the RAPM schemes generally increase at smaller system utilization values since additional static slack provides better slowdown opportunities [58,59,65].

22.5 Dynamic RAPM with Online Slack Reclamation

It is well-known that real-time tasks typically take a small fraction of their worst-case execution times at run-time [20]. Moreover, for the RAPM framework with backward recovery techniques, the recovery tasks are invoked and executed only if the executions of their corresponding scaled tasks fail. Otherwise, the processor time reserved for those recovery tasks can be freed and becomes dynamic slack as well. Therefore, significant amount of dynamic slack can be expected at run-time, which should be exploited to further scale down the processing frequency of selected tasks for additional energy savings or to select additional tasks and enhance system reliability.

22.5.1 Dynamic RAPM for Frame-Based Tasks

For frame-based task systems where the tasks share a common period (and deadline), any dynamic slack generated at run-time is accessible to all remaining tasks within the current frame. Therefore, whenever additional dynamic slack is generated at run-time, one approach would be to redistribute all system slack by solving an RAPM problem of smaller size for the remaining tasks. However, the required computational overhead can be high, especially for systems with many tasks.

Another straightforward approach is to allocate all available dynamic slack to the next task to be dispatched. If the next task already has a statically assigned recovery task, such dynamic slack can be utilized to further scale down the processing frequency for the task as long as the resulting frequency is not lower than the energy-efficient frequency f_{ee}. Otherwise (i.e., if there is no statically assigned recovery task), in case that the dynamic slack is not enough to schedule a recovery for the next task, it cannot be reclaimed by the next task and will be saved for future tasks. For cases where the amount of dynamic slack is large enough, after scheduling a recovery for the next task, the remaining dynamic slack can be utilized to scale down the processing frequency of the task. It has been shown that such greedy slack reclamation is quite effective for additional energy savings [55,56].

22.5.2 Dynamic RAPM for Periodic Tasks

In periodic execution settings, the dynamic slack may be generated at different priorities and may not always be *reclaimable* by the next ready job. A piece of slack is reclaimable for a job only if the slack has higher priority than that of the job [4]. Moreover, possible preemptions that a job could experience *after* it has reclaimed some slack further complicate the problem. This is because, in the RAPM framework, once the execution of a job is scaled through DVFS, additional slack must be reserved for the potential recovery operation to preserve system reliability. Therefore, conserving the reclaimed slack by a job until it completes its execution (at which point the slack may be used for recovery operation if errors occur, or freed otherwise) is essential in reliability-aware power management settings.

Slack management for periodic real-time tasks has been studied extensively (as in the CASH-queue [8] and α-queue [4] techniques) for different purposes. By borrowing and also extending some fundamental ideas from these studies, we discuss the *wrapper-task* mechanism to track and manage dynamic slack, which can guarantee the *conservation* of the reclaimed slack by a job for dynamic RAPM schemes. Essentially, each wrapper task represents a piece of dynamic slack generated at run time. At the highest level, there are three rules for managing dynamic slack with wrapper tasks:

- **Rule 1 (slack generation):** When new slack is generated due to early completion of jobs or unneeded recovery jobs, a new wrapper task is created with the following two timing parameters: a *size* that equals the amount of dynamic slack generated and a *priority* that is equal to that of the job whose early completion gave rise to this slack. Then, the newly created wrapper task is added into a wrapper-task queue (*WT-Queue*), which is used to track available dynamic slack. The wrapper tasks with the same priority can be merged into a larger wrapper task with size equal to the summation of these wrapper tasks' sizes.

- **Rule 2 (slack reclamation):** The slack is reclaimed when: (a) the next dispatched highest priority job is a non-scaled job and its reclaimable slack is larger than the job's WCET (which ensures that a recovery, in the form of re-execution, can be scheduled to preserve reliability); or, (b) the next dispatched job has already been scaled (i.e., CPU time has been already reserved for its recovery) but its scaled frequency is still higher than the energy-efficient frequency f_{ee} and reclaimable slack exists. After reclamation, the corresponding wrapper tasks are removed from the *WT-Queue* and destroyed, which guarantees the conservation of slack for the job under consideration.

- **Rule 3 (slack exchange):** After slack reclamation, the remaining wrapper tasks in the *WT-Queue* compete for the processor along with ready jobs. When a wrapper task has the highest priority and is "scheduled": (a) if there are available ready jobs, the wrapper task will "fetch" the highest priority job and "*wrap*" the execution of that job during the interval when the wrapper task is "executed." In this case, the corresponding slack is actually borrowed to the ready job. When the job returns such slack to the system, the new slack will have a lower priority than that of the job; (b) otherwise, if there is no ready job, the processor becomes idle. The wrapper task is said to "execute no-ops" and the corresponding dynamic slack is consumed/wasted during this time interval.

It has been shown that the wrapper task–based mechanism can effectively manage dynamic slack at run-time for dynamic RAPM schemes [58,59]. In general, higher energy savings can be obtained when more dynamic slack is generated at run-time (with higher workload variability). Also, by potentially managing more tasks, higher system reliability can be achieved by the dynamic RAPM schemes. Interested readers are referred to [58,59] for detailed discussions.

22.6 Reliability Maximization with Energy Constraints

In the RAPM schemes discussed so far, the goal was to minimize the system energy consumption while preserving the system's original reliability by appropriately scheduling the required recovery tasks. In this section, considering the energy-constrained operation settings where the system energy consumption for any given interval must not exceed a hard bound [1,11,35,47], we discuss the schemes that aim at *maximizing system reliability*. Here, the negative effects of DVFS on the system reliability are incorporated but no recovery task is considered/scheduled. That is, we consider the problem of determining task-level frequency assignments to maximize overall system reliability (the probability of completing all tasks successfully) within the given energy budget and without missing the deadlines of tasks. For simplicity, we focus on frame-based task systems with a common deadline; however, the approach can be extended to periodic real-time tasks as well [54].

Let E_{budget} be the energy budget that can be utilized by all tasks within a frame that has the period (deadline) of D. Suppose that the processing frequency for task T_i is f_i ($\leq f_{max}$), which can vary from task to task. The system energy consumption within a frame can be given as follows:

$$E(f_1,\ldots,f_n) = P_s \cdot D + \sum_{i=1}^{n} E_i(f_i)$$

$$= P_s \cdot D + \sum_{i=1}^{n} \left(P_{ind_i} \cdot \frac{c_i}{f_i} + C_{ef} \cdot c_i \cdot f_i^2 \right) \qquad (22.19)$$

Here, $E_i(f_i)$ stands for the active energy consumed by task T_i, which comes from the frequency-independent and frequency-dependent power (see Section 22.2.2). We consider a general case where the frequency-independent power P_{ind_i} can vary from task to task. Note that $E_i(f_i)$ is a strictly convex function and is minimized when $f_i = f_{ee_i}$; which is the energy-efficient frequency for task T_i that can be derived through Equation 22.2.

Given that the rate of soft errors caused by transient faults follows a Poisson distribution as discussed in Section 22.2, the reliability (i.e., the probability of completing a task without having errors caused by transient faults) of task T_i in one frame at the frequency f_i is $R_i(f_i) = e^{-\lambda(f_i)*c_i/f_i}$, where $\lambda(f_i)$ is given by Equation 22.3. Let $\varphi_i(f_i) = \lambda(f_i) \cdot c_i/f_i$. Considering that the correct operation of a system depends on all its tasks, the system reliability within a frame can be represented as

$$R(f_1,\ldots,f_n) = \prod_{i=1}^{n} R_i(f_i) = e^{-\sum_{i=1}^{n} \varphi_i(f_i)} \tag{22.20}$$

Considering the well-known features of the exponential functions, to maximize system reliability $R(f_1,\ldots,f_n)$, we need to minimize $\sum_{i=1}^{n} \varphi_i(f_i)$. Therefore, the *energy-constrained reliability management* (ECRM) problem can be stated as: find the processing frequency f_i $(1 \leq i \leq n)$ so as to

$$\text{minimize} \quad \sum_{i=1}^{n} \varphi_i(f_i) \tag{22.21}$$

Subject to:

$$E(f_1,\ldots,f_n) \leq E_{budget} \tag{22.22}$$

$$\sum_{i=1}^{n} \frac{c_i}{f_i} \leq D \tag{22.23}$$

$$f_{ee_i} \leq f_i \leq f_{max} \qquad (1 \leq i \leq n) \tag{22.24}$$

Here, the first inequality corresponds to the hard energy constraint; the second one encodes the deadline constraint. The last constraint set gives the range of feasible frequency assignments for the tasks.

Let E_{min} be the minimum energy that must be allocated to the given task system to allow their completion before or at the deadline D. Given the task parameters, E_{min} can be computed by the polynomial-time algorithm developed in [3]. As a by-product, the same algorithm yields also the optimal task-level frequency assignments $(fl_1, fl_2, ..., fl_n)$ for the tasks when the total system energy consumption is *exactly* E_{min}. Obviously, if the energy budget E_{budget} is less than E_{min}, there is no solution to the problem as the system would lack the minimum energy needed for timely completion.

Moreover, let $E_{max} = E(f_{max},\ldots,f_{max})$ be the maximum energy consumption of the task set when all tasks run at f_{max}. As another boundary condition, when the given energy budget $E_{budget} \geq E_{max}$, executing all tasks at the maximum frequency is the optimal solution. Note that $R_i(f_i)$ is a strictly concave and increasing function of f_i. Therefore, in what follows, we will focus exclusively on settings where $E_{min} \leq E_{budget} < E_{max}$.

We first explain that in the optimal solution to the ECRM problem, the resulting total energy consumption $E(f_1^{opt},\ldots,f_n^{opt})$ must be equal to E_{budget}. Otherwise, with the assumption that $E_{budget} < E_{max}$, there must be a frequency for task T_i such that $f_i^{opt} < f_{max}$. In this case, it should be possible to increase f_i^{opt} by ε (> 0) such that $f_i' = (f_i^{opt} + \varepsilon) \leq f_{max}$ and $E(f_1^{opt},\ldots,f_i',\ldots,f_n^{opt}) \leq E_{budget}$. It is clear that the deadline and energy constraints are still satisfied after this modification. Further, as $R_i(f_i)$ increases monotonically with the increasing frequency f_i, the overall system reliability can be improved due to the execution of T_i at frequency f_i', which is higher than f_i^{opt}. Hence, in the optimal solution $(f_1^{opt},\ldots,f_n^{opt})$, the resulting total energy consumption $E(f_1^{opt},\ldots,f_n^{opt})$ must be equal to E_{budget}, the given energy budget within a frame.

Therefore, the energy constraint for the original ECRM problem (Equation 22.22) can be rewritten as

$$E(f_1, \ldots, f_n) = E_{budget} \tag{22.25}$$

which leads to a new form of nonlinear (convex) optimization problem ECRM.

The new ECRM problem can be solved, for instance, by *Quasi-Newton* techniques developed for constrained nonlinear optimization [5]. The technique exploits the well-known Kuhn–Tucker optimality conditions for nonlinear programs in an iterative fashion by transforming the original problem to a quadratic programming problem and solving it optimally. While optimal, a theoretical complication with this approach is that it is very difficult to express the maximum number of iterations as a function of the number of unknowns which, in this case, corresponds to the number of tasks n.

The new ECRM problem can also be tackled with an efficient heuristic scheme (denoted as ECRM-LU) that provably runs in polynomial time and satisfies the energy deadline and frequency constraints. ECRM-LU proceeds as follows. We temporarily ignore the deadline constraint (Equation 22.23) and solve the problem only by considering the new energy constraint (Equation 22.25) and frequency range constraints (Equation 22.24). Notice that by excluding the deadline constraint, the problem is transformed to a separable convex optimization problem with n unknowns, $2n$ inequality constraints, and a single equality constraint. This problem, in turn, can be solved in time $O(n^3)$ by iteratively manipulating the Kuhn–Tucker optimality conditions in a way similar to the technique illustrated in algorithm given in [3]. Now, if the resulting solution satisfies also the deadline constraint, obviously it is the solution to the ECRM problem.

Otherwise, we can rewrite the frequency constraint set as

$$fl_i \leq f_i \leq f_{max} \quad (1 \leq i \leq n) \tag{22.26}$$

where fl_i is the frequency assignment to task T_i in the solution where the task set completes *at exactly* $t = D$ and with energy consumption E_{min}. Again, the corresponding $\{fl_i\}$ values can be computed in time $O(n^3)$ [3]. By enforcing the new frequency constraint set given earlier, we make sure that the final frequency assignments satisfy also the deadline constraint. Once again, this version of the problem where the deadline constraint is handled implicitly by enforcing the lower bounds on frequency assignments can be solved in time $O(n^3)$. Hence, the overall time complexity of ECRM-LU is also $O(n^3)$.

The evaluation through extensive simulations show that ECRM-LU yields reliability figures that are extremely close (within 1%) to those achieved by the optimal solution [52,54]. In general, the achieved system reliability increases with increasing energy budget as more energy enables the system to operate at higher processing frequencies, which results in fewer errors due to the reduced number of transient faults. Moreover, the system workload also has an interesting effect on system reliability. Increasing the workload leads to increased execution time, which in turn means increased probability of incurring errors caused by transient faults. However, higher system workload also forces the system to adopt higher frequencies in order to meet the timing constraints of tasks, which instead has a positive impact on the system reliability due to the reduced error rates. Our results show that for low workload (for instance, where the amount of slack is more than the amount of computation of tasks), the first effect dominates. After the workload reaches a certain point, the second effect becomes the dominant factor and leads to better system reliability.

22.7 Other Techniques and Future Directions

In addition to the RAPM framework discussed earlier, there have been other studies on the comanagement of system reliability and energy consumption, with and without the consideration of the negative effect of DVFS on system reliability, respectively. After a brief overview of other reliability and energy comanagement techniques, in this section, we point out some open problems and possible future directions for this line of research.

22.7.1 Comanagement of Energy and Reliability

One of the earliest energy-aware fault tolerance schemes has been proposed by Unsal et al. for a set of independent periodic real-time tasks [43]. Based on the primary and backup fault tolerance model, the scheme postpones as much as possible the execution of backup tasks to minimize the overlap between primary and backup executions and thus to reduce system energy consumption. With the goal of tolerating a fixed number of transient faults, Melhem et al. explored the optimal number of checkpoints, *uniformly* or *nonuniformly* distributed, to achieve the minimum energy consumption for a duplex system (where two hardware processing units are used to run the same software concurrently for fault detection) [30]. To reduce the energy consumption in a traditional TMR system (triple modular redundancy, in which three hardware processing units are used to run the same software simultaneously to detect and mask faults), Elnozahy et al. studied an interesting *Optimistic-TMR* (OTMR) scheme, which allows one processing unit to run at a scaled processing frequency, provided that it can catch up and finish the computation before the deadline if there is a fault [18]. In [63], Zhu et al. further explored the optimal frequency settings for an OTMR system and presented detailed comparisons among Duplex, TMR, and OTMR for reliability and energy consumption. For parallel real-time tasks running on multiprocessor systems, Zhu et al. studied the optimal redundant configuration of the processors to tolerate a given number of transient faults through backward recovery techniques [62,64].

With the assumption that the arrivals of transient faults follow a Poisson distribution with a constant arrival rate, Zhang et al. studied an adaptive checkpointing scheme to tolerate a fixed number of transient faults during the execution of a single real-time task [49]. The scheme dynamically adjusts the checkpointing interval during the execution of a task based not only on the slack time but also on the occurrences of transient faults during task's execution, so as to reduce system energy consumption. The adaptive checkpointing scheme was extended to a set of periodic real-time tasks on a single processor system with the EDF (earliest deadline first) scheduling algorithm [51]. In [50], the authors further considered the cases where faults may occur within checkpoints. Following a similar idea and considering a fixed priority rate-monotonic scheduling (RMS) algorithm, Wei et al. studied an efficient online scheme to minimize energy consumption by applying DVFS with the consideration of the run-time behaviors of tasks and fault occurrences while still satisfying the timing constraints [44]. In [45], the authors extended the study to multi-processor real-time systems. Izosimov et al. studied an optimization problem for mapping a set of tasks with reliability constraints, timing constraints, and precedence relations to processors and for determining appropriate fault tolerance policies (re-execution and replication) for the tasks [24]. However, these studies did not address the negative effects of DVFS on system reliability due to the higher rate of soft errors caused by transient faults at lower supply voltages.

Taking such negative effects into consideration, in addition to the RAPM schemes discussed in this chapter, for real-time periodic tasks that have different reliability requirements, a research effort that proposes the schemes that selectively recover a subset of jobs for each task is given in [66]. For real-time tasks with known statistical execution times, an optimistic RAPM has also been proposed. The scheme deploys smaller size recovery tasks while still preserving the system's original reliability [60].

Ejlali et al. studied schemes that combine the information (about hardware resources) and temporal redundancy to save energy and to preserve system reliability [17]. By employing a feedback controller to track the overall miss ratio of tasks in soft real-time systems, Sridharan et al. [42] proposed a reliability-aware energy management algorithm to minimize the system energy consumption while still preserving the overall system reliability. Pop et al. studied the problem of energy and reliability trade-offs for distributed heterogeneous embedded systems [32]. The main idea is to transform the user-defined reliability goals to the objective of tolerating a fixed number of transient faults by switching to predetermined contingency schedules and re-executing individual tasks. A constrained logic programming-based algorithm is proposed to determine the voltage levels, process start time, and message transmission time to tolerate transient faults and minimize energy consumption while meeting the timing constraints of the application. Dabiri et al. studied the problem of assigning frequency and supply voltage to tasks for energy minimization subject to reliability as well as timing constraints [14].

More recently, following the similar idea in OTMR [18], Ejlali et al. studied a standby-sparing energy-efficient hardware redundancy technique for fault tolerance, where a standby processor is operated at a low-power state whenever possible, provided that it can catch up and finish the tasks in time [16]. This scheme was shown to have better energy performance when compared to that of the backward recovery-based RAPM approach. For frame-based task systems with independent tasks, Qi et al. investigated global scheduling based RAPM problem and studied several individual recovery–based schemes depending on when and how the tasks are selected [34].

22.7.2 Future Research Directions

Among the open problems, the extension of the shared recovery technique to the preemptive periodic execution settings needs to be mentioned. Moreover, with the emergence of multicore processors, extending that technique to multiprocessor real-time systems can also be an interesting direction. With the inherent hardware redundancy, multicore systems provide excellent opportunities to tolerate permanent faults. However, how to effectively integrate the hardware and temporal redundancy to tolerate both permanent and transient faults while reducing energy consumption, especially considering the intriguing interplay between power/energy, temperature, and rate of system failure, is another major open problem.

22.8 Summary

In this chapter, we discussed several reliability-aware power management (RAPM) schemes for real-time tasks running on a single processor. The motivation comes from the negative effect of the widely deployed power management technique, namely *dynamic voltage and frequency scaling* (DVFS), on system reliability due to the increased transient fault rates associated with operation at low supply voltages. The key idea of these RAPM schemes is to recuperate the reliability loss due to DVFS by scheduling proper recovery tasks, while still exploiting the remaining system slack to save energy.

Starting with the case of a single real-time task, we showed that, by scheduling a recovery task before scaling down the execution of the task, the RAPM scheme can preserve the system reliability while still obtaining energy savings. Then, for frame-based task systems where multiple real-time tasks share a common deadline, the RAPM problem is shown to be NP-hard and two efficient heuristics with individual recovery tasks are discussed. Aiming to address the pessimism of the individual recovery–based schemes, the RAPM scheme with a shared recovery task is presented. For general periodic real-time tasks that have different deadlines, we presented a task-level RAPM technique where all jobs of the same periodic task receive the same treatment. Then, for the *earliest-deadline-first* (EDF) scheduling policy, the utilization-based RAPM scheme are discussed, followed by the priority-monotonic RAPM schemes for *rate-monotonic scheduling* (RMS). Dynamic RAPM schemes that exploit dynamic slack generated at run-time are further considered. For energy-constrained systems, we discussed the RAPM scheme that aims at maximizing system reliability. At the end, an overview of other related studies is provided and a few open research problems are identified.

References

1. T.A. AlEnawy and H. Aydin. On energy-constrained real-time scheduling. In *Proceedings of the 16th Euromicro Conference on Real-Time Systems (ECRTS)*, Catania, Italy, pp. 165–174, 2004. Washington, DC: IEEE Computer Society.

2. Advanced Micro Devices. AMD Opteron quad-core processors; http://www.amd.com/us/products/embedded/processors/, 2009.

3. H. Aydin, V. Devadas, and D. Zhu. System-level energy management for periodic real-time tasks. In *Proceedings of The 27th IEEE Real-Time Systems Symposium (RTSS)*, Rio de Jeneiro, Brazil, 2006. Piscataway, NJ: IEEE CS Press.

4. H. Aydin, R. Melhem, D. Mossé, and P. Mejia-Alvarez. Power-aware scheduling for periodic real-time tasks. *IEEE Transactions on Computers*, 53(5):584–600, 2004.

5. M.S. Bazaraa, H.D. Sherali, and C.M. Shetty. *Nonlinear Programming: Theory and Algorithms*. John Wiley & Sons, 2005.

6. E. Bini, G.C. Buttazzo, and G. Lipari. Speed modulation in energy-aware real-time systems. In *Proceedings of the 17th Euromicro Conference on Real-Time Systems (ECRTS)*, Palma de Mallorca, Spain, pp. 3–10, 2005. Washington, DC: IEEE Computer Society.

7. T.D. Burd and R.W. Brodersen. Energy efficient CMOS microprocessor design. In *Proceedings of the 28th Hawaii International Conference on System Sciences (HICSS)*, Kihei, Maui, Hawaii, pp. 288–297, 1995. Washington, DC: IEEE Computer Society.

8. M. Caccamo, G. Buttazzo, and L. Sha. Capacity sharing for overrun control. In *Proceedings of the 21st IEEE Real-Time Systems Symposium (RTSS)*, Orlando, FL, pp. 295–304, 2000. Washington, DC: IEEE Computer Society.

9. X. Castillo, S. McConnel, and D. Siewiorek. Derivation and calibration of a transient error reliability model. *IEEE Transactions on Computers*, 31(7):658–671, 1982.

10. J.-J. Chen and T.-W. Kuo. Multiprocessor energy-efficient scheduling for real-time tasks with different power characteristics. In *Proceedings of the 2005 International Conference on Parallel Processing (ICPP)*, pp. 13–20, 2005. Washington, DC: IEEE Computer Society.

11. J.-J. Chen and T.-W. Kuo. Voltage-scaling scheduling for periodic real-time tasks in reward maximization. In *Proceedings of the 26th IEEE Real-Time Systems Symposium (RTSS)*, Miami, FL, pp. 345–355, 2005. Washington, DC: IEEE Computer Society.

12. J.-J. Chen and T.-W. Kuo. Procrastination determination for periodic real-time tasks in leakage-aware dynamic voltage scaling systems. In *Proceedings of the 2007 IEEE/ACM International Conference on Computer-Aided Design (ICCAD)*, San Jose, CA, pp. 289–294, 2007. Piscataway, NJ: IEEE Press.

13. Intel Corporation Intel embedded quad-core Xeon; http://www.intel.com/products/embedded/processors.htm, 2009.

14. F. Dabiri, N. Amini, M. Rofouei, and M. Sarrafzadeh. Reliability-aware optimization for DVS-enabled real-time embedded systems. In *Proceedings of the Ninth International Symposium on Quality Electronic Design (ISQED)*, San Jose, CA, pp. 780–783, 2008. Washington, DC: IEEE Computer Society.

15. V. Degalahal, L. Li, V. Narayanan, M. Kandemir, and M.J. Irwin. Soft errors issues in low-power caches. *IEEE Transactions on Very Large Scale Integration (VLSI) Systems*, 13(10):1157–1166, October 2005.

16. A. Ejlali, B.M. Al-Hashimi, and P. Eles. A standby-sparing technique with low energy-overhead for fault-tolerant hard real-time systems. In *Proceedings of the Seventh IEEE/ACM International Conference on Hardware/Software Codesign and System Synthesis (CODES)*, Grenoble, France, pp. 193–202, 2009. New York: ACM.

17. A. Ejlali, M.T. Schmitz, B.M. Al-Hashimi, S.G. Miremadi, and P. Rosinger. Energy efficient seu-tolerance in DVS-enabled real-time systems through information redundancy. In *Proceedings of the International Symposium on Low Power and Electronics and Design (ISLPED)*, San Diego, CA, pp. 281–286, 2005. New York: ACM.

18. E. (Mootaz) Elnozahy, R. Melhem, and D. Mossé. Energy-efficient duplex and tmr real-time systems. In *Proceedings of The 23rd IEEE Real-Time Systems Symposium (RTSS)*, Austin, TX, pp. 256–265, 2002. Washington, DC: IEEE Computer Society.

19. D. Ernst, S. Das, S. Lee, D. Blaauw, T. Austin, T. Mudge, N.S. Kim, and K. Flautner. Razor: Circuit-level correction of timing errors for low-power operation. *IEEE Micro*, 24(6):10–20, 2004.

20. R. Ernst and W. Ye. Embedded program timing analysis based on path clustering and architecture classification. In *Proceedings of the International Conference on Computer-Aided Design (ICCAD)*, San Jose, CA, pp. 598–604, 1997. New York: ACM.

21. P. Hazucha and C. Svensson. Impact of CMOS technology scaling on the atmospheric neutron soft error rate. *IEEE Transactions on Nuclear Science*, 47(6):2586–2594, 2000.

22. S. Irani, S. Shukla, and R. Gupta. Algorithms for power savings. In *Proceedings of the 14th Annual ACM-SIAM Symposium on Discrete Algorithms (SODA)*, Baltimore, MA, pp. 37–46, 2003. Philadelphia, PA: Society for Industrial and Applied Mathematics.

23. R.K. Iyer, D.J. Rossetti, and M.C. Hsueh. Measurement and modeling of computer reliability as affected by system activity. *ACM Transactions on Computer Systems*, 4(3):214–237, August 1986.

24. V. Izosimov, P. Pop, P. Eles, and Z. Peng. Design optimization of time-and cost-constrained fault-tolerant distributed embedded systems. In *Proceedings of the Conference on Design, Automation and Test in Europe (DATE)*, Munich, Germany, pp. 864–869, 2005. Washington, DC: IEEE Computer Society.

25. R. Jejurikar, C. Pereira, and R. Gupta. Leakage aware dynamic voltage scaling for real-time embedded systems. In *Proceedings of the 41st Design Automation Conference (DAC)*, San Diego, CA, pp. 275–280, 2004. New York: ACM.

26. T. Juhnke and H. Klar. Calculation of the soft error rate of submicron CMOS logic circuits. *IEEE Journal of Solid-State Circuits*, 30(7):830–834, 1995.

27. A.R. Lebeck, X. Fan, H. Zeng, and C.S. Ellis. Power aware page allocation. In *Proceedings of the Ninth International Conference on Architectural Support for Programming Languages and Operating Systems (ASPLOS)*, Cambridge, MA, pp. 105–116, 2000. New York: ACM.

28. J. Lehoczky, L. Sha, and Y. Ding. The rate monotonic scheduling algorithm: Exact characterization and average case behavior. In *Proceedings of the IEEE Real-Time Systems Symposium (RTSS)*, Santa Monica, CA, pp. 166–171, 1989. Washington, DC: IEEE Computer Society.

29. C.L. Liu and J.W. Layland. Scheduling algorithms for multiprogramming in a hard real-time environment. *Journal of ACM*, 20(1):46–61, 1973.

30. R. Melhem, D. Mossé, and E. (Mootaz) Elnozahy. The interplay of power management and fault recovery in real-time systems. *IEEE Transactions on Computers*, 53(2):217–231, 2004.

31. P. Pillai and K.G. Shin. Real-time dynamic voltage scaling for low-power embedded operating systems. In *Proceedings of the 18th ACM Symposium on Operating Systems Principles (SOSP)*, Chateau Lake Louise, Banff, Canada, pp. 89–102, 2001. New York: ACM.

32. P. Pop, K.H. Poulsen, V. Izosimov, and P. Eles. Scheduling and voltage scaling for energy/reliability trade-offs in fault-tolerant time-triggered embedded systems. In *Proceedings of the Fifth IEEE/ACM International Conference on Hardware/Software Codesign and System Synthesis (CODES+ISSS)*, Salzburg, Austria, pp. 233–238, 2007. New York: ACM.

33. D.K. Pradhan. *Fault Tolerance Computing: Theory and Techniques*. Upper Saddle River, NJ: Prentice Hall, 1986.

34. X. Qi, D. Zhu, and H. Aydin. Global reliability-aware power management for multiprocessor real-time systems. In *Proceedings of the IEEE International Conference on Embedded and Real-Time Computing Systems and Applications (RTCSA)*, Macau, People's Republic of China, pp. 183–192, 2010. Los Alamitos, CA: IEEE Computer Society.

35. C. Rusu, R. Melhem, and D. Mossé. Maximizing rewards for real-time applications with energy constraints. *ACM Transactions on Embedded Computing Systems*, 2(4):537–559, November 2003.

36. S. Saewong and R. Rajkumar. Practical voltage scaling for fixed-priority RT-systems. In *Proceedings of the Ninth IEEE Real-Time and Embedded Technology and Applications Symposium (RTAS)*, Toronto, ON, Canada, pp. 106–115, 2003. Washington, DC: IEEE Computer Society.

37. C. Scordino and G. Lipari. A resource reservation algorithm for power-aware scheduling of periodic and aperiodic real-time tasks. *IEEE Transactions on Computers*, 55(12):1509–1522, 2006.

38. N. Seifert, D. Moyer, N. Leland, and R. Hokinson. Historical trend in alpha-particle induced soft error rates of the Alpha^{TM} microprocessor. In *Proceedings of the 39th IEEE Annual International Reliability Physics Symposium*, Orlando, FL, pp. 259–265, 2001. Washington, DC: IEEE Computer Society.

39. Tezzaron Semiconductor. Soft errors in electronic memory: A white paper. Available at http://www.tachyonsemi.com/about/papers/, 2004.

40. K. Seth, A. Anantaraman, F. Mueller, and E. Rotenberg. Fast: Frequency-aware static timing analysis. In *Proceedings of the 24th IEEE Real-Time System Symposium (RTSS)*, Cancun, Mexico, pp. 200–224, 2003. New York: ACM.

41. P. Shivakumar, M. Kistler, S.W. Keckler, D. Burger, and L. Alvisi. Modeling the effect of technology trends on the soft error rate of combinational logic. In *Proceedings of the International Conference on Dependable Systems and Networks (DSN)*, Bethesda, MA, pp. 389–398, 2002. Washington, DC: IEEE Computer Society.

42. R. Sridharan, N. Gupta, and R. Mahapatra. Feedback-controlled reliability-aware power management for real-time embedded systems. In *Proceedings of the 45th Annual Design Automation Conference (DAC)*, Anaheim, CA, pp. 185–190, 2008. New York: ACM.

43. O.S. Unsal, I. Koren, and C.M. Krishna. Towards energy-aware software-based fault tolerance in real-time systems. In *Proceedings of the International Symposium on Low Power Electronics Design (ISLPED)*, Monterey, CA, pp. 124–129, 2002. New York: ACM.

44. T. Wei, P. Mishra, K. Wu, and H. Liang. Online task-scheduling for fault-tolerant low-energy real-time systems. In *Proceedings of IEEE/ACM International Conference on Computer-Aided Design (ICCAD)*, San Jose, CA, pp. 522–527, 2006. New York: ACM.

45. T. Wei, P. Mishra, K. Wu, and H. Liang. Fixed-priority allocation and scheduling for energy-efficient fault tolerance in hard real-time multiprocessor systems. *IEEE Transactions on Parallel and Distributed Systems (TPDS)*, 19:1511–1526, 2008.

46. M. Weiser, B. Welch, A. Demers, and S. Shenker. Scheduling for reduced CPU energy. In *Proceedings of the First USENIX Symposium on Operating Systems Design and Implementation (OSDI)*, Monterey, CA, 1994. Berkeley, CA: USENIX Association.

47. H. Wu, B. Ravindran, and E.D. Jensen. Utility accrual real-time scheduling under the unimodal arbitrary arrival model with energy bounds. *IEEE Transactions on Computers*, 56(10):1358–1371, October 2007.

48. F. Yao, A. Demers, and S. Shenker. A scheduling model for reduced CPU energy. In *Proceedings of the 36th Symposium on Foundations of Computer Science (FOCS)*, Milwaukee, WI, pp. 374–382, 1995. Washington, DC: IEEE Computer Society.

49. Y. Zhang and K. Chakrabarty. Energy-aware adaptive checkpointing in embedded real-time systems. In *Proceedings of the Conference on Design, Automation and Test in Europe (DATE)*, Munich, Germany, pp. 918–923, 2003. Washington, DC: IEEE Computer Society.

50. Y. Zhang and K. Chakrabarty. Task feasibility analysis and dynamic voltage scaling in fault-tolerant real-time embedded systems. In *Proceedings of IEEE/ACM Design, Automation and Test in Europe Conference (DATE)*, Paris, France, pp. 1170–1175, 2004. Los Alamitos, CA: IEEE Computer Society.

51. Y. Zhang, K. Chakrabarty, and V. Swaminathan. Energy-aware fault tolerance in fixed-priority real-time embedded systems. In *Proceedings of the 2003 IEEE/ACM International Conference on Computer-Aided Design (ICCAD)*, San Jose, CA, pp. 209–214, 2003. Los Alamitos, CA: IEEE Computer Society.

52. B. Zhao, H. Aydin, and D. Zhu. Reliability-aware dynamic voltage scaling for energy-constrained real-time embedded systems. In *Proceedings of the IEEE International Conference on Computer Design (ICCD)*, Lake Tahoe, CA, pp. 633–639, 2008. Piscataway, NJ: IEEE CS Press.

53. B. Zhao, H. Aydin, and D. Zhu. Enhanced reliability-aware power management through shared recovery technique. In *Proceedings of the ACM/IEEE International Conference on Computer Aided Design (ICCAD)*, San Jose, CA, pp. 63–70, 2009. New York: ACM.

54. B. Zhao, H. Aydin, and D. Zhu. On maximizing reliability of real-time embedded applications under hard energy constraint. *IEEE Transactions on Industrial Informatics*, 6(3):316–328, 2010.

55. D. Zhu. Reliability-aware dynamic energy management in dependable embedded real-time systems. In *Proceedings of the IEEE Real-Time and Embedded Technology and Applications Symposium (RTAS)*, San Jose, CA, pp. 397–407, 2006. Piscataway, NJ: IEEE CS Press.

56. D. Zhu. Reliability-aware dynamic energy management in dependable embedded real-time systems. *ACM Transactions on Embedded Computing Systems (TECS)*, 10(2):26.1–26.27, December 2010.

57. D. Zhu and H. Aydin. Energy management for real-time embedded systems with reliability requirements. In *Proceedings of the International Conference on Computer Aided Design (ICCAD)*, San Jose, CA, pp. 528–534, 2006. New York: ACM Press.

58. D. Zhu and H. Aydin. Reliability-aware energy management for periodic real-time tasks. In *Proceedings of the IEEE Real-Time and Embedded Technology and Applications Symposium (RTAS)*, Bellevue, Washington, DC, pp. 225–235, 2007. Piscataway, NJ: IEEE CS Press.

59. D. Zhu and H. Aydin. Reliability-aware energy management for periodic real-time tasks. *IEEE Transactions on Computers*, 58(10):1382–1397, 2009.

60. D. Zhu, H. Aydin, and J.-J. Chen. Optimistic reliability aware energy management for real-time tasks with probabilistic execution times. In *Proceedings of the 29th IEEE Real-Time Systems Symposium (RTSS)*, Barcelona, Spain, pp. 313–322, 2008. Piscataway, NJ: IEEE CS Press.

61. D. Zhu, R. Melhem, and D. Mossé. The effects of energy management on reliability in real-time embedded systems. In *Proceedings of the International Conference on Computer Aided Design (ICCAD)*, San Jose, CA, pp. 35–40, 2004. New York: ACM Press.

62. D. Zhu, R. Melhem, and D. Mossé. Energy efficient configuration for QoS in reliable parallel servers. In *Proceedings of the Fifth European Dependable Computing Conference (EDCC)*, Lecture Notes in Computer Science, Budapest, Hungary, pp. 122–139, April 2005. New York: Springer.

63. D. Zhu, R. Melhem, D. Mossé, and E.(Mootaz) Elnozahy. Analysis of an energy efficient optimistic TMR scheme. In *Proceedings of the Tenth International Conference on Parallel and Distributed Systems*, Newport Beach, CA, pp. 559–568, 2004. Piscataway, NJ: IEEE CS Press.

64. D. Zhu, D. Mossé, and R. Melhem. Energy efficient redundant configurations for real-time parallel reliable servers. *Journal of Real-Time Systems*, 41(3):195–221, April 2009.

65. D. Zhu, X. Qi, and H. Aydin. Priority-monotonic energy management for real-time systems with reliability requirements. In *Proceedings of the IEEE International Conference on Computer Design (ICCD)*, Lake Tahoe, CA, pp. 629–635, 2007. Piscataway, NJ: IEEE CS Press.

66. D. Zhu, X. Qi, and H. Aydin. Energy management for periodic real-time tasks with variable assurance requirements. In *Proceedings of the IEEE International Conference on Embedded and Real-Time Computing Systems and Applications (RTCSA)*, Koushisung, Taiwan, pp. 259–268, 2008. Piscataway, NJ: IEEE CS Press.

67. J.F. Ziegler. Terrestrial cosmic ray intensities. *IBM Journal of Research and Development*, 42(1):117–139, 1998.

68. J.F. Ziegler. Trends in electronic reliability: Effects of terrestrial cosmic rays. Available at http://www.srim.org/SER/SERTrends.htm, 2004.

Energy Minimization for Multiprocessor Systems Executing Real-Time Tasks

Yumin Zhang
Synopsys Inc.

Xiaobo Sharon Hu
University of Notre Dame

Danny Ziyi Chen
University of Notre Dame

23.1 Introduction

Real-time systems must satisfy not only functional requirements but also temporal requirements. That is, certain computations would be considered incorrect if they could not be completed by some given deadlines. Computationally demanding real-time systems are often implemented with multiprocessor systems (or multiprocessors-on-chip [SoCs]). Such systems have wide applications, e.g., in aviation, automobile, and personal electronic industries [23]. Saving energy in such systems is highly desirable due to its impact on cost, weight, reliability, etc.

Almost all processors today are equipped with dynamic voltage scaling (DVS) capabilities as well as different power-off modes. Two main energy saving techniques used for such processors are: voltage selection (VS) (also called voltage scheduling [6]), which selects a processor's supply voltage according to the performance requirement by tasks, and power management (PM), which shuts down a processor when there is no task running [3]. Applying VS judiciously can achieve significant energy saving [18]. There is research demonstrating that the transition of voltages can be done on the fly [16] and some DVS processors can still execute instructions during a transition [5]. However, most processors commercially available today do incur nonnegligible transition overhead both in terms of time and energy. The time overhead can be on the order of tens of microseconds, which may need to be taken into account when considering real-time systems with deadlines on the scale of microseconds. Energy overhead, regardless of its magnitude, should be kept at minimum to reduce its adversary effect on energy saving.

In this chapter, we consider real-time tasks with possible dependencies to be executed on a given number of DVS processors where the voltage transition energy overhead cannot be ignored. The challenge is to find a system implementation that consumes the least amount of energy. Note that such implementation would also reduce the time overhead due to voltage transition. A system implementation is determined by the *task assignment and scheduling* (which task runs on which processor in what order), the *voltage selection* (how many cycles at which voltage), and the *voltage determination* (which voltage each cycle of every task runs on).

There exists a large body of research on saving energy for real-time systems (e.g., [2,21,24]). Most related work concentrates on saving energy of independent tasks or on a single processor [4,12,18], and thus cannot handle the dependencies among tasks on different processors. Tasks in real-world applications usually have control or data dependencies and many systems have multiple processors. Some approaches have been proposed in, e.g., [2,8,15,20] to solve the energy minimization problem for dependent tasks on multiple DVS processors. The following observations can be made through studying these existing approaches. First, task partition/scheduling without consideration of the following VS may not present the best energy saving potential [2,8,15]. Furthermore, approaches that evaluate eligible tasks one by one as in [8,20] fail to recognize the joint effect of slowing down certain tasks simultaneously. The joint effect is critical in finding global optimal voltage settings and an example in Section 23.4.1 explains the effect in detail.

This chapter presents a framework that integrates task scheduling, VS [25], and energy overhead consideration [26] to minimize energy consumption of real-time dependent tasks on a system consisting of multiple DVS-enabled processors. This integrated framework achieves larger energy saving compared with existing work through exploiting the interdependencies between task scheduling and VS. Both formal mathematical programming techniques as well as efficient heuristics are employed in designing the framework. Strategies on how to assign tasks to processors, how to schedule tasks on each processor, and how to select voltage levels will be discussed in detail.

The remaining part of the chapter is organized as follows. Section 23.2 describes preliminary information related to task and processor models. Section 23.3 provides an overview of the framework. The VS approach is presented in Section 23.4, task scheduling is presented in Section 23.5, and energy overhead consideration is presented in Section 23.6. Experimental results are presented in Section 23.7 and the chapter concludes in Section 23.8.

23.2 Preliminaries

This section presents the models and terminology used for real-time tasks as well as microprocessor delay and energy.

23.2.1 Task Model

We use a DAG to represent a task set. In the DAG, each node represents a task, while an edge from node u to v indicates that v can start only after u finishes. Task u may have release time rt_u and deadline dl_u that constrain u's starting time and finishing time. Deadline T_{con} on a set restricts all tasks' finishing time. The DAG for a 5-task set with $t_1, t_2, ..., t_5$, is shown in Figure 23.1a. The number inside each task is the number of time unit the task takes to finish. The time unit can be the cycle time, or it can be any unit that, applying voltage transition, is sensible. To simplify the discussion, we use cycle time as the time unit

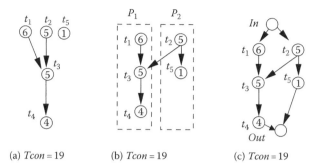

(a) *Tcon* = 19   (b) *Tcon* = 19   (c) *Tcon* = 19

FIGURE 23.1 (a) A 5-task set. (b) Tasks scheduled on P_1 and P_2. Edge (t_2, t_5) is added due to processor sharing between t_2 and t_5. (c) New DAG with IN and OUT nodes added and new edges added.

in the remaining of the chapter. The total execution time of a task is the sum of the time that each cycle takes. A task set after scheduling can still be represented as a DAG with additional edges, if necessary, to capture processor sharing. The DAG after scheduling of tasks in Figure 23.1a to two processors, P_1 and P_2, is shown in Figure 23.1b.

We assume that preemption is not allowed in task execution. That is, once a task starts to be executed, it cannot be interrupted by other tasks. This assumption is used in many practical systems where predictable execution is of prime importance, or the complexity of implementing preemption and the preemption overhead is too costly.

23.2.2 Delay and Energy

The number of cycles, N_u, that task u needs to finish does not change with the supply voltage V_{dd}, while a processor's cycle time CT changes with V_{dd}. The change of CT affects task u's total execution time d_u. Dynamic energy consumption of task u per cycle ET_u also changes with the supply voltage V_{dd}. For current CMOS technology, the dynamic energy consumption is the dominant part of energy consumption. CT and ET_u are functions of V_{dd} and can be computed as

$$CT = \frac{k \cdot V_{dd}}{(V_{dd} - V_{th})^\alpha} \tag{23.1}$$

$$ET_u = C_u \cdot V_{dd}^2 \tag{23.2}$$

where k, α are device-related parameters, V_{th} is the threshold voltage, and C_u is the effective switching capacitance per cycle. The value of α typically ranges from 1.2 to 2. If the supply voltage is the same for N_u cycles, we compute u's delay d_u and energy consumption E_u as $d_u = N_u \cdot CT$ and $E_u = N_u \cdot ET_u$. We depict the relation of E_u and d_u for different α in Figure 23.2 when V_{dd} can change continuously. It is clear that E_u is a convex function of d_u [12] and both are functions of V_{dd}. Note that our work allows tasks to have different power characteristics, such as different effective switching capacitance C_u.

When using a variable voltage processor, different cycles of task u may use different supply voltage to reduce energy consumption. The energy saving of a cycle operating at V_l instead of V_h, $\Delta E_u(V_h, V_l)$, is

$$\Delta E_u(V_h, V_l) = C_u \cdot |V_h^2 - V_l^2| \tag{23.3}$$

A voltage/frequency converter is needed to alter the supply voltage to a variable voltage processor. The energy overhead E_o and time overhead T_o of a typical voltage/frequency converter when voltage switches

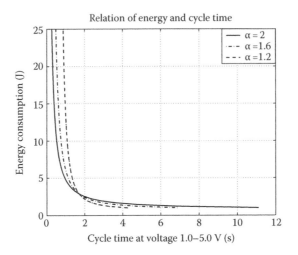

FIGURE 23.2   Convex relation of energy and delay.

between V_h and V_l can be computed as follows [4].

$$E_o = (1 - \eta) \cdot C_{DD} \cdot \left| V_h^2 - V_l^2 \right| \tag{23.4}$$

$$T_o = \frac{2C_{DD}}{I_{max}} \cdot |V_h - V_l| \tag{23.5}$$

where
   η is the efficiency of the DC-DC converter in the voltage/frequency converter
   C_{DD} is the capacitor that stores the charge
   I_{max} is the maximum output current of the converter.

Since E_o and T_o are both proportional to C_{DD}, minimizing C_{DD} will help reduce both the transition energy overhead and time overhead. However, some other system parameters, such as voltage ripple effect, loop stability, and transition efficiency, require that the value of C_{DD} not be too small. The relation of ΔE_u and E_o tells us that energy can be saved after paying the transition overhead if there are enough consecutive cycles running at V_l between transitions.

23.3 Framework Overview

This three-phase approach integrates task scheduling, VS, and energy overhead consideration together into one framework to achieve the maximum energy saving on a given number of variable voltage processors. In the first phase, the task scheduling phase, we use a priority-based deadline driven algorithm to schedule tasks on a single processor. On multiprocessor systems, tasks are scheduled based on a priority function, which is related to a task's release time and deadline, and the availability of processors at scheduling steps. The task with the best priority is assigned to a best-fit processor. In the second phase, the VS phase, the non-overhead VS problem on a given schedule is formulated as a unified IP problem for both single processor and multiple processors, and for continuous and discrete voltage cases. For some cases, we prove that the IP problem can be solved in polynomial time with optimal solutions. When the IP problem is not polynomial time solvable, we use an efficient approximation approach that achieves results close to optimal solutions shown in experiments. In the third phase, the energy overhead consideration

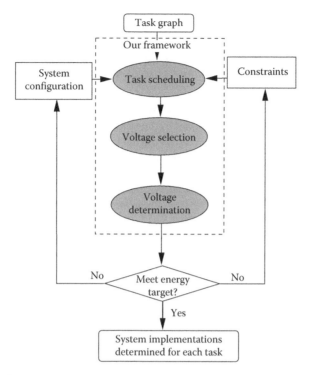

FIGURE 23.3 Design flow with our approach.

phase, we order the cycles with different voltages in such a way that the number of transitions is minimized. The minimization of number of transitions reduces the total transition overhead of both time and energy and increases the energy saving. The EOC phase also eliminates cycles at lower voltages if the energy consumed by involved transitions is more than the energy saving of running those cycles at lower voltage.

For those processors whose transition overhead of time and energy both need to be considered, timing constraints should not be violated when transition time overhead is considered. The amount of time that a task can be slowed down when time overhead is considered will be less than that when overhead is not considered. Our approach that minimizes number of transitions can decrease the time spent by transitions and increase the time that can be spent by slowing down tasks when timing overhead is considered. Moreover, in the third phase of our approach, we can examine the effect of both energy overhead and timing overhead and choose to slow down task cycles only when energy overhead can be overcome and timing constraints can still be met. Our three-phase framework can be used to optimize energy at the system level in a design flow like the one shown in Figure 23.3. The third phase is optional and is not needed if overhead can be ignored.

23.4 Voltage Selection

We present an integer programming (IP) formulation that solves the non-overhead VS problem on a given schedule of dependent tasks on a given number of processors. Assume that a task assignment and schedule are given. (How to obtain these will be discussed in Section 23.5.) A simplified version of the IP formulation can be used to solve the VS problem on a single processor. The IP formulation solves the VS problem exactly and it provides guidance for task scheduling.

23.4.1 A Motivational Example

We use the schedule in Figure 23.1b to show that different VS approaches perform quite differently. This example motivated us to make VS decisions in a global manner. Assume P_1 and P_2 are identical and can operate on two voltages, $V_h = 2$ and $V_l = 1$. For simplicity, based on rough estimation according to (23.1) and (23.2), we set $CT_{V_h} = 1$, $CT_{V_l} = 2$, the energy consumption per cycle at V_h to be 4 units and at V_l to be 1 unit. We summarize the delay and energy per cycle at different voltages in Table 23.1.

Figure 23.4 shows the system implementations by different approaches on the schedule in Figure 23.1b. Cycles in blank are executed at V_h and cycles in shade are executed at V_l. Figure 23.4a shows the execution of tasks on P_1 and P_2 when the two processors are always on V_h. The energy consumption of implementation in Figure 23.4a can serve as a baseline since it does not utilize the variable voltages. The VS decision by the LEneS approach [8] is shown in Figure 23.4b. It fails to recognize that slowing down t_1 and t_2 simultaneously is the best even though slowing down each will take away the potential of slowing down t_3, t_4, and t_5. The VS result by Luo and Jha's approach [15] is shown in Figure 23.4c. It is clear that even distribution of slack is not optimal in saving energy. The best energy saving is achieved in Figure 23.4d and the number of cycles at V_l is 10, which is an optimal solution for the example. The total

TABLE 23.1 Cycle Time and Energy per Cycle at Different Voltages

	V	CT_V	ET_V
V_h	2	1	4
V_l	1	2	1

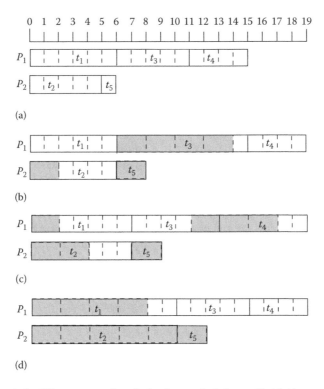

(a)

(b)

(c)

(d)

FIGURE 23.4 VS results by different approaches. Cycles that are shaded are at V_l. (a) Always on V_h. (b) By approach in (8). (c) By approach in (17). (d) Optimal solution.

	N_{V_h}	N_{V_l}	d	E	Saves (%)
(a)	21	0	15	$21 \times 4 = 84$	0
(b)	15	6	19	$15 \times 4 + 6 \times 1 = 66$	21
(c)	14	7	19	$14 \times 4 + 7 \times 1 = 63$	25
(d)	11	10	19	$11 \times 4 + 10 \times 1 = 54$	36

numbers of cycles at V_h and V_l, the delay, and the total energy consumption for the four implementations in Figure 23.4a through d are shown in column N_{V_h}, N_{V_l}, d, and E, respectively, in Table 23.2. The optimal solution saves 36% of energy compared with the baseline consumption, while the other two approaches save 21%–25%. Our VS approach achieves the optimal solution for this example.

23.4.2 Unified IP Formulation for VS

We need to guarantee that after the VS technique is applied, tasks with deadlines still finish before their deadlines and all tasks finish no later than T_{con}. That is, the delay on each path in the DAG should still not be greater than T_{con}. However, the number of paths in a DAG can be exponential to the number of edges and this makes it not practical to do optimization on path exploration. This is one reason VS approaches on single-processor systems or independent tasks cannot be readily extended to solve the VS problem on multiple processors. In a single-processor system, there is only one path. For independent tasks, the number of paths is the number of processors in the system. Here we adopt a model that is used to solve gate resizing problems [7] to formulate timing constraints on a DAG not on paths, but on nodes and edges.

We add two nodes, *IN* and *OUT*, edges from *IN* to the first task on each processor, and edges from the last task on each processor to *OUT* to form a new DAG, as shown in Figure 23.1c. Execution time of task *u at the highest supply voltage* V_h is a constant T_u, its release time rt_u and deadline dl_u are also constants. T_{OUT} and T_{IN} are set to be 0. Besides task's delay d_u, we associate another variable D_u with *u*. Timing constraints on $G(V, E)$ in Figure 23.1c can be modeled as

$$D_{OUT} - D_{IN} \le T_{con} \tag{23.6}$$

$$D_v - D_u - d_u \ge 0 \quad \forall e(u, v) \in E \tag{23.7}$$

$$D_u + d_u \le dl_u \quad \forall u \text{ with } dl_u \tag{23.8}$$

$$D_u \ge rt_u \quad \forall u \text{ with } rt_u \tag{23.9}$$

$$d_u \ge T_u, \ D_u \ge 0, \text{ int}, \ \forall u \in V \tag{23.10}$$

Intuitively, D_u represents *u*'s starting time. D_u and d_u are constrained to be integers since for most systems, delays can only be integral multiples of a timing unit. For a feasible schedule, if D_{IN} is set to 0, the aforementioned constraints guarantee that timing constraints, rt_u, dl_u and T_{con}, are all satisfied.

The objective of VS is to minimize the sum of each task *u*'s energy consumption. To trade the increase of delay for energy saving, we need to establish the relationship between d_u and E_u. If the supply voltage can change continuously, d_u can change continuously from one integer to the next with the step of one. In this case, E_u is a convex function of d_u, as depicted in Figure 23.2, and we have $E_u = f(d_u)$, where $f(\cdot)$ is a convex function. Combining $E_u = f(d_u)$ and the constraints in (23.6) through (23.10), we have the IP formulation that solves the VS problem for the continuous voltage case.

$$\text{Min:} \quad \sum_{u \in V} f(d_u) \tag{23.11}$$

$$\text{S.t.:} \quad D_{OUT} - D_{IN} \leq T_{con} \tag{23.12}$$

$$D_v - D_u - d_u \geq 0 \; \forall e(u, v) \in E \tag{23.13}$$

$$D_u + d_u \leq dl_u \quad \forall u \; \text{with} \; dl_u \tag{23.14}$$

$$D_u \geq rt_u \quad \forall u \; \text{with} \; rt_u \tag{23.15}$$

$$d_u \geq T_u, \; D_u \geq 0, \; \text{int}, \; \forall u \in V \tag{23.16}$$

where

D_u and d_u are variables whose values need to be determined by solving the IP problem
dl_u, rt_u, T_u, and T_{con} are constants

In the discrete voltage case, only a certain number of voltages are available. For example, the Athlon 4 Processor from AMD [1] can operate at five voltage levels. The discreteness of voltage levels implies that a task's delay can only take discrete values. Let the highest voltage be V_h and m other available voltages be $V_1, V_2, \ldots, V_m$, where $V_i < V_h, 1 \leq i \leq m$. We introduce m new variables, $N_{u,i}$, to represent the number of cycles that task u is executed at V_i. For any given V_i, cycle time CT_i and energy consumption per cycle $ET_{u,i}$ are constants as computed in (23.1) and (23.2). The total number of cycles u takes, N_u, remains a constant. d_u and E_u are computed as linear functions of $N_{u,i}$ as follows:

$$d_u = T_u + \sum_{i=1}^{m} N_{u,i}(CT_i - CT_h) \tag{23.17}$$

$$E_u = C_u \cdot \left(\sum_{i=1}^{m} N_{u,i} V_i^2 + \left(N_u - \sum_{i=1}^{m} N_{u,i} \right) \cdot V_h^2 \right) \tag{23.18}$$

Combining the computation of d_u and E_u in (23.17) and (23.18), and adding $\sum_i N_{u,i} \leq N_u$ to constrains in (23.6) through (23.10), we have the Integer Linear Programming (ILP) for the discrete voltage case.

23.4.3 Solving the IP Problem

In general, IP problems are hard to solve. But the IP problem for the continuous voltage case on a given number of processors is polynomial time solvable. The basis for this observation is the fact that an IP problem is polynomial time solvable if the objective function is a separable convex function and the constraint matrix is totally unimodular [10]. The observation is formally stated in the following theorem.

Theorem 23.1 *The VS problem of executing dependent tasks on systems with one or more processors with continuous voltage is polynomial time solvable.*

Proof: The IP formulation of the VS problem on continuous voltage can be transformed to another IP in which the objective function is the sum of separable convex functions and the constraint matrix is totally unimodular.

We transform the IP problem in the following way. First, each D_u in (23.11) through (23.16) is replaced with another variable D_{su}. Then we introduce another variable D_{fu} and $D_{fu} = D_{su} + d_u$. The IP in (23.11) through (23.16) is transformed to the following IP.

$$\text{Min:} \quad \sum_{u \in V} f(d_u) \tag{23.19}$$

$$\text{S.t.:} \quad D_{OUT} - D_{IN} \leq T_{con} \tag{23.20}$$

$$D_{sv} - D_{fu} \geq 0 \quad \forall e(u, v) \in E \tag{23.21}$$

$$D_{fu} = D_{su} + d_u \tag{23.22}$$

$$D_{fu} \leq dl_u \quad \forall u \text{ with } dl_u \tag{23.23}$$

$$D_{su} \geq rt_u \quad \forall u \text{ with } rt_u \tag{23.24}$$

$$d_u \geq T_u, D_{su}, D_{fu} \geq 0, \text{int}, \quad \forall u \in V \tag{23.25}$$

where D_{su}, D_{fu}, and d_u are variables whose values need to be determined by solving the IP problem, and dl_u, rt_u, T_u, and T_{con} are constraints. Intuitively, D_{su} and D_{fu} represent the starting and finishing time of task u.

As shown in Figure 23.2, energy consumption is a convex function of delay. Each task's energy consumption is only related to its own delay. Thus the objective function in (23.19) is a separable convex objective function. The coefficients of variables D_{su}, D_{fu}, and d_u in (23.20) through (23.25) are 1, 0, or -1 and more importantly the constraint matrix is a totally unimodular matrix [19]. An IP problem is polynomial time solvable if the objective function is a separable convex function and the constraint matrix is totally unimodular [9]. The IP formulation in (23.19) through (23.25) has the same form as a polynomial time solvable IP problem. The transformation from our original IP in (23.11) through (23.16) to such a polynomial time solvable IP in (23.19) through (23.25) can be performed in polynomial time. Thus the formulation in (23.11) through (23.16) for the continuous voltage case is also polynomial time solvable. $\qquad \square$

The ILP problem for some special discrete values is also polynomial time solvable. If there is only one lower voltage V_l and $CT_{V_h} - CT_{V_l} = 1$, the ILP problem for the discrete voltage case can also be transformed to an IP that has a totally unimodular constraint matrix. Such an ILP problem is polynomial time solvable [10]. An LP solver, such as *LP_SOLVE* [14], can be used to get the integer solutions in polynomial time.

To solve the ILP problem for general discrete voltages in short runtime, we employ a simple yet effective approximation method. First the integer constraint on variables is relaxed and the corresponding LP problem is solved. Then we use the floor integer of the non-integer solutions for $N_{u,i}$ in the LP as the integer solutions for $N_{u,i}$ in the ILP. In this way, timing constraints that are satisfied by the non-integer solutions are still satisfied by the integer solutions. The objective function value lost due to the approximation is minimal and the approximation is shown to be very efficient by experimental testing.

23.4.4 Simplified IP on a Single Processor

One special feature of schedules on a single processor is that all tasks are on one execution path. We can utilize this feature and simplify the IP to solve the VS problem for tasks that have the same release time on a single processor as follows:

$$\text{Min:} \quad \sum_{u \in V} E_u \tag{23.26}$$

$$\text{S.t.:} \quad \sum_{u \in V} d_u \leq T_{con} \tag{23.27}$$

$$\sum_{v \in B_u} d_v \le dl_u \ \forall u \text{ with } dl_u \tag{23.28}$$

$$d_u \ge T_u, \text{ int}, \ \forall u \in V \tag{23.29}$$

where B_u is the set that includes u and all tasks scheduled before u. The number of variables and constraints are all reduced. The relation between delay and energy for the continuous and discrete voltage can be plugged into the simplified IP in the same way as we did for the general IP that solves the VS problem.

23.4.5 More than Two Voltages Needed

It is not always true that at most two voltages are needed to minimize energy in the discrete voltage case. The number of voltages needed depends on the voltage values available. More than two voltages are needed when the combinations of two voltages cannot produce an execution time that is required to minimize the energy consumption. Denote two voltages as V_h and V_l where $V_h > V_l$, and the corresponding cycle time as CT_{V_h} and CT_{V_l}. For a task that takes N cycles to finish, there are in total N different execution times that the combinations of these two voltages can produce. They are

$$N_1 * (CT_{V_l} - CT_{V_h}) + N * CT_{V_h}, \quad N_1 \le N \tag{23.30}$$

For any given V_h and V_l, CT_{V_h} and CT_{V_l} are constants. Since N is a constant for a given task, the N different execution times change in the step of $CT_{V_l} - CT_{V_h}$. If $CT_{V_l} - CT_{V_h} > 1$, the N different execution times are not consecutive integers, but rather an array of integers with the same increase step. If the target execution time determined for the best energy is in the middle of two of the N values, a third voltage is needed. Otherwise, the task will have to finish earlier and consume more energy than finishing right on time.

Theorem-23.1 in [12] states that if a processor can use only a small number of discrete variable voltages, the voltage scheduling with at most two voltages minimizes the energy consumption under any time constraint. There is a strong condition for this theorem to be applicable and that is the execution time by the combinations of the two voltages can meet timing constraint exactly.

23.5 Task Assignment and Scheduling

Task assignment and scheduling should provide the maximum slowing down potentials for VS to utilize to minimize energy. A large amount of work has been done for task assignment and scheduling. However, the main concern of prior work is timing instead of energy. Recent approaches in [8,15,20] considered energy during scheduling, but they do not consider the joint effect of slowing down certain tasks simultaneously. In this section, we present the assignment and scheduling that is guided by our VS approach. For the simple case of a single processor, we show that a deadline-driven non-preemptive earliest deadline first (NP-EDF) scheduling is optimal in providing slowdown opportunities if tasks have the same release time. For multiple processors, we use a priority-based scheduling in order to provide the best energy saving opportunity.

23.5.1 Scheduling on a Single Processor

It is clear that an individual task's deadline imposes constraints on the design space of delays of tasks that are scheduled before it, as shown by constraints in (23.7). Energy saving is affected by the design space of tasks' delays because the delay will be traded for energy saving. To maximize the design space of delays, tasks with earlier deadlines should be scheduled earlier in order to avoid imposing constraints

Algorithm 23.1 NP-EDF(S, t)

  Input: Task set S and time step t
  Output: Task set S with all tasks' starting time set
  if $S \neq \phi$ **then**
    if $\exists$ u with $rt_u \leq t$ **then**
      if $\exists$ u with dl_u **then**
        Find u with minimum dl_u
      else
        find a $u \in S$
      end if
    else
      Find u with minimum rt_u
      $t = rt_u$
    end if
    $S1 = PRE_u$
    $S2 = S - S1 - \{u\}$
    NP-EDF($S1, t$)
    $s_u = t$
    $t = t + T_u$
    NP-EDF($S2, t$)
  end if

on more tasks. We achieve this goal by incorporating NP-EDF scheduling in the following recursive algorithm NP-EDF(S, t), shown in Algorithm 23.1, to schedule tasks on a single processor. The input to the algorithm is the task set S, and t is initialized to be 0. Denote PRE_u as the set of u's predecessors. The output of the algorithm is the same task set with each task's starting time s_u being assigned.

The NP-EDF algorithm schedules a task with the smallest deadline at each step. The scheduled task's predecessors are scheduled recursively before this task and the rest of the tasks are scheduled after this task recursively. The schedule after NP-EDF specifies the ordering of tasks on the single processor and the ordering is one of the topological sort results. Tasks are ordered by their dependencies and deadlines. The NP-EDF scheduling provides the biggest solution space for the VS problem on a single processor when tasks have the same release time, as stated by the following theorem.

Theorem 23.2 *The NP-EDF scheduling on tasks with the same release time is optimal for the VS process to save energy on a single processor.*

Proof. We prove the theorem by comparing the IP formulation for the VS problem on a NP-EDF ordering, shown in Figure 23.5a, and a non-NP-EDF ordering, shown in Figure 23.5b. A non-NP-EDF algorithm will schedule at least one task with later deadline, denote as x, before a task with smaller deadline, denote

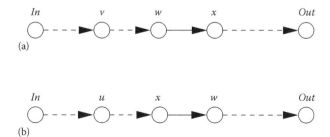

FIGURE 23.5 Different schedules on one processor.

as w, while in the NP-EDF ordering, w is scheduled before x. Assume that w is scheduled exactly one task later in Figure 23.5b than in Figure 23.5a. The constraint w's deadline imposes on the delay of tasks ordered before it in the NP-EDF ordering is

$$\sum_{v \in BE_w} d_v + d_w \le dl_w \qquad (23.31)$$

where BE_w is the set of tasks that are ordered before w in Figure 23.5a. The constraint that w's deadline imposes on tasks in the non-NP-EDF ordering is

$$\sum_{u \in BN_w} d_u + d_w \le dl_w \qquad (23.32)$$

where BN_w is the set of tasks ordered before w in Figure 23.5b. Delay of tasks in BE_w and BN_w should also satisfy

$$d_v \ge T_v \quad \forall v \in BE_w \quad \text{or} \quad \forall v \in BN_w \qquad (23.33)$$

Tasks in Figure 23.5a and b are scheduled in the same order, except w and x. There is an extra task x in BN_w than BE_w. For tasks that are both in BE_w and BN_w, those in BN_w has equal to or tighter constraints than ones in BE_w. For task x that is in BN_w, but not BE_w, its delay is not constrained by dl_w in the NP-EDF ordering in Figure 23.5a, but is constrained by dl_w in the non-NP-EDF ordering in Figure 23.5b. Thus tasks in BN_w have tighter or equal constraint imposed by dl_w. This means that tasks in BN_w have an equal or smaller solution space than tasks in BN_w.

For a schedule that has more tasks not following the NP-EDF order, we can always make a task with smaller deadline to be scheduled before other tasks with longer deadlines and find another ordering that has one more task in the NP-EDF order. By using the same constraints comparison, we can prove that the schedule with one more task in the NP-EDF order has an equal or bigger solution space. We can repeat this process until we find the schedule that follows the NP-EDF order and has the biggest solution space. Thus we prove that the NP-EDF ordering is optimal in providing slowing down opportunity for VS to utilize. If a feasible schedule exists, a NP-EDF schedule is also feasible [13]. This means that we can always find a NP-EDF schedule and get the biggest solution space for the VS problem from a feasible schedule.

The combination of the NP-EDF scheduling and our IP formulation for the VS problem solves the general energy minimization problem of dependent tasks with the same release time on a single variable voltage processor exactly. Scheduling of tasks with different release time is not solved optimally with our NP-EDF approach.

23.5.2 Scheduling on Multiple Processors

The NP-EDF scheduling is not optimal on multiple processors because tasks will be on multiple paths and affect tasks on these paths differently. To provide more energy saving opportunities in the schedule for VS to utilize, we should try to avoid (1) putting tasks on unnecessary paths that might pose tighter constraints, (2) making unnecessary long paths, (3) putting unrelated tasks on the same path, and (4) letting more tasks constrained by smaller deadlines. To achieve these goals, we use a list scheduling algorithm that relies on a priority function to select one task at each scheduling step and find a best-fit processor for the selected task.

The priority relates to a task's deadline, dependencies, and the usage of processors in the system. Tasks are assigned a latest starting time that they must meet for them or their successors to finish by their deadlines. Task u's latest starting time lst_u is defined as

$$lst_u = \min(dl_u - T_u, lst_v - T_u | \forall v, e(u, v) \in E) \qquad (23.34)$$

A leaf task u's latest starting time is the u's deadline dl_u minus the u's delay T_u or T_{con} minus T_u if u does not have a deadline. Starting from leaf tasks and traversing the task set DAG backward, we can assign each task a latest starting time according to (23.34). Denote task u's ready time, which is the maximum of its release time and the time when all u's predecessors have finished, as r_u, its earliest starting time when it is ready and there is a processor available as es_u, the number of processors in the system as N, and processor P_i's available time as a_{P_i}. The values of r_u, es_u, and a_{P_i} are updated during the assignment and scheduling process.

We build the schedule step by step. At each scheduling step, eligible tasks' priorities are evaluated. Tasks are eligible if their release times have passed and their predecessors have been scheduled. The priority is defined as

$$PRI_u = lst_u + es_u \tag{23.35}$$

$$es_u = \max(r_u, \min(a_{P_i} | \forall i \leq N)) \tag{23.36}$$

Task u with the smallest PRI_u is assigned to a best-fit processor. The best-fit processor for u is selected sequentially according to the following three steps and u's starting time s_u, u's finishing time f_u, and a_{P_i} are set accordingly.

1. Select P_i if $a_{P_i} = r_u$. That is to select the processor that is available at u's ready time. In this case, $s_u = r_u = a_{P_i}, f_u = s_u + T_u$, and $a_{P_i} = f_u$.
2. Select P_i if $a_{P_i} < r_u, a_{P_i} > a_{P_j} | \forall j \neq i,\ a_{P_j} < r_u$. That is to select the processor that becomes available the latest among all processors available before u is ready. In this case, $s_u = r_u, f_u = s_u + T_u$, and $a_{P_i} = f_u$.
3. Select P_i if $a_{P_i} \leq a_{P_j},\ \forall j \neq i$. That is to select the processor that is available the earliest. In this case, $s_u = a_{P_i}, f_u = s_u + T_u$, and $a_{P_i} = f_u$.

After the selection and assignment of one task, a new set of eligible tasks is evaluated and the step repeats until all tasks are scheduled. If the priority-based schedule is not feasible, we can increase the weight of the latest starting time in the priority function and try to get a feasible schedule.

23.6 Energy Overhead Consideration

The integer solution obtained directly or approximated from non-integer solutions for the integer programming (IP) formulation determines the numbers of cycles at different voltages for every task. If there is no energy overhead in changing voltage/frequency on a processor, different sequences of voltages do not affect energy consumption. The saving is affected only by the number of cycles at each voltage, and that number is determined by solving the IP for the VS problem. If the energy overhead due to voltage transitions is not negligible, energy consumed by transitions will offset the saving of running tasks on lower voltage levels. To minimize energy consumption, we need to set the voltage for each cycle of every task based on the energy overhead as well.

23.6.1 Motivational Example

The following example motivated us to consider energy overhead while deciding voltage levels for task cycles. Consider the schedule in Figure 23.6a. Assume the two processors have the delay and energy data as summarized in Table 23.1. We will let the energy overhead per transition be one of the three values, 0, 3, and 11, to test how overhead affects overall energy consumption. The three values of overhead may not be very realistic; it does not undermine the purpose of the example so as to show that overhead should be considered.

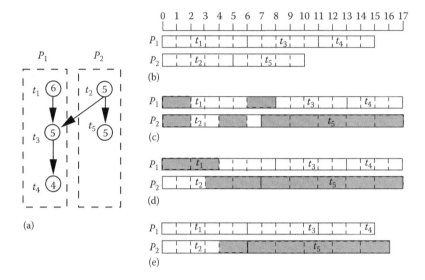

FIGURE 23.6 (a) A given schedule of a 5-task set on two processors. (a) through (e) different voltage settings. (b) Always on V_h (c) Unnecessary transitions (d) Optimal solution when $E_o = 3$ (e) Optimal solution when $E_o = 11$ Cycles in shade are V_l cycles.

TABLE 23.3   Energy of Settings for Different Transition Overhead

	N_{V_h}	N_{V_l}	N_{tr}	$E_0(\%)$	$E_3(\%)$	$E_{11}(\%)$
(b)	25	0	0	100	100	100
(c)	16	9	7	73	94	150
(d)	16	9	2	73	79	95
(e)	19	6	1	82	85	93

Figure 23.6b through e shows four different voltage settings for the schedule in Figure 23.6a. The number of cycles at V_h and V_l, the number of transitions, and the total energy consumption when energy overhead is 0, 3, and 11 for the four voltage settings are shown in columns N_{V_h}, N_{V_l}, N_{tr}, E_0, E_3, and E_{11}, respectively, in Table 23.3. The setting in Figure 23.6b can be viewed as a baseline where the two processors are always on V_h and there are no transitions. The numbers of V_l cycles in Figure 23.6c and d are the same and they can be determined by solving the IP formulation described in Section 23.4. When there is no transition energy overhead, settings in Figure 23.6c and d are optimal solutions and they save 27% of energy over the baseline. However, when the energy overhead per transition is 3, system implementation in Figure 23.6d is the optimal solution that saves 21% energy, while the setting in Figure 23.6b only saves 6%. When the energy overhead increases to 11, Figure 23.6e is the optimal solution that saves 7% of energy. The setting in Figure 23.6d saves 5%, while in the setting in Figure 23.6c consumes 50% more over baseline. This example shows that energy overhead affects overall saving dramatically and must be considered while determining voltage levels. Our EOC phase considers the overhead and can achieve the optimal solution for this example under all three different energy overheads.

23.6.2 Maximizing Saving and Minimizing Transitions

We first find the voltage setting that has the minimum number of transitions, including both intra-task and inter-task transitions. Then each sequence of lower voltage cycles will be examined to decide whether to keep the lower voltage for these cycles.

23.6.2.1 Two Voltages Case

Let us start with the case where only two voltage levels, V_h and V_l, are available. The total energy saving is the difference of saving of V_l cycles and the energy consumed by transitions:

$$E_s = \sum_u N_{l,u} \cdot \Delta E_{s,u} - N_{tr} \cdot E_0 \tag{23.37}$$

where
    $N_{l,u}$ is the number of u's V_l cycles
    N_{tr} is the number of transitions

The first term on the right-hand side of (23.37) is the objective function of the ILP formulation that solves the VS problem for discrete voltages. The objective value obtained after solving the ILP is optimal or close to optimal for systems with negligible overhead. The solutions to the ILP problem serve as a good starting point. We set the minimum number of intra-task transitions of a task to be 0 if there is no V_l or V_h cycle for this task, or 1 if there is at least one V_l cycle and one V_h cycle for this task.

To minimize the number of transitions and save more energy, we also need to minimize the number of inter-task transitions. A minimum inter-task transition setting can be found by a greedy approach that keeps the same voltage across task boundaries whenever possible. For example, in Figure 23.7a, there are two tasks, t_1 and t_2, on the same processor and t_1 is scheduled before t_2. t_1 has two V_l cycles and two V_h cycles, while t_2 has three V_l cycles and four V_h cycles. There are four possible ways of arranging the sequences of these cycles and they are shown in Figure 23.7b through e. Intra-task transitions are already minimized in all four settings. Apparently, always having tasks start with V_l or V_h does not minimize the number of inter-task transitions, as shown in Figure 23.7d and e. Keeping the same voltage across task boundaries minimizes the number of transitions, as shown in Figure 23.7b and c. The greedy approach can be proven to be optimal in finding the minimum number of inter-tasks transitions, as stated in the following theorem.

Theorem 23.3 *Keeping the same voltage across task boundaries whenever possible minimizes the number of inter-task transitions for discrete voltages.*

Proof: We prove the theorem by considering all possible voltage sequences and computing the number of transitions in each sequence to show that having the same voltage across task boundaries always results in the minimum number of transitions.

Tasks with only one voltage do not need to be considered since they do not change the voltage sequence. Let us consider a task u with both V_h cycles and V_l cycles. Assume the sequence before it, S_1, and after it, S_2, already have minimum number of transitions, TR_{S_1} and TR_{S_2}. We compute the number of transitions of the total sequence containing S_1, u, and S_2 and show it in Table 23.4. All possibilities of the voltage at

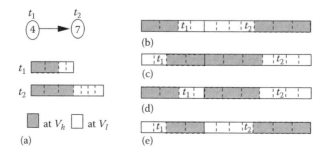

FIGURE 23.7 (a) Schedule of two tasks on one processor. (b) through (e) Different voltage sequences of task cycles.

TABLE 23.4 Number of Transitions for Different Voltage Sequences

	$Vend_{S_1}$	$Vstart_u$	$Vend_u$	$Vstart_{S_2}$	TR_{num}
(a)	V_l	V_l	V_h	V_h	$TR_{S_1} + 1 + TR_{S_2}$
(b)	V_l	V_h	V_l	V_h	$TR_{S_1} + 3 + TR_{S_2}$
(c)	V_l	V_l	V_h	V_l	$TR_{S_1} + 2 + TR_{S_2}$
(d)	V_l	V_h	V_l	V_l	$TR_{S_1} + 2 + TR_{S_2}$
(e)	V_h	V_h	V_l	V_h	$TR_{S_1} + 2 + TR_{S_2}$
(f)	V_h	V_l	V_h	V_h	$TR_{S_1} + 2 + TR_{S_2}$
(g)	V_h	V_h	V_l	V_l	$TR_{S_1} + 1 + TR_{S_2}$
(h)	V_h	V_l	V_h	V_l	$TR_{S_1} + 3 + TR_{S_2}$

the end of S_1, at the start of u, at the end of u, and at the start of S_2 are enumerated in columns $Vend_{S_1}$, $Vstart_u$, $Vend_u$, and $Vstart_{S_2}$, and the number of transitions for each case is computed in column TR_{num}. It is clear that when the same voltage is kept across the boundary of S_1 and u, the number of transitions is minimized. The voltage at the end of S_1 is the same in cases (a) and (b), and the voltage at the start of S_2 is the same in these two cases too. The number of transitions in case (a) is smaller than that in (b) because u starts with the same voltage as S_1 ends with. We can compare cases (c) and (d), (e) and (f), and (g) and (h) in the same way and draw the same conclusion that when the same voltage is kept across task boundaries, the number of transitions is minimized. The greedy approach finds the sequence with the minimum number of transitions by keeping the same voltage across task boundaries whenever possible. □

After finding the settings with the minimum number of transitions, we check each V_l sequence. If the sequence is not long enough to offset the transition overhead involved, these cycles will be changed back to V_h and the number of transitions will decrease by up to two. Thus the solutions for the ILP are changed and both the first and second term in (23.37) are decreased, while the total energy saving E_s is increased. In the example in Figure 23.6e, when the transition overhead is 11, the two V_l cycles of t_1 are changed back to V_h.

The minimization of transitions is beneficial when transition time overhead cannot be ignored. Time that is not spent during transitions can be used to slow down tasks to save energy. If transition time overhead cannot be ignored, each low-voltage sequence can be checked to see whether timing constraints are still being met when time overhead is considered, at the same time when the sequence is checked to see whether it saves energy when energy overhead is considered.

23.6.2.2 Multiple Voltages Case

The multiple voltage case can be handled in the same fashion by first minimizing intra- and inter-task transitions and then eliminating unbeneficial lower voltage cycles. We formulate the problem of finding the minimum transition cost as a shortest path problem. We use one set of nodes to represent all possible settings for each task. In these settings, cycles with the same voltage are grouped together. For a task t_i with cycles on m_i different voltages, there are in total $m_i!$ different settings representing different permutations for the cycle sequences with different voltages. Thus a total of $m_i!$ nodes are introduced for task t_i. Even though m_i is not bounded by 2, it is usually a very small integer. We introduce an edge from each node $n_{i,j}$ in the set for t_i, where $0 < j \le m_i!$, to every node $n_{i+1,k}$ in the set for t_{i+1}, where $0 < k \le m_{i+1}!$. A complete bipartite graph between nodes for consecutive tasks t_i and t_{i+1} on a processor is formed in this way. The transition cost on each node is defined as the sum of the overhead of each intra-task transition in the setting. The cost of every edge in the bipartite graph is defined as the transition overhead between the end voltage of $n_{i,j}$ to the start voltage of node $n_{i+1,k}$. A shortest path of the graph is the setting with the minimum transition cost.

An example of two tasks t_1 and t_2 scheduled on a processor is shown in Figure 23.8a. t_1 has cycles running at three different voltages, while t_2 has cycles on two different voltages. The complete bipartite

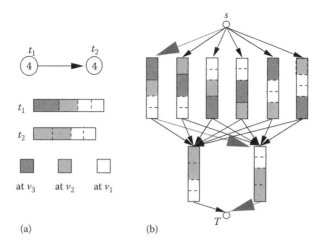

FIGURE 23.8 (a) An example of two tasks with three and two different voltages. (b) A complete bipartite graph. The minimum transition overhead setting is linked by wider edges.

graph for the example is shown in Figure 23.8b. The shortest path that represents the minimum transition cost is marked with wider edges.

For the continuous voltage case, each task has one voltage for all its cycles and there is no intra-task transition. A transition happens between two consecutive tasks with different voltages on the same processor. Inter-task transition is fixed if we keep the solution for the convex IP problem. We need to check whether a sequence of cycles at lower voltages provides more saving than the transition overhead. Two tasks t_a at V_a and t_b at V_b can be treated as one sequence if the saving of running t_a at V_a and t_b at V_b is more than the overhead of the transition between V_a and V_b. Only when the saving is more than overhead, will we allow tasks to be executed with lower voltages and transitions between tasks.

One may point out that the increase of voltage will decrease a task's execution time and if the following task cannot start earlier (constrained by other tasks on another processor with a later finishing time), there will be idle time on a processor. We want to stress that we always try to assign a task to the same processor as its latest immediate predecessor is assigned to, which is available right when the task is ready. Thus it is very likely for a task to start earlier if its immediate predecessor finishes earlier. Otherwise, we can check its other immediate predecessors to decide the starting time for this task.

23.7 Experimental Validation

To evaluate the effectiveness of our framework in saving energy, we implemented it and conducted experiments on various task sets and systems. Nine task sets are generated with *TGFF* [22] where each set consists of 10–500 tasks. Since single-processor systems are a special case of multiprocessor systems and we have proven that the NP-EDF scheduling for tasks with the same release time on a single processor is optimal, we concentrate the experiments on multiprocessor systems. We tested systems consisting of two to eight processors that can operate at two and four different voltages. (Most commercially available variable voltage processors have 2–5 V levels.) First we investigate the effectiveness of our ILP formulation for solving the VS problem by comparing the number of slowdown cycles with a scaling VS approach on the same schedule. Then we test how our scheduling helps saving energy by comparing the VS results on different schedules. The energy savings by our framework on different system configurations are also tested. At last, we investigate how energy overhead affects overall saving.

23.7.1 Testing VS and Scheduling

The first experimental system consists of five processors that can operate at two different voltages, V_h and V_l. For simplicity, we assume that the energy saving per cycle at V_l is the same for all tasks. (Note this is not required by our ILP formulations.) Thus the total number of slowed down cycles SN represents the energy saving, which is proportional to $SN\left(V_h^2 - V_l^2\right)$. The cycle time of processors increases from 1 time unit to 4 time units when voltage is switched from V_h to V_l.

We use our scheduling in the first phase and alternate different VS approaches in the second phase to test how much improvement our ILP approach provides. Since the LEneS approach [8] deals with task ordering in VS and the approach by Luo and Jha [15] has other objectives during its VS process, we feel it is not fair to compare our ILP approach with the two directly. We compare the number of cycles at V_l by our IP approach with a scaling VS approach that is also discussed in [8]. The scaling approach scales down the delay of all tasks by the ratio of timing constraint over critical path length on a given schedule. The approach by Luo and Jha [15] distributes slack evenly and shares the same spirit with the scaling approach. Table 23.5 summarizes the performance of our ILP approach and the scaling approach. Column NT, NC, T_{con}, and T_{cri} show the number of tasks, the number of task cycles, the timing constraints, and the critical path delay at V_h, respectively. Timing constraint T_{con} is set to be 1.5 times of T_{cri}. Column SC and ILP show the number of cycles at V_l by the scaling approach and our ILP approach, and column IM shows the improvement of number of cycles at V_l by our ILP approach (ILP) over the scaling approach (SC), which is computed as $((ILP) - (SC))/(SC)$. It is clear that on average our approach can execute 58% more cycles at V_l and save more energy than the scaling approach.

The approximation method we use to get the integer solutions for the ILP problem is very effective. The objective function value of the ILP, total number of slowdown cycles, after the approximation of integer solutions is within 97% of the objective value achieved with non-integer solutions from solving the LP. The objective of the LP is the upper limit of the objective value that the corresponding ILP can achieve. This means that solutions by the approximation is within at least 97% of the optimal solution of the ILP problem. The whole process including scheduling, solving LP, and approximation finishes within seconds for all task sets.

To test how our scheduling helps to provide more slowdown opportunities, we alternate two scheduling approaches in the first phase, one is our priority-based, non-preemptive ordering and best-fit processor selection approach (NP-PEDF), and another is a baseline approach that uses a non-preemptive EDF ordering and selects the processor that becomes available the earliest (E-EDF). Our ILP formulation is applied in the second phase to minimize energy. The numbers of cycles at V_l for several representative task sets on the same 5-processor experimental system used in the VS comparison are summarized in Table 23.6. In the table, column E-EDF and NP-PEDF show the number of cycles at V_l on the baseline schedule and on our priority-based scheduling, respectively, and column IM shows the improvement of

TABLE 23.5 Numbers of V_l Cycles by Scaling and Our Approach

Set	NT	NC	T_{cri}	T_{con}	SC	ILP	IM (%)
s1	9	81	49	74	11	37	236
s2	50	422	111	167	51	104	104
s3	101	922	243	365	116	225	94
s4	151	1501	464	696	188	413	120
s5	213	2988	722	1083	413	644	56
s6	245	2871	652	978	376	557	48
s7	305	2643	572	858	314	495	58
s8	463	4310	910	1365	541	755	40
s9	514	4633	949	1424	556	808	45
Average	228	2263	519	779	285	449	58

TABLE 23.6 Numbers of V_l Cycles on Different Schedules

Set	NT	NC	T_{con}	E-EDF	NP-PEDF	IM (%)
s2	50	422	167	89	104	17
s3	101	922	365	200	225	13
s7	305	2643	858	432	495	15
Average	152	1329	463	240	275	14

TABLE 23.7 Numbers of V_l Cycles on Different Number of CPUs

Set	NC	T_{con}	2-cpu	5-cpu	8-cpu
s1	81	66	9	31	31
s2	422	266	29	268	397
s3	922	581	67	588	803
s4	1501	989	118	905	1318
s5	2988	1861	208	1939	2910
s6	2871	1760	197	1862	2840
s7	2643	1601	177	1728	2640
s8	4310	2616	287	2832	4310
s9	4633	2789	310	3076	4633
Average	2263	1392	156	1470	2209
%	100		7	65	98

cycles at V_l on our schedule (NP-PEDF) over the baseline schedule (E-EDF). The results show that our scheduling can provide 14% more slowing down opportunities than a baseline scheduling.

It is clear that our first-phase priority-based scheduling and second-phase IP approaches all outperform other respective approaches. We want to further explore how the framework performs on different system configurations. We first investigate how the number of processors affects the energy saving. With fixed timing constraints, systems with more processors should have shorter critical path length and more tasks with more slacks. The slacks can be traded for energy saving. Table 23.7 shows the number of cycles at V_l on systems with different number of processors. Processors can operate at two values and the cycle time changes from 1 unit to 4 units when voltage is switched from V_h to V_l, which is the same as in the first two experiments. The energy saving increases dramatically when the number of processors increases from two to five, but not so much when the number of processors increases from five to eight. This is because the limited parallelism among tasks has been exploited by the increase in number of processors from two to five. Additional processors do not provide much benefit if there are not many tasks that can run in parallel. If tasks are independent, every task can run in parallel and we should expect the increase in number of processors will lead to energy saving increase more evenly.

The number of voltages and voltage values also affect the energy saving. With more voltage levels available, the search space for the VS problem is bigger and energy savings should be higher. For different voltage values, a lower voltage value increases the cycle time more and may prevent other cycles from slowing down if there is not enough slack. However, the energy saving per cycle is more for the lower voltage value. It is not immediately clear how voltage values affect the overall energy saving. We did experiments on the same nine task sets on three five-processor systems. The configurations for the three systems are Sys1: $V_{l1} = V_h/2$, $V_{l2} = V_h/3$, and $V_{l3} = V_h/4$, Sys2: $V_l = V_h/4$, and Sys3: $V_l = V_h/5$. Assume that cycle time increases from 1 to 2, 3, 4, and 5 when voltage changes from V_h to $V_h/2$, $V_h/3$, $V_h/4$, and $V_h/5$. This is a rough estimation but it should not make any conclusion we draw invalid. The number of slowdown cycles and the energy saving in the three systems are shown in the upper and lower part in Figure 23.9. In the figure, the first bar corresponds to data for Sys1, the second for Sys2, and the third for Sys3. The results show that our approach is capable of utilizing the bigger design space provided in Sys1 and achieves more energy savings when more voltage levels are available. In Sys1, where processors

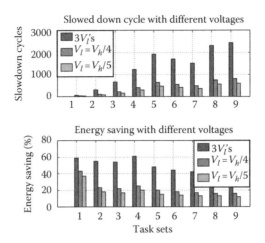

FIGURE 23.9   Energy saving of different voltages.

can operate at three different levels, most cycles are slowed down to operate at V_{l1}, which is the closest to V_h. The numbers of cycles operating at the other two lower voltages are very small. Sys2 with $V_l = V_h/4$ has more cycles slowed down and saves more energy than Sys3 with $V_l = V_h/5$. This is because the closer a V_l is to V_h, the more cycles can be slowed down, and this results in more energy saving even though the saving per cycle is smaller for a V_l that is closer to V_h. Of course, if timing constraint is so loose that every cycle can be executed at the lowest voltage, then the lower the voltage, the more the energy saving. Continuous voltage is the ideal case and should give the best energy saving.

23.7.2 Testing the EOC Phase

In the following, we illustrate the effect of transition overhead on the overall energy saving. We use the data measured by Pouwesle et al. [17] for the highest and lowest voltages of the StrongARM SA-1100 processor [11]. Timing and energy data at $V_h = 1.65$ V and $V_l = 0.79$ V are summarized in Table 23.8. Our testing system consists of five such processors that can operate at V_h and V_l. Assume the saving per V_l cycle is the same for all tasks. We use the parameters of a frequency/voltage converter that is designed for a voltage scalable processor system [4] in our experiments. The overhead is 1 µJ when the capacitor is optimized to be 5 µf and a typical value of the capacitor can be 100 µf, which increases the overhead to 20 µJ per transition. Voltage range in the original system [4] is 1.2–3.8 V and thus overhead per transition in that system is higher.

We tested the EOC phase on nine task sets. The number of tasks, N_t, number of task cycles, N_c, and the critical path delay, T_{cri}, are shown in Table 23.9. Timing constraint on each task set, T_{con}, is set to be $2T_{cri}$ in the following experiments. The baseline consumption is the energy consumed when all tasks are executed at the highest voltage V_h.

Different orderings of task cycles affect the energy consumption when overhead is not negligible. Our approach keeps the same voltage across task boundaries whenever possible and this avoids many unnecessary transitions. We compare with a fixed ordering that lets a task always start from its V_h cycles if the task has cycles on V_h. The number of cycles at V_l can be obtained from solving the ILP problem

TABLE 23.8   SA-1100 Processor Data

	Voltage (V)	Frequency (MHz)	CT (ns)	Power (MW)	Energy/Cycle (nJ)
V_h	1.65	251	3.98	696.7	2.78
V_l	0.79	59	16.9	33.1	0.56

TABLE 23.9 Task Set Parameters and Timing Constraints

Set	N_t	N_c (K)	T_{cri} (µs)
s1	9	81	196
s2	50	422	444
s3	101	922	972
s4	151	1501	1848
s5	213	2988	2880
s6	245	2871	2604
s7	305	2643	2284
s8	463	4310	3636
s9	514	4633	3796
Average	228	2263	2073

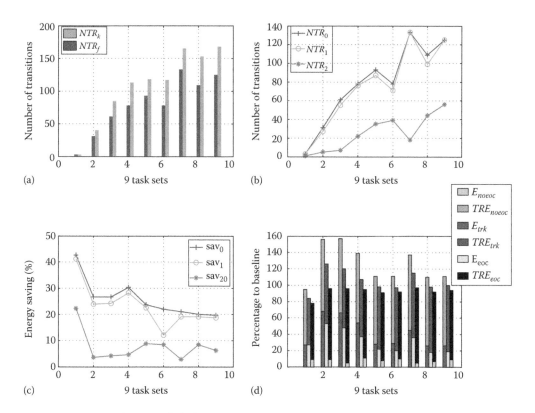

FIGURE 23.10 (a) Different number of transitions by a fixed ordering and our ordering. (b) Number of transitions decreases when overhead per transition increases. (c) Energy saving decreases when overhead per transition increases. (d) Energy consumption without EOC is much higher than with EOC.

and it is not changed by ordering cycles differently. The numbers of transitions by our approach and by the fixed policy for the nine task sets are shown in Figure 23.10a. In the figure, the left bar shows the number of transitions by our approach, NTR_k, and the right bar shows the number of transitions by the fixed policy, NTR_f. By keeping the same voltage across task boundaries, we can decrease the number of transitions by 32% compared with the fixed policy. When transition energy overhead is not negligible, the decrease in the number of transitions translates directly to the increase of energy saving and improvement of timing.

Our approach is not only able to reduce the number of transitions by keeping the same voltages across task boundaries, it can also eliminate non-beneficial V_l sequences to further decrease the number of transitions and the energy consumed by transitions. To measure the effect of energy overhead on the energy saving, we set the overhead per transition to be 0, 1 μJ, and 20 μJ. The number of transitions by our approach decreases when the overhead per transition increases, as shown in Figure 23.10b. In the figure, NTR_0, NTR_1, and NTR_{20} are numbers of transitions by our EOC phase when energy overhead per transition is 0, 1 μJ, and 20 μJ. However, the decrease of transitions still results in more energy consumption because the overhead per transition increases. The energy saving by our approach for the nine task sets on systems with different overhead is shown in Figure 23.10c. In the figure, sav_0, sav_1, and sav_{20} are savings achieved after our EOC phase when energy overhead per transition is 0, 1 μJ, and 20 μJ. We can see when there is no overhead, the average energy saving is 25.8%, but that is decreased to 7.7% when the energy overhead per transition is increased to 20 μJ. This tells us that transition overhead will put a limit on how much energy can be saved through varying supply voltage.

In the following, we use the data on the system where overhead is 20 μJ to show that our EOC phase is very important in reducing energy consumption. In Figure 23.10d, we show the total energy and the energy consumed by transitions for three different cases after the number of cycles for V_h and V_l are decided after solving the ILP problem. In the first case, energy overhead is not considered, tasks always start from their highest voltage cycles and no V_l sequences are eliminated. In the second case, task cycles are ordered to have the same voltage across task boundaries whenever possible. But there is no elimination of V_l cycles. The third case uses our EOC phase that keeps the same voltage across task boundaries and eliminates unbeneficial V_l cycles. Energy consumption is represented in percentage of the baseline consumption. The left bar shows the energy consumed by executing tasks, E_{noeoc}, and energy by transitions, TRE_{noeoc}, of the first case where overhead is not considered and tasks are fixed to always start from their highest voltage cycles. The center bar shows the tasks energy consumption E_k and the energy consumed by transitions TRE_k for the second case where task cycles are ordered to keep the same voltage across task boundaries whenever possible, but no V_l sequences are eliminated. The right bar shows the task energy E_{eoc} and energy consumed by transitions TRE_{eoc} of the third case that uses our EOC phase that keeps same voltage across task boundaries and eliminates unbeneficial V_l cycles.

It is clear that when overhead is not considered at all, eight out of the nine task sets consume more energy than the baseline where all tasks are executed at V_h and there are no transitions. The average energy consumption is 125% and transitions consume 41% of the baseline consumption. When voltage is kept the same across task boundaries whenever possible, the average energy consumption decreases to 105% and energy consumed by transitions decreases to 31% of the baseline. With our EOC phase, the average energy consumption is 92% and the energy consumed by transitions is only 8% of the baseline. The EOC phase is particularly important when overhead per transition is high. If the number of transitions is not decreased wisely, the energy consumption will increase linearly with the increase of overhead per transition and eventually becomes the dominant part and offsets all the benefits of having variable voltages. However, since our approach orders task cycles to minimize the number of transitions and eliminates unbeneficial transitions, we are able to control the energy consumed by transitions to be below 11% for all test cases.

23.8 Conclusion

This chapter presents a three-phase framework that integrates task assignment and scheduling, VS, and transition overhead consideration together to achieve the maximum energy saving of executing real-time-dependent tasks on one or multiple DVS-enabled processors. The first phase strives to provide more slowing down opportunities than a baseline schedule and encompasses an NP-EDF scheduling approach for a single processor and a priority-based scheduling approach and best-fit processor assignment for multiple processors. The second phase uses a novel IP formulation for the VS problem when overhead can be ignored. It can be solved optimally in polynomial time for the continuous voltage and some special discrete voltage cases, or solved efficiently by the simple approximation method for general discrete voltages. The third phase, energy overhead consideration phase, determines the voltage level for each cycle of every task to maximize the energy saving after transition energy overhead is taken into account. The EOC phase is very important in reducing energy consumption when energy overhead cannot be ignored. Due to its high-quality solutions and low computational cost, the framework is a good candidate for design space exploration during the design of multiprocessor real-time systems.

References

1. AMD Athlon 4 Processor, Data sheet reference #24319, Advanced Micro Devices, Inc., 2001.
2. A. Andrei, M. Schmitz, P. Eles, Z. Peng, and B. Al-Hashimi, Overhead-conscious voltage selection for dynamic and leakage power reduction of time-constraint systems, *DATE'04*, pp. 518–523, Paris, France, 2004.
3. L. Benini, A. Bogliolo, and G. De Micheli, A survey of design techniques for system-level dynamic power management, *IEEE Transactions on VLSI Systems*, 8(3):299–316, June 2000.
4. T. Burd, Energy-efficient processor system design, PhD Dissertation, University of California at Berkeley, Berkeley, CA, 2001.
5. T. Burd, T. Pering, A. Stratakos, and R. Brodersen, A dynamic voltage scaled microprocessor system, *IEEE Journal of Solid-State Circuits*, 35:1571–1580, 2000.
6. A. Chandrakasan and R. Brodersen, *Low Power Digital CMOS Design*, Kluwer Academic Publishers, Boston, MA, 1995.
7. W. Chuang, S. Sapatnekar, and I. Hajj, Delay and area optimization for discrete gate sizes under double-sided timing constraints, *CICC'93*, pp. 9.4.1–9.4.4, San Diego, CA, 1993.
8. F. Gruian and K. Kuchcinski, LEneS: Task scheduling for low-energy systems using variable supply voltage processors, *ASP-DAC'01*, pp. 449–455, Yokohama, Japan, 2001.
9. D. Hochbaum, Integer programming and combinatorial optimization, http://queue.IEOR.Berkeley. EDU/hochbaum/#lecture_notes, 1994
10. D. Hochbaum and J. Shanthikumar, Convex separable optimization is not much harder than linear optimization, *Journal of ACM*, 37(4):843–862, 1990.
11. Intel StrongARM SA-1100 microprocessor developer's manual, http://developer.intel.com/design/strong/manuals/278088.htm
12. T. Ishihara and H. Yasuura, Voltage scheduling problem for dynamically variable voltage processors, *ISLPED'98*, pp. 197–202, Monterey, CA, 1998.
13. K. Jeffay, D. Stanat, and C. Martel, On non-preemptive scheduling of periodic and sporadic tasks, *Proceedings of the 1991 Real-time Systems Symposium*, pp. 129–139, San Antonio, TX, 1991.
14. LP_SOLVE, ftp://ftp.es.ele.tue.nl/pub/lp_solve
15. J. Luo and N. Jha, Power-conscious joint scheduling of periodic task graphs and a periodic tasks in distributed real-time embedded systems, *ICCAD'00*, pp. 357–364, San Jose, CA, 2000.
16. N. Namgoong, M. Yu, and T. Meng, A high-efficiency variable-voltage CMOS dynamic DC-DC switching regulator, *ISSCC'97*, pp. 380–381, San Francisco, CA, 1997.
17. J. Pouwelse, K. Langendoen, and H. Sips, Dynamic voltage scaling on a low-power microprocessor, *Proceedings of the 2001 International Symposium on Mobile Multimedia Systems and Applications*, pp. 251–259, Rome, Italy, 2001.
18. G. Quan and X. Hu, Energy efficient fixed-priority scheduling for real-time systems on voltage variable processors, *DAC'01*, pp. 828–833, Las Vegas, NV, 2001.
19. T. Ralphs, Discrete optimization lecture notes, http://www.lehigh.edu/tkr2/teaching/ie418/lectures/Lecture10.pdf

20. M. Schmitz, B. Al-Hashimi, and P. Eles, Energy-efficient mapping and scheduling for DVS enabled distributed embedded systems, *DATE'02*, pp. 514–521, Paris, France, 2002.
21. Y. Shin, K. Choi, and T. Sakurai, Power optimization for real-time embedded systems on variable speed processors, *Proceedings of the 2000 International Conference on Computer-Aided Design*, pp. 365–368, San Jose, CA, 2000.
22. TGFF, http://helsinki.ee.Princeton.EDU/dickrp/tgff
23. W. Wolf, Hardware-software co-design of embedded systems, *Proceedings of IEEE*, 82(7):967–989, 1994.
24. F. Yao, A. Demers, and S. Shenker, A scheduling model for reduced CPU energy, *Proceedings of the 1995 Symposium on Foundations of Computer Science*, pp. 374–382, Milwaukee, WI, 1995.
25. Y. Zhang, X. Hu, and D. Chen, Task scheduling and voltage selection for energy minimization, *DAC'02*, pp. 183–188, New Orleans, LA, 2002.
26. Y. Zhang, X. Hu, and D. Chen, Energy minimization of real-time tasks on variable voltage processors with transition energy overhead, *ASP-DAC'03*, pp. 65–70, Kitakyushu, Japan, 2003.

Energy-Aware Scheduling and Dynamic Reconfiguration in Real-Time Systems

Weixun Wang
University of Florida

Xiaoke Qin
University of Florida

Prabhat Mishra
University of Florida

24.1 Introduction

Various research efforts have focused on design and optimization of real-time systems in recent years. These systems require unique design considerations since timing constraints are imposed on the workloads (i.e., tasks). Tasks have to complete execution by their deadlines in order to ensure correct system behavior. For hard real-time systems, as in safety-critical applications like medical devices and aircrafts, violating task deadlines may lead to catastrophic consequences. Due to these stringent constraints, real-time scheduler must perform task *schedulability analysis* based on task characteristics such as priorities, periods, and deadlines [29]. A task set is considered to be *schedulable* only if there exists a valid schedule that satisfies all task deadlines. As embedded systems become ubiquitous, real-time systems with soft timing constraints, in which missing certain deadlines are acceptable, gain widespread applications including gaming, housekeeping, as well as multimedia equipments. Minor deadline misses may result in temporary service or quality degradation, but will not lead to incorrect system behavior. For example, users of video-streaming on mobile devices can tolerate occasional jitters caused by frame droppings, which does not affect the quality of service.

Energy conservation is a primary optimization objective for embedded systems design since these systems are generally driven by batteries with a limited energy budget. Various low-power computing techniques exist which tune the system during runtime (*dynamically reconfigure*) to meet optimization goals by changing *tunable* system parameters. Research has been targeted at determining *how* and *when* to dynamically reconfigure tunable parameters to achieve higher performance, lower energy consumption, and balance overall system behavior. Dynamic voltage scaling (DVS) [18] of the processor takes advantage of the fact that linear reduction in the supply voltage can quadratically reduce the power consumption while linearly slows down the operating frequency. Many general as well as specific-purpose processors nowadays support DVS [23,32] with multiple available voltage levels. Processor idle time (also know as slack time) also provides a unique opportunity to reduce the overall energy consumption by putting the system into a low-power sleep mode using dynamic power management (DPM) [5]. Research has shown that DVS should be used as the primary low-power technique for processor [21] while DPM could be beneficial after applying DVS. Memory hierarchy, especially the cache subsystem, has become comparable to the processor with respect to the contribution in overall energy consumption [30]. Dynamic cache reconfiguration (DCR) offers the ability to tune the cache configuration parameters at runtime to meet application's unique requirement so that significant amount of memory subsystem energy consumption can be saved [44,47]. Specifically, the working set of the application decides the favored cache capacity, while the spatial and temporal locality reflect the cache line size and associativity requirements, respectively. Research shows that specializing the cache for the application can lead to significant amount of energy reduction [16].

DCR can be successfully used in general-purpose platforms such as desktop-based systems [16,53]. However, there are major challenges in using reconfigurable caches in real-time systems. Given a run-time reconfigurable cache, determining the best cache configuration is difficult. Dynamic and static analysis are two possible techniques. With dynamic analysis, cache configurations are evaluated in the system during runtime to determine the best configuration. However, existing dynamic analysis methods are inappropriate for real-time systems since it either imposes unpredictable performance overhead during exploration or requires power hungry auxiliary data structure and thus can only be operated periodically [16]. With static analysis, various cache alternatives are explored and the best cache configuration is selected for each application in its entirety [15]—application-based tuning, or for each phase of execution within an application [40]—phase-based tuning. Since applications tend to exhibit varying execution behavior throughout execution, phase-based tuning allows for the cache configuration to be specialized to each particular period, resulting in greater energy savings than application-based tuning. Regardless of the tuning method, the predetermined best cache configuration (based on design requirements) could be stored in a look-up table or encoded into specialized instructions. *The static analysis approach, therefore, is most appropriate for real-time systems due to its nonintrusive nature.*

Figure 24.1 shows the distribution of system-wide energy consumption for a typical system-on-chip (SoC) [27]. It can be seen that processor, cache subsystem, memory, and bus are the four main components which make comparable contributions to overall power consumption. Therefore, system-wide energy optimization techniques should consider all of them in order to reflect practical benefits. *Clearly, it will be promising to employ both DVS and DCR simultaneously to conserve both processor and memory energy dissipations.*

In the past, leakage energy was negligible compared to dynamic energy. However, in the last decade, we have observed a continuous CMOS device scaling process in which higher transistor density and smaller device dimension have led to increasing leakage (static) power consumption. This is mainly due to the proportionally reduced threshold voltage level with the supply voltage which decreases along with the

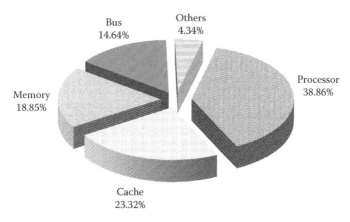

FIGURE 24.1   System-wide power consumption of a typical SoC.

power supply at a speed of 0.85X per generation [12]. Lower threshold voltage results in larger leakage current which mainly consists of subthreshold current and reverse bias junction current. Study has shown that leakage power is increased by about five times in each technology generation [7]. It is responsible for over 42% of the overall power dissipation in the 90 nm generation [25] and can exceed above half of the total in recent 65 nm technology [26]. Static energy is projected to account for near 70% of the cache subsystem's budget in 70 nm technology [26]. Leakage power also constitutes a major fraction of the total consumption for on-chip buses—almost comparable to dynamic part even when leakage control scheme is employed [37]. Memory modules, both with DRAM and SRAM, can also consume significant amount of leakage power (i.e., standby power) [22], especially when DVS is employed so that the standby time of those components increases [54]. *Therefore, decisions should be made judiciously on whether to slow down the system to save dynamic power or to finish execution faster and switch the system into sleep mode to reduce static power.* Extra consideration needs to be taken when DCR is also employed and other system components are considered.

Since energy consumption is converted into heat dissipation, high heat flux increases the on-chip temperature. The "hot spot" on current microprocessor die, caused by nonuniform peak power distribution, could reach up to 120°C [8]. This trend is observed in both desktop and embedded processors [52]. Thermal increase will lead to reliability and performance degradation since CMOS carrier mobility is dependent on the operating temperature. High temperature can result in more frequent transient errors or even permanent damage. Due to the severe detrimental impact, we have to control the instantaneous temperature so that it does not go beyond a certain threshold. Thermal management schemes at all levels of system design are widely studied for general-purpose systems. However, in the context of embedded systems, traditional packaging and cooling solutions are not applicable due to the limits on device size and cost. Therefore, it is extremely important to develop scheduling techniques that are efficient both in terms of energy and temperature.

The remainder of this chapter describes three important topics related to energy-aware dynamic reconfiguration and scheduling for real-time multitasking systems. Section 24.2 describes the model of the reconfigurable cache architecture. It also provides energy and thermal models. In Section 24.3, we introduce how to apply DCR in soft real-time systems. Section 24.4 discusses system-wide energy minimization employing DVS and DCR simultaneously for real-time multitasking systems. An energy- and temperature-constrained scheduling method is presented in Section 24.5. Section 24.6 presents experimental results to demonstrate the usefulness of these approaches for reducing system-wide energy and temperature requirements. Finally, Section 24.7 concludes this chapter.

24.2 Modeling of Real-Time and Reconfigurable Systems

24.2.1 System Model

The system can be modeled as

- A highly configurable cache architecture which supports h different configurations $C\{c_1, c_2, \ldots, c_h\}$
- A voltage scalable processor which supports l discrete voltage levels $V\{v_1, v_2, \ldots, v_l\}$
- A set of m independent tasks $T\{\tau_1, \tau_2, \ldots, \tau_m\}$
- Periodic task $\tau_i \in T$ has known attributes including worst-case workload, arrival time, deadline, and period
- Aperiodic/sporadic task $\tau_i \in T$ has known attributes including workload and inter-arrival time

The runtime overhead of voltage scaling is variable and depends on the original and new voltage levels. The context switching overhead is assumed to be constant.

24.2.2 Reconfigurable Cache Architectures

There are many existing general or application-specific reconfigurable cache architectures. Motorola M*CORE processor [30] provides way shutdown and way management, which has the ability to specify the content of each specific way (instruction, data, or unified way). Settle et al. [39] proposed a dynamically reconfigurable cache specifically designed for chip multiprocessors. The reconfigurable cache architecture proposed by Zhang et al. [53] imposes no overhead to the critical path, thus cache access time does not increase. Furthermore, the cache tuner consists of a small custom hardware or a lightweight process running on a coprocessor, which can alter the cache configuration via hardware or software configuration registers. The underlying cache architecture consists of four separate banks as shown in Figure 24.2a, each of which acts as a separate way. The cache tuner can be implemented either as a small custom hardware or lightweight software running on a coprocessor which changes the cache configuration through special registers. In order to reconfigure associativity, way concatenation, shown in Figure 24.2b, logically concatenates ways together so that the associativity can be changed accordingly without affecting total cache size. The required configure circuit consists of only eight logic gates and two single-bit registers. Varying cache size is achieved by shutting down certain ways, as shown in Figure 24.2c, using gated-V_{dd} technique. An extra transistor is used for every array of SRAM cells. Cache line size is configured by setting a unit-length base line size and then fetching subsequent lines if the line size increases, as illustrated in Figure 24.2d. Therefore, the configurable cache architecture achieves configurability using rather simple hardware and thus requires very minor overhead which makes this architecture especially suitable for embedded systems [53].

24.2.3 Energy Models

24.2.3.1 Cache Energy Model

Cache energy consumption is modeled as the sum of dynamic energy E_{cache}^{dyn} and static energy E_{cache}^{sta}:

$$E_{cache} = E_{cache}^{dyn} + E_{cache}^{sta} \tag{24.1}$$

The number of cache accesses n_{cache}^{access}, cache misses n_{cache}^{misses} and clock cycles CC are obtained from microarchitectural simulation for given tasks and cache configurations. t_{cycle} is decided by the processor

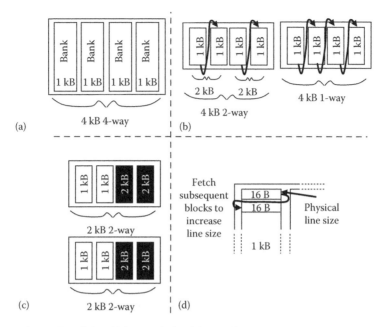

FIGURE 24.2 Cache configurability: (a) base cache bank layout, (b) way concatenation, (c) way shutdown, and (d) configurable line size.

frequency. Let E_{access} and E_{miss} denote the energy consumed per cache access and miss, respectively. Therefore, we have

$$E_{cache}^{dyn} = n_{cache}^{access} \cdot E_{access} + n_{cache}^{misses} \cdot E_{miss} \tag{24.2}$$

$$E_{miss} = E_{offchip} + \underline{E_{\mu P\_stall}} + E_{block\_fill} \tag{24.3}$$

$$E_{cache}^{sta} = P_{cache}^{sta} \cdot CC \cdot \underline{t_{cycle}} \tag{24.4}$$

where

$E_{offchip}$ is the energy required for fetching data from lower levels of memory hierarchy
$E_{\mu P\_stall}$ is the energy consumed when the processor is stalled due to cache miss
E_{block_fill} is for cache block refilling after a miss
P_{cache}^{sta} is the static power consumption of cache.

Values of E_{access}, P_{cache}^{sta}, and E_{block_fill} can be collected from CACTI [19] for different cache configurations.

For system-wide energy minimization as discussed in Section 24.4, since lower-level memory and system buses are all modeled simultaneously, $E_{offchip\_access}$ and $E_{\mu P\_stall}$ in Equation (24.3) are counted during the power estimation of the corresponding components (e.g., for a L2 cache miss, $E_{offchip\_access}$ is incorporated in the energy consumption of off-chip buses and DRAM memory).

24.2.3.2 Processor Energy Model

Since short-circuit power is negligible [43], the energy consumed in a processor mainly comes from dynamic and static power. The dynamic power can be computed as

$$P_{proc}^{dyn} = C_{eff} \cdot V_{dd}^2 \cdot f \tag{24.5}$$

where C_{eff} is the total effective switching capacitance of the processor, V_{dd} is the supply voltage level, and f is the operating frequency. We introduce the analytical processor energy model based on [31], whose accuracy has been verified with SPICE simulation. The threshold voltage V_{th} is presented as

$$V_{th} = V_{th1} - K_1 \cdot V_{dd} - K_2 \cdot V_{bs} \tag{24.6}$$

where V_{th1}, K_1, K_2 are all constants and V_{bs} represents the body bias voltage. Static current mainly consists of the subthreshold current I_{subth} and the reverse bias junction current I_j. Hence, the static power is given by

$$P_{proc}^{sta} = L_g \cdot (V_{dd} \cdot I_{subth} + |V_{bs}| \cdot I_j) \tag{24.7}$$

where

L_g is the number of devices in the circuit
I_j is approximated as a constant
I_{subth} can be calculated by

$$I_{subth} = K_3 \cdot e^{K_4 V_{dd}} \cdot e^{K_5 V_{bs}} \tag{24.8}$$

where K_3, K_4, and K_5 are constant parameters. Obviously, to avoid junction leakage power overriding the gain in lowering I_{subth}, V_{bs} has to be constrained (between 0 and -1 V). Let P_{proc}^{on} be the intrinsic energy needed for keeping the processor on (idle energy). The processor power consumption can be computed as

$$P_{proc} = P_{proc}^{dyn} + P_{proc}^{sta} + P_{proc}^{on} \tag{24.9}$$

The cycle length, t_{cycle}, is given by a modified α power model:

$$t_{cycle} = \frac{L_d \cdot K_6}{(V_{dd} - V_{th})^\alpha} \tag{24.10}$$

where K_6 is a constant. L_d can be estimated as the average logic depth of all instructions' critical path in the processor. The constants mentioned earlier are technology- and design-dependent. Table 24.1 lists the constants for a 70 nm technology processor.

The processor energy consumption becomes

$$E_{proc} = P_{proc} \cdot \underline{CC} \cdot t_{cycle} \tag{24.11}$$

24.2.3.3 Bus Energy Model

The average dynamic power consumption of various system buses can be calculated by [14]:

$$P_{bus}^{dyn} = \frac{1}{2} \cdot C_{bus} \cdot V_{dd}^2 \cdot n_{trans} \cdot f \tag{24.12}$$

TABLE 24.1 Constants for 70 nm Technology

Const	Value	Const	Value	Const	Value
K_1	0.063	K_6	5.26×10^{-12}	V_{th1}	0.244
K_2	0.153	K_7	-0.144	I_j	4.80×10^{-10}
K_3	5.38×10^{-7}	V_{dd}	$[0.5, 1.0]$	C_{eff}	0.43×10^{-9}
K_4	1.83	V_{bs}	$[-1.0, 0.0]$	L_d	37
K_5	4.19	α	1.5	L_g	4×10^6

where

C_{bus} is the load capacitance of the bus

V_{dd} is the supply voltage

f is the bus frequency

n_{trans} denotes the average number of transitions per time unit on the bus line, as shown in the following:

$$n_{trans} = \frac{\sum_{t=0}^{T-1} H(B^{(t)}, B^{(t+1)})}{T} \tag{24.13}$$

where

T is the total number of discretized time units of the system execution time

$H(B^{(t)}, B^{(t+1)})$ gives the Hamming distance between the binary values on the bus at two neighboring time units in T.

Therefore, the total energy consumption of a bus is determined by its dynamic power P_{bus}^{dyn} and static power P_{bus}^{sta}:

$$E_{bus} = \left(P_{bus}^{dyn} + P_{bus}^{sta}\right) \cdot \underline{CC} \cdot t_{cycle} \tag{24.14}$$

24.2.3.4 Main Memory Energy Model

Memory consists of DRAM has three main sources of power consumption: dynamic energy due to accesses E_{mem}^{dyn}, static power P_{mem}^{sta}, and refreshing power P_{mem}^{ref}. Specifically, we have:

$$E_{mem}^{dyn} = n_{mem}^{access} \cdot E_{access} \tag{24.15}$$

where n_{mem}^{access} is the number of memory accesses and E_{access} denotes the dynamic energy required per access. Therefore, we have

$$E_{mem} = E_{mem}^{dyn} + \left(P_{mem}^{sta} + P_{mem}^{ref}\right) \cdot \underline{CC} \cdot t_{cycle} \tag{24.16}$$

Values of E_{access}, P_{mem}^{sta}, and P_{mem}^{ref} can be collected from CACTI [19].

24.2.4 Thermal Model

A thermal RC circuit is normally utilized to model the temperature variation behavior of a microprocessor [52]. Here we introduce the RC circuit model proposed in [41], which is widely used in recent researches [52], to capture the heat transfer phenomena in the processor. If P denotes the power consumption during a time interval, R denotes the thermal resistance, C represents the thermal capacitance, T_{amb} and T_0 are the ambient and initial temperature, respectively, the temperature at the end of the time interval t can be calculated as

$$T = P \cdot R + T_{amb} - (P \cdot R + T_{amb} - T_{init}) \cdot e^{-t/RC} \tag{24.17}$$

where t is the length of the time interval. If t is long enough, T will approach a steady-state temperature $T_s = P \cdot R + T_{amb}$.

24.3 Scheduling-Aware Cache Reconfiguration

This section presents a novel methodology for employing reconfigurable caches in real-time systems with preemptive task scheduling [47]. Since both periodic and sporadic tasks are considered, task characteristics such as arrival time and deadline constraints are unknown. As a result, with preemptive scheduling, one task may be preempted by higher-priority tasks and resume at any time instance. Therefore, for hard real-time systems, the benefit of cache reconfiguration is limited since it may eventually lead to unpredictable system behavior. However, soft real-time systems offer much more flexibility, which can be exploited to achieve considerable energy savings at the cost of minor impacts to user experiences. In this section, we focus on real-time systems with soft timing constraints. Even in soft real-time systems, task execution time cannot be unpredictable or prolonged arbitrarily. The goal is to realize maximum energy savings while ensuring the system only faces an innocuous amount of deadline violations (if any). This methodology—scheduling-aware cache reconfiguration (SACR)—provides an efficient and near-optimal strategy for cache tuning based on static program profiling for both statically and dynamically scheduled systems.

SACR statically executes, profiles, and analyzes each task intended to run in the system. The information obtained in the profiling process is fully utilized at runtime to make reconfiguration decisions dynamically. The remainder of this section is organized as follows. First, we present an overview of SACR using simple illustrative examples. Next, static analysis techniques for cache configuration selection are presented. Finally, we describe how the static analysis results are used during runtime for statically- and dynamically-scheduled real-time systems.

24.3.1 Illustrative Example

This section presents a simple illustrative example to show how reconfigurable caches benefit real-time systems. This example assumes a system with two tasks, $T1$ and $T2$. Traditionally if a reconfigurable cache technique is not applied, the system will use a *base cache* configuration Cache$_{base}$, which is defined in Definition 24.1.

Definition 24.1 *The term* **base cache** *refers to the configuration selected as the optimal one for tasks in the target system with respect to energy as well as performance based on static analysis. Caches in such systems are fixed throughout all task executions and chosen to ensure feasible task schedules.*

In the presence of a reconfigurable cache, as shown in Figure 24.3, different optimal cache configurations are determined for every "phase" of each task. For ease of illustration, we divide each task into two phases: $phase_1$ starts from the beginning to the end, and $phase_2$ starts from the half position of the dynamic instruction flow (midpoint) to the end. The terms Cache$_{T1}^1$, Cache$_{T1}^2$, Cache$_{T2}^1$, and Cache$_{T2}^2$ represent the optimal cache configurations for $phase_1$ and $phase_2$ of task $T1$ and $T2$, respectively. These configurations are chosen statically to be more energy efficient (with same or better performance) in their specific phases than the global base cache, Cache$_{base}$.

Figure 24.4 illustrates how energy consumption can be reduced by using SACR. Figure 24.4a depicts a traditional system and Figure 24.4b depicts a system with a reconfigurable cache. In this example, $T2$ arrives (at time $P1$) and preempts $T1$. In a traditional approach, the system executes using Cache$_{base}$ exclusively. With a reconfigurable cache, the first part of $T1$ executes using Cache$_{T1}^1$. Similarly, Cache$_{T2}^1$ is used for execution of $T2$. Note that the actual preemption point of $T1$ is not exactly at the same place where we precomputed the optimal cache configuration (midpoint) since tasks may arrive at any time. When $T1$ resumes at time point $P2$, the cache is tuned to Cache$_{T1}^2$ since the actual preemption point is closer to the

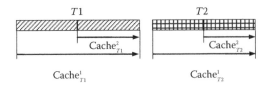

FIGURE 24.3 Cache configurations selected based on task phases.

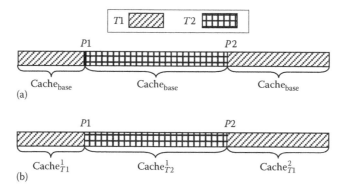

FIGURE 24.4 Dynamic cache reconfigurations for tasks *T*1 and *T*2. (a) Traditional system (b) SACR.

midpoint compared to the starting point (other heuristics can also be employed). Using a reconfigurable cache results in energy savings due to the fact that specific energy optimal caches are used for each phase of task execution compared to using one global base cache in the traditional system. Experimental results suggest that SACR can significantly reduce energy consumption of the memory subsystem with very little performance penalty.

24.3.2 Phase-Based Optimal Cache Selection

This section describes static analysis approach to determine the optimal cache configurations for various task phases. In a preemptive system, tasks may be interrupted and resumed at any point of time. Each time a task resumes, cache performance for the remainder of task execution will differ from the cache performance for the entire application due to its own distinguishing behaviors as well as cold-start compulsory cache misses. Therefore, the optimal cache configuration for the remainder of the task execution may be different.

Definition 24.2 *Phase* *is defined as the execution period between one potential preemption point (also called* **partition points**) *and task completion. The phase that starts at ith partition point is denoted as phase p_n^i, where n is the total number of phases of that task.*

Figure 24.5 depicts the general case where a task is divided into $n - 1$ predefined *potential preemption points* $(P_1, P_2, \ldots, P_{n-1})$. P_0 and P_n are used to refer to the start and end point of the task, respectively. Here, $C_0, C_1, \ldots, C_{n-1}$ represent the optimal cache configuration (either energy or performance) for each phase, respectively. During static profiling, a *partition factor* is chosen that determines the number of potential preemption points and resulting phases. Partition granularity is defined as the number of dynamic instructions between two partition points and is determined by dividing the total number of dynamically executed instructions by the partition factor. Intuitively, the optimal partition granularity should be a single instruction, potentially leading to the largest amount of energy savings. However, such a tiny granularity would result in a prohibitively large look-up table, which is not feasible due to area as well as searching time constraints. Thus, a trade-off should be made to determine a reasonable partition factor based on energy-savings potential and acceptable overheads. The rule of thumb is to find a partition factor minimizing the number of neighboring partitions that share the same optimal cache configuration. It could be a local optimal factor for each task if varying number of table entries for different tasks are allowed or it could be a global optimal factor for the task set. Empirically, a partition factor ranging from 4 to 7 is sufficient to achieve most amount of energy savings [47].

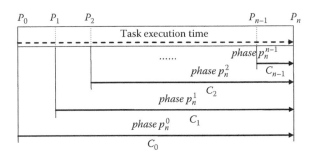

FIGURE 24.5 Task partitioning at n potential preemption points (P_i) resulting in n phases. Each phase comprises execution from the invocation/resumption point to task completion. C_i denotes the cache configuration used in each phase.

Static profiling generates a *profile table* that stores the potential preemption points and the corresponding optimal cache configurations for each task. Next two sections describe how this profile table is used during runtime in statically and dynamically scheduled systems.

24.3.3 Statically Scheduled Systems

With static scheduling, arrival times, execution times, and deadlines are known a priori for each task and this information serves as scheduler input. The scheduler then provides a schedule detailing all actions taken during system execution. According to this schedule, we can statically execute and record the energy-optimal cache configurations that do not violate any task's deadline for every execution period of each task. For soft real-time systems, global (system-wide) energy-optimal configurations can be selected as long as the configuration performance does not severely affect system behavior. After this profiling step, the profile table is integrated with the scheduler so that the cache reconfiguration hardware (cache tuner) can tune the cache for each scheduling decision.

24.3.4 Dynamically Scheduled Systems

With dynamic scheduling (online scheduling), scheduling decisions are made during runtime. In this scenario, task preemption points are unknown since new tasks may enter the system at any time with any time constraint. In this section, we present two versions of SACR based on the nature of the target system.

24.3.4.1 Conservative Approach

In some soft real-time systems where high service quality is required thus time constraints are pressing, only an extremely small number of violations are tolerable. The conservative approach could ensure that given a carefully chosen partition factor, almost every task could meet their deadlines with only very few exceptions. To ensure the largest task acceptance ratio, any reconfiguration decision will only change the cache into the lowest energy configuration whose execution time is no longer than that of the base cache. In other words, only cache configurations with equal or higher performance than the base cache are chosen for each task phase. We denote them as deadline-aware energy-optimal cache configurations. The scheduler chooses the appropriate cache configuration from the generated profile table that contains the energy-optimal cache configurations for each task phase.

During system execution, the scheduler maintains a task list keeping track of all existing tasks. In addition to the static profile table, runtime information such as arrival time, deadline, and number of already executed dynamic instructions are also recorded. This information is stored not only for the scheduler, but also for the cache tuner. As indicated in Section 24.3.2, potential preemption points are pre-decided during the profile table generation process. However, it is highly unlikely that the actual preemptions will occur exactly on these potential preemption points. Hence, a *nearest-neighbor* method is used to determine which cache configuration should be used. Essentially, if the preemption point falls between two partition points, the nearest point will be referred to select the new cache configuration. Details of the conservative approach is available in [47].

24.3.4.2 Aggressive Approach

For soft real-time systems in which only moderate service quality is needed, a more aggressive version of SACR can reveal additional energy savings at the cost of possibly violating several future task deadlines, but remain in an acceptable range. Similar to the conservative approach, a profile table is associated with every task in the system; however this profile table contains the performance-optimal cache configuration (whose execution time is the shortest) in addition to the energy-optimal configuration (the one with lowest energy consumption among all candidates) for every task phase. In order to assist dynamic scheduling, the profile table also includes the corresponding phase's execution time (in cycles) for each configuration. Note that the performance and energy efficiency of a cache configuration may not be opposite. The energy-optimal one does not necessarily have the worst performance. Compared to the base cache, it could have both better energy efficiency and performance.

DCR decisions are made either when a new task with a higher priority than the current executing task arrives or when the current task finishes execution. The cache tuner checks the schedulability of all ready tasks in the system by iteratively examining whether each task can meet its deadline if all the preceding tasks (with higher priorities), including itself, use performance-optimal cache configurations. If not, the task is subject to be discarded. This process is done in the order of tasks' priority (from highest to lowest) to achieve minimum task drop rate. Afterward, the appropriate cache configuration for next executing task is selected based on whether it is safe to use energy-optimal cache configuration. Details of the aggressive approach is available in [47].

24.4 Energy Optimization with DVS and DCR

The previous section considered DCR to reduce energy consumption in the cache subsystem. In this section, we describe how to integrate DVS and DCR together in hard real-time systems to minimize system-wide energy consumption [45]. We examine the correlation among the energy models of all the components (Section 24.2.3) and their impact on the decision making of both DVS and DCR. Energy optimization decisions are made at design time based on static slack allocation, and task procrastination is carried out at runtime to achieve more idle energy savings. For hard real-time embedded systems, off-line analysis is of great importance for energy-efficient scheduling techniques [35]. It is mainly due to the fact that these systems normally have highly deterministic characteristics, e.g., release time, deadline, input set, and workload, which should be fully utilized for energy saving. Furthermore, sophisticated analysis such as memory behavior profiling can only be carried out during design time.

There are major challenges including design space exploration, system-wide energy analysis and configuration selection to reduce overall energy consumption while meeting all task deadlines. Figure 24.6 illustrates the workflow for system-wide energy optimization. Each application in the task set is fed into a simulation process which is driven by a design space exploration heuristic. The total system energy consumption for each simulation is computed by the energy estimation framework which has separate power analyzers for different system components and integrates them systematically. Based on the task set characteristics and the profile tables as well as the scheduling policy, processor voltage level and cache configuration are selected for each task. DVS/DCR assignments and task procrastination algorithm are then used in a one-pass scheduling which produces the total energy consumption of the task set. Section 24.4.1 describes the energy estimation framework based on static profiling outcomes. Section 24.4.2 presents

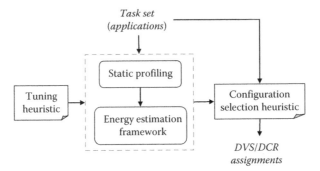

FIGURE 24.6 Workflow of system-wide energy optimization.

important observations in critical speed variation when different system components are taken into considerations. Sections 24.4.3 and 24.4.4 discuss profile table generation and DVS/DCR selection heuristics. Tuning heuristics for multilevel cache reconfiguration design space exploration are available in [44].

24.4.1 Energy Estimation Framework

Since the focus is not to minimize development time and costs, the energy estimation framework, as shown in Figure 24.7, targets at a specific SoC micro-architecture and is able to trade more design time for higher accuracy than the one proposed in [42]. For each application (task) and cache configuration, we run a simulation and collect the statistics, memory access statistics, and bus activity traces. These information, along with the processor voltage levels, are provided to energy models for each system components, based on which the total system energy is computed. Note that in this framework, the inputs to each energy model are all from one single micro-architectural simulation and thus are more comprehensive and systematic, as opposed to the approaches in which the inputs are collected separately using instruction-set simulator, memory trace profiler, cache simulator, and bus simulator [42]. Furthermore, by doing this, the impact on DVS/DCR decisions from other system components as well as

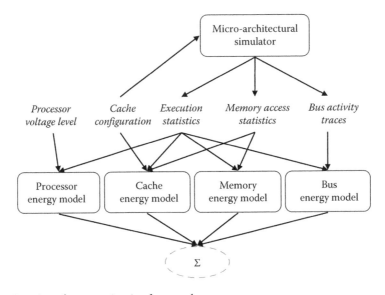

FIGURE 24.7 Overview of energy estimation framework.

their correlations can be reflected accurately. This framework also allows different energy models and analyzers to be used.

24.4.2 Critical Speed

The *critical speed* for processor voltage scaling defines a point beyond which the processor speed cannot be slowed down otherwise DVS will no longer be beneficial [20]. The dynamic power consumption of processor, which is exclusively considered in traditional DVS, is usually a convex and increasing function of the operating frequency. However, since lowering processor speed makes the task execution time longer which leads to higher static energy consumption, the total energy consumed per cycle in the processor will start increasing due to further slowdown.

By taking DCR into consideration, we find that cache configuration has significant impact on the critical speed with respect to the overall system energy consumption. Note that as described in Section 24.2.3, there exists strong correlation among the energy models of processor, cache subsystem and other system components. Since different cache configurations lead to different miss ratios and miss penalty cycles, the number of CC required to execute an application is decided by the cache configuration, which directly affects the energy consumption of other components, as shown in Equations 24.11, 24.14, and 24.16. On the other hand, the length of each clock cycle (t_{cycle}), which is determined by the processor voltage/frequency level, also directly affects the energy consumption of other components, as shown in Equations 24.3, 24.14, and 24.16. In other words, DVS and DCR will affect the overall system energy consumption. On the other hand, due to leakage power, all system components will have impact on decision making of DVS/DCR, especially the critical speed. Specifically, when the processor is slowed down by DVS, increasing static energy consumed by cache hierarchy, bus lines and memory will compromise the benefit gained from reduced processor dynamic energy and thus lead to higher system-wide energy dissipation. Therefore, considering DCR and other system components effects, the critical speed is going to increase drastically, as shown in this section.

24.4.2.1 Processor + L1/L2 Cache

We use a motivating example in which a single benchmark* (*cjpeg* from MediaBench [28]) is executed under all processor voltage levels. It can be observed that in Figure 24.8, when only processor energy is considered, the critical speed is achieved at $V_{dd} = 0.7$ V, which matches the results in [20]. However, as shown in Figure 24.9, with respect to the total energy consumption, combining processor and L1 caches (both configured to 8 kB of capacitance, 32 B line size, and 2-way associativity) increases the critical speed slightly to around $V_{dd} = 0.75$ V, due to the effect from L1 cache's leakage power dissipation. This highlights the importance of considering other system components for accurate analysis when applying DVS. In other words, if L1 caches are incorporated, $V_{dd} = 0.7$ V is no longer a beneficial choice with respect to the overall energy savings. Note that in Figure 24.9, dynamic energy consumption of L1 caches only includes access energy E_{access} and block refilling energy E_{block_fill}. Energy consumed on buses and lower-level memory hierarchy during L1 cache misses will be incorporated when we gradually add the corresponding components into consideration, as shown in following sections.

Figure 24.10 shows the impact on the critical speed if L2 cache (with capacity of 64 kB, line size 128 B, and 8-way associativity) is considered in the overall energy consumption. The critical speed increases to the frequency corresponding to $V_{dd} = 0.85$ V. For L1 caches, as shown in Figure 24.9, dynamic energy dominates and leakage energy becomes comparable only when the processor voltage level drops below 0.6 V. However, in L2 cache, for *cjpeg*, leakage energy dissipation dominates while dynamic energy is almost negligible. It is expected since L1 access rate is much higher than L2 while the capacity, thus leakage power, of L2 cache is much larger than L1. Note that, although some other benchmarks (e.g., *qsort* from MiBench [17]) shows nonnegligible dynamic energy consumption in L2 cache, the leakage part still

---

* Although results for *cjpeg* is shown in this section, similar observations have been made for other benchmarks.

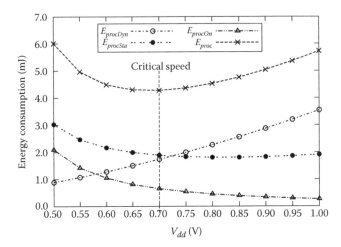

FIGURE 24.8 Processor energy consumption E_{proc} for executing *cjpeg*: $E_{procDyn}$ is the dynamic energy, $E_{procSta}$ is the static energy, and E_{procOn} is the intrinsic energy needed to keep processor on.

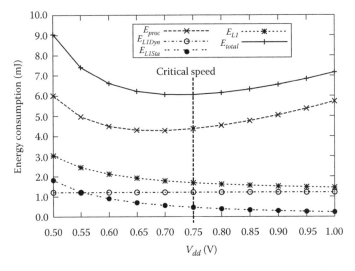

FIGURE 24.9 Overall system energy consumption E_{total} of the processor and L1 caches for executing *cjpeg*: E_{L1Dyn} and E_{L1Sta} are the dynamic and static L1 cache energy consumption, respectively.

dominates when the voltage level goes below a certain point. Therefore, when processor voltage decreases, the total leakage energy consumption increases drastically due to the L2 cache. Generally, when DCR is applied, different cache configurations will lead to different critical speed variations.

24.4.2.2 Processor + L1/L2 Cache + Memory

Figure 24.11 illustrates the fact that memory energy consumption also makes the critical speed increase. The memory is modeled as a common DRAM with size of 8 MB. It can be observed that memory has a similar effect on the critical speed as L2 cache. In fact, for the configurations we considered, the static energy consumptions are comparable for L2 cache and the memory. Although DRAM needs to have its capacitor charge refreshed all the time (which consumes relatively negligible power in 70 nm technology [19]), it requires only one transistor to store one bit. Therefore, it consumes much less leakage power per bit compared to cache, which is smaller but made of more power expensive SRAM.

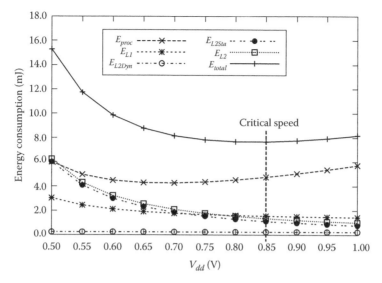

FIGURE 24.10 Overall system energy consumption E_{total} of the processor, L1 caches and L2 cache (configured to 64 kB, 128 B, 8-way) for executing *cjpeg*: E_{L2Dyn} and E_{L2Sta} are the dynamic and static L2 cache energy consumption, respectively.

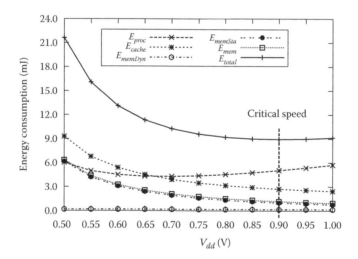

FIGURE 24.11 Overall system energy consumption E_{total} of the processor, L1/L2 caches and memory for executing *cjpeg*: E_{memDyn} and E_{memSta} are the dynamic and static memory energy consumption, respectively; E_{cache} represents the total energy consumption of both L1 and L2 caches.

24.4.2.3 Processor + L1/L2 Cache + Memory + Bus

System bus lines have double effect on the critical speed in overall system energy consumption. On one hand, since on-chip buses should have equal frequency as the processor (which makes them dominate in terms of energy among all system buses), DVS will lead to dynamic energy reduction in them. On the other hand, like other system components, static power dissipation on system buses is also going to increase along with voltage scaling down, which compromises the dynamic energy reduction. As a result, system buses make very minor impact on critical speed, as shown in Figure 24.12.

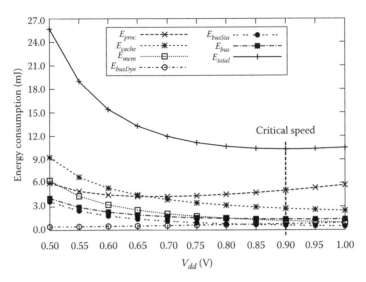

FIGURE 24.12 Overall system energy consumption E_{total} of the processor, L1/L2 caches, memory and system buses for executing *cjpeg*: E_{busDyn} and E_{busSta} are the dynamic and static bus energy consumption, respectively.

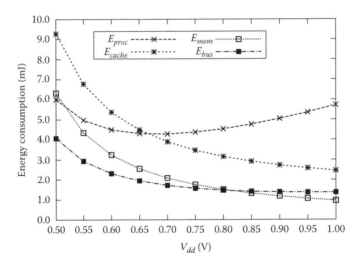

FIGURE 24.13 Processor voltage scaling impact on energy consumption of various system components.

To summarize, we show how energy consumption (both dynamic and static) of each system components vary with voltage scaling in Figure 24.13. When DVS is not applied ($V_{dd} = 1$ V), the processor accounts for over half of overall energy consumption while others also take considerable share. This observation matches with Figure 24.1. When the voltage level decreases, we can see that the energy consumed by the cache hierarchy and memory subsystem increases drastically and, after certain point, it becomes comparable with the processor or even larger. For system buses, due to the reason explained earlier, this effect is less significant compared to cache and memory.

The preceding discussion leads to several interesting observations and important questions. The critical speed is going to change as different system components are considered—increases when leakage energy dominant components are added and decreases when dynamic energy dominant components (DVS-controllable) are added. One would wonder whether DVS is really practically beneficial since the case study shows that the critical speed is at $V_{dd} = 0.9$ V and potentially adding more components may increase it further (possibly close to 1.0 V)? A simple answer is yes but it has to be evaluated using leakage-aware DVS and DCR. It is also important to notice that system properties, application characteristics and reconfiguration decisions together will affect the critical speed, which typically varies between $V_{dd} = 0.65$ and 0.9 V in this case.

24.4.3 Profile Table Generation

We define a *configuration point* as a pair of processor voltage level and cache hierarchy configuration: (v_j, c_k) where $v_j \in V$ and $c_k \in C$. Note that each c_k represents a configuration combination of the L1 caches and L2 cache. For each task, we can construct a *profile table* which consists of all possible configuration points as well as the corresponding total energy consumption and execution time. All points with the voltage level lower than the critical speed are eliminated. Furthermore, non-beneficial configuration points, which is inferior in both energy and time compared to some other points, are also discarded. In other words, only Pareto-optimal trade-off points are considered

An important observation is that cache configurations behave quite consistently across different processor voltage levels. For example, as shown in Figure 24.14, the L1 cache configuration favored by *cjpeg*, 8 kB cache size with 32 B line size and 2-way associativity, outperforms all the other configurations with respect to the total energy consumption. Similar observations can be made when we fix L1 cache configuration while vary L2 cache. Therefore, the profile table for each task actually consists of favored cache configuration combinations with voltage levels equal to or higher than the corresponding critical speed. In fact, we find that only the most energy-efficient cache hierarchy configuration with the voltage level equal to or higher than the critical speed exist in the profile table.

24.4.4 Configuration Selection Heuristic

Most existing DVS algorithms are not applicable when DCR is employed since the energy consumption as well as the impact on task's execution time from system components other than the processor cannot be simply calculated from energy models. SACR discussed in Section 24.3 are also not applicable since it only supports soft task deadlines. Given a static slack allocation, we can assign the most energy efficient configuration point which does not stretch the execution time over the allocated slack. As long as the

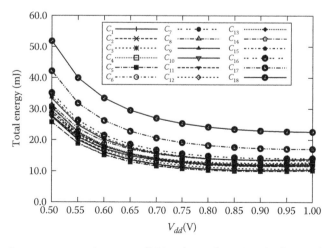

FIGURE 24.14 Total energy consumption across all L1 cache configurations (with L2 cache configured to 64 kB, 128 B, and 8-way) for executing *cjpeg*.

slack allocation is safe, we can always ensure that the schedulability condition is satisfied. Therefore, a simple heuristic motivated by the uniform constant slowdown scheme which is proved to be optimal in continuous voltage scaling [3] is applicable. The optimal common slowdown factor η is given by the system utilization rate. Since we only consider a finite number of discrete configuration points as defined earlier, we select the configuration point with minimum energy consumption but equal or shorter execution time compared to the one decided by the optimal slowdown factor for each task. Note that we use each task's execution under the highest voltage and largest cache configuration as the base case, which is used in the optimal slowdown factor calculation.

24.5 Energy- and Temperature-Constrained Scheduling

So far, we have discussed techniques related to energy optimization. In this section, we describe scheduling techniques that consider both energy and temperature constraints [48]. DVS is acknowledged as one of the most efficient techniques not only for energy optimization but also for temperature management. In existing literatures, *temperature (energy)-constrained* means there is a temperature threshold (energy budget) which cannot be exceeded, while *temperature (energy)-aware* means that there is no constraint but maximum instantaneous temperature (total energy consumption) needs to be minimized. In this section, we introduce a formal method based on model checking for temperature- and energy-constrained (TCEC) scheduling problems in multitasking systems. This approach is also capable of solving other problems including temperature-constrained (TC) scheduling, temperature-aware (TA) scheduling, temperature-constrained energy-aware (TCEA) scheduling, and energy-constrained temperature-aware (TAEC) scheduling.

Figure 24.15 outlines the workflow for solving TCEC problems. The task information describes the characteristics of the tasks running in the system and is fed into the scheduler along with the scheduling policy. Any scheduling algorithm is applicable in this approach. The scheduler executes the task set under

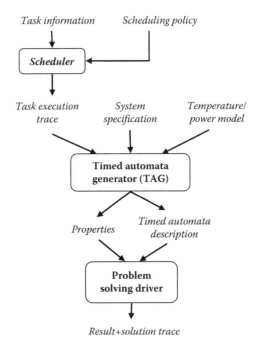

FIGURE 24.15 Overview of the TCEC schedulability framework.

the highest voltage level and produces a trace of *execution blocks*. An execution block is defined as a piece of task execution in a continuous period of time under a single processor voltage/frequency level. Each execution block is essentially a whole task instance in non-preemptive systems. However, in preemptive scheduling, tasks could be preempted during execution hence one block can be a segment of one task. The scheduler records runtime information for each block including its corresponding task, required workload, arrival time, and deadline, if applicable.

The task execution trace, along with system specification (processor voltage, frequency levels, temperature constraints or/and energy budget) and thermal/power models are fed into the timed automata generator (TAG). TAG generates two important outputs. One is the corresponding timed automata model, which will be discussed in Section 24.5.2.2, and the other one is properties reflecting the temperature/energy/deadline constraints defined in system specification. After that, a suitable solver (e.g., a model checker or a SAT solver) is applied to find a feasible schedule of the tasks, or confirm that the required constraints cannot be met.

24.5.1 TCEC Problem Formulation

The methodology described in this section can be applied to both scenarios in which task set has a common deadline and each task has its own deadline. For ease of discussion, the following definition of TCEC problem is constructed for task sets with a common deadline. The general case (tasks with individual deadlines) will be discussed in Section 24.5.2.3.

Given a trace of m jobs $\{\tau_1, \tau_2, \ldots, \tau_m\}$, if tasks are assumed to have the same power profile (i.e., α is constant), the energy consumption and execution time for τ_i under voltage level v_j, denoted by w_{ij} and t_{ij}, respectively, can be calculated based on the given processor model. Otherwise, they can be collected through static profiling by executing each task under every voltage level. Let ψ_{ij} and ω_{ij} denote runtime energy and time overhead, respectively, for scaling from voltage v_i to v_j. Since power is constant during a execution block, temperature is monotonically either increasing or decreasing. We denote $T(i)$ as the final temperature of τ_i. If the task set has a common deadline D, the safe temperature threshold is T_{max} and the energy budget is W, TCEC scheduling problem can be defined as follows.

Definition 24.3 *TCEC problem: Determine whether there exists a voltage assignment $\{l_1, l_2, \ldots, l_m\}^*$ such that:*

$$\sum_{i=1}^{n}(e_i^{k_i} + \psi_{v_{k_{i-1}}, v_{k_i}}) \leqslant \mathcal{E} \tag{24.18}$$

$$T_i \leqslant T_{max}, \forall i \in [1, n] \tag{24.19}$$

$$\sum_{i=1}^{n}(t_i^{k_i} + \omega_{v_{k_{i-1}}, v_{k_i}}) \leqslant D \tag{24.20}$$

where T_i is calculated based on Equation 24.17. Here Equations 24.18 through 24.20 denote the energy, temperature, and common deadline constraints, respectively.

24.5.2 TCEC Modeling with Timed Automaton

24.5.2.1 Timed Automata

A classical timed automaton [2] is a finite-state automaton extended with notion of time. A set of clock variables are associated with each timed automaton and elapse uniformly with time in each state

---

* l_i denote the index of the processor voltage level assigned to b_i.

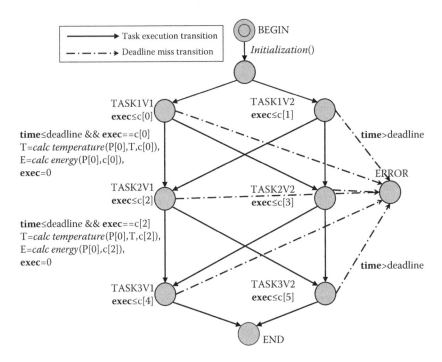

FIGURE 24.16   TCEC problem modeled in extended timed automata.

(i.e., location). Transitions (i.e., edges) are performed instantaneously from one state to another. Each transition is labeled with a set of guards which are Boolean constraints on clock variables and must be satisfied in order to trigger the transition. Transitions also have a subset of clock variables that need to be reset by taking the transition. Formally, it is defined as follows:

Definition 24.4   *A timed automaton $\mathcal{A}$ over clock set $\mathcal{C}$, state set $\mathcal{S}$, and transition set $\mathcal{T}$ is a tuple $\{\mathcal{S}, \mathcal{C}, \mathcal{T}, s_0\}$ where s_0 is the initial state. Transition set is represented as $\mathcal{T} \subseteq \mathcal{S} \times \Phi(\mathcal{C}) \times 2^{\mathcal{C}} \times \mathcal{S}$, where each element ϕ in clock constraint (guard) set $\Phi(\mathcal{C})$ is a conjunction of simple conditions on clocks ($\phi := c \leq t \mid t \leq c \mid \neg\phi \mid \phi_1 \wedge \phi_2$ where $c \in \mathcal{C}, t \in R$). $2^{\mathcal{C}}$ represents the subset of clock variables that will be reset in each transition.*

24.5.2.2 Modeling with Extended Timed Automata

Timed automata are suitable for modeling the TCEC problem based on DVS. Here we need to extend the original automata with notions of temperature and energy consumption. This model supports both scenarios in which task set has a common deadline and each task has its own deadline. For ease of discussion, the terms of *task, job,* and *execution block* refer to the same entity in rest of the discussion. For illustration, an extended timed automata $\mathcal{A}$ generated by TAG is shown in Figure 24.16 assuming that there are three tasks and two voltage levels. Generally, l states[*] are used for each task, forming disjoint sets (horizontal levels of nodes in Figure 24.16) among tasks, to represent different voltage selections. An error state is also introduced, which is reached whenever there is deadline miss. There are also a source state and a destination state denoting the beginning and the end of the task execution. Therefore, there are totally $(n \cdot l + 4)$ states. There is a transition from every state of one task to every state of its next task. In other words, the states in neighboring disjoint sets are fully connected. There are also transitions from every task state to the error state. All the states of the last task have transitions to the end state.

---

[*] There are l voltage levels.

The system temperature and cumulative energy consumption are represented by two global variables, named T and E, respectively. The execution time for every task under each voltage level is pre-calculated and stored in a global array $c[\ ]$. The common deadline D is stored in variable *deadline*. Constants such as processor power values, thermal capacitance/resistance, ambient temperature and initial temperature are stored in respective variables. There are two clock variables, **time** and **exec**, which represent the global system time and the local timer for task execution, respectively. The **time** variable is never reset and elapses uniformly in every state. Both clock variables are initially set to 0.

The transition from the source state carries a function *initialization()* which contains updates to initialize all the variables and constants. Each state is associated with an invariant condition, in the form of **exec** $\leqslant c[\ ]$, which must be satisfied when the state is active. This invariant represents the fact that the task is still under execution. Each transition between task states carries a pair of guard: **time** $\leqslant$ *deadline* && **exec** $==c[\ ]$. The former one ensures that the deadline is observed and the latter one actually triggers the transition, reflecting the fact that the current task has finished execution. Note that the overhead can be incorporated here since we know the start and end voltage level, if they are different. Each transition is also labeled with three important updates. The first one, $T = calcTemperature(P[\ ], T, c[\ ])$, basically updates the current system temperature after execution of one task based on the previous temperature, average power consumption, and the task's execution time. The second one, $E = calcEnergy(P[\ ], c[\ ])$, adds the energy consumed by last task to E. The third update resets clock *exec* to 0. All the transitions to the error state are labeled with a guard in the form of **time** $>$ *deadline*, which triggers the transition whenever the deadline is missed during task execution. Note that not all the transition labels are shown in Figure 24.16.

The extended timed automata's current configuration is decided by valuations of clock variables (**time** and **exec**) and global variables (T and E). Therefore, the system execution now is transformed into a sequence of states from the source state to the destination state. The sequence consists of one and only one state from each disjoint set which represents a task. Solving the TCEC problem as formulated earlier is equal to finding such a sequence with the following properties. First, the final state is the destination state which guarantees the deadline constraint. Next, the temperature T is always below T_{max} in every state. Finally, the energy consumption E is no larger than $\mathcal{E}$. We can write this requirement as a property in computation tree logic (CTL) as

$$\mathbf{EG}((T < T_{max} \ \wedge \ E < \mathcal{E}) \ \mathbf{U} \ \mathcal{A}.end) \tag{24.21}$$

where $\mathcal{A}.end$ means the destination state is reached. Now, we can use the model checker to verify this property and, if satisfied, the witness trace it produces is exactly the TCEC scheduling that we want.

However, it is possible that the model checker's property description language does not support the operator of "until" (**U**), e.g., UPPAAL [4]. In that case, two Boolean variables, *isTSafe* and *isESafe*, are added to denote whether T and E are currently below the constraints. These two Boolean variables are updated in functions *calcTemperature()* and *calcEnergy()*, respectively, whenever a transition is performed. Once the corresponding constraint is violated, they are set to *false*. We can express the requirement in CTL as

$$\mathbf{EF}(isTSafe \ \wedge \ isESafe \ \wedge \ \mathcal{A}.end) \tag{24.22}$$

Note that in the timed CTL that UPPAAL uses, the preceding property can be written as follows, where *Proc* represents the timed automata $\mathcal{A}$, which is called a "Process" in UPPAAL.

$$\mathbf{E} <> (Proc.End \ \mathbf{and} \ Proc.isTSafe \ \mathbf{and} \ Proc.isESafe) \tag{24.23}$$

24.5.2.3 Problem Variants

The model checking-based approach is also applicable to other problem variants by modifying the property and making suitable changes to invocation of the model checker.

24.5.2.3.1 Task Set with Individual Deadlines

In the scenario where each task has its own deadline, e.g., periodic tasks, all the execution blocks have to finish no later than their corresponding task's deadline. A global array, $d[\ ]$, is used to store the deadline constraints of each execution block. If not applicable, i.e., the block does not end that task instance, its entry in $d[\ ]$ is set to -1. Therefore, instead of Equation 24.20, we have

$$\sum_{j=1}^{i} \left(t_j^{k_j} + \omega_{v_{k_{j-1}}, v_{k_j}} \right) \leqslant d[i], \forall d[i] > 0 \tag{24.24}$$

Figure 24.17 shows part of the new timed automata. The difference lies in the guard of transitions. Instead of **time** $\leqslant$ *deadline*, the guard for transitions between task states is in the form of $((d[\ ] > 0 \ \&\& \ \mathbf{time} \leqslant d[\ ]) \ || \ d[\ ] < 0)$. The transition from task state to error state now carries a guard of $(d[\ ] > 0 \ \&\& \ \mathbf{time} > d[\ ])$.

TC: Temperature-constrained scheduling problem is a simplified version of TCEC. It only needs to ensure that the maximum instantaneous temperature is always below the threshold T_{max}. Therefore, the property can be written in CTL as

$$\mathbf{EG}(T < T_{max} \ \mathbf{U} \ \mathcal{A}.end) \tag{24.25}$$

TA: To find a schedule so that the maximum temperature is minimized, we can employ a binary search over the temperature value range. Each iteration invokes the model checker to test the property (24.25) parameterized with current temperature constraint T_{max}. Initially, T_{max} is set to the mid-value of the range. If the property is unsatisfied, we search in the range of values larger than T_{max} in the next iteration. If the property is satisfied, we continue to search in the range of values lower than T_{max} to further explore better results. This process continues until the lower bound is larger than the upper bound. The minimum T_{max} and associated schedule, which makes the property satisfiable during the search, is the result. Note that the temperature value range for microprocessors is small in practice (e.g., 30°C, 120°C). Hence, the number of iterations is typically no more than 7.

TAEC: TAEC has the same objective as TA except that there is an energy budget constraint. Therefore, we can solve the problem by using property (24.21) during the binary search.

TCEA: TCEA can be solved using the same method as TAEC except that the binary search is carried on energy values and temperature acts as a constant constraint. Since energy normally has a much larger value range, to improve the efficiency, we can discretize energy value to make trade-off between solution quality and design time. Since the number of iterations has a logarithmic relationship with the length of energy value range, only moderate discretization is enough.

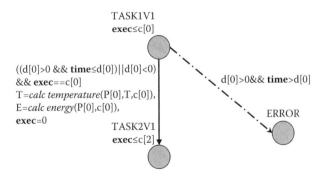

FIGURE 24.17 Problem modeling when every task has own deadline (partial graph).

24.5.3 Using SAT Solver

Timed automata can be used to model the TCEC problem effectively. However, when the number of tasks is large, it can be time consuming to check the properties on the timed automata directly. The reason is that the underlying symbolic model checker like UPPAAL sometimes cannot handle large problems due to the state space explosion problem. Fortunately, this problem can be alleviated by transforming the model checking problem to a Pseudo-Boolean satisfiability problem [9]. Similar to SAT-based techniques [11,34], which are used to reduce the test generation time, Pseudo-Boolean satisfiability solvers usually have better scalability over symbolic model checker and therefore can be used to perform model checking on timed automata more efficiently.

We use Boolean variable x_i^k to indicate whether block i is executed at voltage level v_k. The original model checking problem is equivalent to finding an assignment to $x_i^k, 1 \leqslant i \leqslant n, 1 \leqslant k \leqslant l$, which satisfies

$$\sum_{k=1}^{l} x_i^k = 1, \forall i \in [1, n] \tag{24.26}$$

$$\sum_{i=1}^{n} \sum_{k=1}^{l} x_i^k \cdot \left(e_i^k + \sum_{k'=1}^{l} x_{i-1}^{k'} \cdot \psi_{v_{k'}, v_k} \right) \leqslant \mathcal{E} \tag{24.27}$$

$$T_i \leqslant T_{max}, \forall i \in [1, n] \tag{24.28}$$

$$\sum_{i=1}^{n} \sum_{k=1}^{l} x_i^k \cdot \left(t_i^k + \sum_{k'=1}^{l} x_{i-1}^{k'} \cdot \omega_{v_{k'}, v_k} \right) \leqslant D \tag{24.29}$$

Since there are production terms in the constraints, this problem is a nonlinear Pseudo-Boolean satisfiability problem. It can be solved by normal linear solvers like PBclasp [33] after standard normalization [6]. As shown in Section 24.6, compared to direct model checking of the original timed automata, it is more scalable to perform model checking using Pseudo-Boolean solver. Details of the TCEC framework is available in [48].

24.6 Experiments

24.6.1 Scheduling-Aware Cache Reconfiguration

Selected benchmarks from the MediaBench [28], and EEMBC Automotive [13] benchmark suites, representing typical tasks that might be present in a soft real-time system, are used to evaluate the effectiveness of SACR. All applications were executed with the default input sets provided with the benchmarks suites. We utilized the configurable cache architecture for L1 cache developed by Zhang et al. [53] with a four-bank cache of base size 4 kB, which offers sizes of 1, 2, and 4 bytes, line sizes ranging from 16 to 64 bytes, and associativity of 1-way, 2-way, and 4-way. For comparison purposes, we define the *base cache* configuration to be a 4 bytes, 2-way set associative cache with a 32-byte line size, a reasonably common configuration that meets the needs of the benchmarks studied. The L2 cache is set to a 64 K unified cache with 4-way associativity and 32 bytes line size. The energy model has been described in Section 24.2.3.1. To obtain cache hit and miss statistics, we used the SimpleScalar toolset [10] to simulate the applications. The design space of 18 cache configurations is exhaustively explored during static analysis to determine the energy-, performance-, and deadline-aware energy-optimal cache configurations for each phase of each benchmark.

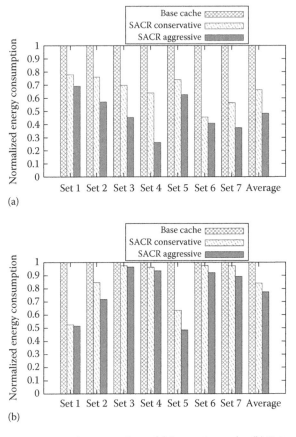

FIGURE 24.18 Cache energy consumption comparisons. (a) Instruction cache. (b) Data cache.

To model sample real-time embedded systems with multiple executing tasks, we created seven different task sets. In each task set, the three selected benchmarks have comparable dynamic instruction sizes in order to avoid behavioral domination by one relatively large task. For system simulation, task arrival times and deadlines are randomly generated. Three different schemes are compared in terms of energy consumption for each task set: a fixed base cache configuration, the conservative approach, and the aggressive approach. Energy consumption is normalized to the fixed base cache configuration. Figure 24.18 presents energy savings for the instruction and data cache. Energy savings in the instruction cache subsystem ranges from 22% to 54% for the conservative approach, while it reaches as high as 74% for the aggressive approach. In the data cache subsystem, energy saving is generally less than that of the instruction cache subsystem due to less variation in cache configuration requirements. In the data cache subsystem, energy savings range from 15% to 47% for the conservative approach, while it reaches as high as 64% for the aggressive approach, and the average are 16% and 22% for the conservative and aggressive approaches, respectively. SACR conservative approach leads to very minor deadline misses (0%–4%) while aggressive approach can lead to slight higher deadline misses (1%–18%).

24.6.2 Energy Minimization Using DVS and DCR

Selected benchmarks from MediaBench [28], MiBench [17], and EEMBC [13] which from four task sets (with each consisting of 5–8 tasks) are used to evaluate the discussed approach in this section. We consider the following techniques across various system utilizations (from 0.1 to 0.9 in a step of 0.1):

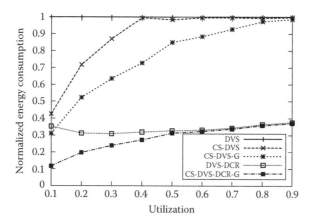

FIGURE 24.19 System-wide energy consumption using different approaches.

- **DVS**: Traditional DVS without DCR which assigns the lowest feasible* voltage level at the aim of minimizing the processor dynamic energy consumption.
- **CS-DVS**: Leakage-aware DVS without DCR which assigns lowest feasible voltage level above the critical speed decided by processor energy consumption.
- **CS-DVS-G**: Leakage-aware DVS without DCR which assigns lowest feasible voltage level above the critical speed decided by system-wide energy consumption[†].
- **DVS-DCR**: Traditional DVS + DCR which assigns the configuration point for minimizing the dynamic energy consumption of processor and cache subsystem.
- **CS-DVS-DCR-G**: Leakage-aware DVS + DCR which assigns the most energy-efficient while feasible configuration point above the critical speed decided by system-wide energy consumption.

All the results are the average of all task sets and are normalized to **DVS** scenario. Figure 24.19 shows the results. The first observation is that, for DVS-only approaches, considering other system components (CS-DVS-G) can achieve 12.8% additional energy savings on average (up to 26.6%) compared with traditional leakage-aware DVS (CS-DVS). Generally, applying DVS and DCR together (DVS-DCR) outperforms traditional DVS (DVS) and CS-DVS-G across all utilization rates by 66.3% and 42.1% on average, respectively. System-wide leakage-aware DVS + DCR (CS-DVS-DCR-G) outperforms CS-DVS-G by 47.6% on average. It can be observed that leakage-aware and leakage-oblivious DVS + DCR approaches behave similarly when the system utilization ratio is beyond 0.5. It is because both of them are inclined to select similar configuration points which have voltage levels above the critical speed (V_{dd} is around 0.8–0.9). In other words, in these scenarios, DVS-DCR does not make inferior DVS decisions, which can lead to dominating leakage power, due to limited available slack. However, when the utilization ratio is low, CS-DVS-DCR-G can achieve around 4.6%–23.5% more energy savings than DVS-DCR since CS-DVS-DCR-G does not lower down the voltage level below the critical speed.

24.6.3 Temperature- and Energy-Constrained Scheduling

This section describes performance evaluation of different methods in solving TCEC problems. A DVS-capable processor StrongARM [32] is modeled with four voltage/frequency levels. The experiments are performed on synthetic task sets, which are randomly generated with each of them having execution time

---

* By saying "feasible," it refers to the configuration points that satisfy the slack allocation described in Section 24.4.4.
† In other words, the difference of CS-DVS-G from CS-DVS is that it considers other system components in determining processor voltage level.

TABLE 24.2 TCEC Results on Different Task Sets

TS	#Blk	TC	EC	DL	Found?	AT	AE	AD
		85	180,000	7000	Y	77	171,612	6865
1	10	85	150,000	8000	Y	77	149,623	7966
		80	140,000	8000	N			
		85	70,000	2500	Y	79	66,375	2499
2	12	85	60,000	2700	Y	76	59,911	2667
		80	60,000	2500	N			
		90	90,000	2600	Y	90	81,287	2540
3	14	85	80,000	2800	Y	79	71,649	2702
		90	80,000	2700	N			

in the range of 100–500 ms. These are suitable and practical sizes to reflect variations in temperature, and millisecond is a reasonable time unit granularity [52]. The ambient temperature and initial temperature of the processor are set to 32°C and 60°C, respectively.

Table 24.2 shows the results on task sets with different number of blocks and constraints. The first and the second column are the index and number of blocks of each task set, respectively. The next three columns present the temperature constraint (TC, in °C), energy constraint (EC, in mJ), and deadlines (DL, in ms) to be checked on the model. The sixth column indicates whether there exists a schedule which satisfies all the constraints. The last three columns give the actual maximum temperature (AT), total energy cost (AE), and time required to finish all blocks (AD) using the schedule found by the model checker (UPPAAL). It can be observed that such an approach can find the solution (if exists) which satisfied all the constraints.

The efficiency of conventional symbolic model checker (UPPAAL) and the Pseudo-Boolean satisfiability-based model checking algorithm (PB approach) are compared on task sets with different number of blocks. The first six columns of Table 24.3 are same as Table 24.2. The last two columns of Table 24.3 shows the results (running time in seconds). Since UPPAAL failed to produce result for task set 4 and 5, only the running time of the PB-based approach are reported. It can be seen that PB-based approach outperforms UPPAAL by more than 10 times on average. Moreover, PB-based approach can solve much larger problems in reasonable running time.

TABLE 24.3 Running Time Comparison on Different Task Sets

TS	#Blk	TC	EC	DL	Found?	UPPAAL	PB
		85	180,000	7,000	Y	9.6	0.1
1	10	85	150,000	8,000	Y	9.9	0.1
		80	140,000	8,000	N	9.4	0.1
		85	70,000	2,500	Y	18.5	0.9
2	12	85	60,000	2,700	Y	106.6	0.1
		80	60,000	2,500	N	17.5	0.8
		90	90,000	2,600	Y	65.1	9.7
3	14	85	80,000	2,800	Y	648.3	3.1
		90	80,000	2,700	N	208.6	14.3
4	50	85	380,000	39,500	Y	—	104.8
5	100	85	720,000	83,800	Y	—	428.2

24.7 Conclusion

In this chapter, we presented three optimization techniques related to energy and temperature for real-time systems. Section 24.3 introduced a novel technique for applying reconfigurable cache in soft real-time systems to reduce cache subsystem's energy. Section 24.4 showed how to minimize system-wide energy consumption using processor voltage scaling and cache reconfiguration. Section 24.5 discussed a temperature- and energy-constrained scheduling technique based on formal methods in real-time systems. There are also a lot of other interesting researches in energy-aware scheduling and reconfiguration for real-time multitasking systems. Techniques for reclaiming dynamic slacks are proposed in [49]. Energy-efficient dynamic task scheduling is studied in [55]. In multi-core systems, cache partitioning is designed at the aim of performance improvement [36], off-chip memory bandwidth minimization [51], as well as energy consumption reduction [38]. Meanwhile, cache partitioning is also beneficial for real-time systems to improve worst-case execution time (WCET) analysis, system predictability, and cache utilization [38]. Energy-aware scheduling based on DVS in multi-core systems are also widely studied in recent years [1,24]. Approximation algorithms which can guarantee to give solutions within a specific bound of the optimal case for DVS are proposed both for uniprocessor [46] as well as multiprocessor systems [50]. Given the increasing importance of low-power systems, energy-aware dynamic reconfigurations is expected to remain in the forefront of research for real-time systems.

References

1. I. Ahmad, S. Ranka, and S.U. Khan. Using game theory for scheduling tasks on multi-core processors for simultaneous optimization of performance and energy. In *Proceedings of the IEEE International Symposium on Parallel and Distributed Processing IPDPS 2008*, Miami, FL, pp. 1–6, 14–18 April 2008.
2. R. Alur and D.L. Dill. A theory of timed automata. *Theoretical Computer Science*, 126(2):183–235, 1994.
3. H. Aydin, R. Melhem, D. Mosse, and P. Mejia-Alvarez. Dynamic and aggressive scheduling techniques for power-aware real-time systems. In *Proceedings of Real-Time Systems Symposium*, London, U.K., pp. 95–105, 2001.
4. J. Bengtsson, K. Larsen, F. Larsson, P. Pettersson, and W. Yi. Uppaal—A tool suite for automatic verification of real-time systems. In *Proceedings of the DIMACS/SYCON Workshop on Hybrid Systems III : Verification and Control*, New Brunswick, NJ, pp. 232–243, Secaucus, NJ, 1996. Springer-Verlag, New York, Inc.
5. L. Benini, R. Bogliolo, and G.D. Micheli. A survey of design techniques for system-level dynamic power management. *IEEE Transactions on VLSI Systems*, 8:299–316, 2000.
6. T. Berthold, S. Heinz, and M.E. Pfetsch. Nonlinear pseudo-boolean optimization: Relaxation or propagation? In *SAT '09: Proceedings of the 12th International Conference on Theory and Applications of Satisfiability Testing*, Swansea, Wales, U.K., pp. 441–446, 2009. Berlin, Heidelberg: Springer-Verlag.
7. S. Borkar. Design challenges of technology scaling. *IEEE Micro*, 19(4):23–29, July 1999.
8. S. Borkar, T. Karnik, S. Narendra, J. Tschanz, A. Keshavarzi, and V. De. Parameter variations and impact on circuits and microarchitecture. In *Proceedings of the Design Automation Conference*, Anaheim, CA, pp. 338–342, 2–6 June 2003.
9. E. Boros and P.L. Hammer. Pseudo-Boolean optimization. *Discrete Applied Mathematics*, 123(1–3): 155–225, 2002.
10. D. Burger, T.M. Austin, and S. Bennett. Evaluating future microprocessors: The simplescalar tool set. Technical report, University of Wisconsin–Madison, Madison, WI, 1996.
11. M. Chen, X. Qin, and P. Mishra. Efficient decision ordering techniques for sat-based test generation. In *Design, Automation Test in Europe Conference Exhibition (DATE 2010)*, Dresden, Germany, pp. 490–495, 2010.

12. B. Doyle, R. Arghavani, D. Barlage, S. Datta, M. Doczy, J. Kavalieros, A. Murthy, and R. Chau. Transistor elements for 30 nm physical gate lengths and beyond. *Intel Technology Journal*, 6:42–54, 2002.

13. EEMBC. The Embedded Microprocessor Benchmark Consortium. http://www.eembc.org/, 2000.

14. W. Fornaciari, D. Sciuto, and C. Silvano. Power estimation for architectural exploration of HW/SW communication on system-level buses. In *Proceedings of the Seventh International Workshop on Hardware/Software Codesign (CODES '99)*, Rome, Italy, pp. 152–156, 3–5 May 1999.

15. A. Gordon-Ross, F. Vahid, and N. Dutt. Fast configurable-cache tuning with a unified second-level cache. In *Proceedings of the International Symposium on Low Power Electronics and Design ISLPED '05*, San Diego, CA, pp. 323–326, 8–10 August 2005.

16. A. Gordon-Ross, P. Viana, F. Vahid, W. Najjar, and E. Barros. A one-shot configurable-cache tuner for improved energy and performance. In *Proceedings of the Design, Automation and Test Conference in Europe*, Nice, France, pp. 755–760, 2007.

17. M. Guthaus, J. Ringenberg, D.Ernest, T. Austin, T. Mudge, and R. Brown. Mibench: A free, commercially representative embedded benchmark suite. In *Proceedings of the IEEE International Workshop on Workload Characterization*, Austin, TX, pp. 3–14, 2001.

18. I. Hong, D. Kirovski, G. Qu, M. Potkonjak, and M.B. Srivastava. Power optimization of variable-voltage core-based systems. *IEEE Transactions on Computer-Aided Design of Integrated Circuits and Systems*, 18:1702–1714, 1999.

19. *CACTI 5.3, HP Laboratories Palo Alto, CACTI 5.3.* http://www.hpl.hp.com/, 2008.

20. R. Jejurikar, C. Pereira, and R.K. Gupta. Leakage aware dynamic voltage scaling for real-time embedded systems. In *Proceedings of the Design Automation Conference*, Paris, France, pp. 275–280, 2004.

21. N. Jha. Low power system scheduling and synthesis. In *Proceedings of the International Conference on Computer-Aided Design*, pp. 259–263, 2001.

22. Y. Joo, Y. Choi, H. Shim, H.G. Lee, K. Kim, and N. Chang. Energy exploration and reduction of SDRAM memory systems. In *DAC '02: Proceedings of the 39th Annual Design Automation Conference*, ACM, New York, pp. 892–897, 2002.

23. J. Kahle, M. Day, H. Hofstee, C. Johns, T. Maeurer, and D. Shippy. Introduction to the cell multiprocessor. *IBM Journal of Research and Development*, 49:589–604, 2005.

24. J. Kang and S. Ranka. DVS based energy minimization algorithm for parallel machines. In *Proceedings of the IEEE International Symposium on Parallel and Distributed Processing IPDPS 2008*, pp. 1–12, 14–18 April 2008.

25. J. Kao, S. Narendra, and A. Chandrakasan. Subthreshold leakage modeling and reduction techniques. In *ICCAD '02: Proceedings of the 2002 IEEE/ACM International Conference on Computer-Aided Design*, ACM, New York, pp. 141–148, 2002.

26. N.S. Kim, T. Austin, D. Blaauw, T. Mudge, K. Flautner, J.S. Hu, M.J. Irwin, M. Kandemir, and V. Narayanan. Leakage current: Moore's law meets static power. *Computer*, 36(12):68–75, 2003.

27. K. Lahiri and A. Raghunathan. Power analysis of system-level on-chip communication architectures. In *CODES+ISSS '04: Proceedings of the Second IEEE/ACM/IFIP International Conference on Hardware/Software Codesign and System Synthesis*, ACM, New York, pp. 236–241, 2004.

28. C. Lee, M. Potkonjak, and W.H. Mangione-smith. Mediabench: A tool for evaluating and synthesizing multimedia and communications systems. In *Proceedings of the International Symposium on Microarchitecture*, pp. 330–335, 1997.

29. J. Liu. *Real-Time Systems*. Upper Saddle River, NJ: Prentice Hall, 2000.

30. A. Malik, B. Moyer, and D. Cermak. A low power unified cache architecture providing power and performance flexibility. In *Proceedings of the International Symposium on Low Power Electronics and Design*, Rapallo, Italy, pp. 241–243, 2000.

31. S.M. Martin, K. Flautner, T. Mudge, and D. Blaauw. Combined dynamic voltage scaling and adaptive body biasing for lower power microprocessors under dynamic workloads. In *Proceedings of the IEEE/ACM International Conference on Computer Aided Design ICCAD 2002*, San Jose, CA, pp. 721–725, 10–14 November 2002.

32. Marvell. StrongARM 1100 processor. http://www.marvell.com/, 1997.

33. M. Gebser, R. Kaminski, B. Kaufmann, M. Ostrowski, T. Schaub, and M. Schieder. Potassco: The potsdam answer set solving collection. *Aicom*, 24(2): 105–124, 2011.

34. X. Qin, M. Chen, and P. Mishra. Synchronized generation of directed tests using satisfiability solving. In *23rd International Conference on VLSI Design, VLSID '10*, Bangalore, India, pp. 351–356, 2010.

35. G. Quan and X.S. Hu. Energy efficient DVS schedule for fixed-priority real-time systems. *ACM Transactions on Design Automation of Electronic Systems*, 6:1–30, 2007.

36. M.K. Qureshi and Y.N. Patt. Utility-based cache partitioning: A low-overhead, high-performance, runtime mechanism to partition shared caches. In *39th Annual IEEE/ACM International Symposium on Microarchitecture, (MICRO-39)*, Orlando, FL, pp. 423–432, December 2006.

37. R.R. Rao, H.S. Deogun, D. Blaauw, and D. Sylvester. Bus encoding for total power reduction using a leakage-aware buffer configuration. *Very Large Scale Integration (VLSI) Systems, IEEE Transactions on*, 13(12):1376–1383, December 2005.

38. R. Reddy and P. Petrov. Cache partitioning for energy-efficient and interference-free embedded multitasking. *ACM Transactions in Embedded Computing Systems*, 9(3):1–35, 2010.

39. A. Settle, D. Connors, and E. Gibert. A dynamically reconfigurable cache for multithreaded processors. *Journal of Embedded Computing*, 2:221–233, 2006.

40. T. Sherwood, E. Perelman, G. Hamerly, S. Sair, and B. Calder. Discovering and exploiting program phases. In *Proceedings of the International Symposium on Microarchitecture*, San Diego, CA, pp. 84–93, 2003.

41. K. Skadron, M.R. Stan, K. Sankaranarayanan, W. Huang, S. Velusamy, and D. Tarjan. Temperature-aware microarchitecture: Modeling and implementation. *ACM Transactions on Architecture and Code Optimization*, 1(1):94–125, 2004.

42. C. Talarico, J.W. Rozenblit, V. Malhotra, and A. Stritter. A new framework for power estimation of embedded systems. *Computer*, 38(2):71–78, 2005.

43. H.J.M. Veendrick. Short-circuit dissipation of static CMOS circuitry and its impact on the design of buffer circuits. *IEEE Journal of Solid-State Circuits*, 19(4):468–473, August 1984.

44. W. Wang and P. Mishra. Dynamic reconfiguration of two-level caches in soft real-time embedded systems. In *Proceedings of the IEEE Computer Society Annual Symposium on VLSI*, Tampa, FL, pp. 145–150, 2009.

45. W. Wang and P. Mishra. Leakage-aware energy minimization using dynamic voltage scaling and cache reconfiguration in real-time systems. In *Proceedings of the IEEE International Conference on VLSI Design*, Bangalore, India, pp. 357–362, 2010.

46. W. Wang and P. Mishra. PreDVS: Preemptive dynamic voltage scaling for real-time systems using approximation scheme. In *Proceedings of the Design Automation Conference*, Anaheim, CA, pp. 705–710, 2010.

47. W. Wang, P. Mishra, and A. Gordon-Ross. SACR: Scheduling-aware cache reconfiguration for real-time embedded systems. In *Proceedings of the IEEE International Conference on VLSI Design*, New Delhi, India, pp. 547–552, 2009.

48. W. Wang, X. Qin, and P. Mishra. Temperature- and energy-constrained scheduling in multitasking systems: A model checking approach. In *Proceedings of the International Symposium on Low Power Electronics and Design*, Austin, TX, pp. 85–90, 2010.

49. C.-Y. Yang, J.-J. Chen, and T.-W. Kuo. Energy-efficiency for multiframe real-time tasks on a dynamic voltage scaling processor. In *CODES+ISSS '09: Proceedings of the Seventh IEEE/ACM International Conference on Hardware/Software Codesign and System Synthesis*, Grenoble, France, pp. 211–220, 2009.

50. C.-Y. Yang, J.-J. Chen, T.-W. Kuo, and L. Thiele. An approximation scheme for energy-efficient scheduling of real-time tasks in heterogeneous multiprocessor systems. In *ACM/IEEE Conference of Design, Automation, and Test in Europe (DATE)*, Nice, France, 2009.
51. C. Yu and P. Petrov. Off-chip memory bandwidth minimization through cache partitioning for multi-core platforms. In *DAC '10: Proceedings of the 47th Design Automation Conference*, New Orleans, LA, pp. 132–137, 2010. New York: ACM.
52. S. Zhang and K.S. Chatha. Approximation algorithm for the temperature aware scheduling problem. In *Proceedings of the International Conference on Computer-Aided Design*, Tempe, AZ, pp. 281–288, 2007.
53. C. Zhang, F. Vahid, and W. Najjar. A highly configurable cache for low energy embedded systems. *ACM Transactions on Embedded Computing Systems*, 6:362–387, 2005.
54. X. Zhong and C. Xu. System-wide energy minimization for real-time tasks: Lower bound and approximation. In *Proceedings of the International Conference on Computer-Aided Design*, Austin, TX, pp. 516–521, 2006.
55. J. Zhuo and C. Chakrabarti. System-level energy-efficient dynamic task scheduling. In *Proceedings of the Design Automation Conference*, San Diego, CA, pp. 628–631, 2005.

Adaptive Power Management for Energy Harvesting Embedded Systems

Jun Lu
Binghamton University, State University of New York

Shaobo Liu
Binghamton University, State University of New York

Qing Wu
Air Force Research Laboratory

Qinru Qiu
Binghamton University, State University of New York

25.1 Introduction

Real-time embedded systems (RTES) have wide applications from sensor networks, personal communication, and entertainment devices to control and management units of unmanned vehicles. *Renewable energy* is energy generated from natural resources such as sunlight, wind, rain, tides geothermal heat, which are naturally replenished. *Energy harvesting*, also called energy scavenging, refers to the process of collecting and converting renewable energy to be utilized by electronic systems. It is a promising technology for overcoming the energy limitations of battery-powered systems and has the potential to allow systems to achieve energy autonomy. Recently, great interest has risen in powering RTES with renewable energy sources.

There are different types of possible energy harvesting sources [1,2] including solar, thermal, vibration, and kinetic energy. Two prototypes, namely, Heliomote [3] and Prometheus [4], have been developed to extract solar energy. Heliomote is a wireless sensor node that uses solar panels as harvesting source to operate autonomously for much longer durations. Prometheus is also a wireless sensor node that uses multistage energy storage system, for example, a supercapacitor [5] and a Li-polymer battery to increase the lifetime of the battery. Both prototypes show that systems may operate perpetually by scavenging solar energy. Beside solar power, other energy sources are also used in environmental powered system. For example, piezoelectric shoes [6], designed by MIT Media Lab, can power radio frequency identification (RFID) tag system; wind energy is used to power wireless sensor node [7].

The distinct characteristics of the energy harvesting system and the intrinsic working behaviors of the embedded system impose new and significant challenges in the design and optimization of energy harvesting real-time embedded systems (EH-RTES). The objective of low power design of EH-RTES is more than minimizing the energy dissipation while trying to meet the performance requirements. With the existence of the energy harvested from the environment, it is more important to develop novel techniques that efficiently utilize this energy for the maximum performance gain.

Many techniques and approaches have been studied in the area of power management [28–32] of EH-RTES. In this chapter, we will discuss the existing challenges in the power management of an EH-RTES system and review the research status in this area. This chapter is organized as follows: Section 25.2 discusses the model of a typical EH-RTES system and introduces the characteristics of each system component. Section 25.3 presents the research challenges in power management of an EH-RTES and the current research status. Section 25.4 evaluates the performance trend of EH-RTES with different harvesting power and storage capacity. We conclude this chapter in Section 25.5.

25.2 Hardware Architecture of EH-RTES

An EH-RTES has three major modules [8,9]: the *energy harvesting module* (*EHM*, e.g., a solar panel), the *energy storage module* (*ESM*, e.g., a battery or ultracapacitor), and the *energy dissipation module* (*EDM*, e.g., an RTES system). The EHM generates energy; the ESM stores the energy. Depending on the energy availability, the RTES will be powered by the EHM or the ESM or both of them. Two *energy conversion modules* (*ECM*, e.g., DC–DC converter) are used to regulate the voltage to the range, which could be used by ESM and EDM. Figure 25.1 shows the basic system block diagram and the energy flow in EH-RTES. In the next, we will introduce some unique characteristics and basic working principles of those system components.

25.2.1 Energy Harvesting Module

The forms of energy that could be harvested by the EH-RTES include mechanical, thermal, photovoltaic, electromagnetic, biological, etc. Different energy harvesting technologies are needed for different forms of the ambient energy. For example, mechanical energy is usually harvested by piezoelectric or vibration harvesting devices [10]. Thermoelectric technologies are used for harvesting thermal energy [10]. However, the most well-known and prevalent energy source is sunlight. Photovoltaic cells, which are the building blocks of solar panels, are the primary harvesting technology for light energy.

In most research, solar panels are chosen to be the primary energy harvesting technology for EH-RTES. Among all the different characteristics of the solar panel, two of them have the most significant impact on the performance, energy efficiency and thus the optimality of power management techniques for the EH-RTES: *harvesting rate* (output power) and *maximum power point* (*MPP*).

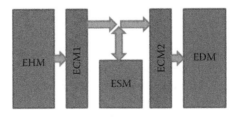

FIGURE 25.1 System diagram and energy flow of an EH-RTES.

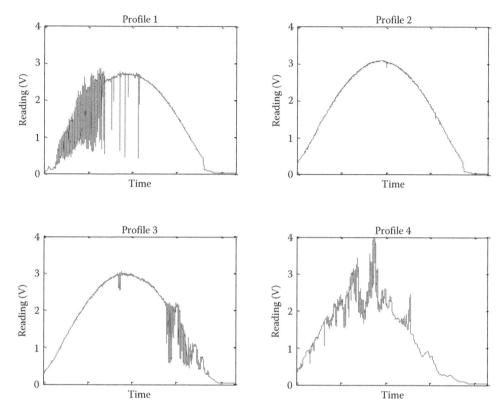

FIGURE 25.2 Four different solar irradiation profiles.

Solar energy harvesting is a process with high uncertainty. The energy harvesting rate is heavily dependent on the operating environment and fluctuates in real time. For example, the output power of a solar panel is usually regarded as a random process affected by the light intensity, temperature, output voltage, manufacturing process variation, etc. Among those, the sun radiation intensity has the first-order impact to the harvesting rate. The sun radiation level again can be affected by many factors, including location, climate, weather, and time. Figure 25.2 shows four different daytime solar radiation profiles collected from 7:00 a.m. to 7:00 p.m. in February in upper New York state [8,9]. As we can see, in general, the sun radiation increases steadily in the morning. It reaches the peak at noon and then starts decreasing. Despite this trend, we can also see dramatic fluctuations in the sunlight intensity caused by weather condition.

In [11–13], the author uses random number generator to model the output power of a solar panel. The model is given by Equation 25.1, where $N(t)$ is a random number that follows a normal distribution with mean 0 and variance 1. Figure 25.3 shows a trace of the harvested solar power generated based on this model. As we can see, the model captures the stochastic and periodic characteristics of the harvesting rate of a solar panel:

$$P_H(t) = \left| 10 \times N(t) \times \cos\left(\frac{t}{70\pi}\right) \times \cos\left(\frac{t}{120\pi}\right) \right|. \tag{25.1}$$

The MPP [14] is defined as a voltage–current combination that maximizes the power output for a given sunlight condition and temperature. Figure 25.4a shows the equivalent circuit of a solar cell that may be used to model the behavior of a solar cell, and Figure 25.4b shows the current–voltage (I–V) and power–voltage (P–V) characteristic of a solar cell for a given sunlight intensity. We can see that there is a single operation point where both the output current and voltage are at their maximum level and the

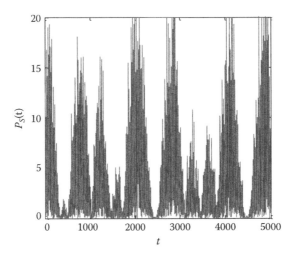

FIGURE 25.3   Energy source behavior by random number generator.

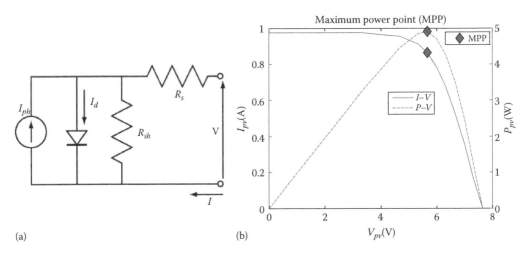

(a)   (b)

FIGURE 25.4   (a) Equivalent circuit and (b) I–V, P–V characteristics of a solar panel.

solar panel yields the maximum output power. This particular point is called the MPP. Before reaching the MPP, solar panel behaves as a current source and its output power increases almost linearly with the increase of the output voltage. After MPP, the internal diode of the solar is turned on and the output current drops drastically. If we keep the solar panel operating at MPP, it can harvest 200%–400% more energy compared to the worst case.

Since MPP makes such a large difference in energy harvesting, much research has been carried out to develop software- and hardware-based techniques that can perform automatic MPP tracking (MPPT) [14,15] for solar panels as well as other energy harvesting technologies [7]. They can generally be divided into two categories: direct and indirect methods.

The indirect methods use empirical data, for example, the recorded solar panels' current, voltage, and power, to estimate the MPP. For example, the authors of [16] develop regression models to approximate MPP. They model the P–V characteristic of a solar panel using Equation 25.2, where a, b, c, and d are coefficients obtained using curve fitting based on sampled power and voltage values. The MPP voltage, whose analytical form is given in Equation 25.3, is then found by maximizing Equation 25.2:

$$P = aV^3 + aV^2 + cV + d, \tag{25.2}$$

$$V_{MPP} = -\frac{b\sqrt{b^2 - 3ac}}{3ac}. \tag{25.3}$$

The direct methods of MPPT do not require any a priori knowledge of the solar panel. They rely on feedback control to dynamically adjust the output voltage to reach the MPP. For example, the authors of [17] propose a gradient search method to solve the following different equations:

$$\frac{dP}{dt} = V, \quad \frac{dI}{dt} + I\frac{dV}{dt} = 0. \tag{25.4}$$

The solution defines the MPP voltage. The key challenge in the direct MPPT is how to perform the search in real time at low cost. For example, to solve different equation (25.4) using the gradient search method, a digital signal processor (DSP) is needed and at least eight calculations are required at each step.

For both direct and indirect MPPT methods, a DC–DC converter must be inserted at the output of the EHM, which controls the output voltage and helps MPPT.

25.2.2 Energy Dissipation Module

In this chapter, we limit out discussion of the EDM only to RTES. An RTES is a computer designed for dedicated applications under real-time constraints. It has wide applications from sensor networks, personal communication and entertainment devices, to control and management units of unmanned vehicles. An RTES usually has very strict requirement in reliability, which is measured by the deadline miss rate (DMR) of its tasks. In a soft real-time system, a deadline miss will cause the system to drop the current task and reduce its quality of service (QoS); while in a hard real-time system, a deadline miss will cause catastrophic failure.

A task τ_m in an RTES is characterized by a triplet (a_m, d_m, and $wcet_m$), where a_m, d_m, and $wcet_m$ indicates the arrival time, the relative deadline, and the worst-case execution time of the task, respectively. In real-time applications, a scheduler controls the execution order of the tasks. A simple example of task scheduling is the earliest deadline first (EDF) [18] policy. The system is considered to be preemptive. The task with the earliest deadline has the highest priority and should be executed first; it preempts any other task if needed. A task that has arrived but not yet been executed is placed in a *ready queue*.

The embedded processor is the primary (in many systems, the only) computational resource in an RTES and consumes most of the energy. Many state-of-the-art embedded processors such as Intel XScale [8,9] and PowerPC 405PL[19] supports run-time *dynamic voltage and frequency scaling (DVFS)*, which is the primary power management technique for embedded systems. The theory behind energy saving by DVFS is quite simple. In CMOS (complementary metal-oxide semiconductor) VLSI (very large scale integration) circuits, the energy dissipation is a quadratic function of supply voltage, while the clock frequency is a linear function of the same variable. Reducing the supply voltage can decrease the energy dissipation quadratically and only increase delay linearly. For example, if we reduce the supply voltage of an embedded processor by $2\times$, we can reduce its energy dissipation by $4\times$, but the execution time of a task will be increased by $2\times$ because clock frequency is reduced by $2\times$. This property allows us to trade-off speed for energy saving when peak performance is not required.

25.2.3 Energy Storage Module

Energy storage is usually required in EH-RTES to maintain a continuous operation even when there is no energy to harvest (e.g., at night for a solar-powered system). As an energy buffer, analogous to the ready

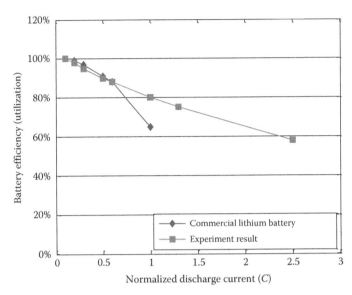

FIGURE 25.5 Efficiency factor versus discharge current. (From Pedram, M. et al., Design considerations for battery-powered electronics, *Proceedings of Design Automation Conference*, 1999.)

queue for tasks, the energy storage unit can provide a large energy-performance trade-off space for power management techniques to obtain better results.

Rechargeable batteries are the predominant choices for energy storage. One of the main characteristics of battery technology is the so called rate-capacity [20] property when charging or discharging the battery. Simply speaking, there is energy overhead when moving energy into or out of the batteries. The magnitude of the overhead is usually dependent on the charging/discharging current of the battery. The efficiency factor of energy storage is defined as actual energy that can be used divided by the amount of energy that is stored. Figure 25.5 shows the efficiency factor versus discharge current curves for a commercial lithium battery. When the discharge current increases from 0.5 to 1 A, the effective battery capacity drops almost 40%. Two simple functions [20] are used to approximate the battery efficiency factor for analytical purpose:

$$\mu = 1 - \beta \times I, \tag{25.5}$$

$$\mu = 1 - \gamma \times I^2. \tag{25.6}$$

where β and γ are positive constant numbers. Equations 25.5 and 25.6 provide good modeling for the efficiency-current relation of batteries as long as the appropriate value of β or γ is chosen.

Ultracapacitors (or supercapacitors) [5] are an emerging and promising technology for energy storage and buffering. Compared to batteries, ultracapacitors do not have the aging and rate-capacity problems. The shortcomings of them are limited energy capacity and higher leakage. There is also energy overhead when charging or discharging the ultracapacitors. However, the efficiency factor of ultracapacitor is usually constant, independent of the current.

25.2.4 Energy Conversion Module

Voltage regulators are designed to automatically maintain a constant voltage level, and they are used to bridge the gap between the supplier and the consumer in energy harvesting systems. As shown in Figure 25.1, this model has two electrical energy conversion units in the EH-RTES. The ECM-1 converts energy from the output of the EHM so that it can be used by the ESM. Depending on the type of energy harvesting technology, ECM-1 can be a voltage regulator, either DC/DC or AC/DC converter with or without MPPT capabilities. ECM-2 is usually a DC/DC converter that regulates the supply voltage level of the EDM. For DVFS-enabled processors, the output voltage of ECM-2 should be controllable. In EH-RTES, the conversion efficiency of the ECMs is dependent on multiple system operation conditions such as input voltage/current, output voltage/current, MPPT working status, DVFS control state, etc.

The regulators used in EH-RTES can generally be divided into two categories, linear regulators and switching regulator. Linear regulators are available as integrated circuits. They output clean, stable power. However, they usually have lower conversion efficiency, associate with a voltage drop, and dissipate more heat. On the other hand, switching regulators have much higher efficiency. They can be further divided into buck, boost, and buck–boost regulators. Buck regulators perform voltage step-down conversion and are efficient, but the input voltage must be higher than the output voltage or else it does not work properly. Boost regulators perform voltage step-up conversion but are less efficient. Buck–boost regulators, as the name implies, can work as a buck or a boost depending on if the input voltage is higher or lower than the output.

25.3 Power Management in EH-RTES

Run-time power management has been widely used in state-of-the-art computing systems. It turns off or slows down idle or underutilized components for energy saving. The objective of the conventional power management is to minimize the system power consumption while meeting the performance constraint. While the conventional power management works for traditional battery power embedded systems by extending the battery cycle time, it is not suitable for an EH-RTES. Because the energy supply of an EH-RTES is replenishable, it is not necessary to overemphasize on energy saving. Instead, it is more important to focus on efficient energy utilization.

In an EH-RTES, both the energy incoming rate and the energy dissipation rate are random variables. The former is determined by the harvesting rate, which exhibits high variations from time to time due to environment impact, while the later is determined by the workload on the embedded processor, which is again a random process. Although the system may eventually gather enough energy to finish a task, if this does not happen before the deadline, the task will be dropped and the QoS of the system is impaired. While the ESM works as low-pass filter to smooth out the abrupt changes in the incoming and outgoing energy, without careful management its benefit is limited. For example, any unused energy harvested from the environment will be wasted when the storage is full.

The power management in an EH-RTES is to control the energy flow in the system to minimize the chances of deadline miss due to insufficient energy. Although we do not have the control of the incoming energy, we do have means to reshape the outgoing energy by task scheduling and DVFS. The former affects the starting time of an energy consumption period while the later determines the magnitude and duration of the energy consumption. Therefore, the objective of power management in an EH-RTES is to maximize performance (i.e., minimize DMR) subject to the given energy constraints (imposed by the environment and the energy harvesting system).

Task scheduling and DVFS are two main approaches used in power management of EH-RTES. Their application should consider the energy status and the workload status of the system. Running a task at lower speed and lower supply voltage reduces its energy dissipation, however, also steals CPU time that could be used to process future tasks and hence may lead to deadline miss in the future due to insufficient processing time. On the other hand, running a task at high speed and higher supply voltage saves CPU time however increases energy usage. Hence, it may still lead to deadline miss in the future due to insufficient energy. A good power management approach finds balance between performance and energy. This must be achieved with the knowledge of the future energy harvesting rate and the current status of the ESM. For example, an energy conservative scheduling and DVFS policy will be selected when the future energy harvesting is predicated to be low and vice versa. The ability of utilizing the predicted information in power management is referred as *proactive power management*. In addition, the power manager must be able to adjust the workload to match the intrinsic resistance of the energy harvesting system to achieve its maximum efficiency [15]. This is referred as *load matching*.

The goal of power management is to minimize DMR, which is a measure of the QoS of the system, subject to the energy constraint. The center of the power manager is scheduling and DVFS techniques. They are supported by energy prediction, proactive power management, and load matching techniques. In the next, we will introduce research status in each of these areas.

25.3.1 Task Scheduling for Minimal Deadline Miss in EH-RTES

For RTES, DMR is one of the most important performance metrics. It is defined as the ratio of the number of tasks missing their deadline to the total number of tasks. In [11,21], the authors proposed task scheduling algorithm, named *LAZY scheduling algorithm* (LSA), to minimize the DMR on EH-RTES.

LSA does not consider the DVFS capability of an embedded processor. It assumes that all tasks are running at full speed and hence has power consumption P_{max}. Its main idea is to delay the task until the time when the system gathered enough energy. First, the LSA order the task based on the ascending order of their deadline. The task with the earliest deadline has the highest priority to be executed. Let $E_c(a_i)$ be the remaining energy in ESM at a_i, and $E_s(a_i, d_i)$ be the harvested energy during time a_i and d_i, where a_i and d_i are arrival time and deadline of task i, respectively. Also, let C denote the maximum capacity of the ESM. It is easy to know that if the ESM does not overflow during the time (a_i, d_i), then total available energy for task i is $E_c(a_i) + E_s(a_i, d_i)$, otherwise it is $C + E_s(s_i, d_i)$, where s_i is the starting time of task i. With this information, the LAZY scheduling determines the starting time s_i for task i using the following three equations:

$$S_i^* = d_i - \frac{E_C(a_i) + E_S(a_i, d_i)}{P_{max}}, \tag{25.7}$$

$$S_i' = d_i - \frac{C + E_S(S_i', d_i)}{P_{max}}, \tag{25.8}$$

$$s_i = max\left(s_i^*, s_i'\right). \tag{25.9}$$

Equation 25.7 calculated the starting time of task i with the assumption that the ESM does not overflow during the time (a_i, d_i), and Equation 25.8 calculated the starting time of task i with the assumption that the ESM will be filled up during the time (a_i, d_i). The actual starting time of the task is determined by the maximum value of Equations 25.7 and 25.8, as shown in Equation 25.9.

It can be proved in [11,21] that the LSA algorithm is optimal under the assumption that the energy harvesting rate is constant and the RTES does not support DVFS. This LSA method greatly reduces the DMR in an EH-RTES, compared with the traditional EDF algorithm [18], which is used in traditional embedded systems.

25.3.2 Harvesting-Aware DVFS for Minimal Energy Dissipation

The DVFS technique has been proven to be effective in reducing the energy dissipation in traditional RTES. A DVFS-enabled processor has N operating voltage and frequency levels $\{(v_n, f_n)|1 \leq n \leq N, f_{min} = f_1 < f_2 < \cdots < f_N = f_{max}\}$. Its power consumption is denoted as p_n when running at (v_n, f_n). It is easy to know that $p_1 < p_2 < \cdots < p_N$. We define a slowdown factor $S(n)$ as the ratio between f_n and the maximum frequency f_{max}, that is, $S(/S_n, where\ n) = f_n/f_{max}$. On a DVFS-enabled processor, the actual execution time of task m at frequency f_n is $wcet_m/S_n$, where $wcet_m$ is the worst-case execution time of the task under maximum frequency.

In research works for traditional RTES, DVFS technique has been proven to be effective in reducing the energy dissipation. [22] addresses the variable voltage scheduling of tasks with soft deadlines, while [23] proposes an online scheduling for hard real-time task. A heuristic DVFS algorithm presented in [24] utilizes the slack of the fixed priority tasks so that energy efficiency is achieved. The work in [18] shows that with DVFS, the system-wide energy consumption is minimized if the workload is allocated evenly among processors.

The authors of [8,9,12,13] extend the tradition DVFS algorithm and apply it to the EH-RTES. The basic idea is to be conservative in energy dissipation as much as possible so that the deadline miss due to insufficient energy can be avoided. They first create an initial schedule using the LAZY scheduling and denote the initial starting time and finish time of a task i as ist_i and ift_i.

The initial schedule puts all tasks at full speed. In the next, tasks will be "stretched" and run as slow as possible for maximum power saving. The algorithm is given in Figure 25.6. This step involves N rounds of DVFS optimization, where N is the number of available operating frequencies to the processor. In each round, the starting time (st_m) of each task in the ready queue is calculated. If task τ_m is the first task (i.e., $m = 1$), then its starting time is the maximum of the current time and its arrival time (i.e., $st_1 = max(a_1, current_{time})$). Otherwise, the starting time of task τ_m, $m > 1$, is the maximum of its arrival time and the finish time of previous task (i.e., $st_m = max(a_m, ft_{m-1})$, $m = 2, \ldots, M$).

$$st_m = \begin{cases} max(a_1, current_{time}), & m = 1 \\ max(a_m, ft_{m-1}), & m = 2, \ldots, M \end{cases} \quad (25.10)$$

The finishing time (ft_m) of task τ_m is determined by the answer of two questions: (1) Can the speed of task τ_m be further reduced? Let SI_m denote the level of frequency and voltage that is used to execute task τ_m, and then slowdown factor of τ_m is $S(SI_m)$. If the following inequality holds, $st_m + wcet_m/S(SI_m - 1) < ift_m$, then the deadline of τ_m can still be met after further slowing down task τ_m. (2) Are the slowdown factors

Algorithm 1: Workload balancing and DVFS
Require: get the initial schedule for M tasks in Q
1. **for** $n = 1 : N$ **do**
2.   **for** $m = 1 : M$ **do**
3.   **if** $m == 1$, **then**
4. $st_m = max(a_m, current\_time)$
5.   **else**
6. $st_m = max(a_m, ft_{m-1})$
7.   **end if**
8.   **if** $st_m + w_m/S(SI_m - 1) < ift_m$ && *no deadline miss for the other tasks*, **then**
9. $SI_m = SI_m - 1$
10.   **end if**
11. $ft_m = st_m + w_m/S(SI_m)$
12. **end for**
13. **end for**

FIGURE 25.6 Algorithm for workload balancing and DVFS.

for tasks indexed from $m + 1$ to M still valid (no deadline misses) if τ_m gets further stretched? If the answers to these two questions are "yes," then the SI_m for task τ_m is decremented by 1, and the operating frequency of task τ_m is reduced to $f_{(SIm-1)}$ from $f_{(SIm)}$. Otherwise, SI_m remains its current value. The finishing time ft_m can be calculated as $ft_m = st_m + wcet_m/S(SI_m)$.

25.3.3 Energy Prediction and Proactive Task Scheduling

Both the LAZY scheduling and the harvest aware DVFS algorithm need to know the energy harvesting rate in the future (i.e., $E_s(a_i, d_i)$.) Traditionally, this information is assumed to be a constant or can be obtained from profiling [11–13,21]. There assumptions are not realistic either. Recently, researchers started to study the harvesting energy prediction methods.

Accurate prediction of the near-future harvested energy is crucial to effective power management of the EH-RTES. It has been acknowledged in references [8,25] that the efficiency of the optimization techniques of EH-RTES largely depends on the accuracy of the energy harvesting profiling and prediction. A good energy prediction model must have the following properties:

1. High accuracy: The predicted future harvested energy has direct impact on the scheduling of the tasks. The prediction accuracy will affect the overall system performance in terms of DMR, etc.
2. Low computation complexity: The prediction technique should be able to be implemented as software or hardware components without incurring significant energy and performance overhead.
3. Low memory requirements: Most EH-RTES have very small memory space for programs and data; thus, we can only store limited number of data points of previously harvested energy. Therefore, any prediction technique that relies on large amount of observed data points will not fit in this application.

The prediction of energy harvesting rate can be divided into two categories: long-term prediction [25] and short-term prediction [8]. Both works target at solar energy prediction. The former predicts the average energy harvesting rate over a relatively long period of time in terms of days or hours. The later gives prediction at fine time resolution in terms of seconds or milliseconds. Long-term energy prediction guides the general task selection and system configuration of the EH-RTES which does not change frequently, while the short-term energy prediction directs the task scheduling and voltage frequency selection, which has fast dynamics during run time.

25.3.3.1 Long-Term Prediction

Two long-term prediction models are presented in reference [25]. They are weather-conditioned moving average (WCMA) model and exponentially weighted moving average (EWMA) model. The WCMA model considers how the weather condition affects the future solar irradiation. The WCMA relies on the recorded sequence of energy that entered the system during the past several days. It predicts the next energy harvesting rate as the weighted sum of current harvesting rate and the average harvesting rate over the past D days at the same time. Equation 25.11 shows how predicted energy value is calculated:

$$E(d, n + 1) = \alpha E(d, n) + (1 - \alpha)\frac{\sum_{i=1}^{D} E_{i,n+1}}{D}, \tag{25.11}$$

where

$E(i, j)$ is the energy in the matrix of the jth sample on the ith day

D is the number of days which is also the number of rows in the matrix

The variable α is a weighting factor that decides how important the actual value is compared to the calculated value.

The principle of the EWMA is to apply an exponentially decreasing weighting factor to the previous value. In this way, an old harvesting rate value will have lower importance in the prediction. At the same time, the prediction takes into account every single value with different weights. The EWMA algorithm is depicted as follows:

$$E_p(x+1) = \beta E_R(x) + (1 - \beta)E_P(x), \tag{25.12}$$

where

$E_P(x + 1)$ is the new predicted value
$E_R(x)$ is the last recorded value
β is a weighting factor between 0 and 1

It predicts the next value as the weighted sum of the last recorded value and the previous predicted value.

25.3.3.2 Short-Term Prediction

In reference [8], three different time series prediction techniques are applied for short-term prediction of energy harvesting rate.

25.3.3.2.1 Regression Analysis

The first is the regression analysis. Regression analysis is a statistical technique for modeling and investigating the relationship among observed variables. It is used to estimate and predict the value of one variable by taking into account the other related. Forecasting using simple regression is based on the following equation:

$$x = b_0 + b_{1Z} + \varepsilon. \tag{25.13}$$

The value of b_0 and b_1 is estimated using the minimal least square method as depicted in Equations 25.14 and 25.15,

$$\widehat{b_1} = \frac{\sum_{i=1}^{n} (z_i - \bar{z}) \sum_{i=1}^{n}}{(x_i - \bar{x}) \sum_{i=1}^{n} (z_i - \bar{z})^2}, \tag{25.14}$$

$$\widehat{b_0} = \bar{x} - \widehat{b_1}\bar{z}, \tag{25.15}$$

where $(x_1, z_1), (x_2, z_2), \ldots, (x_n, z_n)$ are the n observations available. For the energy harvesting prediction problem, x_i and z_i are the sunlight intensity and time at the ith sample period, respectively. And $\bar{x}$ and $\bar{z}$ are their arithmetic means.

The fitted simple linear regression model is

$$\hat{x} = \widehat{b_0} + \widehat{b_{1Z}}. \tag{25.16}$$

Figure 25.7 shows the plots of predicted solar irradiation by regression analysis versus the actual values.

25.3.3.2.2 Moving Average

The second short-term prediction introduced in [8] is the moving average method. The main objective of the *moving average* technique is to predict future values based on averages of the past values. It is useful in reducing the random variations in the observation data. The simple moving average uses the N past observation data to calculate the next prediction time series value, as shown in Equation 25.17.

$$\hat{x} = \frac{x_{(t)} + x_{(t-1)} + \cdots + x_{(t-N+1)}}{N}. \tag{25.17}$$

The quality of simple moving average depends on the number of past observations to be averaged, but it gives equal weight to all past data, of which a large number to be stored and used for forecasts. The actual values in solar irradiation and predicted values using moving average are shown in Figure 25.8.

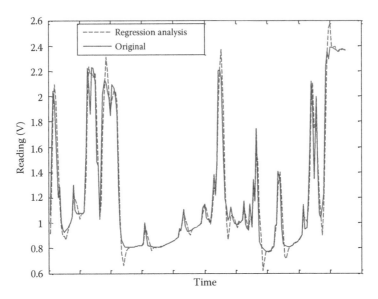

FIGURE 25.7   Regression analysis ($n = 4$).

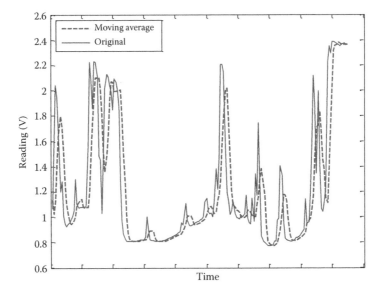

FIGURE 25.8   Moving average ($n = 4$).

25.3.3.2.3 Exponential Smoothing

The last short-term prediction method studied in [8] is the exponential smoothing which is widely used for short-time forecasting. It employs exponentially decaying weighting factors for past values. Simple exponential smoothing can be obtained by the following function:

$$\hat{x} = X_{e(t)} = \alpha x_{(t)} + (1 - \alpha)x_{e(t-1)}, \tag{25.18}$$

where
    $x_{e(t)}$ is called the exponentially smoothed value
    $x_{(t)}$ is the observed value at the same point of time

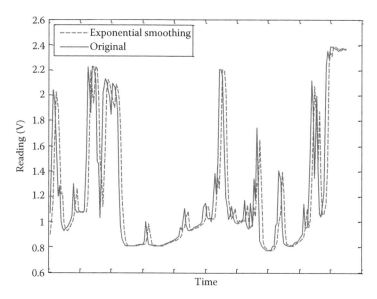

FIGURE 25.9 Exponential smoothing ($\alpha = 0.2$).

The fraction α is called the smoothing constant, and $x_{e(t-1)}$ is the previous exponentially smoothed value. Figure 25.9 shows the actual measurements in solar irradiation and predicted values using exponential smoothing.

The quantitative comparison on these three prediction techniques incorporated with DVFS scheduling methods is given in [8]. The result shows that the regression analysis approach and the moving average approach are slightly superior to exponential smoothing especially when utilization is higher.

25.3.4 Scheduling Adjustment Based on Energy Availability

In addition to use the predicted energy harvesting value to replace the variable $E_S(a_i, d_i)$ in Equations 25.7 and 25.8, this information also helps to predict if there is enough energy to finish a task before it started. A task that is predicted to have insufficient energy will be delayed until the system gathered enough energy or it will be drop even without starting to save energy and CPU time for other tasks.

An algorithm is proposed in [8] to check run-time of energy availability and adjust the scheduling. After selecting the voltage and frequency level of each task using the harvesting-aware DVFS, for the first task τ_m in the ready queue, the following inequality is checked:

$$E_S(st_m) + E_H(st_m, ft_m) < E_D(st_m, ft_m) \tag{25.19}$$

where

$E_S(st_m)$ is the remaining energy in the ESM at time st_m
$E_H(st_m, ft_m)$ is the predicted harvested energy between the starting time (st_m) and finishing time (ft_m)
$E_D(st_m, ft_m)$ is the effective energy dissipation of τ_m

If the inequality does not hold, then there is an energy shortage and the algorithm first tries to delay the start time of task τ_m by dl_m, the selection of dl_m should satisfy the following inequality:

$$E_S(st_m) + E_H(st_m, ft_m + dl_m) = E_D(st_m + dl_m, ft_m + dl_m). \tag{25.20}$$

If the deadline of task τ_m and all tasks after τ_m can still be met after the delay, then the schedule is updated and the task τ_m is executed during time interval $[st_m + dl_m, ft_m + dl_m]$ at the frequency $f_{(SIm)}$. The schedule for tasks with lower priority is updated accordingly. If the task τ_m or any task after it cannot meet its deadline after the delay, then task τ_m is removed from Q. All other tasks after τ_m will be rescheduled using harvesting-aware DVFS or LAZY scheduling to reutilize the energy and CPU time that originally occupied by τ_m.

25.3.5 Load Matching

All the works mentioned previously assume that the EHM is working at the MPP while the efficiency of the ESM and ECM is ignored. Such assumptions again will lead to suboptimal solution. A good power manager should be able to adjust the workload of an EH-RTES so that (1) the EHM can work at its MPP, and (2) the overheads of the ESM and ECM are minimized.

25.3.5.1 Load Matching for Maximum Power Point

In [3,4,7,26], the author designed wireless sensor node with energy harvesting. Some of them [3,4] cannot perform MPPT, while [7,26] perform approximate MPPT with or without microcontroller. It has been found that *load matching* is very important for the solar panel to deliver the MPP.

In [15], the authors proposed a load resistance matching method to maximize the efficiency of solar-powered system. Solar cells can be modeled as an ideal power source, an internal resistor R_i, concatenated with a load resistor R_L [15]. It has a much larger internal resistance and it is a function of the solar output and current drawn. The R_i forms a voltage divider with the load resistance; therefore, the value of R_L directly determines the efficiency of the solar panel. In other words, if the R_L is not matched with R_i, then much of the power can be wasted. If the load is matched, then the system can get the maximum benefit from the available power for longer operation. The authors measured output power and voltage under different load with different sunlight intensity, and computed the load resistance values that maximize the out power of solar panel under certain sunlight intensity. Then, the authors embedded a light sensor into the power source and build a solar model and a control unit into the system, as shown in Figure 25.10. Based on the ambient sunlight intensity, the system is able to dynamically adjust its load to yield the maximum power available for the best system performance.

In [7,26], the similar load matching technique is implemented on a wireless sensor node powered by solar energy. Their load matching strategy achieves high conversion efficiency of harvested energy with lower power overhead and improves system power utilization significantly, compared to a conventional solar powered system without load matching capability.

25.3.5.2 Load Matching for Minimum Overhead

In [9], a load matching algorithm is proposed for adaptive power management in an EH-RTES. The algorithm adjusts the workload to make sure that the EHM always works at MPP and makes decision on the fly when the harvested energy should be used to speed up task execution instead of go to the energy storage to reduce the overhead of the ESM operations.

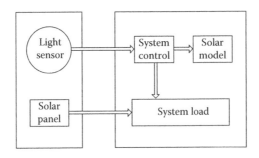

FIGURE 25.10   Block diagram of MPP design.

The EHM is the energy source for EH-RTES. If the EHM does not work efficiently, then the energy efficiency of overall system is degraded. In order to maximize the output power from the EHM, the following equation should hold:

$$P_{H,max} = P_D + P_S, \tag{25.21}$$

where

$P_{H,max}$ denotes the maximum possible harvesting power from the EHM (when it is working at MPP)
P_D and P_S denote the power consumption of the EDM and the power goes into the ESM

Equation 25.21 states that the load for the solar panel should match the MPP for it to deliver $P_{H,max}$.

Note that P_S can be either negative or positive in order that Equation 25.21 holds. At a given time, the sunlight intensity is a constant, and therefore, the maximum output power $P_{H,max}$ is fixed. When the voltage and frequency level is determined, the power consumption of the embedded process P_D is also fixed. In order to make Equation 25.21 hold, the only choice we have is to adjust the charging/discharging power of the ESM. So ESM functions as variable load buffer, it charges or discharges energy to make the EHM work at the MPP. Unfortunately, charging and discharging the energy storage also associate with high overhead, leading to lower system-wide energy efficiency. The basic idea of the load matching algorithm in [9] is to adjust the value of P_D (by changing the frequency and voltage level of the embedded processor) to minimize the charging and discharging of the ESM and hence minimizes its overhead.

Assume that the efficiency of charging and discharging the ESM is η. Let f_n denote the frequency level for task τ chosen by the harvesting-aware DVFS algorithm and $P_D(f_n)$ denote the corresponding power consumption of the embedded processor at this frequency. When the harvested power $P_{H,max}$ is less than $P_D(f_n)$, we must discharge the ESM at power $(P_D(f_n) - P_{H,max})/\eta$ to sustain the operation of the embedded processor because the task cannot be slowed any further without violating the deadline.

When the harvested power $P_{H,max}$ is greater than $P_D(f_n)$, the task can either still run at frequency f_n or run at a higher frequency f'. In order to minimize the charging/discharging overhead, the frequency f' should be selected so that $P_D(f') = P_{H,max}$. Note that the processor has discrete operating frequency levels. Thus, f' may not be among $f_1, f_2, \ldots, f_N$. We can assume that $f_{n1} \leq f' \leq f_{n1+1}$, where $1 \leq n1 \leq N-1$ and n ($n1$. Now we have 3 three possible choices for the operation frequency for the task.

25.3.5.2.1 Option 1

Running the task at frequency f_n. The execution time of τ at f_n is $wcet/S_n$ and the energy storage is charged at rate $P_{H,max} - P_D(f_n)$ during this period of time. The effective energy stored is $\eta(P_{H,max} - P_D)wcet/S_n$. Note that when energy is discharged from the ESM, the system will incur loss again. Thus, the amount of energy usable later is only $\eta^2(P_{H,max} - P_D)wcet/S_n$, and the overhead E_{O1} caused by charging the ESM can be calculated as follows:

$$E_{O1} = \frac{(1 - \eta^2)(P_{H,max} - P_D)wcet}{S_n}, \tag{25.22}$$

E_{O1} indicates the energy "consumed" by the ESM. Let E_{T1} denote the energy dissipation for the embedded processor to finish all tasks in its task queue. The total energy dissipation E_1 needed for supporting all M tasks in Q is the summation of E_{T1} and E_{O1}, which is

$$E_1 = E_{T1} + E_{O1}. \tag{25.23}$$

25.3.5.2.2 Option 2

Running the task at frequency f_{n1}. In this way, the task will finish earlier than its scheduled finishing time, which indicates all other tasks in Q will have more slack time. We reschedule the rest of task with the consideration of this extra slack. The energy dissipation to execute all tasks in the queue based on this new scheduling is calculated and denoted as E_{T2}. Since $f_{n1} \leq (f''$, the energy storage is will be charged at the rate $P_{H,max} - P_D(f_{n1})$, and the overhead E_{O2} is computed as follows:

$$E_{O2} = \frac{(1 - \eta^2)(P_{H,max} - P_D(f_{n1}))wcet}{S_{n1}}. \tag{25.24}$$

The overall energy dissipation for this option is

$$E_2 = E_{T2} + E_{O2}. \tag{25.25}$$

25.3.5.2.3 Option 3

Running the task at frequency f_{n1+1}. Again, we will redistribute the slack and get a new schedule for the rest of the tasks. The energy dissipation under this scheduling is denoted as E_{T3}. Since $f' \leq (f_{n1+1}$, the ESM is discharged at rate $PD(f_{n1+1}) - PH,max$, and the overhead EO3 caused by discharging the energy storage is computed as

$$E_{O3} = \frac{(1/\eta - 1)(P_D(f_{n1+1}) - P_{H,max})wcet}{S_{n1+1}}. \tag{25.26}$$

And the correspondent total energy demand E_3 is

$$E_3 = E_{T3} + E_{O3}. \tag{25.27}$$

The best operating frequency is the one that gives the minimum total energy among those three operations. Figure 25.11 presents the algorithm that adjusts the task scheduling and regulates the charging/discharging power of the ESM in order to improve the energy efficiency of the ESM and the EDM.

Algorithm 2: Load matching and scheduling adjustment
Require: scheduling based on AS-DVFS algorithm
1. **if** $P_{H,max} - P_D > 0$, **then**
2. compute f', and determine f_{n1} and f_{n1+1}
3. compute E_1, E_2, and E_3 using Equations 25.23, 25.25, and 25.27
4. **if** E_1 is the smallest, **then**
5.   execute τ_m at f_n,
6. **else if** E_2 is the smallest, **then**
7.   execute τ_m at f_{n1}, and update scheduling for other $M - 1$ tasks based on AS-DVFS algorithm
8. **else**
9.   execute τ_m **at** f_{n1+1}, and update scheduling for other $M - 1$ tasks based on AS-DVFS algorithm
10. **end if**
11. **else if** $P_{H,max} - P_D < 0$, **then**
12. execute τ_m as scheduled at f_n
13.   **end if**

FIGURE 25.11 Algorithm for load matching and scheduling adjustment.

25.4 Evaluations

Using a large EHM that harvests more energy or a large ESM that provides better buffering for the harvested energy will alleviate the problem of energy fluctuation and improve the system performance. However, this inevitably increases the hardware cost and system size. In this section, the relationship between system performance of EH-RTES and the size of EHM/ESM is evaluated based on a discrete event-driven simulator [8,9] in C++.

25.4.1 Experiment Setup

A DVFS-enabled Intel XScale processor is used as the EDM. Its power, voltage, and frequency levels are shown in Table 25.1. The overhead between the processor switches to different operating voltage/frequency settings is negligible and ignored in the experiment.

The ESM is assumed to be a rechargeable battery or supercapacitor. Without loss of generality, the charging/discharging efficiency of the ESM is fixed to be 0.9, the conversion efficiency of ECM-1 and ECM-2 is 0.9. The capacity of the ESM, E_{cap}, is set to be 1000 J. To speed up the simulation, the ESM has 500 J energy to start each simulation.

The EHM in the simulation is set to be a solar panel with area of 10 cm × 10 cm and conversion efficiency of 10%. Figure 25.2 in Section 25.2 shows four different daytime (7:00 a.m. ∼ 7:00 p.m.) solar irradiation profiles that the authors [8,9] collected during February and March of 2010 at upper state New York. The data are collected every 5 s and the readings are in "*Volts*." The linear interpolation is used to generate more data points to fit the simulation step size. The readings represent the possible power output from the solar panel, that is, P_H. If the reading is denoted as Y, we can convert it to power output of the solar panel using the following equation:

$$P_H = Y \times U \times A \times \varphi, \tag{25.28}$$

where

"U" is a constant value 250 (with unit of W/(m²V)) determined by the pyranometer sensor [27]
"A" is the area of the solar panel (0.01 m²)
"φ" is the conversion efficiency, which is 10%

Similar to most of research work [8,9,11–13,21] in this area, synthetic task sets are used for simulation. Each synthetic task set contains the arbitrary number of periodic tasks. In a specific synthetic task set, the period p_m of a task τ_m is randomly chosen from the set $\{10s, 20s, 30s, \ldots, 100s\}$, and the relative deadline d_m is set to its period p_m. The worst-case execution time w_m is randomly drawn in the interval $[0, p_m]$. Note that the large (in the time magnitude of seconds) tasks are assumed to maintain reasonable CPU time for simulation.

To design different workload categories, the notation of utilization U is introduced, which can be calculated as

$$U = \sum_m \frac{w_m}{p_m}, \tag{25.29}$$

TABLE 25.1 Intel XScale Processor Power and Frequency Levels

Frequency (MHz)	150	400	600	800	1000
Voltage (V)	0.75	1.0	1.3	1.6	1.8
Power (mW)	80	170	400	900	1600
Normalized speed	0.15	0.4	0.6	0.8	1.0

where w_m and p_m are the worst-case execution time and the period of task τ_m, respectively. The utilization U characterizes the percentage of the busy time of the processor if applications are executed at the full speed. Thus, U cannot be greater than 1. In the simulation, different U values are generated by scaling up/down the worst-case execution time of each task in the synthetic task set.

For a given solar profile and utilization setting, the simulation run covers the complete 1 day solar profile. This process is repeated 1000 times, each time with a new random task set. The results presented here are the average of these runs.

25.4.2 Performance Trend

In real applications of EH-RTES, the size of solar panel can be changed due to application and technology, which causes the harvested power from solar panel changes in a wide range; the capacity of ESM can also be changed. Therefore, it is important to study the performance trend of EH-RTES with different harvesting power and storage capacity. Harvesting-aware DVFS scheduling with four different harvesting power settings derived from Profile 3 in Figure 25.2: P_H, $2P_H$, $3P_H$, and $4P_H$. Also, four different capacity of ESM are tested, from 500 to 2000 J with step of 500. The simulation results show that in addition to the utilization ratio U, the harvested power and the ESM capacity (E_{cap}) also have significant impact on the DMR.

Figure 25.5 shows the plots of sweeping both harvest power and E_{cap} for four different utilization ratios of 0.2, 0.4, 0.6, and 0.8. We have the following observations:

1. With increase of the harvest power and/or the storage capacity (E_{cap}), the DMR decreases under all workload settings. The higher the harvest power and/or storage capacity is, the lower the DMR is. It is obvious that tasks are able to be finished before deadline without causing DMR because of more energy coming from the harvested energy or storage or both.

2. DMR reduction increases when the processor utilization ratio increases. At high utilization settings, less slack can be used to slow down the task execution and tasks are executed at high-speed and high-power mode. As a result, the system exhausts the available energy faster and causes more deadline misses.

3. DMR reduction is not linear when harvest power or storage capacity (E_{cap}) increases. Also, it is easy to note that the most significant decreases happen between P_H and $2P_H$, and between E_{cap} of 500 and 1000 J. With further increasing P_H (e.g., from $3P_H$ and $4P_H$) and E_{cap} (e.g., from 1500 and 2000 J,), the improvements in the DMR are not as significant, as shown in Figure 25.12.

25.5 Conclusions

This chapter first describes the hardware architecture of EH-RTES and the major system components of EH-RTES. Then, we presented the research challenges in power management of an EH-RTES and a detailed survey of existing research methods for energy and performance optimization of EH-RTES. Finally, we present some evaluation results that reveal the relation among the size of the EHM, the capacity of the ESM, and the performance (i.e., the DMR) of the EH-RTES. The research in power and performance optimization of the EH-RTES still has long way to go. For example, better energy prediction model that considers the secondary impact such as temperature, process variations, and output voltage should be developed, nonlinear characteristics of energy storage unit should be considered, and better power management policies based on stochastic control or machine learning techniques should be investigated. As the multi-core architecture becomes more and more widely used in embedded systems, the power management of multi-core embedded system with energy harvesting capability is another research challenge to be investigated.

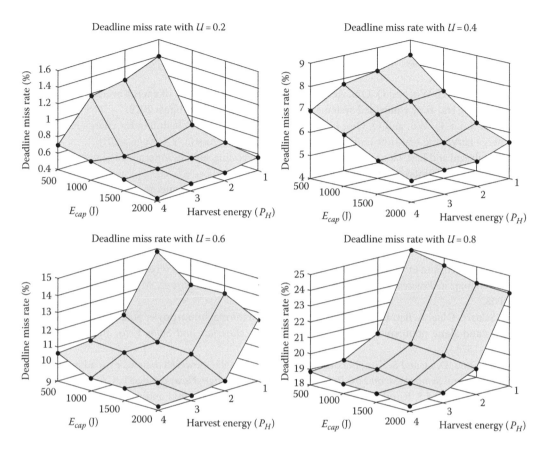

FIGURE 25.12 Sweeping harvest energy and the capacity of ESM. (From Lu, J. et al., Accurate modeling and prediction of energy availability in energy harvesting real-time embedded systems, in *Green Computing Conference, 2010 International*, Chicago, IL, Aug. 15–18, 2010, pp. 469–476.)

References

1. S. Roundy, D. Steingart, L. Frechette, P. K. Wright, and J. M. Rabaey, Power sources for wireless sensor networks, in *Wireless Sensor Networks, First European Workshop, EWSN 2004, Proceedings*, Lecture Notes in Computer Science, Berlin, Germany, January 19–21, 2004, pp. 1–17.
2. J. Rabaey, F. Burghardt, D. Steingart, M. Seeman, and P. Wright, Energy harvesting, a systems perspective, in *IEEE International Electron Devices Meeting*, Washington, DC, December 2007, pp. 363–366.
3. K. Lin, J. Yu, J. Hsu, S. Zahedi, D. Lee, J. Friedman, A. Kansal, V. Raghunathan, and M. Srivastava, Heliomote: Enabling long-lived sensor networks through solar energy harvesting, in *Proceedings of the 3rd International Conference on Embedded Networked Sensor Systems*, San Diego, CA, November 2005.
4. X. Jiang, J. Polastre, and D. E. Culler, Perpetual environmentally powered sensor networks, in *Proceedings of the Fourth International Symposium on Information Processing in Sensor Networks*, Los Angeles, California, April 2005, pp. 463–468.
5. A. M. van Voorden, L. M. R. Elizondo, G. C. Paap, J. Verboomen, and L. van der Sluis, The application of super capacitors to relieve battery-storage systems in autonomous renewable energy systems, in *IEEE Power Tech Conference*, Lausanne, Switzerland, July 2007, pp. 479–484.
6. http://www.media.mit.edu/resenv/power.html

7. C. Park and P. H. Chou, AmbiMax: Efficient, autonomous energy harvesting system for multiple-supply wireless sensor nodes, in *Proceedings of the Third Annual IEEE Communications Society Conference on Sensor, Mesh, and Ad Hoc Communications and Networks* (SECON), Reston, VA, September 2006.

8. J. Lu, S. Liu, Q. Wu, and Q. Qiu, Accurate modeling and prediction of energy availability in energy harvesting real-time embedded systems, in *WIPGC*, Chicago, IL, August 2010.

9. S. Liu, J. Lu, Q. Wu, and Q. Qiu, Load-matching adaptive task scheduling for energy efficiency in energy harvesting real-time embedded systems, in *ISLPED*, Austin, TX, August 2010.

10. S. Chalasani and J. M. Conrad, A survey of energy harvesting sources for embedded systems, *IEEE in Southeastcon*, Huntsville, AL, April 2008, pp. 442–447.

11. C. Moser, D. Brunelli, L. Thiele, and L. Benini, Lazy scheduling for energy-harvesting sensor nodes, in *Working Conference on Distributed and Parallel Embedded Systems*, Braga, Portugal, October 2006.

12. S. Liu, Q. Wu, and Q. Qiu, An adaptive scheduling and voltage/frequency selection algorithm for real-time energy harvesting systems, in *Proceedings of Design Automation Conference*, San Francisco, CA, July 2009.

13. S. Liu, Q. Wu, and Q. Qiu, Dynamic voltage and frequency selection for real-time system with energy scavenging, in *Proceedings of Design, Automation and Test in Europe*, Munich, Germany, March 2008.

14. V. Salas, E. Olias, A. Barrado, and A. Lazaro, Review of the maximum power point tracking algorithms for stand-alone photovoltaic systems, *Solar Energy Materials and Solar Cells*, July 2006, 90(11), 1555–1578.

15. P. H. Chou, D. Li, and S. Kim, Maximizing efficiency of solar-powered systems by load matching, in *International Symposium on Low Power Electronics and Design*, Newport Beach, CA, August 2004.

16. N. Takehara and S. Kurokami, Power control apparatus and method and power generating system using them. Patent US5, 654, 883, 1997.

17. L. T. W. Bavaro, Power regulation utilizing only battery current monitoring, Patent US4, 794, 272, 1988.

18. B. Mochocki, X. Hu, and G. Quan, Transition-overhead-aware voltage scheduling for fixed-priority real-time systems, *ACM Transactions on Design Automation of Electronic Systems*, 12(2), April 2007.

19. G. Zeng, T. Yokoyama, H. Tomiyama, and H. Takada, Practical energy-aware scheduling for real-time multiprocessor systems, *IEEE International Conference on Embedded and Real-Time Computing Systems and Applications*, Beijing, China, August 2009.

20. M. Pedram and Q. Wu, Design considerations for battery-powered electronics, *Proceedings of Design Automation Conference*, New Orleans, LA, June 1999.

21. C. Moser, D. Brunelli, L. Thiele, and L. Benini, Real-time scheduling with regenerative energy, in *Euromicro Conference on Real-Time Systems*, Dresden, Germany, July 2006, pp. 261–270.

22. J. R. Lorch and A. J. Smith, Improving dynamic voltage scaling algorithms with PACE, in *Proceedings of the ACM SIGMETRICS Conference*, Cambridge, MA, June 2001.

23. I. Hong, M. Potkonjak, and M. B. Srivastava, On-line scheduling of hard real-time tasks on variable voltage processor, in *The International Conference on Computer-Aided Design (ICCAD)*, San Jose, CA, November 1998.

24. G. Quan and X. Hu, Energy efficient DVS schedule for fix-priority real-time systems, *ACM Transactions on Embedded Computing Systems*, 6(4), September 2007, pp. 1–31.

25. C. Bergonzini, D. Brunelli, and L. Benini, Algorithms for harvested energy prediction in batteryless wireless sensor networks, *3rd International Workshop on Advances in Sensors and Interfaces*, Trani, Italy, June 2009, pp. 144–149.

26. F. Simjee and P. H. Chou, Efficient charging of supercapacitors for extended lifetime of wireless sensor nodes, *IEEE Transactions on Power Electronics*, 23(3), May 2008, pp. 1526–1536.

27. Silicon-cell photodiode pyranometers, Apogee Instruments, Inc., http://www.apogeeinstruments.com/pyr_spec.htm

28. A. Kansal, J. Hsu, S. Zahedi, and M. B. Srivastava, Power management in energy harvesting sensor networks, in *ACM Transactions on Embedded Computing Systems (in revision)*, May 2006, also available from: NESL Technical Report Number: TR-UCLA-NESL-200605-01.

29. V. Raghunathan, A. Kansal, J. Hsu, J. Friedman, and M. B. Srivastava, Design considerations for solar energy harvesting wireless embedded systems, *International Conference on Information Processing in Sensor Networks*, Los Angeles, CA, April 2005, pp. 457–462.

30. V. Raghunathan and P. H. Chou, Design and power management of energy harvesting embedded systems, *IEEE International Symposium on Low Power Electronics and Design*, Tegernsee, Germany, October 2006.

31. D. Dondi, A. Bertacchini, D. Brunelli, L. Larcher, and L. Benini, Modeling and optimization of a solar energy harvester system for self-powered wireless sensor networks, *IEEE Transactions on Industrial Electronics*, 55(7), July 2008, 2759–2766.

32. C. Moser, L. Thiele, D. Brunelli, and L. Benini, Adaptive power management in energy harvesting systems, in *Design, Automation and Test in Europe*, Nice, France, April 2007, pp. 773–778.

26

Low-Energy Instruction Cache Optimization Techniques for Embedded Systems

Ann Gordon-Ross
University of Florida

Marisha Rawlins
University of Florida

26.1 Introduction

Extensive past research for improving microprocessor performance and power has focused on the instruction cache due to the cache's large impact on those design factors. Proposed optimization techniques typically exploit an instruction stream's spatial and temporal locality. Popular techniques include prefetching, victim buffers [64], filter caches [29,36], loop caches [22,25,38], code compression [9], cache tuning [6,23,27,65,66], and code reordering [34,47].

Most optimization techniques are proposed independently of other techniques, and with so many techniques available, the interplay of those techniques is now important to study as embedded system designers under tight time-to-market constraints cannot be expected to themselves undergo this time-consuming examination. Studying the interplay may demonstrate that one technique dominates other techniques, perhaps making some techniques unnecessary. However, if several techniques are complementary, then finding the most effective combination of the techniques is necessary. In this chapter, we examine the interplay of two major software-based approaches, code reordering and code compression, and two major hardware-based approaches, loop caching and cache tuning.

Code reordering/reorganization/layout at the basic block level is a mature technique developed in the late 1980s to tune an instruction stream to the instruction cache to improve cache hit rates and improve the cache's utilization. This widely researched technique increases performance on average, however decreases performance for some benchmarks. Code reordering places an application's hot path instructions (frequently executed instructions) contiguously in memory, thus moving infrequently executed instructions so that infrequently executed instructions do not pollute the cache (via cache prefetching techniques or large cache line sizes). Code reordering also reduces conflict misses through procedure placement. Initial code reordering techniques were non-cache-aware, meaning these techniques did not consider the specific cache configuration when applying code transformations. Enhanced code-reordering techniques are cache-aware and provide improved performance over non-cache-aware techniques, but cache-aware techniques require a priori knowledge of the target cache configuration. Typically, code reordering is a compile-time or link-time optimization requiring profile information to determine an application's hot path. Runtime/dynamic methods for code reordering also exist [12,30,31,55], resulting in a simpler tool flow but incur some runtime overhead. Several previous works evaluate code reordering impacts, such as the impact of the instruction set [12] and instruction fetch architectures [50] on code reordering, the impact code reordering has on branch prediction [49], code placement for improving branch prediction accuracy [48], and the combined effects of code ordering and victim buffers [6].

Code compression techniques were initially developed to reduce the static code size in embedded systems. However, recent code compression work [10,40] investigated the effects of code compression on instruction fetch energy in embedded systems. In these systems, energy is saved by storing compressed instructions in the level one instruction cache and decompressing these instructions (during runtime) with a low energy/performance overhead decompression unit.

Loop caches are small devices that provide an effective method for decreasing memory hierarchy energy consumption by storing frequently executed code (critical regions) in a more energy efficient structure than the level one cache [22,51]. The main purpose of a loop cache is to provide the processor with as many instructions as possible while the larger, more power hungry level one instruction cache remains idle. The preloaded loop cache (PLC) [22] requires designer-applied static preanalysis to store complex code regions (code with jumps) whereas the adaptive loop cache (ALC) [51] performs this analysis during runtime and requires no designer effort.

Due to new hardware technologies and core-based design methodologies, cache tuning is a more recently developed technique that tunes a cache's parameters, such as total size, associativity, and line size, to an application's instruction stream for decreased energy consumption and/or increased performance. Applications have distinct execution behaviors that warrant distinct cache requirements [66]. Reducing a cache's size or associativity just enough, but not too much, can minimize an application's energy consumption. Tuning the line size, larger for localized applications, smaller for nonlocalized applications, can reduce energy and also improve performance. Caches may be configured in core-based techniques in which a designer synthesizes a customized cache along with a microprocessor [3–5,43,61]. On the other hand, caches in predesigned chips may instead be hardware configurable, with configuration occurring by setting register bits during system reset or during runtime [2,41,65]. To determine the best (lowest energy and/or best performance as defined by the system optimization goals), the cache configuration design space may be explored exhaustively or using a search heuristic.

Studying the interaction of existing techniques reveals the practicality of combining optimization techniques. For example, if combining certain techniques provides additional energy savings but the combination process is nontrivial (e.g., circular dependencies for highly dependent techniques [26]), new design techniques must be developed to maximize savings. On the other hand, less dependent techniques may be easier to combine but may reveal little additional savings. Finally, some combined techniques may even degrade each other. These studies provide designers with valuable insights for determining if the combined savings are worth the additional design effort. This chapter explores the combined effects of code reordering, code compression, loop caching, and cache tuning; analyzes combined benefits; provides technique combination strategies; and provides designer-guided assistance.

26.2 Background and Related Work

26.2.1 Code Reordering

26.2.1.1 Non-Cache-Aware Code Reordering

Much previous research exists in the area of code reordering (also referred to as code layout and code reorganization in literature) at the basic block, loop, and procedure level. Code reorganization at the basic block level dates back to initial work in 1988 by Samples et al. [53]. Early work by McFarling [42] used basic block execution counts to reorder code at the basic block level and exclude infrequently used instructions from the cache. Pettis and Hansen [47] (PH) and Hwu et al. [33] presented similar methods for both basic block and procedural reordering using edge profile information and showed performance benefits of nearly 15%. PH code reordering served as the basis for the majority of all modern code reordering techniques.

Many modern tools implement PH code reordering directly or slightly modified [14,15,19,37,45]. Much research focused on improving the PH code reordering to include cache line coloring to reduce conflict misses [1,28,34,35] by placing frequently executed procedures in memory such that the number of overlapping cache lines was minimized. Also, various production tools offer PH-based code reordering as an optimization option such as Compaq's object modification tool (OM) [59], its successor Spike [14], and IBM's FDPR [57]. Other works provided techniques for applying code reordering to commercial applications [49].

For several of the experiments presented in this chapter, we used the Pentium link-time optimizer (PLTO) [56], which uses an improved PH code reordering algorithm. In PLTO's PH code reordering algorithm, basic blocks are reordered to reduce the number of taken branches and to reduce the number of misses in the instruction cache by increasing instruction locality. For example, loop bodies frequently contain an error condition that is checked in each loop iteration, and the error code is infrequently executed. The error handling code is loaded into the instruction cache, polluting the instruction cache with code that may never be fetched. Code reordering moves infrequently executed code out of the loop body, replacing the code with a jump to the relocated code. Additionally, a jump is inserted at the end of the relocated code to transfer control back to the loop body.

26.2.1.2 Cache-Aware Code Reordering

PH code reordering was designed to exploit a single-level direct-mapped cache. Basic block reordering is performed without any attention to how the ordering may cause contention in a set associative cache or how code reordering affects conflicts in the other levels of the cache hierarchy. More complex algorithms extend code reordering to include cache-aware code placement [8] and multiple levels of cache [19]. For the experiments in this chapter, we used one of the most advanced cache-aware code reordering tools, developed by Kalmatianos et al. [34,35], which performs procedure reordering and cache line coloring. Using an instruction trace file of the application, the tool constructs a conflict miss graph, which is an undirected graph where each node represents a procedure. Every edge between two procedures is weighted with an estimation of the worst-case number of conflict misses between those two procedures. Conflict misses can only occur if the two procedures are simultaneously live—both procedures occupy the cache at the same time. Next, the tool prunes the conflict miss graph to remove unpopular edges and removes procedures with no edges remaining after pruning. After pruning, the tool applies cache line coloring to place the procedures into the cache such that the number of conflict misses between simultaneous live procedures is minimized. Pairs of nodes are processed from the graph by decreasing edge weights. During this step, the tool is conscious of the placement of the procedures in main memory to keep the memory footprint as small as possible.

26.2.2 Code Compression

Several code compression techniques are based on well-known lossless data compression mechanisms. Wolfe and Chanin [63] used Huffman coding to compress/decompress code for reduced instruction set computing (RISC) processors. They also introduced line address tables (LATs), which mapped program instruction addresses to their corresponding compressed code instruction addresses.

Lekatsas et al. [40] incorporated different data compression mechanisms by separating instructions into groups. Codes appended to the beginning of an instruction group identified the group's compression mechanism. This approach achieved system (cache, processor, and busses) energy savings between 22% and 82%.

Benini et al. [10] proposed a low overhead decompression on fetch (DF) (Figure 26.1b) technique based on fast dictionary instructions. The authors noted that since in the DF architecture, the decompression unit was on the critical path (since the decompression unit was invoked for every instruction executed), the unit must have a low decompression (performance) overhead. In their approach, the authors profiled the executable to identify the 256 most frequently executed instructions (denoted as S_N) and replaced those instructions with an 8-bit code if that instruction and its neighboring instructions could be compressed into a single cache line. Results showed average system energy savings of 30%.

26.2.3 Loop Caching

The ALC is the most flexible loop cache (loop cache contents are dynamically loaded/changed during runtime) and can store complex loops (i.e., loops with control of flow [cof] changes such as taken branches and forward jumps). Figure 26.1a shows the loop cache's architectural placement. The ALC [51] identifies and caches loops during runtime using lightweight control flow analysis. The ALC identifies loops when the loop's last instruction (a short backward branch [sbb] instruction) is taken. The ALC fills the loop cache with the loop instructions on the loop's second iteration and from the third iteration onward, the ALC supplies the processor with the loop instructions (i.e., the level one cache is idle). Since loop caches

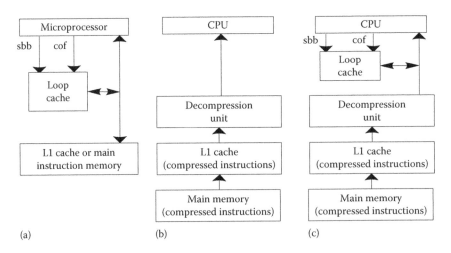

FIGURE 26.1 (a) Architectural placement of the loop cache, (b) the DF architecture, and (c) the DF architecture with a loop cache to store decompressed instructions.

require 100% hit rates, the ALC stores valid bits (exit bits) to determine whether the next instruction should be fetched from the ALC or the L1 cache. (Thus, this transition incurs no additional cycle penalty)

PLC [22] operation is similar to the ALC's (the PLC can store complex loops); however, PLC contents are statically profiled and preanalyzed during design time and loaded during system start-up.

Both the ALC and PLC can reduce instruction memory hierarchy energy by as much as 65% [22,51]. Since instructions stored in the PLC never enter the level one cache, using a PLC affects the locality of the level one cache. Eliminating instructions from the level one cache could affect the overall energy savings of the system since the level one cache (an important source of energy savings) takes advantage of an application's locality for improved performance and energy consumption.

26.2.4 Cache Tuning

Su et al. [60] showed in early work that the memory hierarchy is very important in determining the power and performance of an application. Recently, Zhang et al. [65] showed the vastly different cache configurations required to achieve minimal cache energy consumption. If a cache does not reflect the requirements of an application, excess energy may be consumed. For example, if the cache size is too large for an application, excess energy will be consumed fetching from the large cache. If the cache size is too small, excess energy may be consumed due to thrashing—the working set of an application is constantly being swapped in and out of the cache. Tunable parameters normally include cache size, line size, and associativity; however, other parameters such as the use of a victim buffer, instruction/data encoding, bus width, etc., could also be included as tunable parameters.

Recent advances in research and technology have made possible the configurability of cache parameters. The availability of a tunable cache enables designers to specify cache parameters in a tunable core-based design for custom system synthesis [3–5,43,61]. Additionally, predesigned chips may contain hardware that supports cache tuning during system reset or even during runtime [2,41,65].

Given a configurable cache with many tunable parameters and tunable parameter values, determining appropriate tunable parameter values is nontrivial. Platune [21] used an exhaustive method to search the cache configuration design space for reduced energy consumption. Whereas an exhaustive method produces optimal results, the time needed to exhaustively search the design space may not be available. To decrease exploration time, heuristic methods exist to explore the design space. Palesi et al. [46] presented an extension to Platune that explored the design space using a genetic algorithm and produced near-optimal results in a fraction of the time. Ghosh et al. [20] presented a method for directly computing cache parameter values given design constraints. Balasubramonian et al. [7] presented a runtime method for redistributing the cache size between the various cache levels. Zhang et al. [66] presented a prototyping methodology for level one cache tuning exploration for Pareto-optimal points trading off energy and performance. Further work by Zhang et al. [65] showed a methodology for runtime cache tuning for reduced energy consumption resulting in energy savings of 45%–55% on average. Gordon-Ross et al. [27] developed a heuristic for quickly searching a two-level configurable cache hierarchy for separate instruction and data caches showing energy savings averaging 53%.

In this chapter, we use the configurable cache tuning methodology described by Zhang [65,66] and extended by Gordon-Ross et al. [27] for single-level cache tuning. The cache tuning heuristic utilized in this chapter efficiently explores the cache parameters based on their impact on total system energy and miss rate and explores cache parameters having a larger impact on system energy and miss rate before parameters having a smaller impact. The heuristic is summarized in the following steps: (1) Holding the cache line size and associativity at their smallest values, determine the cache size yielding the lowest energy consumption; (2) fixing the cache size at the size determined in the previous step and the associativity at the smallest value, determine the cache line size yielding the lowest energy consumption; and

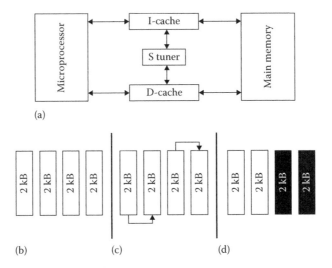

FIGURE 26.2   (a) Configurable cache architecture, (b) 8 kB 4-way base cache with for 2 kB sub-banks, (c) 8 kB 2-way cache using way concatenation, and (d) 4 kB 2-way cache using way shutdown.

(3) fixing the cache size and the cache line size at the values determined in the previous steps, determine the cache associativity yielding the lowest energy consumption. Since previous work and our experimental observations found that this heuristic determined the optimal cache configuration in most cases, we will refer to the heuristically determined cache configuration as the optimal cache configuration.

To enable the cache tuning heuristic to tune during runtime, a tunable cache is necessary. Zhang [66] describes tunable cache hardware and presents verification of the hardware layout of this configurable cache. Figure 26.2a shows the configurable cache architecture. Figure 26.2b shows the base cache consisting of four separate banks that may be turned on, concatenated with another bank (Figure 26.2c), or turned off (Figure 26.2d) via a configuration register. Due to the use of configurable banks, certain cache configurations are not possible. For instance, for an 8 kB base cache size, four 2 kB banks are used, providing 8 kB direct-mapped, 2-way, and 4-way set associative caches. To reduce the cache size to 2 kB, three of the banks must be shut down leaving one remaining 2 kB bank. Since banks are used to increase associativity and there is only one bank available in a 2 kB cache, only a direct-mapped cache is available for 2 kB. Further details are available in [66].

26.3 Interplay of Code Reordering and Cache Tuning

Code reordering and cache tuning can be applied at different times during application design. Code reordering is typically carried out during design time as a designer-guided step, while cache tuning can be applied at design time or runtime. For code reordering, the designer must compile and profile the code and then generate an optimized executable by either recompiling the code or using a link-time code optimizer. Sanghai et al. [54] presented a framework to provide code layout assistance for embedded processors with configurable memories. Dynamic procedure placement [55] is possible but has received little attention due in part to potentially significant runtime overhead. Cache tuning can also be applied as a designer-guided step; however, recent research focuses on cache tuning during runtime to eliminate the need for designer intervention [65] with little to no runtime overhead. Designer-guided optimization steps increase the complexity of the design task, whereas runtime optimization requires no special design efforts and also ensures that optimizations use an application's real data set.

26.3.1 Non-Cache-Aware Code Reordering and Cache Tuning

First, we explored the interplay of non-cache-aware code reordering and cache tuning. In this section, we describe our evaluation framework (which is similar to the framework used for our next study using cache-aware code ordering) and evaluate our results.

26.3.1.1 Evaluation Framework

To determine the combined effects of code reordering and cache tuning, we used 26 benchmarks: 12 benchmarks from the Powerstone benchmark suite [41], 3 benchmarks from the MediaBench benchmark suite [39], and 11 benchmarks from the Embedded Microprocessor Benchmark Consortium (EEMBC) benchmark suite [17]. For each benchmark suite, we report data for every benchmark that successfully ran through the compilation and simulation tools. Some benchmarks would not compile, would not run through the tools, or would not execute correctly after code reordering. For all benchmarks, we used the provided input vectors.

We used PLTO [56] to perform code reordering on the applications. PLTO is similar to the popular ALTO [45] tool but works with the ×86 architecture instead of the Alpha architecture. We performed the following steps to produce code reordered executables: (1) Compile the code with flags to include the symbol table and relocation information, not to patch any of the instructions, and statically link all libraries; (2) invoke PLTO to instrument the executable to gather edge profiles; (3) run the instrumented executable to produce a file containing the edge counts; and (4) rerun PLTO with edge profiles and perform code reordering.

Since PLTO offers many other link-time optimizations, we ensured that only code reordering was applied by turning off all other optimizations at the command line. Additionally, for comparison purposes, we created executables without code reordering using the same steps as described above except that in step 4, we turned off the code reordering optimization.

We used Perl scripts to drive the cache tuning heuristic along with an instruction cache simulator to determine cache statistics. Most ×86 cache simulators are trace driven, requiring an instruction trace file for execution. Due to the long execution time of some of the benchmarks studied, trace driven cache simulation was not feasible. To alleviate the need for instruction traces, we obtained a trap-based profiler from the University of Arizona to perform execution driven cache simulation [44]. The trap-based profiler combines the trace cache simulator Dinero IV [16] and PLTO to create an execution driven cache simulation. The trap-based profiler executes the application using PLTO, traps instruction addresses, and passes the instruction addresses to Dinero.

We determined energy consumption for a cache configuration for both static and dynamic energy using the energy model depicted in Figure 26.3. We used Cacti [52] to determine the dynamic energy consumed by each cache fetch for each cache configuration using 0.18-micron technology. The trap profiler provided us with the cache hits and misses for each cache configuration. Miss energy determination is difficult because it depends on the off-chip access energy and the CPU stall energy, which are highly dependent on the actually system configuration used. We could have chosen a particular system configuration and obtained hard values for the *CPU_stall_energy*; however, our results would only apply to one particular system configuration. Instead, we examined the stall energy for several microprocessors and

$$total\_energy = static\_energy + dynamic\_energy$$
$$dynamic\_energy = cache\_hits * hit\_energy + cache\_misses * miss\_energy$$
$$miss\_energy = offchip\_access\_energy + miss\_cycles * CPU\_stall\_energy + cache\_fill\_energy$$
$$miss\_cycles = cache\_misses * miss\_latency + cache\_misses * memory\_bandwidth$$
$$static\_energy = total\_cycles * static\_energy\_per\_cycle$$
$$static\_energy\_per\_cycle = energy\_per\_Kbyte * cache\_size\_in\_Kbytes$$
$$energy\_per\_Kbyte = ((dynamic\_energy\_of\_base\_cache * 10\%)/base\_cache\_size\_in\_Kbytes)$$

FIGURE 26.3 Energy model used for evaluating code reordering and cache tuning.

estimated the *CPU_stall_energy* to be 20% of the active energy of the microprocessor. We obtained the *offchip_access_energy* from a standard low-power Samsung memory. To obtain miss cycles, the miss latency and bandwidth of the system are required. We estimated a cache miss to take 40 times longer than a cache hit to transfer the first block (16 bytes) and subsequent blocks (each additional 16 bytes) would transfer in 50% of the time it took to transfer the first block. Previous work [27] showed that cache tuning heuristics remained valid across different configurations of miss latency and bandwidth. We determined the static energy per kilobyte as 10% of the dynamic energy of the base cache divided by the base cache size in kilobytes.

We chose cache parameters to reflect those available in typical embedded processors. We explored cache sizes of 2, 4, and 8 kB, cache line sizes of 16, 32, and 64 bytes, and set associativities of direct-mapped, 2-way, and 4-way.

We generated cache statistics for every cache configuration for every benchmark, with and without code reordering, to determine the optimal cache configurations. To determine cache energy savings due to cache tuning, normally a large cache is used as a base cache for comparison purposes. This cache size reflects a common configuration likely to be found in a platform to accommodate a wide range of target applications. However, research shows that code reordering is most effective for small to medium cache sizes [42] because an application may entirely fit into too large of a cache—only in a small cache do we see large numbers of conflict misses. To best show the benefits of code reordering, we have chosen the smallest cache as our *base cache configuration*—a 2 kB direct-mapped cache with a line size of 16 bytes. This small cache size is not too small as to be dominated by capacity misses. Using the smallest cache possible is also a goal of many cost-constrained embedded systems.

26.3.1.2 Energy and Performance Evaluation

We explored the interplay of code reordering and cache tuning using four energy and performance results for each benchmark. The results included energy and performance values for the base cache configuration for each benchmark without code reordering and for each benchmark after code reordering has been performed. Additionally, we applied cache tuning to each benchmark without code reordering and for each benchmark after code reordering has been applied.

Figure 26.4 shows the energy savings and performance impact of cache tuning both with and without code reordering. All values have been normalized to the base cache configuration without code reordering for each benchmark. Overall, results showed a similar trend for both energy and performance as expected since code reordering simply reduces the number of cache misses. For code reordering alone, average energy savings and execution time reduction were approximately 3.5% over all benchmarks. However, the averages included two benchmarks, *CACHEB* and *CANRDR*, where code reordering performed very poorly. Removing these two benchmarks from the average increased the average energy savings and execution time reduction to approximately 9%, closely reflecting results obtained in previous research [45]. (Ultimately, a designer would be able to detect the decreased performance of code reordering on an application and then choose to not apply code reordering).

When cache tuning is applied to the benchmarks, Figure 26.4 shows that on average both the energy savings and performance benefits were nearly identical for cache tuning without code reordering and cache tuning with code reordering. Energy savings obtained due to cache tuning were on average 15% without code reordering and 17% with code reordering over all benchmarks. Likewise, execution time reduction with cache tuning averaged 17% without code reordering and 18.5% with code reordering over all benchmarks. From these results, we might conclude that the benefits due to code reordering are nearly negated when cache tuning is used. A designer can thus eliminate the special tools, profiling setup, and time required to perform code reordering if runtime cache tuning is available. The additional savings due to adding code reordering to cache tuning are nominal and probably not worth the extra design effort required by the designer. Runtime cache tuning produces the benefits without designer effort.

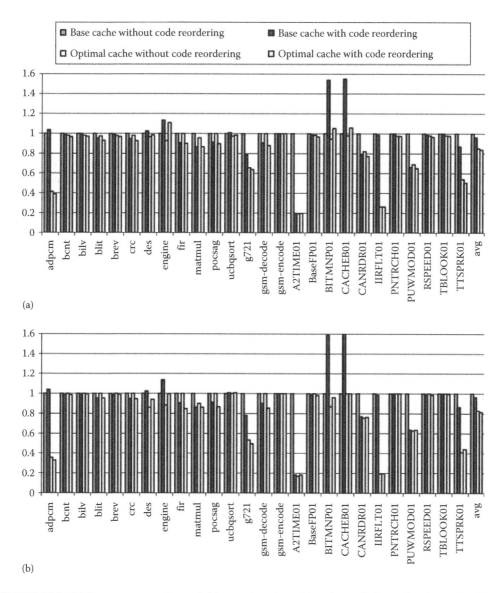

FIGURE 26.4 (a) Energy consumption and (b) execution time with code reordering and cache tuning. Energy consumption and execution time for each benchmark are normalized to the energy consumption and execution time, respectively, of the base cache without code reordering.

Additionally, we observed a very interesting trend across all benchmarks. As Figure 26.4 shows, in a few benchmarks, both energy and execution time increased when code reordering was applied. However, when cache tuning was applied along with code reordering, there was no execution time degradation for any benchmark. Execution time for each application was either as good or better than the base cache configuration with no code reordering, thus alleviating some of the negative performance impacts some applications incur due to code reordering.

26.3.2 Cache-Aware Code Reordering and Cache Tuning

Next, we explored the interplay of cache-aware code reordering and cache tuning. We present our evaluation framework, associated design exploration heuristics, and evaluate our results.

26.3.2.1 Evaluation Framework

Since our evaluation framework is similar to that used for non-cache-aware code reordering, we limit our discussion here to the differences. Due to the use of a different tool suite, our benchmark suite is slightly different and included 30 embedded system benchmarks: 11 benchmarks from the Powerstone benchmark suite [41], 3 benchmarks from the MediaBench benchmark suite [39], and 16 benchmarks from the EEMBC benchmark suite [17]. We could only explore three benchmarks from the MediaBench benchmark suite because the other benchmarks produced trace files too large to process in a reasonable amount of time by the code reordering tool.

The code reordering tool we used [34,35] required as input an instruction trace of the application to gather profile information, the location of all procedures in the application, a list of all active procedures in the application, and the cache configuration. To obtain the instruction trace file, we modified the sim-cache portion of SimpleScalar [11] to output each instruction address during execution. We obtained location information of all active and inactive procedures using Looan [62], a loop analysis tool that takes as input an application binary and the instruction trace and outputs the location of all loops and procedures in the application along with execution frequencies. A header file provides the cache configuration information to the code reordering tool.

The code reordering tool first performs code reordering for the cache configuration and then simulates the instruction cache to obtain cache statistics. For tests that involved reordering the code for a configuration other than the current cache configuration being explored, we modified the tool, so that both the cache configuration to reorder for and the actual cache configuration to simulate could be different.

26.3.2.2 Design Exploration Heuristics to Combine Cache-Aware Code Reordering and Cache Tuning

We also sought to develop the best heuristic for combining the two methods and to compare that heuristic to each method applied alone. We considered three possible exploration heuristics: *reorder-tune*, *tune-reorder-tune*, and *reorder-during-tuning*.

26.3.2.2.1 Reorder-Tune

Figure 26.5a shows the reorder-tune heuristic. This heuristic first performs code reordering and then applies cache tuning to the reordered code. However, since cache-aware code reordering must have the cache configuration as input, difficulty arises in choosing the cache configuration to reorder for before the best cache configuration is known. We chose to perform code reordering for the base cache configuration since our results will be compared to the base cache configuration. After code reordering, the cache tuning

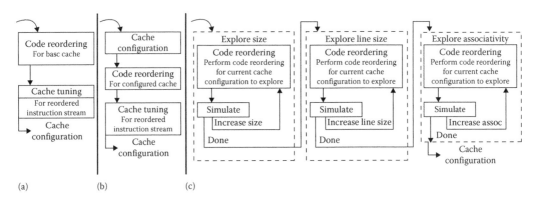

FIGURE 26.5 Flow charts depicting optimization methods for the (a) reorder-tune heuristic, (b) the tune-reorder-tune heuristic, and (c) the reorder-during-tuning heuristic.

heuristic configures the cache for optimal energy consumption given the reordered instruction stream. With one code reordering round and one cache tuning round, this heuristic is the fastest to apply.

26.3.2.2.2 Tune-Reorder-Tune

The drawback of the reorder-tune heuristic is that the cache configuration must be arbitrarily chosen for cache-aware code reordering. The second heuristic we developed addresses this issue. Figure 26.5b shows the tune-reorder-tune heuristic. Instead of applying code reordering first, the tune-reorder-tune heuristic configures the cache given the original instruction stream. This cache configuration is then used as input to the code reordering phase, and the code is reordered for the optimal cache. However, after the code reordering phase, the cache may need to be tuned to the new optimized instruction stream. To account for this, the tune-reorder-tune heuristic applies another round of cache tuning to tune the cache to the optimized instruction stream. With one round of code reordering and two rounds of cache tuning, the tune-reorder-tune heuristic takes nearly twice as long to apply as the reorder-tune heuristic.

26.3.2.2.3 Reorder-During-Tuning

The final heuristic we developed represents the near-optimal search where code reordering is performed during the cache tuning process. This heuristic does not perform an exhaustive search of all possible code reordering situations for all possible cache configurations but searches only the most interesting cases. Figure 26.5c shows the reorder-during-tuning heuristic. This heuristic incorporates code reordering into the cache tuning process by performing code reordering for each cache configuration before the configuration is simulated. Since the reorder-during-tuning heuristic performs code reordering for each cache configuration, this heuristic may take as much as seven times longer to simulate than the reorder-tune heuristic (on average, the heuristic explores seven cache configurations, and code reordering is applied to all seven configurations).

26.3.2.3 Experiments

26.3.2.3.1 Cache-Aware Code Reordering versus Cache Tuning

Figure 26.6a shows the energy consumed by each benchmark for the code reordering method and the cache tuning method. The x-axis shows each benchmark studied and the y-axis represents the energy consumption of the instruction cache for each heuristic normalized to the energy consumption of the base cache configuration without any code reordering or cache tuning (shown as 1.0). The first bar shows the energy consumption of the base cache configuration with code reordering. The second bar shows the energy consumption of the optimal cache configuration with no code reordering applied.

The results show that code reordering yielded average improvements of 6.5% and increased energy consumption in some examples, by as much as 12% in *rawcaudio*. Previous work [45] also observed such increases in energy consumption. Eliminating code reordering for these benchmarks improved the average only by a few percent. On the other hand, for some benchmarks code reordering obtained good energy savings, up to 60% savings for *RSPEED01*.

Cache tuning, on the other hand, yielded energy savings of 36.5% on average. This number closely matches results obtained by Zhang [65], even though their base cache was a high-performance configuration rather than a small-sized configuration. The results showed that cache tuning yielded energy savings as high as 90% for *bliv*. Cache tuning also yielded nearly equal or better savings than code reordering for every benchmark. Stated another way, for applications for which code reordering obtained substantial energy savings, cache tuning achieved nearly equal or better savings. Thus, one may conclude that, among the two methods, the hardware approach of cache tuning is superior, assuming either method is possible.

Figure 26.6b shows similar data comparing benchmark performance results, yielding similar conclusions.

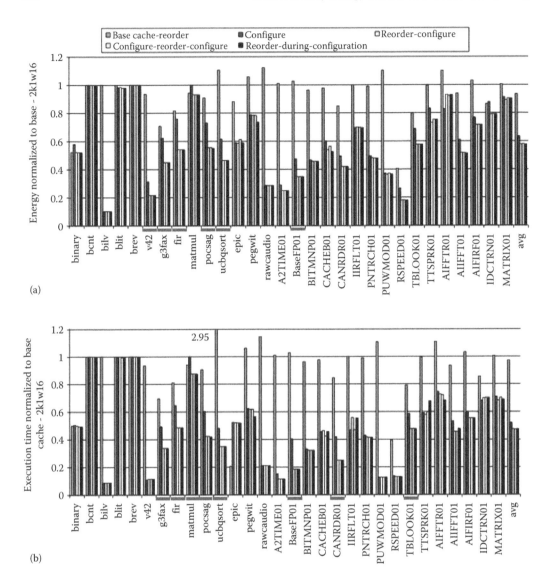

FIGURE 26.6 Results for each benchmark showing (a) energy consumption and (b) execution time normalized to the base cache configurations. Underlined benchmarks highlight those that benefit exceptionally well from cache tuning combined with code reordering.

26.3.2.3.2 *Combined-Methods Heuristic*

Although cache tuning is clearly superior to code reordering alone, combining the two methods may still yield improvement for some benchmarks. We therefore compared the energy savings for each of three heuristics on the 30 benchmarks, with results shown in Figure 26.6a. The third, fourth, and fifth bars show the energy resulting from the reorder-tune, tune-reorder-tune, and reorder-during-tuning heuristics, respectively.

The results revealed that the three combined-methods heuristics performed about the same on average. More significantly, the results showed that combining code reordering and cache tuning only slightly improved the average energy compared to cache tuning alone, by about 6%. Thus, if a designer is concerned with reducing average energy savings across a variety of benchmarks, the hardware solution of a configurable cache may be sufficient.

The results also showed that for particular benchmarks, the combined-methods heuristics did obtain additional 10%–15% energy savings compared to cache tuning alone. Thus, combining the methods may still be useful for applications with tight energy constraints.

The performance data in Figure 26.6b revealed similar results, yielding similar conclusions.

With regard to comparing the three heuristics, we see they performed about the same on average, and that all three heuristics performed equally well for nearly every benchmark. These results were rather surprising to us, as we expected the "optimal" search performed by the reorder-during-tuning heuristic to achieve better results. Instead, the results show that the far more computationally efficient reorder-tune heuristic obtained near-optimal results.

We explain these results by first pointing out that, in most cases, the majority of energy and performance benefits come from cache tuning alone. Even if code reordering is applied to a suboptimal configuration, cache tuning will still determine the lowest energy cache for the instruction stream. Second, for both the reorder-tune and the tune-reorder-tune heuristics, code reordering is applied to wisely chosen cache configurations. For the reorder-tune heuristic, code reordering is applied to the base cache configuration, which is the smallest available cache configuration (code reordering is most effective at small cache sizes [42]). For the tune-reorder-tune method, code reordering is applied to the cache determined by applying cache tuning to the original instruction stream, again, a wise cache for which to perform reordering because that cache closely reflects the needs of the application. The goal of code reordering is twofold—to increase the spatial locality of code by placing hot paths through the application in contiguous memory, thus increasing the benefits of a larger line size, and to reduce conflict misses through cache line coloring. We observed that in the cases where code reordering was beneficial, the only change from the configured cache without code reordering to the configured cache with code reordering was a larger line size.

26.3.2.3.3 Non-Cache-Aware Code Reordering versus Cache-Aware Code Reordering

In order to highlight the benefits gained by cache-aware code reordering, we compare the results for both non-cache-aware and cache-aware code reordering. We note that the instruction set used for the non-cache-aware code reordering results was the ×86 instruction set, whereas the instruction set used for the cache-aware code reordering was the PISA instruction set. Thus, direct normalization between the results is not possible (as even the instruction counts for the base caches differ due to the different instruction set architectures); however, magnitudes can be compared. In addition, due to different tool chains and thus different benchmark sets, we can only compare benchmarks that successfully completed both tool chains.

Figure 26.7 compares (a) energy savings and (b) performance impacts for non-cache-aware code reordering with cache tuning and cache-aware code reordering with cache tuning for the best average heuristic as determined in Section 26.3.2.3.2. Results revealed the importance of considering the actual cache configuration while performing code reordering. Non-cache-aware code reordering with cache tuning revealed a 17% reduction in average energy consumption and a 19% reduction in average execution time. Cache-aware code reordering with cache tuning revealed a 47% reduction in average energy consumption and a 56% reduction in average execution time. Cache-aware code reordering with cache tuning increased the energy savings and reduced the execution time by 2.8× and 2.9×, respectively. However, when looking at individual benchmarks, results revealed that cache-aware code reordering increased energy consumption and execution time as compared to non-cache-aware code reordering with cache tuning for 6 benchmarks and 5 benchmarks, respectively, out of 18 benchmarks. Out of these benchmarks, only two benchmarks, *TTPPRK01* and *IIRFLT01*, showed a large increase in energy consumption and execution time (all other benchmarks showed a negligible difference). We attribute these differences to the general uncertainty of code reordering benefits with respect to application particulars; however, overall, cache-aware code reordering with cache tuning helped to alleviate most of this uncertainty.

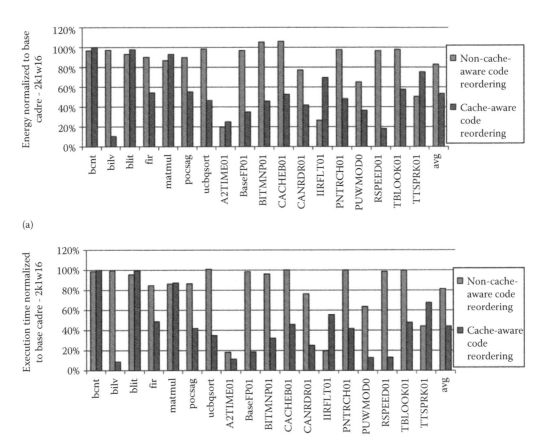

(a)

(b)

FIGURE 26.7 Comparing non-cache-aware code reordering and cache-aware code reordering: (a) energy consumption and (b) execution time normalized to the base cache configuration without code reordering.

26.3.2.3.4 Exploration Speedup

The reorder-tune heuristic we present provides near-optimal results and provides significant design space exploration speedup over an exhaustive method. Since code reordering is typically a link-time optimization, a designer applying both code reordering and cache tuning would likely perform cache tuning exploration in a simulation-based environment. In a simulation environment, the application is executed for each cache configuration to gather cache statistics. Simulation of complex systems can be quite time-consuming, easily requiring many hours or perhaps days to run just a single cache configuration, making development of an efficient heuristic essential to reducing design space exploration time.

The exhaustive reorder-during-tuning approach explored 18 possible cache configurations, applying code reordering to each of the 18 configurations before simulation. The reorder-tune heuristic explored only seven cache configurations, and only once performed code reordering—reducing the number of cache configurations explored by 62% and reducing the number of code reorderings by 95%.

26.4 Interplay of Loop Caching and Cache Tuning

Since loop caches store the most frequently executed instructions (an application's working set and main source of temporal locality), a loop cache effectively removes most temporal locality from the level one cache, and thus may change an application's optimal energy level one cache configuration. In addition, the loop cache itself can benefit from configuration, as working set sizes are application dependent. In this section, we evaluate the combined benefits of a configurable loop cache and configurable level one cache.

26.4.1 Experimental Setup

To determine the combined effects of loop caching and cache tuning, we determined the optimal (lowest energy) loop cache and level one cache configurations for systems using the ALC and the PLC for 31 benchmarks from the EEMBC [17], MiBench [18], and Powerstone [58] benchmark suites (all benchmarks were run to completion, however, due to incorrect execution not related to the loop caches, we could not evaluate the complete suites).

We used the energy model and methods in [51] to calculate energy consumption for each configuration. For comparison purposes, we normalize energy consumption to a *base system* configuration with an 8 kB, 4-way set associative level one instruction cache with a 32 byte line size (a configuration shown in [66] to perform well for a variety of benchmarks on several embedded microprocessors) and with no loop cache. We implemented each loop cache design in SimpleScalar [11]. We varied the level one instruction cache size from 2 to 8 kB, the line size from 16 to 64 bytes, and the associativity from direct-mapped to 4-way [65,66], and varied the loop cache size from 4 to 256 entries [51]. In our experiments, we searched all possible configurations to find the optimal configuration; however, heuristics (such as in [24,27,65]) can also be applied for dynamic configuration.

Our experiments evaluated three different system configurations. In the first experiment, we tuned the level one cache with a fixed 32-entry ALC for the EEMBC and MiBench and a fixed 128-entry ALC for Powerstone (denoted as *tuneL1 + ALC*) ([51] showed that these sizes performed well on average for the respective benchmark suites). In the second experiment, we quantified additional energy savings gained by tuning both the level one instruction cache and the ALC (denoted as *tuneL1 + tuneALC*). In our final experiment, we tuned the level one cache while using a fixed 128-entry PLC (denoted as *tuneL1 + PLC*). For thorough comparison purposes, we also report energy savings obtained by tuning the ALC using a fixed level one base cache configuration (denoted as *tuneLC + base*) and tuning the level one cache in a system with no loop cache (denoted as *noLC*).

26.4.2 Analysis

Figure 26.8 depicts energy savings for all experiments normalized to the base system. In summary, these results compare the energy savings for combining loop caching and level one cache tuning with the energy savings for applying loop caching and cache tuning individually.

First, we evaluated energy savings for each technique individually. Level one cache tuning alone achieved average energy savings of 53.62%, 59.61%, and 37.04% for the EEMBC, Powerstone, and MiBench benchmark suites, respectively. ALC tuning in a system with a base level one cache achieved average energy savings of 23.41%, 45.55%, and 26.04% for the EEMBC, Powerstone, and MiBench benchmark suites, respectively. These results revealed that in general, ALC tuning alone did not match the energy savings of level one cache tuning alone. In this case, a smaller optimal level one cache saved more energy than the ALC combined with the (much larger) base cache. For example, tuning the ALC with a fixed base level one cache achieved 33.57% energy savings for *IDCTRN01*. However, when level one cache tuning was applied, the 8 kB, 4-way, 32 byte line size base level one cache is replaced with a smaller 2 kB, direct-mapped, 64 byte line size level one cache, resulting in energy savings of 55.07%.

However, loop cache tuning alone can save more energy than level one cache tuning without a loop cache when the optimal level one cache configuration is already similar to the base cache (e.g., *dijkstra*). Also, when ALC loop cache access rates are high, ALC cache tuning alone is sufficient, such as with *PNTRCH01*, *blit*, and *CRC32*, which all have loop cache access rates greater than 90%.

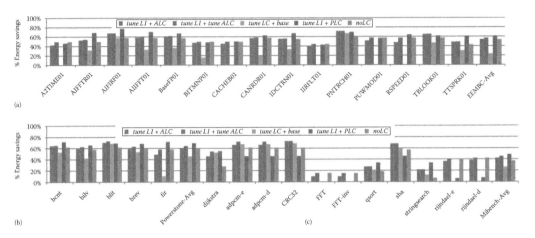

(a)

(b)   (c)

FIGURE 26.8 Energy savings (compared with the base system with no loop cache) for loop caching and cache tuning for the (a) EEMBC, (b) Powerstone, and (c) MiBench benchmark suites.

Next, we evaluated the combined effects of a fixed sized ALC with level one cache tuning (*tuneL1 +
ALC* in Figure 26.8). Additional energy savings were minor as compared to level one cache tuning
alone (average energy savings were 54.24%, 60.59%, and 42.06% for the EEMBC, Powerstone, and
MiBench benchmark suites, respectively). Although the average improvement in energy savings across
the benchmark suites was approximately 1%, adding a fixed sized ALC improved energy savings by as
much as 14.91% for *stringsearch*. Also, in cases where loop caching alone resulted in negative energy
savings (benchmarks with less than a 10% loop cache access rate), this negative impact was offset using
level one cache tuning. For example, even when the ALC caused a 9% increase in energy consumption,
the overall energy savings were still 47.83% for *RSPEED01* since level one cache tuning dominated the
overall energy savings.

Our next experiment investigated the effects of tuning both the level one cache and the loop cache
(*tuneL1 + tuneALC* in Figure 26.8). Although there were improvements in energy savings, the resulting
average energy savings of 56.88%, 64.36%, and 45.56% for the EEMBC, Powerstone, and MiBench
benchmark suites, respectively, were not significantly better than the energy savings achieved by level one
cache tuning alone. Even though the improvement in average energy savings across all benchmark suites
was approximately 4%, additional energy savings reached as high as 26.30% for *dijkstra*.

Additionally, when comparing a system with a tuned level one cache and a fixed sized ALC, the
improvement in average energy savings was minor, averaging only 2% over all benchmark suites, however,
improvements reached as high as 10% for *RSPEED01*. The reason for the minor additional energy savings
was because the energy savings for the optimal ALC size were very close to the savings for the fixed
sized ALC for two reasons: (1) the optimal ALC size was typically similar to the fixed sized ALC size
(the ALC's size was chosen because it performed well on average for each particular suite) and (2) loop
cache access rates leveled off as the loop cache size increased [51]. This finding is significant in that it
reveals that level one cache tuning obviates ALC tuning. If a system designer wishes to incorporate an
ALC, simply tuning the level one cache and adding an appropriately sized ALC is sufficient. This finding's
significance is also important for dynamic cache tuning since using a fixed sized ALC decreases the design
exploration space by a factor of 7 since we eliminate the need to combine each level one configuration
with 7 ALC sizes.

The results presented thus far suggest that, in general, in a system optimized using level one cache
tuning, an ALC can improve energy savings, but it is not necessary to tune the ALC since level one cache
tuning dominates the energy savings. We observed that since the optimal ALC configuration does not
change the optimal level one cache configuration, there is no need to consider the ALC during level one

cache tuning. The level one cache configuration remains the same regardless of the presence of the ALC because using an ALC does not remove any instructions from the instruction stream, nor does the ALC prevent those instructions from being cached in the level one cache and therefore, does not affect the spatial locality. In fact, the level one cache supplies the processor with instructions during the first two loop iterations to fill the ALC [51]. The additional energy savings achieved by adding an ALC to the optimal level one cache configuration results from fetching instructions from the smaller, lower energy ALC [51]. The tradeoff for adding the ALC is an increase in area, which can be as high as 12%. However, this area increase is only a concern in highly area-constrained systems, in which case the system designer should choose to apply level one cache tuning with no ALC.

Since the ALC does not change the actual instructions stored in the level one cache (the ALC only changes the number of times each instruction is fetched from the level one cache), our final experiment involved combining level one cache tuning with a fixed sized PLC, since the PLC actually eliminates instructions from the level one cache, thus changing the spatial locality. Tuning the level one cache and using a fixed sized PLC resulted in average energy savings of 61.04%, 69.33%, and 48.91% for the EEMBC, Powerstone, and MiBench benchmark suites, respectively. On average, adding the PLC to level one cache tuning revealed an additional energy savings of 9.64% as compared to level one cache tuning alone (with no loop cache) with individual additional savings ranging from 10% to 27% for 12 of the 31 benchmarks. Furthermore, since the PLC is preloaded and the preloaded instructions never enter the level one instruction cache, using a PLC can change the optimal level one cache configuration, especially when PLC access rates are very high. Adding the PLC changed the optimal level one cache configuration for 14 benchmarks, which resulted in area savings as high as 33%. Whereas these additional savings may be attractive, we reiterate that these additional savings come at the expense of the PLC preanalysis step and require a stable application.

26.5 Interplay of Code Compression, Loop Caching, and Cache Tuning

Using a loop cache can decrease decompression overheads (performance and energy) for DF techniques by storing/caching uncompressed instructions (Figure 26.1c) in a smaller, more energy efficient loop cache. The magnitude of this overhead reduction is dependent on an application's temporal and spatial locality. In addition, code compression reduces the level one cache requirements. In this section, we quantify the overhead reduction afforded by introducing a loop cache as an instruction decompression buffer, in addition to cache tuning for both the level one and loop caches.

26.5.1 Experimental Setup

To determine the combined effects of code compression with cache tuning, we determined the optimal (lowest energy) level one cache configuration for a system using a modified DF architecture (Figure 26.1c) for the same 31 benchmarks and experimental setup as described for the loop cache and cache tuning experiments. For comparison purposes, energy consumption and performance was normalized to a base system configuration with an 8 kB, 4-way set associative level one base cache with a 32 byte line size (with no loop cache). Based on [51], we used a 32-entry ALC and a 64-entry PLC for our experiments.

We used Huffman encoding [32] for instruction compression/decompression. Branch targets were byte aligned to enable random access decompression and a LAT (loop address table) translated uncompressed addresses to corresponding compressed addresses for branch and jump targets.

We modified SimpleScalar [11] to include the decompression unit, LAT, and loop cache. The energy model used for the loop cache and cache tuning experiments was modified to include decompression energy. We also measured the performance (total number of clock cycles needed for execution).

The performance measured was normalized to the performance of the base system with uncompressed instructions and no loop cache.

26.5.2 Analysis

Figure 26.9 depicts the (a) energy and (b) performance of the optimal level one cache configuration for a system that stores compressed instructions in the level one cache and uncompressed instructions in a loop cache (ALC or PLC) normalized to the base system with no loop cache. For brevity, Figure 26.9 shows average energy and performance for each benchmark suite and selected individual benchmarks that revealed interesting results.

Figure 26.9a shows that, on average, for the EEMBC benchmarks, the optimal level one cache configuration combined with the 32-entry ALC did not result in energy savings. However, on average, the Powerstone and MiBench benchmarks achieved energy savings of 20% and 19%, respectively, (Figure 26.9a) for the system with an ALC.

Analysis of the benchmarks' structure revealed that both Powerstone and MiBench benchmarks contain only a few loops that iterate several times (several Powerstone and MiBench benchmarks stay in the same loop for hundreds of consecutive iterations) resulting in energy savings and a lower performance overhead. EEMBC benchmarks, however, contain many loops that iterate fewer times than the Powerstone and MiBench benchmarks (several EEMBC benchmarks stay in the same loop for less than 20 consecutive iterations). EEMBC benchmarks spend a short time fetching uncompressed instructions from the ALC before a new loop is encountered and the decompression unit is invoked again resulting in low energy savings and a large performance overhead. However, EEMBC benchmarks with a high loop cache access rate achieved energy savings (e.g., *PNTRCH01* with a 97% loop cache access rate [51] achieved 69% energy savings (Figure 26.9a) with only a small decompression overhead (Figure 26.9b).

Figure 26.9a shows that, on average, both the Powerstone and MiBench benchmark suites achieved energy savings of 30% for the system with an optimal level one cache configuration combined with a 64-entry PLC. An additional 10% average energy savings was gained by eliminating the decompression overhead, which would have been consumed while filling the ALC. Figure 26.9a shows that MiBench's *dijkstra* and *adpcm-e* benchmarks saved 56% and 38% more energy, respectively, when using the PLC instead of the ALC. Results for Powerstone's *blit* benchmark highlight the impact of the decompression overhead. For *blit*, the loop cache access rate for the 32-entry ALC is higher than the loop cache access rate for the 64-entry PLC (80% compared with 30% [51]) but by removing the decompression energy consumed during the first two iterations of the loop, the system with the PLC saved almost as much energy as the system with the ALC (Figure 26.9a).

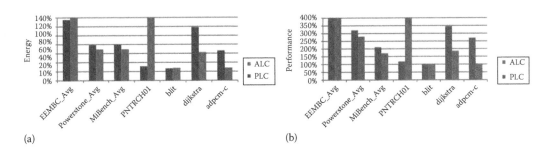

(a)   (b)

FIGURE 26.9 (a) Energy and (b) performance (energy and performance normalized to the base system with no loop cache) for the lowest energy cache configuration averaged across the EEMBC benchmarks (EEMBC-Avg), Powerstone benchmarks (Powerstone-Avg), MiBench benchmarks (MiBench-Avg), and for selected individual benchmarks (*PNTRCH01*, *blit*, *dijkstra*, and *adpcm-e*).

Figure 26.9a also shows that, on average, for the EEMBC benchmarks, using the PLC did not result in energy savings and that the ALC outperformed the PLC. This result is expected since, for the EEMBC benchmarks, the PLC only outperformed the ALC for the 256-entry loop cache [51].

Figure 26.9b shows that, on average, the performance of the system increased for both the ALC and the PLC because of the large decompression overhead (the loop cache does not affect system performance since it guarantees a 100% loop cache hit rate). The average increase in performance due to decompression overhead ranged from as much as 4.7× for EEMBC benchmarks with a PLC to 1.7× for MiBench benchmarks with a PLC (Figure 26.9b). We also observed that using the PLC instead of the ALC reduced the decompression overhead by approximately 40% for Powerstone and MiBench benchmarks. Individual results showed that, for most benchmarks, the PLC reduced the decompression overhead but increased system performance as compared to a system with no PLC. As shown in Figure 26.9, for the system with the PLC, MiBench's *adpcm-e* achieved 73% energy savings (38% more than the system with the ALC) and reduced performance overhead to only 2% more than the performance of the base system.

For our experiments, we tuned the level one cache while keeping the loop cache size fixed to find the optimal level one cache and loop cache combination. We compared these new level one cache configurations to the lowest energy level one cache configurations for a system with uncompressed instructions and no loop cache. We found that for 12 out of 31 benchmarks, the new level one cache configurations were smaller for the systems using compression compared with the level one cache configurations for the systems not using compression. These benchmarks were able to use smaller level one cache configurations since the level one cache stored compressed instructions, and effectively increased the cache size. However, we did not observe a change in level one cache configuration for systems with low loop cache access rates and no energy savings. Additionally, for some benchmarks, the optimal level one configuration for the uncompressed system was already the smallest size (2 kB), so adding a loop cache did not result in a smaller level one cache configuration.

We calculated the area savings gained by replacing the level one cache storing uncompressed instructions with the smaller level one cache storing compressed instructions combined with the loop cache for the 12 benchmarks with new optimal level one configurations. The benchmarks that replaced an 8 kB level one cache with a 2 kB level one cache and loop cache achieved approximately 50% area savings. The benchmarks that replaced an 8 kB level one cache with a 4 kB level one cache and loop cache and replaced a 4 kB level one cache with a 2 kB level one cache and loop cache achieved approximately 30% and 20% area savings, respectively. For the remaining benchmarks, the level one cache configuration did not change, and thus adding a loop cache increased the area of the system. Some benchmarks achieved energy savings but not area savings. For example, *PNTRCH01* had a loop cache access rate of 97% and achieved 69% energy savings with the ALC, but the level one configuration was the same for both the uncompressed and compressed system, which resulted in an increase in area of approximately 14%.

26.6 Conclusions

This chapter provided a detailed study of the interplay between several instruction cache optimization techniques: code reordering, code compression, loop caching, and cache tuning. When comparing code reordering and cache tuning, we found that cache tuning yielded superior average energy and performance benefits, and is always better or nearly equal to code reordering on every benchmark examined. For applications where combining the two methods yielded additional savings, the additional savings reached as high as 15%. We found that a simple heuristic that applies code reordering once, followed by cache tuning, yielded near-optimal results. Furthermore, we showed the importance of cache-aware code reordering. Cache-aware code reordering with cache tuning revealed an addition 2.8× and 2.9× increase in energy savings and reduction in execution time, respectively, as compared to non-cache-aware code reordering with cache tuning.

Next, we investigated the effects of combining loop caching with level one cache tuning and found that in general, cache tuning dominated overall energy savings indicating that cache tuning is sufficient for energy savings. However, we observed that adding a loop cache to an optimal (lowest energy) cache increased energy savings by as much as 26%. Finally, we investigated the possibility of using a loop cache to minimize runtime decompression overhead and quantified the effects of combining code compression with cache tuning. Our results showed that a loop cache effectively reduced the decompression overhead, resulting in energy savings of up to 73%. However, to fully exploit combining cache tuning, code compression, and loop caching, a compression/decompression algorithm with lower overhead than the Huffman encoding technique is required.

Acknowledgments

This research was supported in part by the National Science Foundation (CNS-0953447, CCR-0203829, and CCR-9876006). Any opinions, findings, and conclusions or recommendations expressed in this material are those of the author(s) and do not necessarily reflect the views of the National Science Foundation.

References

1. Aydin, H. and Kaeli, D. Using cache line coloring to perform aggressive procedure inlining. *ACM SIGARCH News*, 28 (1), 62–71, March 2000.
2. Albonesi, D. H. Selective cache ways: On demand cache resource allocation. *Journal of Instruction Level Parallelism*, vol. 2, May 2002.
3. Altera. Nios embedded processor system development, http://www.altera.com/corporate/news_room/releases/products/nr-nios_delivers_goods.html, 2010.
4. Arc International. www.arccores.com, 2010.
5. ARM. www.arm.com, 2010.
6. Bahar, I., Calder, B., and Grunwald, D. A. Comparison of software code reordering and victim buffers. In *3rd Workshop of Interaction between Compilers and Computer Architecture*, San Jose, California, October 1998.
7. Balasubramanian, R., Albonesi, D., Buyuktosunoglu, A., and Dwarkadas, S. Memory hierarchy reconfiguration for energy and performance in general-purpose processor architecture. In *33rd International Symposium on Microarchitecture*, Monterey, California, December 2000.
8. Bartolini, S. and Prete, C. A. Optimizing instruction cache performance of embedded systems. *ACM Transactions on Embedded Computer Systems*, 4 (4), 934–965, 2005.
9. Benini, L., Macii, A., Macii, E., and Poncino, M. Selective instruction compression for memory energy reduction in embedded systems. In *International Symposium on Low Power Embedded Systems*, San Diego, California, August 1999.
10. Benini, L., Macii, A., and Nannarelli, A. Cached-code compression for energy minimization in embedded processors. In *Proceedings of the 2001 International Symposium on Low Power Electronics and Design*, Huntington Beach, California, August 2001.
11. Burger, D., Austin, T., and Bennet, S. Evaluating future microprocessors: The simplescalar toolset. University of Wisconsin–Madison. Computer Science Department technical report *CS-TR-1308*, July 2000.
12. Chen, J. and Leupen, B. Improving instruction locality with just-in-time code layout. In *Proceedings of the USENIX Windows NT Workshop*, Seattle, Washington, August 1997.
13. Chen, Y. and Zhang, F. Code reordering on limited branch offset. *ACM Transactions on Architecture and Code Optimizations (TACO)*, 4 (2), June 2007.

14. Cohn, R., Goodwin, P., Lowney, G., and Rubin, N. Spike: An optimizer for Alpha/NT executables. In *USENIX Windows NT Workshop*, Seattle, Washington, August 1997.

15. Cohn, R. and Lowney, P. G. Design and analysis of profile-based optimization in Compaq's compilation tools for Alpha. *Journal of Instruction Level Parallelism*, vol. 2, May 2000.

16. Dinero IV. http://www.cs.wisc.edu/~markhill/DineroIV/

17. EEMBC. The Embedded Microprocessor Benchmark Consortium, www.eembc.org

18. Guthaus, M. R., Ringenberg, J. S., Ernst, D., Austin, T. M., Mudge, T., and Brown, R. B. MiBench: A free, commercially representative embedded benchmark suite. In *IEEE Fourth Annual Workshop on Workload Characterization*, Austin, Texas, December 2001.

19. Gloy, N., Blackwell, T., Smith, M. D., and Calder, B. Procedure placement using temporal ordering information. In *Proceedings of the 30th Annual ACM/IEEE Intl. Symposium on Microarchitecture*, pp. 303–313, Research Triangle Park, North Carolina, December 1997.

20. Ghosh, A. and Givargis, T. Cache optimization for embedded processor cores: An analytical approach. In *International Conference on Computer Aided Design*, November 2003.

21. Givargis, T. and Vahid, F. Platune: A tuning framework for system-on-a-chip platforms. *IEEE Transactions on Computer Aided Design*, 21 (11), Pages 1317–1327, November 2002.

22. Gordon-Ross, A., Cotterell, S., and Vahid, F. Exploiting fixed programs in embedded systems: A Loop cache example. *Computer Architecture Letters*, vol. 1, January 2002.

23. Gordon-Ross, A., Lau, J., and Calder, B. Phase-based cache reconfiguration for a highly-configurable two-level cache hierarchy. In *Proceedings of the 18th ACM Great Lakes Symposium on VLSI (GLSVLSI)*, Orlando, Florida, May 2008.

24. Gordon-Ross, A., Viana, P., Vahid, F., Najjar, W., and Barros, E. A one-shot configurable-cache tuner for improved energy and performance. In *Proceedings of the Conference on Design, Automation and Test in Europe (DATE)*, Nice, France, April 2007.

25. Gordon-Ross, A. and Vahid, F. Dynamic loop caching meets preloaded loop caching—A hybrid approach. In *International Conference on Computer Design*, Freiburg, Germany, September 2002.

26. Gordon-Ross, A., Vahid, F., and Dutt, N. A first look at the interplay of code reordering and configurable caches. In *Proceedings of the 15th ACM Great Lakes Symposium on VLSI (GLSVSLI'05)*, Chicago, IL, April 17–19, 2005.

27. Gordon-Ross, A., Vahid, F., and Dutt, N. Fast configurable-cache tuning with a unified second-level cache. In *IEEE Transactions on Very Large Scale Integration (TVLSI)*, 17 (1), 80–91, January 2009.

28. Hashemi, A., Kaeli, D., and Calder, B. Efficient procedure mapping using cache line coloring. In *Proceedings of the International Conference on Programming Language Design and Implementation*, Las Vegas, Nevada, June 1997.

29. Hines, S., Whalley, D., and Tyson, G. Guaranteeing hits to improve the efficiency of a small instruction cache. In *IEEE/ACM International Symposium on Microarchitecture*. Chicago, Illinois, December 2007.

30. Huang, X., Blackburn, S., Grove, D., and McKinley, K. Fast and efficient partial code reordering: Taking advantage of a dynamic recompiler. *International Symposium on Memory Management*, 2006.

31. Huang, X., Lewis, T., and McKinley, K. Dynamic code management: Improving whole program code locality in managed runtimes. *ACM International Conference on Virtual Execution Environments*, Ottawa, Canada, June 2006.

32. Huffman, D. A. A method for the construction of minimum-redundancy codes. In *Proceedings of the IRE*, 4D, 1098–1101, September 1952.

33. Hwu, W. W. and Chang, P. Achieving high instruction cache performance with an optimizing compiler. In *Proceedings of the 16th Annual Intl. Symposium on Computer Architecture*, Jerusalem, Israel, June 1989.

34. Kalamatianos, J. and Kaeli, D. Code reordering for multi-level cache hierarchies. Northeastern University Computer Architecture Research Group, November 1999.

35. Kalamatianos, J. and Kaeli, D. Accurate simulation and evaluation of code reordering. In *Proceedings of the IEEE International Symposium on the Performance Analysis of Systems and Software*, Austin, Texas, April 2000.

36. Kin, J., Gupta, M., and Mangione-Smith, W. The filter cache: An energy efficient memory structure. In *IEEE Micro*, December 1997.

37. Lee, D., Baer, J., Bershad, B., and Anderson, T. Reducing startup latency in web and desktop applications. In *Windows NT Symposium*, Boston, Massachusetts, July 1999.

38. Lee, L. H., Moyer, W., and Arends, J. Low cost embedded program loop caching—Revisited. University of Michigan technical report CSE-TR-411-99, December 1999.

39. Lee, C., Potkonjak, M., and Mangione-Smith, W. H. MediaBench: A tool for evaluating and synthesizing multimedia and communication systems. In *Proceedings of the 30th Annual International Symposium on Microarchitecture*, North Carolina, December 1997.

40. Lekatsas, H., Henkel, J., and Wolf, W. Code compression for low power embedded system design. In *Design Automation Conference*, Los Angeles, California, USA, June *2000. Proceedings 2000.* 37, 294–299, 2000.

41. Malik, A., Moyer, W., and Cermak, D. A low power unified cache architecture providing power and performance flexibility. In *International Symposium on Low Power Electronics and Design*, Rapallo, Italy, July 2000.

42. McFarling, S. Program optimization for instruction caches. In *Proceedings of the Third International Conference on Architectural Support for Programming Languages and Operating Systems* (ASPLOS III), Boston, Massachusetts, April 1989.

43. MIPS Technologies. www.mips.com, 2010.

44. Moseley, P., Debray, S., and Andrews, G. Checking program profiles. In *Third IEEE International Workshop of Source Code Analysis and Manipulation*, Amsterdam, The Netherlands, September 2003.

45. Muth, R., Debray, S., Watterson, S., and de Bosschere, K. Alto: A link-time optimizer for the Compaq Alpha. *Software Practice and Experience,* 31 (6), 67–101, January 2001.

46. Palesi, M. and Givargis, T. Multi-objective design space exploration using genetic algorithms. In *International Workshop on Hardware/Software Codesign,* May 2002.

47. Pettis, K. and Hansen, R. Profile guided code positioning. In *Proceedings of the ACM SIGPLAN conference on Programming Language Design and Implementation*, White Plains, New York, June 1990.

48. Ramirez, A. Code placement for improving dynamic branch prediction accuracy. In *ACM SIGPLAN Conference on Programming Language Design and Implementation (PLDI)*, Chicago, Illinois, June 2005.

49. Ramirez, A., Larriba-Pey, J., and Valero, M. The effect of code reordering on branch prediction. In *International Conference on Parallel Architectures and Compilation Techniques (PACT)*, Philadelphia, Pennsylvania, October 2001.

50. Ramirez, A., Larriba-Pey, J., and Valero, M. Instruction fetch architectures and code layout optimizations. *Proceedings of the IEEE,* 89, 11, November 2001.

51. Rawlins, M. and Gordon-Ross, A. Lightweight runtime control flow analysis for adaptive loop caching. In *Proceedings of the 20th Symposium on Great Lakes Symposium on VLSI (GLSVLSI'10)*, Providence, RI, May 16–18, 2010.

52. Reinman, G. and Jouppi, N. P. Cacti2.0: An integrated cache timing and power model. *COMPAQ Western Research Lab,* 1999.

53. Samples, A. D. and Hilfinger, P. N. Code reorganization for instruction caches. Technical Report UCB/CSD 88/447, University of California, Berkeley, October 1988.

54. Sanghai, K., Kaeli, D., Raikman, A., and Butler, K. A code layout framework for embedded processors with configurable memory hierarchy. In *Workshop on Optimizations for DSP and Embedded Systems (ODES)*, San Jose, California, March 2007.

55. Scales, D. Efficient dynamic procedure placement. Technical report WRL-98/5, Compaq WRL Research Lab, May 1998.

56. Scharz, B., Debray, S., Andrews, G., and Legendre, M. PLTO: A link-time optimizer for the Intel IA-32 architecture. In *Proceedings 2001 Workshop on Binary Translation (WBT-2001)*, Barcelona, Spain, September 2001.

57. Schmidt, W. J., Roediger, R. R., Mestad, C. S., Mendelson, B., Shavit-Lottem, I., and Bortnikov-Sitnitsky, V. Profile-directed restructuring of operation system code. *IBM Systems Journal*, 37 (2), 270–297, 1998.

58. Scott, J., Lee, L., Arends, J., and Moyer, B. Designing the low-power M~CORE architecture. In *International Symposium on Computer Architecture Power Driven Microarchitecture Workshop*, Barcelona, Spain, pp. 145–150, July 1998.

59. Srivastava, A. and Wall, D. W. A practical system of intermodule code optimization at link-time. *Journal of Programming Languages*, 1, (1), 1–18, December 1992.

60. Su, C. and Despain, A. M. Cache design trade-offs for power and performance optimization: A case study. In *International Symposium on Low Power Electronics and Design*, Dana Point, California, April 1995.

61. Tensilica. Xtensa Processor Generator, http://www.tensilica.com/, 2010.

62. Villarreal, J., Lysecky, R., Cotterell, S., and Vahid, F. Loop analysis of embedded applications. UC Riverside technical report UCR-CSR-01-03, 2001.

63. Wolfe, A. and Chanin, A. Executing compressed programs on an embedded RISC architecture. In *Proceedings of the 25th Annual International Symposium on Microarchitecture*, Portland, OR, December 1–4, 1992. IEEE Computer Society Press, Los Alamitos, CA, pp. 81–91.

64. Zhang, C. and Vahid, F. Using a victim buffer in an application-specific memory hierarchy. In *Design, Automation and Test (DATE) Conference in Europe*, Paris, France, February 2004.

65. Zhang, C. and Vahid, F. A self-tuning cache architecture for embedded systems. *Design, Automation and Test (DATE) Conference in Europe*, Paris, France, 2004.

66. Zhang, C., Vahid, F., and Najjar, W. A highly-configurable cache architecture for embedded systems. In *The 30th Annual International Symposium on Computer Architecture*, San Diego, CA, June 2003.

Index

Printed and bound by CPI Group (UK) Ltd, Croydon, CR0 4YY

21/10/2024

01777040-0017